W9-BHU-267

J. M. Thomas, W. J. Thomas

Principles and Practice of Heterogeneous Catalysis

Distribution:
VCH, P.O. Box 101161, D-69451 Weinheim (Federal Republic of Germany)
Switzerland: VCH P.O. Box, CH-4020 Basel (Switzerland)
United Kingdom and Ireland: VCH (UK) Ltd., 8 Wellington Court, Cambridge CB1 1HZ
USA and Canada: VCH, 333 7th Avenue, New York, NY 10001 (USA)
Japan: VCH, Eikow Building, 10-9 Hongo 1-chome, Bunkyo-ku, Tokyo 113 (Japan)

ISBN 3-527-29288-8 (Hardcover)
ISBN 3-527-29239-X (Softcover)

J. M. Thomas, W. J. Thomas

Principles and Practice of Heterogeneous Catalysis

Weinheim · New York · Basel · Cambridge · Tokyo

Professor Sir J. M. Thomas
Peterhouse
University of Cambridge
Cambridge CB2 1QY
UK
and
The Royal Institution, London

Professor W. J. Thomas
School of Chemical Engineering
University of Bath
Claverton Down
Bath B1Z 7AY
UK

Published jointly by
VCH Verlagsgesellschaft mbH, Weinheim (Federal Republic of Germany)
VCH Publishers Inc., New York, NY (USA)

Editorial Director: Dr. Michael Bär
Production Manager: Dipl.-Wirt.-Ing. (FH) Bernd Riedel
Copy Editing: Diana Boatman

Cover illustration: The Background shows a high-resolution electron micrograph of the zeolitic catalyst ZSM-5. The diameter of each large white patch is *ca* 5.5 Å. The left-hand inset presents a series of X-ray absorption spectra, recorded in situ, of a mesoporous titano–silica catalyst (right-hand inset) for the epoxidation of cyclohexene in the presence of H_2O_2. The bottom inset shows a catalytic monolith for converting automobile exhaust gases: the monolith is housed within a metal cluster canister the cone ends of which are connected to the exhaust manifold.

Colour graphics by D. W. Lewis and R. G. Bell

Library of Congress Card No. applied for.

A catalogue record for this book is available from the British Library.

Deutsche Bibliothek Cataloguing-in-Publication Data:

Thomas, John Meurig:
Principles and practice of heterogeneous catalysis / J. M. Thomas; W. J. Thomas. – Weinheim; New York; Basel; Cambridge; Tokyo: VCH, 1996
ISBN 3-527-29288-8 Gb.
ISBN 3-527-29239-X brosch.
NE: Thomas, W. John:

Printed on acid-free and chlorine-free paper

Composition: Asco Trade Typesetting Limited, Hong Kong
Printing: Strauss Offsetdruck, D-69509 Mörlenbach
Bookbinding: Wilh. Osswald & Co., D-67433 Keustadt/Weinstr.

Printed in the Federal Republic of Germany

Preface

Catalysis occupies a pivotal position in the physical and biological sciences. As well as being the mainstay of the chemical industry, it is the means of effecting many laboratory syntheses and the root cause of all enzymatic processes. Nowadays it is increasingly implicated in the suppression of atmospheric pollution, in the design of environmentally compatible new technologies, and in the pursuit of new ways of generating energy and materials either by harnessing solar radiation or through selected chemical conversions of abundant hydrocarbons such as methane, and small molecules such as the oxides of carbon. More importantly from the viewpoint of the academic scientist, catalysis also serves as a nexus that brings together numerous disciplines, each with its own identity: surface and solid-state chemistry, solid-state physics, biochemistry, materials science and engineering, chemical engineering, organometallic and theoretical chemistry. As a consequence, catalysis is a topic conducive of cross-disciplinary activity and debate.

Writing a single unified text on the subject is, therefore, no mean task. Indeed, it is often argued that the subject is now so diversified and expansive that only through a continuing series of monographs, chronicling selected advances, can we adequately cope with its burgeoning growth. Regular monographs charting the progress of catalysis certainly have their place. But they cannot rival the educational advantages provided by unified texts that are targeted at senior undergraduates and fresh researchers. In 1967 we published our *Introduction to the Principles of Heterogeneous Catalysis*. It was well received; and we were urged, after stocks of its fourth printing had been exhausted, to prepare a second edition. This book began as such. But it soon became apparent that a comprehensive re-writing was called for. Less than 10% of our earlier effort has been incorporated, in a radically reorganized state, into this text, which is based largely on courses given to undergraduate chemists at the University of Cambridge and chemical engineers at the University of Bath.

It is instructive to identify some of the major changes in heterogeneous catalysis since the appearance of our first book. Immense strides have been taken in characterizing the surface and bulk structure of catalysts: photoelectron and Auger electron spectroscopy permit the ready identification of submonolayer amounts of material, and the nature of their bonding; and two-dimensional surface structures can be gleaned from low-energy-electron diffraction and X-ray spectroscopy. It has been recognized that surface reconstruction can be both ubiquitous and extensive; and that catalysed reactions can often exhibit an oscillatory, even chaotic character. Incommensurate phases, non-stoichiometric, intergrown and structurally adaptable as well as multicomponent and multiphasic solid catalysts are a reality. Laboratory-scale 'smart catalysts' are also a reality (see Hill and Zhang *Nature* (*London*), **1995**, *373*, 324), and much is expected on the catalytic front from recent advances in combinatorial chemistry. Moreover, many new procedures, entailing the use of synchrotron radiation for X-ray absorption spectroscopy, neutron beams and laser-based spectroscopies, can now probe the behaviour of catalysts under reaction conditions. Again, so dramatic has been our increased expertise in solid-state chemistry that many new types of catalytically active solids can be fashioned, some capable of releasing constituent atoms to gaseous reactants. This is particularly true in the case of selective oxidation catalysts and in the area of shape-selective catalysts, where well-defined microporous and mesoporous solids are of the essence. To a lesser extent, it is also true in the tailoring of enantioselective catalysts by prior modification of the active solid surface. There is a continuing trend to design catalysts from materials that are plentiful or readily synthesized, such as layered clays or zeolites; and to make even more efficient use of supported rare-metal catalysts (platinum and rhodium), especially in the commercially important processes of auto-emission control and in the reforming of hydrocarbons so as to produce better fuels. Solid acid catalysts as well as immobilized enzymes and catalytic antibodies are all in the ascendant; and photocatalysis, which holds the key to more efficient utilization of solar energy, not to mention specialized organic syntheses, is of increasing importance. There is hope that, just as various electronic devices can be expertly designed from a knowledge of bonding and band structure, so also will it prove possible to design better photocatalysts on similar principles. Some of the concepts and ideas that, 20 years ago, were seen as notable milestones have now receded into the perspective of a vanishing road. Some have been replaced and supplanted; still others, a few of relatively ancient lineage, have survived, even flourished. It has become apparent that, in formulating mechanisms and quantum-mechanical theories of heterogeneous catalysis, there are advantages in employing localized pictures, akin to those used in homogeneous catalysis. Active sites and ensembles remain in vogue; structure-sensitive and structure-insensitive catalysed reactions have become part of our vocabulary, as have spillover and turnover frequencies.

Arguably the single most dramatic change that has occurred in science generally since the mid-1970s has been the enormous growth in computational power which experimentalists and theoreticians alike now have at their disposal. This means that chemical engineers can cope with problems that hitherto seemed quite intractable, and can solve, by numerical methods, fiendishly complex sets of equations such as those appropriate for the operation of catalytic reactors. Nowadays, chemical engineers can operate a refinery entirely by computer control: they also handle non-linear phenomena or coupled equations with consummate computational ease. Fast computers likewise make it possible for protein engineers, pharmaceutical chemists and quantum pharmacologists to lay down the semi-empirical principles of molecular recognition, and these could well figure increasingly in the design of catalysts for the future and in the molecular simulation of their mode of action. For the computational chemist, the formalisms of Monte Carlo and molecular-dynamics simulations have brought with them their own power and seductiveness. So long as one remains alert to the pitfalls inherent in simulation divorced from reality, these developments add power to the elbow of catalyst designers.

In preparing this book, which concentrates on fundamental principles and which explains the vocabulary, grammar and literature of catalysis from the laboratory-orientated model study through to the operating plant, we have been fortunate to have conferred on a multiplicity of topics with experts from far and wide. Individuals and research groups worldwide have entered into discussions with us, and we are grateful for their contributions.

As with many other authors coping with a rapidly changing vista, the thoughts expressed by Dr Johnson (1709–1784) in the preface to his famous *Dictionary* spring to mind:

When the mind is unchained from necessity, it will range after convenience; when it is left at large in the fields of speculation, it will shift opinions; as custom is disused, the words that express it must perish with it; as any opinion grows popular, it will innovate speech in the same proportion as it alters practice. No dictionary of a living tongue ever can be perfect, since while it is hastening to publication, some words are budding and some falling away.

J. M. Thomas
Davy Faraday Research Laboratory
The Royal Institution, London
and
Peterhouse
University of Cambridge
UK

W. J. Thomas
School of Chemical Engineering
University of Bath
UK

Contents

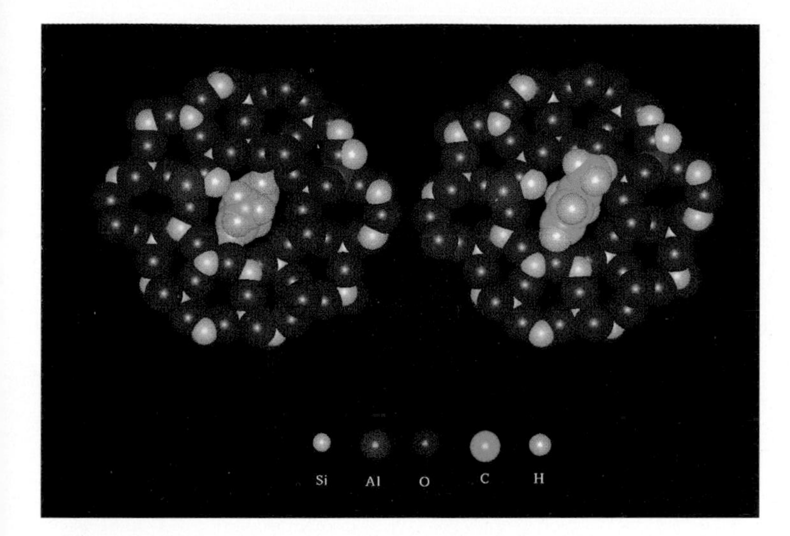

Colour plate 1 Computer graphics reveal that *p*-xylene fits sedately within the pores of the zeolitic catalyst ZSM-5 and diffuses smoothly along them (left), whereas *m*-xylene, because of its size and shape, does not fit inside the pores and cannot diffuse along them (right).

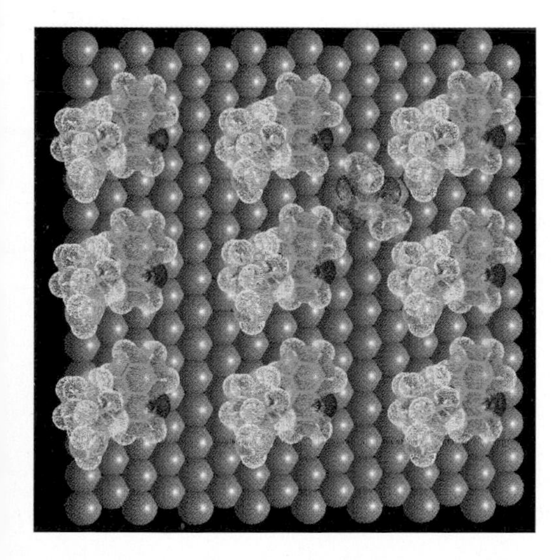

Colour plate 2 Schematic illustration of the way in which cavities in the ordered monolayer of cinchonidine (on a platinum surface) may serve as centers where a prochiral reactant may be stereopreferentially adsorbed. This preordains the production of an enantiomeric excess of the methyl lactate formed by catalytic reduction on the modified metal surface.

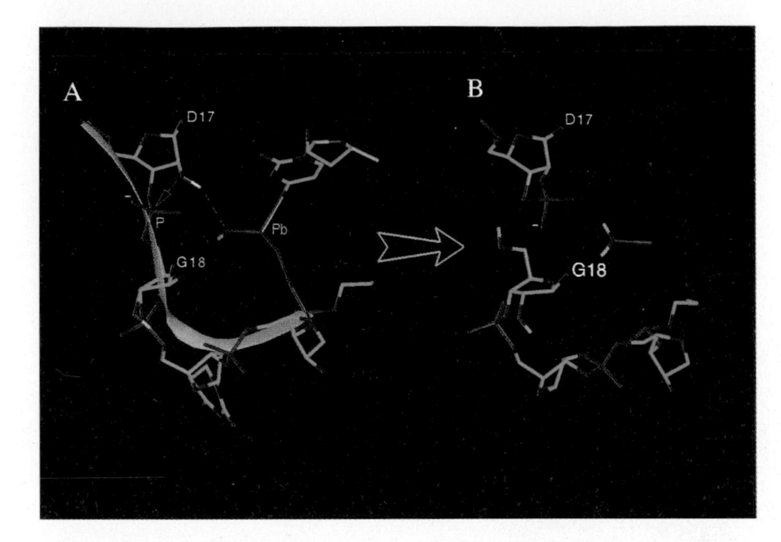

Color plate 3 Representation of the proposed mechanism for the catalysed cleavage of the sugar-phosphate backbone of tRNA. The precise siting of the Pb, and its role as a catalyst, were inferred from the results of in-situ X-ray studies. With permission from R. S. Brown, J. C Dewan, A. Klug, *Biochemistry* **1985**, *224*, 4785.

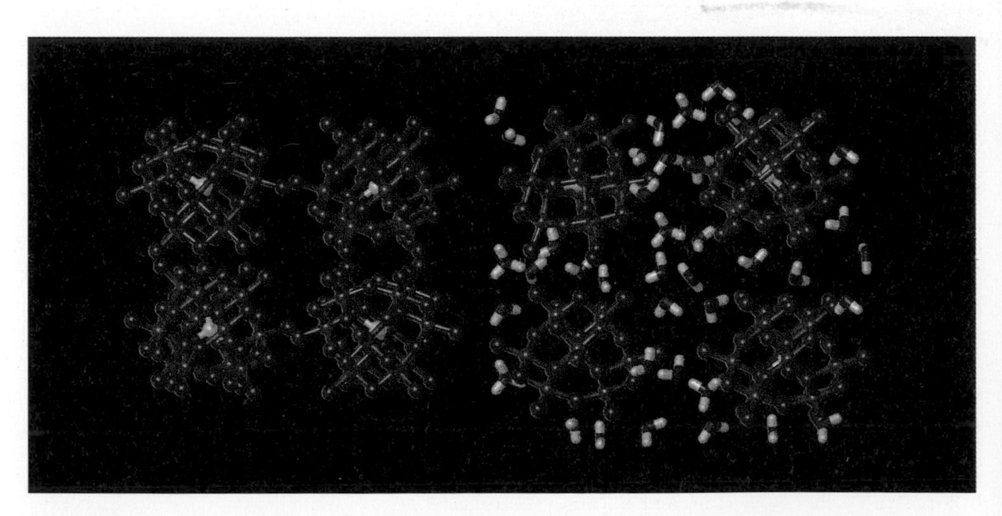

Colour Plate 4 Four Keggin ions (left) are composed of a central atom of Si or P (tiny yellow) surrounded by O (red) and Mo or W (larger yellow). Structures made up of Keggin ions absorb polar molecules (red and white), which have a slight electrical charge, to become swollen quasi-liquids (right) with a large surface area and high acidity (After J. M. Thomas, *Sci. American*, **1992**, *266*, 85).

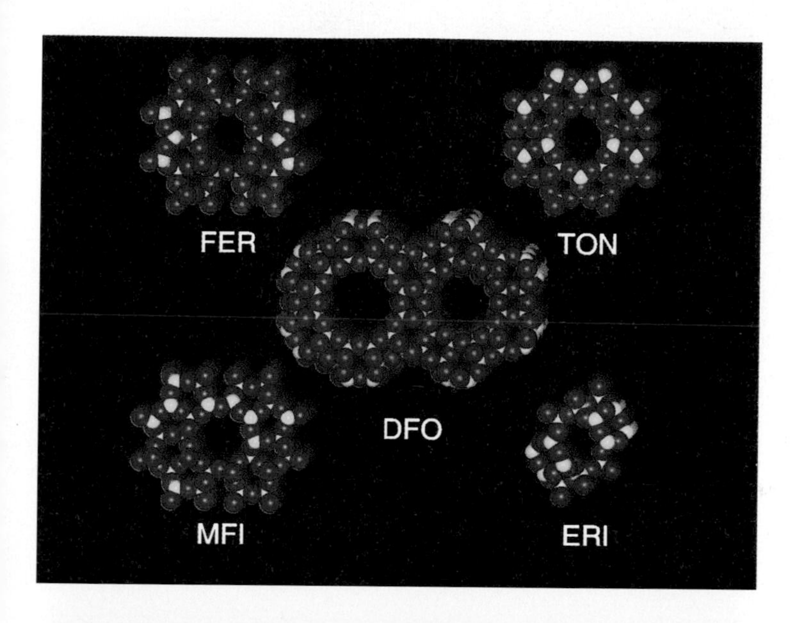

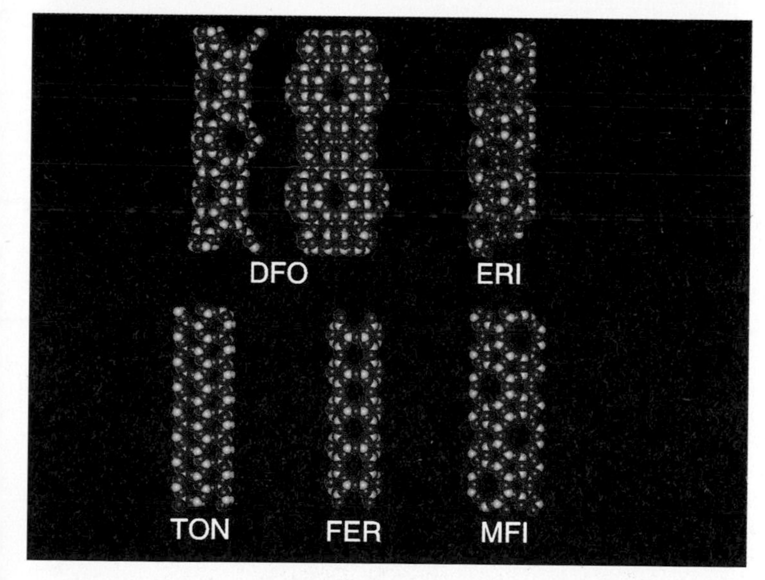

Colour Plate 5 Plan view (top) and elevation (bottom) of five distinct types of shape-selective molecular-sieve solids that function as solid acid catalysts for the skeletal isomerization of 1-butene. FER ≡ H^+-ferrierite. TON ≡ H^+-*T*heta *one*. MFI ≡ H^+-ZSM-5 (*M*obil *F*ive). ERI ≡ H^+-*eri*onite and DFO ≡ H^+-*D*avy-*F*araday *one*. (DFO (DFI) has two distinct kinds of pores.) (Reprinted from J. M. Thomas, *Angew. Chem., Int. Ed. Engl.* **1994**, *33*, 913.)

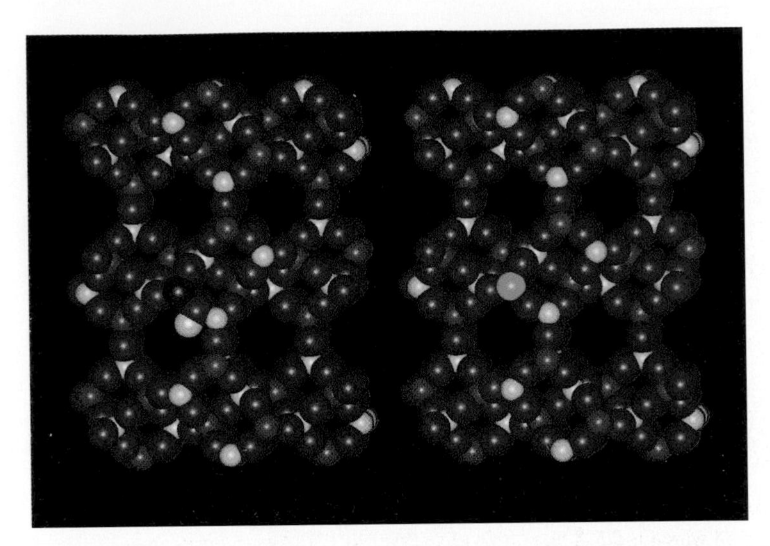

Colour Plate 6 Computer graphic representation of cobalt-containing ALPO-18 structure. The framework is closely similar to that of the aluminosilicate mineral chabazite. The actual colours of the calcined and reduced forms of the catalyst are green and blue respectively. (Colour code: red, oxygen; blue, Co^{II}; green, Co^{III}; white, hydrogen; purple, phosphorus; yellow, aluminium). (Based on J. M. Thomas, G. N. Greaves, C. R. A. Catlow, *Nucl. Inst. Meth.* **1995**, *B97*, 1.)

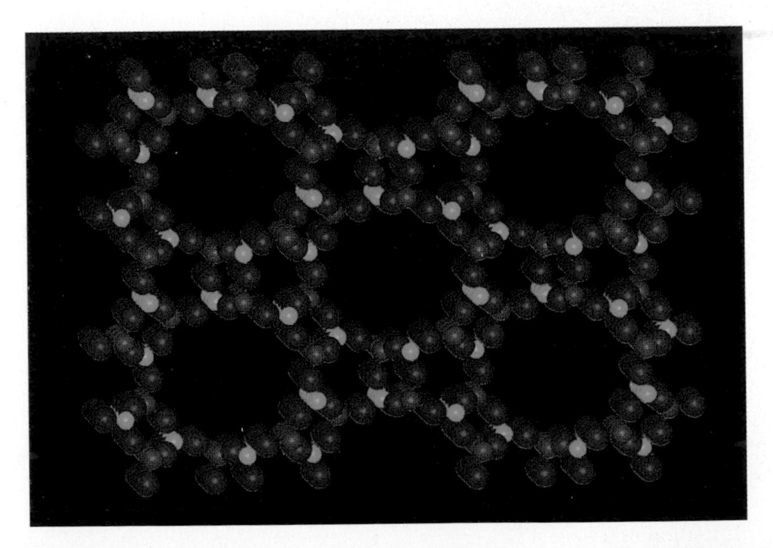

Colour Plate 7 The energy-minimized ALPO structure for MeALPO-36, Me = Mg, Zn, etc. Note that Al and P alternate in the tetrahedra. Red = oxygen, green = phosphorus, violet = aluminium. Reprinted from P. A. Wright, S. Natarajan, J. M. Thomas, R. G. Bell, P. L. Gai-Boyes, R. H. Jones, J. Chen, *Angew. Chen., Int. Ed. Engl.* **1992**, *31*, 1473.

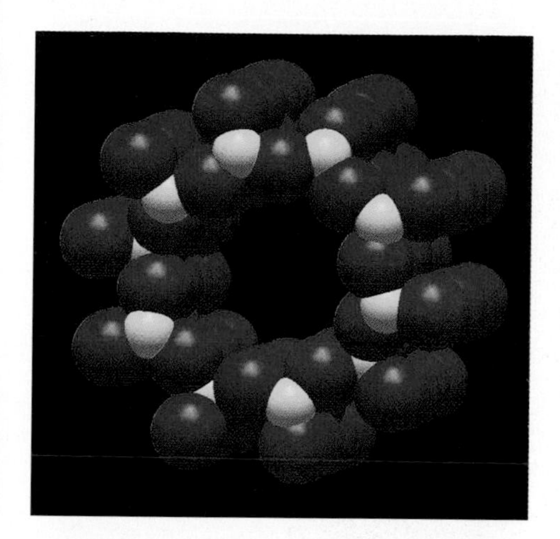

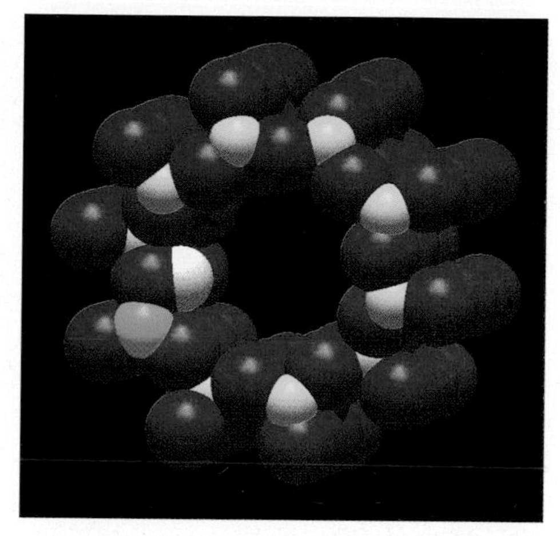

Colour Plate 8 Scalar representation (left) of the pure, siliceous molecular seive known as silicalite (SiO_2), the pore diameter of which is 5.5 Å as in ZSM-5 (right). H^+-ZSM-5 is a protonated form of the siliceous analogue but with 1 in 10 to 1 in 50 or so of the Si atoms in the framework replaced by Al atoms. There is one proton loosely attached to framework O atoms for every Al^{3+} ion (or other trivalent ion such as Fe^{3+}) that substitutes for a Si^{4+}.

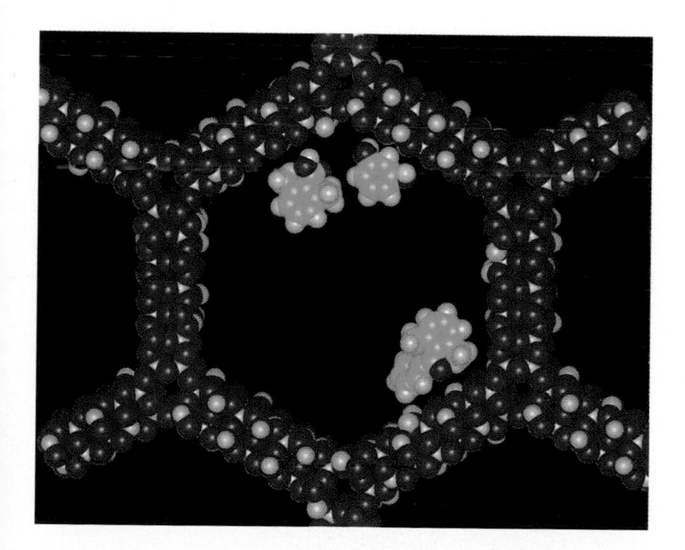

Color Plate 9 Graphic illustration of how van Bekkum et al. have used individual members of the newly developed MCM family of mesoporous silica to inject shape-selectivity into organic processes such as the Fries reaction. Reactants such as 2,6-dimethylbenzoic acid and resorcinol, which may be anchored by pendant Si–OH groups to the inner walls of the silica, undergo the Fries reaction to yield benzophenone derivatives. (After J. M. Thomas, *Angew. Chem., Int. Ed. Engl.* **1994**, *33*, 913.)

1 Setting the Scene

1.1 Introduction

Our objective in this chapter is to trace the emergence, application, study, and interpretation of heterogeneous catalysis. The phenomenon of catalysis is so intricately woven into the fabric of chemistry, and the recognition of its nature and importance is so intimately associated with the pioneering endeavours of the founders of modern chemistry and physics – Berzelius, Davy, Faraday, Nernst, Kirchhoff and Ostwald – that it hardly seems necessary to recall what the term 'catalysis' signifies. To be sure that there is no misunderstanding, we define a catalyst as a substance that increases the rate of attainment of chemical equilibrium without itself undergoing chemical change. We shall consider later whether the catalyst suffers any change, transitory or permanent. But it follows from this definition that if the rate of a forward reaction, e.g. hydrogenation, is speeded up in the presence of a particular catalyst, the reverse reaction, dehydrogenation, will likewise be facilitated to the same degree.

Heterogeneous catalysis is perennially relevant, it is endlessly fascinating, and it continues to be deeply enigmatic. More than 90% of the chemical manufacturing processes in use throughout the world utilize catalysts in one form or another: much of the food we eat and the medicines we take, many of the fabrics and building materials that keep up warm, and almost all the fuels that transport us by road, sea or air are produced by heterogeneously catalysed reactions. The current world production of ammonia, required principally as an agricultural fertilizer, is *ca* 140 million tons per annum. Owing to the expanding need to feed mankind, this figure is increasing at some 3% per annum. The science and technology of catalysis are therefore of central practical importance. However, when we recall that, until recently, the majority of commercially significant catalysts were discovered and developed principally by empirical methods, we appreciate how much more remains to be learned about the principles and manifestations of catalysis. Rational design of certain kinds of catalyst and chemical engineering process, thanks to very recent advances, is now a reality.

But it is not the industrial scientist alone who responds to the challenge of catalysis: the academic is also profoundly aroused. How is it that molecules impinging upon certain (catalytic) surfaces at velocities of typically 1600 km h^{-1} can be converted at that surface, with high efficiency and often with spectacular selectivity, into a desired product, whereas the same species impinging upon other (inert) surfaces merely rebound with more or less retention of translational, vibrational, and rotational energy? This is one of the key questions we endeavour to answer in this book. In doing so, we invoke many other considerations, which span thermodynamics, kinetics, and theories of bonding and crystal structure, and draw upon the panoply of techniques that have deepened our understanding of electron transfer and atomic transformations at solid surfaces. There is no single, all-embracing theory of heterogeneous catalysis in the sense that there are theories of evolution, relativity or electromagnetism. There is, however, a corpus of principles which serves as an interpretive and predictive framework for coping with heterogeneous processes, especially those that occur at the gas–solid interface.

1.1.1 The Selectivity of Catalysts

A good catalyst must possess both high activity and long-term stability. But its single most important attribute is its selectivity, S, which reflects its ability to direct conversion of the reactant(s) along one specific pathway.

When a reactant A can be transformed to either B or C at rates R_1 and R_2 (Scheme 1.1), respectively, the selectivity S is calculated as shown in scheme 1.1. It is also sometimes convenient, when R_3 is zero, to define another index of selectivity, $p = R_1/R_2$.

Over a silver catalyst, ethylene (ethene) is selectively converted to ethylene oxide – an important precursor in the manufacture of ethylene glycol which, in turn, is required for the production of polyester fibres and antifreeze agents – in preference to the other two possibilities represented in Scheme 1.2. Finely dispersed platinum, as in an auto-exhaust catalyst, selectively favours the total combustion of ethylene. The oxidation of ethylene to acetaldehyde is best performed homogeneously in the presence of aqueous palladium chloride and copper ions, this being the essence of the industrial Wacker process.

Depending upon the solid catalyst employed, one or other of the products shown in Scheme 1.3 can be selectively generated from propylene (propene). From a mixture of carbon monoxide and hydrogen, also known as 'synthesis gas' or 'syn-gas',

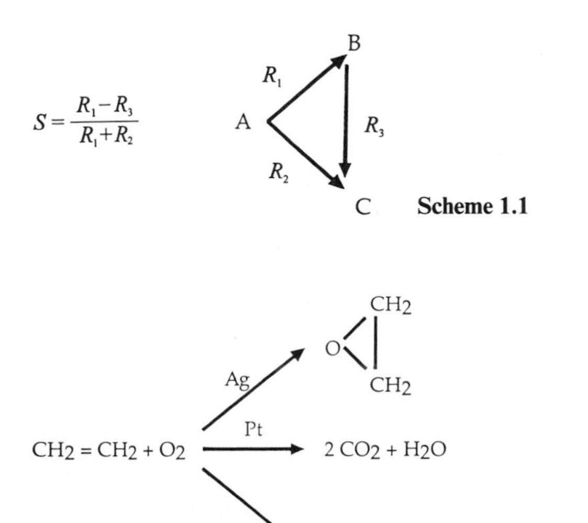

Scheme 1.1

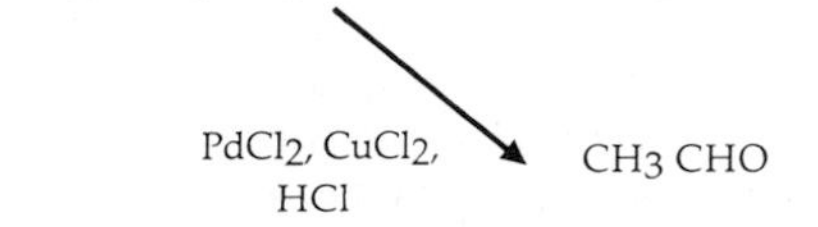

Scheme 1.2

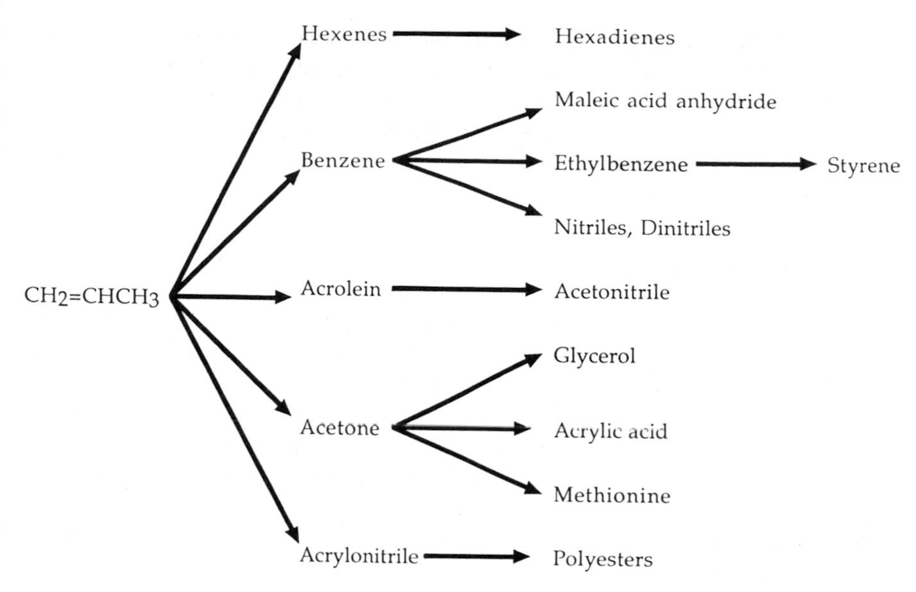

Scheme 1.3

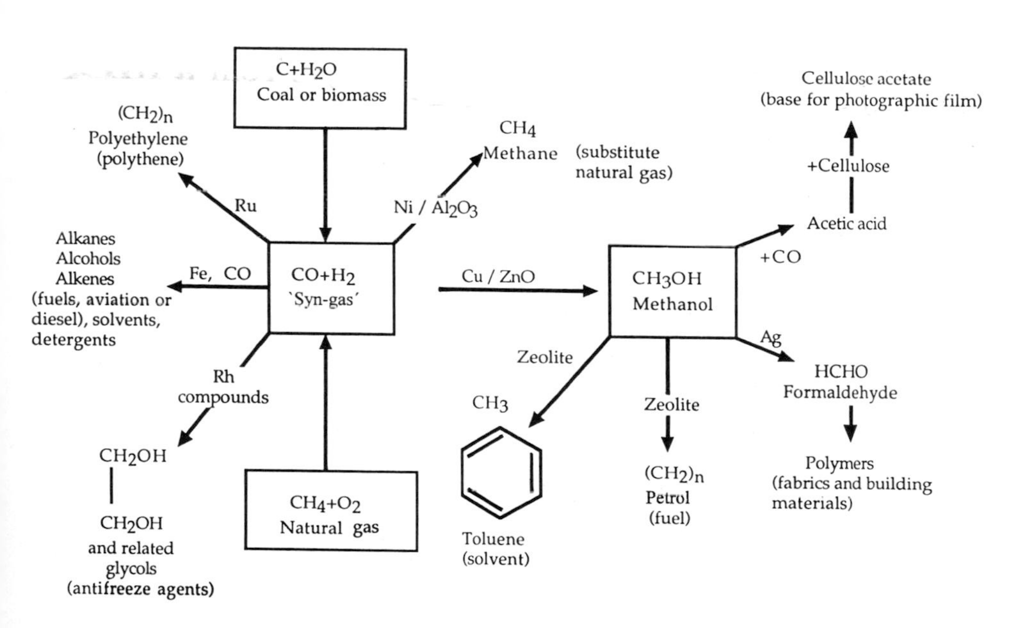

Scheme 1.4 Synthesis gas (abbreviated 'syn-gas') is a mixture of carbon monoxide and hydrogen and may be produced either by oxidising coal with steam, or by burning methane. A wide diversity of products can be generated from 'syn-gas', using the appropriate catalyst, as shown here.

produced either by the older method of gasification of coal by steam ($C + H_2O \rightarrow CO + H_2$) or by the newer method of partial oxidation and 'steam reforming' of methane from natural gas ($2\,CH_4 + O_2 \rightleftharpoons 2\,CO + 4\,H_2$; and $CH_4 + H_2O \rightleftharpoons CO + 3\,H_2$), many commercially important products can be prepared with high efficiency (see Scheme 1.4).

1.2 Perspectives in Catalysis: Past, Present, Future

Berzelius, who in 1836 first introduced the terms 'catalysis' (meaning 'loosening down' in Greek), endowed catalysts with some mysterious quality – he talked of a recondite catalytic force – and only in comparatively recent times has the aura of the occult been finally exorcized from discussions of the subject. The phenomenon of catalysis has been extensively studied since the early decades of the 19th century, and used unconsciously for a much longer period: it may not, like Melchizedek, have existed from eternity, but it was certainly harnessed by the ancients in the pursuit of some of their primitive arts.

In 1814, Kirchhoff noted the catalysed hydrolysis of starch by acids, a classic example of homogeneous catalysis, Then in 1817 Humphry Davy discovered that the introduction of hot platinum into a mixture of air and coal gas led to the metal becoming white hot. In 1824, Henry reported the first example of poisoning of a catalyst: ethylene inhibited the reaction between hydrogen and oxygen on platinum. He also noted selective oxidation in the reaction between oxygen and a mixture of hydrogen, carbon monoxide and methane. At about that time Döbereiner introduced his remarkable 'tinderbox' (*Feuerzeug*), which consisted of a miniature generator of hydrogen and spongy platinum. The box, which was commercialized for the purpose of lighting fires and smoking pipes, contained a small Kipp's apparatus with zinc and dilute sulphuric acid. A jet of hydrogen produced by this means was directed on to the supported platinum where it catalytically combined with oxygen to yield a gentle flame. It has been said that over a million tinderboxes were sold in the 1820s.

In 1834, Michael Faraday examined the power of platinum plate to effect the recombination of gaseous hydrogen and oxygen, which he had produced by the electrolysis of water. Grove in 1845 demonstrated that a hot platinum filament was equally good as a catalyst for the decomposition of water vapour into hydrogen and oxygen. Like Henry, Faraday observed that certain other gases, notably carbon monoxide and ethylene, suppressed the catalytic activity of the metal. A few decades later, another English scientist, Phillips, patented the use of platinum for oxidizing sulphur dioxide to sulphur trioxide with air. He also noted that the loss of catalytic activity arose because the surface of the metal was poisoned by other reactants. In 1871, an industrial process (the Deacon process) for the oxidation of hydrochloric acid to chlorine (used for making bleaching powder) was developed. The catalyst was a clay brick impregnated with cupric salts.

A few years later (in 1877), Lemoine demonstrated that the decomposition of hydriodic acid to hydrogen and iodine reached the same equilibrium point (19%) at 350 °C, irrespective of whether the reaction was carried out homogeneously and slowly in the gas phase or rapidly and heterogeneously in the presence of platinum sponge. Bertholet, in 1879, working on the acid-catalysed esterification of organic acids and hydrolysis of esters, confirmed that the catalyst did not influence the position of equilibrium, a key observation in our understanding of the nature of catalysis.

The introduction of catalytic hydrogenation is rightly associated with the work of the French chemist Paul Sabatier, whose insight into the nature of surface phenomena was exceptional. A German contemporary, Wilhelm Normann, also played a crucial role in establishing catalytic hydrogenation, using finely divided nickel powder, as a means of converting oils, fats, and waxes into edible foodstuffs and other products. Normann, in 1901, described how he converted oleic acid (*cis*-9-octadecenoic acid, $C_{17}H_{33}COOH$), a liquid, into its saturated analogue, stearic acid ($C_{17}H_{35}COOH$), which is crystalline. Reactions of this kind carry an echo of the remarkable accomplishment of Hipployte Mège-Mouries, the Frenchman who invented margarine (in 1869 he won a prize offered by Napoleon III for a satisfactory substitute for butter). Catalytic hydrogenation of oils and fats is still of major importance for the production of foodstuffs, vitamins, medicines, soap, perfumery, paints, varnishes, lubricants, polishes, candles, and chocolate. Remarkably, metallic nickel remains the preferred catalyst in these hydrogenations.

A landmark in the history of applied catalysis was 2 July 1909. On that day in Karlsruhe Fritz Haber succeeded in preparing copious quantities of ammonia from nitrogen and hydrogen in the presence of a reduced magnetite (Fe_3O_4) catalyst using a high-pressure apparatus. This catalyst was to be perfected by Bosch and Mittasch at the laboratories of Badische Anilin & Soda Fabrik (BASF) in Oppau, Germany, in the years preceding World War I (see Section 8.3). But already, in 1903, Ostwald had shown that ammonia could be catalytically oxidized over a platinum gauze to yield oxides of nitrogen which, in turn, were converted to nitric acid. The first synthetic methanol plant, also utilizing high-pressure reactors, was commissioned by BASF in 1923. The process operated at *ca* 400 °C and 200 bar using a zinc oxide–chromium oxide catalyst (see Section 8.1). Shortly afterwards, also in Germany, the Fischer–Tropsch process, which converts syn-gas to hydrocarbons and alcohols, became operational using cobalt or iron catalysts (see Section 8.2). In due course, it became feasible to produce on an industrial scale formaldehyde, phthalic anhydride, and maleic anhydride by the selective oxidation of methanol, naphthalene, and benzene, respectively. By 1937, the Union Carbide company had commercialized the selective, silver-catalysed oxidation of ethylene to ethylene oxide.

In the late 1930s, catalytic cracking, which refers to the rupture of C–C bonds in order to convert large petroleum molecules, such as those that occur in gas oil, into small hydrocarbons of the kind found in fuel, first came into prominence. Alkanes are 'cracked' to give alkenes and smaller alkanes; alkenes yield smaller alkenes; and alkyl aromatics undergo dealkylation (see Section 8.3). For cracking, the most popular catalyst was initially acid-treated clay of the montmorillonite type (see Section 8.4), although many years earlier Friedel–Crafts catalysts, consisting of aluminium trichloride ($AlCl_3$), had been used for this purpose. As a consequence of ingenious chemical engineering, in particular the optimization and regeneration of deactivated catalysts, Houdry devised fixed-bed catalytic cracking reactors; and from such units came most of the aviation fuel (gasoline) consumed by the Allies in the Battle of Britain. In 1941, fluid-bed or fluidized catalytic cracking (FCC) became a commercial reality, the culmination of revolutionary engineering design work by the American workers Lewis and Gilleland. In FCC units, fine particles of catalyst are maintained in suspension in a stream of vaporized (heavy) hydrocarbon that is blown through the so-called transfer-time reactor and subsequently passed through the regenerator. Thus contact times can readily be adjusted, thereby offering greater control in optimizing product yield. In fluidized reactors, catalyst lifetime depends upon the mechanical or attritional resistance of the fine particles.

Another important development that gained rapid momentum in the 1930s was the work of Ipatieff and Pines in the oligomerisation of gaseous alkenes with 'silicophosphoric' acid, sometimes designated 'solid phosphoric acid' (SPA). This solid acid catalyst is made by taking phosphoric acid, a liquid which is awkward to handle on a commercial scale because of its corrosiveness, and mixing it with kieselguhr (diatomaceous earth) to form a plastic composite calcinable at 200–300 °C. The final acid (composition 60% P_2O_5 and 40% SiO_2) can be readily fashioned into granular or cylindrical pieces of acceptable physical strength at the temperatures required for catalysis (300–500 °C). Iso-octane was produced by oligomerisation of lower olefins, followed by hydrogenation. This became an industrial reality following Ipatieff's work. Ipatieff also discovered paraffin alkylation in which isobutane was reacted with butenes or propylene, the initial catalyst being those first used by F. C. Whitmore for alkene isomerisation: mineral acids.

1.2.1 Applied Catalysis since the 1940s

After World War II, acid-treated clays for catalytic cracking became unpopular because of their lack of long-term stability. They were gradually supplanted by amorphous synthetic silica–alumina catalysts, which were more stable under regeneration conditions: they also gave better product distributions.

Synthetic zeolite catalysts were first reported to be especially active and selective for isomerizing hydrocarbons by Rabo et al. in 1960. Shortly after Milton (USA) and Barrer (UK) had independently shown that zeolites could readily be synthesized, considerable effort was made to explore the reactions they catalyse. We now know that a bewildering variety of reactions are catalysed to a greater or lesser degree by zeolites. But it was Plank and Rosinsky who demonstrated the remarkable performance of synthetic zeolites as cracking catalysts in 1964. Synthetic zeolites were later used also for hydrocracking, and for shape-selective conversions, thanks to the pioneering work of Weisz et al. at the Mobil Company Laboratories. In shape-selectivity, advantage is taken of the convenient fact that the intracrystalline space available to reactant and product molecules has dimensions comparable with those of the molecules themselves. Highly branched hydrocarbons cannot, therefore, enter the internal volume of the zeolite, where most of the 'surface' area resides and where the active sites (see Fig. 1.1, Colour Plate 1, Section 8.8) are situated. Monobranched or linear molecules like the *n*-alkanes, on the other hand, can readily diffuse into zeolite catalysts, which, consequently, can be used in selective oxidation of linear hydrocarbons without converting the branched or aromatic ones. Since 1974,

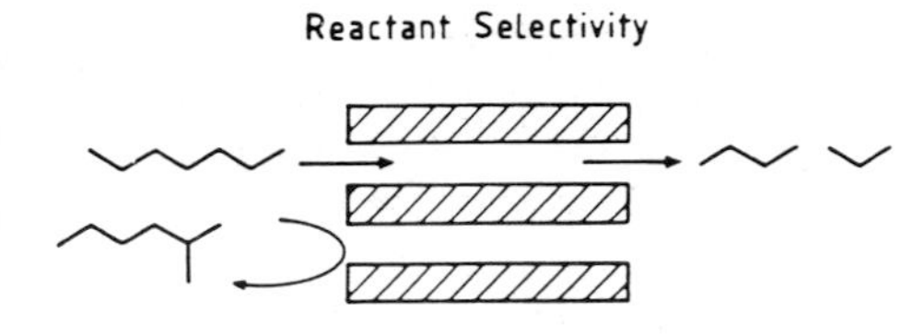

Figure 1.1 A shape-selective catalyst, such as a synthetic zeolite with cylindrical pores of *ca* 5.5 Å diameter, permits ready ingress of straight-chain reactant alkanes (e.g. *n*-heptane) but not of branched isomers (e.g. 2-methylhexane). The acid centres lining the pores can therefore catalyse the cracking of the *n*-heptane but not of the branched hexane. For similar reasons, in the acid-catalysed disproportionation of toluene to xylene and benzene, production of the *p*-xylene is favoured (see Colour Plate 1).

a series of novel catalytic processes based on the unique properties of a synthetic zeolite called ZSM-5 has been introduced. This catalyst has pore openings of *ca* 5.5 Å (see Fig. 1.2) and exhibits shape-selectivity as well as acid activity, an unusual resistance to coking, and freedom from poisoning. ZSM-5 is nowadays the catalyst used to convert *m*ethanol *t*o petrol (*g*asoline) – the MTG process – and to effect xylene isomerization (so as to maximize production of *p*-xylene) and a number of other key industrial processes.

Catalytic reforming, in particular naphtha reforming, which entails the isomerization of alkanes, the dehydrogenation of cyclohexanes, the dehydroisomerization of methylcyclopentanes, as well as the aromatization of some alkanes and the hydrocracking of other hydrocarbons, serves to enhance the octane number of a fuel. In

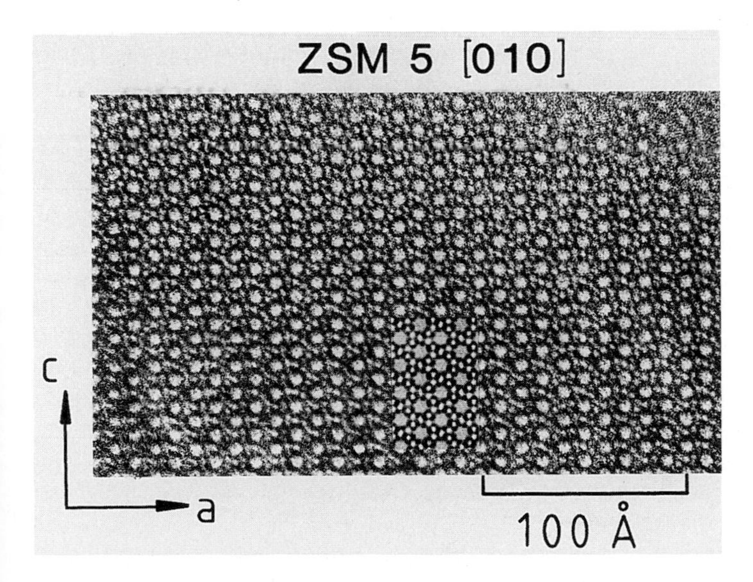

Figure 1.2 This high-resolution electron micrograph shows up the aperture openings (large white spots) of 5.5 Å diameter present in the shape-selective zeolitic catalyst, ZSM-5. (The rectangular inset shows the calculated image.)

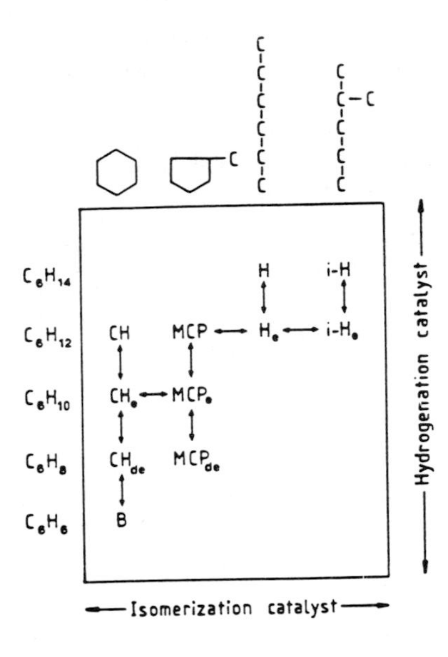

Figure 1.3 In a dual-function catalyst, such as finely dispersed platinum on an acidic Al_2O_3 support, the role of the acid support is principally to facilitate isomerization (e.g., methylcyclopentane (MCP) from *n*- or isohexane (H_e or i-H_e)) whereas the metal facilitates the dehydrogenation (of cyclohexane, CH, to cyclohexadiene, CH_{de}, to benzene, B). With permission from H. Heinemann, G. A. Mills, J. B. Hattmann, F. W. Kirsch, *Ind. Eng. Chem.* **1953**, *45*, 130.

1949, a new generation of reforming catalysts was introduced by Haensel at the United Oil Products Company. Such catalysts consist of finely dispersed platinum on an acidic support, generally γ-alumina (γ-Al_2O_3). The dual-function character of these reforming catalysts, possessing hydrogenative–dehydrogenative capabilities as well as acidic properties, was appreciated by Mills et al., whose summarizing interpretation of the essence of the reforming (or 'platforming') process is shown in Fig. 1.3. There is no doubt that alkenes play an important role here. In the period 1967–1971, two major improvements in reforming catalysts, one from the Chevron Oil Company, the other from the Exxon Company, were announced. The first used platinum–rhenium bimetallic particles as the catalyst, the second platinum–iridium. The improvement results because the second element (e.g., iridium) is much more active than platinum in effecting the hydrogenolysis (rupture of C–C bonds) of carbonaceous residues that tend to accumulate on the catalyst and poison its surface.

Hydrodesulphurization and, more recently, hydrodenitrification, which consists of removing the organic sulphur- and nitrogen-containing components from crude oil or from the products of cracking (such as hydrogen sulphide or ammonia, respectively) have become progressively more important catalytic reactions since about the early 1960s because of the increasing supply of 'high-sulphur' and 'high-nitrogen' crude oils, which tend to poison the catalysts that effect reforming, and the pressing environmental need to eliminate sulphur- and nitrogen-containing pollutants from the atmosphere. The most common catalysts are Co/MoS_2 and Ni/WS_2, which are often prepared on alumina supports.

Production of hydrogen, the preliminary step in so many major commercially significant catalytic systems, typified by the synthesis of ammonia and methanol and by the Fischer–Tropsch process, has undergone many changes since the early 1960s. Hydrogen currently comes chiefly from two sources: from naphtha (hydrocarbon)

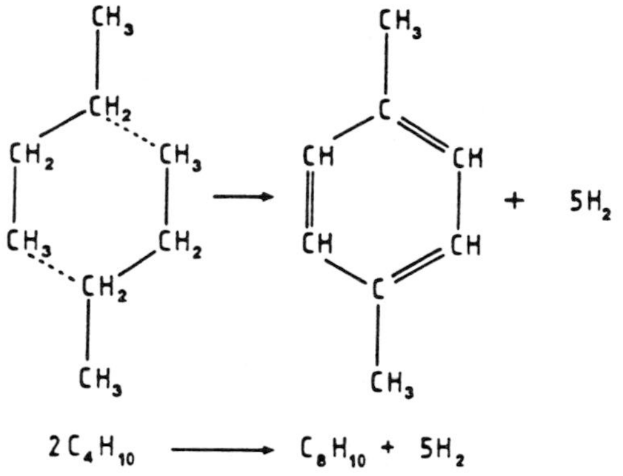

$$2C_4H_{10} \longrightarrow C_8H_{10} + 5H_2$$

Scheme 1.5 A recently developed catalyst converts *n*-butane, present in natural gas, to xylene and hydrogen, both of which are desirable products.

reforming mentioned above, and from the steam–hydrocarbon reaction. The second of these is also called steam-reforming; and 'syn-gas' ($CO + H_2$ mixture) is often produced by the steam-reforming reaction of natural gas (chiefly methane, CH_4). An important technical advance in the industrial production of hydrogen, and hence in the emergence of a new generation of ammonia plants, hinged upon the fact that the carbon–steam reaction ($C + H_2O \rightleftharpoons CO + H_2$) is greatly accelerated by the presence of alkali or alkaline earth elements added to the nickel catalyst normally used for this conversion. Much hydrogen in future could well come from natural gas and from propane and butane, especially in those parts of the world where liquefied petroleum gas (LPG) is plentiful. A high-activity zeolite catalyst recently developed by British Petroleum and United Oil Products converts propane and butane to monocyclic aromatics plus hydrogen (Scheme 1.5): shape-selective zeolites impregnated with tellurium or other elements are good for this purpose.

Other important large-scale catalysed reactions that gained prominence in the mid-1960s included dehydrogenation of butane to butenes and/or butadienes, and of ethylbenzene to styrene monomer:

$$CH_3CH_2CH_2CH_3 \longrightarrow CH_3CH_2CH{=}CH_2 \longrightarrow CH_2{=}CHCH{=}CH_2$$

$$C_6H_5CH_2CH_3 \longrightarrow C_6H_5CH{=}CH_2$$

Both of these conversions are oxidative dehydrogenations, and are often quoted as examples of selective oxidation, which has an important commercial manifestation in the conversion of propylene to acrolein over a bismuth molybdate catalyst. Closely related to the latter is the production of acrylonitrile, $CH_2{=}CHCN$, from a mixture of propylene, ammonia and air, in the so-called 'ammoxidation' process of the Sohio Company. Multicomponent catalysts, containing iron, potassium, and manganese and other additives to bismuth molybdate, are used for this purpose.

Disproportionation and polymerization of alkenes have also assumed industrial significance since the 1960s. Propylene, for example, can be converted to ethylene and butene in the renowned triolefin process associated with the Phillips Company, a typical metathesis reaction using $Mo(CO)_6$ or $W(CO)_6$ supported on alumina:

```
    H       CH3          H         H    H         CH3
     \     /              \       /      \       /
2     C=C       ---->      C=C        +    C=C
     /     \              /       \      /       \
    H       H            H         H   CH3         H
```

The best examples of polymerization are Ziegler–Natta conversions, which permit the production of polyethylene and crystalline stereoregular polypropylene from the respective monomers. This is achieved using a mixture of aluminium trialkyls and, as the key component of the Ziegler–Natta catalyst, titanium (III) chloride ($TiCl_3$). Nowadays, $TiCl_3$ supported on solid magnesium chloride ($MgCl_2$) is the catalyst of choice.

1.2.2 Current Trends in Applied Catalysis

Since the mid-1980s, several other major themes have emerged in applied catalysis. Some of these reflect the growing commitment to protecting the natural environment, others represent the logical extension of pure research, whilst still others have arisen from a desire to produce foodstuffs and other useful products from precursors that are plentiful.

1.2.2.1 Auto-Exhaust Catalysts

In developing effective automobile catalytic converters, the fundamental question is how one may transform undesirable species such as carbon monoxide (CO), nitric oxide (NO) and small hydrocarbons (Fig. 1.4) into harmless products such as carbon dioxide, nitrogen, and water? Effective catalysts consist of fine platinum–rhodium bimetallic particles supported on high-area ceric oxide–alumina (CeO_2/Al_2O_3) mixed

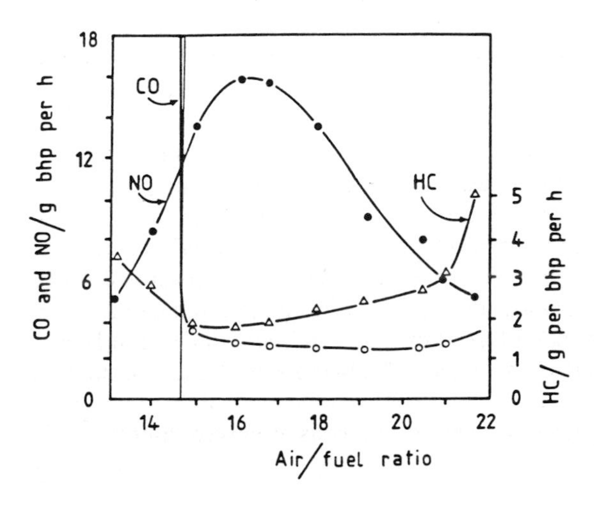

Figure 1.4 The effects of air/fuel ratio on the emission in auto-exhausts of hydrocarbon (HC), NO, and CO. With permission from L. L. Hegedus, J. J. Gumbleton, *Chemtech.* **1980**, 630.

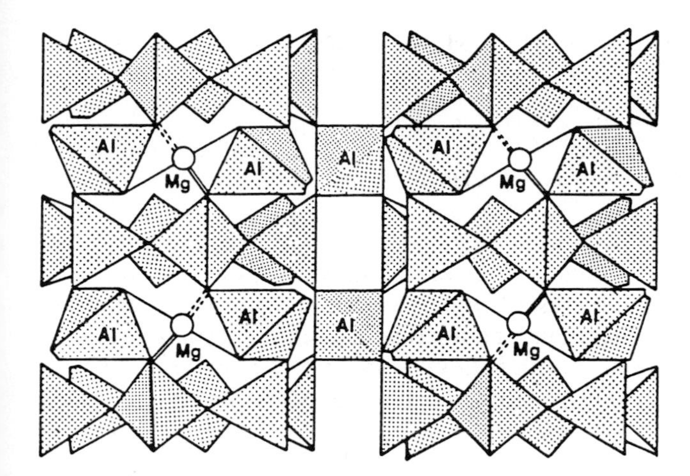

Figure 1.5 Thermal and mechanical stability are important properties required in support material for the metals used (rhodium and platinum) in auto-exhaust catalysts. Synthetic cordierite ($Mg_2Al_4Si_5O_{18}$) has a structure closely akin to that of the very strong mineral and gemstone beryl. The interpenetrating rings and chains of corner-sharing tetrahedra (AlO_4 and SiO_4) confer these qualities upon the cordierite.

oxides which, in turn, are distributed on to a monolith of cordierite ($Mg_2Si_5Al_4O_{18}$), an aluminosilicate which, because of its structure (see Fig. 1.5), stands up well to thermal shock. Platinum alone is effective, since it coadsorbs CO and atomic oxygen; but rhodium as a 'promoter' improves the uptake of NO, thereby facilitating the conversion

$$CO\,(ad) + NO\,(ad) \longrightarrow N_2\,(g) + CO_2\,(g)$$

The γ-Al_2O_3 maintains thc high arca of the metal particles which it supports, but the added presence of CeO_2 is an extra advantage. It not only stabilizes the Al_2O_3 on to the cordierite, but, in view of its solid-state properties, it possesses the twin advantages of releasing and mopping up oxygen under reducing and oxidizing conditions respectively. (In CeO_x, which has the fluorite structure, x may take the values $1.72 \leqslant x \leqslant 2$, while retaining its skeletal atomic structure.)

NO and NO_2, collectively termed NO_x, lie at the heart of smog, acid rain and the greenhouse effect. For as long as the internal combustion engine, employing high-temperature oxidation, is used the production of NO_x is inevitable. NO_x would not be produced if an effective catalyst for the low-temperature 'cool' (and clean) combustion of hydrocarbon fuel were developed. Alternatively, the development of an efficient electrocatalyst that would enable electricity to be produced directly from fuel combustion – in the modern version of a Grove fuel cell first described in 1839 – would have the extra bonus of a much higher overall efficiency in the use of the fuel for propulsion.

More practical details pertaining to auto-exhaust catalysts are given later (Section 8.6).

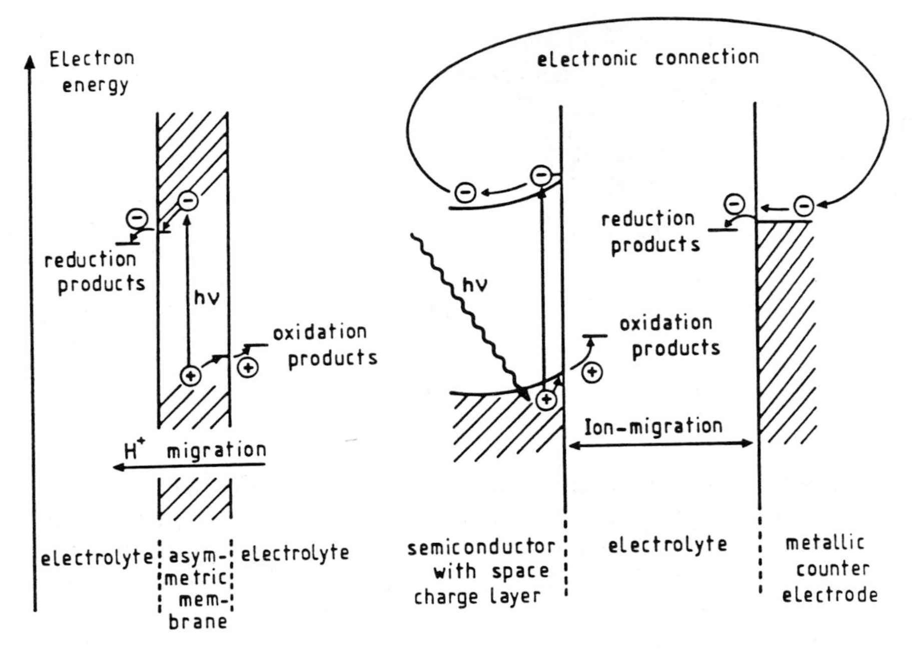

Figure 1.6 Photoelectrolysis (right) and photosynthesis (left) have much in common. In each case light is used to create electrons and holes which then serve to effect reduction and oxidation.

1.2.2.2 Catalysts in Electrochemistry and Photoelectrochemistry

The efficient production of fuels from inexpensive precursors by utilization of solar energy with cheap and stable chemical systems is nowadays the target of much pure and applied research. Desirable reactions are, typically, the reduction of water to hydrogen (H_2) and of carbon dioxide to methanol, each driven by the absorption of light. Since absorption of light creates electron–hole pairs – a 'hole' is simply that which is left behind in an orbital or band of orbitals when an electron is promoted to higher energy – the fuel-producing reaction must be accompanied by an oxidation reaction. Ideally, this oxidation reaction should consume a plentiful material, e.g., water, thereby generating oxygen (O_2) or, alternatively, produce a chemical of commercial value such as chlorine (Cl_2) from Cl^- ions. There is an analogy here with photosynthesis, which employs light absorption to produce vital organic materials and O_2 (see Fig. 1.6). A number of photochemical schemes have been formulated with the aim to harness solar energy. To be effective, it is necessary to engineer solids with band gaps, i.e. energy separations of the highest filled and lowest unfilled orbitals (see Section 5.9), of around 2 eV, so as to take good advantage of the solar spectrum.

The engineering of semiconductor solids that absorb light leads us, in turn, to design 'dual-function catalysts' of a kind different from those discussed in Section 1.2.2.1, where we considered hydrocarbon reforming. What is meant here is that there should be (as in Fig. 1.6) a semiconductor catalyst possessing the appropriate

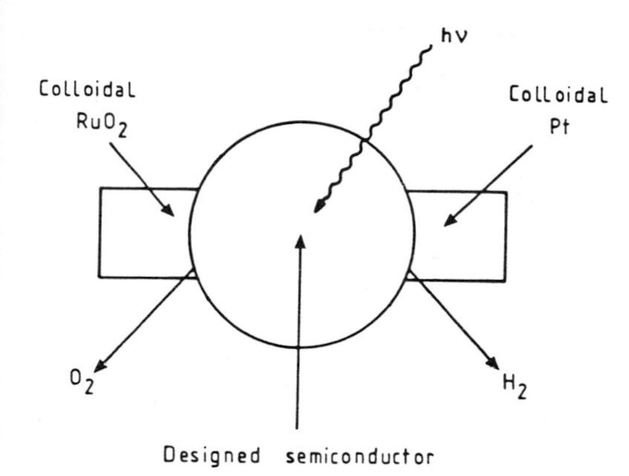

Figure 1.7 It should be feasible to design a semiconductor catalyst microcapsule for the photocleavage of water and other abundant materials (see text).

electronic properties so that electrons and holes can be used to effect reduction at one electrode (say colloidal platinum) and oxidation at another (colloidal ruthenium dioxide, RuO_2) (Fig. 1.7). Extending the ideas of the Swiss worker Grätzel, the entire catalyst 'capsule' or 'microcapsule', in which all three components cohere, would then be so arranged so as to effect continuous photocleavage of water.

The principles summarized in Figs 1.6 and 1.7 are also used in the solar-driven clean-up of environmentally harmful chemical by-products, especially chlorohydrocarbons, using titanium dioxide (TiO_2) as a photocatalyst. Insofar as destruction of C_1 and C_2 chlorohydrocarbons are concerned, however, catalytic combustion with platinum-group metals is still the preferred option.

1.2.2.3 Asymmetric Sites on Heterogeneous Catalysts

A third lively theme in present-day applied catalysis is the quest for solids the surfaces of which can be modified to function enantioselectively, just as certain asymmetric homogeneous catalysts do. An example is the asymmetric hydrogenation of α-(acetamino)acrylic acid in which, depending upon the enantioselectivity of the optically active catalyst used, different proportions of the enantiomers of acetylalanine are formed:

$$\mathrm{(H)(H)C{=}C(NHAc)(COOH)} + H_2 \longrightarrow \mathrm{(H)(H)(H)C{-}C(NHAc)(H)(COOH)}$$

One of the major triumphs of modern homogeneous catalysis is the Monsanto synthesis of the L-isomer of 3,4-dihydroxyphenylalanine, otherwise known as L-dopa, the drug used in the treatment of Parkinson's disease. The key feature for selective synthesis of one optical isomer of a chiral molecule in preference to another is an

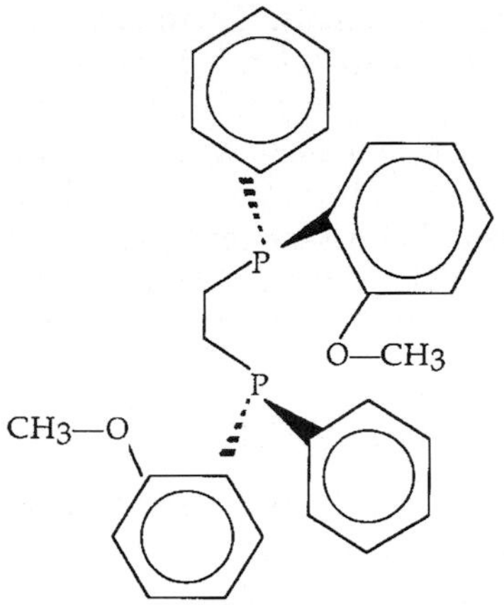

(**1**) DiPAMP

asymmetric catalyst site that will anchor a prochiral olefin or ketone in one conformation. This 'recognition' of the preferred conformer can be accomplished nicely in homogeneous metal-complex catalysts by the use of an appropriately tailored chiral ligand. Thus, in the L-dopa synthesis a 'chiral hole' is in effect created within the coordination sphere of the rhodium atom in a variant of the well-known Wilkinson catalysts, chlorotris (triphenylphosphine) rhodium, $Rh(PPh)_3Cl$. The triphenylphosphine is replaced by an asymmetric ligand such as the chelating diphosphine (consisting of a *d*imer of *p*henyl *o*-anisyl *m*ethyl *p*hosphine (PAMP), known as DiPAMP (**1**). The prochiral olefin that is to be hydrogenated in a stereoselective fashion to produce L-dopa is a substituted cinnamic acid (**2**), which coordinates to the rhodium atom both through its olefinic bond and through the carbonyl group of the acetamide function. Situated in this rigid fashion at the rhodium atom in the chiral hole circumscribed by the DiPAMP, only one 'face' of the olefin is accessible to the hydrogen in the crucial catalytic act.

Homogeneous catalysts, effective as they are for hydrogenating double bonds, are not yet demonstrated to be as versatile as their heterogeneous counterparts for

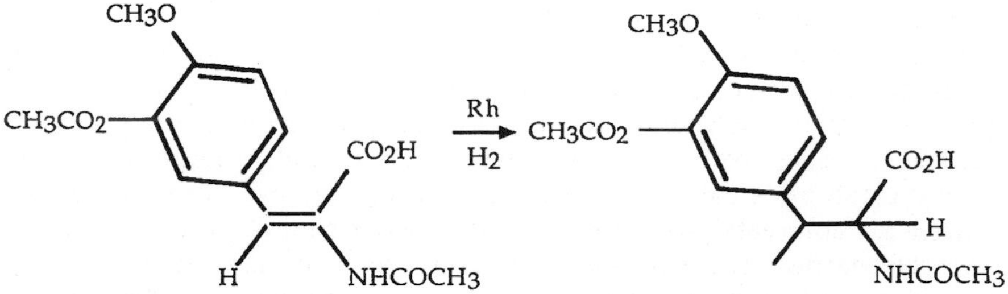

(**2**)

selectively converting heterocycles and other important reactants. There has consequently been renewed interest in early experiments which demonstrated that metal catalysts, after suitable modification of their surfaces by asymmetric reagents, can then yield certain products in enantiomeric excess. In 1956, Japanese workers, using palladium supported on natural silk, obtained optical yields of *ca* 36% in the hydrogenation of oximes and oxazolones. More recently, Izumi and Sachtler have shown that Raney nickel catalysts, when first modified so as to effect corrosive chemisorption (see Chapter 2) of (*R*,*R*)-tartaric acid, can then smoothly hydrogenate methyl acetoacetate to 3-methyl-3-hydroxybutyrate:

$$CH_3{-}CO{-}CH_2COOCH_3 + H_2 \longrightarrow CH_3{-}C(OH)H{-}CH_2COOCH_3$$

American workers using a catalyst similarly modified with a chiral tartaric acid 'template', hydrogenated 2-methylpyridine to optically active D-2-methylpiperidine; and workers at the Zelinskii Institute in Moscow showed that metallic copper, cobalt, and palladium, when modified with chiral amino acids, are also good catalysts for a number of enantioselective hydrogenations. In effect, the role of the modifier or molecular template is to generate at the metal surface so-called enantioselective sites, at which the transition states for the two possible configurations of the half-hydrogenated state possess significantly different free energies of activation.

The scope for tailoring metal catalysts by pre-adsorbing upon their surfaces bulky organic molecules which create prochiral cavities for the incoming reactant is obviously considerable. Thus platinum, modified by the presence of presorbed cinchonidine (a member of the cinchona alkaloids), becomes active for the stereoselective hydrogenation of an α-ketoester (e.g. methyl pyruvate) to an optically active product (e.g. (*R*)-(+)-methyl lactate). One may explore, by computer graphics (see Colour Plate 2) the way in which this stereoselective process might proceed. The plausible assumption here is that cinchonidine forms an ordered, well-spaced-out array of interstices in the sorbed layer. The interstices permit ingress of the methyl pyruvate in a preferred fashion so as to pre-ordain the formation of a given stereo product.

1.2.2.4 Immobilized Transition Metals

Modifying a metal catalyst by first introducing a chiral template is just one manifestation of the much wider practice of surface derivatization (or heterogenizing) that constitutes another major theme in present-day applied catalysis. A universally recognized way of combining the best features of homogeneous and enzyme catalysts on the one hand, and heterogeneous catalysts on the other, is to immobilize the former using an appropriate adsorbent, so that the resulting surface complex rivals or surpasses the performance of an analogous heterogeneous catalyst. Sometimes the adsorbent is an organic polymer, sometimes an inorganic gel or crystalline solid. This expedient is designed to take advantage of the normally high selectivity of homogeneous catalysts, while at the same time ensuring that one of the key advantages of heterogeneous catalysts – relative ease of separation of products from reactants – is safeguarded. There are various ways available for derivatizing surfaces, and in particular for depositing highly dispersed metal atoms, metal clusters, or ions, in an immobilized fashion on silica-rich surfaces. One of the difficulties in this kind of work is finding out the chemical state and degree of dispersion of the tethered catalyst, but great progress has recently been made in this regard using some of the powerful new techniques of catalyst characterization, such as solid-state nuclear magnetic resonance (NMR) and extended X-ray absorption fine structure (EXAFS) which are described in Chapter 3. There seems little doubt, for example, that the silanol groups on a silica surface can be utilized to tether single metal atoms (M = Zr, Hf, Nb, Cr, Mo, W, Re, Ni, Pd, Pt, Rh) by the procedure:

$$[(O)_3Si-O]\begin{matrix}-OH\\-OH\end{matrix} + M(allyl)_x \longrightarrow [(O)_3Si-O]\begin{matrix}-O\\-O\end{matrix}\!\!>M(allyl)_{x-2} + CH_3CH{=}CH_2$$

Some of the metals laid down on high-area siliceous surfaces in this fashion function as good catalysts in the conversion of synthesis gas, others in the hydrogenation of toluene. Anchored osmium clusters prepared from $Os_3(CO)_{12}$, $Os_6(CO)_{18}$, and $H_2Os_{10}(C(CO)_{24})$ on silica, alumina and titania are catalytically active in the hydrogenolysis of ethane. New generations of anchored catalysts, in which a transition-metal complex is tethered via phosphido linkages or σ-bonds between the metal and a phosphinated polymer, montmorillonite clay or mesoporous silica are under investigation. So too are catalysts in which enzymes are bound to high-area solids so that advantage can be taken both of efficient separation of products from reactants and of the full scope of the exceptional selectivity of biological catalysts.

1.2.2.5 Immobilized Enzymes and Cells: Present and Future

Several tens of thousands enzymes are used for life on earth, but no more than about 3000 enzymes have been identified. Several hundreds have been isolated, purified, and characterized; but the structures of only some 300 are known in atomic detail. They effect, under the mildest of conditions, highly specific reactions at rates that far exceed those typical of ordinary catalysis (homogeneous or heterogeneous). The idea of intentionally immobilizing an enzyme on the accessible surface of a high-area carbon support was first demonstrated in the 1920s by Nelson and Griffin. Their strategy was later imaginatively exploited by Katchalkski and his colleagues at the Weizmann Institute in Israel. Nowadays polymeric supports (Fig. 1.8) are favoured: but solids such as pillared clays might prove even more effective, since they are more rugged.

The activity of an enzyme covalently bonded to its polymeric conjugate (the support) can vary from zero to a value even greater than that of the native enzyme. One cannot, in general, predict the activity of the bound enzyme but it is usually lower than that of the dispersed parent. The first commercial process utilizing an immobilized enzyme catalyst was the resolution of amino acids with an aminoacylase adsorbed on an anion-exchange resin. Resolution of racemic mixtures was effected

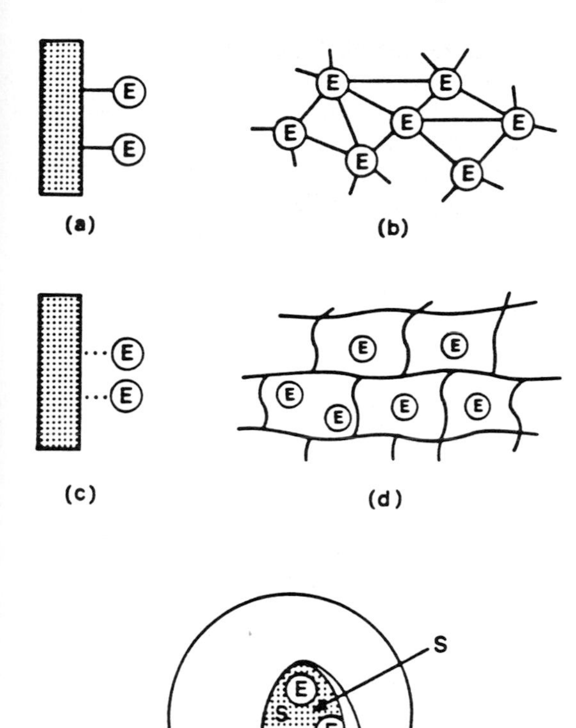

Figure 1.8 Schematic representations of immobilized enzyme systems. E, S and P stand for enzyme, substrate (i.e. reactant) and product molecules respectively. (a) Covalently bonded enzyme–polymer conjugate; (b) covalently bonded intermolecularly crosslinked enzyme conjugate; (c) adsorbed enzyme–polymer conjugate; (d) polymer lattice-entrapped enzyme conjugate; and (e) microencapsulated enzyme. With permission from O. R. Zaborsky, *Methods in Enzymology* (Ed. K. Mosbach). Academic Press, New York 1976, p. 317.

by the hydrolysis of *N*-acetyl-DL-amino acids by the immobilized enzyme in a packed-bed reactor; this was accomplished in Japan by the Tanabe Seiyaku Company. The enzyme amidase, which cleaves penicillin G or V to the penicillin nucleus, has also been successfully immobilized using techniques evolved by the pharmaceutical companies Beecham (in the UK) and Bayer (in Germany), in association with Lilly and Dunnill at University College, London. Another important process is the isomerization of glucose to fructose with the immobilized enzyme, glucose isomerase. The scale of this process is reflected by the fact that in the mid-1980s over a billion (10^9) kilograms of fructose were produced (for the soft drink market) from corn syrup in the USA alone (see Table 1.1)

Immobilized enzymes also play a central role in modern-day sensor technology. This again is an example of heterogeneous catalysis in which biological processes and electrochemical principles merge. Several automated systems for assaying enzymes in blood or other fluids rely on immobilized enzymes. The Enzymax Analyser (Leeds and Northrup Co.) uses an immobilized glucose oxidase to convert the glu-

Table 1.1 Some current industrial-scale applications of immobilized biocatalysts.

Application	Enzyme	Microorganism
Optical resolution of α-amino acids	Amino acylase	*Aspergillus oryzae*
Glucose isomerisation	Xylose (glucose) isomerase	*Actinoplanes sp.*
Starch saccharification	Glucoamylase	*Aspergillus niger*
Steroid conversions	11-β-Hydroxylase	*Curvilarice lunata*
6-Aminopenicillanic acid from benzylpenicillin	Penicillin acylase	*Escherichia coli*

cose to δ-gluconolactone and hydrogen peroxide (H_2O_2), the latter being quantitatively monitored by an electrochemical probe.

Immobilized enzymes are also prominent for their role in correcting metabolic disorders. Urease, for example, in its immobilized state is the centrepiece of an artificial kidney. The blood to be processed is fed in at the top (Fig. 1.9). Urease converts urea to CO_2 and NH_4^+ ions, and the latter, along with other contaminants, such as creatine and uric acid, are adsorbed on the adsorbents. The CO_2 is eliminated through the lungs.

Major possible future applications of immobilized enzymes include the conversion of lignin to useful organic products, the modification of non-edible proteins to edible foodstuffs, the synthesis of ammonia from molecular nitrogen by nitrogenase (or some appropriatelty genetically modified enzyme), and the synthesis of important medicinal and pharmaceutical compounds. It is timely to reflect that, armed with the added refinement made available by chemical modification of recombinant DNA (which, in turn, leads to the production of 'new' enzymes) and hence our ability, in principle, to modify the active sites of any enzyme by site-directed mutagenesis, one can foresee an enormous expansion in the use of immobilized tailored enzymes for effecting highly specific reactions in near-ambient conditions. It seems reasonable to suppose that microporous microcrystalline solids of the kind that exhibit shape-selectivity (see Figs. 1.1 and 1.2 and Colour Plate 1), can profitably be used as the support for the enzyme. The apertures of zeolites (see Section 8.3) are probably too small to allow access to the enzyme before the act of immobilization. But the new family of mesopore structures synthesized in 1992 by Beck and co-workers in the Mobil Company and the two-dimensional zeolites generated by pillaring clays could be fashioned for the particular enzyme and reactants that are under consideration.

In the final analysis, the design of optimal systems involving immobilized enzymes will merge with the topic of designing artificial enzymes *de novo* using, for example, cyclodextrins, as has been done by Bender, Breslow, Tabushi, and others.

Among the current industrial processes using immobilized cells, that for the production of fructose-enriched glucose is a major one. ICI (now Zeneca) plc uses *Anthrobacter* cells flocculated by polyelectrolyte. Each kilogram of catalyst converts more than 2000 kg of glucose syrup. The best-known industrial immobilized-cell process is that of the Tanabe–Seiyaku company for the stereoselective conversion of fumaric acid to L-aspartic acid. It employes *Escherichia coli* cells trapped in poly-

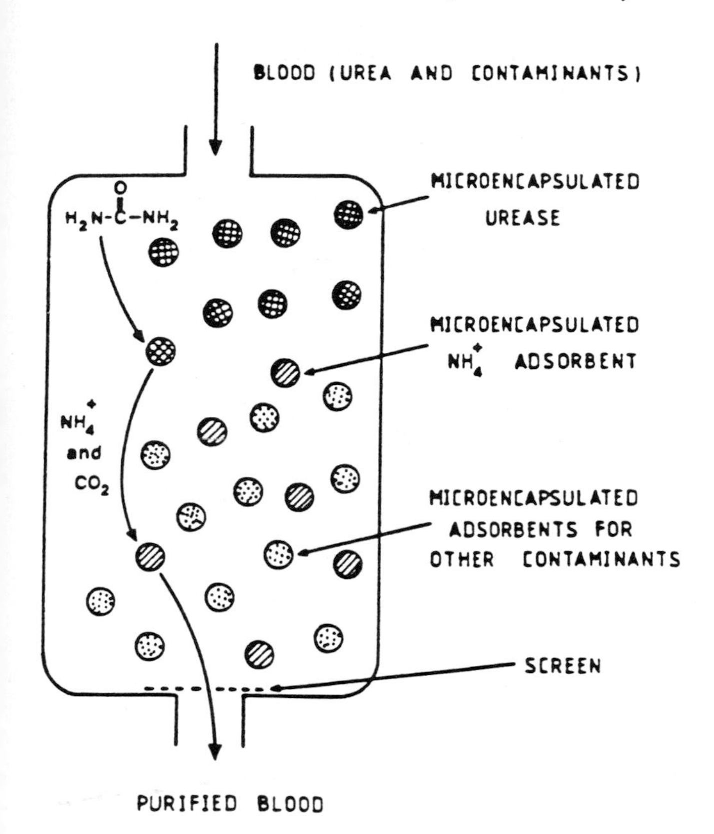

Figure 1.9 Artificial kidney based on the use of an immobilized enzyme catalyst, urease. This extracorporeal 'shunt' contains microencapsulated enzyme and adsorbents. With permission from O.R. Zaborsky, *Methods in Enzymology* (Ed. K. Mosbach). Academic Press, New York 1976, p. 317.

acrylamide and subsequently treated so as to destroy the normal cell permeability barrier. The half-life of immobilized cell catalysts is 120 days, and the cost of production of aspartic acid by this method has been estimated to be 40% less than that by the conventional process employing freely suspended cells.

It is increasingly appreciated that immobilized enzyme or cell catalysis need not be restricted to water-soluble materials. For example, the interesterification of fats in organic solvents and the dehydrogenation of steroids in mixed aqueous/organic solvents have both been achieved by means of immobilized enzymes.

With the growing range of highly selective organic conversions that may now be effected via biocatalysts (enzymes or cells) in aqueous or non-aqueous media there is a pressing need to evolve a rationale so as to evaluate the characteristics of the biotransformation and thence to deduce the constraints imposed by these on the reaction medium, by the resulting kinetics and by the properties of the catalyst. Lilly and co-workers in London have indeed constructed an intelligent knowledge-based system (IKBS) for reactions involving poorly water-soluble reactants. The value of the resulting IKBS is twofold. First, on the basis of the generic rules and experience gained already, one may eliminate many possibilities at an early stage so that a few key selections may then be considered in more detail. Second, it highlights very early what information is missing and needs to be determined before a rational choice can be made. In both the food and pharmaceutical industries, where biocatalysis is of central importance, this is a great step forward as it introduces rational design in place of case-by-case study as the platform for future advance.

1.2.2.6 Catalytic Antibodies

Nature has the remarkable property, manifest in the immunoglobulin protein of the immune system, of creating spontaneously within the folds of the polypeptides (the immunoglobulin antibodies) active sites (cavities) to recognize and bind a vast array of chemical structures that are exhibited by foreign invaders, the antigens. It is thought that nature utilizes no more than about 30–40 thousand enzymes to effect all the bewildering range of coupled processes responsible for the teeming manifestations of life. So far as practical applications, in a biotechnological sense, are concerned, the limitation of enzymatic catalysis is that there are so few enzymes compared with the enormous numbers of scientifically important chemical reactions which one seeks to catalyse. Antibodies, like enzymes, are proteins. Their structural features complement, in a literal sense, those of the antigenic sites of intruder species. They have the ability to recognize foreign substances and are of breathtaking diversity. The immune system, which is where antibodies are generated, is able to make over 100 million different kinds of antibodies.

Both antibodies and enzymes exert their influence in much the same way: by binding to other molecules. The question therefore naturally arises: can one capitalize upon the similarity of antibodies and enzymes (as molecular recognizers) to make the antibodies act as biocatalysts? If so, the door is opened to an almost limitless range of new catalysts. But antibodies do not, in their normal role, catalyse chemical transformations. The way in which they have been recently 'cultured' to function as catalysts is by feeding the immune system with a stable structural analogue of the transition state through which a particular reactant molecule must pass to be transformed. The antibody will then bind that chosen 'starter' molecule – the 'hapten' – so as to produce the appropriate disposition of catalytic groups within its created cavity. Effectively, this enables the free energy with which antibodies bind to their targets to be harnessed in the catalytic process. This kind of 'abzyme' chemistry – the term is a hybrid of *a*nti*b*odies and an en*zyme* – is still in its infancy.

The first reports of antibody-catalysed chemical reactions appeared in 1986. Since then considerable progress has been achieved. Among the transformations catalysed by antibodies are carbonate, ester, and amide hydrolysis reactions as well as the corresponding reverse reactions, β-eliminations, *cis–trans* isomerizations, Claisen rearrangements, Diels–Alder reactions, light-induced cycloadditions, redox processes, and metal-chelating reactions. The problem in generating catalytic antibodies is the design of the hapten which must be chemically stable and at the same time resemble the structural and electronic properties of the transition state of the

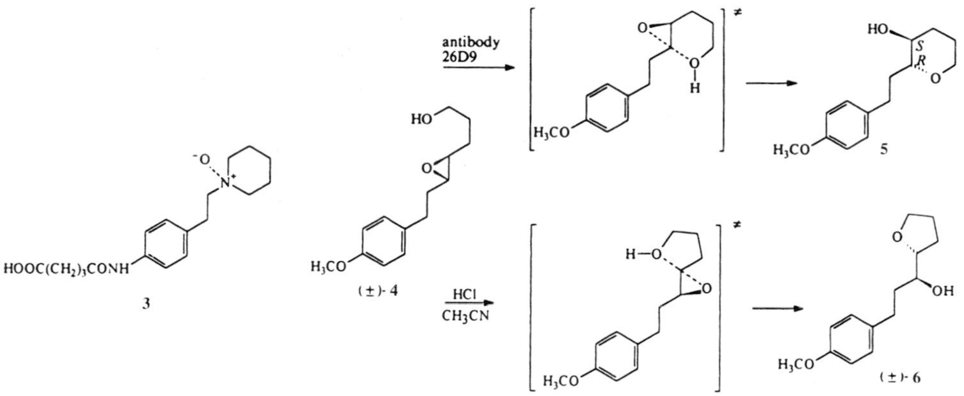

Scheme 1.6 With permission from C. Leumann, *Angew. Chem. Int. Ed. Engl.* **1993**, *32*, 1291.

reaction to be catalysed. For this purpose, the immune system uses complexes between the appropriate haptens and carrier proteins as targets. The antibodies produced in this way bind to the transition state of the target reaction and thus accelerate the reaction. These catalysts may show similar catalytic efficiency and substrate selectivity as natural enzymes.

Scheme 1.6 shows how the stereochemical course of a substitution reaction can be controlled by antibodies. The antibody 26D9 against hapten **3** catalyses the intramolecular substitution of the hydroxyepoxide **4** to give the tetrahydropyran derivativc **5**. Conventional acid catalysis without the antibody leads to the formation of the tetrahydrofuran derivative **6**. The antibody is also substrate-specific, transforming only one enantiomer of **4** into enantiomerically pure **5**. The reason for the observed course of the antibody-catalysed reaction is the close resemblance between the structurc of thc haptcn **3** and the transition state of the reaction in the upper part of Scheme 1.6. Thus, the reaction is not only accelerated but also forced onto a specific path that does *not* correspond to the inherent reactivity of the free substrate.

Another example for regioselectivity induced by antibody catalysis is the reduction of the diketone shown in Scheme 1.7. The antibody 37B.39.3 against the hapten **7** (which again resembles the transition state of the desired reaction) catalyses the enantioselective reduction of the *p*-nitrobenzyl ketones **8–10** to the corresponding alcohols **11–13** (with 86–96% *ee*(*S*) configuration in **11**). In the antibody-catalysed reduction of **14**, of the three possible products **(15–17) 15** is formed selectively in 95% yield (57% conversion) with 96% *ee*. Without the antibody, virtually no preference for the reduction of one of the two keto groups can be observed. The catalytic antibody thus stimulates the keto group next to the nitrobenzyl residue to react about 75 times as fast as the second one. The remarkable conclusion is that while the antibody will tolerate a range of residues R in the ketones **8–10** it discriminates sharply between methoxybenzyl and nitrobenzyl groups. This again indicates the relative insensitivity of antibodies with respect to structural variations near the conjugation site of the hapten-protein conjugate.

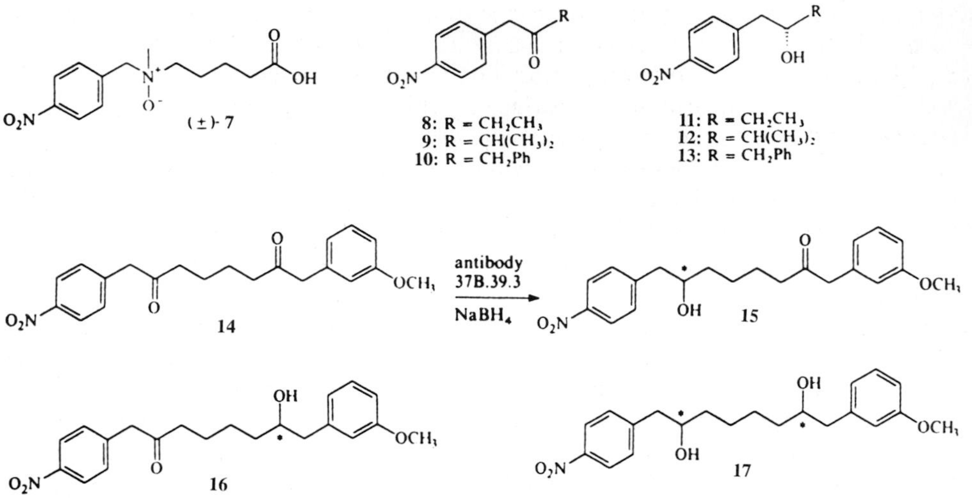

(±)-7

8: R = CH_2CH_3
9: R = $CH(CH_3)_2$
10: R = CH_2Ph

11: R = CH_2CH_3
12: R = $CH(CH_3)_2$
13: R = CH_2Ph

14 —antibody 37B.39.3, $NaBH_4$→ 15

16

17

Scheme 1.7 With permission from C. Leumann, *Angew. Chem. Int. Ed. Engl.* **1993**, *32*, 1291.

While a few years ago most investigations aimed at extending the number of reactions catalysed by antibodies or to elucidate mechanistic details, today's research concentrates mainly on practical applications. A good example is the development of an antibody catalysing the hydrolysis of cocaine to (–)-ecgonine (see Scheme 1.8). The motivation behind this work was the wish to develop a new cure for drug addiction. Self-administration of the drug is reinforced by a number of factors such

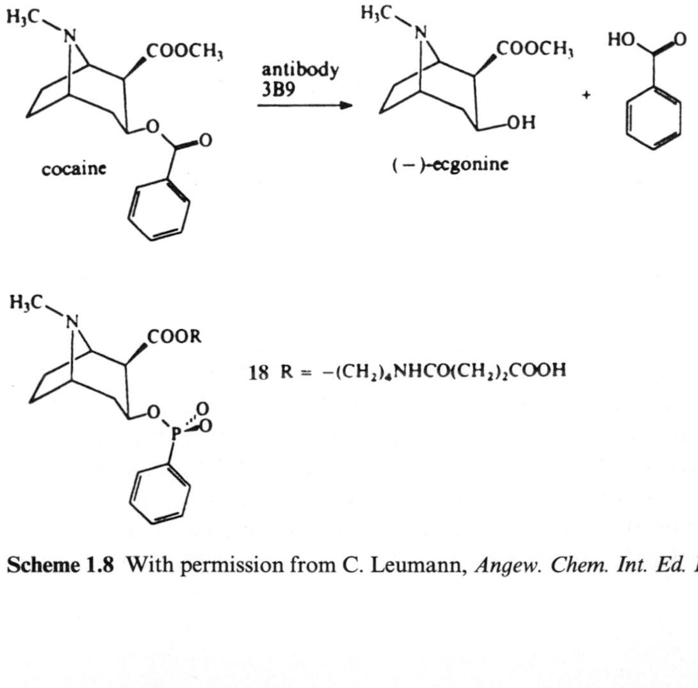

cocaine —antibody 3B9→ (–)-ecgonine + benzoic acid

18 R = $-(CH_2)_4NHCO(CH_2)_2COOH$

Scheme 1.8 With permission from C. Leumann, *Angew. Chem. Int. Ed. Engl.* **1993**, *32*, 1291.

as peak concentration, positive concentration gradients, and changes in the level of cocaine in the blood. Thus the idea was to develop a drug that degrades cocaine in the blood before it can be transported to the central nervous system. An antibody which catalyses the hydrolysis of cocaine to yield (physiologically inactive) ecgonine could fulfill this hope. The catalytic activity of the antibody against the phosphonate **18** is similar to that of the main enzyme responsible for the breakdown of cocaine in the blood, butyrylcholinesterase. To be therapeutically useful, an antibody would need to show much higher activity, however. This pioneering work may therefore be regarded as a first step towards future applications of catalytic antibodies in drug development.

In spite of recent advances, the preparation of catalytic antibodies is still a problem. Not only the immunization process but also the use of test animals for the induction and production of antibodies are time-consuming and expensive. However, the tools and techniques of molecular biology have made in vitro production and affinity screening of antibodies possible. For example, the 're-definition' of an anti-tetanus-toxoid antibody into an fluorescein-binding antibody was achieved by replacing the part of the gene coding for one of the hypervariable regions of the heavy chain of the original antibody with a synthetically prepared mixture of oligodeoxynucleotides of the same length. The resulting large number of antibodies was then expressed on the surface of phages. These were selected according to their ability to bind to fluorescein, and after amplification and isolation the molecular nature of the individual antibodies was determined. Some of the antibodies produced in this way show a similar affinity to fluorescein as those obtained by direct immunization of mice with fluorescein derivatives.

It is still too early to judge whether catalytic antibodies will find widespread use as drugs or diagnostic tools. Their success will not only depend on further rate enhancements (which today typically range from two to six orders of magnitude), but also on the development of methods that avoid the immunization procedure and the use of test animals. The first promising steps in this direction have already been taken in Cambridge and elsewhere.

1.2.2.7 Ribozymes

One of the surprises to emerge from the work of biological chemists during the mid-1980s was the discovery made by Cech and Altman that RNA could function as a catalyst. Catalytic RNAs, or ribozymes (which is a term applied to both metal-free and metal-containing RNAs), are of profound significance so far as the fundamental principles of biocatalysis in general are concerned, since they are ideal molecules for 'evolution' experiments *in vitro*. A large, heterogeneous pool of RNAs can be subjected to multiple rounds of selection, amplification and mutation, leading to the development of variations that have some desired phenotype (genetic characteristics acquired as a result of interaction of the inherited characteristics with its environment). Such experiments allow the investigator to correlate specific genetic changes with quantifiable alterations of the catalytic properties of the RNA.

The work of Lehman et al., at the Scripps Institute in California, began with a pool of 10^{13} variants of the *Tetrahymena* ribozyme: they carried out *in vitro* evolution experiments that led to the generation of ribozymes with the ability to cleave an RNA substrate in the presence of Ca^{2+} ions, an activity that does not exist for the wild-type molecule. Over the course of 12 generations, a seven-error variant emerged that has substantial Ca^{2+}-dependent RNA-cleavage activity.

Bartel and Szostak, at Massachusetts General Hospital, used an interactive *in vitro* selection procedure to isolate a new class of catalytic RNAs (ribozymes) from a large pool (some 10^{15}) of different random-sequence RNA molecules. These ribozymes ligate two RNA molecules that are aligned on a template by catalysing the attack of a 3′-hydroxyl on an adjacent 5′-triphosphate – a reaction similar to that employed by the familiar protein enzymes that synthesize RNA. The corresponding uncatalysed reaction also yielded a 3′,5′-phosphodiester bond. *In vitro* evolution of the population of new ribozymes led to improvement of the average ligation activity and the emergence of ribozymes with reaction rates that were seven million times as fast as the uncatalysed one.

Whereas the full biotechnological significance of ribozymes is not yet apparent, it is clear that the *in vitro* evolution experiments that they allow make it possible to elucidate important aspects of both evolutionary biology and structural biochemistry on modest and accessible timescales. Ribozymes, like catalytic antibodies, constitute the other side of the coin so far as creating biological macromolecules that catalyse chemical reactions are concerned. On the one side there is the rational design of enzymes – either by assembling miniature, artificial, 'mimicking' enzymes or by use of site-selective mutagenesis – whilst on the other there is the selection from a large pool of randomly generated biomolecules a few that present an effective array of catalytic functional groups to a reactant (substrate).

1.2.2.8 Catalytic Oxidation of Methane: the Centrepiece of Future Power Sources

Methane, the simplest and most abundant of saturated hydrocarbons, has four C–H bonds, which are much more difficult to activate and to sever sequentially than most other chemical bonds. It is not, however, difficult to sever all four C–H links more or less simultaneously: the ease with which this complete and relatively uncontrolled oxidation is accomplished thermally in air is what makes natural gas a fuel. Ways of partially oxidizing CH_4 to produce 'syn-gas' are well known, as we saw in Section 1.2.1, these being the bases of the industrial production of methanol, hydrogen, and many other chemicals, principally by Fischer–Tropsch catalysis. In view of the superabundance of methane (CH_4) and the dwindling reserves of oil, scientists are looking afresh at the rationale that lies behind present-day methods of producing petrol (gasoline). In brief, the CH_4 is partially oxidized to $CO + H_2$ mixtures, which are then converted either to methanol (CH_3OH) and onward (via a zeolite catalyst, the MTG process; see Section 1.2.1) to the aromatics that comprise transport fuel or, alternatively, via Fischer–Tropsch to the hydrocarbons which, after further catalytic reorganization (such as dehydrocyclodimerization, Scheme 1.5). Neither of these routes would remain commercially viable if CH_4 could be directly converted to petrol. Equally, many of the speciality chemicals, including vinyl and acetate monomers and other precursors, are nowadays prepared from syn-gas. How much more effective it would be if CH_4 could be converted directly into these molecules!

In the mid-1970s new approaches were pursued, especially by Bhasin and Keller in the USA, to convert CH_4 to C_2-hydrocarbons such as ethylene and ethane by oxidative coupling:

$$CH_4 \xrightarrow[\text{oxide}]{O_2} CO + CO_2 + H_2O + C_2H_4 + C_2H_6$$

A range of oxide-based solid catalysts that yield promising conversions and selectivities have been reported, and strenuous efforts are now being made to improve the performance of these catalysts, notably by the US workers Lunsford and Klier and their associates and by numerous groups in Europe (Hutchings, Burch, Joyner, Baerns, van Santen, Lambert) and Japan (Ueda and Otsuka).

Organometallic chemists have realized that one of the reasons why the first of the two reactions shown below for gently oxidizing CH_4 proceeds so much more readily than the second is attributable to the fact that M–C bonds are weaker by some 90 kJ mol^{-1} than are M–H bonds:

$$M + H-H \longrightarrow M\langle^{H}_{H}$$

$$M + -\overset{|}{\underset{|}{C}}-H \longrightarrow M\langle^{C-}_{H}$$

Using higher alkanes, Crabtree set out to effect multiple C–H bond rupture at a metal in the presence of a hydrogen acceptor such as *t*-butylethylene (TBE). The thermodynamics are therefore so modified as to make reactions such as the following one feasible:

$$R^1CH_2{-}CH_3 + R^2CH{=}CH_2 \longrightarrow R^1CH{=}CH_2 + R^2CH_2{-}CH_3$$

Cyclopentane can indeed by dehydrogenated by a homogeneous iridium complex at 80 °C in the presence of TBE. It is thought that both in metal complexes and at metal surfaces rich in monatomic steps, C–H–M bridges such as those shown in Fig.1.10 are implicated.

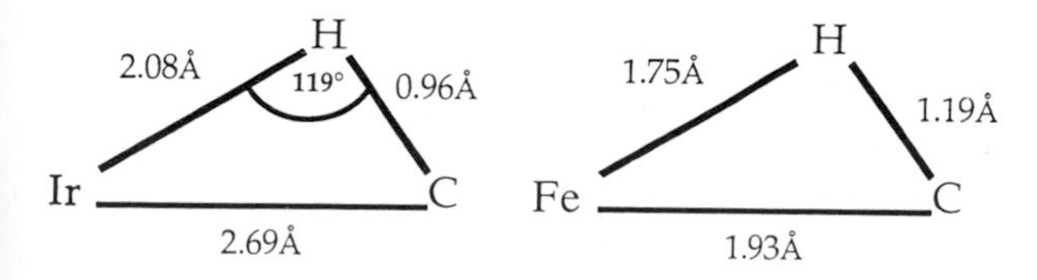

Figure 1.10 C–H–M bridges.

A deeper understanding of electronic charges accompanying the attachment of small molecules to metal surfaces encourages the belief that better transition-metal catalysts as well as families of various mixed oxides can be developed in due course for the selective conversions of methane.

A fundamentally new approach to the conversion of plentiful gas (methane) to desired liquid (hydrocarbon fuels and feedstocks) entails the use of an 'autothermal reactor' developed in the UK by British Petroleum. Practical details are not available, and it is not yet clear to what extent catalysis plays a part. But through adroit design of the methane reactor, liquid products are obtained in good yields at modest temperatures. Great efforts are also expended in perfecting reactors which secure complete oxidation of methane for more efficient power stations, and which achieve operating temperatures at which only minute amounts of undesirable NO_x are produced. For these reactors, which are also of great concern in the better utilization of fuel (both domestic and transport), mixed metal oxide catalysts are of central importance.

The 'sweetening' of methane (i.e., the elimination of H_2S impurities which often coexist in natural gas) can be beautifully achieved by catalysis (as demonstrated at the Boreŝkov Institute in Siberia) whereby H_2S is selectively oxidized to sulphur and the hydrocarbon remains unconverted.

1.3 Definition of Catalytic Activity

It was essential at the outset to define catalytic selectivity in order to appreciate the subtleties involved in the phenomenon of catalysis (see Section 1.1.1). Having referred frequently to 'catalytic activity', we must now specify precisely what is meant by this term.

In general, the rate of any gas–solid or liquid–solid-catalysed reaction can, as we shall discuss in detail in Chapter 2, be expressed as the product of the apparent rate coefficient k and a pressure- (or concentration-) dependent term:

$$\text{rate} = kf(p_i) \tag{1}$$

where p_i is the partial pressure of the reactant i. The rate coefficient for the overall catalytic reaction may incorporate the rate coefficients of many of the elementary reaction steps that precede the rate-determining step. For several reasons this rate coefficient will change as the prevailing conditions of the reaction (temperature, pressure, surface concentrations, etc.) vary, and it is operationally convenient to use the Arrhenius equation:

$$k = A' \exp(-E'/RT) \tag{2}$$

where A' is a temperature-independent pre-exponential factor and E' is the *apparent* activation energy of the catalytic reaction. E' cannot be expected to be the true activation energy, even if the catalyst structure remains unchanged with varying temperature, because the concentration of reactant at the catalyst surface will, in general, be temperature-dependent. For this and other reasons it is best not, as seems first logical, to define catalytic activity in terms of activation energy. Far more convenient is the use of the concept of *turnover frequency* or *turnover number*.

The *t*urn*o*ver *f*requency (often designated TOF) is simply the number of times n that the overall catalytic reaction in question takes place per catalytic site per unit time for a fixed set of reaction conditions (temperature, pressure or concentration, reactant ratio, extent of reaction). In words,

$$\mathrm{TOF} = \frac{\text{number of molecules of a given product}}{(\text{number of active sites}) \times (\text{time})}$$

or

$$\mathrm{TOF} = \frac{1}{S}\frac{\mathrm{d}n}{\mathrm{d}t} \tag{3}$$

where S is the number of active sites. When the number of active sites S is known, as is generally the case with enzymatic processes and almost invariably with homogeneously catalysed reactions, the turnover frequency can be specified quantitatively. In heterogeneous catalysis, however, it is sometimes difficult to determine the number of active sites. For such situations, S is often replaced by the total, readily measurable, area A of the exposed catalyst. Clearly $(\mathrm{d}n/\mathrm{d}t)/A$ sets a lower limit to the turnover frequency.

As well as in terms of the unit total area, the turnover frequency can also be expressed per mass or per volume of the catalyst, or, in chemical engineering contexts, per volume of packed reactor. The IUPAC recommendation is that TOF, expressed per unit total area, be termed the 'areal rate of reaction' but this usage is not yet in vogue. Note that TOF is a rate, not a rate coefficient, so that it is necessary to specify all the prevailing conditions of the catalytic reaction.

Notwithstanding what was said earlier about the difficulties of determining the number of active sites in a heterogeneous catalyst, the use of TOF as a measure of catalytic activity is sensible, partly because with some such catalysts (e.g. zeolites and enzymes) it *is* possible accurately to specify the number of active sites; and even with finely dispersed supported-metal catalysts it is increasingly possible, using the techniques described in Chapter 3, to count the number of surface atoms. Comparisons can therefore profitably be drawn between catalytic activities of single-crystal model catalysts studied in the laboratory and of real-life catalysts used in industrial plants.

1.3.1 Magnitude of Turnover Frequencies and Active Site Concentrations

For most heterogeneous reactions involving the catalytic transformation of small molecules in the temperature range 100–500 °C and pressures of up to a few bars, turnover frequencies fall between 10^{-2} and $10^{2}\ \mathrm{s}^{-1}$. These values are to be compared with those associated with well-known enzymatic reactions: 10^{3} for chymotrypsin, 10^{4} for urease and acetylcholinesterase, and 10^{7} for catalase. It is striking how much larger the frequencies are for enzymes than for their inorganic analogues. The differences can be much less between enzymes and the immobilized catalysts discussed above and also between enzymes and zeolitic catalysts. Indeed, since it is possible to

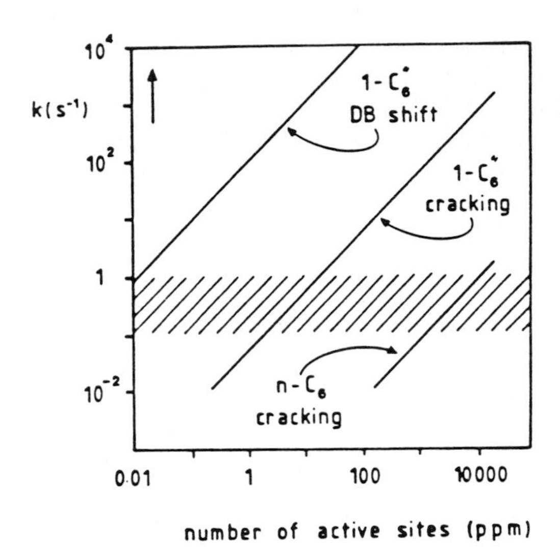

Figure 1.11 Plot of first-order rate coefficients versus concentration of active sites in ZSM-5 catalysts for the cracking of *n*-hexane (n-C_6), for the cracking of 1-hexene (1-C_6'') and for the shift of double bond in 1-hexene (DB shift of 1-C_6''). With permission from W. O. Haag, R. M. Lago, P. B. Weisz, *Nature* **1984**, *309*, 589.

determine the precise number of active sites in zeolites such as ZSM-5 (see Colour Plate 1), TOF can be quantitatively specified. It transpires that the activity of this catalyst at *ca* 450 °C, for certain reactions, rivals that of an enzyme under ambient conditions: for the cracking of 1-hexene, the turnover frequency is *ca* 5×10^2; for the isomerization of this alkene it is *ca* 10^7.

To link turnover frequencies with reaction rate coefficients and other kinetic parameters used to describe catalytic activity, it is instructive to cite the work of Haag et al. on ZSM-5 catalysts in further detail (Fig. 1.11). We note that the first-order rate coefficient for the cracking of the 1-hexene is 800 times as large as that of the *n*-hexane. Many commercial processes occur with first-order rate coefficients in the range 0.1–1.0 s^{-1}. Thus the active site concentration needed to achieve such technically relevant reaction rates – a magnitude cross-hatched in Fig. 1.11 – would be *ca* 1500 ppm for the cracking of *n*-hexane and *ca* 10 ppm for that of 1-hexene. It is noteworthy that kinetic principles require some 10^{18} active sites per cm^3 of catalyst volume to obtain technically useful catalytic conversion rates in a process operating at 500 °C and requiring an activation energy of 125 kJ mol^{-1}.

It is often advantageous to quote reaction probabilities, R_p, instead of turnover frequencies. R_p reveals the overall efficiency of the catalyst and is defined thus:

$$R_p = \frac{\text{rate of formation of product}}{\text{rate of incidence of reactant}} \tag{4}$$

It is simply the turnover frequency divided by the flux of reactant incident upon the catalyst (see Eq. (1), Chapter 2). Figure 1.12 illustrates how the turnover frequency and reaction probability compare for platinum-catalysed hydrocarbon reactions. Somorjai has compiled a comprehensive set of R_p values and other relevant kinetic parameters for reactions such as hydrogenation, dehydrogenation, cracking, ring-opening, dehydrocyclization and isomerization.

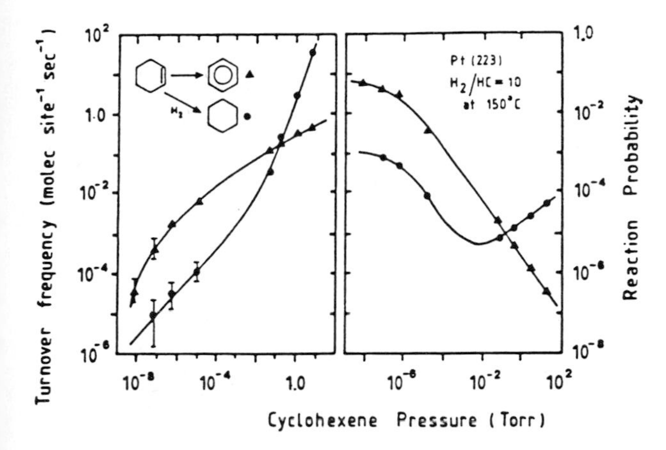

Figure 1.12 Correlation of reaction rates and reaction probabilities over a pressure range of ten orders of magnitude for the conversions of cyclohexene. The dehydrogenation and hydrogenation were catalysed by the (223) plane of a platinum crystal. With permission from S. M. Davis, G. A. Somorjai, *J. Catalysis* **1980**, *65*, 78.

1.3.2 Volcano Plots

Another quite widely used index of catalytic activity is the temperature required for the reaction under consideration to attain an arbitrary degree (or rate) of conversion. This approach can be misleading. Relative activities of a series of catalysts will vary with the degree of conversion chosen unless all the catalysts exhibit the same activation energy, a situation not likely always to prevail. There is something to be said, therefore, for using as an index of catalytic activity the relative efficiencies of different catalysts at the same temperature, although, ideally, one ought to relate activity to energies of activation and pre-exponential factors. But fixing a certain temperature may turn out to be just as arbitrary as fixing a certain degree of conversion – quite different orders of catalytic efficiency may be obtained if a different standard temperature is fixed. On balance, the criterion of temperature at fixed conversion is preferred over conversion at fixed temperature, chiefly on the practical consideration that temperatures of equal conversions are measured quantities, whereas conversions at equal temperatures are partly extrapolations.

The type of curve that results when this particular index of catalytic activity (temperature at fixed conversion for a series of related metals) is plotted against some enthalpic function of those metals is illustrated in Fig. 1.13. This is known as a Balandin volcano plot. It is the enthalpy of formation of the metal formate that is plotted on the abscissa in this case, where we focus on metal-catalyzed decomposition of formic acid. If the data on the ordinate of Fig. 1.13 are plotted against the enthalpy of formation per metal atom of the highest oxide, the resulting curve is known as a Tanaka–Tamaru plot, and when the enthalpy of formation per mole of oxygenation of the most stable oxide is plotted along the abscissa, the curve is known as a Sachtler–Fahrenfort plot. All three curves are volcano-shaped; as we shall see

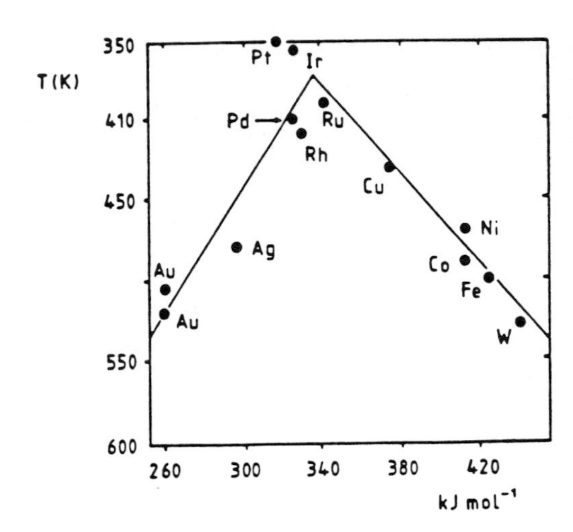

Figure 1.13 A typical volcano plot of the kind discussed by Balandin. The ordinate gives the temperature for a specified conversion of the reactant (formic acid) on a metal catalyst, and the abscissa gives the enthalpy of formation of the metal formate. With permission from J. Fahrenfort, L. L. van Reijen, W. M. H Sachtler, *The Mechanism of Heterogeneous Catalysis* (Ed. J. H. de Boer). Elsevier, Amsterdam 1960.

later, the significance of this fact is that it offers broad support to Sabatier's idea that an intermediate compound is formed at the surface of a catalyst. Too high an enthalpy of formation would not facilitate catalysis, and neither would too low a value as, in this case, there would be little propensity for the intermediate compound to form. An optimum value of the enthalpy would balance out these two conflicting tendencies, and this is why the peak of the 'volcano' corresponds to the highest catalytic activity.

1.3.3 The Evolution of Important Concepts and Techniques in Heterogeneous Catalysis

In catalysis, as in most other subjects, concepts and techniques are inextricably mingled. Theoretical insight naturally prompts experiment; correspondingly, technical virtuosity engenders fresh perception, which, in turn, stimulates ever more ambitious experiment. Nowadays, thanks to the repertoire of delicate techniques at our disposal (see Chapter 3), less than one-hundredth of a monolayer of adsorbed material – on a solid surface of area less than 1 cm^2 – can be identified. Moreover, the nature of its bonding before and after attachment to the catalyst can be probed both by direct experiment and by theoretical computation. Oxidation states of surface atoms – their steady-state concentration as well as their spatial distribution – can likewise be determined, at least for model catalysts though not usually for their real-life analogues. However, as with other processes involving chemical transformations, we cannot probe the transition states directly, in view of their exceptionally short lifetimes. Experience teaches us, as we shall see, that the lifetime of species implicated in heterogeneous catalysis exceeds 10 μs; the crucial acts of electronic and atomic rearrangement generally involve a much more rapid timescale, of the order of pico- or femto-seconds.

It is prudent to trace the emergence of some of the key concepts and experimental advances in catalysis since the early 20th century. By 1920, thanks largely to the work of Sabatier, it had been appreciated that a metal, such as nickel, which catalysed hydrogenations possessed its activity because it could readily form an intermediate hydride which, in turn, decomposed to regenerate the free metal. Langmuir, probably influenced by W. H. Bragg, had almost completed his landmark demonstration of the inadequacy of the Nernst theory – which had satisfactorily explained the kinetic features of the dissolution of solids in liquids – to account for gas uptake and reactions at surfaces. In due course, after Langmuir, Rideal, Hinshelwood and their associates had studied the kinetics of many heterogeneously catalysed reactions, it became possible to formulate some generalized principles to account for the various rate–pressure relationships that had been observed experimentally. Thus, the Langmuir–Hinshelwood mechanism (see Chapter 2) for catalysed processes postulated that the rate of a heterogeneous reaction is controlled by the reaction of the adsorbed molecules, and that all adsorption and desorption processes are in equilibrium. The Rideal–Eley mechanism (see Chapter 2), on the other hand, envisaged that a heterogeneous reaction could take place between strongly adsorbed atoms (that is, those chemisorbed) and molecules held to the surface only by weak, van der Waals, forces (that is, those physically adsorbed).

In the next decade. H. S. Taylor advanced cogent reasons for believing that preferential adsorption on a catalyst surface would take place at those atoms situated at peaks, fissures, and other topographical discontinuities. Moreover, it was implied that such atoms would have greater catalytic activity than those on flat surfaces. This was the genesis of the idea of 'active sites' and 'active centres', terms which, along with 'catalyst poisoning' or 'deactivation' to which they are related, are still widely used, but with rather wider meanings. Most present-day authors use the term 'active site' to describe the locus of catalytic conversion; but often, especially in discussions on chemisorption *per se*, this same term is used to specify that site at which adsorption is strongest. It will emerge in Chapter 2 that strong adsorption is inimical to catalysis, so that the most 'active' site for chemisorption is by no means the most favourable site for facile chemical conversion. In his classic paper, Taylor said, with remarkable perspicacity, that 'the amount of surface which is catalytically active is determined by the reaction catalysed', and also wrote about mechanisms 'whereby both the constituents of a (catalysed) hydrogenation process may be attached to one and the same (surface) atom'. The first of these quotations is most relevant to modern interpretations of kinetic oscillations in catalytic oxidations such as the burning of carbon monoxide (CO) on platinum studied by Ertl and others. The second foreshadows some of the views expressed four decades later in formulating modes of interconversion of various surface intermediates which are σ- or π-bonded as transient ligands on the same metal atom or ion. This second notion also contains the germ of the ideas overtly expressed by Nyholm, Burwell, Rooney and others in the early 1960s about the kinship between homogeneous and heterogeneous catalysis.

Taylor also suggested that the process of chemisorption frequently involves an activation energy, an idea which was soon to receive theoretical support when Lennard-Jones introduced potential-energy diagrams as an interpretive framework for discussing adsorptive and catalytic phenomena. Another important contribution by Taylor and his school in Princeton and by Farkas in Cambridge was to employ D_2 and deuterated reactants for studies of hydrogenation. This strategy, and others

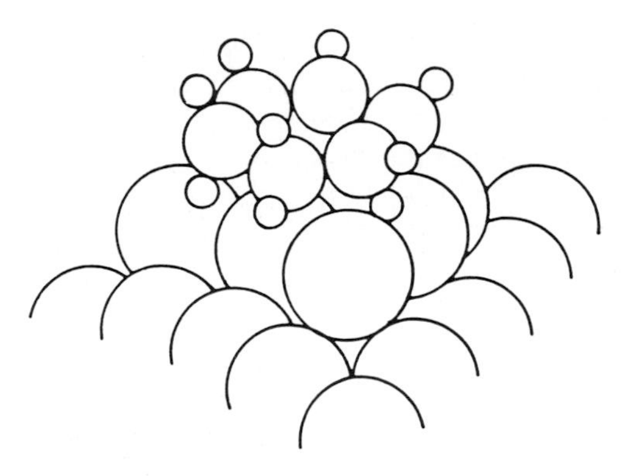

Figure 1.14 Schematic illustration of the Balandin concept of multiplets involved in the bonding to certain metals that are good catalysts for cyclohexane–benzene interconversions. This particular multiplet is a sextet (see text).

based on it, was to prove exceptionally fruitful in ensuing years in mechanistic studies in the hands of numerous workers with access to mass spectrometric analysis.

In the late 1930s significant advances came in the wake of Brunauer, Emmett and Teller's theory of physical adsorption (see Chapter 2), which offered, for the first time, a moderately reliable method of measuring surface areas of catalysts. This meant that comparisons of activity in a family of related catalysts could thereafter be put on a quantitative basis. It was soon demonstrated that the pore-size distribution, another important characterizing property of a catalyst, could be retrieved from adsorption isotherms (see Chapter 4).

Significant contributions also emerged in the mid-1940s from the (then) USSR. Balandin formulated his multiplet theory, the prime feature of which hinged on the postulate that the activity of a catalyst depends to a large degree on the presence on the surface of correctly spaced groups (or multiplets) of atoms to accommodate the various reactant molecules; Kobozev proposed the idea that 'ensembles' – the smallest group of catalytically active atoms – might be a helpful concept in surface phenomena. The computations of Twigg and Rideal at Cambridge in 1940, in addition to earlier ones by Eyring in the USA and the later experiments of Beeck, also in the USA, lent support to the view that a 'geometric factor' could be important in catalysis. Balandin, Krylov and their co-workers used their multiplet theory to predict that metals with interatomic distances ranging from 2.48 to 2.77 Å should exhibit catalytic activity for the hydrogenation of benzene and the dehydrogenation of cyclohexane, since, for these reactions, the metal spacings match the interatomic distances in the cyclic molecules. This prediction has been verified experimentally. It is interesting to note that Balandin's schematic picture of his so-called 'sextet complex' (Fig. 1.14) is strikingly confirmed by recent studies of benzene on nickel, where the aromatic molecule is found to lie flat on the metal surface; by the beautiful X-ray crystallographic studies of Gallop et al. (see Fig. 1.15), who find evidence of π-bonded and σ-bonded modes of attachment of benzene in the cluster compounds $[Os_3(CO)_9(C_6H_6)]$ and $[Os_3H_2(CO)_9(C_6H_4)]$ respectively; and in the elegant low-energy-election diffraction (LEED) studies by Somorjai and van Hove of benzene adsorbed on single-crystal faces of rhodium (Fig. 1.16). It is noteworthy that, both from LEED and X-ray studies, there is some evidence (from C–C bond distances)

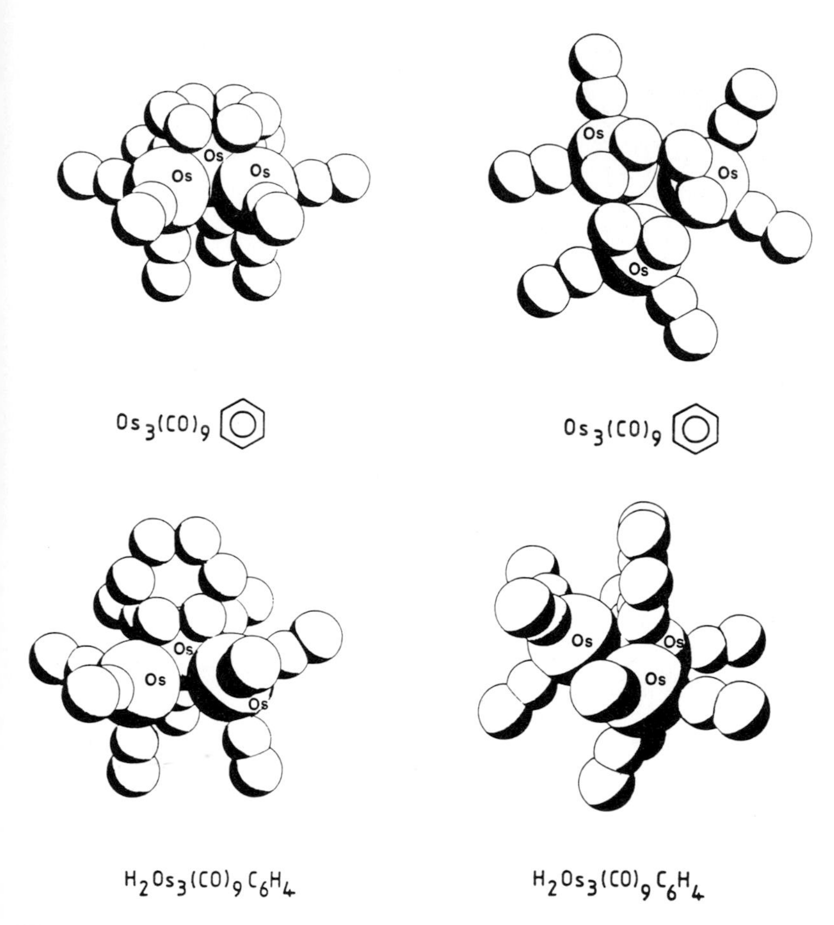

Figure 1.15 Experimental evidence from metal carbonyl cluster chemistry of σ- and π-bonded benzenoid groups to tri-osmium skeleton. With permission from M. A. Gallop, M.-P. Gomez-Sal, C. E. Housecroft, B. F. G. Johnson, J. Lewis, S. M. Owen, P. R. Raithby, A. H. Wright, *J. Am. Chem. Soc.* **1992**, *114*, 2502.

of a triene structure in which the benzene sits flat on top of an array of metals as substratum.

By the mid-1940s, arguments based on the electronic band structure of the bulk catalyst became fashionable; and in 1950 Dowden published his famous classification of catalytic solids into metals, semiconductors and insulators. The idea that catalytic activity could be directly related to bulk electronic properties, although at first promising in interpreting the behavioural trends within the compositional variations in a fixed alloy system, turned out to be oversimplified, especially when it was later realized that surface compositions could be very different from bulk ones. With solid oxides and chalcogenides, however, some progress was made by Hauffe, Weisz, Wolkenstein, Stone and others in predicting degrees of uptake of reactants in terms of the fundamental properties of the semiconductor. (Band theory and band struc-

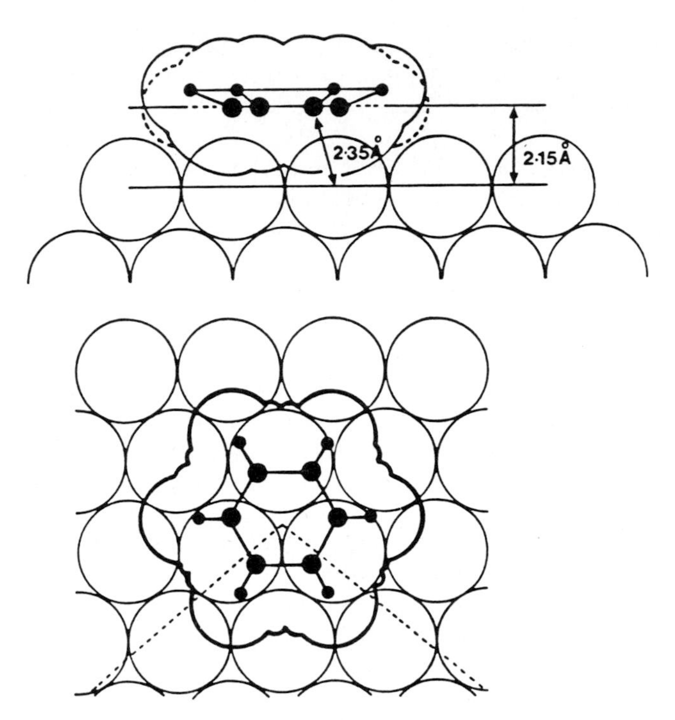

Figure 1.16 Elevation and plan view of the benzene molecule bound to a close-packed surface of rhodium. From the bond distances it is thought that the adsorbed benzene has a triene structure. With permission from G. A. Somorjai, *Chem. Soc. Rev.* **1984**, *13*, 332.

ture are still of great relevance in numerous contexts in heterogeneous catalysis, as we discuss in Chapter 5.)

Two significant landmarks in the mid-1950s were the discovery by Ziegler and Natta of stereoregular polymerization, and the full-blooded introduction, first by Eischens, Terenin, and Sheppard and their co-workers, of spectroscopic methods for probing adsorbed species attached to supported catalysts. Other spectroscopic techniques, along with diffraction methods, began to be widely used to their fullest advantage in the early to mid-1960s. Turkevich, Kazansky, and Lunsford separately pursued the use of electron spin resonance to probe the status of transition metal ions (in zeolites) and of unusual valences of oxygen in oxides. Germer and his school were prominent in the application to surface studies of LEED discovered much earlier. Photoelectron spectroscopy, pioneered by Siegbahn, Price and Turner, was demonstrated by Thomas et al. to be sensitive enough to detect adsorbed monolayers and, in particular, by Roberts and Joyner, to establish beyond doubt that CO on certain metals is dissociatively adsorbed. Auger electron spectroscopy (AES) was found to be a very sensitive technique for detecting light elements at surfaces; and electron-energy-loss spectroscopy (EELS) was resurrected and shown to be a powerful means of probing the vibrational modes of chemisorbed entities. By the early 1970s it was realized that combined experimental approaches, incorporating dynamic mass spectrometry for monitoring temperature-programmed desorption, LEED, AES, X-ray and UV-induced photoelectron spectroscopy, as well as other probes for monitoring the electronic and atomic states of surfaces, were feasible for the study of model catalysts.

So far as recent experimental advances are concerned, the current preoccupation is with the development of techniques suited to the in-situ studies of real-life catalysts. Infrared and laser Raman spectroscopy have their merits in this connection. So also do the inelastic or elastic scattering of neutrons; the absorption and elastic scattering of X-rays, with the rich information that emerges about crystal phase, oxidation state, coordination number, and site symmetry of surface atoms from the fine structures of the X-ray spectra; and the use of isotopic traces in so-called transient-response kinetic studies, which can determine both the lifetime and fractional coverage of intermediates at catalyst surfaces under operating conditions. The use of rather well-established techniques such as NMR and X-ray diffraction, hitherto regarded as unpromising for in-situ studies of catalysis, are also currently in course of development.

Several key concepts invoked nowadays in discussions of heterogeneous catalytic phenomena have emerged since the early 1970s. We shall elaborate these in later chapters, but it is helpful to adumbrate at this stage points of general interest relating to these concepts. First we recognize that numerous surface entities may be present on reacting catalysts. Some of these, however, are no more than 'spectator species' which are not implicated in the critical steps of catalysis, Ethylidyne (CH_3C), for example, contrary to earlier suggestions (see Scheme 1.9), plays no significant role in the hydrogenation of ethylene on supported palladium catalysts. Unusual states of oxide ions (O^-, O_2^-, O_3^- and O_4^-) on oxide catalysts have been identified, and the first of these (O^-) when bound to a Li^+ ion as a substitutional dopant in MgO, is directly implicated in the catalytic oxidative dimerization of CH_4.

Second, Boudart's subdivision of catalysts into 'structure-sensitive' and 'structure-insensitive' categories has proved to be a profitable concept. A catalytic reaction is said to be structure-sensitive if its rate changes markedly as the particle size of the (supported) catalyst is changed, or as the crystallographic face of a (single-crystal) catalyst is altered. Conversely, the rate of a structure-insensitive reaction is not significantly modified by such changes. The hydrogenolysis of ethane ($C_2H_6 + H_2 \longrightarrow$

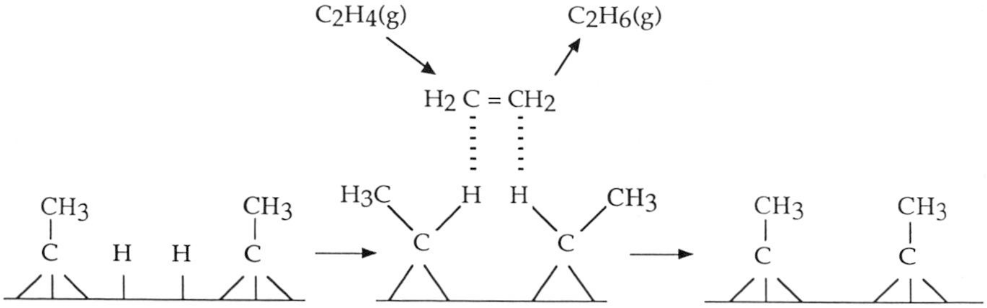

Scheme 1.9

$2\,CH_4$) and the synthesis of ammonia are good examples of structure-sensitive reactions: the metal catalysts used for these reactions are particularly susceptible to poisoning. The hydrogenation of ethylene or benzene and the oxidation of CO, on the other hand, are examples of structure-insensitive reactions, in which the effect of poisons on the metal catalyst is relatively minor.

Third, there is the notion of bifunctional or multifunctional catalysis. The recognition that supported metal catalysts (e.g., palladium or platinum on Al_2O_3 or zeolites) behave in a clearly identified dual fashion has helped enormously in the design of catalytic reactors. In the hydroprocessing of petrochemicals, for example, the metal serves to dissociate H_2 as well as to facilitate the equilibration of alkanes, alkenes and alkynes, whereas the acid support serves to catalyse the build-up of vital carbonium ion (alkylcarbenium ion) intermediates (see Fig. 1.3). We note that the support functions not only to stabilize the highly dispersed catalyst (thereby retaining high surface area), and to activate the metal by electron transfer between it and metal, but that it can also be directly involved in crucial elementary steps in overall reactions.

Implicit in the sketch in Fig. 1.17 is the 'spillover' of hydrogen atoms that are generated by dissociation of H_2 at the metal particle. It has been demonstrated by Khoobiar and others that hydrogen atoms which spill over in this fashion, and are mobile on the support, play an important role in many heterogeneously catalysed hydrogenations.

We note also that the Brønsted acidity of an oxide or mixed-oxide catalyst is of major importance in many catalytic processes. It was long ago recognized by Bernadskii in Russia and Pauling in the USA that aluminosilicates possess pronounced (Brønsted) acidic properties. This fact was skillfully harnessed by Rabo and others in developing the zeolitic catalysts mentioned earlier and discussed further in Section 8.8. A good acidic oxide or mixed oxide is one in which the surface OH groups at the solid A dissociate such as to yield $A{-}O^- + H^+$ rather than $A^+ + OH^-$. It is advantageous, therefore, to incorporate a second oxide BO (a basic one) to provide a corrdination shell in which the oxide ion attached to A can be accommodated, thus increasing the acid dissociation constant. On this score, we expect, and do indeed find, that the following oxide 'solid solutions' are good Brønsted acids: $SiO_2{-}Al_2O_3$, $P_2O_5{-}Al_2O_3$, $SiO_2{-}ZrO_2$, $B_2O_3{-}Al_2O_3$ and Ipatieff's 'solid phosphoric acid'.

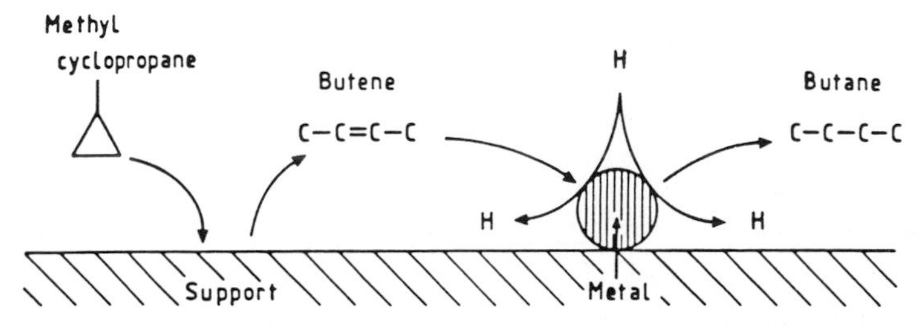

Figure 1.17 A bifunctional catalyst such as platinum on Al_2O_3 facilitates the isomerization of methylcyclopropane to 2-butene as well as the hydrogenation of 2-butene to butane (after Boudart).

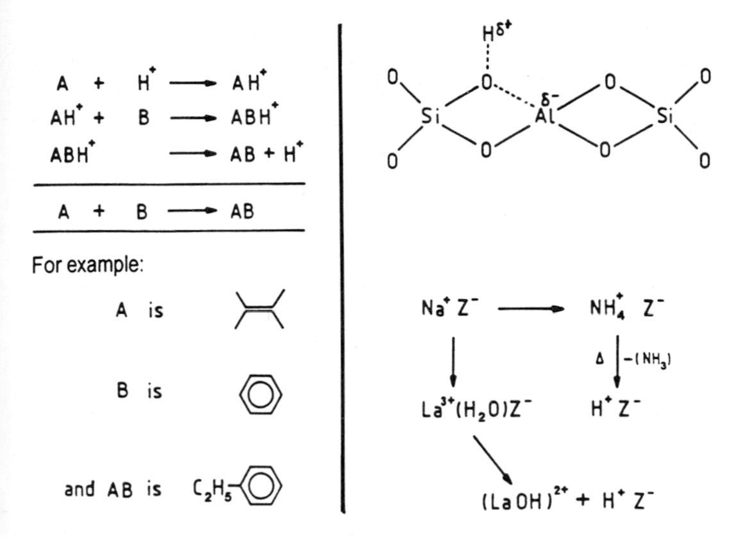

Figure 1.18 A Brønsted acid catalyst, such as a zeolite or an exchanged clay, functions through its ability to release and accept protons (top left). There is one acid hydrogen for every tetrahedrally bonded aluminium in the zeolite (top right). Zeolitic catalysts are often rendered acidic (H^+Z^-) by exchange with NH_4^+ ions followed by heating; or by simple exchange with polyvalent ions (e.g. La^{3+}) which then hydrolyse the bound water (bottom right).

It cannot be overemphasized how important a role Brønsted acidity plays in such processes as the catalytic cracking, alkylations and isomerizations of hydrocarbons. The gradual progression from acid-washed clays and silica–aluminas of the 1950s to the crystalline, well-defined zeolite catalysts of the present for effecting these processes underlines this fact. In simple mechanistic terms, we see (Fig. 1.18) the role of Brønsted catalysts in generating the carbonium (i.e. alkylcarbenium) ions which, as demonstrated in different contexts by Koptyug in Siberia and by Olah in the USA, are the versatile intermediates that serve to propagate the desired chemical changes. The development by Weisz et al. of shape-selective inorganic catalysts took advantage, as we shall see in Chapter 8, of the convenient fact that the active sites (which are here synonymous with the loci of the detachable protons) are situated largely inside the pores and cavities of the solid, and are accessible only to those species possessing the requisite shape and diffusive characteristics. By taking advantage of the ability of neutron-scattering procedures to solve the structure of powdered catalysts, we now have a picture, in atomic detail, of the active and hence the acid site in La^{3+}-exchanged zeolite-Y, where the 'active' proton is initially generated as a result of intracrystallite cation hydrolysis: $La^{3+} + H_2O \rightarrow (LaOH)^{2+} + H^+$.

The period from the mid-1960s to the late 1970s witnessed much conceptual activity and turmoil, prompted chiefly by the increasing pace with which new types of experimental information were uncovered. Quite apart from the discoveries that surfaces often had atomic structures different from equivalent sheets in the bulk solids, that substantial amounts of ordered impurities were present at exterior surfaces of solids, and that electronic reorganization was associated with these impurities, several other important bodies of facts and concepts became apparent or took on a new significance. The concepts of ensembles as well as geometric factors were resuscitated; intermediate compound formation and electronic factors received further attention; there was widespread use of isotopic labelling to elucidate reaction mechanisms and the nature of surface intermediates; the ideas of organometallic chemistry and of solid-state physics and the greater appreciation of the nature of orbital interactions and bonding were assimilated into surface chemistry; major advances were registered in our understanding of enzyme catalysis, largely as a result of the brilliant work of D. C. Phillips and C. A. Vernon, who interpreted the crystal structure of lysozyme with and without the polysaccharides it catalytically attacks.

The concept of ensembles has gained considerable prominence because, without it, it seems impossible to interpret the catalytic behaviour of alloys or the poisoning of single-component surfaces. At the root of this concept is the notion that, for many metal-catalysed reactions, a family or "ensemble" of several contiguous surface atoms can form bonds with a molecule, or be in some loose sense implicated in the formation of a transition state. Ponec, and Sachtler and Sinfelt, working with alloys such as Pt–Au, Ru–Cu and Ni–Cu, came independently to the conclusion that hydrogenolysis requires the largest ensemble of surface nickel atoms. The ensemble requirement for a given reaction can be studied by diluting the metal in an alloy with a chemically inert metal, thus reducing the concentration of large ensembles of the active metal on the surface. It seems that as many as 12 contiguous nickel atoms are required for the splitting of C_2H_6 to CH_4. For CO dissociation on metal, large ensembles seem to be required: for the dehydrogenation of propane on Pt–Au alloys, one exposed metal atom suffices. The role of ensembles in the poisoning of catalysts was discussed elegantly by Rideal. He showed, as have others in greater detail subsequently, that the deposition of species of poison (e.g. sulphur or phosphorus atoms) on uniform transition-metal catalysts will selectively inhibit the reactions that demand the larger ensemble sizes. For adspecies (poisons) which are distributed randomly, the surface concentration of ensembles of size n (i.e., θ_n) will vary as $(1-\theta_p)^n$, where θ_p is the fraction of the surface covered by the poisonous species. The situation at the catalyst surface can be much more complicated if there are interactions between the species of poison. Thus, when these are attracted to one another, the situation is computed (by Monte Carlo methods) to be as shown in Fig. 1.19(a); whereas when they repel one another, a surface such as that depicted in Fig. 1.19(b) results.

The geometric factor re-emerged as a useful concept in catalysis when analogies were drawn between the functioning of certain homogeneous catalysts on the one hand and hydrodesulphurization catalysts (MoS_2) on the other. On the grounds that MoS_2 catalysts function best when the metal centre is coordinatively unsaturated, so as then to be in a position to accommodate bound hydrogen atoms and the organic reactant as ligands, it is evident that the nonbasal faces of the layered sulphide will (and indeed do) exhibit greater catalytic activity than the basal surfaces, in which the metal atoms are submerged below the sulphur atoms (Fig. 1.20).

Important as geometric effects undoubtedly are, they seldom dominate all other factors. We are reminded of this in contemplating the sharp contrast in behaviour between metallic platinum (generally a good catalyst) and metallic gold (generally a

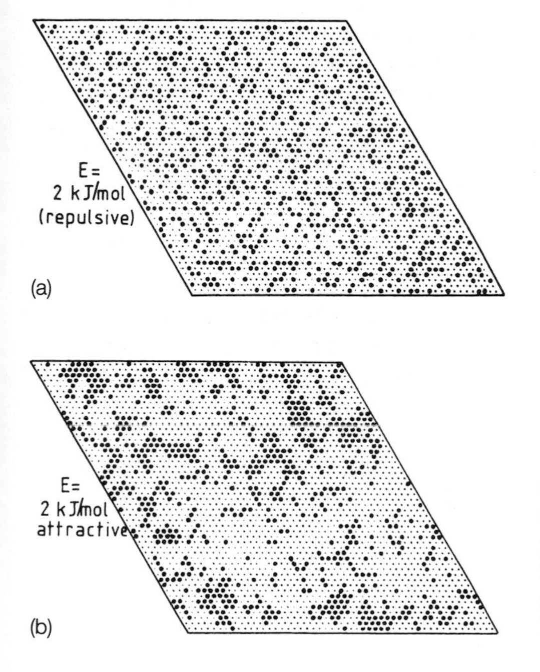

Figure 1.19 Monte Carlo simulation methods can be used to investigate the extent of adsorption of a given species (the probe) on a simple substratum as the surface concentration of the poison is altered. In these two simulations small dots denote the centres of the atoms of the substratum (the adsorbent); open and black circles represent probe and poison species, respectively when $\theta_p = 0.3$. In (a) there is a (pairwise) attractive interaction between poison species of 2 kJ mol^{-1}; in (b) there is a repulsive interaction of 2 kJ mol^{-1}. With permission from J. S. Foord, D. Tildsley, A. E. Reynolds, *Surf. Sci.* **1986**, *166*, 13.

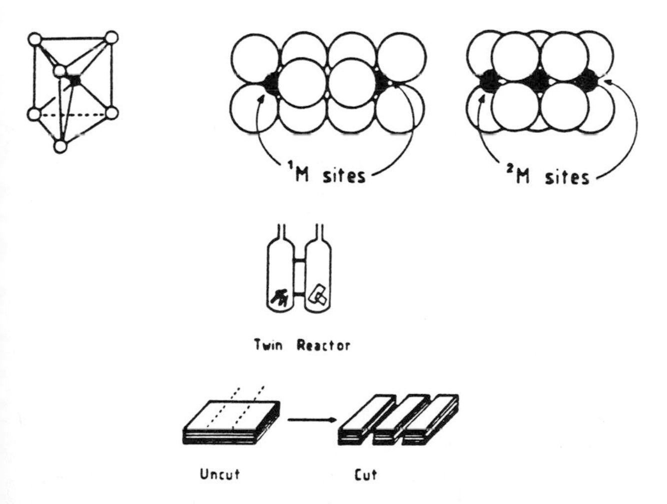

Figure 1.20 At the edges of crystals of transition-metal chalcogenides (e.g. MoS_2) there are some singly (^1M) and doubly (^2M) coordinatively unsaturated sites. For a given mass of crystalline catalysts, their surface concentration can be increased by cutting perpendicular to the basal planes. Tanaka probed the catalytic performance of such sites using a twin reactor, one limb of which contained uncut crystals, the other cut crystals of MoS_2.

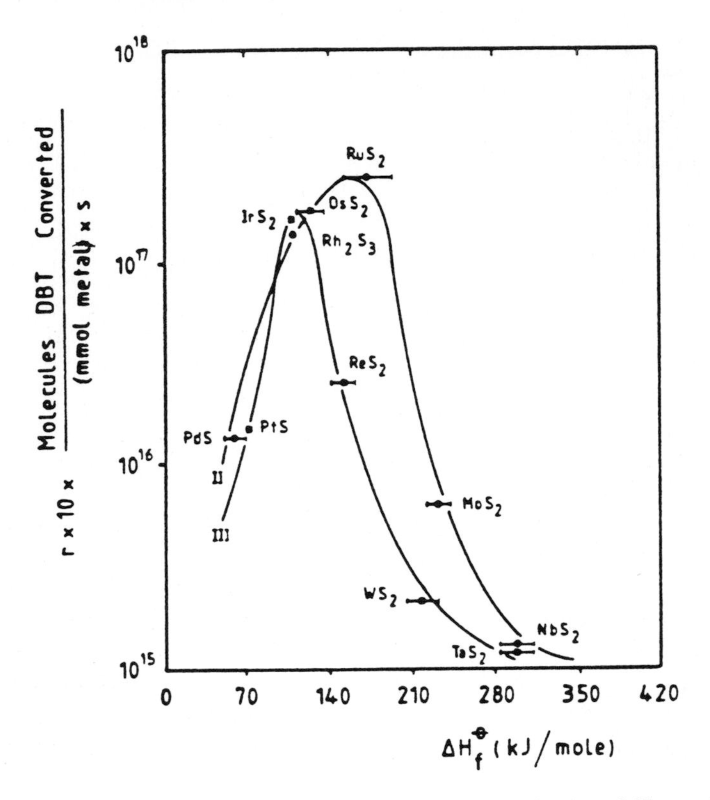

Figure 1.21 Catalytic activity in the hydrodesulphurization of dibenzothiophene varies in a 'volcano' fashion (see text) as a function of the enthalpy of formation of the bulk metal sulphide. With permission from R. R. Chianelli, T. A. Pecoraro, *J. Catalysis* **1984**, *86*, 226.

poor catalyst). These metals have exactly the same structure (face-centred cubic – see Chapter 5) and their interatomic spacings differ by only a few per cent. The same can be said of silver and gold. Even in the case of hydrodesulphurization, electronic factors can be very important, as demonstrated by Chianelli and Topsøe, who showed that catalytic activity can vary by several powers of ten in proceeding along the sulphides of a series of transition metals (Fig. 1.21). From this volcano plot we deduce that the enthalpy of formation for maximal activity takes an intermediate value: the most effective catalysts are those metal sulphides which have the ability to form and regenerate sulphur vacancies, required to create coordinative unsaturation at the metal centre.

The re-emergence of the intermediate compound theory of catalysis first propounded by Sabatier occurred amongst Dutch workers. As early as 1954, Mars and van Krevelen concluded that the catalytic oxidation of hydrocarbons took place in two steps: a reaction between the oxide and the hydrocarbon, in which the latter is oxidized and the former reduced, followed by the reaction of the reduced oxide with

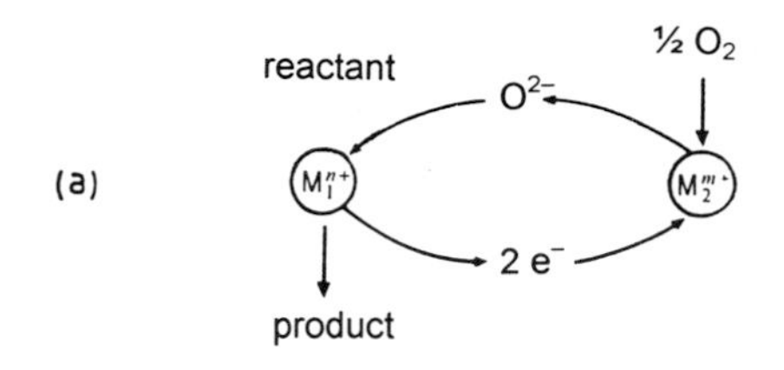

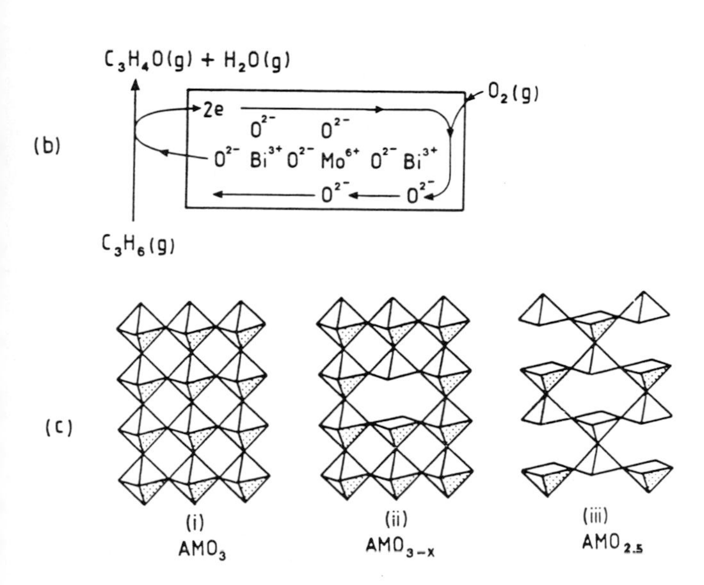

Figure 1.22 (a) In certain catalytic oxidations the so-called Mars--van Krevelen mechanism (an extension of Sabatier's original view) operates in which the hydrocarbon uproots oxygen from the oxide and the loss from the solid is subsequently made good by incorporation of gaseous oxygen. (b) Bismuth molybdate catalysts function in this way in the conversion of propylene to acrolein. With permission from R. K. Grasselli, J. D. Burrington, *Adv. Catalysis* **1981**, *30*, 133. (c) Some perovskite catalysts (general formula ABO_3; see Chapter 5), simplified in this schematic representation, are good selective oxidation catalysts. On releasing structural oxygen, octahedra become square pyramids, but the skeletal structure remains intact.

O_2 to restore the initial state. Obviously the tendency of an oxide (or a mixed oxide) to donate its structural oxygen is of key importance in governing its efficacy as a selective oxidation catalyst. If reduction of the oxide catalyst is too facile, it may be active, but it ceases to be selective; however, if it is not facile it may be active, but it ceases to be selective; if it is not facile enough the catalyst inevitably is of low activity. These generalizations echo our earlier remarks about volcano plots (Figs 1.13 and 1.21). Again, it is a balance between ease of loss of oxygen from the solid and the ability to convert gaseous oxygen into the bulk oxide, as well as the mobility of the entities (electrons and ions) that hold sway in the solid (Fig. 1.22).

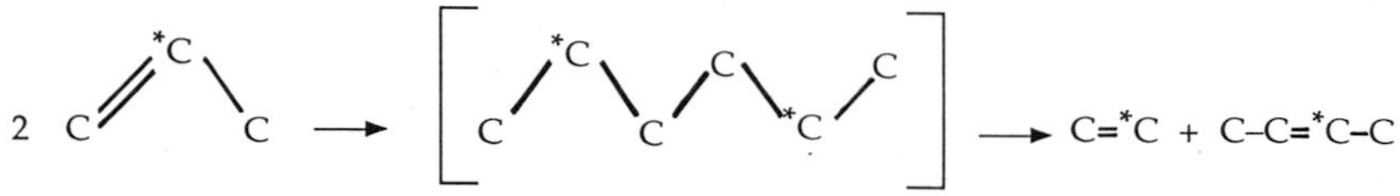

2 C=*C–C → [C–*C–C–C–*C–C] → C=*C + C–C=*C–C

Scheme 1.10

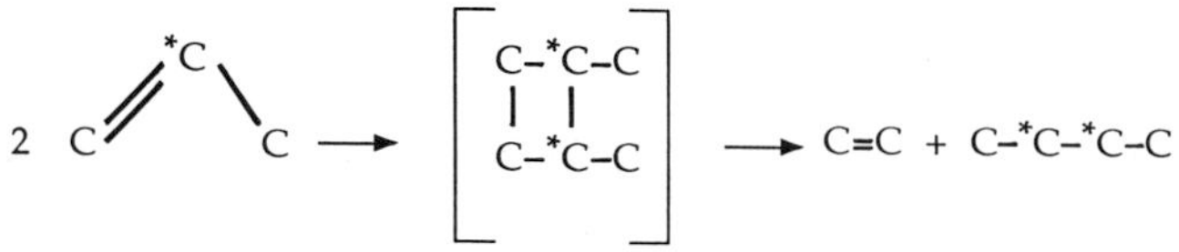

2 C=*C–C → [C–*C–C / C–*C–C] → C=C + C–*C–*C–C

Scheme 1.11

1.3.3.1 Mechanistic Insights from Isotopic Labelling

Experiments employing isotopically labelled reactants have greatly clarified our understanding of the mechanisms of catalysed reactions. As stated earlier, H. S. Taylor foresaw the value of using deuterium as a probe shortly after its discovery. Decades later, with the greater availability and ease of detection of radio-isotopes, many ingeneous investigations were made of 'live' catalyst surfaces, especially by Thomson and his school at Glasgow. In the early 1960s, they showed that, in general, only a fraction of chemisorbed species participate in catalysis, and that the surfaces themselves are energetically heterogeneous. They also pinpointed how much carbonaceous material is present at the surfaces during the conversion of hydrocarbons on metal catalysts. A good example of how ^{14}C-labelling elucidates reaction mechanism concerns the disproportionation of propylene: $2\,C_3H_6 \longrightarrow C_2H_4 + CH_3CH{=}CHCH_3$. Does this reaction, catalysed by ReO_3/Al_2O_3, proceed via a linear or a cyclic intermediate? The two alternatives are represented by Schemes 1.10 and 1.11. By labelling the propylene in the 2-position, the Dutch workers Mol and Moulijn found that the ethylene showed no radioactivity, whereas the butene had twice that of the starting gas. The first mechanism is, therefore, not valid, and a 4-membered intermediate, as shown in Scheme 1.11, must be implicated. Using both 1-*C- and 3-*C-labelled propylene it was also established that the identity of the terminal CH_3 group is retained during the disproportionation. Over bismuth molybdate catalysts Sachtler, working with 1-, 2-, and 3-labelled propylene, showed that the terminal CH_3 does lose its identity in the course of its coversion to acrolein:

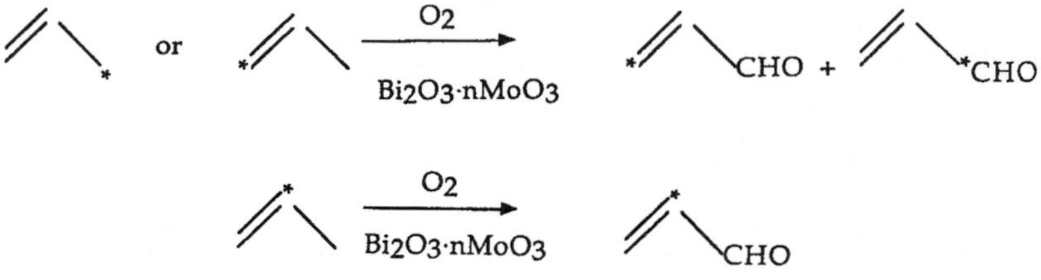

$$\text{(3-*C or 1-*C propylene)} \xrightarrow[Bi_2O_3 \cdot nMoO_3]{O_2} \text{*C=C–CHO} + \text{C=C–*CHO}$$

$$\text{(2-*C propylene)} \xrightarrow[Bi_2O_3 \cdot nMoO_3]{O_2} \text{C=*C–CHO}$$

19

This suggests that a π-allyl intermediate **19** is formed. Approaches such as these have established the identity of numerous intermediates at oxide and other catalyst surfaces. Propylene alone gives rise to five distinct intermediates, depending upon the nature of the oxide catalyst used (Table 1.2). We see that the hapticity of the ligand (the number of carbon atoms in the ligand directly bonded to the surface) can range from one to three.

Isotopic labelling has proved invaluable in numerous other catalytic contexts, e.g., Ponec's proof of the dissociation of CO as a primary step in Fischer–Tropsch syn-

Table 1.2 Intermediates identified when propylene reacts with oxide catalysts (after Kemball, *Chem. Soc. Rev.* **1984**, *13*, 375).

Catalyst	Intermediate	Name and hapticity
$Bi_2O_3 \cdot nMoO_3$	$CH_2 \cdots CH \cdots CH_2$ (CH bonded to surface)	π-Allyl (tri)
Ga_2O_3; Al_2O_3	CH_2(–surface)–CH=CH_2	σ-Allyl (mono)
Al_2O_3	CH_2=C(–surface)–CH_3	Propen-2-yl (mono)
	and	
	CH(–surface)=CH–CH_3	Propen-1-yl (mono)
Zeolites, ZrO_2	CH_3–^{+}CH–CH_3	Carbenium ion (zero)

thesis; the demonstration by Lambert of the comparative unimportance of adsorbed dioxygen in the silver-catalysed synthesis of ethylene oxide; and the separate experiments of Keulks, of Adams and of Ashmore which demonstrated that structural oxygen from solid catalysts was incorporated into the products of selective oxidation. These days, stable isotopes are more convenient to monitor (by mass spectrometry and NMR) than radioactive ones. They are almost invariably used in the transient response method of probing the surfaces of fine catalysts, described in Chapter 3.

1.3.3.2 Concepts from Organometallic Chemistry

So far as the incorporation of the ideas of organometallic chemistry into the concepts of catalysis is concerned, an important factor has been the realization that metal atoms attached to organic ligands can be bound in a multiplicity of ways (involving σ and/or π bonds) with more or less donation to, or back-donation from, the metal centre. Combining the results derived from difference sources, we see in Fig. 1.23 how the bonding and energetic situation for an isolated metal atom attached to CO differs from that pertaining to a metal surface.

In 1966 Bond drew attention to the usefulness of conceptually isolating the d-orbitals of individual atoms at the surface of a transition-metal catalyst. By so doing, it follows logically why both terminal and bridge-bonded CO should exist at catalyst surfaces. With a fuller appreciation of band theory and the language of solid-state physics (see Chapter 5), we can more fully appreciate that interactions between the orbitals of an isolated CO molecule can be favoured with 'localized' orbitals in the metal.

In addition, the demonstrated mobility of ligands on the surfaces of small organometallic clusters vindicated the interpretations of others who saw evidence for similar phenomena at the surfaces of bulk catalysts. Another important set of insights that organometallic chemistry has given to the student of catalysts is into the manner in which atoms of carbon can attach themselves to metals. Carbon is now reckoned to be ubiquitous on the surfaces of many transition metal and alloy catalysts, but comparatively little is known about the way in which it is bonded. So far, several non-reactive cluster carbides have been discovered since Dahl's remarkable discovery in 1962 of the compound $Fe_5C(CO)_{15}$. Examples of non-reactive 'cage' and peripheral carbide clusters are shown in skeletal outline in Fig. 1.24.

Figure 1.23 (a) The molecular orbitals formed when oxygen and carbon combine to yield CO (the antibonding 2π-orbital is sometimes referred to as $2\pi^*$). (b) The orbital limits of the various σ- and π-orbitals in CO. In (c) the arrows denote the direction of electron transfer when CO is attached to a metal centre. As well as donation of σ-electrons from the CO to the metal, there is back-donation of electrons from the d-orbitals of the metal to the antibonding orbitals of the CO. (d) The orbital interactions involved in bridge-bonding of CO. (e) The relative energies of the various energy levels of the bonding and antibonding orbitals of CO, as well as the highest occupied orbitals (HOMO) of the isolated metal (E_M) and the bulk metal (E_F). With permission from E. Shustorovich, R. C. Baetrold, *Science* **1985**, *227*, 376.

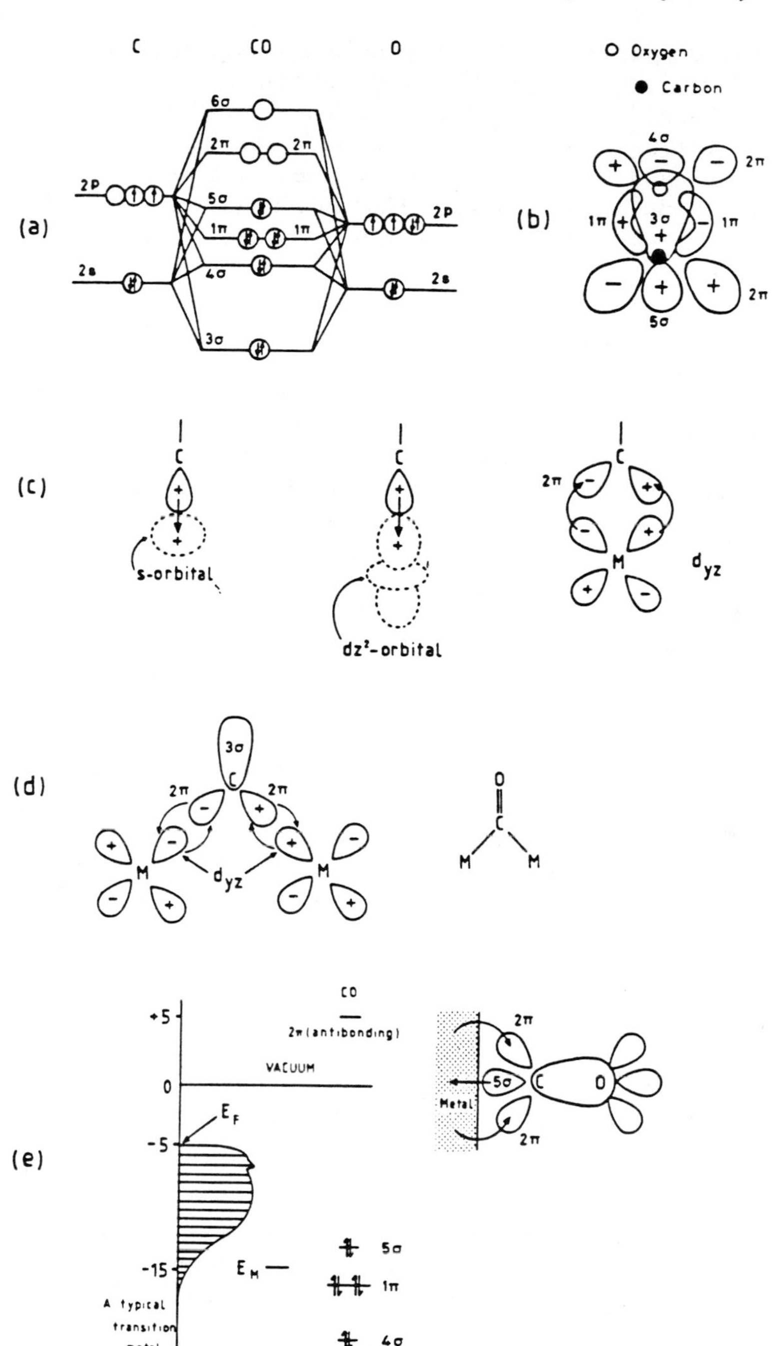
C
CO
O
O Oxygen
Carbon
6σ
2π
2π
2P
5σ
1π
1π
2P
2s
4σ
2s
3σ
(a)
(b)
4σ
2π
1π
3σ
1π
2π
5σ
(c)
s-orbital
dz²-orbital
2π
M
d_{yz}
(d)
3σ
2π
2π
d_{yz}
M
M
CO
2π (antibonding)
VACUUM
E_F
(e)
E_M
5σ
1π
4σ
A typical transition metal
Metal
2π
5σ
C
O
2π

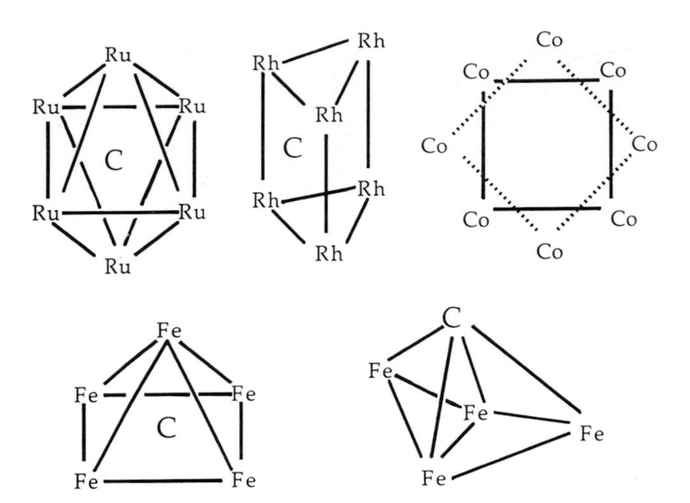

Figure 1.24 Nonreactive 'cage' and peripheral carbide clusters.

The recognition of the so-called 'agostic' interaction (in the terminology of Brookhard and Green), C–H···M, is also an important step forward in organometallic chemistry, and likely to be of future value in catalysis. The agostic interaction results because, in effect, transition metals can form stable complexes in which the electronic requirements of the metal are satisfied by interaction with the electrons of the C–H bond. Such interactions might be considered to resemble those present in transitory species in the processes of β-elimination/olefin insertion (Scheme 1.12).

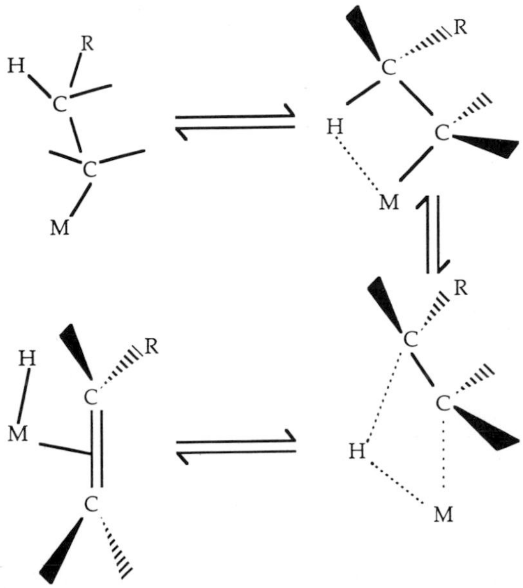

Scheme 1.12

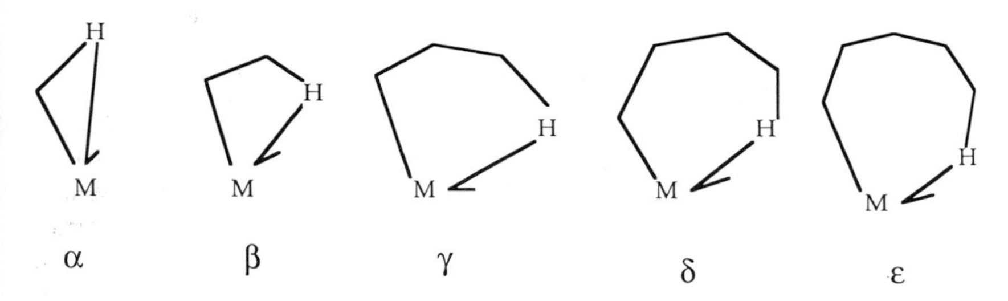

Figure 1.25 C–H environments for agostic interactions in organometallic complexes.

So far, agostic interactions have been identified with some 15 or so different metals, and they encompass a range of C–H environments, with the carbon ring being α-, β-, γ-, δ- or ε- to the metal involved in the three-centre interaction (Fig. 1.25). N–H···M interactions have also been identified.

Yet a further advantage of incorporating the concepts of organometallic chemistry into catalysis is seen in the formulation of reaction mechanisms. In this regard Clarke and Rooney have made valuable contributions, especially in invoking the role of metallocyclobutanes in metal-catalysed homologation reactions (the formation of the next-higher *n*-alkane). The mechanism they proposed consists of addition of surface methylene (for the existence of which there is good evidence) to the unsubsituted vinylic carbon of an α-olefin, which is formed initially by dehydrogenation. The resulting metallocyclobutane may then hydrogenate to yield the next-higher homologue or isomerize to the corresponding α-olefin which, in turn, may repeat the same reaction (Scheme 1.13).

The language of organometallic chemistry has also been of value in identifying the

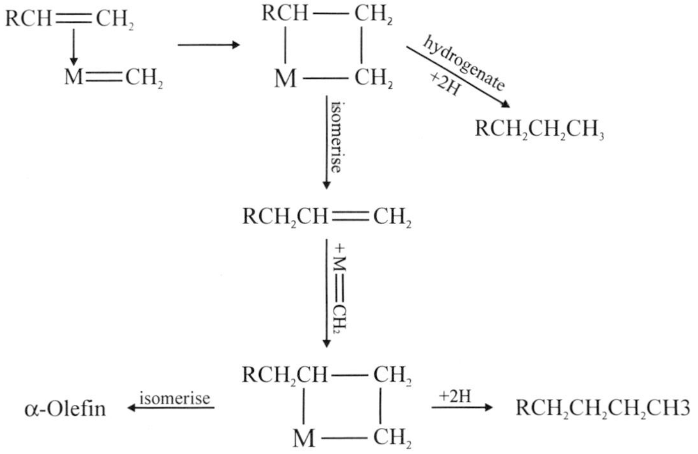

$RCH{=}CH_2$ ↓ $M{=}CH_2$ ⟶ $RCH{-}CH_2$ / $M{-}CH_2$ (metallocyclobutane) —hydrogenate, +2H⟶ $RCH_2CH_2CH_3$

↓ isomerise

$RCH_2CH{=}CH_2$

↓ $+M{=}CH_2$

$RCH_2CH{-}CH_2$ / $M{-}CH_2$ —+2H⟶ $RCH_2CH_2CH_2CH3$

α-Olefin ⟵isomerise— $RCH_2CH{-}CH_2$ / $M{-}CH_2$

Scheme 1.13

nature of certain metal–support interactions. The work of Henrici-Olive and Olive illustrates this well. Finely divided chromium behaves quite differently on SiO_2 and Al_2O_3 supports. From the observed magnetic properties, there is little doubt that the chromium is in the chromium(II) state; whereas the SiO_2 support provides a strong ligand field, thereby favouring the low-spin configuration for the octahedrally coordinated ($3d^4$) Cr^{II} ion, the Al_2O_3 support provides a weak ligand field, thereby favouring the high-spin state of the Cr^{II}. Metal–support interactions, and the way they can be influenced by appropriate pretreatment, have been given much attention of late. The reality of such interaction had been recognized for some time by Schwab and Solymosi before being further investigated by Tauster et al. The catalytic activity for alkane hydrogenolysis, for example, can be suppressed if the supported catalyst is subjected to high-temperature pretreatment in H_2. The probable cause of the diminution in catalytic activity is the migration of species from the oxide support on to the exposed surface of the otherwise active metal.

1.3.3.3 Contributions from Theoretical and Computational Chemistry

In the 1980s, some of the most conspicuous endeavours in heterogeneous catalysis focused on two major themes: the integration of the concepts and procedures of theoretical and computational chemistry into the mainstream of surface phenomena; and applying the principles of solid-state chemistry and cognate subjects to the design of new catalysts. Both of these themes, and especially the former, are discussed in Chapter 5. One of the key points about theoretical chemistry is that, possibly alone among all other techniques, it can in principle cope with the structural elucidation of surface intermediates, the lifetime and concentration of which are so small (e.g., 10^{-12} s and 10^{-3} monolayer, respectively) as to make them well-nigh inaccessible to direct experimental study. But theoretical chemistry has made its impact in other ways, especially in offering unifying interpretive frameworks for rationalizing the performance of known catalysts and in attempting to design new ones. Molecular orbital symmetry conservation in transition-metal catalysis is a case in point. The ideas of Woodward, Hoffmann, Longuet-Higgins, Abrahamson, and, in an earlier era, of M. G. Evans have proved remarkably fruitful in the domain of concerted reactions in organic chemstry generally. They have already been profitably incorporated into many situations in heterogeneous catalysis. Thus the importance (see Fig. 1.23) of the siting and availability of antibonding orbitals has been recognized as a result of theoretical enquiries into the nature of the charge transfer associated with adsorbate–adsorbent interaction. Again, it is recognized that concerted reactions are favoured whent the highest occupied molecular orbital (HOMO) of one reactant and the lowest unoccupied MO (LUMO) of another meet certain symmetry requirements. MO theory and orbital conservation rules enable us to ascertain which reactions are 'allowed' and which 'forbidden'. The addition of H_2 to an olefin is forbidden. But Mango realized that, through the intermediacy of a complex, such as that shown in Fig. 1.26, a symmetry-forbidden transformation, such as cyclobutane formation from two alkenes, can be switched to a symmetry-allowed one. Similarly Waugh showed that, during the selective oxidation of benzene to

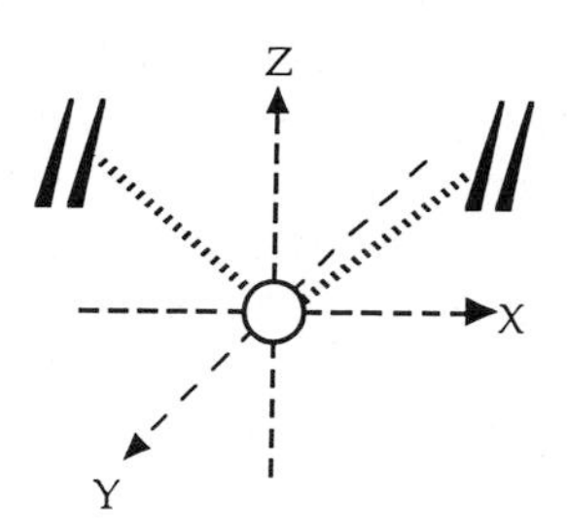

Figure 1.26 Intermediate complex for transition from symmetry-forbidden to symmetry-allowed concerted reaction.

maleic anhydride on an oxide catalyst, the concerted rearrangement of the benzene molecule is allowed.

Progress in the rational design of catalysts has, for certain categories of catalyst, been most encouraging. Much has already been acheived in the design of enzymes and catalytic antibodies (which we discuss briefly below), and there is growing confidence that the rational design of inorganic, especially monophasic, molecular-sieve catalysts will be increasingly possible. We postpone until Chapter 5 discussions on the strategies available for the design of inorganic catalysts.

1.4 Some Intellectual and Practical Challenges for Catalysis in the 21st Century

Catalysis lies at the heart of both the chemical and the petrochemical industries; it is also of pivotal importance in the production of fuel, foodstuffs, pharmaceuticals, and numerous other manufactured goods and is a vital component of a number of critical technologies in this globally industrialized age. The milestones reached (Table 1.3) in industrial catalysis in the 20th century testify to its central importance in the creation of wealth. But the general public is rightly becoming more concerned with environmental issues, and decisions are now being made that are not primarily based on science and technology. Public concern is a potent influence; and for many industries where catalysis already looms large it is becoming increasingly difficult to obtain permits, eliminate waste, construct incinerators, and receive and transport toxic or otherwise hazardous materials.

In the 21st century it will become increasingly necessary to devise processes with close to 100% yields, to effect catalyst recovery, to regenerate and to recycle on a routine basis, and in all this catalysis will play an increasingly dominant role. Elimination of by-products and process waste is becoming a major issue and will demonstrably determine the viability of future chemical processes. It is a sign of the times that the State of California has already decreed that, by AD 2000, 2% of the motor vehicles on its roads must have zero deleterious emission, even though ways of securing this end are at present by no means clear.

Hazardous and toxic materials such as HF, HCN, HCl, H_2SO_4, H_3PO_4, Cl_2, acrylonitrile, formaldehyde, ethylene oxide and phosgene, for example, are more-or-less essential building reagents in the chemical industry since they often possess

Table 1.3 Milestones in the introduction of industrial heterogeneous catalysis during the 20th century.

Decade	Process	Catalyst (Prime constituent)
1900	Methane from $CO + H_2$ (syn-gas)	Ni
	Hydrogenation of fat	Ni
1910	Liquefaction of coal	Fe
	Synthesis of ammonia from $N_2 + H_2$	Fe/K
	Oxidation of ammonia to nitric acid	Pt
1920	High-pressure synthesis of methanol from $CO + H_2$	(Zn, Cr) oxide
	Fischer–Tropsch synthesis of alkanes, alkenes and alkanols from $CO + H_2$	Co, Fe
	Oxidation of SO_2 to SO_3 (sulphuric acid production)	V_2O_5
1930	Catalytic cracking of oil (fixed-bed, Houdry process)	Montmorillonite clays
	Epoxidation of ethylene	Ag
	Oxidation of benzene to maleic anhydride	V
1940	Catalytic reforming of hydrocarbons (gasoline)	Pt/Al_2O_3
	Hydrogenation of benzene to cyclohexane	Ni, Pt
	Inversion of sucrose (and decolorization of golden syrup)	Immobilized enzyme (invertase) on charcoal
1950	Polymerization of ethylene to polyethylene:	
	Ziegler–Natta	Ti
	Phillips process	Cr
	Production of polypropylene and polybutadiene (Ziegler–Natta)	Ti
	Hydrodesulphurization	(Co, Mo) sulphides
	Hydrotreatment of naphtha	$Co–Mo/Al_2O_3$
	Oxidation of naphthalene to phthalic anhydride	(V, Mo) oxides
1960	Oxidation of butene to maleic anhydride	(V, P) oxides
	Oxidation of propylene to acrolein	(Bi, Mo) oxides
	Ammoxidation of propylene to acrylonitrile	(Bi, Mo) oxides
	Improved means of reforming hydrocarbons	Pt–Ir on Al_2O_3 Pt–Re on Al_2O_3
	Metathesis of alkenes	(W, Mo or Re) oxides
	Improved means of cracking of hydrocarbons	Zeolites (Faujasite-based)
	Production of vinyl acetate from ethylene	Pd/Cu
	Oxychlorination of ethylene to vinyl chloride	Cu chloride
	Triolefin process (propylene to butene and ethylene)	$Mo(CO)_6$ or $W(CO)_6$ on Al_2O_3
	Production of *ortho*-xylene from phthalic anhydride	V_2O_5 on TiO_2
	Hydrocracking	$Ni–W/Al_2O_3$
	Water-gas shift (at high temperatures)	$Fe_2O_3/Cr_2O_3/MgO$
	Water-gas shift (at low temperatures)	$CuO/ZnO/Al_2O_3$

Table 1.3 (cont.)

Decade	Process	Catalyst (Prime constituent)
1970	Xylene isomerization (shape-selective formation of *para*-xylene)	H-ZSM-5 (zeolite)
	Methanol from $CO + H_2$	$Cu–Zn/Al_2O_3$
	Disproportionation of toluene to benzene and *para*-xylene	H-ZSM-5
	Catalytic dewaxing	H-ZSM-5
	Auto-exhaust catalyst	Pt, Rh, Pd on oxide
	Conversion of benzylpenicillin to 6-aminopenicillanic acid (production of semisynthetic penicillins)	Immobilized enzyme (penicillin amidase)
	Isomerization of glucose to fructose	Immobilized glucose isomerase
	Hydroisomerization	Pt/zeolite
	Selective reduction of NO with NH_3	V_2O_5 on TiO_2
	Production of MTBE (methyl *t*-butyl ether) from methanol and 2-methylpropylene	Acidic ion-exchange resins
	Fructose-enriched from glucose (from glucose syrup)	Immobilized cells (*Arthrobacter*)
	Hydrolysis of raffinose	Immobilized cells (*Mortierella vinacea*)
1980	Conversion of ethylene and benzene to ethylbenzene	H-ZSM-5
	Methanol to gasoline (petrol) (MTG)	H-ZSM-5
	Conversion of ethylene and acetic acid to vinyl acetate	Pd
	Oxidation of *t*-butanol to methyl methacrylate	Mo oxides
	Improved means of liquefying coal	(Co, Ni) sulphides
	Production of diesel fuel from $CO + H_2$ (syn-gas)	Co
	Hydrotreatment of hydrocarbons	Pt/zeolite Ni/zeolite H^+-Ferrierite
	Catalytic distillation (in MTBE production)	Acidic ion-exchange resins
	Vitamin K_4 production	Pd membrane
	Dehydrocyclization ("Cyclar") of alkanes } Conversion of light alkanes to aromatics }	Ga–ZSM5
	Oxidation of methacrolein } Hydration of isobutene }	Mo–V–P (hetero-polyacid)
	Polymerization of tetrahydrofuran	Phase-transfer catalysis
1990	Production of dimethyl carbonate from acetone	Cu chloride
	Conversion of phenol to hydroquinone and catechol	Ti–silicalite
	Isomerization of but-1-ene to 2-methylpropylene	H^+-Ferrierite } acidic H^+-Theta-1 } zeolites
	Isomerization of oxime of cyclohexanone to ε-caprolactam	Silicoalumino-phosphate molecular sieve (SAPO-11)
	Ammoxidation of cyclohexanone to its oxime using H_2O_2	Ti–silicalite

Table 1.3 (cont.)

Decade	Process	Catalyst (Prime constituent)
1990	Production of acrylamide from vinyl cyanide	Immobilized nitrile hydratase
	Complete combustion of natural gas (at *ca* 1300 °C)	Noble metals and/or mixed oxides
	"Sweetening" of natural gas by selective oxidation of H_2S to S	Mixed oxides
	Oxidation of benzene to phenol via cyclohexene	Zeolite
	Methanol to light alkenes	Silicoalumino-phosphate molecular sieve
	Olefin oligomerization (Shell polygasoline and kerosene process)	Zeolite
	Production of L-aspartic acid and L-analine from ammonium fumarate	Immobilized micro-organisms
	Conversion of toluene to toluene *cis*-glycol	*Pseudomonas putida*
	Production of 2,6-diisopropylnaphthalene using propylene as alkylating agent	Acidic zeolite (mordenite)
	Decomposition of hypochlorite	NiO
	Dehydration of alkanols	Heteropolyacid salts

reactivity or functionality required for further chemical reactions. Future business practices which entail the inventory and transportation of these materials will almost certainly have to be avoided or drastically diminished.

Methyl isocyanate (MIC) is familiar as it was at the centre of the tragic incident that occurred in Bhopal, India, some years ago. At that time and in that place MIC was produced by the phosgenation of methylamine:

$$CH_3NH_2 + COCl_2 \rightarrow CH_3NCO + 2\,HCl$$

It is no longer acceptable to store MIC as was done in Bhopal, and the use of reagents such as phosgene is greatly discouraged. The consequence of this is that industrial companies have sought, and succeeded in producing, safer alternatives. The DuPont Company, for example, now use the catalytic dehydrogenation process (although by 1993 they had not yet disclosed the nature of the catalysts that they developed for these reactions):

$$CH_3NH_2 \xrightarrow{CO} CH_3NHCHO \xrightarrow{O_2} CH_3NCO$$

The DuPont strategy enables them to make MIC and to convert it in situ to an important agrichemical product. In-situ manufacture will become progressively more prominent in the 21st century, since it greatly reduces the risk of unwanted exposure.

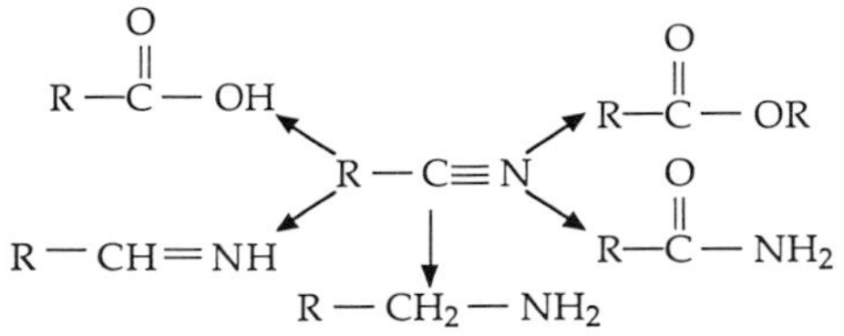

$R{-}C(=O){-}OH$ ← $R{-}C{\equiv}N$ → $R{-}C(=O){-}OR$

$R{-}CH{=}NH$ ← $R{-}C{\equiv}N$ → $R{-}C(=O){-}NH_2$

$R{-}C{\equiv}N$ → $R{-}CH_2{-}NH_2$

Scheme 1.14

This example, entailing a re-analysis of the entire mode of manufacture, could be replaced many times over. The Enichem Company in Italy, in its novel catalytic method of manufacturing the useful monomer dimethyl carbonate, which formerly used massive quantities of phosgene, now follows the following 'safe' process:

$$CH_3COCH_3 + O_2 \rightarrow (CH_3)_2CO_3$$

Yet another example concerns hydrogen cyanide (prussic acid), HCN, which is the starting point for introducing the versatile nitrile functionality in a wide range of organic commodities. Amides, acids, amines, esters, etc., are all readily generated from the corresponding nitrile (Scheme 1.14).

All this underlies the recognition that catalysis is of central importance, and is likely to remain so into the forseeable future, in securing environmentally benign products. Catalysis likewise holds the key to the safer and cheaper manufacture of desirable products. Typical processes are shown in Schemes 1.15 and 1.16.

An attractive method for the utilization of solar energy is its direct conversion into chemical fuels (e.g., syn-gas), thereby making it available as a source either of chemical feedstock, or of heat (by catalytically effecting a reverse reaction – see Fig. 1.27), or both. The Russian ADAM–EVA cycle depicted in Fig. 1.27 is already functioning smoothly. (See Problem 10 in Section 1.5.)

Conceptually it is convenient to divide the future challenges into broad categories such as environmental (Table 1.4), technological (including biotechnological) (Table 1.5) and fundamental. And although it is possible to draw up a list of targets as is done in these tables, it must never be forgotten that these three categories are quite

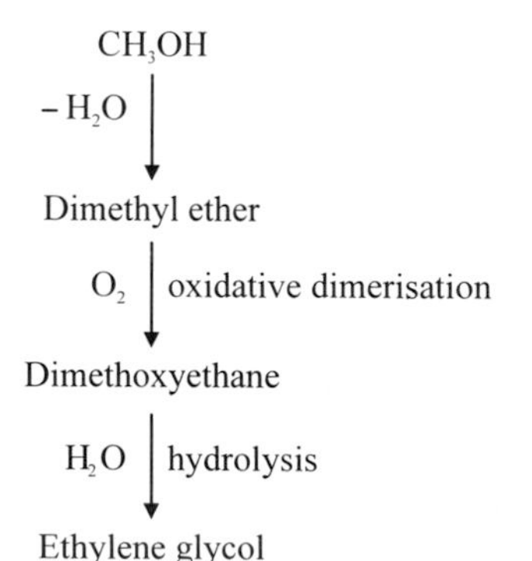

Scheme 1.15 Production of ethylene glycol from methanol.

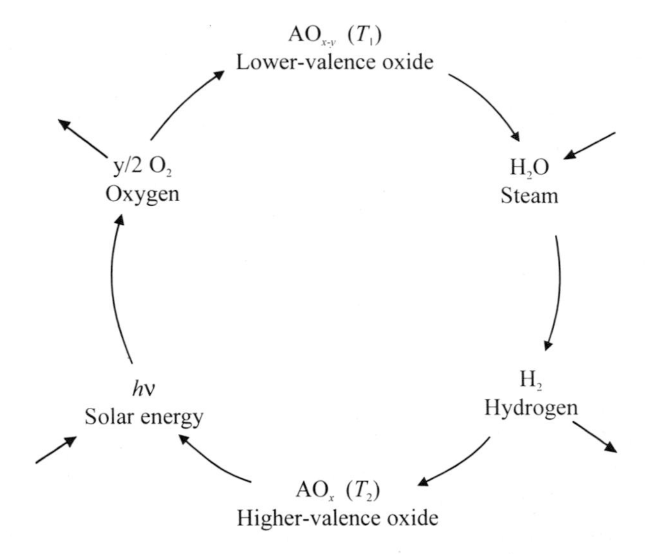

Scheme 1.16 Thermochemical cycles with metal oxide for the production of hydrogen from water. Ideally the 'higher' oxide needs to decompose (catalytically) very rapidly at a high temperature T_2 and the steam to react with the 'lower' oxide at a low temperature T_1.

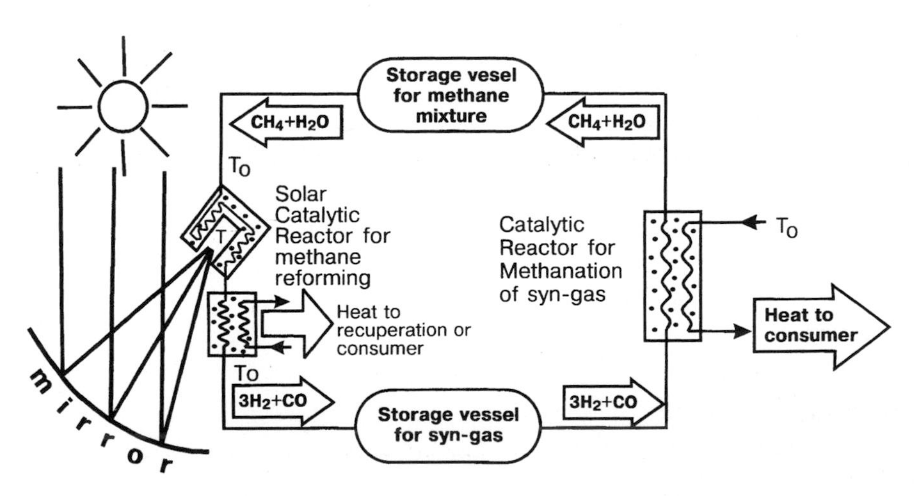

Figure 1.27 Scheme of thermocatalytic conversion of solar energy based on the closed thermochemical 'ADAM–EVA' cycle. T is the temperature inside the cavity of the solar catalytic reactor; T_0 is the temperature of environment. With permission from K. I. Zamaraev, *Topics in Catalysis*, **1996**, *3*, 1.

Table 1.4 A selection of environmental challenges.

Development of 'zero-waste' processes
Minimization of hazardous products and 'greenhouse' gases
Replacement of corrosive liquid acid catalysts by benign solid acid catalysts
Evolution of sustainable systems
Reduction in volume of by-products
Development of processes requiring less 'consumption' of catalysts
Elimination of voluminous by-products

Table 1.5 A selection of technological challenges.

Reformulated transport fuels (containing lower amounts of aromatics and volatile components, and larger amounts of more completely combustible additives)
Development of (catalytic) automobiles operating on methanol dissociation[a]
Better catalysts for hydrodesulphurization, hydrodenitrification of light oils and coals and hydrotreatment of heavy oils and tars
Single-step synthesis of desirable products:
acetaldehyde from ethane
aromatics from ethane
phenol from benzene
acrolein from propane
acrylonitrile from propane by ammoxidation
acetic acid from methanol
2-methylpropylene from syn-gas
Better methods for isomerizing linear alkanes into branched-chain ones
Functionalization of light alkanes, especially methane (e.g., by use of heterogenized metalloporphyrins, P450s, or cytochromes)
Efficient routes to cheaper feedstocks for the chemical and pharmaceutical industries
Development of robust, re-usable, chiral catalysts
New shape-selective catalysts, e.g., for nonthermodynamic ratios of mono- and di-methylamines from methanol and ammonia
Development of processes using CO_2 as reactant
Cheaper and safer methods of generating hydrogen
New catalytic membranes
Better electrocatalysts for fuel-cell consumption of plentiful hydrocarbons
Development of modified enzymes, organisms or transgenic plants for 'natural' production of polymers[b]
Families of solid catalysts for 'tunable' conversion of methanol to either ethylene or propylene
Fischer–Tropsch catalysts for sharply defined reaction products
Efficient, safe methods of generating hydrogen peroxide (from H_2 and O_2)[c]

Table 1.5 (cont.)

Designed solids capable of controlled release of structural elements such as oxygen or hydrogen
Uniform, molecular-sieve catalysts (of redox or Brønsted type) possessing well-defined larger pores (40–100 Å diameter)
'Targetable' antibody catalysts as therapeutic products[d]
Engineered proteins for pharmaceutical use

[a] A vehicle running on a fuel in which methanol dissociates to CO and H_2 on board has many attractions. First, such a fuel is more efficient than undissociated methanol. Second, the heat required for endothermic dissociation of the alcohol can be supplied by the engine exhaust gas. This recovers waste heat and increases the heating value of the fuel. Third, an engine functioning in this manner can be operated with excess air (i.e., under lean burning conditions), which facilitates the complete combustion of CO and hydrocarbons. Lastly, problems associated with formaldehyde emission are appreciably reduced, and emission of the oxides of nitrogen are likewise greatly diminished.

[b] Zeneca plc (formerly ICI) produced 100 tons of the natural polymer Biopol in 1993. Biopol is a bacterial storage polymer, produced by fermentation of an alkaligenous strain. It consists of polyhydroxybutyrate or a copolymer with polyhydroxyvalerate. It is both biocompatible and biodegradable.

[c] In 1992, several companies including Mitsubishi, DuPont and Interox announced palladium-or platinum-based catalysts for this synthesis.

[d] Zeneca plc have also recently illustrated the potential power of biotechnology in general, and biocatalysis in particular, in a medical context: they developed a toxin–antibody conjugate (D0490) aimed at colorectal cancer. The idea in using antibody conjugates this way is to link a 'warhead' to a sophisticated guidance system in a modern version of the so-called 'magic bullet'. As the warhead, Zeneca chose ricin, a natural plant toxin with two protein strands. Ricin depends on its carrier B-chain to convey the toxic A-chain to its site of action. Recombinant DNA technology enables the A-chain of the toxin to be produced independently by fermentation.

arbitrary; the distinctions between them are blurred. Purists rightly argue that, in most circumstances, the solutions to practical problems almost invariably demand a deeper understanding of fundamentals. Moreover, distinctions between the technological and biotechnological can themselves be fuzzy. Recall that hitherto, for example, the source of C_{10}–C_{14}-hydrocarbons for use in, say, detergency was petrochemical. Nowadays, thanks to major advances in the production of transgenic plants and other consequences of the molecular biological revolution, it is possible to develop oilseed rape hybrids rich in C_{12}-oils. And when it comes to biocatalysts, and the use of immobilized cells, modified organisms, or cloned genes (to generate spectacularly stereoselective and active catalysts), the distinction between heterogeneous and homogeneous systems also becomes less well defined.

Returning to fundamentals, it is almost a truism that the perennial quest has to be understanding the factors that enhance specificity, increase activity and improve lifetime. A better catalyst must, by definition, do one or all of these things. To understand better how we may achieve these ends is the pervasive theme of this book.

1.5 Problems

These problems focus largely on thermodynamic and a little on strategic aspects of catalysis. They are intended to consolidate the reader's assumed acquaintance with thermodynamic and related considerations. In each subsequent chapter the problems set are linked directly to the topics discussed in that chapter, and serve to illustrate or extend the relevant principles.

1 Using only thermodynamic principles, show that for any given temperature and total pressure the maximum conversion of hydrogen to ammonia occurs when the ratio of hydrogen to nitrogen in the feed gas to an ammonia synthesis reactor is 3:1.

2 Natural gas consists chiefly of methane. It has been suggested that one way of avoiding dependence on oil for the production of important chemicals is for benzene to be produced according to the following reaction:

$$6\,CH_4\,(g) \rightarrow C_6H_6\,(g) + 9\,H_2\,(g)$$

Before embarking on a search for a suitable catalyst to effect this conversion, we must first determine whether this reaction is feasible. Pursue this question quantitatively given that ΔC_p for the above reaction is:

$$\begin{aligned} &42.0 - 32.1 \times 10^{-3}T + 3.83 \times 10^{-6}T^2 \text{ cal K}^{-1}\text{ mol}^{-1} \\ &= 176 - 134 \times 10^{-3}T + 16.0 \times 10^{-6}T^2 \text{ JK}^{-1}\text{ mol}^{-1} \end{aligned}$$

[Hint: You should begin by consulting thermodynamic tables in other text books, or in National Bureau of Standards compilations, and work out $\Delta H^\ominus_{298}$ and $\Delta S^\ominus_{298}$ for the proposed reaction.]

3 Investigate the effect of temperature on the disproportionation of toluene, given the accompanying data.

T[°C]	50	100	150	200	250
K_o	0.059	0.065	0.070	0.074	0.078
K_m	0.206	0.208	0.209	0.210	0.211
K_p	0.0837	0.0875	0.0904	0.0928	0.0949

K_o, K_m, K_p are thermodynamic equilibrium constants for the disproportionation reaction yielding benzene and *ortho*-, *meta*- or *para*-xylene respectively.

What temperature would you choose to effect (a) the best conversion to *m*-xylene, and (b) the best conversion to benzene? What effect would total pressure have on the product distribution?

4 Choose an approximate operating pressure for the industrial preparation at 500 °C of methanol from synthesis gas (a mixture of carbon monoxide and hydrogen), given the following data on free energies ($\Delta G^{\ominus}$) and enthalpies ($\Delta H^{\ominus}$) of formation:

	CO (g)	CH_3OH (g)
$\Delta G^{\ominus}_{298}$/kJ mol^{-1}	−7836	−9241
$\Delta H^{\ominus}_{298}$/kJ mol^{-1}	−6310	−11483

5 The composition of a gas emerging from the secondary reformer of a modern ammonia plant has the following composition:

$$39.5\%\ H_2, 16.3\%\ N_2, 28.3\%\ H_2O, 4.5\%\ CO_2, 10.7\%\ CO, 0.2\%\ A, 0.5\%\ CH_4$$

Assuming that equilibrium is established, what would the CO content of this gas be if it were fed directly to the first stage of a shift converter employing an iron oxide/chromium oxide catalyst and operating at 450 °C? The equilibrium constant for the reaction

$$CO + H_2O \rightleftharpoons CO_2 + H_2$$

is 7.337 at 450 °C. Why is it normal practice to have a low-temperature (ca 250 °C) shift converter as a second stage? Would the same catalyst as used for the high-temperature (first) stage be an appropriate choice for the second stage?

6 With the phase-down in the use of lead additives in petrol (gasoline), there is a growing need to obtain blending agents which have high octane numbers. One such material is methyl *tert*-butyl ether, MTBE (2-methyl-2-methoxypropane). It can be synthesized, using clay catalysts (see Ballantine et al., *Chem. Lett.* **1985**, *6*, 763) from isobutene (2-methylpropylene). An attractive route for the production of isobutene, which is comparatively scarce in contrast to methanol, of which there is a glut, is to take *n*-butane (from natural gas), isomerize it to isobutane (iB) and then to dehydrogenate the latter to 2-methylpropylene (2MP), as shown in Scheme 1.17. Shape-selective zeolitic catalysts are good for the first and platinum on alumina for the second of these two steps. Of considerable importance in designing a feasible operating system are the equilibrium fractions of the various

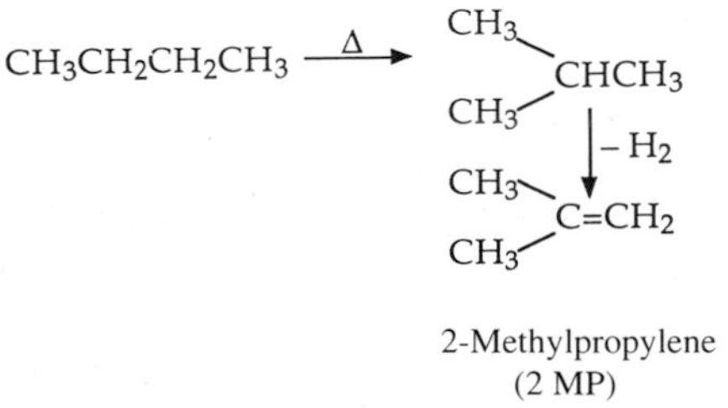

$$CH_3CH_2CH_2CH_3 \xrightarrow{\Delta} (CH_3)_2CHCH_3 \xrightarrow{-H_2} (CH_3)_2C{=}CH_2$$

2-Methylpropylene
(2 MP)

Scheme 1.17 Production of 2-Methylpropene from *n*-butane.

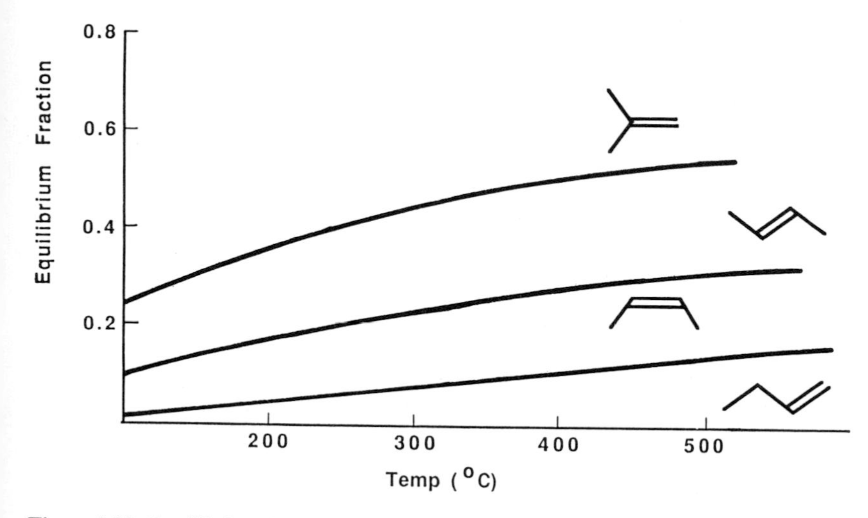

Figure 1.28 Equilibrium fraction of various alkenes in the thermal dehydrogenation of isobutane:

$$\begin{matrix} CH_3 \\ CH_3 \end{matrix}\!\!>CH\,CH_3 \rightleftharpoons C_4H_8 + H_2$$

possible isomers. These equilibrium fractions are shown in Fig. 1.28 for a total pressure of 1 bar.

Explain how such quantitative information is derived; and discuss which temperature ranges are the most appropriate for overall conversion. Consult recent literature (e.g. S. Natarajan et al., *J. Chem. Soc., Chem. Commun.* **1993**, 1861) for an alternative catalytic route to 2MP.

7 A popular method of producing formaldehyde on an industrial scale is to pass a mixture of methanol vapour and air at a total pressure of 1 bar over a metallic silver catalyst at 550 °C. During the course of this process, the silver slowly loses its lustre and gradually disintegrates. Using the following data, examine whether this might be due to the formation of silver oxide:

$$\Delta G^{\ominus}_{298}\,(Ag_2O) = -10.826\,\text{kJ mol}^{-1}$$

$$\Delta H^{\ominus}_{298}\,(Ag_2O) = -30.556\,\text{kJ mol}^{-1}$$

The relevant heat capacities [$\text{J K}^{-1}\,\text{mol}^{-1}$] are: Ag, 26.75; Ag_2O, 65.63; O_2, 31.35.

8 Benzaldehyde can be generated from carbon monoxide and benzene over an appropriate catalyst at 50 °C and 500 bar total pressure. Given the information below, describe the calculations by which you would estimate an upper limit to the fraction of the benzene converted to benzaldehyde.

(a) Heat capacities and standard free energies and enthalpies of the compounds in question at 298 K,

(b) The densities of benzene and benzaldehyde at 298 K; and
(c) p–V data for carbon monoxide at 20 °C and 30 °C in the pressure range up to 500 bar.

9 Nitric oxide leads to the depletion of the ozone layer ($NO + O_3 \rightarrow NO_2 + O_2$). Since NO is liberated in vast quantities from power stations currently in use, efforts are now underway to reduce its concentration by catalytic ammoxidation using V_2O_5–TiO_2. One such reaction is: $4\,NH_3 + 3\,O_2 \rightarrow 2\,N_2 + 6\,H_2O$. Write down other feasible reactions leading to N_2 or N_2O and H_2O as sole products. From thermodynamic data, estimate the equilibrium constants of these reactions.

10 Several strategies are available for the use of catalysts in harnessing solar energy. One of these, illustrated in Fig. 1.27, is the so-called ADAM–EVA cycle operated by the Boreskov Institute; another, by the Weizmann Institute in Israel, has been described (I. Dostrovsky, Scientific American **Dec. 1991**, *265*, 50). Both these thermocatalytic converters entail the cycle of 'syn-gas' production (from the endothermic reaction $CH_4 + H_2O$) followed by methanation ($CO + 3\,H_2 \rightarrow CH_4 + H_2O$) with liberation of heat. But several other strategies are available, e.g., production of syn-gas and production of hydrogen.

(a) Indicating which catalyst and the reactor conditions you would use, draw up a minimum of eight specific syntheses of products (such as alkanes, alkenes, alkanols and gasoline) that could be produced from a solar furnace.
(b) Starting from either Fe_2O_3 or Fe_3O_4 and water, describe a feasible set of reactions and conditions (including appropriate catalysts) for the solar production of hydrogen and oxygen.

1.6 Further Reading

1.6.1 General Background

G. C. Bond, *Heterogeneous Catalysis and Applications*, 2nd Edition, Oxford University Press, Oxford, **1987**.

M. Boudart, G. Djega-Mariadassou, *Kinetics of Heterogeneous Catalytic Reactions*, Princeton University Press, Princeton, NJ, **1984**.

R. L. Burwell (Ed.), *Heterogeneous Catalysis: Selected American Histories*, ACS Symp. Series No. 211, American Chemical Society, Washington, DC, **1983**.

B. C. Gates, *Catalytic Chemistry*, Wiley, New York, **1992**.

B. C. Gates, L. Guczi, H. Knözinger (Eds.), *Metal Clusters in Catalysis*, Elsevier, Amsterdam, **1986**.

D. W. Goodman, J. Phys. Chem. **1996**, *100*, 13090.

G. L. Haller, D. E. Resasco, *Catalysis*. In *Encyclopedia of Applied Physics* (Ed. G. L. Trigg), Vol. 3, VCH, Weinheim, **1992**, p. 67.

G. L. Haller, R. S. Weber, *Concepts in Heterogeneous Catalysis*. In *Metal–Ligand Interactions* (Eds. D. R. Salahub, N. Russo). Kluwer, Amsterdam, **1992**, p. 71.

L. L. Hegedus (Ed.), *Catalyst Design, Progress and Perspectives*, Wiley, New York, **1987**.

R. Hoffmann, *Solids and Surfaces: A Chemist's View of Bonding in Extended Structures*, VCH, Weinheim, **1988**.

J. A. Moulijn, P. W. N. M. van Leeuwen, R. A. van Santen (Eds.), *Catalysis: An Integrated Approach to Homogeneous, Heterogeneous and Industrial Catalysis*, Elsevier, Amsterdam, **1993**.

R. Prins, G. C. A. Schuit (Eds.), *Chemistry and Chemical Engineering of Catalytic Processes*, NATO-ASI No. 39, Sijthoff and Noordholt, Alphen aan de Rijn, **1980**.

M. W. Roberts, C. S. McKee, *Chemistry of the Metal–Gas Interface*, Clarendon, Oxford, **1978**.

T. N. Rhodin, G. Ertl (Eds.), *The Nature of the Surface Chemical Bond*, North-Holland, Amsterdam, **1979**.

C. N. Sattersfield, *Heterogeneous Catalysis in Practice*, 2nd edition, McGraw-Hill, New York, **1991**.

D. F. Shriver, M. J. Sailor, *Acc. Chem. Res.* **1988**, *21*, 374.

J. H. Sinfelt, *Bimetallic Catalysts*, Wiley, New York, **1983**.

G. A. Somorjai, *Principles of Surface Chemistry and Catalysis*, Wiley, New York, **1994**.

K. Tamaru, *Dynamic Heterogeneous Catalysis*, Academic Press, London, **1978**.

J. M. Thomas, W. J. Thomas, *Introduction to the Principles of Heterogeneous Catalysis*, Academic Press, London, **1967**.

J. M. Thomas, K. I. Zamaraev (Eds.), *Perspectives in Catalysis*, IUPAC/Blackwells, Oxford, **1992**.

H. Topsøe, B. S. Clausen, F. E. Massoth, *Hydrotreating Catalysts*, Springer, Berlin, **1996**.

R. A. van Santen, *Theoretical Heterogeneous Catalysis*, World Scientific, Singapore, **1991**.

1.6.2 Short Reviews

A series of review volumes edited by D. A. King and D. P. Woodruff (*The Chemical Physics of Solid Surfaces and Heterogeneous Catalysis*, Elsevier, Amsterdam), along with another on *Catalysis* (published by the Royal Society of Chemistry, London) and three well-established review journals that appear more or less annually – edited by A. T. Bell (*Catalysis Reviews: Science and Engineering*, Plenum, New York), by J. R. Anderson and M. Boudart (*Catalysis*, Springer, Berlin), and by W. O. Haag, D. D. Eley and B. C. Gates (*Advances in Catalysis*, Academic, New York) – are invaluable sources of information.

A collection of short reviews, dealing with new developments in alternatives for chlorofluorohydrocarbons (by G. Webb), enantioselective hydrogenations (P. B. Wells), the molecular basis of catalytic activity (D. A. King), exhaust catalysts in lean burn engines (R. W. Joyner, M. Bowker), new uses for natural gas (R. Burch), zeolitic catalysts (G. J. Hutchings), and photocatalysts for detoxifying aqueous organic pollutants (R. I. Bickley), appear *Chemistry in Britain* **1992**, 28, 989–1014.

An excellent short review entitled *Catalysis Looks to the Future*, prepared by a group of US experts under the chairmanship of A. T. Bell, was published by the National Academy Press, Washington, **1992**, for the US National Research Council.

Another invaluable source of information is the comprehensive *Handbook of*

Heterogeneous Catalysis edited by Ertl, Knözinger and Weitkamp (VCH, Weinheim 1997) which covers almost every aspect of heterogeneous catalysis in nearly 200 articles of varying length.

1.6.3 General Articles

Recent publications, intelligible to the advanced undergraduate, include:

A. J. Bard, Electro- and photo-catalysis, *J. Phys. Chem.* **1982**, *86*, 172.

G. Centi, Selective heterogeneous oxidation of light alkanes, *Catal. Lett.* **1993**, *22*, 53.

A. Corma, Isoparaffin-olefin alkylations, *Catal. Rev. Sci. Technol.* **1992**, *34*, 481.

J. A. Cusumano, New technology and the environment, *Chemtech* **1992**, 482 (see also *Perspectives in Catalysis*, J. M. Thomas, K. I. Zamarev (Eds.), IUPAC/Blackwells, Oxford, **1992**, p. 1).

B. Delmon, Hydrotreating catalysts: new challenges, *Catal. Lett.* **1993**, *22*, 1.

I. Dostrovsky, Chemicals from solar light, *Scientific American*, **1991**, *265*, 50.

G. Ertl, Oscillatory Kinetics in Reactions at Solid Surfaces, *Science*, **1991**, *254*, 1750.

C. M. Friend, Catalysis on surfaces, *Scientific American* **1993**, *267* (April), 42.

F. Küber, Dreams of the perfect plastic, *New Scientist*, **1993** (Aug. 14), 28.

M. D. Lilly, Advances in biotransformation processes, *Chem. Eng. Sci.* **1994**, *49*, 151.

J. H. Lunsford, Oxidative coupling of natural gas, *Langmuir* **1989**, *5*, 12.

R. J. Madix, Molecular transformations on single crystal metal surfaces, *Science* **1986**, *233*, 1159.

L. E. Manzer, V. N. M. Rao, Towards catalysis in the 21st century chemical industry, *Adv. Catal.* **1993**, *39*, 329.

P. M. Maitlis, A. Haynes, G. J. Sonley, M. J. Howard, Methanol carbonylation revisited: thirty years on. *J. Chem. Soc. Dalton Trans.*, **1996**, 2187.

I. E. Maxwell, M. E. Davis, Solid Catalysts and Porous Solids, *Current Opinion in Solid State and Materials Science*, **1996**, *1*, 55.

Y. Ono, Transformation of lower alkanes into aromatic hydrocarbons, *Catal. Rev. Sci. Technol.*, **1992**, *34*, 179.

V. N. Parmon, Z. R. Ismagilov, M. A. Kerzhentsev, Catalysis for energy production, in *Perspectives in Catalysis*, J. M. Thomas, K. I. Zamarev (Eds.), IUPAC/Blackwells, Oxford, **1992**, p. 37.

J. F. Roth, Industrial catalysis: poised for innovation, *Chemtech* **1991**, 357.

R. J. Schmidt, P. L. Bogdan, N. L. Gilsdorf, Meeting the challenge of reformulated gasoline, *Chemtech* **1993**, 41.

K. C. Taylor, Auto-exhaust catalysts, *Catal. Sci. Technol.* **1984**, *5*, 119 (see also K. C. Taylor, *Catal. Rev. Sci. Technol.*, **1993**, *35*, 457).

J. M. Thomas, New microcrystalline catalysts, *Philos. Trans. R. Soc. London Ser. A* **1990**, *333*, 173.

J. M. Thomas, Solid acid catalysts, *Scientific American* **1992**, *267*(April), 112.

J. M. Thomas, Turning Points in Catalysis, *Angew. Chem., Int. Ed. Engl.* **1994**, *33*, 913.

J. M. White, C. T. Campbell, Surface chemistry in heterogeneous catalysis, *J. Chem. Educ.*, **1980**, *57*, 471.

P. B. Wells, Some Aspects of Catalyst Characterization and Activity, *Farad. Disc. Chem. Soc.* **1989**, *87*, 1.

J. T. Yates, Surface chemistry, *Chem. Eng. News* **1992**, *70*(13), 22.

K. I. Zamaraev, Catalysis and new technologies for sustainable development, *Chemistry for Sustainable Development* **1993**, *1*, 133.

1.6.4 Homogeneous and Heterogeneous Catalysis

For short articles on the topic of correlations and differences between homogeneous and heterogeneous catalysis, see:

J. M. Bassett et al., in *Perspectives in Catalysis*, J. M. Thomas, K. I. Zamarev (Eds.), IUPAC/Blackwells, Oxford, **1992**, p. 125.

M. Che, *Proc. 10th Int. Congr. Catalysis* (Budapest, July 1992), L. Guczi, F. Solymosi, P. Tetanyi (Eds.), Akademia Kaido, Budapest, **1993**, p. 31.

W. S. Knowles, *Acc. Chem. Res.*, **1983**, *16*, 106.

T. J. Marks, *Acc. Chem. Res.* **1992**, *25*, 57; and J. J. Rooney, *J. Mol. Catal.* **1985**, *31*, 147.

G. Wilke, *Angew. Chem., Int. Ed. Engl.* **1988**, *27*, 186.

K. I. Zamaraev, *Topics in Catalysis* **1996**, *3*, 1.

1.6.5 Role of Catalysis

The crucial role of catalysis is exemplified in the US National Research Council's report on *Critical Technologies: The Role of Chemistry and Chemical Engineering*, L. L. Hegedus (Ed.), National Academy Press, Washington, DC, **1992**.

1.6.6 Catalytic Antibodies

Several admirable short reviews of the burgeoning field of catalytic antibodies, which symbolises the merging of chemistry and immunology, are contained in a special issue, edited by P. G. Schultz and R. A. Lerner, of *Accounts of Chemical Research* **1993**, *26*(8). The article by D. R. Burton on 'Monoclonal antibodies from combinatorial libraries' in the same issue (p. 405) is among the best currently available as an introduction to the language and the explanation of the terms of catalytic antibodies. See also *Accounts of Chemical Research* **1996**, 29(3).

1.6.7 Enzyme Catalysis

For synoptic short reviews, see:

W. P. Jencks, Destabilization is as important as binding, *Philos. Trans. R. Soc. London Ser. A*, **1993**, *345*, 3.

J. R. Knowles, Not different, just better, *Nature (London)*, **1991**, *350*, 121.

M. F. Perutz, What are enzyme structures telling us? *Faraday Discuss. Chem. Soc.* **1992**, *93*, 1.

1.6.8 Progress in Biotechnology

Two highly readable short reviews of progress in biotechnology and the use of immobilized cells, enzymes, and biocatalysts for enantiselective synthesis of chiral nonracemic compounds are:

P. Dunnill, Immobilized cell and enzyme technology, *Philos. Trans. R. Soc. London Ser. B*, **1980**, *290*, 409.

P. Doyle, Biotechnology in action (The Zeneca Lecture, October 1993), *Science in Public Affairs* (Royal Society, London) **1993**, 20.

2 The Fundamentals of Adsorption: Structural and Dynamical Considerations, Isotherms, and Energetics

2.1 Catalysis Must Always be Preceded by Adsorption

At least one of the reactants in all heterogeneously catalysed processes must be attached for a significant period of time to the exterior surface of the solid catalyst. We can right away appreciate that two distinct mechanistic situations can arise in the surface-catalysed transformation of gas-phase species A and B to a product C (Fig. 2.1): either both species are attached to the surface, and atomic reorganization takes place in the resulting adsorbed layer (the so-called Langmuir–Hinshelwood mechanism of heterogeneous catalysis); or only one of them is bound, and is converted to product when the other impinges upon it from the gas phase (the Eley–Rideal mechanism). We shall discover later that the first of these mechanisms holds good far more frequently than the second, that the distinction between them is sometimes blurred, and that other subtleties can arise.

It is helpful at the outset to distinguish adsorption from absorption. In principle, *ad*sorption, being by definition the preferential accumulation of material – termed the adsorbate – at a surface, is very different from *ab*sorption, which is a bulk phenomenon. We note that if the amount of gas taken up by a solid at a fixed pressure and temperature is proportional to its surface area, and not to its volume, then adsorption, not absorption, predominates. The distinction is lost, however, when solids either are capable of intercalating guest species (like many layered solids such as clay and graphitic catalysts which generate more internal area by the process of assimilation of guest) or when they are highly microporous (like the zeolites and pillared clays which have cavities and channels of molecular dimensions) so that most of the surface area resides inside the solid.

2.1.1 Physical Adsorption, Chemisorption and Precursor States

It is also helpful to distinguish physical adsorption from chemisorption. The former involves the forces of molecular interaction which embrace permanent dipole, induced dipole, and quadrupole attraction. For this reason it is often termed van der Waals adsorption. Chemisorption, on the other hand, involves the rearrangement of the electrons of the interacting gas and solid, with consequential formation and rupture of chemical bonds. By its very nature, physical adsorption is characterised by enthalpy changes that are small, typically in the range -10 to $-40\,\mathrm{kJ\,mol^{-1}}$ (heats of adsorption of 10–$40\,\mathrm{kJ\,mol^{-1}}$), whereas heats of chemisorption are rarely less than $80\,\mathrm{kJ\,mol^{-1}}$ and often exceed $400\,\mathrm{kJ\,mol^{-1}}$. Other distinctions between these two extreme types of adsorption devolve upon the temperature at which adsorption takes place and its specificity. Physical adsorption occurs, in general, only at temperatures

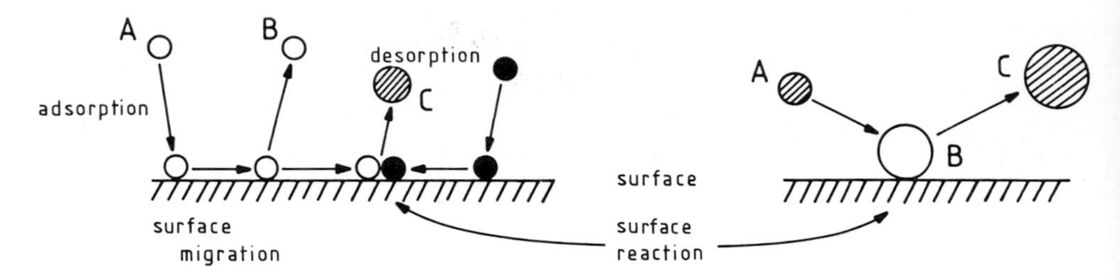

Figure 2.1 Two possible ways in which heterogeneous catalysis proceeds at a surface: the Langmuir–Hinshelwood mechanism (left) and the Eley–Rideal mechanism (right).

close to the boiling point of the adsorbate at the operative pressure, whereas chemisorption has no such restriction. Since chemisorptions are chemical reactions confined to the surfaces of solids, and since chemical reactions are specific, so also are chemisorptions in the sense that, if a species is chemisorbed by a given solid under certain conditions, it does not follow that the same gas will be chemisorbed by another solid of comparable cleanliness under identical conditions. Indeed, it does not follow that because one particular crystallographic face of a given solid is capable of chemisorbing a gas-phase species so also will another face of the *same* solid. Such specificity is quite marked for many metals, non-metals, oxides, and chalcogenides. At low coverages, a certain degree of specificity is found in some processes of physical adsorption, but at higher coverages, as the equilibrium pressure of adsorbate rises, physical adsorption becomes almost indistinguishable from condensation or liquefaction, and there is no crystallographic or material specificity.

At this juncture it is instructive to consider the events which lead to the formation of a chemisorbed overlayer when a gas molecule impinges upon a solid surface. Let us focus on a clean surface and ask what could in principle happen following a collision:

1. The gas molecule may be elastically scattered (i.e. without loss of energy) back into the gas phase.
2. The molecule may lose sufficient translational energy to the solid to become trapped in a physically adsorbed state.
3. If a chemisorbed state can be formed in the vicinity of the site of incidence, the molecule may – if it possesses adequate energy – pass directly to the chemisorbed state (see Fig. 2.2 below), without being trapped in the physically adsorbed state.
4. If the molecule is trapped in the physically adsorbed state at the site of incidence, it may (i) become chemisorbed, (ii) be inelastically (i.e. with loss of energy) scattered back into the gas phase, or (iii) hop to a neighbouring site, in which case pathways (i) and (ii) are again open.
5. During formation of the chemisorbed species, the molecule, or its dissociated constituents, may (a) lose sufficient chemical energy, released to the solid by the exothermic transfer from the physically adsorbed to the chemisorbed state, and become localized at the original site; (b) lose insufficient energy and hence make a limited number of diffusive hops until the excess energy is dissipated; or (c)

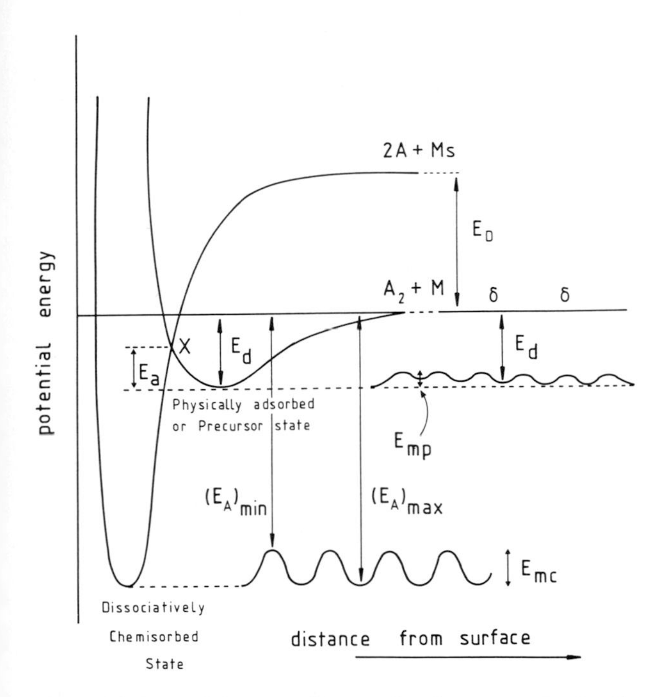

Figure 2.2 Potential-energy diagram for the approach of a diatomic molecule A_2 toward a surface M. E_{mc} and E_{mp} are the activation energies for surface migration in the dissociatively chemisorbed and molecularly, physically adsorbed states. E_{mc} is the difference between the maximum and minimum energy of adsorption. E_D is the bond dissociation energy and E_d the energy of desorption of A_2.

undergo continual migration, depending upon the ratio of the activation energy of migration and the thermal energy kT.

With the experimental techniques now at our disposal (see Chapter 3) we can ascertain which of these many options are taken up by a molecule when it collides with a particular solid surface under a defined set of catalytically relevant circumstances. To appreciate the subtleties involved we take advantage of potential-energy diagrams, of the kind introduced by Lennard-Jones in 1932 explicitly to describe the energetics associated with the approach of a molecule to a solid surface. The first shallow potential-energy minimum (Fig. 2.2) describes the physically adsorbed state. If, as the molecule proceeds nearer to the surface, chemical bonding can occur, then a deeper well is created. There is ample evidence for the importance of this physically adsorbed, or precursor, state in chemisorption, the most important being the fact that, for a wide range of gas–metal systems, chemisorption is a non-activated process. (Note in Fig. 2.2 that whenever the crossover, X, of the physical adsorption and chemisorption curves lies below the potential-energy zero, dissociative chemisorption from the molecular state is non-activated). In the original Lennard-Jones model of adsorption, and in many subsequent ones, the surface was regarded as a continuum, so that there was no periodic variation in energy across it. In reality, however, there are energy maxima and minima to be traversed when both weakly (physically adsorbed or precursor state) and strongly bound adlayers are involved. E_{mp} and E_{mc} (Fig. 2.2) are, respectively, the activation energies for migration in the physically and chemically bound states. We shall have more to say about the magnitude of the respective energies of these processes shortly.

In so far as the nature of the precursor state is concerned, frequently this is synonymous with the physically adsorbed state. In such weakly bound conditions a molecule may have several potential sites for chemisorption during its residence on the surface. In general, precursor states encompass molecules physically adsorbed both on a bare surface and on top of other adsorbed layers. We would expect a molecule that collides with a pre-bound species at a surface to 'stick' less effectively than those that collide with the bare surface. Indeed, a great deal of insight into adsorption mechanism can be gleaned from measurements of sticking probability, s, defined as the ratio:

$$s = \frac{\text{rate of adsorption}}{\text{rate of bombardment}}$$

Typically, the initial sticking probability, s_0, for the adsorption of reactive gases such as hydrogen, carbon monoxide, oxygen and nitrogen on to a clean metal falls in the range 0.1–1.0. Occasionally s_0 for certain gases on some metals (e.g. O_2 on poly-

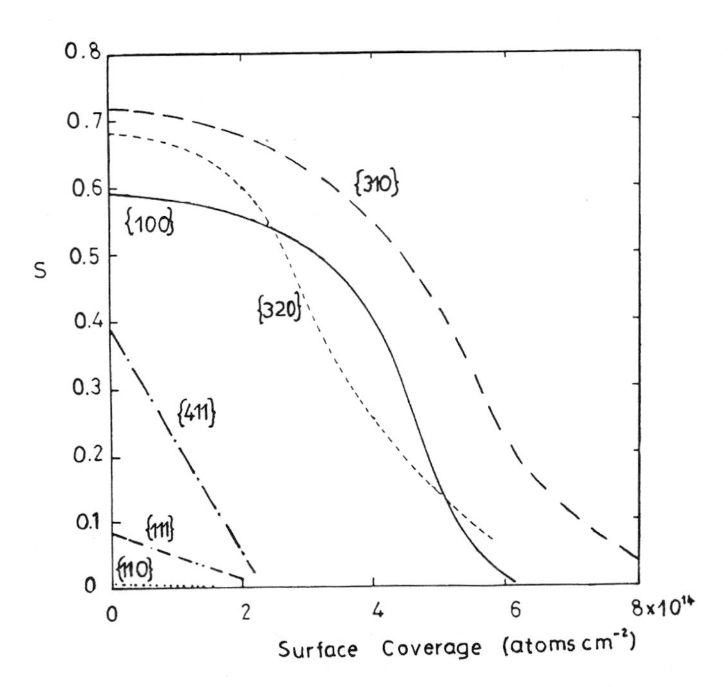

Figure 2.3 Sticking coefficient as a function of surface coverage for nitrogen on some of the single-crystal planes of tungsten at 300 K. With permission from S. P. Singh-Bopari, M. Bowker, D. A. King, *Surf. Sci.* **1975**, *53*, 55.

crystalline silver) lies in the range 10^{-3}–10^{-4}. For O_2 on the basal surface {0001} of graphite s_0 has been estimated to be lower than 10^{-15}. Several systems show extreme crystallographic specificity, as seen in Fig. 2.3; and the existence of such widespread variation in sticking probabilities among the different crystallographic faces of a particular metal, tungsten, foreshadows the phenomenon of catalytic variability as a function of crystal face, a topic to which we shall return often in this text.

The disposition of atoms at the exterior surface of a catalyst is, therefore, a quintessential feature in our understanding of catalysis in general and adsorption in particular. Using the techniques to be described in Chapter 3 we are capable of deducing the precise interatomic distances and coordination numbers both at bare surfaces and within the adsorbed layers. We next summarise some salient aspects of this wealth of structural information, since, without it, we cannot properly appreciate the factors that come into play in our subsequent discussions of adsorption isotherms, of rates of adsorption and desorption, and of the associated energetics.

2.2 The Surfaces of Clean Solids are Sometimes Reconstructed

Chiefly as a result of applying low-energy electron diffraction (LEED) and other scattering techniques (see Chapter 3), we now know that the symmetry and interatomic distances at the surfaces of some solids are the same as those we would expect of an ideal crystal plane obtained by conceptually cleaving perfect crystals of the solids in question. There are, however, some solids – a number of metals and semi-

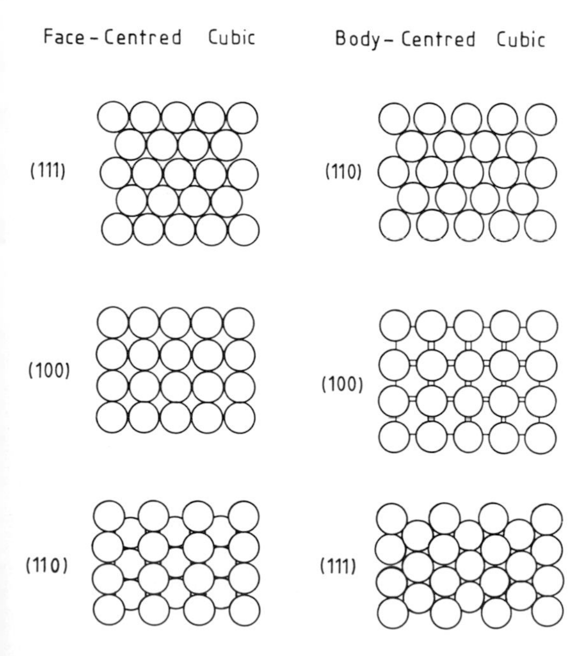

Figure 2.4 Representations of high-symmetry planes of face-centred cubic (fcc) and body-centred cubic (bcc) metals.

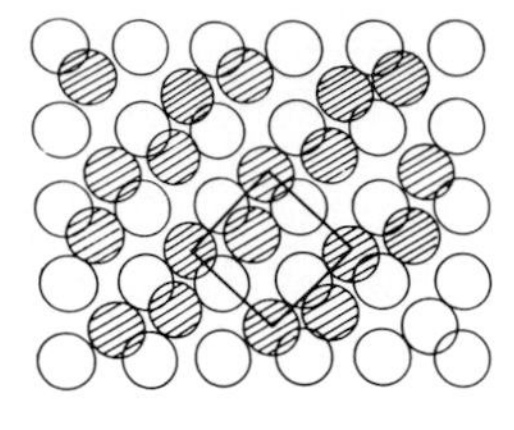

Figure 2.5 Model of the c(2 × 2) structures on W(001) illustrating the displacements in the plane of the surface. The shaded circles signify W atoms in the top layer; the open circles represent the second layer. With permission from M. K. Debe, D. A. King, *J. Phys. Chem.* **1977**, *10*, L303.

conductors among them – which, even in their state of highest surface cleanliness, adopt a structure different from that of the corresponding plane in the bulk.

The ideal surface structures of the so-called high-symmetry planes of face-centred cubic (fcc) and body centred cubic (bcc) metals are shown in Fig. 2.4. Platinum, silver and nickel have fcc structures; iron, tungsten and molybdenum have bcc structures. For most transition metals it is found that the actual surface structures are nearly perfect and deviate little from the structures shown in Fig. 2.4. In the notation used to describe surface structures, we say that unreconstituted patterns such as those shown here possess (1 × 1) periodicity. (The use of indices to described crystal planes and directions is explained in Section 5.2.1.1, and the notation for surface meshes in Sections 3.6.)

However, a clean surface of W(100) upon cooling to below room temperature adopts a supermesh structure designated c(2 × 2) (see Fig. 2.5). Here we see that atoms in the outermost layer have shifted from the positions occupied in the ideal structure. Tungsten is not alone in this respect; molybdenum (100) surfaces behave likewise, and the iridium (100) face adopts a (5 × 1) mesh as symbolized in Fig. 2.6. Gold, too, adopts a reconstructed surface; and recent high-resolution electron micrographs (Fig. 2.7) show direct proof of the occurrence of the two-fold repeat in what is believed to be a (2 × 1) supermesh.

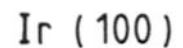

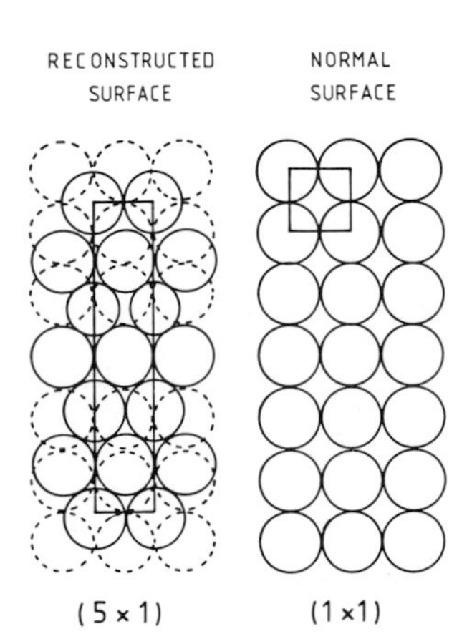

Figure 2.6 Schematic representation of the outer atomic layers of an Ir(001) surface in its reconstructed (5 × 1) structure (left) and in its non-reconstructed structure (right). The outermost atomic layer is indicated by solid circles whereas the broken circles give the positions of the atoms in the second layer for the reconstructed surface.

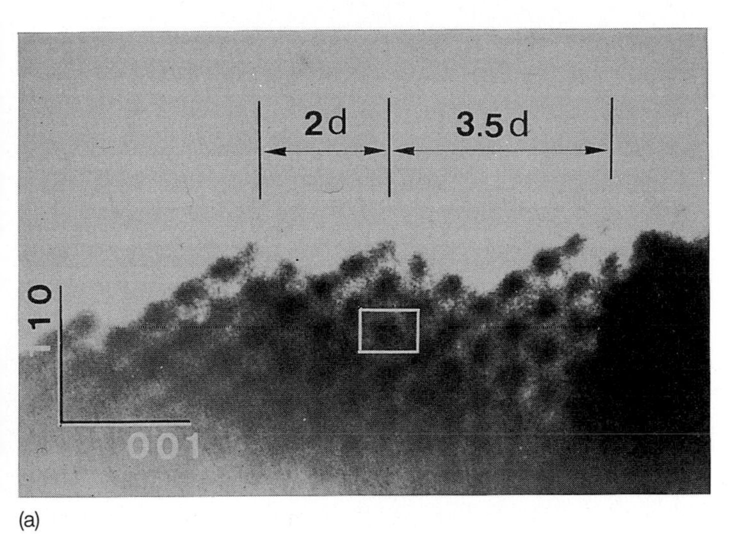

(a)

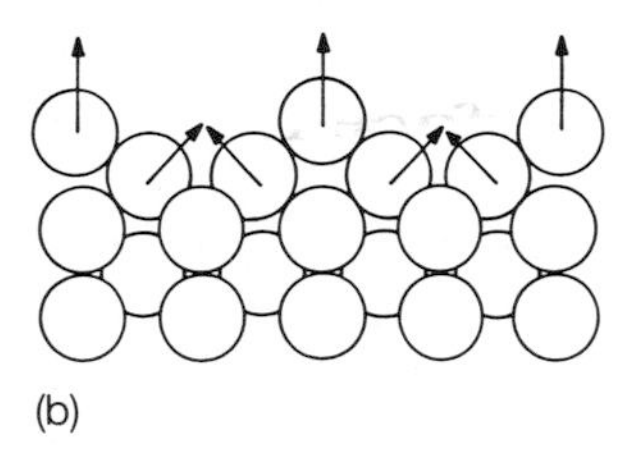

(b)

Figure 2.7 (a) High-resolution electron micrograph (with permission from L. D. Marks, *Surf. Sci.* **1984**, *139*, 281) showing an elevated view of the (110) surface of gold which has undergone reconstruction to a (2 × 1) supermesh, as schematized in (b). In (b) the arrows indicate the directions of displacement of the surface atoms.

Here we need not pursue why such reconstructions take place; such matters are discussed briefly in Chapter 5. It is important to note, however, that the uptake of a fraction of a monolayer of adsorbate can also sometimes give rise to a rearrangement of the outermost layers of the adsorbent (catalyst). Thus, displacements of W atoms from the 'ideal' positions on the (100) face to the positions shown in Fig. 2.8 occur when H_2 is dissociatively chemisorbed by the metal.

We must not regard the surface of any solid as being totally rigidified into a particular structure or mesh where the exposed atoms are fixed in a static fashion at their lattice points. At any finite temperature the individual atoms execute motion about their time-averaged position, the amplitudes of such displacements increasing with increasing temperature. And when sufficient thermal energy is endowed to the crystal and the mean-square displacement exceeds a certain value (which is about one-tenth of the interatomic distance for a metallic or covalent solid) the surface melts. The collective thermal motions of the constituents of a solid are discussed in terms of phonons, the quantum of collective vibration. Surface phonons are implicated in a number of catalytic reactions.

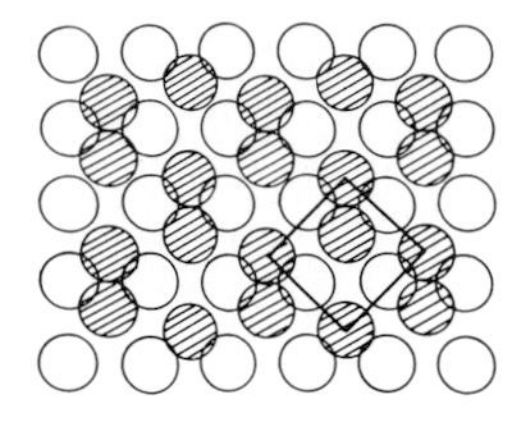

Figure 2.8 The c(2 × 2) structure on W(001) induced to form as a result of hydrogen chemisorption (compare Fig. 2.5). The H atoms are not shown, but they are thought to be located on top of the dimers. (With permission from P. J. Estrup, R. A. Barker, in *Ordering in Two Dimensions* (Ed.: S. K. Sinha), North-Holland, Amsterdam **1980**, p. 39.).

2.3 There are Many Well-Defined Kinds of Ordered Adlayers

If the activation energy for surface migration is small by comparison with the thermal energy (kT) (see Fig. 2.2), the adsorbed species is freely mobile across the surface and its behaviour will be that of a gas molecule restricted to two dimensions. If adsorbate–adsorbent and/or adsorbate–adsorbate interactions are significant we can expect the adlayer to take up an ordered structure which, as in an ordered crystal, reflects a balance of the internal energy and entropic terms so as to minimize the overall free energy.

The conventional model (known as the lattice gas model) for describing adlayers is based on the assumption (not always valid as we saw in the preceding section, Fig. 2.8) that the solid surface can be treated as a rigid matrix of discrete sites for adsorbed species. In general, especially for physical adsorption, this assumption is reasonably valid: it leads to a convenient method of defining surface coverage θ as:

$$\theta = \frac{\text{number of adsorbed species}}{\text{number of atoms in the outermost layer of the substratum}}$$

Experiment shows that well-defined supermesh (often called superlattice) structures are taken up as the surface is gradually populated: the limits of stability of these various structures are shown in simple phase diagrams (temperature–surface coverage plots). For oxygen dissociatively chemisorbed on tungsten (110) faces, the situation is summarized in Fig. 2.9, from which we see that four distinct kinds of surface phases (p(2 × 2), p(2 × 1), p(1 × 1) and a disordered one) exist in this system. For the oxygen/Ni(111) system, five distinct phases, including several ordered and one disordered adlayers, have been identified.

On the basis of the lattice gas model we can understand why ordered adlayer structures are formed. There are rigorous quantitative ways, using Monte Carlo simulations for example, of calculating the phase diagram for a given set of assumptions and parameters relating to particular adlayer–adlayer and adlayer–substrate interaction energies and related terms. But we can qualitatively illustrate how p(2 × 2) and c(2 × 4) structures are adopted at surface coverages of $\theta = 0.25$ on the (100) face of a cubic crystal in terms of the following arguments. The formation of a p(2 × 2) structure (Fig. 2.10(a)) is explained in the lattice gas formalism by the existence of adlayer–adlayer interactions which give a repulsion between atoms on nearest-neighbour sites ($\varepsilon_1 > 0$) and on next-nearest sites ($\varepsilon_2 > 0$), combined with

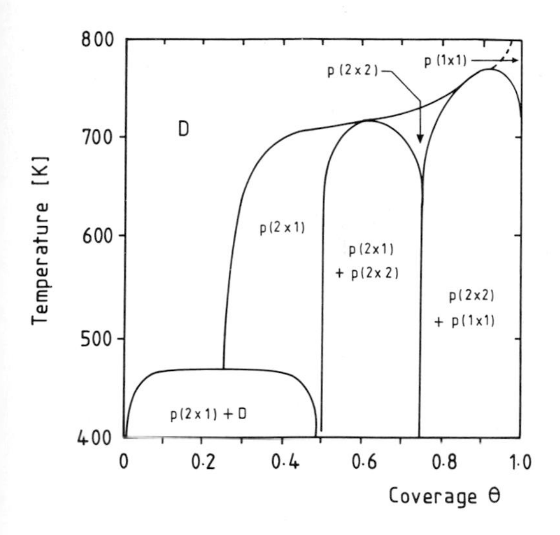

Figure 2.9 Phase diagram for the ordered adlayers formed when oxygen is dissociatively chemisorbed on a W(110) surface. D stands for disordered surface phase. (With permission from L. D. Roelafs, in *Chemistry and Physics of Solid Surfaces IV* (Eds.: R. Vanselow, R. Howe) Springer, Berlin **1983**, p. 209.

attraction at greater distances. If, however, the repulsion extended beyond next-nearest neighbours, a c(2 × 4) supermesh (Fig. 2.10(b)) would be favoured at this coverage. (In principle, the magnitude of ε_1, ε_2, etc., can be found by calculating the phase diagram (compare Fig. 2.9) for different sets of these parameters until a fit with the experimental phase diagram is achieved.)

We now proceed to quote a few examples which illustrate how widespread is the occurrence of ordering, both in cases where the adsorption forces are relatively weak and in those where adsorption is strong.

When gases such as xenon, krypton and methane are adsorbed at low temperatures on a graphite surface a $(\sqrt{3} \times \sqrt{3})$ $R30°$ structure (Fig. 2.11), commensurate with the underlying graphite structure, is formed. For methane this structure dominates at temperatures up to 20 K and coverages less than 0.9 monolayers. For krypton, the classic early work of Thomy and Duval has recently been extended, and we now have an interesting picture that reveals ordering in second and subsequent adsorbed layers on graphite. This is schematized in Fig. 2.12, which refers to the kryp-

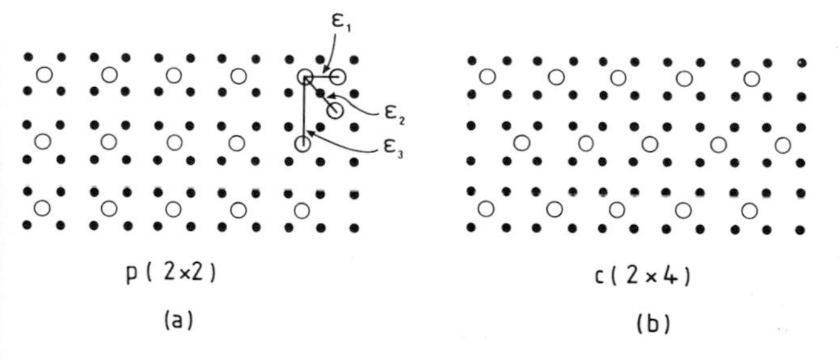

Figure 2.10 Large circles represent adatoms forming a two-dimensional lattice gas at a solid surface. In (a) there is a p(2 × 2) and in (b) a c(2 × 4) structure; ε_1 and ε_2 denote the interaction (repulsion) between nearest neighbour and next-nearest neighbours in the adlayer.

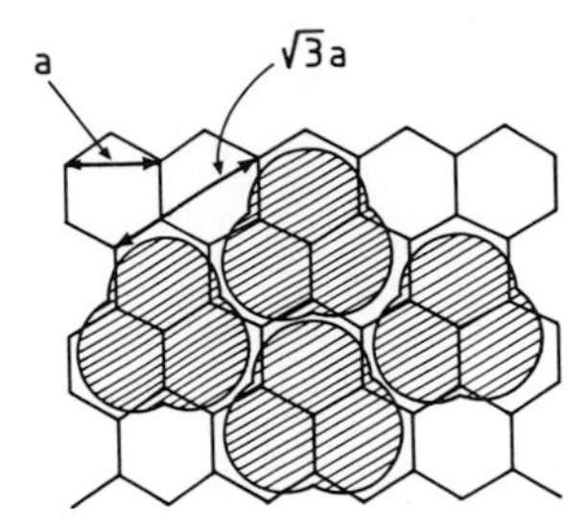

Figure 2.11 A $\sqrt{3} \times \sqrt{3}$ commensurate structure is formed when CH_4 (and several other gases) are adsorbed on the basal surface of graphite at low temperature. The repeat distance for graphite is *a*.

ton adsorption isotherm at 77 K. The main features here are the large vertical steps, which are interpreted to be due to layer-by-layer condensation (physical adsorption). Note that, sandwiched between the first- and second-layer steps, there is a small feature, which we now know marks the transition from a commensurate $(\sqrt{3} \times \sqrt{3})$ $R30°$ monolayer to an incommensurate monolayer with a lattice constant closer to the value of the krypton bulk crystal. An incommensurate structure is defined as one

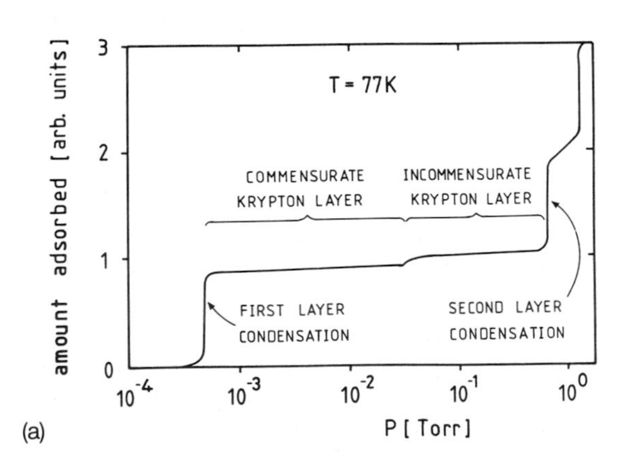

(a)

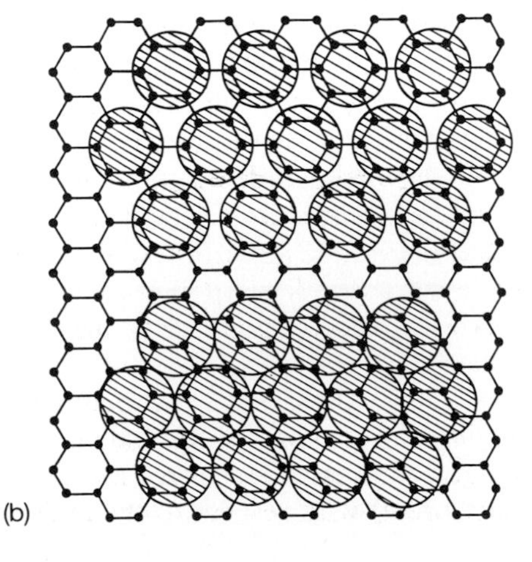

(b)

Figure 2.12 (a) A 'multilayer' adsorption isotherm for krypton on graphite at 77 K. First-, second-, and third-layer condensation is visible and the commensurate and incommensurate adlayer structures (see text) are shown. With permission from S. C. Fain, in *Chemistry and Physics of Solid Surfaces IV* (Eds.: R. Vanselow, R. Howe), Springer, Berlin **1982**, p. 203. (b) Representation of the $\sqrt{3} \times \sqrt{3}$ commensurate phase (top) and of an incommensurate denser monolayer structure (bottom).

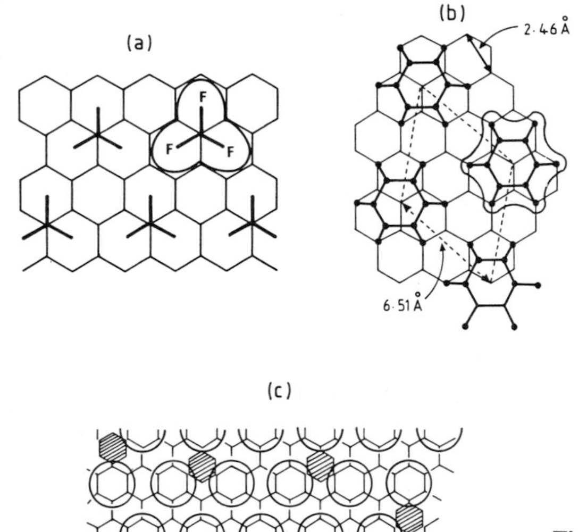

Figure 2.13 CF_4 adsorbed on graphite adopts a 2×2 supermesh (a), whereas benzene forms a $\sqrt{7} \times \sqrt{7}$ structure (b). When *sym*-tribromobenzene is adsorbed, the Br atoms form a 2×2 supermesh, but the benzene skeletons to which they are attached are disordered (c).

in which the ratio of the number of adsorbate molecules to the number of substrate adsorption sites is not a simple fraction, such as the 1/3 value for the $(\sqrt{3} \times \sqrt{3})\ R30°$ structure (see Fig. 2.12b). Very many other types of supermesh structures have been reported for a variety of species adsorbed on to graphite: CF_4 (Fig. 2.13a) can form a 2×2, benzene (Fig. 2.13b) a $\sqrt{7} \times \sqrt{7}$ and GeI_4 a 3×3 structure. An unusual situation, discovered by Lander, arises with *sym*-tribromobenzene where the bromine atoms take up a commensurate 2×2 structure, but with the consequence that the benzene framework of the molecule is disordered (Fig. 2.13c).

Turning to ordered chemisorbed layers, the situation is even more intricate as one might have imagined owing to there being several possible distinct adsorption sites (Fig. 2.14) or sets of sites which may be preferred for a given gas–solid system under certain prescribed conditions of coverage and temperature. Again we see, as with physically adsorbed species, well-defined two-dimensionally ordered structures of the commensurate type. But with some bound species, especially CO which is frequently in its molecular state when adsorbed, incommensurate structures often exist. There is yet a further subtlety, however: the so-called coincidence structure, where some of the ad-species take up well-defined sites with respect to atoms in the substrate, while others do not.

Pritchard's work (1976) illustrates nicely the various kinds of order that may arise when CO is chemisorbed at the (100) and (111) faces of copper. At one time it was thought that adlayers of CO for the high-coverage compressed state on these faces had the ad-species out of registry (i.e., incommensurate) with respect to the under-

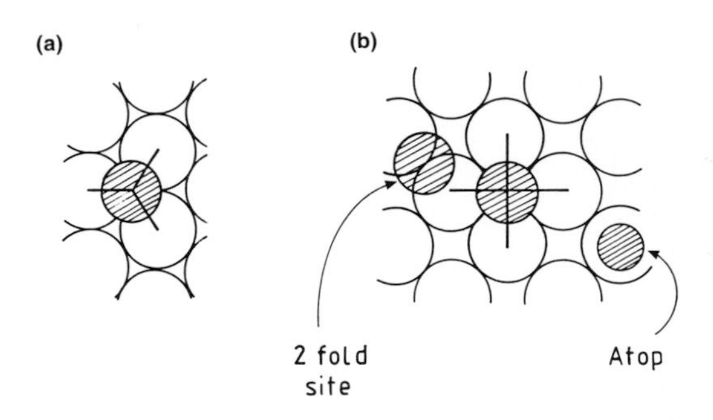

Figure 2.14 Illustration of a three-fold site on the (111) surface of an fcc metal (a) and a four-fold site on the (100) surface (b); two-fold and atop sites also exist on each of these surfaces.

lying metal atoms. Pritchard has drawn attention to the fact that as coverage of the metal gradually increases, structures such as those shown in Fig. 2.15 (a) and (c) [for the c(2 × 2) mesh] are likely. In (a), fourfold hollow sites are favoured; in (c) the CO sits atop (onefold) in a terminally bonded fashion. And in the two coincidence structures, Fig. 2.15 (b) and (d), fourfold sites in (b), or terminal sites in (d), are occupied to an appreciable degree along with others of lower symmetry. Another interesting structure adopted by CO molecules chemisorbed on to Cu(100) is shown in Fig. 2.16, where arrays of bridge-bonded (B) and linearly bonded (L) sites (the latter being synonymous with terminally bonded CO) are arranged in an ordered recurrent fashion across the surface. Two other examples of the adoption of well-defined sites in systems that are of relevance to catalysis are shown in Fig. 2.17, where in (a) we see that CO is adsorbed at a coverage of $\theta = 0.5$ only at bridge sites; and in (b) we see that, owing to mutual interactions between the adsorbed species,

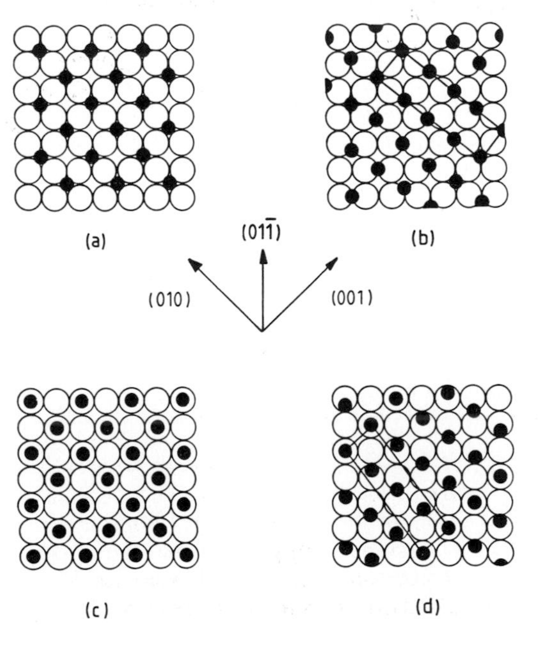

Figure 2.15 Possible sites for CO molecules (filled circles) in the c(2 × 2) supermesh, (a) and (c), and in the coincidence structure (b) and (d). Adsorbent is Cu(100). With permission from J. Pritchard, *Surf. Sci.* **1979**, *79*, 231.

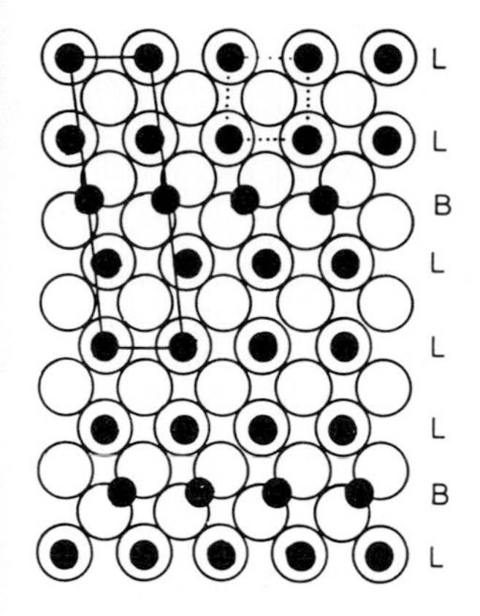

Figure 2.16 An adlayer structure for CO consisting of alternating rows of linear (atop or ten-fold) and bridged (two-fold) sites on a Cu(100) surface. With permission from J. Pritchard, *Surf. Sci.* **1979**, *79*, 231.

oxygen atoms occupy threefold hollow sites on Pt(111) faces, the resulting 2×2 mesh corresponding to saturation at $\theta = 0.25$. Finally, we note that in the uptake of Cl_2 by a Cr(100) surface it is possible to track the sequential formation of ordered phases [$c(2 \times 2) \rightarrow c(2 \times 4) \rightarrow p(2 \times 5)$] with increasing coverage (Fig. 2.18). It is interesting to note that, with the addition of Na atoms to the chlorided chromium surface, an ordered layer of NaCl, in registry with the substratum, forms (Fig. 2.19). This is an example of *epitaxy*, and the (4×4) NaCl layer is said to be in epitactic relationship to the underlying chromium. It has frequently been surmised that the special catalytic behaviour of one oxide spread as a veneer on another (e.g. V_2O_5 on anatase is an excellent catalyst for the selective oxidation of *o*-xylene to phthalic anhydride) is a consequence of epitaxy. Supporting this view is the fact that V_2O_5 on rutile, another polymorphic form of TiO_2 with a different surface structure from that of anatase, is less catalytically efficient.

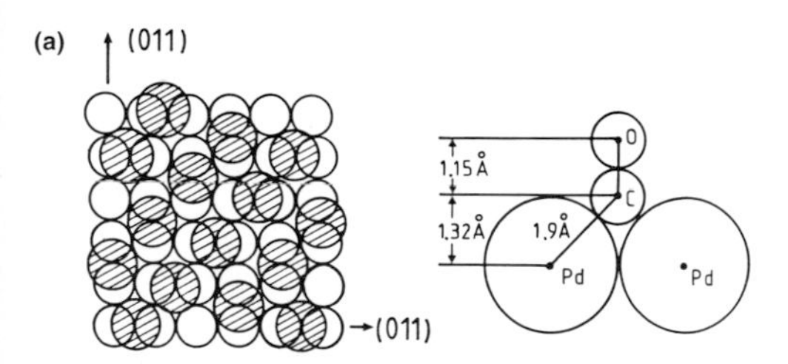

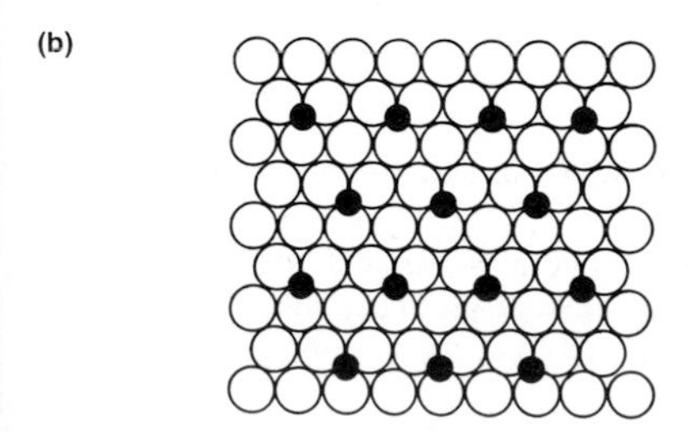

Figure 2.17 (a) Structure of CO adsorbed on a Pd(100) surface at a coverage of $\theta = 0.5$. (b) The open 2×2 structure formed when O_2 is dissociatively chemisorbed (at a 'saturation' value of $\theta = 0.25$) on Pd(111). With permission from G. Ertl, in *Catalysis: Science and Technology*, Vol. 4 (Eds.: J. Anderson, M. Boudart), Springer, Berlin, **1983**, p. 209.

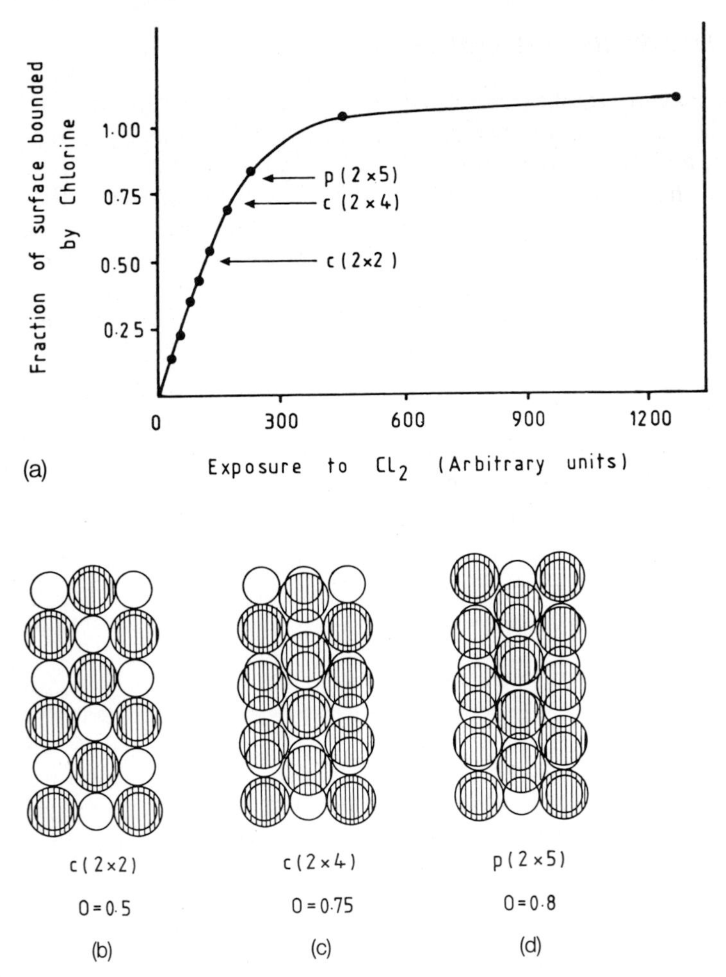

Figure 2.18 (a) Ordered structures are formed as the (100) surfaces of chromium are covered with adsorbed chlorine atoms. In the schematic illustrations (b), (c), and (d), open circles represent Cr atoms, shaded ones Cl. With permission from J. S. Foord, R. M. Lambert, *Surf. Sci.* **1982**, *175*, 141.

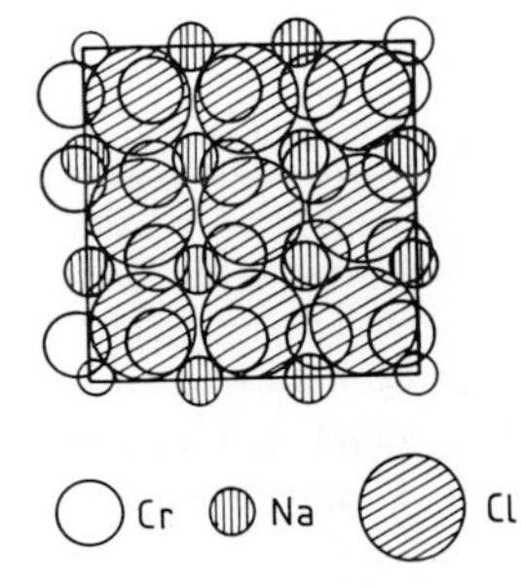

Figure 2.19 Proposed model for the Cr(100) (4 × 4) NaCl epitactic (or epitaxial) overlayer. A square coincidence mesh is outlined (thick lines). With permission from J. S. Foord, R. M. Lambert, *Surf. Sci.* **1982**, *175*, 141.

2.4 Adsorption Isotherms and Isobars

Experience has shown that the equilibrium distribution of adsorbate molecules between the surface of the adsorbent and the gas phase is dependent upon pressure, temperature, the nature and area of the adsorbent, and the nature of the adsorbate. An adsorption isotherm shows how the amount adsorbed depends upon the equilibrium pressure of the gas at constant temperature: an adsorption isobar, how the amount adsorbed varies with temperature at constant pressure. An isostere relates the equilibrium pressure to the adsorption temperature for a stipulated amount of gas adsorbed. Figure 2.18(a) is an adsorption isotherm for a single-crystal surface. High-area solids such as zeolites when they adsorb small molecules (H_2, O_2, N_2 or CO) are typified by the isotherms shown in Fig. 2.20(a). An adsorption isobar typical of those obtained when gas is introduced at low temperature (*ca* 100 K) to a polycrystalline metal or oxide and the system subsequently raise, keeping the pressure constant, to higher temperatures (*ca* 1000 K) is shown in Fig. 2.20(b).

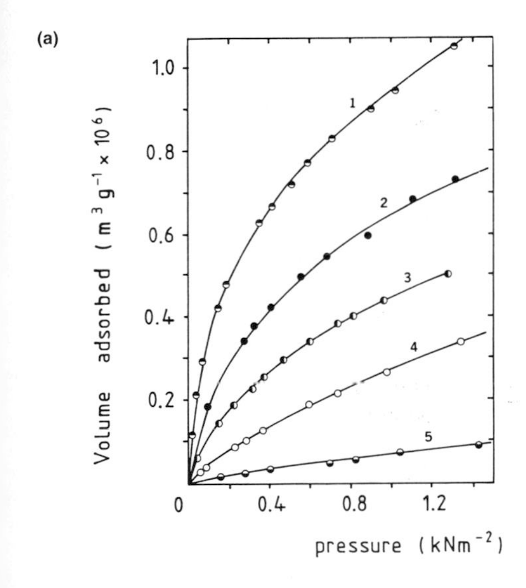

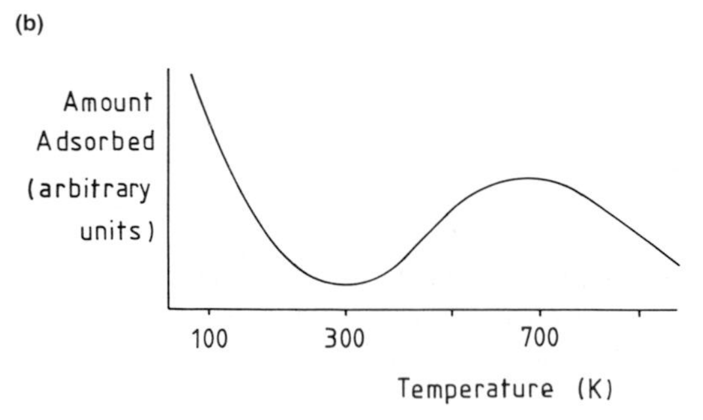

Figure 2.20 (a) A series of adsorption isotherms for the uptake of CO by Ca^{2+}-exchanged zeolite-Y. 1, 273 K; 2, 290.3 K; 3, 304.4 K; 4, 332.6 K; 5, 349 K. With permission from T. A. Egerton, F. S. Stone, *Trans. Faraday Soc.* **1970**, *66*, 2364. (b) Schematic illustration of the kind of adsorption isobar obtained on heating a solid previously exposed to gas at low temperature.

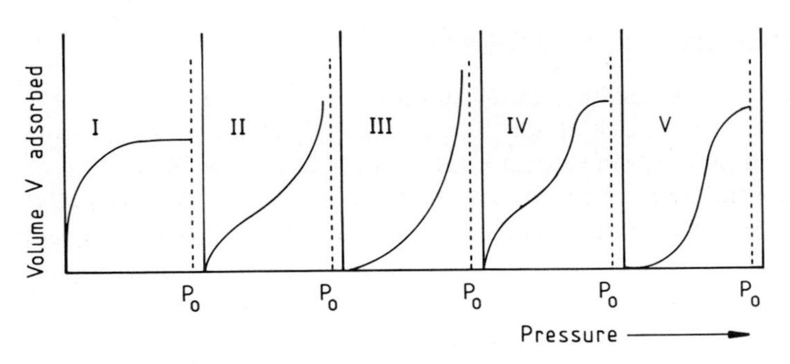

Figure 2.21 The five types of adsorption isotherms in the Brunauer classification.

Table 2.1 A selection of adsorption isotherms.[(a)]

Name	Isotherm equation	Applicability	Equation no.
Langmuir[(b)]	$\frac{V}{V_m} = \theta = \frac{bp}{1+bp}$	Chemisorption and physical adsorption	(41)
Henry[(c)]	$V = k'p$	Chemisorption and physical adsorption at low coverages	
Freundlich	$V = kp^{1/n} (n > 1)$	Chemisorption and physical adsorption at low coverages	(49)
Temkin	$\frac{V}{V_m} = \theta = A \ln Bp$	Chemisorption	(50)
Brunauer–Emmett–Teller (BET)	$\frac{p}{V(p_0 - p)} = \frac{1}{V_m c} + \frac{c-1}{V_m c} \cdot \frac{p}{p_0}$	Multilayer, physical adsorption	(51)
Polanyi[(d)]	$\varepsilon = RT \ln(p_0/p)$	Physical adsorption	(53)
Dubinin–Radushkevich[(e)]	$\ln x = \ln(W_0 \rho) - D[\ln(p_0/p)]^2$	Multilayer formation in microporous solids	(58)
Dubinin–Kaganer–Radushkevich (DKR)	$x = x_m \exp(-B\varepsilon^2)$	Physical adsorption up to a monolayer	(62)
Virial	$\frac{p}{RT} = x(1 + a_1 x + a_2 x^2 + \ldots)$	Multilayer formation in micropores	(63)

(a) Amounts adsorbed at pressure p are represented either by volume V or mass x. Unless otherwise specified, all other symbols in these equations are constants.
(b) V_m (and x_m) correspond to monolayer coverage.
(c) This equation is the limiting form of the Langmuir equation.
(d) The adsorption potential ε is defined by this equation.
(e) W_0 stands for the total volume of all the micropores in a solid.

2.4.1 The Empirical Facts

Viewed empirically, for all manner of solids and adsorbates there are but five types of adsorption isotherms (Fig. 2.21); this is the so-called Brunauer classification. Most isotherms, however, are of type I in this system. Equations describing observed isotherms are readily formulated empirically and many of them can, as we show later (Section 2.6.2), be derived theoretically. Table 2.1 summarizes those isotherms that figure most commonly in discussions of catalysis.

2.4.2 Information that can be Gleaned from Isotherms

At a glance one may tell whether multilayer adsorption is likely to be taking place. Thus, type II and type III isotherms (Fig. 2.21) leave little doubt that uptake continues beyond a monolayer, which is what one would normally (but not invariably) deduce from the occurrence of a type I isotherm for, say, a low-area solid. There are quite reliable methods of deducing from experimental isotherms the quantitative uptake of adsorbate and monolayer coverage. These are discussed in Chapter 4. A rough guide to the achievement of monolayer saturation is the point at which the knee occurs in type I or type II isotherms.

When true adsorption–desorption equilibrium is established, that is when the system strictly conforms to thermodynamics reversibility, the heat of adsorption ($-\Delta H_a$) for a coverage θ may be obtained from adsorption isotherms measured at different temperatures using the Clausius–Clapeyron equation:

$$\left.\frac{\mathrm{d}\ln p}{\mathrm{d}(1/T)}\right|_{\theta=\mathrm{const}} = \frac{-\Delta H_a}{R}$$

A plot of $\ln p$ versus $1/T$ at constant coverage (or a quantity which is directly proportional to coverage, such as Auger electron emission intensity, or change in work function – see Chapter 3) yields the isosteric heat of adsorption at the respective coverage. Thus from Fig. 2.20(a) we construct the plot shown in Fig. 2.22(a), which yields a heat of adsorption for CO on a partially exchanged Ca–Y zeolite of $44\,\mathrm{kJ\,mol^{-1}}$. Proceeding likewise for the family of adsorption isotherms of CO reversibly adsorbed on Pd(111) in the temperature range 295–563 K, we obtain the plot of isosteric heat of adsorption shown in Fig. 2.22(b). The rapid drop in heat of adsorption at coverages beyond $\theta \approx 0.5$ reflects the effective size of the adsorbed species. The extent of orbital repulsion may be estimated by empirical means using appropriate functions that depend upon separation distances in the adlayer. With the system CO/Pd(100), the pairwise repulsion interaction potential U_{ij}, of the form

$$U_{ij} = -A/r_{ij}^{6} + B/r_{ij}^{12} \text{ (Lennard-Jones)}$$

may be derived as a function of the CO $\cdots$ CO distance, r, from the variation of heat of adsorption with coverage. The data (Fig. 2.23) are in reasonable agreement with the interaction potential between gaseous CO molecules and point to the dominance of orbital overlap at higher coverages.

From the magnitude of the heat of adsorption we may draw conclusions about the nature of the adsorbed link. For example, when isopropanol is adsorbed on a nickel

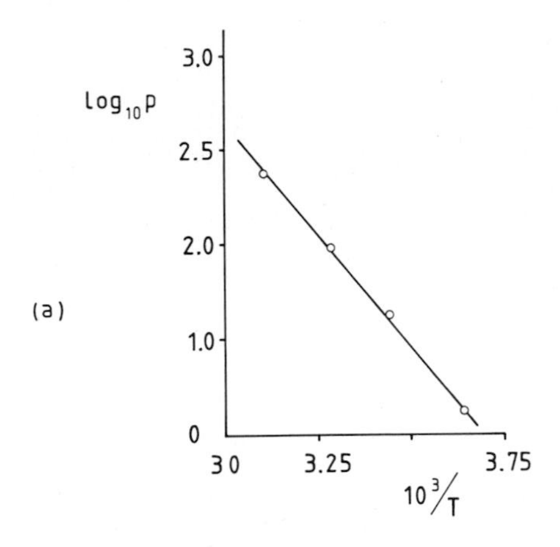

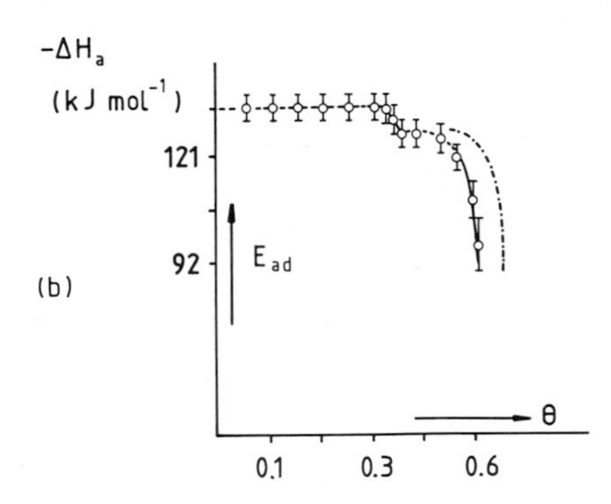

Figure 2.22 (a) The slope of this plot yields, via the Clausius–Clapeyron equation (Section 2.4.2), the isosteric heat of adsorption of 44 kJ mol^{-1} at a coverage of 0.10×10^{-6} m^3 of CO per g of hydrated zeolite. With permission from T. A. Egerton, F. S. Stone, *Trans. Faraday Soc.* **1970**, *66*, 2364. (b) Experimentally derived interaction potential for CO on Pd(100) compared with gas-phase data. With permission from G. Ertl, J. Koch, *Z. Natusforsch., Teil A* **1970**, *25*, 1906.

oxide catalyst the heat of adsorption (66.6 kJ mol^{-1}) surpasses the heat of liquefaction of the adsorbate by 22.9 kJ mol^{-1}. This signifies that the alcohol is hydrogen-bonded to the catalyst surface. Pursuing this line of quantitative argument, it can be shown that when CO and O_2 are co-adsorbed at transition-metal oxide surfaces, a CO_3 complex, distinct from the carbonate ion, is formed in the adlayer. Co-adsorption of NH_3 and HCl (as NH_4Cl) on a mordenite catalyst shows isotherm behaviour distinctly different from that of either HCl or NH_3 alone on the same zeolite (Fig. 2.24). The steeply rising part of the isotherm (Brunauer type IV) is explained by the strong interaction of HCl and NH_3 within the zeolite. A similar phenomenon is observed when xenon is adsorbed by zeolites and when atoms of Hg enter a Ag^+-exchanged zeolite: with increasing pressure of mercury a sharp rise in uptake signifies the formation of clusters of Hg atoms around either Ag atoms or Hg ions (generated by the reaction: $2Ag^+ + Hg \rightarrow 2Ag + Hg^{2+}$).

2.4.3 Adsorption is Almost Invariably Exothermic

It is a commonly held view that adsorption processes are *always* exothermic, the justification for the statement being thermodynamic. Since the Gibbs free energy *G*

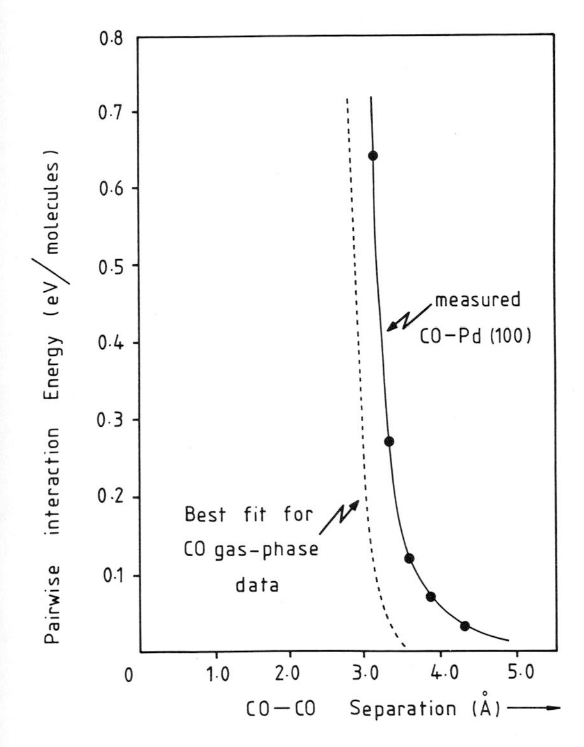

Figure 2.23 Experimentally derived interaction potential for CO on Pd(100) compared with gas-phase data. With permission from J. C. Tracy and P. W. Palmberg, *J. Chem. Phys.* **1969**, *51*, 4852.

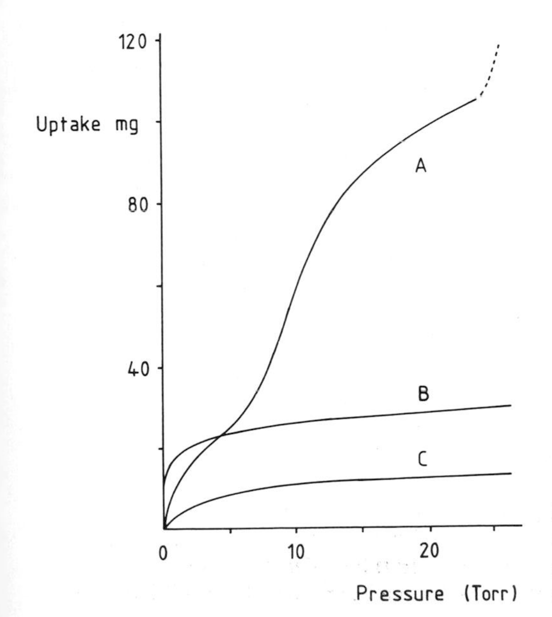

Figure 2.24 Adsorption of NH_3, HCl and (NH_3 + HCl) on the H^+ form of mordenite at 230 °C. A, $NH_4Cl \rightleftharpoons NH_3 + HCl$; B, NH_3 alone; C, HCl alone.

must decrease for any spontaneous event, and since adsorption is one such event which is always accompanied by a decrease in entropy – the number of degrees of freedom in the adsorbed state being less than the number in the gaseous state – then, from:

$$\Delta G = \Delta H - T\Delta S$$

ΔH, the enthalpy change accompanying adsorption, must be negative. For the overwhelming majority of adsorptions, this argument is valid. But consider, for heuristic purposes, the following hypothetical system. A diatomic gas molecule G_2 is dissociatively chemisorbed on the surface of a solid, S. Suppose the strength of the S–G bond is equal to half the strength of the G–G bond. Now, if the adsorbed atoms G have complete two-dimensional mobility, it would then follow that a positive entropy change ΔS would result, corresponding to a net gain of one degree of freedom, and the free-energy change associated with this thermally neutral process ($\Delta H = 0$) would be equal solely to the $T\Delta S$ term. Moving beyond the hypothetical, positive values for the entropy of adsorption can originate from another source: the change in entropy of the adsorbent itself which would happen if the struture of the adsorbent altered on adsorption. Thus, during chemisorption, even if, as is usual, the entropy of the species adsorbed decreases on adsorption owing to a loss in the number of degrees of freedom, this decrease may be exceeded by a concomitant increase in the entropy of the adsorbent itself. This, in turn, would lead to an endothermic process if the $T\Delta S$ term were greater numerically than the ΔG term. Although no detailed thermodynamic data are yet available, there is now irrefragable proof form solid-state, magic-angle NMR, that when some organic species are adsorbed by certain highly siliceous zeolitic catalysts (for example, ZSM-5/silicalite), dramatic changes in the structure of the solid, evidenced by significant movements of ^{29}Si resonances, ensue.

Energetically, it is not difficult to envisage endothermic adsorptions. Figure 2.25 shows two potential-energy curves, one of which refers to the chemisorption of

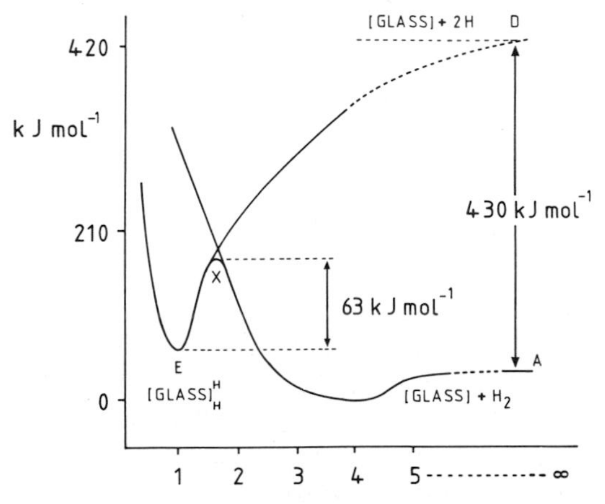

Figure 2.25 The endothermic chemisorption of hydrogen on glass: the heat of adsorption is about $-63\,\mathrm{kJ\,mol^{-1}}$.

atomic hydrogen on glass, the other to the physical adsorption of molecular hydrogen on the same adsorbent. The energy of the two hydrogen atoms adsorbed on glass is *above* that of gaseous molecular hydrogen: for exothermic adsorptions the situations is reversed (compare Fig. 2.2).

2.5 Dynamical Considerations

At 293 K the average velocities of molecules such as H_2, O_2, and H_2O – irrespective of the pressure – are, respectively, 1.76×10^3, 4.4×10^2, and 5.87×10^2 m s^{-1}. These facts are derived from the kinetic theory of gases:

$$\bar{c} = \left(\frac{8kT}{\pi m}\right)^{1/2}$$

where the average velocity $\bar{c}$ is related to the temperature T, the mass of the molecule m, and the Boltzmann constant k: they remind us that the speeds of molecules that impinge upon catalyst surfaces under working conditions, being in the vicinity of 1600 km h^{-1} (1000 mph), exceed the speed of sound under normal conditions (3.30×10^2 m s^{-1}). In general, only a fraction s, as we saw earlier (Section 2.1.1), stick to the surface on impact. But even if the sticking probability is small, a bare surface subjected to such bombardment soon becomes covered owing to the immensity of the flux, F (molecules cm^{-2} s^{-1}) that strikes it. To arrive at F, we again invoke kinetic theory. At a pressure p we have:

$$F = \frac{p}{(2\pi mkT)^{1/2}} \tag{1}$$

which is known as the Hertz–Knudsen formula. We note that, for O_2 at pressures of 10^{-6} bar and 10^{-13} bar at 300 K, the numbers of collisions per cm^2 of surface per second are 2.7×10^{17} and 2.7×10^{10} respectively. With a sticking probability of only 0.01, for example, it would take only about 0.4 s to cover a surface with an adlayer of oxygen at 10^{-6} bar, there being roughly 10^{15} sites for adsorption per cm^2.

2.5.1 Residence Times

Obviously the surface concentration, N_a, of adsorbed species is the product of the flux and the so-called residence time, i.e. the time τ during which the adsorbed entity lingers on the surface:

$$N_a = F\tau \tag{2}$$

Attempts were made many decades ago by Dutch physicists to measure τ directly. A stream of atoms or molecules was directed against a rapidly rotating plate. If the molecules stay for a time τ on the plate before being desorbed, they will travel a short distance with the rotating plate before they escape. The desorbing molecules emerge in all directions, so that if these molecules are then all condensed on a stationary plate which is extremely cold, the centre of the spot of condensed molecules will not

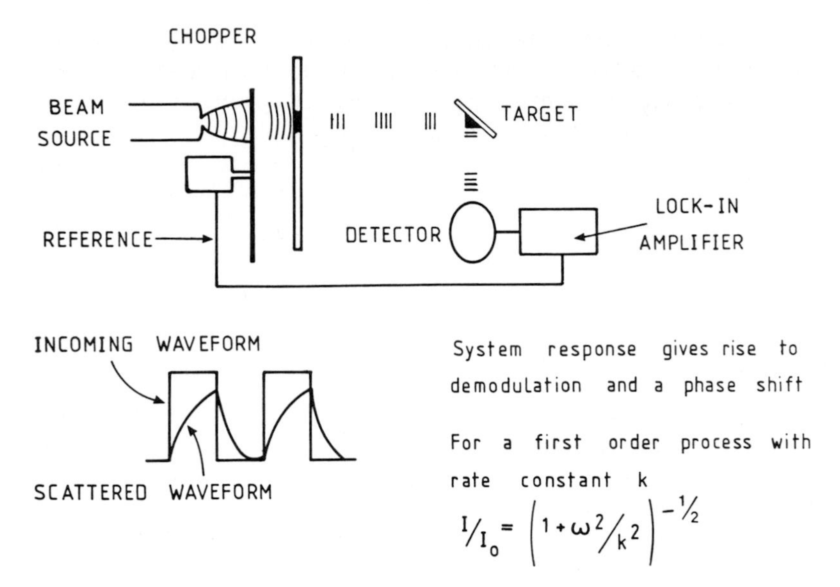

Figure 2.26 Principle of the modulated molecular beam technique. Using this technique the phase lag ϕ between the primary and scattered beam is recorded. For a first-order rate process this is related to the rate constant k by $\tan \phi = \omega/k$, where ω is the chopping frequency. With permission from C. T. Campbell, G. Ertl, J. Signer, *Surf. Sci.* **1982**, *115*, 309.

coincide with the opening through which the original beam of molecules emerged. It will be displaced by a certain length in the direction of the rotation of the rotating plate, and this length is governed by the magnitude of τ (and the speed of the rotating plate). These attempts yielded values which were, for cadmium impinging upon glass at 200 K, somewhere between 10^{-12} and 10^{-6} s. Taking advantage of the panoply of electronic and vacuum techniques now at the disposal of the surface chemist, this early experiment can nowadays be performed using modulated molecular beams (Fig. 2.26). The phase lag ϕ between the primary and scattered signals yields the residence time. For an adlayer that desorbs according to first-order kinetics, the rate constant k_d equals $1/(p\tau)$. Ertl and co-workers found that simple molecules adsorbed on metal surfaces had residence times that could be expressed as:

$$\tau = \tau_0 \exp\left(\frac{-E_d}{RT}\right) \tag{3}$$

where E_d is the activation energy of desorption. Specifically for CO on Pd(111),

$$\tau = 3 \times 10^{-13} \exp\left(\frac{-16\,600}{T}\right) \tag{4}$$

i.e. the activation energy of desorption is 138 kJ mol^{-1}, and residence times over the range 580–700 K fall between 0.1 and 10 ms.

Equation (3) is synonymous with that derived theoretically by the Russian physicist Frenkel, using statistical mechanics, and working on the notion that vibrations executed by adsorbed species perpendicular to the surface are of the essence. In his theory τ_0 is equated to the period of this perpendicular vibration, which should be of the order of 10^{-13} s. To be precise, Frenkel's equation, derived in 1924, was

$$\tau = \tau_0 \exp\left(\frac{Q}{RT}\right) \tag{3a}$$

where Q is the heat of adsorption. We shall see later under which circumstances E_d equals Q.

2.5.2 Rates of Adsorption

The kinetics of non-dissociative adsorption, for a clean surface, are readily formulated in terms of the expressions quoted above. The rate, r_a, of uptake of a gas such as CO is simply the product of the flux and the sticking coefficient, assuming that the process is non-activated.

$$r_a = \frac{dN_a}{dt} = Fs = sp/(2\pi mkT)^{1/2} \tag{5}$$

We have seen, however, that the sticking coefficient varies with coverage (Fig. 2.3), and some consideration needs to be given to the functional dependence of s upon N_a. The simplest model for the variation of s with coverage is the one underlying the derivation of the Langmuir adsorption isotherm (see Section 2.6). If a particle strikes an empty site it becomes adsorbed with a probability s_0, otherwise it is reflected so that

$$s = s_0(1 - \theta) \tag{6}$$

Sometimes s is seen to vary in this linear fashion with coverage, but frequently it drops non-linearly, as seen in Fig. 2.3. One explanation why s does not fall off linearly with coverage is the occurrence of a precursor state in the second layer. Even when an incoming species strikes a species already bound to the surface, there will be a finite interaction leading to a shallow minimum which then serves to trap the incoming species. In this second layer the mobility will be rather high so that the trapped particle will eventually find an empty site for chemisorption. However, with increasing coverage the mean diffusion length required to reach an empty site will continuously increase so that the sticking coefficient will, in turn, decrease.

Dissociative chemisorption of diatomic and other molecules at surfaces is frequently one for which negligible activation energy is required. The energy expended in breaking the bond of the parent diatomic is generally more than compensated by the formation of two new bonds with the surface. Reverting to the potential energy curves of Figs 2.2 and 2.27, we see that a molecule approaching a surface first becomes attracted by a relatively flat potential minimum for non-dissociative physical adsorption (precursor state in the first layer), but can then be readily carried, through an adiabatic transition, into the dissociated state. If the crossing point is below the line of zero potential energy, the overall process is non-activated: if above, the overall process requires activation. Sometimes a metastable chemisorbed state is trans-

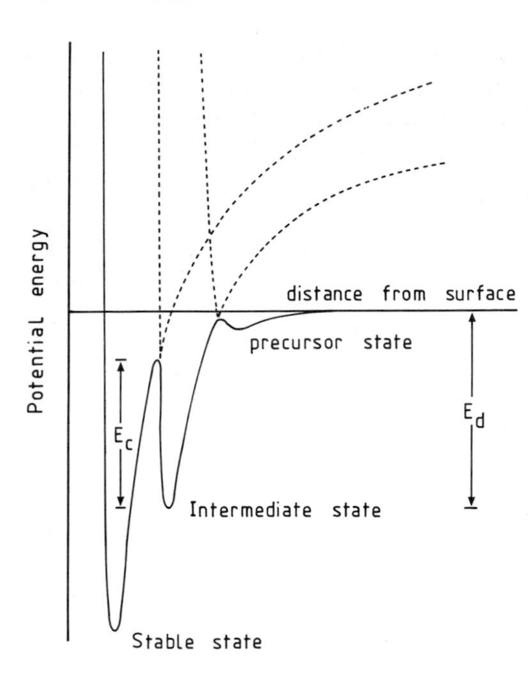

Figure 2.27 Potential-energy diagrams illustrating the relative energies of the various kinds of adsorbed states formed when a gaseous species approaches a surface. E_c is the activation energy for conversion from the intermediate state to the stable state and E_d is the activation energy of desorption from the intermediate state.

versed as an intermediate prior to the formation of the final, stable dissociated state (Fig. 2.27).

With the aid of Fig. 2.28, we see how the energy varies as a function of the interatomic distance as well as of the molecule's distance from the surface. Depending upon whether the saddle point (i.e., the transition state) is located in the entrance or the exit channel, either translational or vibrational excitation is required for

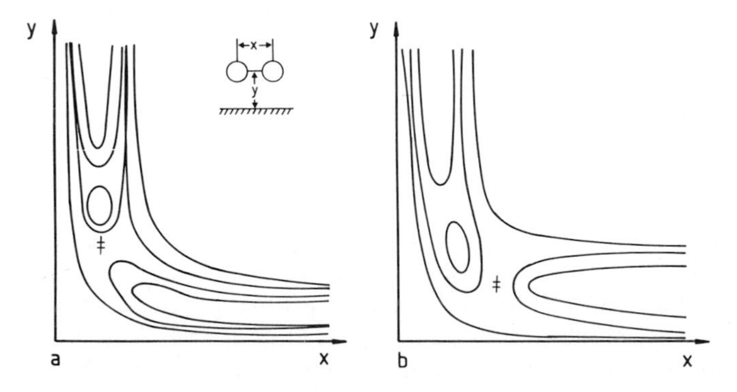

Figure 2.28 Two-dimensional potential-energy diagrams illustrating dissociative chemisorption: (a) with translational energy activation; (b) with vibrational activation. With permission from G. Ertl, in *Catalysis: Science and Technology*, Vol. 4 (Eds.: J. R. Anderson, M. Boudart), Springer, Berlin **1983**, p. 209.

convulsion into the dissociated state. Dissociative adsorption proceeds *directly* (see Section 2.1.1) if the incoming particle has adequate energy to surmount the activated state without previously being trapped in the molecular state. (It has been demonstrated experimentally by Balooch and co-workers that the probability for dissociative chemisorption rises sharply once a critical value of the velocity normal to the surface is reached.)

Dissociative adsorption takes place *indirectly* if the gaseous molecule is first trapped in the (first-layer) precursor state from which it is then thermally activated into the metastable or stable dissociated state. Modern techniques of surface analysis (Chapter 3) can unambiguously identify molecularly chemisorbed precursor species (such as dioxygen or dinitrogen) which subsequently fragment into their bound, dissociated state. This distinction between the direct and indirect mechanisms of dissociative chemisorption is not always clear-cut, and these mechanisms are to be regarded as the extreme ends of a continuum.

In assessing the kinetics and energetics of activated dissociative chemisorption, where precisely we identify the occurrence of the actual process of activation is to some extent arbitrary. We could argue that the sticking coefficient is temperature-dependent. We could also interpret the experimental data by assuming that the trapping probability of impinging molecules is temperature-independent, and attribute the exponential term to the activation required to convert trapped species into their final dissociative state. In general we may write for the rate of dissociative chemisorption:

$$r_a = \frac{p}{(2\pi mkT)^{1/2}} s' \exp\left(\frac{-E_a}{RT}\right) f(\theta) \tag{7}$$

where the first term is the Hertz–Knudsen formula (Eq. (1)); s', the condensation coefficient, is the fraction of those molecules with energy greater than E_a that is adsorbed, E_a is the activation energy of chemisorption and $f(\theta)$ specifies the fraction of the surface uncovered. For dissociation into two or three bound fragments $f(\theta)$ is $(1-\theta)^2$ or $(1-\theta)^3$ respectively; and this picture tacitly assumes that collisions between incoming molecules and adsorbed species do not lead to adsorption. We know from experience that the activation energy for dissociative chemisorption can vary appreciably with surface coverage. Indeed, the Elovich equation (Section 2.5.4) is obeyed when E_a varies linearly with coverage.

2.5.3 Applying Statistical Mechanics to Adsorption

Transition-state theory yields further insights into the factors that govern rates of adsorption. This theory postulates that reaction, in this case adsorption, proceeds via an energetically activated transition state (the saddle point in Fig. 2.27 if we are considering dissociative chemisorption) that is intermediate in structure between the reactants and products. It exists at the top of a potential energy barrier, the height of which is the activation energy of adsorption. Passage over the energy barrier occurs by motion along a path called the reaction coordinate (‡ in Fig. 2.28) that describes the molecular configuration of the reactants and products. It is assumed that the activated complex at the top of the barrier exists in low concentration in equilibrium with the reactants, thereby enabling us to apply statistical theory to the situation.

For non-dissociative adsorption on a uniform surface, let N^* be the equilibrium concentration of activated complexes (molecules cm^{-2}), N_s the number of adsorption sites per cm^2, and N_g the number of gas-phase molecules per cm^3. From statistical mechanics we may write an equilibrium constant, K^*, as:

$$K^* = N^*/(N_g N_s) = f^{*\prime}/(f_g f_s) \tag{8}$$

where the three f terms are the *complete* partition functions for the two species and the surface site ($f = \sum_i g_i \exp(-\varepsilon_i/kT)$ with g_i being the degeneracy of the quantum state ε_i). Hence,

$$N^* = f^{*\prime} N_g N_s/(f_g f_s) \tag{9}$$

Now the rate of adsorption is equal to the concentration N^* of the activated complex multiplied by the frequency of crossing of the barrier. If it is assumed that the activated complex exists in a region of length δ along the reaction coordinate at the top of the barrier, then the average velocity $\bar{v}$ for passage over the barrier is determined from Maxwell–Boltzmann statistics for a one-dimensional problem:

$$\bar{v} = (kT/2\pi m^*)^{1/2} \tag{10}$$

where m^* is the mass of the activated complex. The average time τ' of crossing the barrier is:

$$\tau' = \delta/\bar{v} = \delta(2\pi m^*/kT)^{1/2} \tag{11}$$

The rate of transmission over the barrier is, therefore, $\mathscr{K} N^*/\tau'$, where $\mathscr{K}$ is a transmission coefficient reflecting the probability that the activated complex will surmount the potential barrier to the final adsorbed state. Thus:

$$r_a = \frac{-dN_a}{dt} = \mathscr{K} N^*/\tau' = (\mathscr{K}/\tau)(f^{*\prime} N_g N_s/f_g f_s) \tag{12}$$

It is convenient to factorize $f^{*\prime}$, the complete partition function of the activated complex, into two components: $f^{*\prime} = f^* f^*_{trans(1D)}$, the second term being the one-dimensional translation partition function corresponding to motion over the barrier along the reaction coordinate. The quantum mechanics of a particle in a one-dimensional box of length δ,

$$f^*_{trans(1D)} = (2\pi m^* kT)^{1/2} \delta/h \tag{13}$$

Hence Eq. (12) becomes:

$$r_a = \frac{-dN_a}{dt} = (\mathscr{K} kT/h)(f^* N_g N_s/(f_g f_s)) \tag{14}$$

It is convenient to extract from the partition function f^* the zero-point energy of the initial state of the system by making this energy the arbitrary zero reference energy, and redefining f^* on this basis. Thus if, as is plausible, the transmission coefficient is unity,

$$\frac{-\mathrm{d}N_{\mathrm{g}}}{\mathrm{d}t} = \frac{kT}{h}\frac{f^*}{f_{\mathrm{g}}f_{\mathrm{s}}} N_{\mathrm{g}}N_{\mathrm{s}} \exp\left(\frac{-\varepsilon_{\mathrm{a}}^{\ominus}}{kT}\right) \tag{15}$$

where $\varepsilon_{\mathrm{a}}^{\ominus}$ is the difference in zero-point energy for the reactant and the activated complex, i.e. the activation energy per molecule for non-dissociative adsorption. Expressed differently:

$$r_{\mathrm{a}} = \frac{-\mathrm{d}N_{\mathrm{g}}}{\mathrm{d}t} = \frac{kT}{h} N_{\mathrm{g}}N_{\mathrm{s}} \frac{f^*}{f_{\mathrm{g}}f_{\mathrm{s}}} \exp\left(\frac{-E_{\mathrm{a}}^{\ominus}}{RT}\right) \tag{15a}$$

where $-E_{\mathrm{a}}^{\ominus}$ is the activation energy for adsorption per mole at zero Kelvin (0 K). This result from transition-state theory emerges as a consequence of envisaging the activated complex as vibrating with no restoring force perpendicular to the surface.

By analogy with the above, we may write an equation akin to Eq. (15) for the rate of dissociative chemisorption. If two species are produced upon adsorption, the term N_{s} is replaced by N_{s_2}, the number of bare adjacent sites per cm^2. We shall see later that similar arguments enable us to derive from statistical mechanics equations (Eq. (29) below) for the rate of desorption under a variety of conditions.

2.5.4 Adsorption Kinetics Can Often be Represented by the Elovich Equation

The literature on heterogeneous catalysis often contains an equation of the form

$$\frac{\mathrm{d}q}{\mathrm{d}t} = a\exp(-\alpha q) \tag{16}$$

where q is the amount absorbed at a time t. It was first proposed to describe the kinetics of oxidation and later recognized by Elovich and others as being useful in the description of the adsorption of gases. The constant α is pressure- and temperature-independent and a embodies terms for the rate at which molecules strike the surface as well as the number N_{s} of (active) adsorption sites per unit area available initially and the probability of molecules possessing sufficient activation energy to form chemical bonds at the surface. Thus, a in its full form is

$$a = \frac{s'pN_{\mathrm{s}}}{(2\pi mkT)^{1/2}} \exp\left(\frac{-E_{\mathrm{a}}}{RT}\right) \tag{17}$$

where s' is the condensation coefficient and E_{a} is the activation energy for adsorption (see Sections 2.5.2 and 2.5.3). For a linear dependence of activation energy on amount adsorbed, we have:

$$E_{\mathrm{a}} = E_0 + \beta q \tag{18}$$

where E_0 is independent of the amount adsorbed. Substitution gives:

$$\frac{\mathrm{d}q}{\mathrm{d}t} = \frac{s'pN_{\mathrm{s}}}{(2\pi mkT)^{1/2}} \exp\left(\frac{-E_0}{RT}\right) \tag{19}$$

a more fundamental formulation of the Elovich equation, where N_s, sometimes expressed as the fraction $(1 - \theta)$ of free surface, depends now on the amount adsorbed:

$$N_s = N_s^{\ominus} \exp(-\alpha q) \tag{20}$$

It is important to recognize that, whereas in our previous discussions (see Sections 2.5.2. and 2.5.3) N_s was a fixed quantity, being the number of sites per cm^2, here N_s can vary during reaction. An alternative way of writing Eq. (16), especially when adsorption kinetics are discussed in terms of fractional surface coverage, is

$$r_a = k_a p_{N_2} \exp(-g\theta) \tag{21}$$

This form is utilized later in the Temkin–Pyzhev description for the kinetics of ammonia synthesis (Section 8.3), when the rate of adsorption of N_2 is identical to Eq. (20) but with $p = p_{N_2}$. In its integrated form the Elovich equation may be written

$$q = \frac{1}{\alpha t_0} \ln(t + t_0) \tag{22}$$

where $t_0 = 1/a\alpha$. Clearly a plot of either q against $\ln(t + t_0)$, or of $\log(dq/dt)$ against q, is linear when the Elovich equation is obeyed.

Slow chemisorption processes, sometimes preceded by an initial rapid uptake of the adsorbate, can often be described by the Elovich equation. An example is the slow chemisorption of H_2 onto the mixed oxides of manganese(II) and chromium(VI). Such discontinuities are thought to be caused by the presence of contaminants, which immediately vitiates an elementary description of a single-rate process. When distinct discontinuities occur in Elovich plots for surfaces, the cleanliness of which is established by the absence of spurious X-ray photoelectron spectra

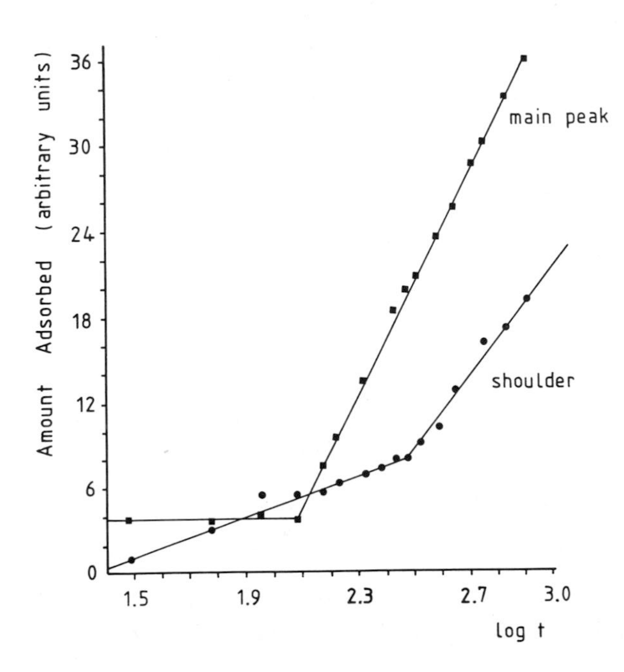

Figure 2.29 Elovich plots representing the growth of a chemisorbed oxygen on a graphite (basal) surface. After M. Barber, E. L. Evans, J. M. Thomas, *Chem. Phys. Lett.* **1973**, *18*, 423.

(XPS) (see Section 3.5) there can be no doubt that a change of slope in the Elovich plot is indicative of distinct rate processes. Figure 2.29 shows Elovich plots relevant to the growth of a peak and shoulder of the XPS of atomic oxygen adsorbing at a clean basal face of graphite. The implication is that there are four different activation energies associated with the process and apparently attributable to two types of bound oxygen. Many other systems exhibit characteristics of this kind.

The basis of most models formulated to explain the Elovich equation is the concept that the rate of adsorption is a functon of the partial pressure p of the gaseous adsorbate and the concentration N_s of available adsorption sites at a given moment. This is the starting point leading to a variety of interpretations. At constant temperature and partial pressure, Eq. (19) may be written

$$\frac{dq}{dt} = kN_s \tag{23}$$

but to obtain the Elovich equation from Eq. (23) further assumptions are necessary. These fall into two classes: models based on a variation in the number of adsorption sites, and those based on an activation energy which varies with coverage. Both classes of theory have been used successfully to explain experimentally observed adsorption kinetics. A theoretical interpretation of the linear portion of a plot of q (or θ) against $\ln(t + t_0)$ thus rests on assumptions concerning the constancy, or otherwise, of the number of adsorption sites.

Cimino et al., in an analysis of the adsorption of H_2 on ZnO, supposed that a surface bond between the gas and solid could arise by thermal excitation of electrons from the valence band of the adsorbent. Positive holes are thus created and these serve as active centres at which bond formation may occur. The very act of chemisorption at a positive hole, however, generates another active centre and there is thus a first-order creation of active sites described by Eq. (23). The free electrons can also combine with positive holes simultaneously formed in the excitation process. These moieties therefore become mutually annihilated, the rate of destruction of active sites being represented by a second-order process:

$$\frac{-dN_s}{dt} = \frac{kN_s^2}{N_s^{\ominus}} \tag{24}$$

where $N_s^{\ominus}$ is the number of active centres initially present. Division of Eq. (23) by Eq. (24) and subsequent integration yields

$$\frac{dq}{dt} = kN_s^{\ominus} \exp\left(\frac{-q}{N_s^{\ominus}}\right) \tag{25}$$

identical in form to the Elovich equation. The crucial assumption involved in manipulating Eqs. (23) and (24) is that k is constant. This, in turn, implies that the activation energy of adsorption, E_a is constant and independent of coverage. In this particular model then, an energetically homogeneous surface, or patches of homogeneity and invariable activation energy for adsorption within those patches, is assumed.

But we know that the activation energy of chemisorption can vary with surface coverage [see Eq. (18)]. The equation describing the kinetics of adsorption on a non-uniform surface with linear increase of activation energy of adsorption with coverage was first deduced by Brunauer et al. In this model the surface is divided into ensembles of adsorbed molecules occupying elements of area, each area constituting a uniform element of surface associated with an activation energy of adsorption ($E_0 + \beta q$), where E_0 is the activation energy for the bare surface. The rate of adsorption is then found by integrating over the entire surface. The result is

$$\frac{\mathrm{d}q}{\mathrm{d}t} = \frac{kRT}{\beta} \exp\left(\frac{-E_0}{RT}\right) \exp\left(\frac{-\beta q}{RT}\right) = a \exp(-\alpha q) \tag{26}$$

where k in this instance is an effective collision number. This equation is also identical in form to the Elovich equation. In principle any type of activation energy distribution function could be used to describe the variation in activation energy of adsorption. Most distribution functions, when integrated, give rise to an equation in which q is related to a logarithmic function of time, as is Eq. (22) when integrated with respect to q at constant temperature.

The Elovich equation is also valid for a situation in which there is (i) an exponential variation of the number of ensembles of molecules requiring a given activation energy for adsorption and (ii) a linear dependence of activation energy on the number of ensembles present. It is clear then that theory will accommodate a number of variations in the type of surface non-uniformity that may exist, yet lead to forms of the Elovich equation which are identifiable with Eq. (16) or are closely related to it. The obedience of experimental results to the Elovich equation clearly does not, of itself, assist in establishing the mechanism of adsorption or the particular dynamics of the rate process. Nevertheless the Elovich equation is a most convenient way of representing the relatively slow chemisorptions typical of a wide variety of gas–catalyst interactions.

2.5.5 Rates of Desorption

If only because desorption is the final step in any catalytic reaction, it is necessary to be able to calculate the rate parameters of desorption for different operating conditions. But there are other reasons for understanding desorption in quantitative terms, not least because mechanistic insights are obtained from the magnitudes of the four important kinetic parameters in the so-called Polanyi–Wigner equation for the rate of desorption, which is:

$$r_\mathrm{d} = \frac{-\mathrm{d}N_\mathrm{a}}{\mathrm{d}t} = v\mathrm{N_a}^m \exp\left(\frac{-E_\mathrm{d}}{RT}\right) \tag{27a}$$

or

$$r_\mathrm{d} = v'\theta^m \exp\left(\frac{-E_\mathrm{d}}{RT}\right) \tag{27b}$$

if we write θ for the fractional coverage, v' now being different from v by a numerical factor.

The four parameters are:

m, the kinetic order of the desorption process;
E_d, the activation energy of desorption;
ν, the pre-exponential factor of the desorption rate coefficient;
N_a, and the number of binding sites of the admolecules.

The kinetic order suggests the nature of the elementary step(s) governing desorption. Thus zero-order kinetics often indicate desorption from a multilayer where the rate of desorption is independent of coverage. First-order kinetics may be indicative of the presence of a single surface species, whereas second-order kinetics are an indication of recombination of adsorbate atoms leading to the production of a diatomic molecule that is then evolved. For a system in which adsorption is non-activated, the activation energy of desorption is the same numerically as the heat of adsorption, so that, under these circumstances, desorption kinetics lead to thermodynamic data. The magnitude of the pre-exponential factor tells us a good deal (see below) about the nature of, and in particular the degree of freedom within, the adsorbed state, or about the occurrence of precursor states. Knowledge of the magnitude of the concentration of surface sites, N_a, is self-evidently of fundamental importance.

The common occurrence of multiple desorption processes (see Section 3.9.1) in temperature-programmed desorption may indicate that a mixture of different kinds of adsorbed species coexist on the surface, or that the act of desorption induces an interconversion from one species (or desorption process) to another as coverage is depleted. For illustrative purposes, we show the results of a study of thermal desorption from a Cr(111) surface previously exposed to different doses of oxygen (Fig. 2.30). There are clearly two types of bound states of oxygen, designated α and β. The total area of the desorption peaks yields the corresponding N_a values which enables the absorption isotherms (Fig. 2.30b) to be constructed. Additional experiments using a gas mixture of 70:30 $^{36}O_2/^{32}O_2$ for the initial dosing and subsequent mass-spectrometric analysis of the 32, 34 and 36 amu peaks reveal that there is no isotope scrambling associated with the lower-temperature desorption peak (from the α-state) but there is for the higher temperature one. The α-state represents non-dissociated oxygen and the β-state dissociatively adsorbed oxygen.

We derive in Section 3.9.1 useful equations, first discussed by Redhead for deducing the activation energy of desorption E_d for processes that are kinetically either first- or second-order ($m = 1$ or 2 respectively). These equations are applicable when temperature rises are linear ($T_{(t)} = T_0 + \beta t$) and they devolve upon measuring the temperature T_p at which the maximum desorption rate occurs as a function of β:

$$\frac{E_d}{RT_p^2} = \frac{\nu}{\beta} \exp\left(\frac{-E_d}{RT}\right) \qquad \text{for } m = 1 \tag{28a}$$

and

$$\frac{E_d}{RT_p^2} = 2(N_a)_p \frac{\nu}{\beta} \exp\left(\frac{-E}{RT_p}\right) \qquad \text{for } m = 2 \tag{28b}$$

where $(N_a)_p$ is the surface coverage at $T = T_p$. We note that for first-order desorption kinetics the temperature at which the maximum rate of desorption occurs is invariant with surface coverage, whereas for second-order kinetics the temperature at

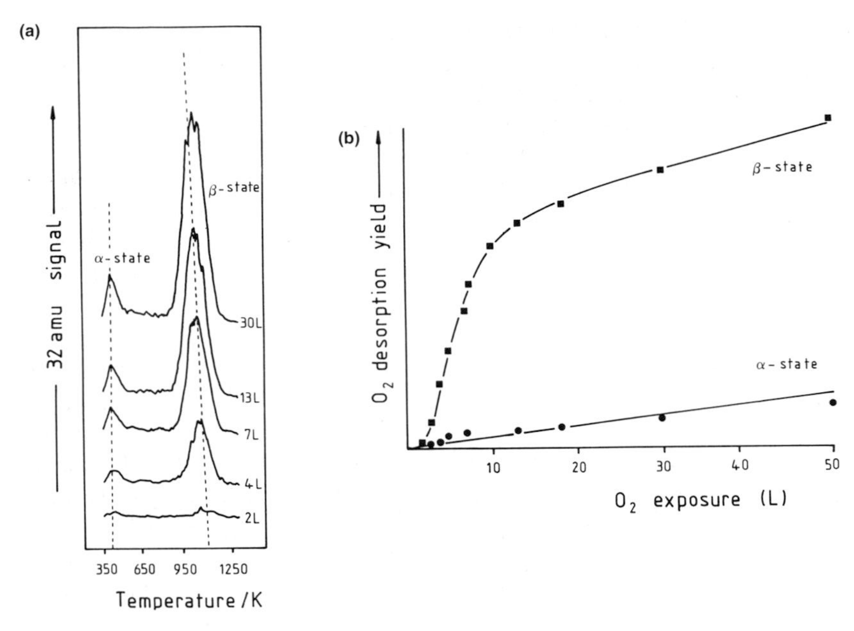

Figure 2.30 (a) Oxygen thermal desorption curves from Cr(111). The α- and β-states represent, respectively, the non-dissociatively and dissociatively chemisorbed oxygen. The α-state desorbs according to first-order kinetics, the maximum temperature of desorption being invariant with coverage. The β-state desorbs according to second-order kinetics and the peak temperature decreases with increasing coverage (see text). (b) Coverage increases with exposure [1 Langmuir (L) = 1×10^{-6} Torr s) in a linear fashion (Henry's Law) for the α-state, but follows a type I isotherm (see Fig. 2.21) for the β-state. The amount adsorbed at a given exposure is computed from the area of the corresponding peak in Fig. 2.30(a). With permission from J. S. Foord, R. M. Lambert, *Surf. Sci.* **1982**, *175*, 147.

which the maximum rate occurs shifts to lower values as surface coverage increases. Reverting to Fig. 2.30, the molecularly adsorbed state desorbs according to first-order kinetics, and the dissociatively adsorbed oxygen according to second-order kinetics, conclusions altogether unsurprising and reassuring.

2.5.6 Applying Statistical Mechanics to Desorption

Using arguments similar to those given in Section 2.5.3, transition-state theory yields for first-order desorption via an activated complex

$$\frac{-\mathrm{d}N_\mathrm{a}}{\mathrm{d}t} = \frac{kT}{h}\frac{f^*}{f_\mathrm{a}} N_\mathrm{a} \exp\left(\frac{-E_\mathrm{d}^\ominus}{kT}\right) \tag{29}$$

(compare Eq. (15) where the transmission coefficient was also taken as unity). Here f_a is the partition function for the adsorbed species, and $E_\mathrm{d}^\ominus$ is the activation energy for desorption per mole, referred to the zero-point energy of the adsorbed state.

For simple first-order kinetics of desorption in which both the adsorbate and the activated complex are immobile, the ratio $f^*/f_a \approx 1$, so that the pre-exponential factor (excluding N_a; i.e., ν in Eq. (2.28)) is kT/h, which is *ca* 10^{13} s^{-1}. To be exact, in the temperature range 50–2000 K, ν for first-order kinetics is:

$$10^{12} \leqslant \nu\left(=\frac{kT}{h}\right) \leqslant 4 \times 10^{13}\ \mathrm{s}^{-1} \tag{30}$$

There are many instances known in which the pre-exponential factors are of this magnitude.

Since motions in the adsorbed state will generally be more restricted than in the activated complex which represents a bound species just in front of the surface, f^*/f_a can clearly be greater than unity. Estimates for limiting cases may be made (Yates, 1985). If the adsorbed state is completely immobile and the activated complex is described by a two-dimensional gas, then, for a molecule of 30 amu at 300 K, f^*/f_a is *ca* 10^3 and ν becomes 10^{16} s^{-1}. If internal degrees of freedom and possible ordering phenomena in the adlayer are taken into account it has been shown by Menzel and co-workers (1978) that ν can increase to about 10^{18} s^{-1}. If $E_a^{\ominus}$ varies with coverage this will influence the partition function of the adsorbed state f_a and, through Eq. (29), the pre-exponential term ν also.

Analogous arguments to those used for first-order desorption kinetics can be applied to desorption phenomena that are kinetically of second order. It has been shown (compare Eq. (30)) that the pre-exponential factor ν *for second-order kinetics* in the temperature range 50–2000 K is:

$$10^{-3} \leqslant \nu\left(=\frac{kT}{hN_s}\right) \leqslant 4 \times 10^{-2}\ \mathrm{cm}^2\ \mathrm{s}^{-1} \tag{31}$$

Moreover, if it is assumed that the activated compelx has complete two-dimensional mobility (i.e. two degrees of translational freedom) and that the adsorbate itself is immobile, there will be a further increase in ν by a factor close to 10^3.

Both for first-order and second-order desorptions, additional enhancement of the pre-exponential factor will occur if other degrees of freedom are permitted in the activated complex for desorption but denied in the chemisorbed state. For example, a diatomic molecule may freely rotate in the activated complex but is restrained from rotation in the adsorbed state.

Although transition-state theory has proved a valuable framework for interpreting the magnitude of pre-exponential factors in desorption, there is a disconcerting number of results where these factors are substantially larger than expected on the basis of the theory even after due allowance for motional freedom. Goddard and co-workers (1983), using the classical stochastic diffusion theory in which the motion of the adsorbed particle is described by classical Newtonian equations, derived an equation for the rate of desorption that appears superior to earlier equations, especially those based on transition-state theory. The pre-exponential term in Goddard's equation is inversely proportional to absolute temperature and fits a broad range of experimental results for the desorption of both atoms and molecules.

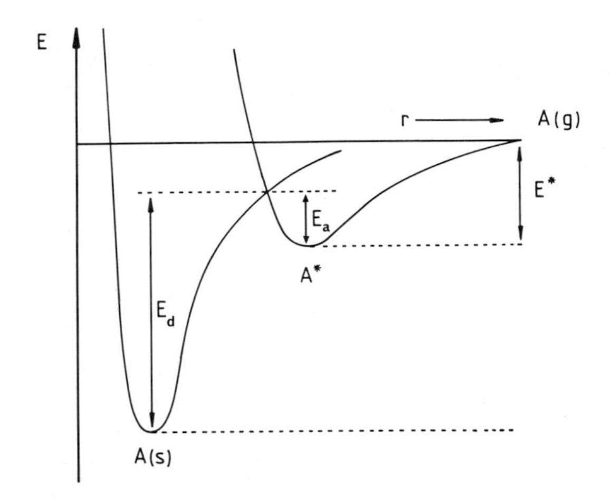

Figure 2.31 Schematic, one-dimensional potential-energy diagram for a simplified precursor-mediated desorption process. The precursor state is at an energy E^* below that of infinitely separated surface and gas-phase species.

2.5.7 The Influence of a Precursor State on the Kinetics of Desorption

We have mentioned earlier that there is evidence for adsorbed species that may enter a mobile precursor state to adsorption, and that this state can then sample both filled and empty adsorption sites as it migrates before entering the stable, chemisorbed state. It is instructive to consider what happens when the process of desorption entails passage through such a mobile precursor state. We recall below the salient features of a treatment by Gorte and Schmidt for a simplifed precursor-mediated first-order desorption process (Fig. 2.31); similar considerations are applicable to more complex situations. Desorption of the adsorbed molecule A(s) proceeds through the following steps:

$$\mathrm{A(s)} \underset{k_\mathrm{a}}{\overset{k_\mathrm{d}}{\rightleftharpoons}} \mathrm{A}^* + \mathrm{site} \xrightarrow{k^*} \mathrm{A(g)} \tag{32}$$

where each k represents a first-order rate constant. Provided the precursor A^* is in equilibrium with A(s), at coverage θ_s, and is at a low, steady-state concentration during desorption, we may write:

$$\frac{\mathrm{d}\theta_{\mathrm{A}^*}}{\mathrm{d}t} = k_\mathrm{d}\theta_\mathrm{s} - k_\mathrm{a}\theta_{\mathrm{A}^*}(1-\theta_\mathrm{s}) - k^*\theta_{\mathrm{A}^*} = 0 \tag{33}$$

and hence

$$\frac{\mathrm{d}\theta_\mathrm{s}}{\mathrm{d}t} = -k^*\theta_{\mathrm{A}^*} = -k^*k_\mathrm{d}\theta_\mathrm{S}/[k^* + k_\mathrm{a}(1-\theta_\mathrm{S})] \tag{34}$$

For $k^* \gg k_\mathrm{a}$ normal first-order kinetics are obeyed, i.e.

$$\frac{\mathrm{d}\theta_\mathrm{s}}{\mathrm{d}t} = -k_\mathrm{d}\theta_\mathrm{s} \tag{35}$$

But when $k^* \ll k_a$, then

$$\frac{d\theta_s}{dt} = -k^* k_d \theta_s / [k_a(1-\theta_s)] \tag{36}$$

The analogous reaction rate at any temperature T is:

$$\frac{d\theta_s}{dt} = -(\nu^* \nu_d/\nu_a)[\theta_s/(1-\theta_s)] \exp[-(E^* + E_d - E_a)/RT] \tag{37}$$

The exponential term in this equation is the barrier height in going from A(s) to A(g) and is synonymous with the heat of adsorption.

We note that, in effect, the precursor-mediated desorption results in the insertion of a $\theta_s/(1-\theta_s)$ term into Eq. (37). Had we considered dissociative chemisorption and a *second-order* desorption process (via a mobile precursor state), the extra term in the equation analogous to Eq. (37) would have been $\theta^2/(1-\theta)^2$. In any event, the influence of the precursor on the corresponding thermal desorption curves (such as those shown in Fig. 2.30) is to broaden them significantly as well as shifting them to lower temperatures.

2.6 Deriving Adsorption Isotherms from Kinetic Principles

Taking the general form of Eq. (7) above for the rate of adsorption r_a and recognizing that $f(\theta) = (1-\theta)$ for non-dissociative adsorption, $f(\theta) = (1-\theta)^2$ for dissociative adsorption into two fragments, $f(\theta) = (1-\theta)^3$ into three, etc.; and taking the general form of Eq. (27a) for the rate of desorption r_d (and writing $\phi(\theta)$ for $N_a{}^m$ such that $\phi(\alpha)$ is replaced by θ for non-dissociative adsorption, and by θ^2 or θ^3 for dissociative adsorption into two or three fragments respectively), we may readily derive an equation for the adsorption isotherm simply by equating r_a and r_d. Thus:

$$\nu\phi(\theta) \exp\left(\frac{-E_d}{RT}\right) = \frac{p}{(2\pi mkT)^{1/2}} s' f(\theta) \exp\left(\frac{-E_a}{RT}\right) \tag{38}$$

For the simplest case of non-dissociative adsorption ($f(\theta) = (1-\theta)$ and $\phi(\theta) = \theta$), and noting that $E_d - E_a = -\Delta H$ (the heat of adsorption), we have:

$$p = (2\pi mkT)^{1/2} \frac{\nu}{s'} \left(\frac{\theta}{1-\theta}\right) \exp\left(\frac{\Delta H}{RT}\right) \tag{39}$$

or

$$p = \frac{\theta}{b(1-\theta)} \tag{40}$$

where

$$\frac{1}{b} = \frac{\nu}{s'} (2\pi mkT)^{1/2} \exp\left(\frac{\Delta H}{RT}\right) \tag{40a}$$

If the heat of adsorption remains constant with coverage, b is a function only of temperature, so that we have:

$$\theta = \frac{bp}{1 + bp} \tag{41}$$

the well-known Langmuir isotherm (Table 2.1). If, on adsorption, each molecule dissociates into two entities and each entity occupies one site, then the Langmuir isotherm equation becomes:

$$\theta = \frac{(bp)^{1/2}}{1 + (bp)^{1/2}} \tag{42}$$

for, under these circumstances, $f(\theta)$ and $\phi(\theta)$ in Eq. (38) are replaced by $(1 - \theta)^2$ and θ^2, respectively. When the adsorbed molecule dissociates into n entities each of which occupies a surface site, then:

$$\theta = \frac{(bp)^{1/2}}{1 + (bp)^{1/2}} \tag{43}$$

The Langmuir isotherm for two gases adsorbed simultaneously and non-dissociatively is of considerable utility in the interpretation and modelling of catalyzed reactions (see Chapter 7). It follows readily that, if θ_A and θ_B refer to the fractions of the sites covered by molecules of type A and B, respectively, then

$$\theta_A = b_A p_A / (1 + b_A p_A + b_B p_B) \tag{44a}$$

and

$$\theta_B = b_B p_B / (1 + b_A p_A + b_B p_B) \tag{44b}$$

Catalysis offers many instances where there is simultaneous chemisorption of different molecules or atoms. For example, the crucial phenomena of poisoning and promotion devolve in part upon simultaneous, competitive adsorption. And, by combining Eqs. (40a), (41), (44a) (44b) we have

$$\frac{\theta_A}{\theta_B} = \frac{k_A\, p_A}{k_B\, p_B} \exp\left(\frac{\Delta H_B - \Delta H_A}{RT}\right) \tag{45}$$

where k_A and k_B are the respective pre-exponential terms for species A and B in Eq. (40a), and the ΔHs are the respective changes in enthalpy on adsorption. To fix our ideas, we cite a specific example. For the simultaneous chemisorption of acetylene and ethylene on a nickel catalyst at 50 °C we deduce from Eq. (45) and the measured heats of adsorption for C_2H_2 and C_2H_4 that:

$$\frac{\theta_{C_2H_2}}{\theta_{C_2H_4}} \approx 10^6$$

(taking $k_A \approx k_B$, as is reasonable). The surface of the catalyst is evidently dominated by adsorbed acetylene, which means that when a gas mixture of C_2H_2 and C_2H_4 is introduced to the nickel and H_2, reduction of the alkyne precedes that of the alkene.

2.6.1 Using the Langmuir Isotherm to Estimate the Proportions of Non-dissociative and Associative Adsorption

Although it is recognized that the Langmuir isotherm equation cannot be expected to be universally valid – because the heat of adsorption often decreases with increasing surface coverage – it enables us to make valuable semi-quantitative estimates of the relative proportions of dissociated and molecular species present at a catalyst surface when they are exposed to such species as H_2 and CO. The treatment below is an abbreviated version of that given by Benzinger, and is likely to be more valid at low, overall surface coverages since heats of adsorption are essentially constant under those circumstances.

Consider the two equilibrium steps:

$$\mathrm{AB(g)} \rightleftharpoons \mathrm{AB(a)}$$

$$\mathrm{AB(g)} \rightleftharpoons \mathrm{A(a)} + \mathrm{B(a)}$$

where (a) represents the adsorbed state. If it is assumed that molecular adsorption requires one adsorption site and dissociative adsorption requires two, then the fractional coverages of molecular (non-dissociated) and dissociated adsorbed species, θ_M and θ_D respectively, are:

$$\theta_\mathrm{M} = \frac{k_\mathrm{M} p}{(1 + 2K_\mathrm{D}^{1/2} p^{1/2} + k_\mathrm{M} p)} \tag{46a}$$

$$\theta_\mathrm{D} = \frac{k_\mathrm{D}^{1/2} p^{1/2}}{(1 + 2K_\mathrm{D}^{1/2} p^{1/2} + K_\mathrm{M} p)} \tag{46b}$$

The Ks are equilibrium constants synonymous with b in Eqs. (40)–(42), and can be expressed, using statistical mechanics, in terms of the respective complete partition functions for the internal degrees of freedom, $f'_{\mathrm{a,M}}$ and $f'_{\mathrm{a,D}}$ and the heats of adsorption at absolute zero, $-\Delta H_\mathrm{M}^\ominus$ and $-\Delta H_\mathrm{D}^\ominus$, respectively:

$$K_\mathrm{M} = \left(\frac{h^2}{2\pi mkT}\right)^{3/2} \frac{1}{kT} \frac{f'_{\mathrm{a,M}}}{f'_\mathrm{g}} \exp\left(\frac{-\Delta H_\mathrm{M}^\ominus}{RT}\right) \tag{47a}$$

$$K_\mathrm{D} = \left(\frac{h^2}{2\pi mkT}\right)^{3/2} \frac{1}{kT} \frac{f'_{\mathrm{a,D}}}{f'_\mathrm{g}} \exp\left(\frac{-\Delta H_\mathrm{D}^\ominus}{RT}\right) \tag{47b}$$

(Note that f_g in Eqs. (9) and (14) is different from the f'_g used here in that the former stands for the complete partition function for the gaseous species, not just for the internal degrees of freedom). The major contribution to the internal degrees of freedom of the gaseous species comes from the rotational ones, since the vibrational contributors are negligible for most diatomic molecules in the range 100–1000 K. We may therefore write:

$$\frac{f'_{\mathrm{a,M}}}{f'_\mathrm{g}} = \frac{f'_{\mathrm{a,D}}}{f'_\mathrm{g}} = \frac{\theta_\mathrm{r}}{T} \approx \frac{2.5}{T} \tag{48}$$

where θ_r is the rotational temperature, which for species such as CO, N_2 and CO, falls in the range 2–3 K.

We next look at some general consequences of the competitive adsorption of molecular (non-dissociated) and dissociated species. First, we note that, at low temperatures, molecular adsorption is preferred whenever the molecular adsorption enthalpy is greater than half the enthalpy for dissociative adsorption. This follows from the fact that the surface is capable of adsorbing twice as many molecular species as dissociated ones, so the energy of the system is minimized by molecular adsorption. However, as the temperature increases, entropy effects (the $T\Delta S$ term in Section 2.4.3) become important. There is an entropy change of *ca* $-200\,\mathrm{J\,mol^{-1}\,K^{-1}}$ upon adsorption of both molecular and dissociated species. But as dissociative adsorption results in only half as many molecules being adsorbed as molecular adsorption, the entropy change per adsorption site is less for dissociated chemisorption. So signif-

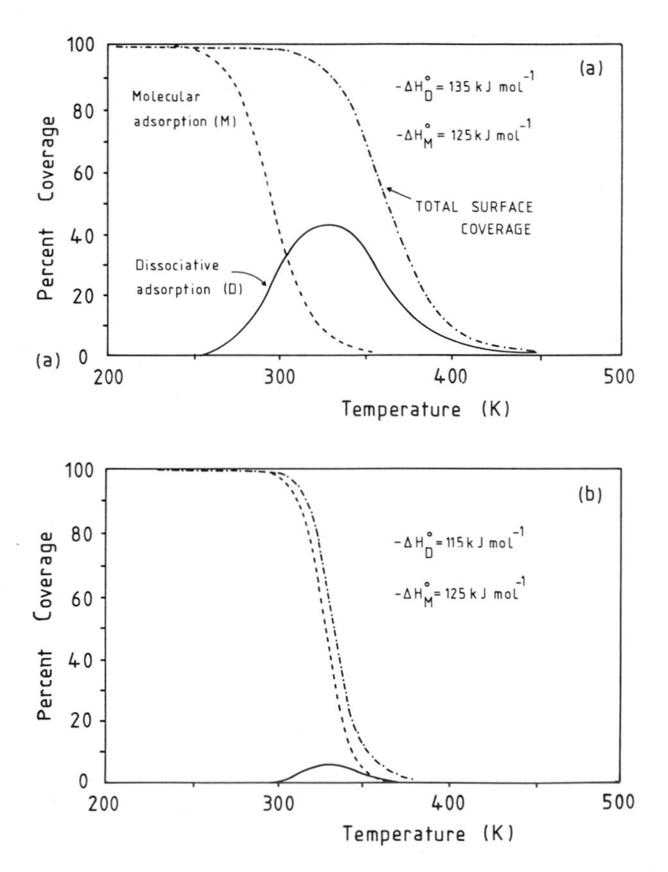

Figure 2.32 Computed (Eq. (2.46)) adsorption isobars ($p = 10^{-8}$ Torr) for competitive dissociative (D) and non-dissociative (molecular, M), adsorption. (a) and (b) refer to two distinct sets of values for the respective heat of adsorption at absolute zero, $-\Delta H^\ominus$. With permission from J. B. Benzinger, *Appl. Surf. Sci.* **1980**, *6*, 121.

icant is the entropy contribution that, for temperatures close to 300 K, it is large enough to overcome an energy difference of 125 kJ mol^{-1} between adsorption of two molecules and one dissociated diatomic molecule. For illustrative purposes adsorption isobars are shown in Fig. 2.32 for some typical situations. It is seen that, when the heats of adsorption for dissociated and molecular species are -135 and -125 kJ mol^{-1}, respectively, no dissociative adsorption occurs below 250 K, even though the enthalpy of dissociative adsorption is more exothermic. Between 250 and 325 K the conversion from molecular to dissociative adsorption occurs, driven, as it were, by the entropic term, $T\Delta S$. Above 450 K neither molecular nor dissociated species remain on the surface at a pressure of 10^{-8} Torr. Pressure also plays a role in governing the extent of molecular and dissociative chemisorption. Increasing the pressure at constant temperature shifts the equilibrium to favour molecular adsorption, as may be deduced from Eqs. (46a) and (46b), which yield $\theta_M/\theta_D \propto p^{1/2}$. Figure 2.33 shows the adsorption isotherms at 300 K when the heats of adsorption for both molecular and dissociated adsorption are 125 kJ mol^{-1}. It is interesting to note in Fig. 2.32 that, for low surface coverage, dissociative adsorption is thermodynamically favoured.

It is to be emphasized that the above arguments take for granted the validity of the major assumption of the Langmuir theory of adsorption, i.e., the constancy of the heat of adsorption with coverage. But already we have seen, and we shall be reminded of it frequently in this text, that the heat of chemisorption often decreases with increasing coverage. Lateral interactions between species in the adlayer manifest themselves as coverage increases; consequently the peaks in a thermal desorption spectrum (e.g. Fig. 2.30a) broaden, sometimes leading to the appearance of multiple peaks. Such peaks are sometimes assigned to different adsorption states. To be sure, for some systems, these states are indeed attributable to genuinely different adsorption sites (terminal, bridge, etc.), but for other systems the multiple states arise from molecules desorbing from sites with identical local geometry but variable occupancy

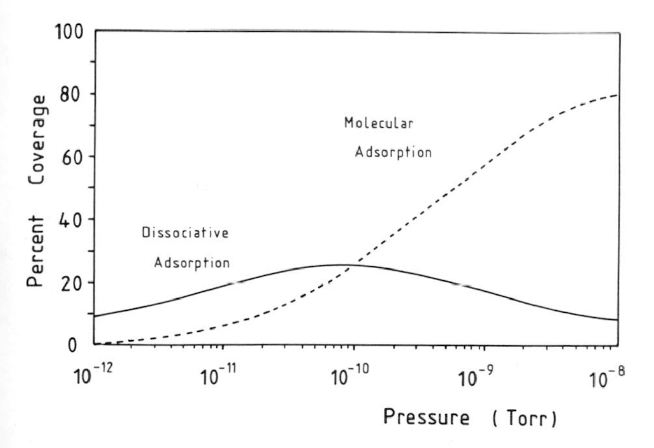

Figure 2.33 Adsorption isotherms (at 300 K) and $-\Delta H^{\ominus}_{M} - \Delta H^{\ominus}_{D}$ values of 125 kJ mol^{-1}. With permission from J. B. Benzinger, *Appl. Surf. Sci.* **1980**, *6*, 121.

of neighbouring sites. A further case for the drop in heat of adsorption with increasing coverage is the progressive population of different sites on an intrinsically heterogeneous surface – terraces, kinks, emergent dislocations, etc.

2.6.2 Other Adsorption Isotherms

2.6.2.1 Henry's Adsorption Isotherm

Henry's adsorption isotherm (Table 2.1) – which is effectively Henry's law applied to a two-dimensional solution instead of a bulk solution – emerges as a special case (low-pressure limit) of Langmuir's isotherm (see Eq. (41) with $bp \ll 1$).

2.6.2.2 The Freundlich Isotherm

At one time it was thought that the Freundlich equation was simply a convenient form of representing the Langmuir equation at intermediate values of θ. But Zeldowitch, in his derivation of an adsorption isotherm for a heterogeneous surface, showed that the Freundlich equation emerged from the basic idea that adsorption sites are distributed exponentially with respect to the heat of adsorption. In Zeldowitch's derivation the surface sites are subdivided into several types, i, each possessing a characteristic constant heat of adsorption (that is, the Langmuir model is assumed to be valid for the sites within one type, there being no repulsion or any other mutual interaction in the portion of the adlayer). Using equations of the form:

$$\frac{\theta_i}{1-\theta_i} = b_i p \quad \text{and} \quad \theta = \frac{\sum\limits_i \theta_i N_i}{\sum\limits_i N_i}$$

where there are N_i sites of the ith kind, and natural extensions of these equations, one arrives at

$$\ln\theta = \frac{RT}{A}\ln p + B \tag{49}$$

where A and B are constants, and this can be recast into $\theta = kp^{1/n}$, the form in which the Freundlich equation is normally written. From this derivation we deduce that Eq. (49) is likely to be valid only at low coverages. It is therefore erroneous to argue, as is sometimes done, that the Freundlich equation predicts progressively increasing coverage with increasing pressure.

2.6.2.3 The Temkin Isotherm

The Temkin isotherm (sometimes designated the Slygin–Frumkin isotherm) relates the fraction covered, θ, to the logarithm of the pressure, p:

$$\theta = A\ln Bp \tag{50}$$

where the constant A is dependent on temperature and B is related to the heat of adsorption. It can be shown that the Temkin isotherm follows from an assumption that the heat of adsorption drops linearly with increasing coverage.

2.6.2.4 The Brunauer–Emmett–Teller Isotherm

Of the isotherms that are appropriate for the description of multilayer (physical) adsorption, the Brunauer–Emmett–Teller (BET) equation is the most widely used. Its merits and utility for the study of catalysts are discussed fully in Sections 4.2. We note here the assumptions on which the BET theory rests. It argues that the first layer of adsorbate is taken up with a fixed heat of adsorption (H_1) whereas the second and subsequent layers are all characterized by heats of adsorption equal to the latent heat of evaporation (H_L). By considering a dynamic equilibrium between each layer and the gas phase, the BET equation is arrived at:

$$\frac{p}{V(p_0 - p)} = \frac{1}{V_m c} + \frac{c-1}{V_m c} \cdot \frac{p}{p_0} \tag{51}$$

where V is the volume of gas adsorbed, p is the pressure of gas, p_0 is the saturated vapour pressure of the liquid at the operating temperature and V_m is the volume equivalent to an adsorbed monolayer. The BET constant c is given by:

$$c = \exp\left(\frac{H_1 - H_L}{RT}\right) \tag{52}$$

A plot of $p/V(p_0 - p)$ versus p/p_0 is usually linear in the range of p/p_0 from 0.05 to 0.35, and the slope and intercept of this plot yield both c and the monolayer capacity V_m.

2.6.2.5 Developments from Polanyi's Adsorption Theory

In dealing with microporous catalysts such as zeolites, high-area carbons, and pillared clays, it is sometimes an advantage to revert to one of the earlier theories of adsorption, that proposed by Polanyi in 1914. According to Polanyi's treatment, the 'adsorption space' in the vicinity of a solid surface is characterized by a series of equipotential surfaces. In Fig. 2.34, ABDC represents a section through the adsorption space associated with unit mass of solid and ABba represents an equipotential surface. When the space (of volume W) between CDcd and ABab is filled with adsorbate, the equilibrium pressure being p, the adsorption potential ε at the surface ABba is, by definition, given by:

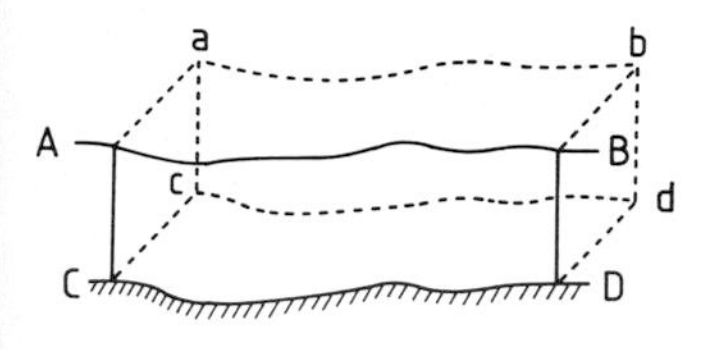

Figure 2.34 Diagrammatic representation of the adsorption space in the vicinity of an adsorbent. ABba is an equipotential surface which mirrors the topography of the surface of the adsorbent, CDdc.

$$\varepsilon = RT \ln p_0/p \tag{53}$$

Polanyi pictured the adsorbate in intimate contact with the solid to be in liquid form so that $W = x/\rho$, with x the mass adsorbed and ρ the density of the liquid. Extending this idea, the Russian workers Dubinin and Radushkevich proceeded to derive a new adsorption isotherm. The adsorption potential, resulting as it does from the dispersion and polar forces between the solid and the adsorbate molecules, though independent of temperature, varies according to the nature of the adsorbate as well as that of the solid. Each of these forces are functions of the polarizability α of the adsorbed molecule. Thus, the adsorption potential of two different vapours at the same value of W on a given solid will bear a constant ratio to one another:

$$\frac{\varepsilon_1}{\varepsilon_2} = \frac{\alpha_1}{\alpha_2} = \beta \tag{54}$$

The constant, β, is known as an affinity coefficient. If another adsorbate is taken as an arbitrary standard, the Eq. (54) becomes

$$\frac{\varepsilon}{\varepsilon_0} = \frac{\alpha}{\alpha_0} = \beta \tag{54a}$$

where the symbols with the suffix zero refer to the standard vapour and those without to the other vapour. Dubinin and Radushkevich argued that the volume of the adsorption space may be expressed as a Gaussian function of the corresponding adsorption potential. For the standard vapour we thus have

$$W = W_0 \exp(-A\varepsilon_0{}^2) \tag{55}$$

where W_0 is the total volume of all the micropores and A is a constant characteristic of the pore size distribution. From Eqs. (54a) and (55) we have:

$$W = W_0 \exp\left[-A\left(\frac{\varepsilon}{\beta}\right)^2\right] \tag{56}$$

which, substituting from Eq. (53), yields:

$$\frac{x}{\rho} = W_0 \exp\left\{-\frac{A}{B^2}\left(RT \ln\left(\frac{p_0}{p}\right)\right)^2\right\} \tag{57}$$

which rearranges to:

$$\ln x = \ln(W_0\rho) - D\left[\ln\left(\frac{p_0}{p}\right)\right]^2 \tag{58}$$

where D is $A(RT/\beta)^2$. Thus, a plot of $\ln x$ versus $(\ln(p_0/p))^2$ should give a straight line of slope D and intercept $\ln(W_0\rho)$.

Dubinin and others have shown that Eq. (58), which, for example, yields a linear plot over the relative pressure range $1 \times 10^{-5} \leqslant p/p_0 \leqslant 0.2$ for benzene and other hydrocarbons on activated carbons, is a convenient method of evaluating the micropore volume, W_0.

2.6.2.6 Kaganer's Isotherm and the DKR Equation

Kaganer's isotherms emerges by modifying Dubinin's argument so as to yield a method of evaluating surface areas from equilibrium uptake curves. In confining attention to the monolayer region and assuming that the distribution of adsorption potential over the sites on the surface is Gaussian, Kaganer wrote, by analogy with Eq. (56):

$$\theta = \exp(-A_1\varepsilon^2) \tag{59}$$

where A_1 is a constant which charaterizes the Gaussian distribution and ε is defined by Eq. (53). Hence:

$$\theta = \exp\left\{-A_1\left(RT\ln\left(\frac{p_0}{p}\right)\right)^2\right\} \tag{60}$$

Recalling that $\theta = x/x_m$, where x_m is the mass of a monolayer of adsorbate, Eq. (60) becomes:

$$\ln x = \ln x_m - A_1(RT)^2\left(\ln\frac{p_0}{p}\right)^2 \tag{61}$$

or

$$x = x_m\exp(-B\varepsilon^2) \tag{62}$$

with the constant B replacing $A_1R^2T^2$. Equation (62), for obvious reasons, is known as the Dubinin–Kaganer–Raduschkevich (DKR) equation, and has been widely used to explore energetic heterogeneity of solid surfaces at low coverages.

2.6.2.7 The Virial Equation of State

It is sometimes convenient, again for microporous solids, to interpret the experimentally observed adsorption isotherm in terms of the virial equation of state. When adsorption is regarded as a process of volume-filling of the adsorbent, then the mean hydrostatic pressure p of the adsorbate molecules inside the solid is related to the amount adsorbed, x, by the virial equation:

$$\frac{p}{RT} = x(1 + a_1x + a_2x^2 + a_3x^3 + \cdots) \tag{63}$$

where the a terms are the virial coefficients. The advantage of using a virial equation approach to adsorption is that it allows other thermodynamic properties to be subsequently calculated, irrespective of any assumptions concerning the mechanism of adsorption or the nature of adsorbate–adsorbate interactions. Equally, theoretical approaches to the computation of adsorption energies using quantum-mechanical procedures, and the derivation of reliable potentials describing the interaction of a species within an adlayer, are advantageously carried out via the agency of the two-dimensional virial isotherm which can be written

$$\frac{x}{p} = \frac{Z_s}{kT} + x\left(\frac{-2B}{A}\right)\frac{Z_s}{kT} + \frac{x^2}{2}\left(\frac{2B}{A^2} - \frac{3C}{A^2}\right)\frac{Z_s}{kT} + \cdots \text{ (higher terms)} \tag{64}$$

where x is the amount of gas adsorbed at a pressure p, Z_s is the configuration integral for a single species on the surface, A is the specific area of the adsorbent and B and C are, respectively, the second and third two-dimensional virial coefficients.

2.7 Energetics of Adsorption

There are many reasons why we need to know, or to be able to estimate, the strength of binding of species to a surface. In the first place, recognizing (Fig. 2.35) that what, in essence, is achieved in heterogeneous catalysis is the circumvention of the 'high-energy', homogeneous path (i), we require to know more about route (ii). We are aware that – for a given set of reactants – depending upon the catalyst used, or the crystallographic phase or the predominantly exposed surface of that phase, quite different values of the activation energy barriers (E_1^*, E_2^*, etc.) or of the energy minima A(ad), I(ad), etc., are possible. The degree of catalytic efficiency gained in following path (ii) is governed by the energetics of these various intermediates, which encompass adsorbed reactant (A(ad)), the activation energy E_1^* required to convert this bound reactant into a surface intermediate (I(ad)) of different structure, the activation energy E_2^* that needs to be surmounted to yield adsorbed products (P(ad) and O(ad)) and the activation energy of desorption of products E_3^*. Another reason why we need to know the energies involved in adsorption is that they clarify the preferred structure of the bound species. Is a small molecule such as CO dissociated? And does a larger molecule such as butane lie flat – and, if so, in what orientation – on a catalyst surface? If we compute how the binding energy varies at different sites on the surface, we can thereby estimate the activation energy of surface diffusion, a factor of considerable importance in catalysis (Section 2.7.1).

2.7.1 Estimating the Binding Energies of Physically Adsorbed Species

When an isolated species approaches the surface of a solid, several interactions come into play, each of which contributes to the heat or energy of physical adsorption. In general, we may write:

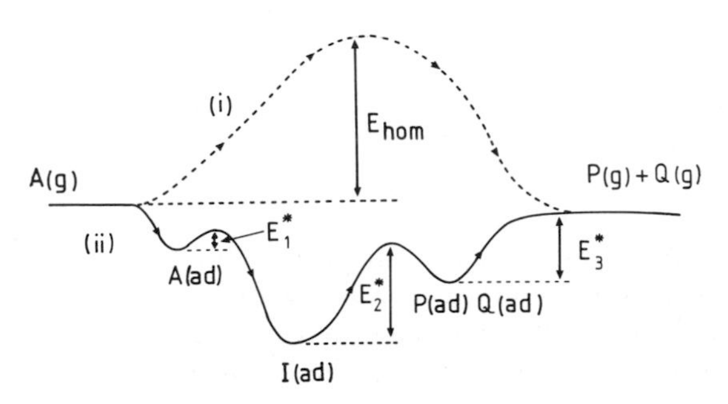

Figure 2.35 In proceeding from gaseous reactant A(g) to gaseous products P(g) and Q(g), route (i) (homogeneous gas-phase reaction) requires greater activation than the alternative (ii) which entails various surface intermediates (see text).

$$U = U_d + U_r + U_p + U_{fd} + U_{fq} + U_{sp} \tag{65}$$

where U, the interaction potential, varies as a function of the position of the adsorbed species with respect to the crystal, and

U_d is the attractive (dispersion) potential,
U_r is the close-range repulsion term,
U_p is the polarization energy,
U_{fd} is the field–dipole interaction,
U_{fq} is the field-gradient–quadrupole interaction, and
U_{sp} is a self-potential which takes into account adsorbate–adsorbate interactions.

The first three terms on the right-hand side of Eq. (65) are always present, irrespective of the nature of the adsorbate and adsorbent. The next two, U_{fd} and U_{fq}, depend upon the presence or absence of permanent dipoles or quadrupoles respectively in the adsorbate; and U_{sp} vanishes for small uptakes. Rigorous calculations of, say, the binding energy of molecules in complicated, catalytically active solids such as zeolites require the evaluation of all terms in Eq. (65). For simpler situations, however, we may proceed satisfactorily using either the Lennard-Jones equation (Eq. (66); see also Section 2.4.2), or a more adaptable variant such as Eq. (67), which expresses the potential U of an atom in terms of its distance r_i from another atom in the solid.

$$U = \sum_i U(r_i) = \sum_i -\frac{A}{r_i^6} + \frac{B}{r_i^{12}} = 4\varepsilon \sum_i \left[\left(\frac{\sigma}{r_i}\right)^{12} - \left(\frac{\sigma}{r_i}\right)^{6} \right] \tag{66}$$

$$U = \sum_i \left[-\frac{A}{r_i^6} + B\exp(-Cr_i) \right] \tag{67}$$

The constants A and B in the attraction and repulsion terms of the Lennard-Jones equation are calculable in terms of the (gas-phase) polarizabilities and ionization energies of the atoms involved; B is proportional to A. From the right-hand side of the Lennard-Jones equation (Eq. 66), we see that the new parameters ε and σ are related by $4\varepsilon\sigma^6 = A$ and $4\varepsilon\sigma^{12} = B$. Expressed in this form ε represents the energy of interaction or well-depth (as may be verified by putting $\mathrm{d}U/\mathrm{d}r = 0$, which gives the condition for the minimum value of the energy). Equations (66) and (67), or further extensions of them such as:

$$U = \sum_i \left[-\frac{A}{r_i^6} + B\exp(-Cr_i) + \frac{D}{r_i^8} \right] \tag{68a}$$

$$U = \sum_i \left[-\frac{A}{r_i^6} + \frac{B\exp(-Cr_i)}{r_i^D} + \frac{K_{q_1 q_2}}{r_i} \right] \tag{68b}$$

in which A, B, C, D, etc. are parameterized values extracted from a data bank of information incorporating enthalpies of vaporization and compressibilities of a large body of materials, have proved particularly useful in computing potential-energy curves for a variety of instructive situations. Each atom in the adsorbed species is

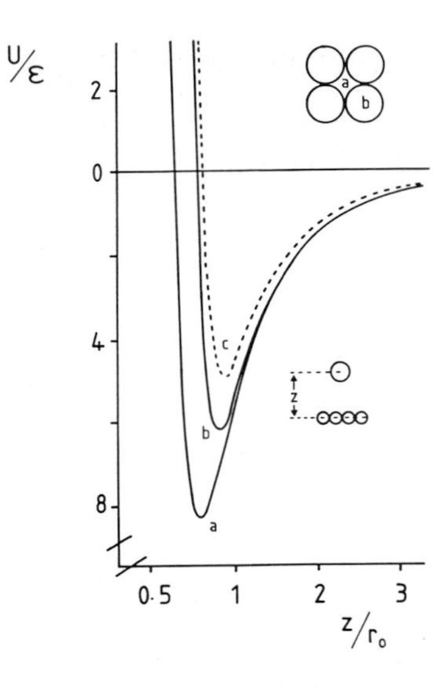

Figure 2.36 Potential-energy curves for the adsorption of xenon at a Pd(100) surface. (a) Xe approaches the four-fold site; (b) Xe approaches an on-top site; (c) adsorption at a structureless homogeneous adsorbent. The unit of distance z above the surface, is r_0, defined by $r_0 = r_{Xe} + r_{Pd}$, where r_{Xe} and r_{Pd} are the radii of the Xe and Pd atoms, respectively. With permission from J. Kuppers, U. Seip, *Surf. Sci.* **1982**, *119*, 297.

'summed' individually. Figure 2.36 shows that the maximum in adsorption energy when a Xe atom approaches a palladium surface occurs at the fourfold site on Pd(100). It is found that the maximum energy coincides with the threefold site on Pd(111). And when a butane molecule approaches the surface of a graphitic carbon, the disposition of the molecule which yields the strongest binding energy is that shown in Fig. 2.37, a fact which is in line with the experimental observations of Groszek, who studied experimentally the uptake of aliphatic hydrocarbons on graphite. Kiselev and his co-workers in Russia have made extensive contributions to the computation and measurement of gas adsorption on carbons and on zeolites (Table 2.2); by calculating the interaction energy of molecules as they are transported (on computer) along a channel in a shape-selective zeolitic catalyst, other workers have evaluated the variations in total energy of interaction experiences of a molecule in its path within the catalyst pores (Fig. 2.38).

The advent of powerful computational facilities has enabled other approaches to be applied to derive heats of adsorption and preferred location of sorbed species.

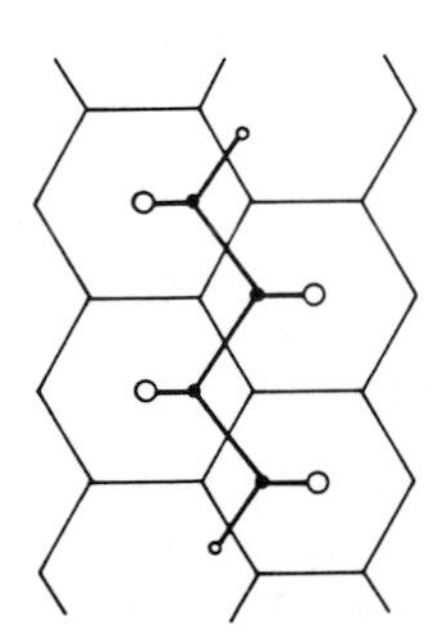

Figure 2.37 Computations based upon Eq. (67) reveal that this is the most favourable orientation, energetically, for a butane molecule when adsorbed on the basal surface of a graphite crystal. With permission from A. J. Goszek, *Proc. R. Soc. London, Ser. A* **1970**, *314*, 473.

Table 2.2 Energies of adsorption ($kJ\,mol^{-1}$) on the basal surface of graphitized carbon black at half coverage. (After A. V. Kiselev et al. *J. Chem. Soc., Faraday Trans. 2* **1978**, *74*, 367.)

Adsorbate	Energy of adsorption	
	Calculated from Eq. (68a)	Observed
Hydrogen	3.7	3.8
Nitrogen	10.9	11.7
Propane	28.4	27.2
n-Hexane	51.9	52.3
n-Octane	67.4	66.9
Benzene	43.1	41.8

These approaches include: energy minimization, Monte Carlo methods, and molecular dynamics, and will be fully illustrated in Chapter 5.

The value of energy minimization procedures in modeling the docking of molecules in zeolites was illustrated by Cheetham and one of the present authors (JMT) in the mid-1980s, when the position of pyridine inside zxeolite L model catalysts was located by computation and found to agree well with the site actually determined by

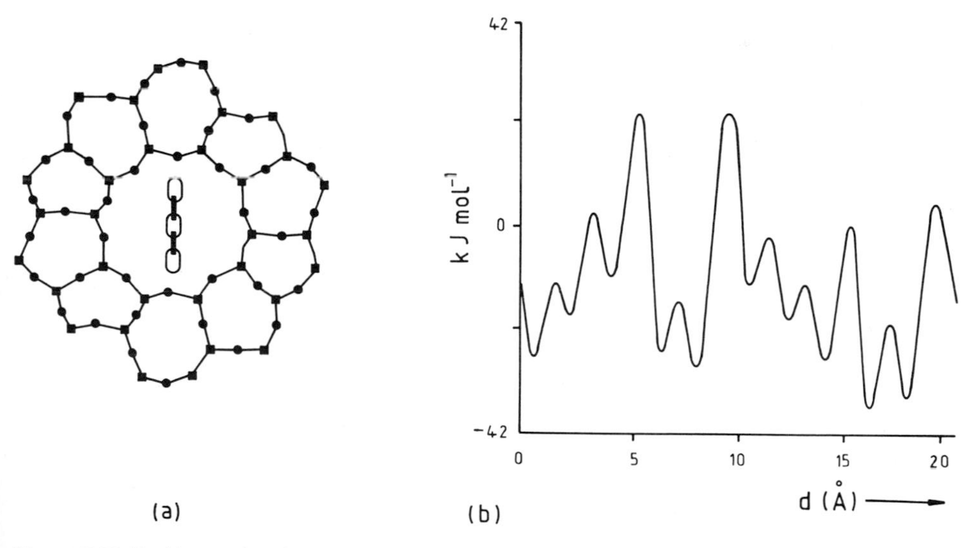

Figure 2.38 In (a) a molecule of benzene, with its plane perpendicular to the paper, is placed at an angle of 30° with respect to the maximum diameter of the large pore in a model zeolitic catalyst. The variation in interaction energy between the benzene and the oxygens lining the pore of the catalyst is shown in (b) as a function of distance traversed along the pore, keeping the orientation of the benzene constant. With permission from S. Ramdas et al., *Angew. Chem., Int. Ed. Engl.* **1984**, *23*, 671.)

neutron scattering. In the Monte Carlo method, ensemble averages are computed by numerical procedures which entail the generation of an ensemble of configurations by a series of random moves: the probability of a configuration being accepted into the ensemble is normally dependent on its Boltzmann factor. Such methods have a useful role in the study of the distribution (and its temperature dependence) of a sorbed molecules over the available sites within a solid and on its surface. A recent elegant extension of this approach, known as the 'computation-bias Monte Carlo technique', is particularly well suited to simulate quantitatively the adsorption of long-chain alkanes (of relevance in catalytic cracking) in zeolites. Molecular dynamics, in which kinetic energy is included explicitly in the simulations, solves numerically the classical equations of motion of the system simulated. This technique yields detailed dynamical behaviour of the system and the simulation of the diffusion of the sorbed species. In this work, classical microcanonical ensemble techniques are used on an ensemble to which periodic boundary conditions are applied so as to be able to cope with an infinite system. Combinations of molecular dynamics (MD) and Monte Carlo (MC) and energy minimization (EM) methods have also been utilized to great effect in, for example, ascertaining which particular microporous solid acid (zeolitic) catalyst is best suited for the skeletal isomerization of 1-butene to 2-methylpropylene (isobutene) – see Freeman *et al.* (1991).

In summary, we note that there are useful semi-empirical methods of calculating the heats of adsorption of physically adsorbed entities. In particular, we see that long-chain hydrocarbons or porphyrins, because their interaction potentials are roughly proportional to the number of atoms in the molecule, will have quite large values (70 $kJ\,mol^{-1}$) for their heats of physical adsorption. These molecules occur in oils such as those that are 'cracked' by zeolitic and clay-based catalysts.

2.7.2 The Binding Energies of Chemisorbed Species

Armed with the wealth of quantitative information pertaining to the strengths of binding of species chemisorbed to surfaces, we conclude that there are no simple rules that connect a single property of the solid with its chemisorptive behaviour. Even for metals, where we would expect more readily interpretable facts, the situation is far from simple, as the data in Table 2.3 reflect. It is not surprising that the general trend so far as the strengths of bonding is concerned is in the order $N > O > H > CO \geqslant NO$. But locating the precise cause of the variations along a series of metals with a given adsorbate is not easy (see below). If, however, we compare the bond dissociation energies of diatomic metal hydrides and the strengths of the corresponding M–H chemisorption bonds and the appropriate energies given in Table 2.3 with the average M–CO bond energies in well-known metal carbonyls (such as $Ni(CO)_4$, $Fe(CO)_5$, where the M–C bond strength is, respectively, 146 and 121 $kJ\,mol^{-1}$), we conclude that localized, covalent bonds could well be formed when molecules are chemisorbed by metals. The fact that the photoelectron spectrum (see Section 3.5) of the species $Fe(PF_3)_5$ is closely similar to that exhibited by a Fe(110) surface exposed to PF_3 vapour reinforces this view.

The notion that covalent bonds are formed when simple molecules are chemisorbed at metal surfaces first gained currency with the work of Eley. For the dis-

Table 2.3 Strengths of metal–adsorbate bonds (kJ mol^{-1})
(a) Adsorption at densely packed planes of transition metals

	N	H	CO	NO
Fe(110)	585	268		
Ni(111)	564	263	113	105
Cu(111)		234	50	
Pd(111)	543	259	142	130
Ir(111)	531	263	142	84
Pt(111)	531	238	134	113

(b) Bond dissociation energies of M–H bonds

	Ni–H	Cu–H	Ag–H	Pt–H
Diatomic molecule	250	276	221	347
Chemisorbed link	263	234	217	238

(c) Strength of the W–H bond at crystallographically different surfaces of tungsten

(100)	(110)	(111)	(211)	(123)	(144)
146	138	153	167	163	142

sociative uptake of H_2 by a metal M:

$$2\,M + H_2 \rightarrow 2(M\text{–}H)$$

Eley assumed that the heat of adsorption at zero coverage is given by:

$$-\Delta H = 2D_{MH} - D_{HH} \tag{69}$$

the Ds being the respective bond energies. Tacit in Eley's work is the assumption that no metal–metal bonds are severed during adsorption. He wrote for D_{MH}, following Pauling,

$$D_{MH} = (D_{MM} + D_{HH})/2 + 23.06(x_M - x_H)^2 \tag{70}$$

where $(x_M - x_H)$ is the difference in electronegativity between M and H. Combining Eqs. (69) and (70) and taking $\Delta H_v/6$ as the magnitude of D_M, where ΔH_v is the enthalpy of vaporization of the metal (each atom is surrounded by 12 neighbouring ones, so that the average M–M bond energy is $2\Delta H_v/12$), Eley calculated D_{MH} for a number of metals. The computed values agreed, to within about 10%, with the experimental values for chemisorption, at zero coverage, on tungsten, tantalum, iron, nickel, chromium, and cobalt, thus lending credence to the view that the bond formed is predominantly covalent.

It is also possible to estimate the heat of adsorption that would correspond to an ionic chemisorptive bond such as M^+H^- or M^-H^+. So enormous is the difference between the observed and computed values that we may safely discount the likelihood of there being predominantly ionic links formed when H_2 is dissociatively chemisorbed at metal surfaces.

When, however, sodium or caesium is bound to a metal surface such as tungsten, there is little doubt that the bond formed is largely ionic:

$$Na(g) + W(s) \rightarrow W^{-}Na^{+}$$

The magnitude of the heat of adsorption may be deduced from:

$$-\Delta H = e\phi - eI + \frac{e^2}{4\pi\varepsilon_0(4R)} \tag{71}$$

where eI is the energy required to remove an electron from the highest occupied level of an isolated atom (in this case Na) to infinity, $e\phi$ is the energy liberated when an electron is transferred to the lowest unoccupied energy level of the metal (tungsten), and the third term, containing the permittivity of free space ε_0, is the electrostatic image energy arising from the presence of the positive ion at a distance R from the surface.

Recall that the electrostatic force between an electron and its image in this situation is $-e^2/(2R)^2$, and also that, for a metal, the electron affinity and the ionization energy are numerically equal to one another and to the so-called work function, each being the energy difference between an electron at the Fermi level and an electron at rest in the vacuum just outside the metal; see Fig. 2.39.

That there is some degree of charge separation in the covalent bonds formed when molecules are adsorbed at metal surfaces is unmistakably indicated by the changes that are observed in the magnitude and sign of the work function in the course of gas uptake (Section 3.8). When electrons flow from the metal to the adsorbate upon formation of the chemisorbed link, there is an increase in the work function and a

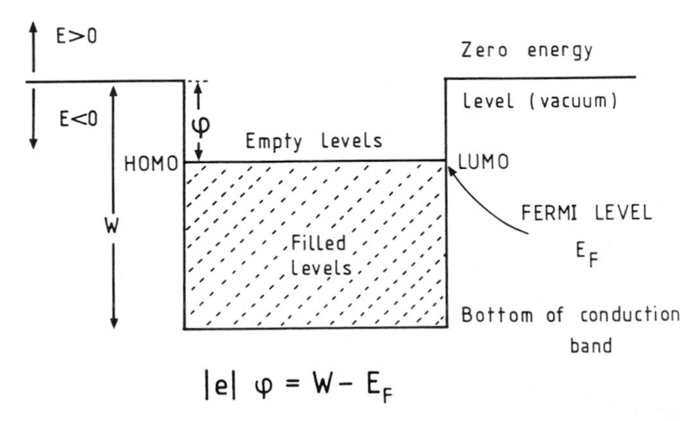

Figure 2.39 For a metal the Fermi level E_F coincides with the highest occupied molecular orbital (HOMO) and it is infinitesimally lower than the lowest unoccupied molecular orbital (LUMO). W is the potential well bonding the conduction electrons to the solid and φ is the work function (see text).

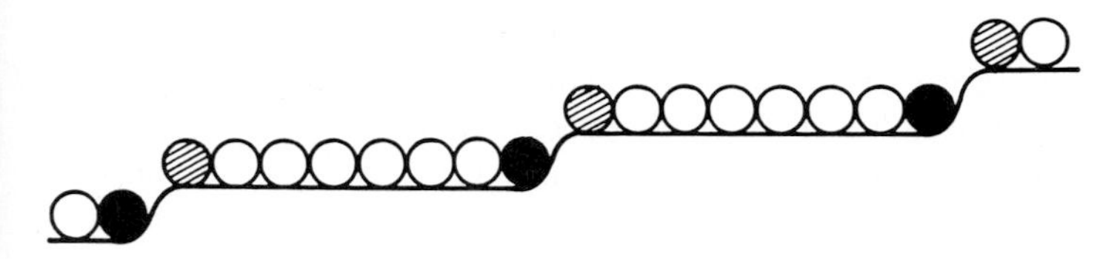

Figure 2.40 Model of monolayer of adsorbed hydrogen atoms (circles) on a stepped surface of platinum. The hydrogens adjacent to the steps have slightly different positive dipoles pointing outwards, whereas the others have negative dipoles pointing outwards.

decrease when electrons are donated to the metal. In general, the surface M–H bonds, irrespective of the nature of the metal M, have the negative end of the dipole pointing outwards. When, however, stepped surfaces of platinum and other metals are exposed to H_2, three distinct types of covalent M–H bonds, differing very slightly in binding energy (by less than 7 kJ mol^{-1}), are formed (Fig. 2.40). Whereas the hydrogen atoms bound to the flat surface have their negative dipoles pointing outwards, those attached on either side of a monatomic step in the platinum surface have the dipole pointing inwards and to a different degree depending upon whether the hydrogen atoms are at the top or bottom of the step.

A number of quantum mechanical procedures are available for estimating the strength of the chemisorptive link; these are outlined in Section 5.4.7. Such procedures reveal the degree of variation we should expect for the heat of adsorption across a single crystal surface. When, for example, CO is 'chemisorbed' on a model

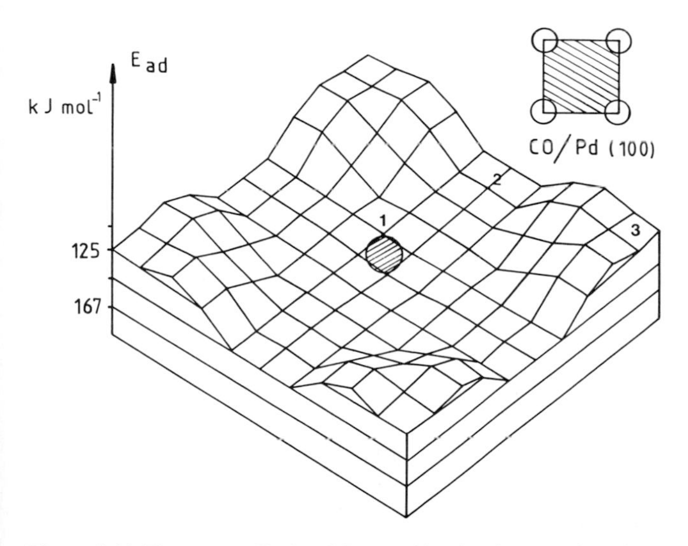

Figure 2.41 Energy profile for CO on Pd(100). The energies of adsorption at sites 1 and 2 (four-fold and two-fold sites, respectively) are almost equal, not so at the atop site (3), where it is less. Surface migration from sites 1 to 2 is, therefore, facile. With permission from G. Ertl, in *The Nature of the Surface Chemical Bond* (Eds.: T. N. Rhodin, G. Ertl), North-Holland, Amsterdam **1979**, Chapter 5.)

surface of nickel (composed of clusters of from 5 to 15 metal atoms) it is found that the most favourable sites are those with the highest coordination of the adsorbate by surface atoms (as found experimentally by LEED; see Section 3.6). Doyen and Ertl (1977) carried out detailed semi-empirical calculations for CO, N_2 and NO adsorbed on nickel, copper, palladium and silver. They found that, on the Pd(100) surface, the adsorption energy varies by only some $10\,\mathrm{kJ\,mol^{-1}}$ across the surface, bridge and four-fold sites being energetically nearly equivalent (Fig. 2.41). This agrees with the structural model shown in Fig. 2.17(a), and also with the observation that the adsorption energy drops by about $25\,\mathrm{kJ\,mol^{-1}}$ above about $\theta = 0.5$, where the molecules of CO are displaced from their symmetric bridge positions. The information conveyed by Fig. 2.41 helps us to understand the phenomenon of surface migration, which we discuss in Section 2.8.

2.7.3 Estimating Heats of Adsorption from Thermodynamic Data

The compilations given in Table 2.3 have emerged largely from direct experimental studies of the systems in question by determination of isosteric heats from adsorption isotherms, or from the appropriate thermally stimulated desorption spectra. Often, however, we may estimate heats of adsorption from already available thermodynamic data such as the enthalpies of formation of bulk compounds.

That useful empirical relationships exist was realized by Trapnell and co-workers (1960) who drew attention to the fact that the heats of adsorption of O_2 on evaporated metal films were similar to the corresponding enthalpies of formation of the bulk oxides. We would not expect these two values to be exactly the same because there are a number of factors associated with the surface which might be expected to affect profoundly the enthalpy of formation of a single layer and render it different from that for the bulk compounds. Nevertheless there are strong similarities electronically between two-dimensional and three-dimensional compounds, because photoelectron spectroscopy tells us (Section 3.5) that, for example, dissociated CO on a metal is structurally similar to a surface carbide and a surface oxide. Roberts showed that the enthalpy of formation of many metal oxides, nitrides, hydrides and sulphides vary smoothly with the heat of adsorption of the corresponding simple diatomic gases on the clean, polycrystalline metal surfaces, and Tanaka and Tamaru have formulated a general rule according to which the initial heats of chemisorption (ΔH_a) of gases such as O_2, N_2, H_2, NH_3 and C_2H_4 on various metal surfaces are empirically expressed by the equation:

$$\frac{\Delta H_\mathrm{a}}{\mathrm{kJ\,mol^{-1}}} = a[(-\Delta H_\mathrm{f}) + 155] + 83.6$$

where $-\Delta H_\mathrm{f}$ is the enthalpy of formation of the highest oxide per metal atom, and a is a constant dependent upon the nature of the gas. Thus, if the heat of adsorption of a particular gas on a metal surface is known, all the values of the heats of chemisorption at zero coverage of the gas on different metal surfaces can be estimated.

Pursuing the analogy between surface and bulk compounds, Benzinger (1980) arrived at a convenient method of estimating the heats of dissociative chemisorption of CO, NO and N_2 on metals in the first, second and third transition series. The

Table 2.4 Estimated heats (kJ mol^{-1}) of dissociative adsorption (after Benzinger – see text)

	Ti	V	Cr	Mn	Fe	Co	Ni
CO	572	410	335	279	136	90	88
NO	882	725	589	603	368	321	333
N_2	672	434	246	256	22	−16	−2

argument states that the heat of dissociative adsorption ΔH_D is given approximately by the difference in enthalpy between the gas-phase molecule and the enthalpies of formation of the metal oxides, carbides, and nitrides:

$$\Delta H_D(CO) = \Delta H_f(MC) + \Delta H_f(MO) - \Delta H_f(CO(g))$$

$$\Delta H_D(NO) = \Delta H_f(MN) + \Delta H_f(MO) - \Delta H_f(NO(g))$$

$$\Delta H_D(N_2) = 2\Delta H_f(MN)$$

A selection of the estimate made by Benzinger is given in Table 2.4.

Apart from their intrinsic usefulness, the fact that the numerical magnitudes of the heats of adsorption closely parallel those of the enthalpies of formation of the corresponding bulk compounds, adds justification to the increasing practice these days of seeking correlations between enthalpies of formation, on the one hand, with catalytic activity, on the other. In the catalysed decomposition of formic acid over a series of metals, for instance, there is a correlation between the enthalpy of formation of the metal formate and the activity of the corresponding metal (see Section 3.2).

2.7.4 Decline of the Heat of Adsorption with Increasing Coverage

For polycrystalline solids, or for single-crystal material containing very many distinct sites for adsorption, we would expect the heat of adsorption to fall as surface coverage increases, solely because the strongest sites will be populated first. This is the main cause of the decrease in heat of adsorption of gases on polycrystalline materials such as those shown in Fig. 2.42(a). We can also expect quite significant decreases in heats of adsorption when one type of chemisorbed link gives way to another at a step, kink, or emergent dislocation. But on single-crystal faces, where there is a uniform, if periodically varying, potential as in Fig. 2.41, we expect (and do indeed find) the heat of adsorption to remain sensibly constant as coverage increases. And by the mere fact that ordered adlayers are frequently formed (Section 2.3, Fig. 2.18 especially), we conclude that there are significant interactions between adsorbed particles. At certain critical values of coverage (see Figs. 2.22 and 2.42), sharp drops in heats of adsorption are attributable to the onset of adsorbate–adsorbate repulsions. There are at least three sources of these repulsions.

First, dipole–dipole interactions obviously become more pronounced as coverage increases – picture the mutual repulsion between the negatively charged chemisorbed hydrogen atoms in the monolayer shown in Fig. 2.40. The pairwise interaction energy between two dipoles separated by a distance r is equal to μ^2/r^3, where μ is the

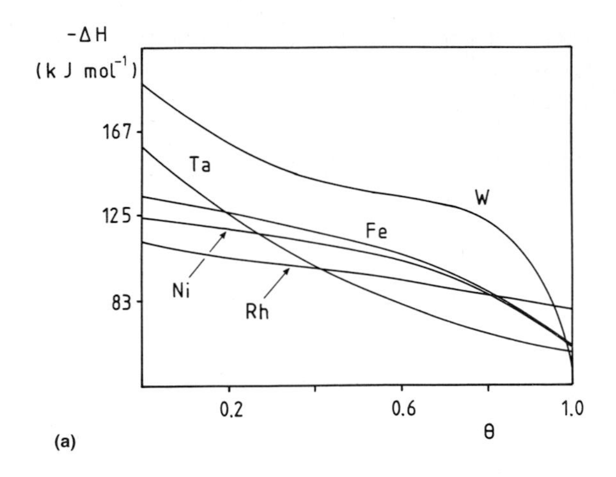

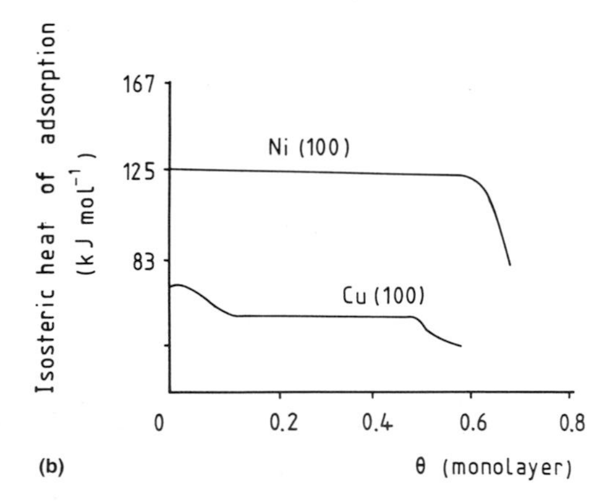

Figure 2.42 (a) Heat of adsorption, $-\Delta H$, determined calorimetrically, plotted against coverage, θ for hydrogen on a series of evaporated films. (b) Variation of isosteric heat of adsorption of CO on the (100) planes of nickel and copper. (Compare Fig. 2.22(b).) With permission from J. C. Tracy, P. W. Palmberg, *J. Chem. Phys.* **1969**, *51*, 4852.

dipole moment. Numerical estimates of this effect for a monolayer of adsorbate show that the total fall hardly amounts to more than about $5\,\text{kJ mol}^{-1}$, even for highly dipolar adsorbates (such as Na^+ on a tungsten surface).

Second, the effective size of an adsorbate species is dictated by the dimensions of its atomic or molecular orbitals. As coverage approaches a maximum, orbitals on neighbouring species will interact, and we could invoke an interaction potential U much as in Eqs. (66) and (67) above to describe the situation. As we have seen earlier (Fig. 2.23) there is strong evidence that this type of interaction dominates chemisorption of CO on palladium and other transition metals at high coverage.

The continuous decrease in heat of adsorption at low coverage on single-crystal surfaces (Figs 2.23 and 2.42b) signifies the operation of a third, long-range, type of interaction. Accompanying this continuous fall in adsorption energy is a concomitant change the C–O stretching frequency of the adsorbed CO. It is thought that the operating interactions are of an indirect ('through-bond') kind. That is, they act via the valence electrons of the metallic adsorbent. Moreover, these interactions exhibit oscillatory character in that they may be repulsive as well as attractive, depending upon separation distance (compare Fig. 2.10). Their magnitude amounts to about one-tenth of the strength of the chemisorption bond. These interactions, which fade out to below the magnitude of the thermal energy RT beyond a few lattice repeat distances, are responsible for the ordering within adlayers, discussed in Section 2.3.

2.8 Mobility at Surfaces

The potential-energy diagram shown in Fig. 2.2 depicts a periodic variation with distance parallel to the crystal surface. Obviously, a surface species, be it a bound adsorbate or a loosely held adsorbent atom, has to surmount a potential energy barrier (E_m) in order to migrate from one lattice site to a neighbouring one. If, after a time t, the mean-square displacement is $\langle R^2 \rangle$, it can be shown that

$$\langle R^2 \rangle = Dt \tag{72}$$

Moreover, the diffusion coefficient D, usually expressed in $cm^2\, s^{-1}$, is temperature-dependent:

$$D = D_0 \exp\left(\frac{-E_m}{RT}\right) \tag{73}$$

Many experimentalists have devised ingenious methods of recording D as a function of temperature, thereby arriving at E_m. It is found that atomic hydrogen diffuses over a tungsten surface with values of E_m ranging from 25 to 70 $kJ\, mol^{-1}$, depending upon surface coverage and crystallographic orientation. It transpires that, in general, the activation energy barrier for surface migration (i.e. E_m) lies between 10 and 20% of the magnitude of the bond energy holding the chemisorbed species to the surface, but it can be larger. Reverting to Fig. 2.41, we immediately see that there is likely to be anisotropy associated with surface migration: energetically it is much more favourable for a bound CO molecule to move from position 1 to 2 (from the fourfold to the twofold site) than to migrate along the diagonal direction via 3.

It is instructive to enquire how frequently an adsorbed entity 'hops' along a catalyst surface. By analogy with Eqs. (3) and (3a) for the residence time of an adsorbate before it desorbs, we may write:

$$\tau' = \tau_0' \exp\left(\frac{E_m}{RT}\right) \tag{74}$$

where, for the reasons given earlier (Section 2.5.1) τ_0' may be taken as 10^{-13} s. Now if E_m is very small (and less than RT), the adsorbed species will move freely over the surface and it behaves as a two-dimensional gas. In many catalytic reactions, especially those operating at high temperatures, this situation is realized. At the other extreme we may envisage a value of E_m of some 20 $kJ\, mol^{-1}$ so that the adsorbed species remains, anchored, at room temperature, at a particular surface site for *ca* 5×10^{-10} s. If the residence time (before desorption) at the same temperature is, typically, a microsecond or so – corresponding to a quite low activation energy of desorption of some 40 kJ mol^{-1} – we see that the adsorbed species will execute an average of some 2000 hops from one site to another during its sojourn on the surface. We note, in passing, that, if the hopping distance is *ca* 300 pm the adsorbed species will have covered a total distance of 6×10^{-5} cm before it is desorbed.

Direct observation of individual adsorbate species using field-ion microscopy and scanning tunneling microscopy (STM) vividly brings home the reality of surface migration. It is, however, easier experimentally to follow (self-) diffusion on the surface of the metal, or diffusion of one metal on another, than to trace the surface meanderings of a light element on a metal although the recent work of Zambelli *et al.* (1996), using STM, has succeeded in recording the surface diffusion and hopping of nitrogen atoms on ruthenium. The values of E_m for the self-diffusion on different crystallographic planes of rhodium are (in kJ mol^{-1}): (111), 15; (110), 58; and (100), 84. This indicates that some surfaces of a particular metal are intrinsically more mobile than others. In common with the surface motion of light adsorbates, self-diffusion can be markedly anisotropic and may sometimes be restricted to one dimension as, for example, on the (110) planes of Rh.

The 'visualization' of the diffusion of sorbed species within zeolitic catalysts is readily accomplished by computer graphics. It is a relatively straightforward matter to 'track' the diffusion paths of sorbed molecules within microporous or mesoporous catalysts (see Chapter 5) on feeding into the system under coonsideration the appropriate potential functions (for atom–atom or three-centre interactions) as well as the bond bending and torsion terms coupled to the interaction parameters (see Eqs. (66)–(68), be they for individual atoms or so-called 'united-atom' descriptions (where, for example, methyl and methylene groups are considered as a single interaction centre).

2.9 Kinetics of Surface Reactions

Traditionally, interpretations pertaining to the kinetics of heterogeneous catalytic reactions have made use of Langmuirian concepts that relate fractional coverage to partial pressure via equations such as Eq. (47) above. But in the foregoing sections we have shown that either because of intrinsic surface heterogeneity, or because of induced surface interactions, the heat of adsorption of a species often varies with coverage, thereby rendering the Langmuir relationship invalid (except over narrow ranges of θ, where the resulting deviations can be ignored). Moreover, when two or more reactants are co-adsorbed, we find that the individually adsorbed species are seldom randomly distributed over the surface (see Fig. 2.43). When, therefore, we consider a reaction such as $A + B \rightarrow C$ taking place at a surface according to either the Langmuir–Hinshelwood or Eley–Rideal mechanisms of catalysis (Fig. 2.1), we are immediately forbidden from writing the rate of reaction as:

$$r = k\theta_A\theta_B \quad \text{(Langmuir–Hinshelwood)} \tag{75}$$

and

$$r = k'\theta_B p_A \quad \text{(Eley–Rideal)} \tag{76}$$

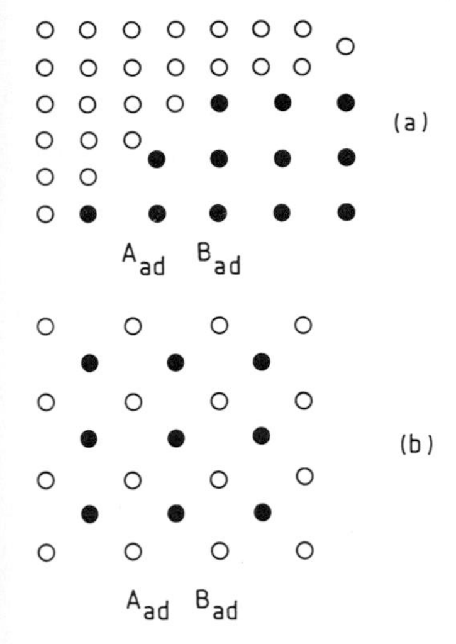

Figure 2.43 Two reactants, A and B, distinguished by filled and empty circles, will form an adlayer structure such as (a) if adsorption is competitive and (b) if it is cooperative. Only at the regions where empty and filled circles meet is a Langmuir–Hinshelwood surface reaction possible in (a), whereas, in principle, it may occur anywhere on the surface in (b). The sequence in which the two reactants are introduced to the surface governs the nature and composition of the adlayer. Thus little oxygen can be chemisorbed on a palladium surface on to which CO has been previously adsorbed, as shown in Fig. 2.17(a).

In effect we cannot straighforwardly correlate coverages with the respective partial pressures. We have seen that even when a single species is adsorbed at a single-crystal surface many subtleties such as variation in sticking coefficients and heats of adsorption with coverage are observed. A real catalyst surface with its a-priori energetic heterogeneity will obviously be much more complicated in so far as relating a steady-state concentration of a surface intermediate to a particular partial pressure is concerned.

Nevertheless there are a few circumstances when (judging by the final results!) it seems that the crude assumptions of Langmuir adsorption can be used. Typically, for the reaction A + B → C, we have:

$$\theta_{\mathrm{A}} = \frac{b_{\mathrm{A}} p_{\mathrm{A}}}{1 + b_{\mathrm{A}} p_{\mathrm{A}} + b_{\mathrm{B}} p_{\mathrm{B}}} \tag{44a}$$

$$\theta_{\mathrm{B}} = \frac{b_{\mathrm{B}} p_{\mathrm{B}}}{1 + b_{\mathrm{A}} p_{\mathrm{A}} + b_{\mathrm{B}} p_{\mathrm{B}}} \tag{44b}$$

Tacit in our model is the notion that product molecule C is so weakly adsorbed that its steady-state concentration is negligible, and that an adsorption site may be occupied only by either A or B in a competitive sense.

For a Langmuir–Hinshelwood (LH) mechanism we therefore rewrite Eq. (75) as:

$$r_{\mathrm{LH}} = k\theta_{\mathrm{A}}\theta_{\mathrm{B}} = \frac{b_{\mathrm{A}} B_{\mathrm{A}} p_{\mathrm{A}} p_{\mathrm{B}}}{(1 + b_{\mathrm{A}} p_{\mathrm{A}} + b_{\mathrm{B}} p_{\mathrm{B}})^2} \tag{77}$$

We note that for a constant temperature and constant p_{B}, the rate will pass through a maximum as p_{A} is varied. The maximum occurs when the two reactants, A and B, have equal surface concentrations: $\theta_{\mathrm{A}} = \theta_{\mathrm{B}}$.

By contrast, with an Eley–Rideal (ER) mechanism we have

$$r_{\mathrm{ER}} = \frac{k' b_{\mathrm{B}} p_{\mathrm{A}} p_{\mathrm{B}}}{1 + b_{\mathrm{B}} p_{\mathrm{B}}} \tag{78}$$

which reveals that the rate will always be of first order with respect to p^{A}, whereas the order with respect to p_{B} passes from unity to zero.

Turning to surface reactions of the type: A(ad) → B + C which involve the rupture of at least one chemical bond, we have

$$r = \frac{kb_{\mathrm{A}}p_{\mathrm{A}}}{1 + b_{\mathrm{A}}p_{\mathrm{A}}} \tag{79}$$

There are two limiting cases: if $b_{\mathrm{A}}p_{\mathrm{A}} \ll 1$ then $r = kb_{\mathrm{A}}p_{\mathrm{A}}$ and the reaction is kinetically of the first order; but if $b_{\mathrm{A}}p_{\mathrm{A}} \gg 1$, then $r = 1$ and the reaction is zero-order in gas pressure. For a strongly adsorbed species b is, by definition, large and $b_{\mathrm{A}}p_{\mathrm{A}} \gg 1$ is likely to be valid for all accessible values of p_{A} so that the observed reaction is always zero-order. At intermediate values of b the reaction order will itself vary with pressure.

2.9.1 The Influences of Precursor States on the Kinetics and Energy Distribution of Catalysed Reactions

At the outset of this chapter we remarked that the conceptual distinction between the Langmuir–Hinshelwood and Eley–Rideal mechanism was clearcut. We also commented that sometimes the distinction between them is blurred. The blurring tends to occur whenever non-thermal processes are a consequence of the existence of hot precursors akin to those discussed earlier (see Sections 2.1.1 and 2.5.6).

On theoretical grounds, the initial collision of a gas-phase molecule with a catalyst can rarely be likened to that of a golf ball with the surface of a sandpit. Subsequently to collision, the molecule either scatters back into the gas (elastically or inelastically), or is trapped in the region close to the surface. If the latter, it then performs a complicated 'precursor motion' for a period of time during which energy is exchanged between the various degrees of freedom of the system, including internal and translational degrees of freedom of the molecule. Parallel transport over short or large distances is feasible, depending upon the nature of the surface–adsorbate interaction and the efficiency of the energy transfer. If the precursor collides with a chemisorbed species a direct reaction may ensue, followed by immediate desorption of the product. The likelihood that such a reaction may take place may be appreciably greater than the probability that the same species diffusing thermally along the surface will react. The reason for this is that the precursor may have high translational energy or be in a state of high internal excitation, the root cause of such excitation being the chemisorption energy that is invested into the precursor species.

The key point here is that the energy distributions of reactants and products at catalyst surfaces may be highly non-thermal. The precursor is never fully chemisorbed in a thermally equilibrated sense so that part of the chemisorption energy is not dissipated by the solid, but appears as kinetic, vibrational, rotational or electronic energy of the product. (The veracity of this statement has been amply demonstrated in recent years in a series of experiments and theoretical assessments involving collisions of beams of molecules with surfaces). As a consequence we would not expect, and we do not find, simple obedience to LH and ER kinetics of the kind encapsulated by Eqs. (75) and (76). An elegant illustration of how neither the LH nor the ER mechanism holds for the catalysed production of H_2O from H_2(g) and O(ad) at a platinum surface has been provided by Harris et al. (1981). H_2 molecules incident at metal sites break apart, liberating H atoms with high kinetic energy parallel to the surface. These atoms collide with chemisorbed O to form OH radicals, and with OH radicals to form H_2O. There is no straightforward dependence of rate of reaction on the concentration of the H atoms!

2.9.2 Comparing the Rates of Heterogeneous and Homogeneous Reactions

One of the most striking ways in which we can illustrate why certain reactions proceed more rapidly on a solid surface than in the gas phase is by utilizing transition-state theory, just as we did earlier for rates of adsorption (Section 2.5.3) and desorption (Section 2.5.6).

By analogy with Eqs. (15) and (29), we may write for the rate of the homogeneous gas-phase reaction between species A and B:

$$r_{\mathrm{hom}} = \frac{kT}{h}\frac{f^*_{\mathrm{hom}}}{f_{\mathrm{A}}f_{\mathrm{B}}} N_{\mathrm{A}}N_{\mathrm{B}} \exp\left(\frac{-E^{\ominus}_{\mathrm{hom}}}{RT}\right) \tag{80}$$

where N_{A} and N_{B} are, respectively, the numbers of molecules of A and B per cm^3 of gas, and $E^{\ominus}_{\mathrm{hom}}$ is the activation energy for the reaction at zero Kelvin. For the heterogeneous process we make the assumption that the reaction is between surface species, the concentration of which is directly proportional to the gas-phase concentration, and that reaction takes place at adjacent pairs of surface sites S_2, the concentration of which is N_{S_2}. We are, in other words, assuming that the classic Langmuir–Hinshelwood mechanism is valid. Hence, the rate, r_{het}, is:

$$r_{\mathrm{het}} = \frac{kT}{h}\frac{f^*_{\mathrm{het}}}{f_{\mathrm{A}}f_{\mathrm{B}}} N_{\mathrm{A}}N_{\mathrm{B}}N_{S_2} \exp\left(\frac{-E^{\ominus}_{\mathrm{het}}}{RT}\right) \tag{81}$$

where f^*_{het} and $E^{\ominus}_{\mathrm{het}}$ stand, respectively, for the partition function for the transition complex formed at the surface and the activation energy of the heterogeneous reaction at zero Kelvin, Hence,

$$\frac{r_{\mathrm{het}}}{r_{\mathrm{hom}}} = N_{S_2}\frac{f^*_{\mathrm{het}}}{f^*_{\mathrm{hom}}} \exp\left(\frac{\Delta E^{\ominus}}{RT}\right) \tag{82}$$

where $\Delta E = E^{\ominus}_{\mathrm{hom}} - E^{\ominus}_{\mathrm{het}}$.

For an essentially smooth surface, most solids have a value of N_{S_2} close to $10^{15}\ \mathrm{cm}^{-2}$. The value of f^*_{hom} ranges from 10^{24} to 10^{30} and, for an immobile transition state at the catalyst surface, $f^*_{\mathrm{het}} \simeq 1$. Equation (82) therefore becomes

$$\frac{r_{\mathrm{het}}}{r_{\mathrm{hom}}} \approx 10^{-12\pm3} \exp\left(\frac{\Delta E^{\ominus}}{RT}\right) \tag{83}$$

Bearing in mind that, in these estimates, the heterogeneous rate refers to $1\ \mathrm{cm}^2$ of suface and to $1\ \mathrm{cm}^3$ of gas phase, and the homogeneous rate to $1\ \mathrm{cm}^3$ of the gaseous reactants, it follows that the heterogeneous reaction would be very much slower if the activation energies for the two reactions were the same. Clearly, the heterogeneous reaction will tend to become more significant if

1. E^{1}_{het} is less than $E^{\ominus}_{hom}$, and
2. the surface area is increased.

By far the more important is the difference in activation energy – for a surface area of $100\,m^2\,g^{-1}$, there would still remain a pre-exponential factor of $10^{-6\pm3}$. It follows from Eq. (83) that, for the rates of the heterogeneous and homogeneous reactions to be equal at room temperatures, the activation energy of the heterogeneous reaction must be less by *ca* $70\,kJ\,mol^{-1}$. Obviously, as the reaction temperature increases, the disparity in activation energy must increase in order to maintain parity of rate. There is ample evidence that bears out these conclusions; and it has consequently become one of the shibboleths of surface chemistry that a catalyst lowers the activation energy of a reaction, just as we saw in Fig. 2.35.

2.10 Autocatalytic, Oscillatory, and Complex Heterogeneous Reactions

Our discussion of the kinetics of catalysed reactions would not be complete without reference to oscillatory phenomena, which are as intriguing and important here as they are in other branches of science (astronomy, hydrodynamics, meteorology and many aspects of biology, including population dynamics, nerve impulses, morphogenesis and enzyme reactions).

Sustained oscillations in catalyst temperature, in concentration of surface species, or most commonly – because of its accessibility – in the time variation of the concentrations of reactants and products in contact with the catalyst, have been described since the early 1970s. Reports from Germany, Russia, the USA and the UK show that, depending upon the temperature range and reaction in question (oxidation of CO or hydrocarbon or NH_3), the cycle time can range from seconds to hours. Both damped and sustained oscillations in concentration for homogeneous reactions in solution have been known since much earlier; and the renowned Belousov–Zhabotinskii (BZ) reaction, now well understood, involves temporal variation in the ratio of cerium(IV)/cerium(III) during the cerium-ion-catalysed oxidation of malonic acid by bromate in aqueous sulphuric acid.

Oscillatory phenomena observed during catalytic reactions may be a consequence of either physical or chemical effects and, indeed, may occur by interaction between the physical and chemical states of a system. We shall see in the following sections of this chapter that innate instabilities within a catalytically reacting system may, by perturbation of conditions such as temperature or concentration, either lead to a new steady state or, alternatively, precipitate regular oscillations in temperature and concentration. It is also clear that oscillatory phenomena may be induced purely as the result of chemical interactions. In a heterogeneous catalytic reaction, such oscillations may be the result of interactions between the concentration of reactants in the fluid phase and the concentration of surface species. On the other hand, oscillations may occur solely because of autocatalytic effects. Each of these examples will be discussed in the following sections, but because autocatalysis is unfettered by physical phenomena superimposed upon chemical effects, we commence our discussion by reference to autocatalysis and, after providing a brief historical background, deal with some specific details concerning instabilities and oscillations.

2.10.1 An Outline of Autocatalysis

Whenever the products of a reaction serve to increase its rate, we talk of autocatalysis. Here, as the reaction ensues and more product is formed, the rate increases. The BZ reaction mentioned above (and now known to involve 18 elementary steps and 21 distinct chemical species) has several steps which are autocatlytic. Thus the $HBrO_2$ formed by:

$$2\,BrO_2\,Ce(III) + 2\,H_3O^+ \rightarrow 2HBrO_2 + Ce(IV) + 2H_2O$$

is a reactant in the first step. In general, for a reactant R yielding a product P in an autocatalytic step, we have $d[P]/dt = k[R][P]$.

If $[R]_0$ and $[P]_0$ are the concentrations of reactant and product at $t = 0$, and x is the amount decomposed at time t, then:

$$dx/dt = k([R]_0 - x)([P]_0 + x) \qquad (84)$$

Integrating by partial fractions using

$$\frac{1}{([R]_0 - x)([P]_0 + x)} = \left\{\frac{1}{([R]_0 + [P]_0)}\right\}\left\{\frac{1}{([R]_0 - x)} + \frac{1}{([P]_0 + x)}\right\} \qquad (85)$$

yields

$$\left\{\frac{1}{([R]_0 + [P]_0)}\right\} \ln \left\{\frac{([P]_0 + x)[R]_0}{([R]_0 - x)[P]_0}\right\} = kt \qquad (86)$$

so that:

$$x/[R]_0 = \frac{(\exp(at) - 1)}{(1 + b\exp(at))} \qquad (87)$$

with $a = ([R]_0 + [P]_0)k$ and $b = [P]_0/[R]_0$.

Armed with the information contained in Eq. (87), it is possible, for realistic values of a and b, to maximise the rates of autocatalytic heterogeneous reactions by optimising the conditions under which stirred reactors (see Chapter 7) operate.

2.10.2 Background to Oscillating Reactions

Much is known, thanks to the efforts of Noyes, Prigogine, and others, about the theory and phenomenology of chemical oscillations. But, interestingly enough, the conditions required for undamped oscillations were arrived at much earlier by A. J. Lotka (1910, 1920), working in Brooklyn on animal populations. The enigmatic Turing, working on morphogenesis, had also encountered oscillatory behaviour.

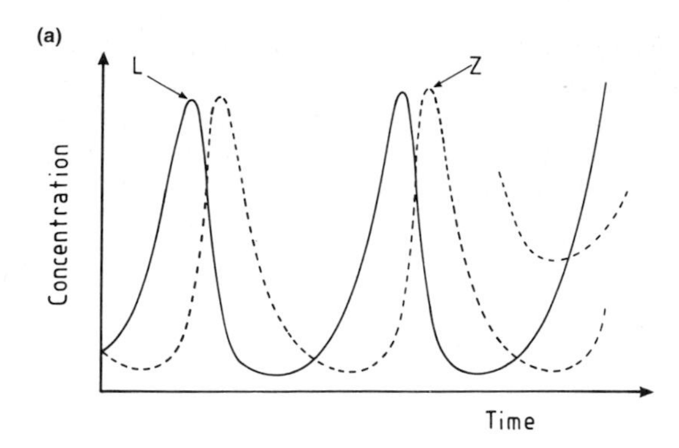

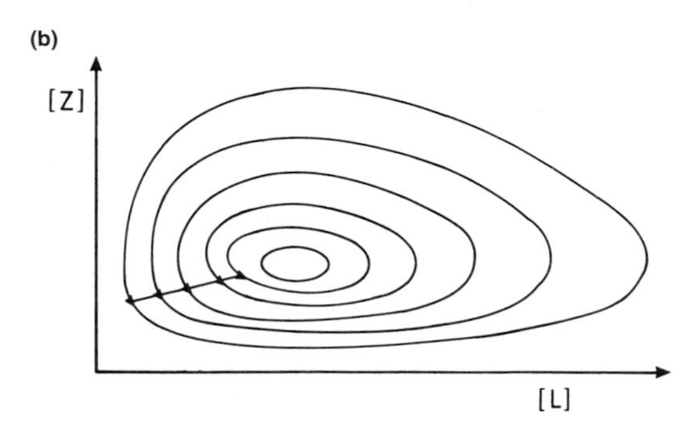

Figure 2.44 (a) Illustration of the temporal variation of the concentrations of the intermediates L and Z in a Lotka reaction sequence (Eqs. (88)). (b) The analogue of (a) in which the concentration of L is plotted versus that of Z. The overall system is represented by closed orbits, different ones being obtained from different starting conditions.

To appreciate the Lotka mechanism, consider the following reaction sequence:

$$(a)\ \mathrm{A} + \mathrm{L} \longrightarrow 2L \qquad \mathrm{d}[\mathrm{A}]/\mathrm{d}t = -k_a[\mathrm{A}][\mathrm{L}] \tag{88a}$$

$$(b)\ \mathrm{L} + \mathrm{Z} \longrightarrow 2Z \qquad \mathrm{d}[\mathrm{L}]/\mathrm{d}t = -k_b[\mathrm{L}][\mathrm{Z}] \tag{88b}$$

$$(c)\ \mathrm{Z} \longrightarrow B \qquad \mathrm{dB}/\mathrm{d}t = k_c[\mathrm{Z}] \tag{88c}$$

in which the first two steps are autocatalytic. We hold the concentration of A constant by adding more of this substance to the reactor. The concentrations [L] and [Z] of the intermediates, however, are variables. We may solve the rate equations (88) for constant [A] by numerical methods; and the results can be displayed in two ways (Fig. 2.44). The meaning of the temporal variations shown in Fig. 2.44(a) are explained as follows. With an initial small value of L, more of it will be formed (from Eq. (88a)), and this reaction, being autocatalytic, produces even more L, and the concentration of this species will increase. But as it is formed, step (b) can commence. After a slow build-up of Z, the autocatalytic step (b) leads to a surge in its concentration. But this step also consumes L, so that step (a) will now slow down, and less of L will be produced. In turn, since less of L is now available, step (b) will slow down. And as less Z is available to consume L, there will be a further burgeoning in the concentration of L. An alternative way of representing periodic variation is shown in Fig. 2.44(b), where the concentrations of L and Z are plotted against one another.

We are now in a position to consider specific aspects of oscillations in heterogeneous catalysis, including those induced by interaction between chemical and physical effects. Before doing so, we should note that, as well as having autocatalytic steps, there has to be considerable chemical complexity (i.e. many intermediates and many steps, as in the BZ reaction) and also the reactions themselves must be far from equilibrium. Under these conditions, supplemented by a few others, even greater complexity can reign. Chaos is a well-defined term reserved for the description of, *inter alia*, extremely sensitive situations which, in a manner of speaking, are synonymous with superoscillations.

2.10.3 Instabilities and Transient Phenomena in Heterogeneous Catalysis

It is natural that the accumulation of experimental data from heterogeneous catalytic reacting systems is sought, normally, from systems which are in a stable steady state, for it is only by this means that we can be sure that experiments are reproducible and the data not spurious. There are, of course, occasions when a deliberate perturbation from the steady state (such as a step change in concentration or temperature) is advantageous in securing kinetic information about the approach to a steady operating state. However, some circumstances give rise to unstable conditions which encourage the reacting system to converge upon a more stable operating state. In rather different circumstances a perturbation may lead to oscillation between steady states of the system.

Heterogeneous catalytic reactions which display more than a single steady state, at least one of which is metastable, are usually exothermic processes in which heat and mass transport between the fluid phase and the solid catalyst play a dominant role. A multiplicity of steady states can, however, arise by virtue of the chemical kinetics of the catalytic system and is certainly not confined to that class of problems identified by the effects of physical transport processes. We shall describe, in turn, the observed phenomena which are a result of system instabilities and which result in transient states.

2.10.4 Multiple Steady States

To illustrate how more than one steady state can arise in a reacting catalytic system, we consider a very exothermic reaction such as, for example, ammonia oxidation, which is discussed in greater detail in Chapter 8. Whether the reaction occurs on a metal gauze (as in the example in Chapter 8) or in a porous material supporting the active metal, heat generated by virtue of reaction at the catalyst surface is dissipated by convection, conduction, and (for some conditions) radiation processes. A given portion of gauze or single individual catalyst pellet will be in a steady state of thermal equilibrium if the rate of heat generation Q_g in the pellet volume V_p is balanced by the rate of heat loss Q_l from its exterior surface area a. In the simplest case, when one can ignore intraparticle diffusion effects, the rate of heat generation due to, say, a first-order reaction is

$$Q_g = V_p(-\Delta H)kc_i \tag{89}$$

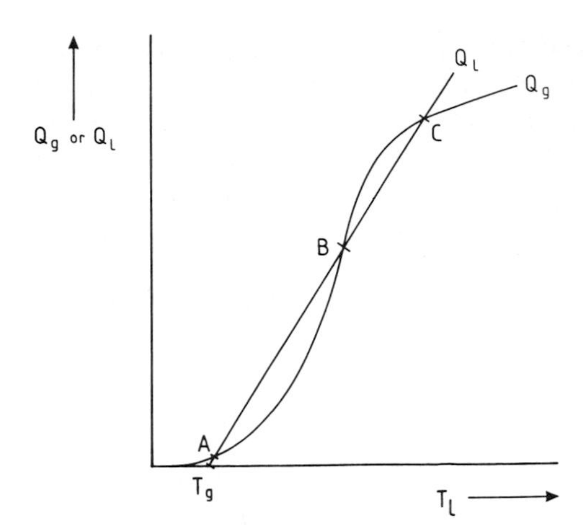

Figure 2.45 Thermal instabilities (see text).

where k is the chemical rate constant (strongly dependent on temperature), $(-\Delta H)$ is the heat of reaction and c_i the interface concentration. For a steady state, the rate of chemical reaction must be balanced by the mass flux of reactant from gas phase to particle, so

$$V_p k c_i = h_D a(c_g - c_i) \tag{90}$$

where h_D is the mass transfer coefficient and c_g is the bulk gas concentration of reactant. Q_g may therefore be expressed as a function of temperature and gas concentration only by elimination of c_i. As a function of temperature, $Q_g(T)$ is a sigmoid curve as shown in Fig. 2.45. One may superimpose upon such a curve (for a given gas concentration) the straight line representing the rate of heat loss from the particle

$$Q_l = ha(T_i - T_g) \tag{91}$$

where h is the solid-to-gas heat transfer coefficient.

It is evident from Fig. 2.45 that as many as three steady states A, B, and C are possible (compare the temperature-increase and heat-exchange curves of Fig. 7.20). States A and C are stable steady states. This can be understood if we analyse what would happen if there were a temperature perturbation in the system. Consider the operating point C (high temperature and high conversion). Any upward perturbation in temperature would result in less heat generation than heat loss and any downward perturbation would result in more heat generation than heat loss. At this particular operation point C, the system is therefore self-compensating in a thermal context. Similar arguments would indicate A to be a stable steady state (though at low temperature and hardly any conversion – a state equivalent to extinction of the reaction). On the other hand, an upward perturbation from the steady operating point B would result in more heat release than heat loss and the system temperature would adjust until the stable operating point C was reached. A downward temperature perturbation at B, however, would reduce the rate of heat generation below the rate of heat loss and the reaction would be extinguished. Here one can draw the analogy between these exothermic catalytic processes and conditions within a diffusion flame: a sudden increase in gas velocity might lift the flame off the burner port, whereas sufficient reduction in gas flow causes the flame to strike back along the gas supply line. The operating point B then is a metastable state. It is not difficult to imagine the violent temperature perturbations which catalyst particles, located at various positions in a packed tubular reactor, might undergo as a reactor assembly is either started up or shut down. Each particle in the bed is likely to have a different temperature history because the extent of reaction will vary along the bed length, as also will the gas-phase temperature.

Different states of chemical stability might also exist for a given catalytic system. Thus Wicke et al. not only reported thermal instabilities during the platinum-catalysed oxidation of hydrogen (which were of the kind described above and attributed to mass- and heat-transfer effects between catalyst particle and gas phase), but also concentration instabilities for isothermal conditions during the palladium-catalysed oxidation of carbon monoxide. Now the isothermal reaction rate for carbon monoxide (CO) oxidation can be represented by an equation of the form

$$R = \frac{kc_i}{(1 + kc_i)^2} \tag{92}$$

CO actually inhibits the oxidation at relatively high interfacial concentrations ci of CO. For a steady state this must be balanced by the rate of mass transport from gas to catalyst particle,

$$R = h_D a(c_g - c_i) \tag{93}$$

where c_g is the bulk gas-phase concentration of CO and $(h_D a)$ is the product of the mass transfer coefficient and the particle external surface area per unit volume. In Fig. 2.46, the curve displaying a maximum represents the reaction rate as a function of bulk gas concentration C (the unknown interfacial concentration can be elimi-

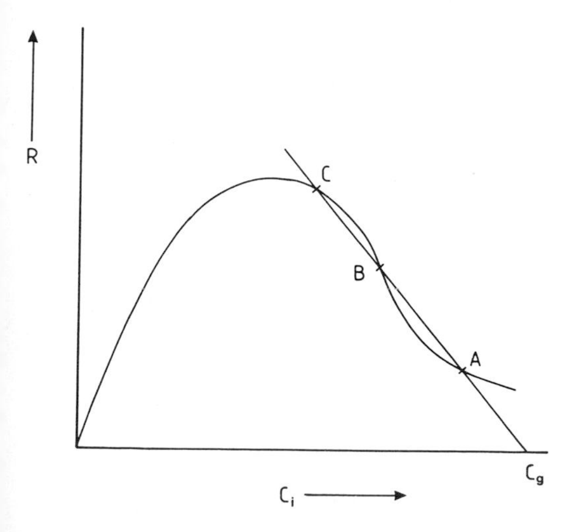

Figure 2.46 Concentration instabilities (see text and Eq. (92)).

nated between these equations and the rate reformulated in terms of the bulk gas concentration): the straight line (of slope $h_D a$) represents the mass transfer rate from bulk gas to interface. At steady-state conditions these rates will be equal. Again we note three possible steady states, the point B being metastable and points A and C stable. Clearly then, concentration instabilities may also exist in heterogeneous reacting systems, but have only been reported for a limited number of systems. It is also possible for multiple steady states to arise by virtue of a balance between intraparticle diffusive fluxes and a reaction rate described by Langmuir–Hinshelwood kinetics.

A proposal that purely chemical kinetic effects might account for the multiplicity of CO oxidation in the presence of platinum has recently been confirmed. Sufficiently small catalyst particle sizes were employed so as to eliminate intraparticle diffusion and, furthermore, a recycle reactor (see Chapter 7) was used to ensure that fluid-to-particle mass- and heat-transfer effects were absent. The experimental results obtained display multiple steady-state phenomena which are explained solely on the basis of adsorption, surface reaction and desorption rates. Apparently the interaction between these three chemical rate processes is sufficient to lead to a multiplicity of steady states under isothermal conditions.

2.10.5 Transient Phenomena

We have just shown that it is possible for some catalytically reacting systems to switch from one steady state to another. In adjusting to such a change, the path followed by the system depends on the direction of the perturbation and leads to the phenomena of hysteresis. For example, consider the heating of a catalyst particle which catalyses an exothermic reaction. We will superimpose upon the sigmoid heat generation curve a number of heat loss lines, each one of which represents the rate of heat loss for a given gas-phase temperature. Figure 2.47 illustrates how the steady-state interface temperature of the particle changes as the gas-phase temperature is increased through the sequence T_1, T_2, T_3, T_4, to T_5. The states of stable equilibrium corresponding to these temperatures are at the points of intersection A, B, F, G, H

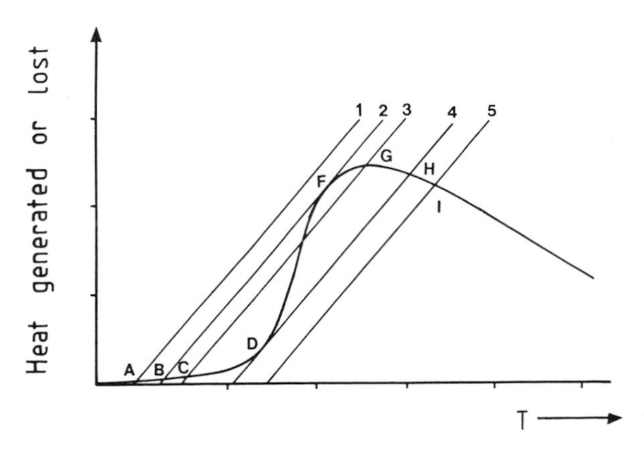

Figure 2.47 Interface temperature changes with change in gas temperature (see text). Adapted from R. Aris, Introduction to the analysis of chemical reactors. Prentice Hall, Englewood Cliffs, 1965.

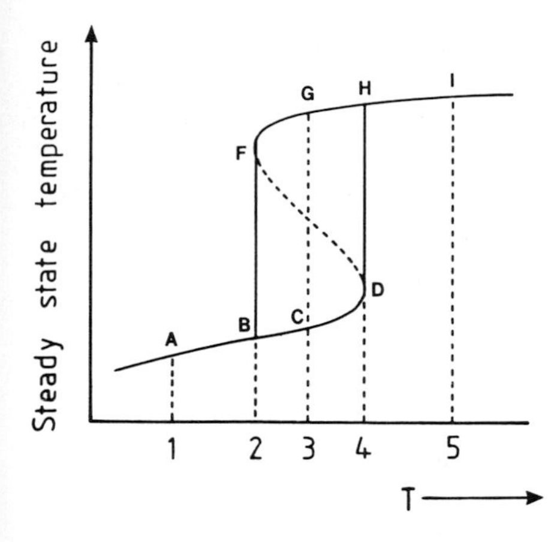

Figure 2.48 Thermal hysteresis (see text). Adapted from R. Aris, Introduction to the analysis of chemical reactors. Prentice Hall, Englewood Cliffs, 1965.

and I of the straight lines with the heat generation curve. On the other hand, if the temperature were decreased from T_5 to T_1 a different path I, H, D, C, B, A would be traced. This is evident from Fig. 2.48, which is a plot of the steady-state interface temperature as a function of gas temperature; it is clear that the equilibrium path that is followed when increasing the temperature differs from that followed when decreasing the temperature.

Evidence that such hysteresis effects occur during the platinum-catalysed oxidation of hydrogen was provided by Wicke and co-workers (1972). Figure 2.49 shows the stepwise transition between lower and upper steady states which were recorded on increasing the concentration of hydrogen. Decreasing the concentration from a

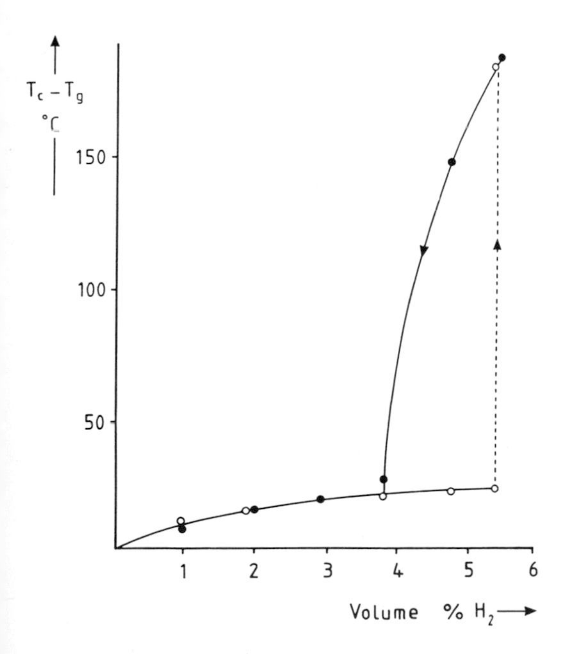

Figure 2.49 Stepwise transition between upper and lower steady states during oxidation of H_2 over platinum (see text and H. Beuch et al., *Adv. Chem. Ser.* **1972**, *109*, 615). The temperature increase which occurs when the concentration of H_2 is increased follows a different path from that observed when decreasing the concentration. A hysteresis effect between upper and lower steady states is thus apparent.

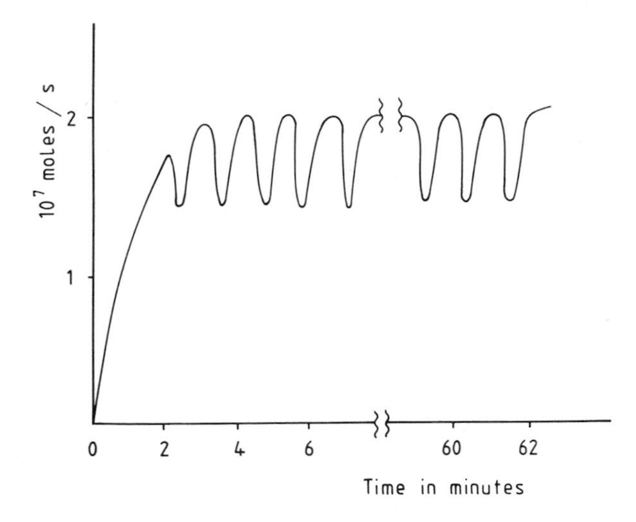

Figure 2.50 Concentration oscillations observed during oxidation of H_2 over nickel foil. The oscillations occur at a relatively low value of gas temperature (*ca* 180 °C) and have a time period of about 1 min.

stable upper steady-state operating condition, however, did not have an identical effect and the system displayed hysteresis, the stable lower steady states being approached along a gradually decreasing concentration path.

During the platinum-catalysed oxidation of hydrogen, sustained oscillations in both the product concentration and the catalyst pellet temperature were observed. These oscillations occurred at a relatively low value of the gas temperature and had a rather long time period (*ca* 1 min). The coincidence of the temperature and concentration maxima and the long, but regular, time interval between maxima suggest that the oscillatory behaviour does not arise as a result of transitions between stable thermal states of the catalyst particle and gas phase. Oscillatory behaviour was also noted during the catalysed oxidation of hydrogen over nickel foil for isothermal conditions (Fig. 2.50). Both the amplitude and frequency of oscillation increase with increase in reaction temperature. Carberry (1975), Schmidt (1993) and co-workers observed little difference (*ca* 0.001 °C) between gas and catalyst temperature at the maxima and minima of reaction rate oscillations during the platinum-catalysed oxidation of CO. Contact potential difference measurements at the surface of a nickel foil catalyst during hydrogen oxidation indicate that the composition of the catalyst surface must also be periodically changing. All of these observations point to the probability that chemical, rather than physical, phenomena are the principal causes of oscillatory behaviour.

Sheintuch (1981) studied the development of oscillatory states occurring when CO is oxidized in the presence of platinum foil in a well-mixed spinning-basket reactor (see Chapter 7). Simultaneous with the observation of the oscillatory behaviour (Fig. 2.51a) he measured the instantaneous derivative of the time-smoothed detector output (conversion versus time) and was therefore able to construct a phase plane portrait, represented in Fig. 2.51(b), depicting the approach to the oscillatory state. It is of interest to note that Fig. 2.51(b) is similar in character to the portrayal of the Lotka reaction given in Fig. 2.44(b). This so-called 'limit cycle' is typical of the

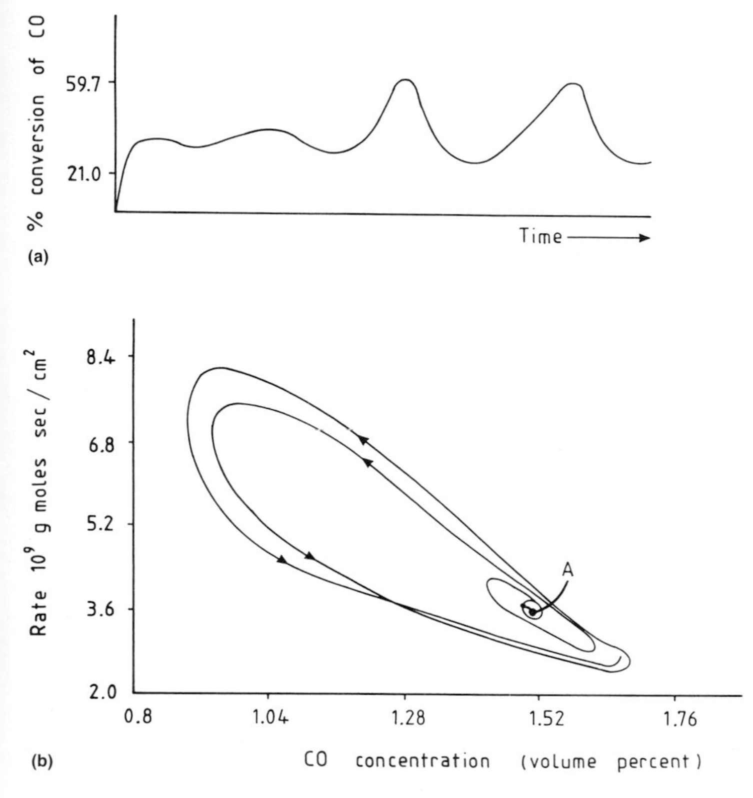

Figure 2.51 (a) Oscillations observed during oxidation of CO over platinum foil. (b) Phase plane portrait indicating a limit cycle during oxidation of CO over platinum foil. With permission from M. Scheintuch, *AICHE J.* **1981**, *27*, 20.

behaviour of systems with a multiplicity of steady states. The limit cycle diagram provides a useful indication of how a system re-establishes either steady or oscillatory behaviour when the system is perturbed. In the case cited, oscillations in the concentration of product grow in amplitude until regular concentration maxima and minima are established: the trajectory therefore commences from the internal point A (Fig. 2.51b) and spirals outwards toward the stable limit cycle B, which becomes established after the elapse of a sufficient period of time (Fig. 2.51a). It should be noted that phase portraits of limit cycles may also be presented as diagrams of the concentration of any appropriate system variable as a function of another variable. Figure 2.52(a), for example, shows the phase plane portrait of a hypothetical catalytic reaction in which gaseous reactants A and B form a product AB by the catalytic action of a surface S. The portrait shows how the surface concentrations of both reactants (X and Y, respectively) vary and approach, from various starting locations, a stable limit cycle. The corresponding time dependence of the surface concentrations in the stable limit cycle is shown in Fig. 2.52(b).

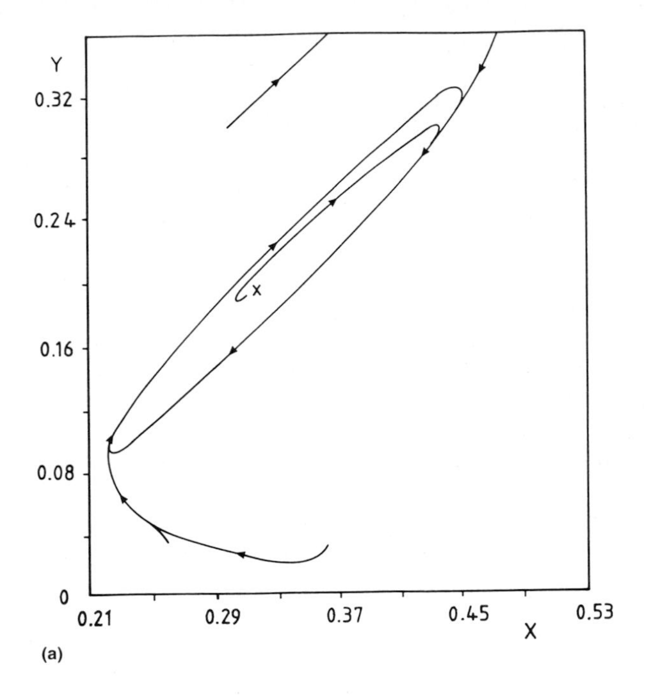

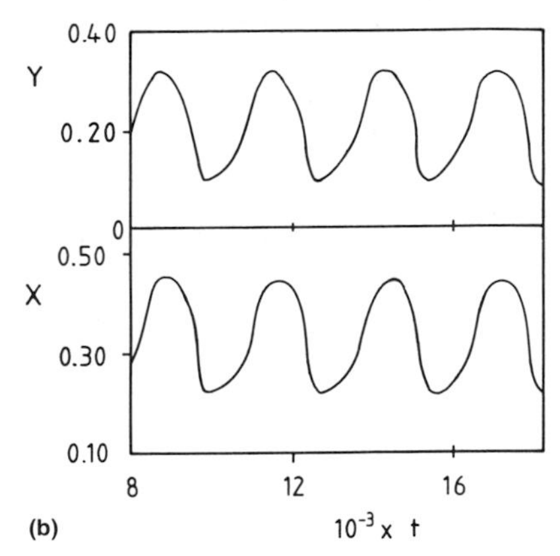

Figure 2.52 (a) Phase plane portrait for a hypothetical surface reaction

$$A(g) + S \rightarrow A\text{–}S$$

$$B(g) + S \rightarrow B\text{–}S$$

$$A\text{–}S + B\text{–}S \rightarrow AB(g) + 2\,S$$

Surface concentrations of A and B are designated x and y, respectively. (b) Sustained oscillation in the surface concentrations of A (designated x) and B (designated y). With permission from C. A. Pikos, D. Luss, *Chem. Eng. Sci.* **1982**, *37*, 375.

Pikios and Luss (1977) computed the phase plane portrait described (Fig. 2.52a) by assigning estimated parameter values to the dynamic interaction of gaseous A and B with the surface S. The catalysis was supposed to occur by the following overall reaction steps:

$$A(g) + * \rightleftharpoons \begin{matrix} A \\ | \\ * \end{matrix}$$

$$\mathrm{B(g)} + * \rightleftharpoons \begin{array}{c}\mathrm{B}\\ |\\ *\end{array}$$

$$\begin{array}{c}\mathrm{A}\\ |\\ *\end{array} + \begin{array}{c}\mathrm{B}\\ |\\ *\end{array} \longrightarrow \mathrm{AB(g)} + * + *$$

where the symbol $*$ denotes an active site at the catalyst surface. The surface was assumed to be energetically heterogeneous, the activation energy of the surface reaction varying linearly with the fraction of surface occupied by one of the reactants. A mathematical stability analysis of the two ordinary differential equations describing the dynamics of the reacting system showed that only a single unstable steady state existed which led to sustained oscillations in the system. Clearly then, a chemical description of the catalytic system is sufficient in this case to account for sustained oscillations.

Hlavacek and Rathousky (1982) have reported both regular and chaotic oscillations in the extent of CO oxidation in a multichannel catalytic monolith (see Chapter 8) whose tubular channels were coated with alumina and impregnated with palladium. These oscillations appear to occur about a unique unstable steady state, the character of system and the initial conditions. The experimental observations are explained qualitatively by a surface memory effect which can be interpreted as a time lag between the condition of the surface at any moment and the response of the reactants to the surface condition.

Recent reviews (see Gray and Scott (1990), Ertl (1991), Schüth et al. (1993) and Thomas (1994), under Further Reading at the end of this chapter) amply confirm that chaos, instability, and oscillatory phenomena of various kinds are an intrinsic feature of heterogeneous catalysis. From 'simple' reaction of diatomic molecules – encompassing the combustions of hydrogen, the hydrogenation of CO, the reaction of NO and H_2 (to yield N_2 and H_2O, as on a rhodium catalyst), and numerous other processes – autocatalysis, with its production of chemical waves, gives rise to a variety of spatio–temporal patterns.

2.11 Problems

The first eight problems are linked to the content of this chapter. To tackle the remainder, the reader is recommended to consult either some later chapters or the relevant references cited.

1 Discuss, with the aid of a potential-energy diagram, the process of dissociative chemisorption of a diatomic gas at a metal surface. What relationship exists between the enthalpy of chemisorption and the activation energy of desorption?

Thermal desorption spectra of atomic oxygen bound to the surface of a ruthenium catalyst show symmetrical peaks with a characteristic shift to lower temperatures at higher initial coverages, suggesting second-order kinetics. Using the data given in the Table, confirm the reaction order and estimate the activation energy of desorption.

Relative desorption rate [peak height]	Coverage	Temperature [°C]
0.83	1.00	448
1.60	0.92	468
2.07	0.80	476
2.93	0.66	494
2.40	0.32	522
0.89	0.14	540
0.13	0.04	548

(Source: Cambridge University Tripos, 1986)

2 At a temperature of 190 K and a pressure of CO of 48 Torr, one-tenth of a monolayer of adsorbate is present at a noble metal surface. At 250 K a partial pressure of 320 Torr is required to achieve the same surface coverage. Estimate the molar enthalpy of adsorption, and comment upon whether the gas is likely to be physically adsorbed or chemisorbed. What other measurements would you perform to identify the nature of the adsorption?

3 It is alleged that the desorption of xenon from a graphite surface obeys zero-order kinetics (as does the desorption of certain metals from the surface of silicon). How would you: (i) establish experimentally that this is indeed so; and (ii) interpret such a result?

4 Estimate how long it would take to cover an atomically clean surface of a nickel catalyst with a monolayer of oxide if it were exposed to O_2 at a pressure of (a) 7.60×10^{-4} Torr or (b) 1.33×10^{-8} Pa, or (c) to a 'good' laboratory vacuum.

5 What information can you glean from the following sets of data?

(a) Acidic mordenite takes up the following amounts x of n-hexane as a function of pressure p at a temperature of 100°C:

x [mmol g^{-1}]	p [Torr]
0.531	4.4
0.555	10.8
0.587	21.2
0.600	41.5
0.607	54.6
0.612	57.4

(b) A pillared clay catalyst takes up the following amounts of N_2 at 78 K:

$[10^2 g\,g^{-1}]$	p [Torr]
7.63	21.1
8.75	130.9
9.05	195.0
9.46	287.2
10.00	422.4
11.00	490.3
9.60	276.6
9.10	159.0
7.95	31.8
7.19	9.4

6 It is known that the enthalpies of adsorption of an atom bound to sites at (a) a flat surface, (b) a ledge site and (c) a kink site are, respectively, 10, 20 and 30 kcal mol^{-1} (42, 84 and 126 kJ mol^{-1}). Estimate the residence time of the atom on each of these sites at 300 K.

7 From the information in the Table pertaining to the adsorption of CO on Ca^{2+}-exchanged zeolite-Y, determine the isosteric heats of adsorption at each of the coverages.

T [K]	Pressure $[N\,m^{-2}]$				
	Amount adsorbed $[10^6\,m^3 g^{-1}]$				
	0.05	0.10	0.20	0.40	0.60
273.0		15.2	34.2	116.2	310.5
290.3		40.0	100.0	356.2	834.3
304.4	28.6	80.0	247.6	800.0	
322.6	91.4	247.6	685.7		

(*Note*: This table has been compiled from Fig. 4 of T. A. Egerton, F. S. Stone, *Trans. Faraday Soc.* **1970**, *66*, 2364. These workers establish that CO is a sensitive probe molecule for investigating the location of unshielded multivalent cations in zeolitic catalysts.)

8 Supported ruthenium catalyst now promise to displace iron-based ones for the synthesis of NH_3 (see Section 8.3). Accordingly, much effort is expended in establishing rates of dissociative chemisorption of N_2 at, and the desorption of surface species from, ruthenium surfaces. The curves in Fig. 2.53, taken from the work of F. Rosowski, O. Hinrichsen, M. Muhler, and G. Ertl, *Catal. Lett.* **1996**, *35*, 195, are TPD data obtained by dosing N_2 at 573 K for 14 h, subsequently cooling in N_2 to 298 K and then using (in separate experiments) three different heating rates: 1 K min^{-1} (trace A), 5 K min^{-1} (B) and 15 K min^{-1}(C). What conclusions can you draw from these data?

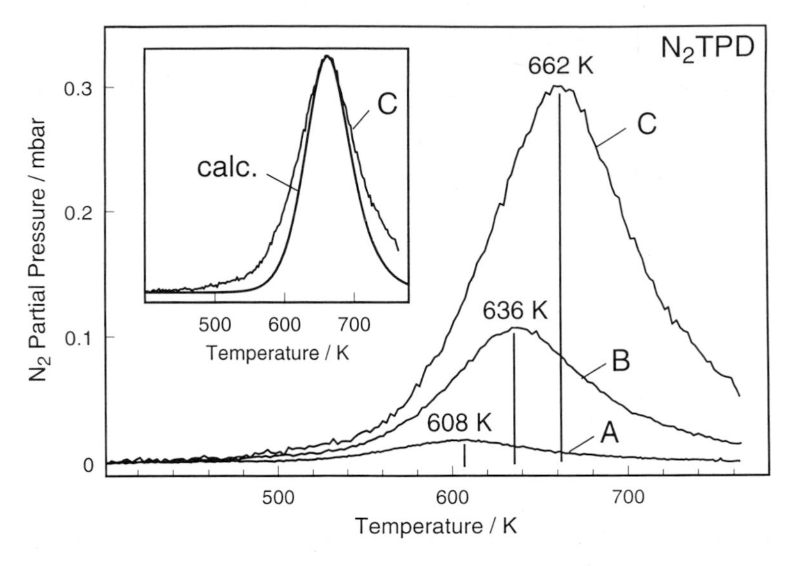

Figure 2.53

9 When NO is adsorbed at low temperatures on metal surfaces, monolayers of NO are followed by multilayer $(NO)_2$ formation. Use an appropriate database to locate papers on this system by M. W. Roberts, M. H. Matloob, **1978**; R. J. Behm, C. R. Brundle, **1984**; and W. A. Brown, P. Gardner, D. A. King, **1995**.

Upon heating the system N_2O is formed. Describe which experiments you would perform to establish whether N_2O is produced via dissociative adsorption:

$$2\,NO \rightarrow N_{ad} + O_{ad} + NO_{ad}$$
$$\rightarrow N_2O_{ad} + O_{ad}$$

or via the formation of dimers:

$$2\,NO \rightarrow (NO)_{2ad}$$
$$\rightarrow N_2O_{ad} + O_{ad}$$

10 The work of Henrich and Freund suggests that a perfect oxide monolayer (NiO) is unreactive towards H_2O. But if the surface is first exposed to O_2 its reactivity towards H_2O is significant. After consulting appropriate sections of Chapter 3 and the work of G.U. Kulkarni, C. N. R. Rao, M.W. Roberts, *Langmuir*, **1995**, *11*, 2572, devise a set of experiments that would elucidate the surface chemistry of this system.

11 Attempt both parts:

(a) The oxygen uptake curve for the surface of a single crystal of rhodium, previously outgassed in ultrahigh vacuum, is shown in Fig. 2.54. Explain how this curve is obtained from:

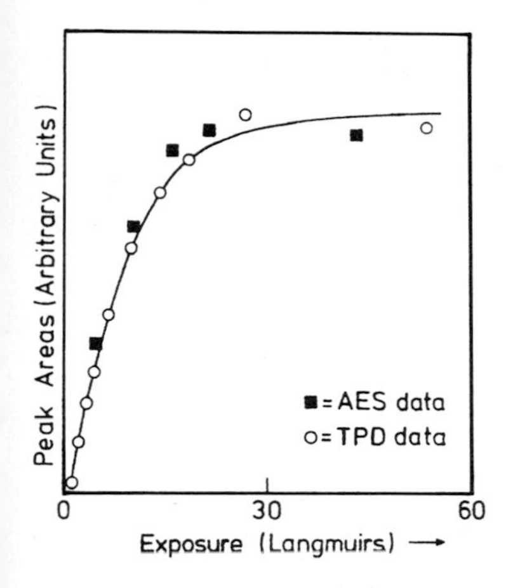

Figure 2.54

(i) temperature-programmed desorption (TPD) data;
(ii) Auger electron-spectroscopic (AES) data.

(b) Derive an expression for $-dN_a/dT$ where N_a is the number of species adsorbed at a given surface, and v and E_d are, respectively, the frequency factor and activation energy for desorption. (Assume that the temperature rises linearly with time: $T_t = T_0 + \beta t$).

(Source: Cambridge University Tripos, 1985.)

12 When surfaces of either porous silicon or its H_2-passivated analogue react with methanol, O–H and Si–OCH_3 as well as Si–H species are formed (see J. T. Yates, Jr, *Surf. Sci.* **1995**, *338*, 125). Which of the kinetic techniques discussed in this chapter, and the spectroscopic ones in the next, would you use to elucidate the nature of these functionalities?

13 The rate of conversion of propane (to pentanes, butanes, propylene and other C_2- and C_1-products) over a sulphated zirconia acid catalyst is extremely slow. B. C. Gates et al. (*Catal. Lett.* **1995**, *34*, 351) have proposed that the initial events include the reactions:

$$\mathrm{C_3H_8 + H^+(ad) \longrightarrow H_3C{-}\underset{H}{\overset{H\ +\ H}{C}}{-}CH_3\ (ad)}$$

$$\mathrm{H_3C{-}\underset{H}{\overset{H\ +\ H}{C}}{-}CH_3\ (ad) \longrightarrow CH_4 + H_3C{-}\underset{H}{\overset{+}{C}}{-}H\ (ad)}$$

Pursuing this line of argument, explore the tenability of their view that the production of butanes and pentanes from C_3H_8 is consistent with Olah superacid solution chemistry.

14 The desorption kinetics of butanes and butenes from silver catalysts may be explored by TPD, XPS and NEXAFS. It is thought that alkane desorption shows evidence of weak attractive intermolecular interactions, whereas alkenes show evidence for moderately strong repulsive interactions in desorption. Explain how spectroscopic techniques, NEXAFS in particular, may clarify the origins of these differences. (You are advised to consult the work of R. J. Madix, J. Stöhr et al., *Surf. Sci.* **1995**, *339*, 8–40, before attempting your answer.)

15 After first consulting the paper by D. A. Butler, B. E. Hayden, *Surf. Sci.* **1995**, *337*, 67, who studied the dynamics of dissociative adsorption of H_2 on a tungsten surface, explain what is meant by direct and indirect channels to dissociation. In particular, consider (i) why the initial sticking probability via the indirect channel decreases remarkably slowly with increasing incident-beam energy, and (ii) the role of a dynamical precursor in the indirect channel to dissociation.

16 Weinberg et al. (see *J. Chem. Phys.* **1994**, *101*, 10997) reported that the adsorption of gas-phase atomic hydrogen on the Ru(001) surface results in a saturation coverage of 1.42 H atoms per primitive unit mesh, which may be compared with a saturation coverage of one H adatom for the dissociative chemisorption of H_2. Consider whether such an ostensibly anomalous result may arise from the operaion of an Eley–Rideal mechanism of surface reaction.

17 The EuroPt-1 Pt/SiO_2 catalyst has been widely used as a model 'test' supported catalyst in a variety of comparative studies. On the basis of the ^1H NMR studies of chemisorbed H_2 (see M. A. Chesters, K. J. Packer et al., *J. Chem. Soc., Faraday Trans.* **1995**, *91*, 2203, critically review the evidence for the fast-exchange, reversible adsorption model.

18 The methods used to treat the kinetics of elementary processes at surfaces (e.g. adsorption, diffusion, desorption, reaction, ordering and reconstruction) fall into two broad categories: analytical and numerical. The analytical approach, entailing the time dependence of such properties as surface coverage, has been employed in this chapter. By reading relevant sections of Chapter 5 and the reviews of H. C. Kang, W. H. Weinberg (*Acc. Chem. Res.* **1992**, *25*, 253; *Chem. Rev.* **1995**, *95*, 667), elaborate upon how the numerical approach, which relies heavily on Monte Carlo techniques, may handle such phenomena as surface diffusion, precursor-mediated adsorption and the study of the compensation effect in Langmuir–Hinshelwood catalysed reactions. (*Note*: For a lucid account of how to compute adsorption isotherms, for Xenon on rutile, consult F. Rittner, D. Paschek, B. Boddenberg, *Langmuir* **1995**, *11*, 3097).

19 'When a Rh surface is first exposed to D_2 then to CO, there is partial mixing and partial segregation in the surface phase. When, however, CO is introduced first followed by D_2 there is no detectable mixing.' Apart from thermal desorption spectroscopy, which other techniques would you use to test the veracity of this claim? (*Hint*: LEED, see Section 3.6, has been used by Y. Kim, H. C. Peebles, J. M. White, *Surf. Sci.* **1982**, *114*, 363 to address this question.)

20 By reference to the early work of R. J. Madix and J. L. Falconer (*Surf. Sci.* **1974**, *46*, 473) and the later work of M. Bowker et al. (*Catal. Lett.* **1993**, *21*, 321, write a coherent account of the root cause of so-called 'surface explosions' involving acetate species at the surfaces of metals such as copper, nickel, palladium, iridium and rhodium.

21 (a) 'The energy at which an incident molecule collides with a surface can be a key factor in determining its reactivity with or on the surface.' By reference to the use of molecular-beam techniques, illustrate the truth of this statement.
(b) How do you account for the fact that the dissociation probability of CH_4 (to produce CH_3(ad) and H(ad) on a nickel surface increases steeply as a function of the normal component of the translation energy? (Consult S. T. Ceyer, *Science* **1990**, *249*, 133.)

22 The following quotations are from C. M. Friend (*Surface Reactions*, (Ed.: R. J. Madix), Springer, Berlin, **1994**, p. 55):

'Catalytic desulphurization reactions are of extreme industrial importance due to the need for upgrading fossil fuel feedstocks with high sulphur content' ... 'The facility for S–H bond breaking is consistent with its rather weak bond strength and the large thermodynamic driving force to form strong metal–sulphur bonds.'

Discuss. (You are advised first to read the later chapters and especially Section 8.7.)

23 The activation energy measured for H_2 adsorption on copper has been studied intermittently since the mid-1930s. The reported values range from 9.6 to 23.1 kcal mol^{-1} (40.2 to 50.5 kJ mol^{-1}) (see. H. A. Michelson, C. T. Rettner, D. J. Auerbach, *Surface Reactions* (Ed.: R. J. Madix), Springer Berlin, **1994**, p. 185). How do you account for this variation?

2.12 Further Reading

J. M. Allison, Y. Zeiri, A. Redordo, W. A. Goddard III, *Chem. Phys. Lett.* **1983**, *97*, 387.

J. B. Benzinger, R. J. Madix, *Surf. Sci.* **1980**, *94*, 119.

H. Beuch, P. Fieguth, E. Wicke, *Adv. Chem. Ser.* **1972**, *109*, 615.

D. Brennan, D. O. Hayward, B. W. M. Trapnell, *Proc. R. Soc. London, Ser. A* **1960**, *256*, 81.

C. T. Campbell, G. Ertl, J. Segner, *Surf. Sci.* **1982**, *115*, 309.

M. K. Debe, D. A. King, *J. Phys. Chem.* **1977**, *10*, L303.

G. Doyen, G. Ertl, *Surf. Sci.* **1977**, *69*, 157.

D. D. Eley, *Disc. Faraday Soc.* **1950**, *8*, 34.

D. Eltinger, K. Horma, M. Keil, J. C. Polanyi, *Chem. Phys. Lett.* **1983**, *78*, 4245.

G. Ertl, *Science* **1991**, *254*, 1750.

G. Ertl, J. Koch, *Z. Naturforsch., Teil A* **1970**, *25*, 1906.

P. J. Estrup, R. A. Barker, in *Ordering in Two Dimensions* (Ed.: S. K. Sinha), North-Holland, Amsterdam **1980**, p. 39.

R. J. Field, E. Körös, R. M. Noyes, *J. Am. Chem. Soc.* **1972**, *94*, 8649.

J. S. Foord, R. M. Lambert, *Surf. Sci.* **1982**, *115*, 141.

C. M. Freeman, C. R. A. Catlow, J. M. Thomas, S. Brode, *Chem. Phys. Lett.*, **1991**, *186*, 137.

P. Gray, S. K. Scott, *Chemical Oscillations and Instabilities*, Clarendon, Oxford, **1990.**

J. Harris, B. Kasemo, E. Tornqvist, *Surf. Sci.* **1981**, *105*, L288.

V. Hlavacek, J. Rathousky, *Chem. Eng. Sci.* **1982**, *37*, 375.

R. Imbihl, G. Ertl, *Chem. Rev.*, **1995**, *95*, 697.

D. A. King, Crit. Rev. *Solid State Mater. Sci.* **1978**, *7*, 1.

A. V. Kiselev, A. G. Bezus, A. A. Lopatkin, Pham Quang Du, *J. Chem. Soc., Faraday Trans. 2* **1978**, *74*, 367.

G. D. Kubiak, G. O. Sitz, R. N. Zare, *J. Chem. Phys.* **1984**, *81*, 6387.

A. J. Lotka, *J. Phys. Chem.* **1910**, *14*, 271; A. J. Lotka, *J. Am. Chem. Soc.* **1920**, *42*, 1595.

E. McCarthy, J. Zahradnik, G. C. Kuczynski, J. J. Carberry, *J. Catal.* **1975**, *39*, 29.

F. Mertens, R. Imbihl, *Nature (London)* **1994**, *370*, 124.

J. Misewich, G. Blyholder, P. L. Houston, R. P. Merrill, *J. Chem. Phys.* **1983**, *78*, 4245.

D. Mukesh, M. Goodman, C. N. Kenney, W. Morton, *Catalysis* **1983**, *6*, 1.

C. A. Pikios, D. Luss, *Chem Eng. Sci.* **1977**, *32*, 191.

H. Pfnür, P. Feulner, H. A. Englehardt, D. Menzel, *Chem. Phys. Lett.* **1978**, *59*, 481.

S. Ramdas, J. M. Thomas, P. W. Bettridge, A. K. Cheetham, E. K. Davies, *Angew. Chem., Int. Ed. Engl.* **1984**, *23*, 671.

H. H. Rotermund, G. Haas, R. U. Franz, R. M. Tromp, G. Ertl, *Science*, **1995**, *270*, 608.

E. Schüth, B. E. Henry, L. D. Schmidt, *Adv. Catal.* **1993**, *39*, 51.

M. Sheintuch, *A. I. Ch. E. J.* **1981**, *27*, 20.

J. M. Thomas, *Angew. Chem., Int. Ed. Engl.* **1994**, *33*, 913.

R. P. Thorman, D. Anderson, S. L. Bernasek, *Phys. Rev. Lett.* **1980**, *44*, 743.

T. L. Tsitovskaya, O. V. Altshuler, O. V. Krylov, *Dokl. Akad. Nauk SSSR* **1973**, *212* (6), 1400.

R. A. van Santen, J. W. Niemantsverdriet, *Chemical Kinetics and Catalysis*, Plenum, New York, **1995**.

J. T. Yates, *Methods Exp. Phys.* **1985**, *22*, 425.

T. Zambelli, J. Trost, J. Wintterun, G. Ertl, *Phys. Rev. Lett.*, **1996**, *76*, 795.

3 Characterizing Catalysts and their Surfaces

3.1 Model Systems and Real-Life Catalysts

Almost without exception, the solid catalysts used industrially possess surface areas that are seldom less than 10 and occasionally more than 300 $m^2 g^{-1}$; they are generally microcrystalline and are often multicomponent. Some important catalysts consist of several distinct phases, and in such materials there may be as many as ten different elements in various states of oxidation and combination. Moreover, the functional groups exposed at the exterior surface can be many and varied. The internal areas of catalysts and their supports can often greatly exceed their external areas; and the pore apertures have diameters anywhere from *ca* 4 to 4000 Å. Model catalysts, from a study of which much of our recent understanding of surface structure and of the dynamics and energetics of heterogeneous reactions has emerged, are, on the other hand, usually single crystals with surface areas of no more than about 1 cm^2. These can be generated in a state of bulk and surface ultrapurity, and are capable of withstanding rigorous methods of eliminating surface contaminants by such procedures as ion-beam bombardment, cleavage with a diamond knife, or cycles of chemical and thermal treatments such as high-temperature oxidation, reduction, outgassing, and annealing in ultrahigh vacuum (UHV) at residual pressures less than 10^{-12} Torr. In view of this dichotomy, surfacc scientists often display an ambivalence in describing or introducing the range of techniques available for the characterization of catalysts. What is appropriate for, say, a single crystal of iron or platinum prepared under stringent laboratory conditions, may be quite inapplicable for the actual, real-life situation in a catalytic reactor, where pressures commonly exceed by a factor of 10^{10} those used in more clinical assessments on laboratory specimens prepared under UHV.

Until recently, hardly any technique devised by the physical chemist seemed capable of probing the surface properties of catalysts under the rather severe conditions that apply in the real-life situation. This is one of the main reasons why so little is known, compared with our knowledge of homogeneously catalysed reactions, in which the gas pressure may be typically 100 bar and the temperature 450 °C. When, as is often the case with homogeneous reactions, conversions occur at atmospheric pressure and modest temperatures, the whole panoply of spectroscopic methods can be brought to bear to elucidate the nature of intermediates and various reaction pathways involved in the catalysis.

Since the mid-1980s, electron-based techniques (see later), in which either electron beams are used as surface probes for diffraction and spectroscopic purposes or photon-induced electron emission studies are carried out, have enormously enlarged our knowledge of the surface chemistry of solids in general and of catalysts in particular. However, because of the small mean free paths of electrons in gases at modest pressure, such studies are necessarily confined to model systems. Fortunately, powerful additional methods, employing, *inter alia*, high-energy photons and/or high-flux photon or neutron beams, as well as certain other means (see later), are suitable for the in-situ study of catalysts under actual reactor conditions. It would be wrong, however, to conclude that electron-based techniques, of which there are many, have only a minor part to play in the study of real-life catalysts. On the contrary, one of the key problems facing the process engineer is how to monitor the degree of deactivation of a catalyst. Usually specimens are removed from the reactor and examined ex situ. This is where very many of the sophisticated methods of the surface scientist prove invaluable.

In summarizing the more important techniques for catalyst characterization, we deal first with those methods that are applicable to model systems. Some of these techniques can be adapted for study of real-life catalysts. But we shall also focus on methods that are uniquely suited for probing catalysts under in-situ conditions. To begin with, however, we set the scene concerning the multiplicity of new techniques that have proved most useful in recent years.

3.2 A Portfolio of Modern Methods: Introducing the Acronyms

Popular methods of surface analysis traditionally employed for the study of catalysts are enumerated in Table 3.1. Adsorption as a phenomenon has an obvious applicability. For example, total accessible area of an iron catalyst used in ammonia synthesis (Chapter 8) is obtained from N_2 adsorption isotherms. The fraction of this surface covered by the promoter K_2O is determined from the uptake of CO_2, and the fraction covered by Fe from the amount of dissociative N_2 chemisorption or uptake of oxygen from N_2O. By difference, one can also deduce the fraction of the surface composed of the other promoter, Al_2O_3. Using a range of adsorbates covering a spectrum of molecular dimensions (Table 3.2), it is possible to determine the fractions of the surface area accessible to an adsorbate species of given size and shape.

Table 3.1 Traditional methods of catalyst characterisation.

Principle used	Information obtained
Gas adsorption:	
Physisorption	Surface area (surface roughness)
	Pore volume
	Pore-size distribution
	Aperture dimension
Chemisorption	'Active' surface area
	Degree of dispersion of a supported catalyst
Changes in work function and electrical conductivity	Direction of flow of electrons to or from surface

Table 3.2 The dimensions of 'probe' molecules used as adsorbates.

Species	Kinetic diameter [Å]	Species	Kinetic diameter [Å]
He	2.6	SF_6	5.5
O_2	3.5	Benzene	5.9
N_2	3.6	CCl_4	5.9
CH_4	3.8	Cyclohexane	6.0
C_2H_4	3.9	Neopentane	6.2
Xe	4.0	$(C_2H_5)_3N$	7.8
C_3H_8	4.3	$(C_4H_9)_3N$	8.1
Isobutane	5.0	$(C_4F_9)_3N$	10.2

(Surface areas as well as pore volumes are considered at length in Chapter 4.) Work function measurements also continue to be used in characterizing catalyst surfaces and specific examples are cited later.

Many of the numerous methods of characterization now in use may be classified (see Table 3.3) according to their ability to absorb, emit, or scatter photons, electrons, neutrons or ions. Whereas some of these methods (e.g., LEED, SIMS and XPS) are rather sophisticated and suitable only for specialized model systems, others (EXAFS, XRD, XRE and PIXE) are more or less universally applicable. There are some important methods (e.g., those based on calorimetric or thermal measurements) which cannot be classified as in Table 3.3.

Table 3.3 Methods of catalyst characterisation generated by combinations of photons, electrons and ions.

Primary beam	Outgoing beam		
	Photons	Electrons	Ions
Photons	Nuclear magnetic resonance (NMR)	UV-induced photoelectron spectroscopy (UPS)	Laser microprobe mass spectrometry (LMMS)
	Electron-spin resonance (ESR)	X-ray induced photo-electron spectroscopy (XPS or ESCA[a])	
	Fourier transform IR (FTIR)	X-ray induced photo-electron diffraction (XPD)	
	Raman spectroscopy (RS)		
	Mössbauer spectroscopy (Moss. S)	Microscopy (PEEM)	Photoelectron emission
	X-ray absorption (near-edge) spectroscopy (XANES)	Conversion-electron Mössbauer spectroscopy (CEMS)	
	Extended X-ray absorption fine structure (EXAFS)		

Table 3.3 (cont.)

Primary beam	Outgoing beam		
	Photons	Electrons	Ions
	X-ray fluorescence (XRF)		
	X-ray diffraction (XRD)		
	Ellipsometry		
	Optical microscopy		Photoemission microscopy
	Surface plasmon microscopy		
Electrons	X-ray emission spectroscopy (XRE)	Low-energy electron diffraction (LEED)	
	Cathodoluminescence (CL)	Auger-electron spectroscopy (AES)	
		Scanning-Auger microscopy (SAM)	
		High-resolution electron-energy-loss spectroscopy (HREELS)	
		Electron-energy-loss spectroscopy (EELS)	
		Scanning electron microscopy (SEM)	
		Transmission electron microscopy (TEM)	
		Electron diffraction (ED)	
		High-resolution electron microscopy (HREM)	
		Scanning transmission electron microscopy (STEM)	
		Scanning tunnelling microscopy (STM)	
Ions	Proton-induced X-ray emission (PIXE)		Ion-scattering spectroscopy (ISS)
	Positron annihilation spectroscopy (PAS)		Secondary-ion mass spectroscopy (SIMS)
			Ion-microprobe microanalysis (IMM)
			Rutherford back-scattering (RBS)

[a] ESCA stands for electron spectroscopy for chemical analysis.

3.3 Which Elements and Which Phases are Present?

Given that most catalysts are quite complicated, multicomponent, multiphasic systems, this question must be answered at the outset. Until recently the bulk composition of a catalyst could always be deduced from a combination of wet chemical (titrimetric, potentiometric and spectrophotometric) methods and their automated successors (atomic absorption and atomic emission spectroscopy). Nowadays, the average elemental composition and overall stoichiometry can be rapidly determined by X-ray emission or by inductively coupled plasma mass spectrometry (ICPMS).

3.3.1 X-Ray Fluorescence (XRF), X-Ray Emission (XRE) and Proton-Induced X-Ray Emission (PIXE)

The principles upon which the three related methods, XRF, XRE, and PIXE, are based are similar (Fig. 3.1). Electrons are first liberated from inner core levels; X-rays are then emitted when the vacancies so created are filled by transitions from outer levels. The intensity I_x of the emitted X-ray is proportional to the concentration of the atoms N_x responsible for the characteristic emission, provided that: (1) there is no thickness-dependent or energy-dependent absorption of the emitted X-rays; (2) all, or a constant fraction, of the created vacancies are filled by a process that leads to X-ray emission. Since the intensity of emission is usually dependent upon the atomic number (Z) and on the subsequent absorption or fluorescence of the X-rays, materials of known composition, preferably similar to the catalyst under study, are first used as standards for calibration.

Commercially available XRF instruments, which have high fluxes of primary X-rays, generally employ wavelength-dispersive X-ray detectors: the wavelength λ and hence the energy (hc/λ) of the X-ray are obtained from the Bragg equation:

$$n\lambda = 2d \sin\theta \tag{1}$$

To encompass all likely values of θ for the X-rays under study, a range of d values is chosen from a set of interplanar distances in a collection of analyzer crystals. Such

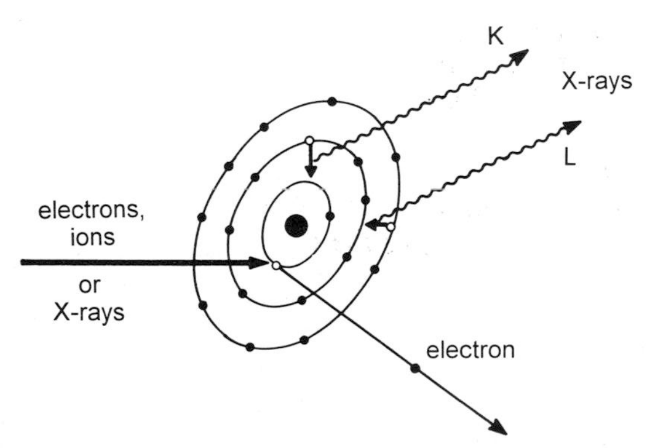

Figure 3.1 Schematic illustration of X-ray production by irradiation with primary particles (electrons, ions or X-rays).

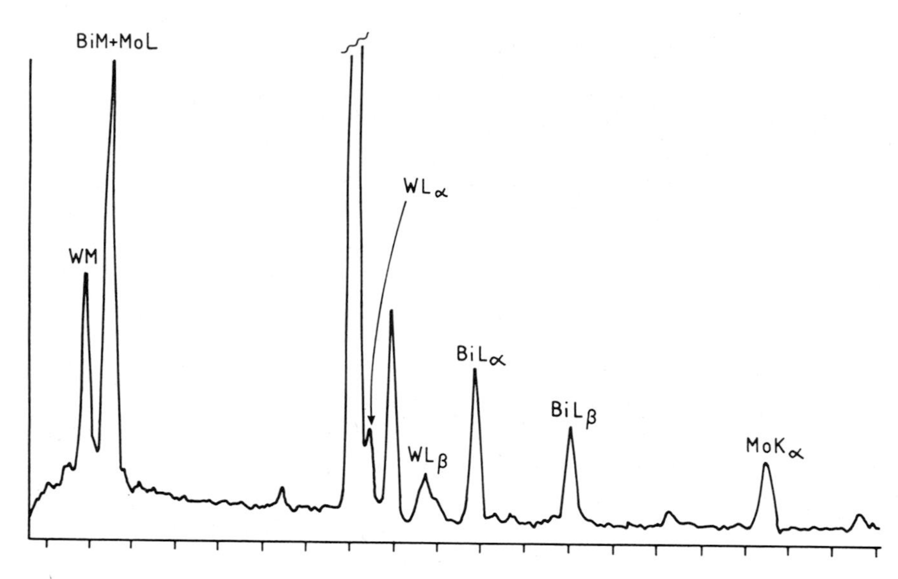

Figure 3.2 A characteristic X-ray emission spectrum, recorded with an electron microscope, from a model Bi–W–Mo oxide model catalyst (D. J. Buttrey, D. A. Jefferson, J. M. Thomas, unpublished work).

instruments are capable of handling milligram quantities of material, and can reliably detect the equivalent of one-tenth to one monolayer of a given element. (Bear in mind that each square centimetre of solid contains about 10^{15} atoms and that the mass of each atom is around 6×10^{-23} g. A monolayer of platinum has a so-called areal density of some 7×10^{-7} g cm^{-2}.)

Energy-dispersive instruments have tended to replace wavelength-dispersive ones, and when high-energy proton beams are available as primary sources, PIXE comes into its own. PIXE spectra are quantitatively reliable; and the technique can detect even trace amounts of heavy elements in catalyst specimens weighing only a few micrograms.

X-ray emission (XRE) spectra are conveniently recorded in an electron microscope. The spectrum shown in Fig. 3.2, for example, was taken a few minutes after insertion of a copper grid (as mount) containing a bismuth molybdate catalyst, deliberately doped so as to modify its performance in the ammoxidation of propylene. The Si(Li) detector, placed close to the electron-irradiated specimen in the microscope usually has a thin beryllium window to protect the detector and yet permit X-rays to pass through it. The soft X-rays liberated by the lighter elements from Li to F (Z from 3 to 9) cannot reach the detector. 'Windowless' detectors have been developed and it is now possible to identify the location, and semi-quantitatively to determine the distribution, of important light elements (C, N, O and F) in catalyst preparations.

3.3.2 Developing Techniques: ICPMS

All three variants of X-ray emission are best suited for determination of bulk composition. To determine the surface composition, electron-based methods are pre-eminent (see Section 3.5). We shall see that X-ray-induced photoelectron spectroscopy (XPS) and Auger electron spectroscopy (AES) are especially good for this purpose, fractions of a monolayer (of any element with $Z > 3$) being routinely detectable. There are other methods available for determining composition. Spark-source mass spectroscopy, which is a highly sensitive but destructive technique, is one. Magic-angle-spinning NMR (MASNMR), a non-destructive technique (see Section 3.7.3) is another. It, too, has quite good sensitivity, being capable of detecting 1 part of aluminum in 1000 parts of silicon in a highly siliceous zeolite known as silicalite. Inductively coupled plasma mass spectrometry (ICPMS) is a powerful new aid for evaluating the composition of a material that can be homogeneously dissolved (in strong acid, for example) prior to being atomized on entry into a plasma for subsequent ionization and ion selection (into a mass spectrometer). This technique has a linear response of peak intensity for six orders of magnitude of concentration.

3.3.3 X-Ray Diffraction (XRD) and Electron Diffraction (ED)

Of these two, electron diffraction is generally less universally applicable as a means of determining which phases are present. For thin-film specimens of thickness less than about 1000 Å, electron diffraction patterns are readily recorded. From the symmetry and intensities of the patterns we discover whether the material is highly crystalline, quasi-crystalline or non-crystalline.

Bulk catalysts, however, are usually and fruitfully subjected to analysis by X-ray diffraction. So comprehensive are the libraries of available characteristic *d*-spacings and intensities of previously studied solids – the ASTM and JCPDS indexes contain over 60 000 entries, and can be accessed by appropriate computer search routines – that the constituents of a catalyst can be quickly deduced from diffractograms such as that shown in Fig. 3.3. X-ray diffractograms from a few hundred milligrams of catalyst can be accumulated or repeatedly scanned over long periods of time (typically 100 h), using computer-linked digital systems, so as to maximise the detection of minority phases.

X-ray diffractograms reveal several important properties. First, they signify whether the catalyst, or a component of it, is non-crystalline or quasi-crystalline; second, they yield an estimate of the size of the microcrystallites that may be present; third, because XRD patterns yield *d*-spacings and unit cell dimensions, we gain insights into the atomic constituents of the unit cell; and, last, we can tell, in favourable circumstances, from in-situ experiments what influence reactant gas mixtures exert upon the internal structure as well as the crystalline order of the exterior surface of the catalyst. Thus, non-crystalline catalyst (typically a silica–alumina gel for hydrocarbon cracking) shows no sharp diffraction peaks, merely broad features. This is what one expects from a material with no-long-range, translational order. Microcrystallites yield diffraction peaks which are broadened because the fewer the planes

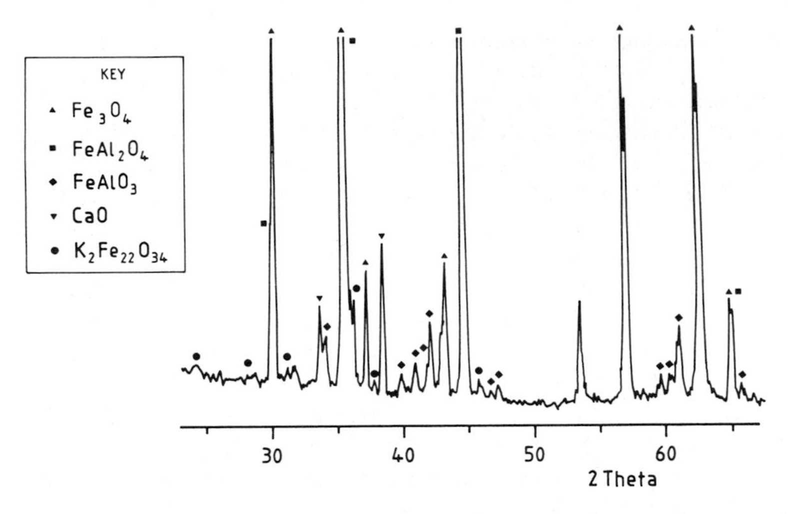

Figure 3.3 X-ray diffraction (XRD) pattern of a promoted magnetite ammonia-synthesis catalyst, taken by slow digital accumulation of the scattered X-rays. Peaks from the various phases identified from the JCPDS file (T. Rayment, R. Schlögl, J. M. Thomas, G. Ertl, *Nature (London)* **1988**, *315*, 311).

that give rise to Bragg diffraction, the less sharp is the peak. If β is the full width at half maximum (FWHM) of the broadened peak, λ is the X-ray wavelength, and t is the thickness of the crystal in a direction perpendicular to the diffracting planes, then we have:

$$t = \frac{K\lambda}{\beta \cos\theta} \tag{2}$$

where θ is the Bragg angle, and K is a constant which to some degree depends on the shape of the peak. Figure 3.4 illustrates the utility of Eq. (2). Here, the peak shapes

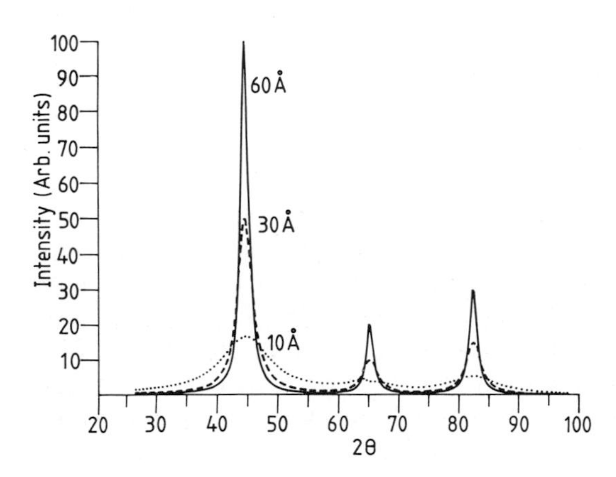

Figure 3.4 Computed diffractograms (see text) for minute particles of α-iron (diameters quoted on curves).

for particles of α-Fe of diameter 60, 30 and 10 Å have been computed from an equation related to Eq. (2). More sophisticated uses of the broadening of X-ray peaks have led to the determination of size distributions of supported metals.

Particle sizes can be determined by X-ray diffraction by quite another means: using small-angle scattering (SAXS). If the particles in a catalyst preparation fall in the size range 500–5000 Å (diameter) they will scatter X-rays in much the way that atoms do. Just as the size of minute water droplets in a mist can be inferred from the maximum angle of scattering suffered by a beam of visible light – in both instances, the maximum angle is given by the ratio λ/D where λ is the wavelength of the radiation and D is the particle diameter – so also can the size of small particles of catalysts (or supported catalysts) be deduced from X-ray scattering. It is evident that the central peak of scattered intensity gets broader as the particle size decreases. With wavelengths of about 1 Å, the usual range of sizes which can be measured is 10–1000 Å; but this range can be expanded by operating at other X-ray wavelengths. For Cu K_α radiation ($\lambda = 1.542$ Å), for example, the scattering occurs predominantly at values of 2θ less than 2°.

3.3.3.1 Mean Size, Surface Area and Particle-Size Distribution from SAXS

Irrespective of whether the geometry of the SAXS experiment relies on pinhole or slit collimation, a function s, known as the scattering vector, needs to be defined:

$$s = 2\sin(\theta/\lambda) \simeq 2\theta/\lambda \tag{3}$$

Guinier showed that, when the particles giving rise to the SAXS are identical, in random orientation, and far removed from one another, $I(s)$, the intensity as a function of s, can be represented approximately by:

$$I(s) = (\Delta\rho_e)^2 V^2 \exp(-4\pi^2 s^2 R_G{}^2/3) \tag{4}$$

where $\Delta\rho_e$ is the difference in electron density between the particles and the surrounding medium, V is the volume of the particle and R_G is the radius of gyration of the particle. Using Eq. (4), R_G is obtained from the slope of the linear plot (the Guinier plot) of $\ln[I(s)]$ versus s^2. (For a sphere of radius R, $R_G = (3/5^{1/2}R)$. If the particles are not identical, R_G is an average radius, and the larger particles are weighted more heavily than the smaller ones. The Guinier plot should be made in the central part of the SAXS curve where the exponential approximation is valid. It was shown by Porod that the intensity of the SAXS in the wing of the curve is such that:

$$\lim\,(I(s)s^4) = [(\Delta\rho_e)^2/8\pi^3]S \tag{5}$$

for point (i.e. pinhole) collimation. Here, at the asymptotic limit, S is the total area of the interface between two phases; so, for a solid–air interface, S is simply the area of the solid phase. Porod's law is valid whenever the product $I(s)s^4$ (or $I(s)s^3$ for the case of linear (i.e. slit) collimation) tends asymptotically to a constant at large s. Under these circumstances, S is deduced using Eq. (5); and it is noteworthy that Porod's law can be applied whatever the dimension, shape, and porosity of the particles responsible for the X-ray scattering. If, specifically, the solid phase is composed of spherical, non-porous particles, then an equivalent mean diameter $\bar{D} = 6/\rho S$ (where ρ is the density of the particle) can be calculated.

When the particles are assumed to scatter independently, the scattering intensity is proportional to the integral:

$$I(s) = \int_0^\infty P(D)\, I(D, s)\, \mathrm{d}D \tag{6}$$

where $I(D, s)$ is the intensity scattered by a particle of diameter D, and $P(D)$ is a distribution function such that $P(D)\,\mathrm{d}D$ equals the probability that the particle has a diameter between D and $D + \mathrm{d}D$. There are several ways of evaluating $P(D)$ by numerical analysis, without restriction as to the form of the distribution. The Fourier transform of $I(s)$ gives the function $\gamma(r)$, a characteristic function of the particles under discussion, or, alternatively, the so-called Patterson function $P(r)$, such that $\gamma(r) = P(r)/\rho^2 V$, where ρ is the density of the particles. $\gamma(r)$ is often used to calculate various kinds of diameter distribution on the assumption of a given particle shape.

To exemplify the utility of the SAXS experiment, we cite the work of Renouprez et al., who found the surface areas (using Eq. (5)) of samples of SiC, SiO_2 and NiO to be, respectively 0.34, 42 and 107 $m^2\,g^{-1}$, in good agreement with the corresponding BET values of 0.35, 41 and 119 m^2g^{-1}.

3.3.3.2 In-Situ Studies by X-Ray Diffraction

With the advent of improved instrumentation for X-ray experiments, it is now possible to obtain good-quality diffractograms with sample temperatures up to 1000 °C and in a gas pressure of up to 50 bar. With the aid of a solid-state X-ray detector (Li–Si crystals), it is also feasible to carry out energy-dispersive diffraction: the detector, set at a given angle, picks up the reflections of those wavelengths in the continuous spectrum (provided by a rotating anode X-ray generator or a synchrotrons source) that satisfy the Bragg law (Eq. (1)). This approach is likely to be used increasingly for high-temperature, high-pressure catalytic studies in the future, even though with solid-state detectors the resolution is not as good as with conventional X-ray cells, and geometry and intensity measurements are more difficult.

Using the conventional geometry, but with a mica, beryllium or Kapton window fitted to the cell, it is possible to record the *d*-spacings of pyrochlore and other solid catalysts for partial oxidation of methane under realistic conditions.

There have been other examples of in-situ XRD studies of supported catalysts, among the first being the work of Ratnasamy and Fripiat et al., who found that minute particles of platinum (15–30 Å diameter) on SiO_2–Al_2O_3 supports retained their face-centered cubic (fcc) structure in a hydrogen atmosphere but tended towards structural disorder in oxygen. Gallezot, in a series of elegant studies, has explored the way in which the structures of 10–20 Å platinum particles resident inside the cages of a zeolite are modified when ‘exposed’ via some of the apertures in a zeolite (see Fig. 3.5) to reacting gases such as benzene or butane. To understand the principle of Gallezot’s approach, we first recall that very small particles, even if crystalline,

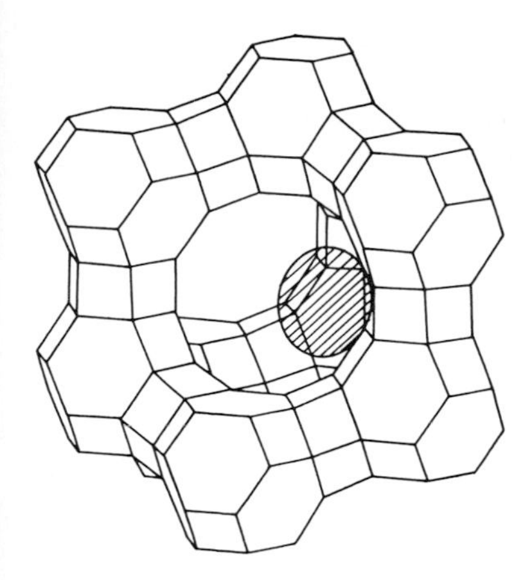

Figure 3.5 Small particles of platinum located inside the supercage of a zeolite-Y catalyst may be reached by small molecules of hydrocarbon, via the large apertures (dia. *ca* 7.4 Å), thereby modifying the structure of the particles (see text). (Based on work of P. Gallezot.)

yield broader diffraction peaks. This broadening is well understood from Debye's equation,

$$I(s) = N \sum_{m} \sum_{n} f_m f_n \sin(2\pi s r_{mn})/2\pi s r_{mn} \tag{7}$$

which refers to N particles composed of atoms of types m and n separated by a distance r_{mn}, the scattering factors for the atoms being f_m and f_n, and s is the scattering vector (Eq. (3)). If we take N identical fcc clusters of platinum each containing 14 atoms, the diffraction pattern computed from Eq. (7) is as shown in Fig. 3.6. These

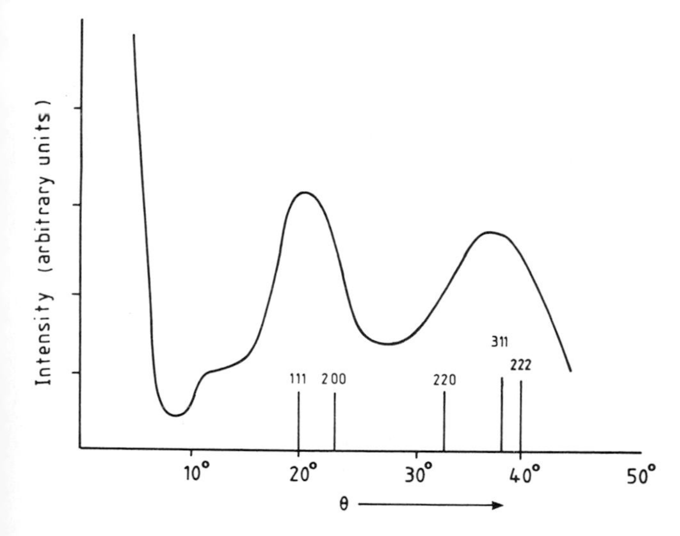

Figure 3.6 Interference function, calculated using the Debye equation (Eq. (7); see text) for a 14-atom cubic cluster of platinum. The positions of the prominent diffraction peaks for the bulk metal are marked on the abscissa.

broad peaks (only two!) differ from the series of five sharp lines (marked on the θ axis) that are obtained in this region of theta space from a bulk fcc sample of platinum. To extract quantitative information from profiles as broad as those in Fig. 3.6 we employ the radial electron distribution (RED) method. Following on from Eq. (7) we arrive at the radial distribution function (RDF) for the case of N identical *atoms* (note the change from definition of N in Eq. (7)):

$$4\pi r^2 \rho(r) = 4\pi r^2 \rho_0 + 8\pi r \int_0^\infty s_i(s) \sin(2\pi rs)\, ds \tag{8}$$

Here, $4\pi r^2 \rho(r) dr$ is the average number of atoms situated between two spheres of radius r and $r + dr$, and ρ_0 is the average atom density; also, $i(s)$ is related to the experimentally measured intensity $I(s)$ in absolute units (electron^2) – suitably corrected for obtrusive artefacts such as polarization and inelastic scattering by $i(s) = [I(s)/Nf^2] - 1$. Thus we see how, from Eq. (8) and with the aid of experimental data that yield I and s (and from the *International Tables of X-Ray Crystallography*, which give the scattering factors and ways of allowing for the artefacts), we may plot $4\pi r^2 \rho(r)$ versus r to obtain the so-called radial electron distributions.

Gallezot and co-workers, using platinum particles even smaller than those visible in electron micrographs, have recorded REDs extracted from diffraction data

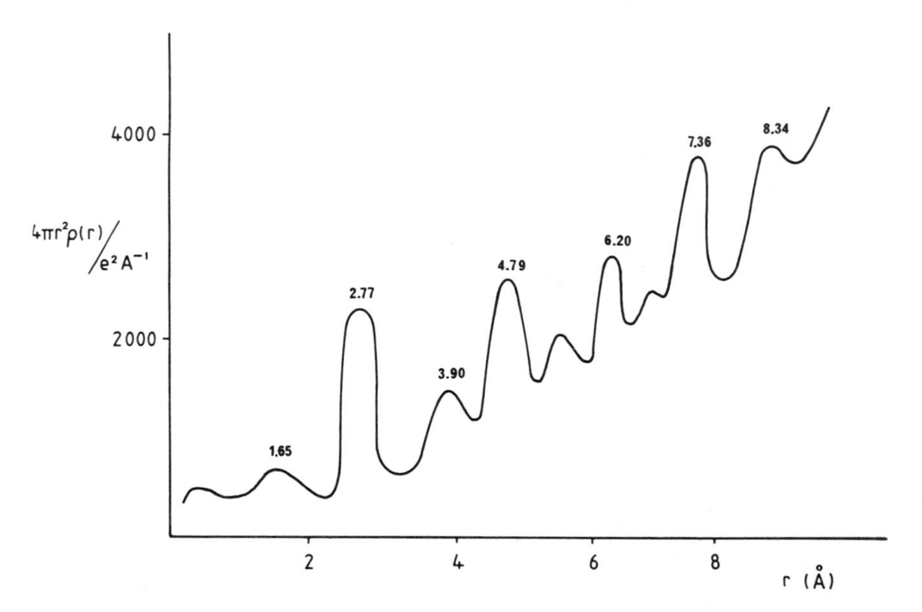

Figure 3.7 Radial electron distribution of 10 Å (average diameter) Pt particles exposed to H_2. (Based on work of P. Gallezot and G. Bergeret in B. Innelik, J. C. Vedrine (eds.) *Catalyst Characterisation: Physical Techniques for Solid Materials*, Plenum, New York, **1994**.

obtained under a series of different catalytic reactor (in-situ) conditions. Thus, in 10 Å platinum particles entrained within a specimen of zeolite-Y, the Pt–Pt separation peaks stand out clearly (Fig. 3.7) when hydrogen or benzene is adsorbed. It is significant that the first peak (Pt–Pt separation of 2.75 Å) exhibited by this catalyst under actual conditions of benzene hydrogenation (a 1:3 mixture of C_6H_6 and H_2) is intermediate between the extreme values (2.77 and 2.71 Å) in H_2 and in C_6H_6. We return to the broader issues of in-situ studies in Section 3.10 below where, *inter alia*, the supreme advantages of synchrotron radiation sources are illustrated.

3.4 Detecting, Identifying, and Counting the Atoms at Solid Surfaces: Sub-monolayer Amounts can be Measured

Most of the techniques we discuss next apply principally to model systems, but some of them can be used for the surface analysis of typical catalysts. So far as the identification and counting of the atoms present at or on the exterior surface of the solids is concerned, among the most impressive methods are those based on the use of charged particles as primary beams.

3.4.1 Ion-Scattering Spectroscopy (ISS): A Detection Limit of 10^{-4} Monolayers

When ions of the noble gases (He^+, Ne^+ or Ar^+) collide with a solid surface there is a high probability that they will be neutralized after the first collision. Those ions that are not neutralized will lose energy to the surface atoms during the scattering process. With primary ions of relatively low energy (e.g. 1–10 keV), the collision process is so rapid that it is much shorter than the time required by the surface atoms to execute a single vibration. It is, therefore, legitimate to picture the process as a mere binary event, involving the primary ion and the 'free' surface atoms only. Under these conditions there is a direct relation between the energy and mass of the primary ion on the one hand and the energy of the scattered ion on the other.

For a scattering angle of 90°, the relation between the energy of the primary ion, E_0, that of the scattered ion, E, and the masses of the surface atom and the primary ion, M_2 and M_1 respectively, is given by:

$$E/E_0 = (M_2 - M_1)/(M_2 + M_1) \tag{9}$$

Clearly, ISS is restricted in analysis to the first layer of atoms at the exterior surface of the solid. It is sensitive to all elements, from lithium to lawrencium, and tests have shown its limit of detection to be 10^{-4} monolayer.

ISS has the ability to detect light elements (C, O, F, etc.) on a catalyst surface; and although the resolution for the heavier elements is not so good it is a very powerful method of monitoring compositional changes at the surface of alloys and other catalysts, especially those used in automotive three-way devices (see Section 8.6).

3.4.2 Nuclear Microanalysis (NMI): A Sensitive Means of Detecting Specific Elements, Including Hydrogen, at Surfaces

Using projectiles such as He^+ or deuteron beams at energies close to 1 MeV one may determine the surface concentrations of various adsorbed or subsurface species, such as H, O, or CO. This can be achieved accurately via nuclear reactions of the following kind:

$$d + {}^{16}O \rightarrow p + {}^{17}O$$

$$d + {}^{12}C \rightarrow p + {}^{13}C$$

$${}^{3}He + {}^{2}H \rightarrow p + {}^{4}He$$

Working with UHV systems capable of attaining residual pressures of *ca* 5×10^{-11} Torr, and using a 2.5 MeV van de Graaff accelerator as source, all three reactions can be calibrated. Thus, for the ${}^{16}O(d,p){}^{17}O$ reaction an anodised Ta_2O_5 film of known thickness was used as a primary standard. It could then be demonstrated that surface coverage of oxygen was measurable to an accuracy of *ca* 2%. Thin, condensed films of CO_2 and D_2O ice were used as standards for the other two of the three reactions given above. The uptake of hydrogen by clean single-crystal faces of platinum is measurable using the third of the above reactions.

By selecting the energy and the projectile carefully, very high sensitivity (down to 10^{12} atoms cm^{-2}) can be achieved using the appropriate prompt nuclear reaction as the basis of the microanalysis. Furthermore, the spatial distributions of particular elements – sulphur on a poisoned reforming catalyst, and carbon or nitrogen laid down on a pellet of cracking catalyst – may be obtained by focusing the projectile beam down to a fine spot (*ca* 2 μm dia.). A few relevant nuclear reactions for a number of light elements, with the appropriate ion energy are:

${}^{10}B$ (d, p) 1.5 MeV

${}^{12}C$ (d, p) 1.0 MeV

${}^{14}N$ (d, p) 1.7 MeV

The second of these reactions permits the determination of the spatial distribution of carbon across the surface of a fouled catalyst.

3.4.3 Rutherford Back-Scattering (RBS)

This technique, also known as nuclear back-scattering spectrometry, involves bombarding a target with a beam of α-particles or protons (from an accelerator) and detecting (usually with a Si(Li) solid-state detector) those primary particles that have been back-scattered through very large angles by the target nuclei. Helium ions are usually employed at an energy below that which would cause nuclear transmutation. When protons are used as primary beams their energy, for this reason, seldom exceeds 200–400 keV.

In RBS, an incoming projectile of low mass, when it undergoes a close encounter with a heavy target atom, transfers only a small fraction of its initial energy, so that the returning particle enters the detector with little loss in energy. However, if the projectile suffers a close encounter with a light nucleus it will transfer a larger proportion of its initial energy. The RB spectrum consists of a series of peaks which represent the residual energy of ions back-scattered from nuclei of different masses. Particular advantages of the technique are that it is nondestructive, it can be rendered quantitative, and it may be effected rapidly.

3.5 Identifying the Atoms at a Surface and Probing their Immediate Environment: SIMS, IR, HREELS, AES and XPS

Knowing the atomic composition of a solid surface, or the species which are attached to it, is important; but it is not enough. The state of combination of the surface atoms and the functional groups that are present also need to be known. We recall that there are *ca* 10^{15} atoms per cm^2 of essentially all solid surfaces, which implies that sensitive techniques are required to characterize fractions of a monolayer. Many of the techniques discussed so far, especially ISS and NMI, can readily meet this requirement; there are others, notably SIMS, which are even more sensitive.

3.5.1 Secondary-Ion Mass Spectroscopy (SIMS): A Detection Limit of 10^{-5} Monolayers

In discussing ISS (Section 3.4.1) we focused our attention on the event schematized in the left-hand side of Fig. 3.8, the reflection after transfer of energy of the primary

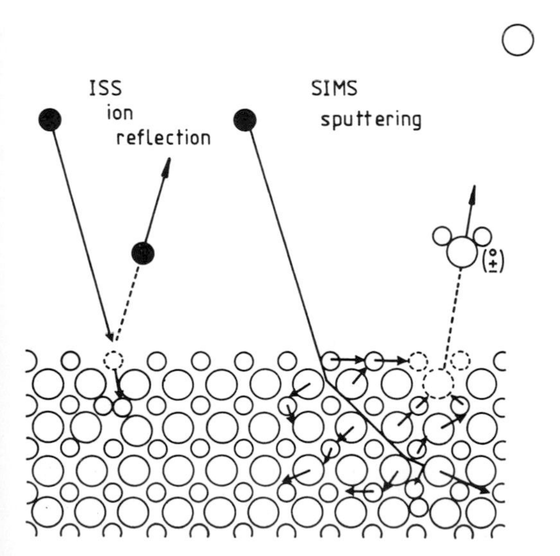

Figure 3.8 Diagrammatic representation of ion-scattering spectroscopy (ISS) and of secondary-ion mass spectroscopy (SIMS).

ion from the surface under investigation. But the primary ion can also penetrate a certain distance into the surface, and thus be implanted in the sub-surface region. During penetration, the energy of the primary ion is dissipated into the surrounding zone; and a small fraction of this energy can reach the surface, with a momentum direction pointing outwards from it. This is why, as an inevitable consequence of impingement of high-energy primary ions, secondary particles are emitted. The latter may be atomic or molecular, in a charged or uncharged state (Fig. 3.8). The use of this phenomenon as an ultrasensitive tool for detecting all the elements at a solid surface is largely a consequence of Benninghoven's work. Typically noble gas ions (e.g., 3 keV Ne^+ or Ar^+) are used, the incident beam current being so small (10^{-11} to 10^{-8} A) that the erosion of the surface can be as low as 1 Å h^{-1}!

Although SIMS is the most sensitive of surface analytical techniques it is, as yet, not the most quantitative. Difficulties arise because of several factors. First, there is in general no direct correlation between intensity of secondary ions (positive or negative) and the surface concentration corresponding to that stoichiometry. Second, there are at least as many neutral secondaries emitted as there are ions. As these neutral species are not conveniently detectable, one obtains a quantitatively unrepresentative picture of the aggregates and clusters that may be present at the surface. Third, the sputtering yield, and hence the limit of detection, are dependent upon the atom in question and the structure of the solid to which it belongs as well as upon the prehistory of that sample. Yields of vanadium are enhanced by factors of 100–1000 if the clean surface is first exposed to oxygen.

Notwithstanding these difficulties, significant progress has been made in establishing relationships between intensities of SIMS spectra and the adsorbate structures of CO on single crystals of nickel, ruthenium, platinum and palladium. By invoking well-authenticated information derived using parallel techniques (IR, HREELS and LEED, all discussed later), a direct correlation has been established by Vickerman between the proportions of secondary ions MCO^+, M_2CO^+, and M_3CO^+ and the fractions of the metal surfaces covered by linearly bonded, bridge-bonded and triply bridged CO (Fig. 3.9). One of his spectra (Fig. 3.10), recorded for a Pd(100) surface on to which CO had been adsorbed, neatly shows evidence for the structures depicted in Fig. 3.9.

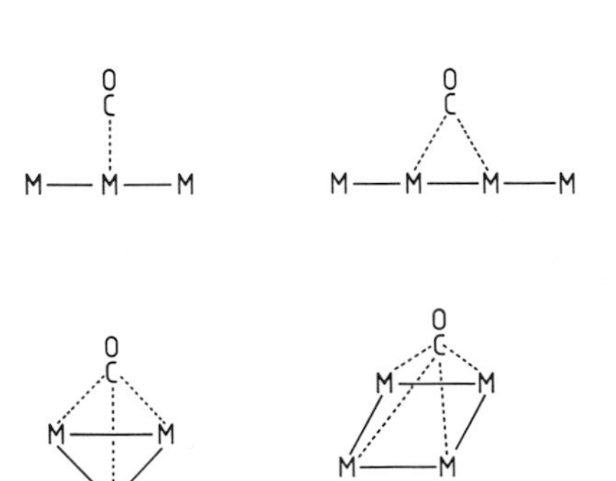

Figure 3.9 Linear (μ_1), bridged (μ_2), triply bridged (μ_3) and quadruply bridged (μ_4) forms of CO bonded to a metal surface.

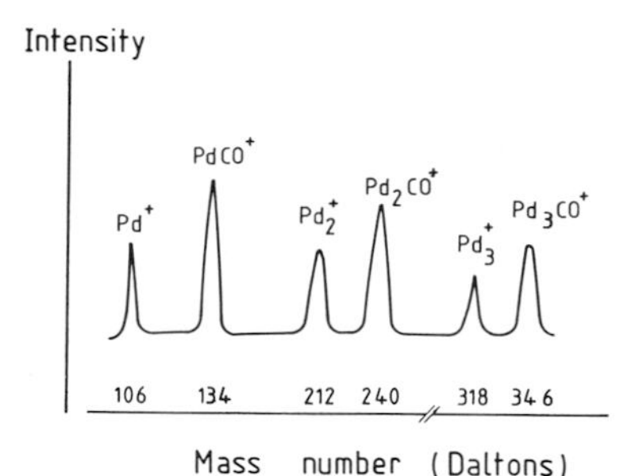

Figure 3.10 SIMS spectrum of a Pd(100) surface covered with CO. (Based on work of J. C. Vickerman.)

3.5.2 Infrared Spectroscopy (IR): A Non-destructive Technique Usable on Catalysts Exposed to High Pressure

In the mid-1950s, Eischens and Sheppard and co-workers, building on earlier foundations laid by Rodebush and Terenin but recognizing new possibilities, initiated studies of seminal importance. They probed, by IR absorption measurements, the nature of the bonding that holds small molecules, such as CO, to the surfaces of transition metals. The metals themselves were finely divided and dispersed on high-area supports such as silica, alumina, or silica–alumina gels. Non-dissociative adsorption of CO, bound via the carbon to the metal, was proved directly and, quite early on, at least two distinct modes of attachment (linear and bridged – see Fig. 3.9) were identified. Such measurements, carried out using straightforward transmission geometry, were later complemented by studies conducted on rigorously outgassed single-crystal surfaces, by the so-called 'reflection–absorption method' pioneered by Pritchard and Greenler (Fig. 3.11). Still later, phase- and amplitude-detection of reflected, plane-polarized IR radiation became possible, a combination known as IR ellipsometry. Photo-acoustic reflection–absorption spectroscopy also arrived in due course. Here the sample is irradiated periodically with an IR laser pulse. It experiences oscillations in temperature as a result of the absorption of radiation and these oscillations are picked up by a microphone or a piezoelectric transducer.

Since Fourier transform instruments incorporating diffuse reflectance are now commercially available, the examination by FTIR of even intensely coloured and

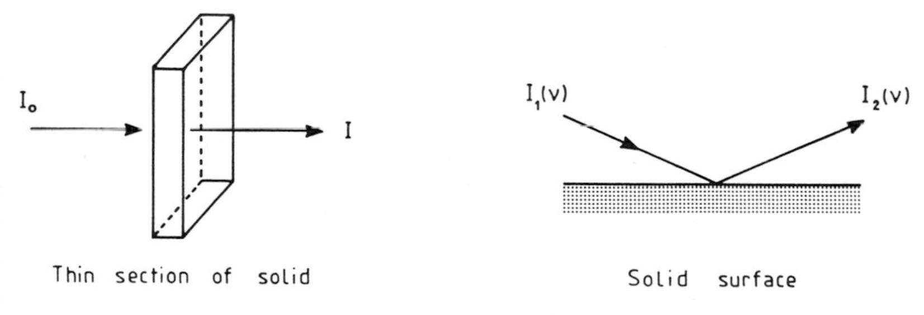

Figure 3.11 The transmission and reflection–absorption methods for the IR study of the adsorbed phase.

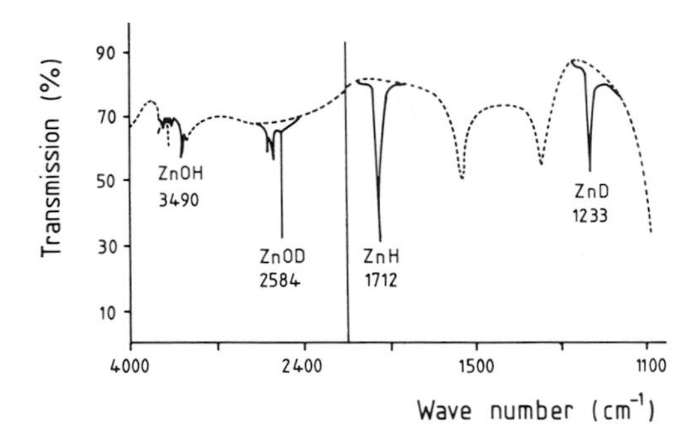

Figure 3.12 IR spectra of H_2 (and D_2 adsorbed on ZnO. Both M–H (M–D) and O–H (O–D) bonds are identifiable. With permission from K. Tamaru, *Dynamic Heterogeneous Catalysis*, Academic Press, London 1978.

powdered solids can be readily undertaken. Moreover, computational accessories make the recording of 'difference' or accumulated spectra, i.e. either before or after exposure to an adsorbate or during the course of a catalytic reaction, often a routine matter. With the advent of 'mirage' detectors, further instrumental developments, and the prospect of remotely sensing the progress of a catalytic reaction by IR, are likely. Diffuse reflectance IR Fourier transform spectroscopy (DRIFTS) is of great value for in-situ studies (see Section 3.10).

Compared with its rival technique (HREELS) for detecting vibrational frequencies of adsorbed and transitory species, IR spectroscopy is inferior: the cross-section for interaction of the oscillating dipole is less, by a factor of 10^6, with IR radiation than with a beam of low-energy electrons. Nevertheless, it is a gentle, non-invasive technique, well equipped to monitor surface changes in catalysts under operating conditions.

Some of the triumphs of IR as a characterizing technique are that:

1. It shows a loss of aromatic character in benzene when the latter is adsorbed on certain catalysts.
2. When H_2 is adsorbed on ZnO, dissociation leads to the production of Zn–H and O–H surface linkages, a fact further vindicated by the detection of $\sqrt{2}$-shifted absorption frequencies when D_2 is used in place of H_2 (Fig. 3.12).
3. When formaldehyde contacts a rutile (TiO_2) surface, no CH_2 links are detected, only CH.
4. When amines and a range of other guests are intercalated by clay catalysts, protonated species are formed.
5. When HCOOH is brought into contact with a wide range of metal surfaces, a monomolecular layer of formate is formed, just as when formaldehyde is adsorbed on rutile.
6. When pyridine is bound to cracking catalysts (SiO_2–Al_2O_3, for example), the various Brønsted and Lewis acid sites can be identified.
7. Linear 'end-on' attachment to some metals, best designated in the 'valence bond' formalism as M^-–N^+≡N (where M is typically Ni), is unmistakably indicated.

So far as the last conclusion is concerned, it is interesting that this was the first-ever identification of an 'end-on' dinitrogen species. Later this type of bonding was discovered for ligands in coordination chemistry. Transmission IR studies of chem-

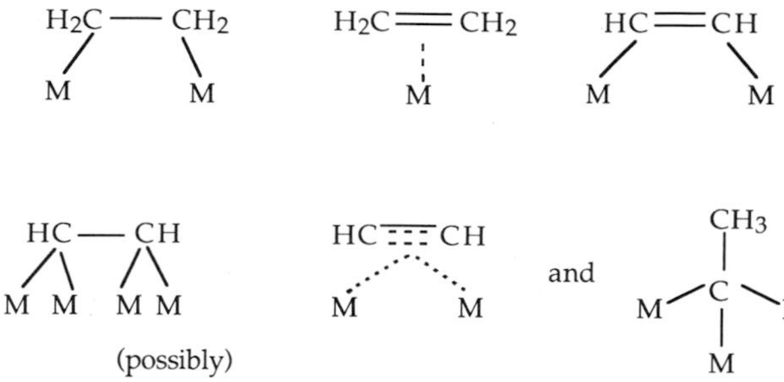

H_2C—CH_2 / M M; H_2C=CH_2 / M; HC=CH / M M

HC—CH / M M M M (possibly); HC≡CH / M M; and CH_3 / C / M M M

Figure 3.13 Species the presence of which has been detected by IR spectroscopy when ethylene is absorbed on oxide-supported platinum or palladium.

isorption of finely divided particles provide evidence for the presence of numerous kinds of different adsorbed species. Thus, for ethylene adsorbed on oxide-supported platinum or palladium, IR studies show or imply the presence of the species shown in Fig. 3.13, many of which coexist on the same catalyst preparation (M ≡ metal atoms).

When propylene is chemisorbed at ZnO surfaces, there is little doubt that the predominant adsorbed species is in the form of a symmetrical allyl group (CH_2–CH–CH_2) formed by the detachment of a hydrogen from the parent alkene and the concomitant formation of a surface OH group. In the context of heterogeneous catalysis, this is a significant result for it is believed that, in the conversion of propylene to acrolein, CH_3CH–CHO or to acrylonitrile, CH_2≡CH–CN, by ammoxidation using molybdates as catalysts, the allylic species is a crucial intermediate.

It has recently been recognized that IR frequencies exhibited by CO or NO groups bonded to organometallic clusters can be a misleading guide to the nature of CO bonding on extended metals. It is nevertheless undeniable that IR has clarified much of our understanding of the way in which CO and H_2 (which are of such central importance in syn-gas chemistry) are attached to metals.

As outlined earlier (Fig. 3.9), CO can be bonded to one metal atom, or it can bridge over two or more atoms. The former, linear or terminal bonding mode, is common on most Group VII metals, but on palladium the multisite situation is predominant. Sheppard and Nguyen give the following assignment of ranges of IR wave numbers to the structures of the adcomplexes of CO:

1650–1800 cm^{-1}	μ_4-bridged, $M_4(CO)$
1800–1920 cm^{-1}	μ_3-bridged, $M_3(CO)$
1860–2000 cm^{-1}	μ_2-bridged, $M_2(CO)$
2000–2130 cm^{-1}	Linear, M–CO

IR adsorptions in the range 2130–2200 cm^{-1} signify a linearly adsorbed CO on a cationic site. These assignments are more likely to hold good for very finely divided than for extended metals, i.e. real-life supported catalysts are likely to conform better

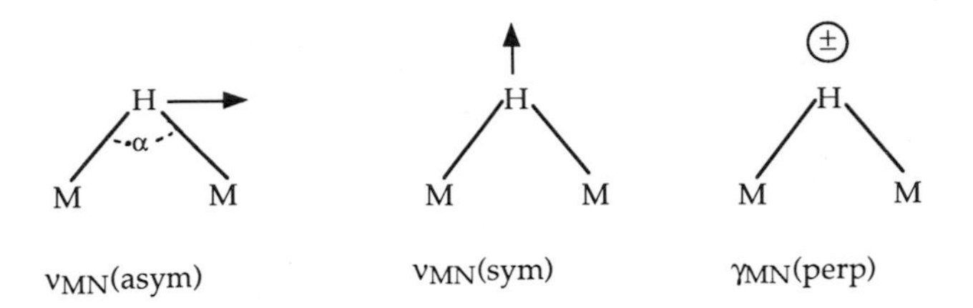

Figure 3.14 Vibration modes of bridging hydrogen species M–H···M.

with these rules than model, single-crystal catalysts. So far as chemisorption of hydrogen on metals and its attachment in organic complexes are concerned, IR spectroscopy has elucidated the nature of bridging hydrogen species of the type shown in Fig. 3.14 (compare μ_2-bridges for CO). Such a grouping has two modes of vibration associated with the stretching of MH bonds, and one 'angle-bending' mode in which, in the presence of other ligands or metal atoms, the H atom vibrates perpendicular to the plane of the M_2H triangle (i.e. perpendicular to the paper). The three vibrations are, therefore as shown in Fig. 3.14. It transpires that, in cluster compounds where bridging hydrogens of this kind can be pinpointed by neutron crystallography, there are variable angles, α. But because the H atom is uniquely light compared with the metal atoms, a simple central force-field treatment yields a theoretical ratio:

$$\nu_{MH(asym)}/\nu_{MH(sym)} = \tan(\alpha/2) \tag{10}$$

This relationship fits the observed IR frequencies very well; and indeed, by identifying the two frequencies (ν_{asym} and ν_{sym}) in a new cluster compound or surface complex, the value of α may be inferred, thereby giving the distance of the H atom above the M···M line. Sheppard et al., armed with this knowledge, were able to reinterpret some earlier, erroneously assigned, HREELS vibrations observed for H atoms bound to tungsten surfaces (W(100)).

3.5.3 High-Resolution Electron-Energy-Loss Spectroscopy (HREELS): The Most Sensitive Tool for Identifying Surface Vibrational Modes

When a beam of low-energy electrons, usually with energy less than 10 eV (energy spread of *ca* $\pm$10 meV), with a well-defined energy of incidence, strikes a solid surface, some of the scattered electrons lose energy because, on impingement, they succeed in exciting various kinds of vibrational modes at the surface. These modes, which typically arise from M–C, M–O, M–H, C–C, C=O, C–H bonds (where M is a metal atom), fall in the range 100–300 meV (1 eV $\triangleq$ 8067 cm^{-1}) and are readily detectable using a high-resolution electron spectrometer. The setup employed is shown schematically in Fig. 3.15, where the detector is situated at the angle of specular reflection. In this direction of the scattered beam, the selection rule governing which vibrational modes are active is the same as for the case of IR reflection, i.e., there is strong absorption only if there is a net change of dipole moment perpendicular to the surface. However, in HREELS, when off-specular beams are monitored

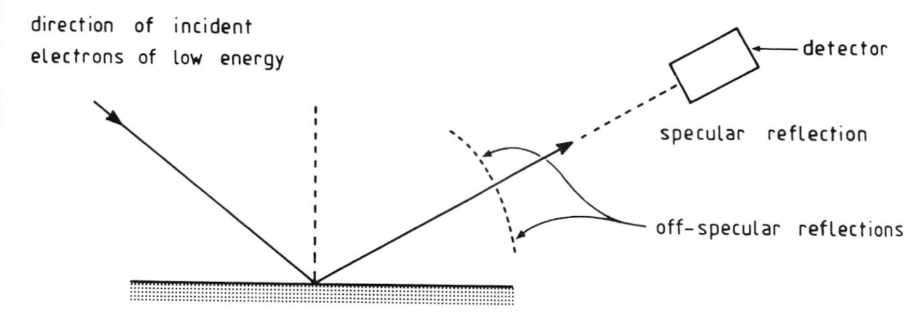

Figure 3.15 Set-up used for recording HREEL spectra of surfaces.

this 'dipole' selection rule breaks down and vibrational modes parallel to the surface are also observed. HREELS is more sensitive than IR, but both methods under their own optimal conditions are capable of detecting 10^{-3} monolayers. Unlike IR, however, HREELS cannot be used for studying catalyst surfaces under real-life conditions, because of the necessity to utilize long mean free paths for the electrons.

A powerful feature of this technique is the relative ease with which it detects adsorbed hydrogen and hydrogenous moieties (C–H, CH_2, –C–CH_3, etc.). Moreover, it can probe changes in hybridization of organic species during adsorption; and when dioxygen is bound to Ag(110) or Pt(111) surfaces the diminution in molecular vibrational frequency in the adsorbate signifies that donation of electrons from the metal to the $2p\pi^*$ antibonding state of the O_2 molecule has occurred (Fig. 3.16). Another important example, taken from the work of Somorjai, underlines the value of model catalyst studies, using HREELS as a tool to ascertain the role of electron additives in heterogeneous catalysis. Thus, in the absence of potassium, two well-defined CO stretching frequency 'losses' (at 1875 and 2120 cm^{-1}) are observed when this molecule is adsorbed on Pt(111) surfaces. From the assignments given in Section 3.5.2 we see that the lower-frequency peak arises from the μ_2-bridged state, the higher

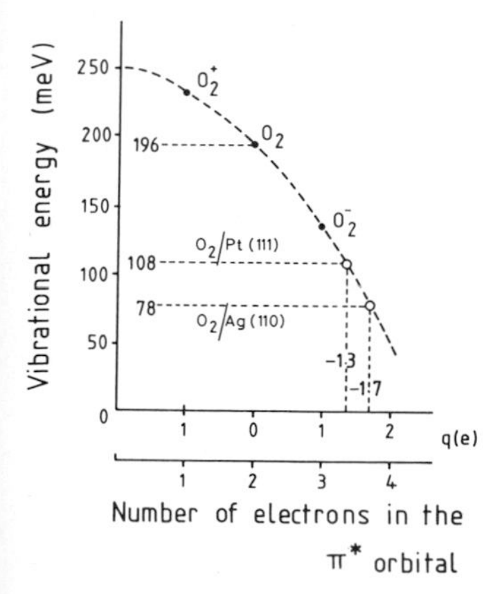

Figure 3.16 Proof, from HREELS, that electrons flow from the metal to bound O_2 molecules on Pt(111) and Ag(110). By extrapolation, we estimate that, on the former, the bound species are $O_2^{1.3-}$, on the latter $O_2^{1.7-}$. (Courtesy Dr C. Backx, Shell Co., Amsterdam.)

from the linear (terminal) state M(CO); in both cases the CO is perpendicular to the surface. As K is added to the surface, more of the bridge sites become occupied and the stretching frequency at this site decreases by more than $300\,cm^{-1}$. This, in effect, signifies a gradual change of bond order, from 2 to 1.5, with increasing coverage of K. The interpretation is that electrons are transferred from K to the transition metal and, from the Fermi level of the latter, electrons are fed into the antibonding molecular orbitals of the CO. Concomitant with the weakening of the C–O bond is a significant strengthening of the Pt–C bond. It is tempting to argue that the reason for potassium's role as an additive in Fischer–Tropsch catalysis (Chapter 8) in increasing the molecular weights of hydrocarbon products stems from the weakening of the C–O link, which, in turn, facilitates dissociation.

Two further examples of the usefulness of HREELS in the study of metal–hydrocarbon interactions merit comment. First, it is found that when cyclohexane is adsorbed on both nickel and platinum there are strong indications, from the observed broadened HREELS band at 2600–2700 cm^{-1}, that 'hydrogen bonds' involving C–H···M are present. Second, Ibach and coworkers identified the decomposition product of ethylene on the Pt(111) surface as an unexpected ethylidyne complex. The totally symmetric modes of the ethylidyne surface species are all identifiable in the 'specular' HREEL spectrum and the 'off-specular' spectrum provides some of the frequencies of the asymmetric modes.

3.5.4 Electron Spectroscopy: The Ability to Probe Composition and Bonding at Surfaces

Some surface techniques, typified by SIMS, are ultrasensitive, but, as a means of quantifying the composition of surface layers, they are not reliable; others, such as ISS, are capable of yielding quantitative surface compositions, but they do not reveal much about surface bonding. The great merit of the three variants of electron spectroscopy, apart from their admirable sensitivity – capable of detecting 10^{-3} monolayer on surface areas of less than $0.2\,cm^2$ – is that they also disclose a good beal about surface bonding. UV-induced photoelectron spectroscopy (UPS) is especially good in probing the surface bonding, less so in 'counting' surface atoms; the reverse is true of Auger electron spectroscopy (AES); and X-ray induced photoelectron spectroscopy (XPS) is good at both.

Figure 3.17 summarizes the various processes involved in electron emission. Photoemission, as rationalized by Einstein's equation, can be described in its broadest terms as:

$$A + h\nu \rightarrow A^{+*} + e$$

where a species A is ionized to A^{+*}, an excited charged state, by a photon of radiation, $h\nu$. If E_{ph} is the energy of the stimulating photon, the relationship between the kinetic energy of the liberated electron, E_{kin}, and the binding energy, E_b, that held it to the atom, molecule or solid from which it was ejected is:

$$E_{kin} = E_{ph} - E_b \tag{11}$$

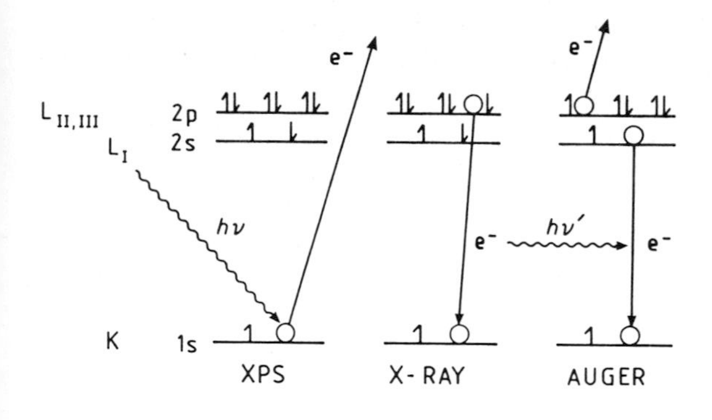

Energetics

XPS : $T_e = h\nu - E_B(1s)$

X-Ray : $h\nu' = E_B(1s) - E_B(2p)$

Auger : $T_e = E_B(1s) - E_B(2s) - E_B'(2p)$

Figure 3.17 Illustrative summary of the processes involved in photoelectron spectroscopy (XPS and UPS) and AES. With permission from D. M. Hercules, S. H. Hercules, *J. Chem. Educ.* **1984**, *61*, 402.

When the energy of the photon is less than *ca* 40 eV, we speak of *U*V-induced *p*hoto-emission *s*pectroscopy (UPS). Generally UPS is carried out with a helium resonance lamp ($E = 21.2\,\text{eV}$) and the electrons emitted emanate from the valence orbitals of the solid or surface under investigation *X*-ray-induced *p*hotoelectron *s*pectroscopy (XPS), on the other hand, entails emission from the core electrons of the solid, the stimulating sources being usually the so-called Al K_α line (E_{ph} =1486.6 cV, for rea-

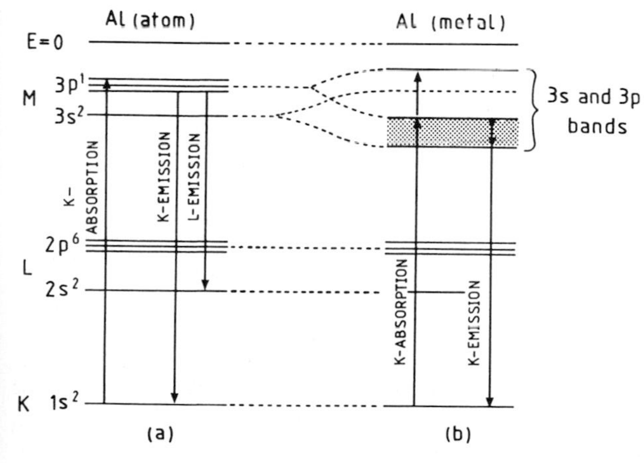

Figure 3.18 Energy levels for aluminium in the isolated atomic (a) and metallic (b) state. Note formulation of energy bands in the latter. Al K_α X-rays are emitted when electrons fall from the s–p band to a core hole in the K-shell.

sons that are apparent from Fig. 3.18) or Mg K_α (E_{ph} =1253.6 eV). The vacancy or hole left in the inner core level after photoemission can be reoccupied by one of two distinct processes: either by reorganization such that when an electron 'falls' from a higher level or band to occupy this hole there is a concomitant emission of an electron from a level of higher energy; or, alternatively, an X-ray photon is emitted when an electron descends from a higher level to fill the hole. The former event is the Auger process, the latter, X-ray fluorescence (XRF). Light atoms have a greater tendency to occupy an inner hole by the Auger event, whereas heavier atoms favour the fluorescence emission. In summary, Auger electron spectroscopy is represented by:

$$A^{+*} \rightarrow A^{2+} + e$$

and X-ray fluorescence by:

$$A^{+*} \rightarrow A^{+} = h\nu'$$

$h\nu'$ being a photon of smaller energy (longer wavelength) than that which gave rise to the initial ionization.

The symbols used in describing AES and XPS spectra often appear confusing. This confusion arises because the nomenclatures of X-ray practitioners evolved separately from those of other (including photoelectron) spectroscopists.

It is important to bear in mind that the *jj* coupling plays a prominent part in AES, where j, the total angular momentum of a particular electron, is the vector sum of the individual electronic spin and angular moments: $j = 1 + s$; and j can take values of $\frac{1}{2}$, $\frac{3}{2}$, $\frac{5}{2}$, etc. Strictly speaking, *jj* coupling is an apt description of electronic interaction in atoms of high Z (>70); but in practice the nomenclature based on *jj* coupling is often used for both Auger and other spectroscopic processes for all parts of the Periodic Table.

We recall that states with principal quantum number $n = 1, 2, 3, 4, \ldots$ are designated K, L, M, N, ... respectively, whereas states with various combinations of $l = 0, 1, 2, 3, \ldots$ and $j = \frac{1}{2}, \frac{3}{2}, \frac{5}{2}, \ldots$ are represented by suffixes $1, 2, 3, \ldots$. Now a KL_1L_2 Auger transition, by definition, means that a hole in the K shell is filled by an electronic transition (down) from L_1 and accompanied by an emission of another electron (out) from L_2. We therefore see that the omnibus description KLL for an Auger event in reality embraces the six transitions: KL_1L_1, KL_1L_2, KL_1L_3, KL_2L_2, KL_2L_3, KL_3L_3. These are typical of the descriptions used in AES; but sometimes, when the 'down' and 'out' electrons are valence-shell electrons (as the L_2 and L_3 levels are, for example, in carbon), the notation then is to replace L_2 and L_3 by V: i.e. a KL_2L_3 Auger event becomes KVV.

The spectroscopic descriptions used in XPS involve, first, the principal quantum number; then come the value of l, such that when $l = 0$, 1, 2, 3, ..., etc., we write s, p, d, f, ... respectively; and, last we quote the appropriate j value as a suffix. Clearly the L_2 state ($n = 2$, $l = 1$, $j = \frac{1}{2}$) in X-ray notation is written $2p_{1/2}$ in XPS notation.

The deeper the core level (i.e. the greater the binding energy of the electron) the less sensitive is the precise value of the binding energy of such an electron to the chemical environment. Therefore we see that, to all intents and purposes, values of

E_b as defined in Eq. (11) are characteristic of the atoms from which the Auger electron emerges. Thus, for a KL_1L_2 Auger electron we write

$$E_{KL_1L_2} = E_K - E_{L_1} - E^*_{L_2} \tag{12}$$

where E_i are the binding energies of the ith atomic energy levels. The 'starring' of E_{L_2} is necessary to signify that we are now referring to the L_2 level in the presence of a hole in level L_1. It is not easy to determine the energy of such excited-state levels, and hence one cannot be too exact in calculating the kinetic energy of an Auger electron. A good empirical equation is available, however:

$$E_{ABC}(Z) = E_A(Z) - \frac{1}{2}[E_B(Z) + E_B(Z+1)] - [E_C(Z) + E_C(Z+1)] \tag{13}$$

where E_{ABC} refers to the Auger transition ABC in an atom of atomic number Z, $E_i(Z)$ refers to the binding energy of the ith level in the same element and $E_i(Z+1)$ the corresponding level in the next element along the Periodic Table. Compilations based on Eq. (13) are available, as are tables of core energy levels for all accessible elements.

We are now fully equipped to understand the simple quantitative principles upon which atom identification and atom counting of elements at surfaces are based. In XPS, with a fixed primary energy, E_{ph}, we can tell from measured peaks E_{kin} in an electron spectrometer which atoms are present (Eq. (13)). With Al K_α radiation ($E_{ph} = 1486.6$ eV), for example, peaks would appear at *ca* 832, 859 and 978 eV (all from oxygen KLL Auger transitions) and *ca* 956, 1954, 1334, 1368, 1384 and 1413 eV if we were examining a SiO_2–Al_2O_3 cracking catalyst. If Mg K_α radiation ($E_{ph} = 1253.6$ eV) were used, each of these except the first three would be shifted by 253 eV to lower kinetic energies. Since Auger electron energies are independent of the energy of the stimulating radiation, one reliable way of identifying an Auger peak in a photoelectron spectrum is by noting those that are not shifted by a change of stimulating radiation. In both XPS and AES the peaks are so widely separated that element identification is straightforward.

AES as a sensitive surface spectroscopic technique is, however, much more readily carried out using primary electrons rather than soft X-rays to create the holes in the core levels. For quantum mechanical reasons the energy E_p of the primary electrons has to be about five times as large as the bonding energy of the electrons that are to be ejected. For the study of C, O, and N overlayers, therefore, E_p needs to be about 2000–3000 eV (for K-shell ionization). A convenient way of recording AE spectra is by adapting the grid optics required to record low-energy electron diffraction patterns; see Section 3.5.4.2 and 3.6.1.

3.5.4.1 The Realization that Electron Spectroscopy is Sufficiently Sensitive to Detect Fractions of a Monolayer

The achievements and detection limits of surface analyses are so impressive that we tend to forget that until the 1970s we lacked the wherewithal to probe monolayers if the surface areas were only a few square centimeters. The turning point came in the mid-1960s with the work of Siegbahn et al. in Sweden on electron spectroscopy for chemical analysis (ESCA – now taken to be synonymous with XPS) and also with that of Harris, Palmberg and Rhodin in the USA on AES.

So far as appreciating the sensitivity of XPS was concerned, an important observation was that involving the $I_{3d_{5/2}}$ signal from successively thicker bilayers of iodostearic acid. The fact that XPS signal intensity leveled off as the thickness of the film increased beyond a certain value implied that the escape depth of the photoemitted electrons was small. Definitive proof that monolayer amounts of adsorbed species could be detected by XPS came from Thomas et al.'s experiments on single crystals of graphite. Enough was known about the sticking coefficients of oxygen on the two main crystallographic faces of graphite to enable a clear-cut XPS experiment to be carried out, even though the vacuum conditions were relatively poor (10^{-6} Torr). The basal {0001} surface of graphite does not adsorb oxygen. (When gaseous dioxygen species collide with this surface the sticking coefficient is estimated to be less than 10^{-15}, which means that a freshly cleaved graphite crystal remains free of bound oxygen even in a vacuum of 10^{-6} Torr for *ca* 10^5 h). The prismatic $\{10\bar{1}0\}$ faces, on the other hand, are very reactive and pick up oxygen immediately on exposure to air. This oxygen cannot be removed except by heating to very high temperature. A flux of Al K_α soft X-rays played on to the basal faces shows no peak at an E_{kin} of 956 eV, whereas when the crystal is rotated so as to let the X-ray beam impinge upon the prismatic faces a peak does appear, proving that a monolayer of adsorbed oxygen could be readily detected by XPS.

In establishing the surface sensitivity of AES, the crucial experiment entailed depositing monatomic layers of one metal upon another and observing that AES signals characteristic of depths of only 5–10 Å could be quite easily recorded. In due course it was found that electrons with energies in the range 10–500 eV are ideally suited for surface analysis, because the plot of the mean distance of travel of an electron in solids as a function of electron energy exhibits a broad minimum in this range. As the mean free path corresponding to this minimum lies between 4 and 10 Å, it follows that electron emission from solids with energies of 10–500 eV must originate from the top few atomic layers.

When electrons are emitted from valence orbitals (or bands) in UPS (or by exposure to synchrotron radiation), their energies also fall in the range 5–30 eV, depending upon the precise value of the chosen source of stimulating light. Thus we see why UPS is also a surface-sensitive tool.

3.5.4.2 Auger Electron Spectroscopy (AES) and Scanning Auger Microscopy (SAM)

Only a small fraction of the electrons in a monoenergetic beam that is allowed to strike a solid surface are scattered elastically. This is because there are so many pathways for inelastic events available in such a situation – the excitation of oscillations in the valence electrons (plasma oscillations), ionization, promotion to higher levels and so forth. The so-called secondary-electron energy distribution is, in general, as sketched in Fig. 3.19. The minute features in the descending curve between the peak at low energies and the small elastic peak are manifestations of the Auger process. Although these features are weak, they are relatively sharp in energy and the peaks for the various light elements (C, N and O) are very well separated from one

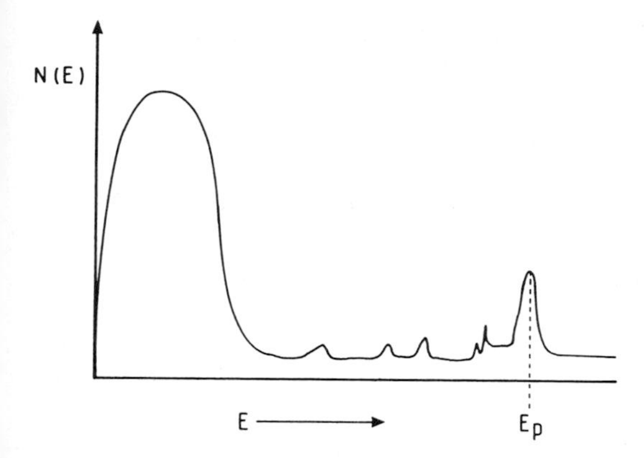

Figure 3.19 Schematic drawing showing the number of scattered electrons $N(E)$ of energy E when a primary beam of energy E_p strikes a solid surface. The small peaks at intermediate energies signify pressures of specific elements.

another. AES is now one of the most widely used of all vacuum-based surface analytical techniques.

Values of absolute stoichiometry for the topmost layers of a solid are not determined with great accuracy by AES, but the stoichiometry of atoms bound to the exterior surface is determined accurately. Moreover, changes in composition and the stoichiometry of adsorbed layers are conveniently and reliably quantified by AES. It has the additional merit of being able to cope with polycrystalline as well as single-crystal material.

Several compilations of Auger transitions are available. In these the Auger peaks are labelled with their respective kinetic energies, thereby making element identification straightforward. It is useful to realize that there are slight differences between the values of energy given in these compilations and those that we would compute from equations such as Eq. (13). This is because in the recorded spectra it is conventional to quote the energy of the Auger peak in the differentiated mode ($dN(E)/dE$ versus E) and it is the position of the minimum in the high-energy negative excursion that is cited. The 'conventional' energy clearly is not synonymous with the peak in the undifferentiated spectrum.

It is convenient, especially in postmortem (or prenatal) studies of catalysts, that scanning Auger microscopy (SAM) is also available. This is achieved by adapting scanning-electron microscopes or by designing, de novo, special equipment for the purpose. Spatial resolutions are such that a volume of $(500\,\text{Å})^2 \times 10\,\text{Å}$ can be probed, so that SAM is well suited to pinpoint element segregation to, or depetion at, grain boundaries, and to monitor element distribution over the surfaces of activated or spent catalysts, in much the same way that element-mapping is accomplished in conventional scanning-electron microscopy (SEM) (Section 3.7.5).

3.5.4.3 X-Ray Photoelectron Spectroscopy (XPS)

On exposing a typical metal catalyst to a monochromatic soft X-ray source such as AlK_α, electrons are liberated from the various sharply defined core levels and from the broader valence bands in the manner schematized in Fig. 3.20. A multicomponent catalyst would yield an XP spectrum correspondingly more complicated; but,

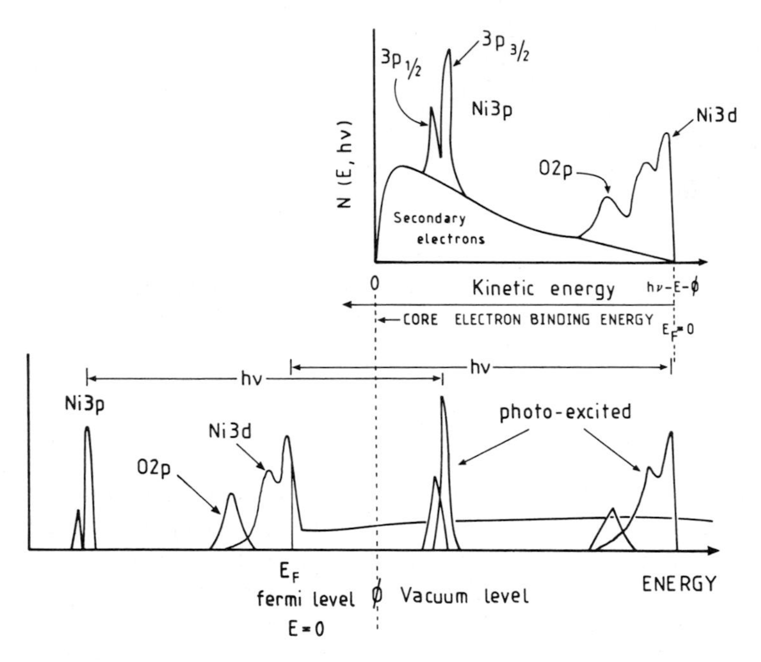

Figure 3.20 In the schematic diagram (top) of the photoelectron spectrum (XPS) of a nickel surface on which oxygen is adsorbed (below), we see how well separated the core electron levels are and also the direct relation between observed kinetic energy and the core electron binding energies.

from tables and atlases of binding energy and Auger transitions, each peak in the resulting spectrum can readily be identified. The seasoned practitioner soon learns to spot the additional features such as plasmon-loss peaks or shake-up peaks that sometimes accompany well-known XPS transitions. Element identification, therefore, presents no problem; and there are well-known tests to ascertain whether trace quantities are present at the exterior surface. A simple expedient is to tilt the sample with respect to the angle subtended by the source and detector sample so as to maximize the fraction of electrons photoemitted from the surface versus those from the sub-surface region.

To extract information about local chemical environment and bonding, we need first to determine the binding energies of the core and valence electrons. Figure 3.21 explains how this is done for the case of a conducting sample in good electrical contact with the electron spectrometer. The kinetic energy of a photo-emitted electron is measured with respect to the zero level of the spectrometer, which is the energy corresponding to the situation where an electron is at rest just outside the material of the spectrometer. (Recall the definition of work function, Section 2.7.) The kinetic energy E_{kin} with respect to the zero binding energy level of the sample is not measured; instead, its counterpart, E'_{kin}, is recorded. This arises because there is a contact potential, ϕ_c, between the conducting sample and the spectrometer, this being set up as a result of the equalization of the Fermi levels in sample and spectrometer. Thus

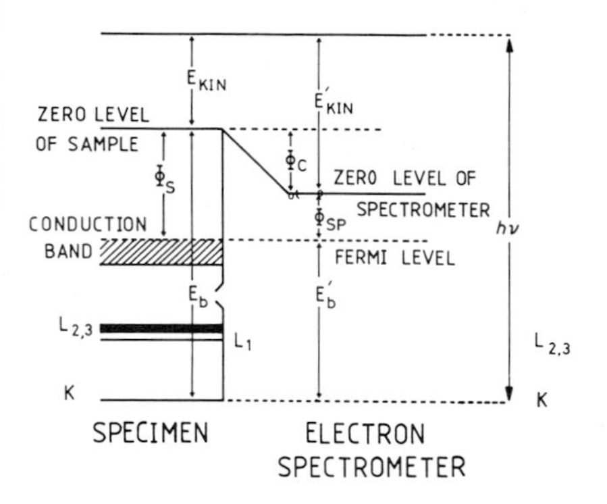

Figure 3.21 Representation of the relationship between the primary energy $h\nu$ and the core energy levels of a conducting sample (work function ϕ_s) in good electrical contact with the spectrometer (cf. Fig. 3.8).

$$E'_b = h\nu - E'_{kin} - \phi_{sp} \tag{14}$$

where ϕ_{sp} is the work function of the spectrometer, and E'_b is the binding energy with respect to the Fermi level of the spectrometer. As the sample and spectrometer are in electrical contact, their Fermi levels coincide, and E'_b is the binding energy with respect to the Fermi level of the sample also. The merit of this procedure is that E'_{kin} (as distinct from E_{kin}) is what is actually measured: moreover, ϕ_{sp} is constant and depends only on the spectrometer, in contradistinction to ϕ_s, the work function of the sample, which is liable to change for a given material upon chemical treatment and which, in any case, varies from sample to sample. The disadvantage, however, is that the reference level is not the actual (vacuum) zero of the sample. Consequently, comparisons of E'_b between samples must take account of differences in their respective work functions, ϕ_s.

Many heterogeneous catalysts are insulating, so that Eq. (14) has to be modified. Procedures for doing this are available; they take into account the fact that, under X-ray irradiation, the sample develops a positive charge which exerts an additional retarding influence on the emitted electrons. In practice, catalysts give rise to at least one peak (e.g. that of $C_{1s_{1/2}}$), which can be used as a subsidiary standard.

What is it that governs the magnitude of the shift in binding energy for core levels of atoms of the same element that are non-equivalent in a given sample? To answer this question we first picture an atom as a sphere of radius r_v, on the surface of which the valence charge q_i resides; the potential inside the sphere is the same at all points and is given by q_i/r_v. If there is a change Δq_i in this valence-electron charge, the potential inside the sphere will suffer a corresponding change $\Delta q_i/r_v$. Let E_i be the binding energy of a particular core level on atom i, and $E_i^{\ominus}$ be an energy reference (e.g., the measured energy in that neutral atom in the gas phase). Recognizing that the atom is embedded in a molecule, a solid, or a surface and surrounded by other atoms j, we may write:

$$E_i = E_i^{\ominus} + mq_i + \sum_{i \neq j} \frac{q_i}{r_{ij}} \tag{15}$$

where m is a proportionality constant that depends on the core level in question and the summation term, known as the Madelung contribution, allows for the potential

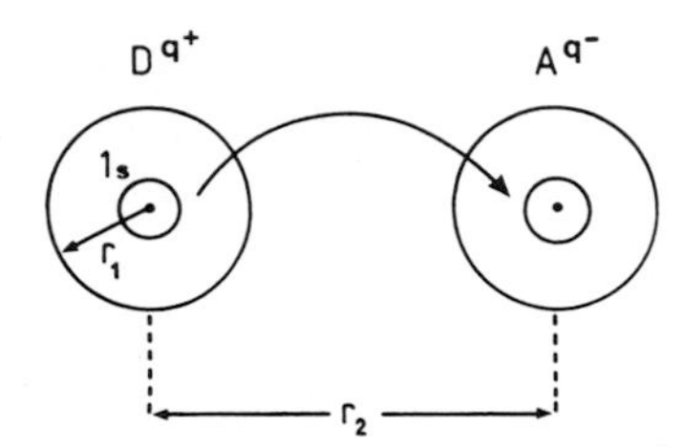

Figure 3.22 A simple model for explaining how charge transfer from or to a central atom changes the core binding energy. Donation of q electrons from D to A shifts the 1s binding energy by $(1/r_1 - 1/r_2)q$. In an ionic crystal the shift is $(1/r_1 - \alpha/r_2)q$, where α is the contribution to the Madelung constant for the atom in question.

exerted on atom i by point charges on surrounding atoms j. If ΔE_i is the shift in binding energy for a given core level of atom i in two distinct environments, we see from Eq. (15) that

$$\Delta E_i = m\{(q_i)_1 - (q_i)_2\} + ((M_i)_1 - (M_i)_2) \tag{16}$$

where M_i is the Madelung term. The first term of this equation accounts for the increase in binding energy that accompanies a decrease in valence-electron density on atom i (it is an observed fact – see Fig. 3.22 – that this is indeed the case). The second term works in opposition to the first; and this is why, depending upon the particular chemical or crystallographic environment that prevails, the magnitude of ΔE_i in going from an element in a lower to a higher oxidation state is very small. Usually, however, there are significant changes in E_i when the oxidation state of the metal is changed; and Ertl took advantage of this fact to follow the conversion of the magnetite precursor catalyst in the ammonia synthesis to its elemental state by recording the ΔE_i for the $Fe_{2p_{3/2}}$ level.

Consider the relationship between intensity of photoemission and the so-called inelastic mean free path λ, the latter being the average distance a photoelectron travels before suffering an inelastic collision. Clearly the probability of an electron escaping in the vertical direction from a depth x is proportional to $\exp(-x/\lambda)$ (see Fig. 3.23). Provided there is no attenuation in the X-ray flux as it passes through the sample, the total intensity I is:

$$I = a\sigma N \int_0^d \exp(-x/\lambda)\,dx$$

$$= a\sigma N\lambda[1 - \exp(-d/\lambda)] \tag{17}$$

where σ is the photoelectric cross-section for the core level under consideration, N is the number of atoms containing that core level per unit volume, and a is a proportionality constant. If, for purposes of calibration, we had an infinitely thick specimen and the intensity for that specimen were I_0 Eq. (17) would become:

$$I = I_0[1 - \exp(-d/\lambda)] \tag{18}$$

which serves as a basis for an experimental method of determining λ. An exemplary application of these principles is found in a recent characterization of oxidized surfaces of platinum. Oxides such as PtO_2, $PtO_2 \cdot mH_2O$, $PtO(OH)_2$, and $Pt(OH)_4$ are present in the overlayer that forms when platinum metal is severely oxidized. We

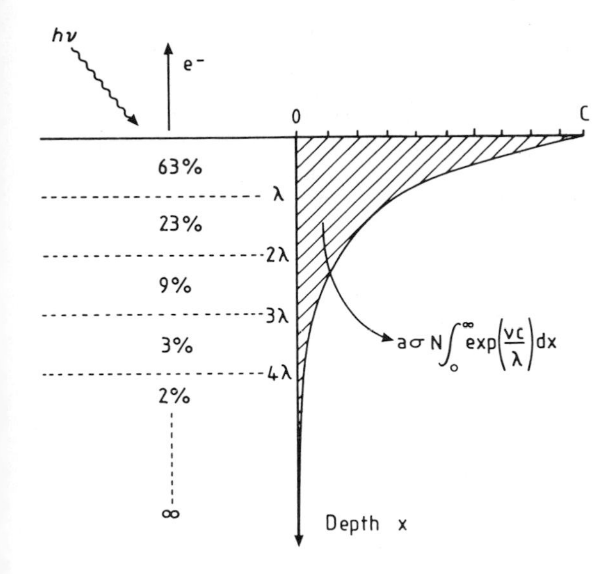

Figure 3.23 The percentage contribution of successive layers (in units of mean free path λ) to the total XPS intensity is shown on the left for electrons liberated in the vertical direction. The curve represents Eq. (17).

determine the thickness d of a given oxide film from the measured total intensity I_{Pt} of the Pt_{4f} signal and $I^{\ominus}_{Pt(metal)}$ and $I^{\ominus}_{Pt(oxide)}$ which are, respectively, the Pt_{4f} signals of pure metal and pure oxide. Clearly

$$I_{Pt} = I^{\ominus}_{Pt(oxide)}[1 - \exp(-d/\lambda_1)] + I^{\ominus}_{Pt(metal)} \exp(-d/\lambda_1) \tag{19}$$

where λ_1 is the inelastic mean free path (otherwise known as the escape depth) of the Pt_{4f} photoelectrons (separately found to be 12 Å). Likewise the build-up of the oxide layer could be monitored from the $O_{1s_{1/2}}$ XP signal intensity I_{ox}, knowing $I^{\ominus}_{ox}$, the corresponding signal for the pure oxide, and assuming that the inelastic mean-free path was also λ_1:

$$I_0 = I^{\ominus}_{ox}[1 - \exp(-d/\lambda_1)] \tag{20}$$

Roberts, Madix, and others have taken full advantage of the complementary aspects of XPS and UPS, especially in the study of the retention, or otherwise, of the integrity of small molecules such as CO, HCOOH and CH_3OH at catalyst surfaces. Some specific examples are cited in the next section.

3.5.4.4 UV-Induced Photoelectron Spectroscopy (UPS)

Whereas the broad principles of UPS and XPS are the same, there are some important differences. The obvious diminution in energy of the stimulating source (21.2 or 40.8 eV for He I and He II radiation) brings with it a better degree of resolution than in XPS; and this stems in part from the greater monochromaticity of far-UV compared with X-ray light sources. But UPS is necessarily restricted to electrons from valence shells.

Consider the salient points concerning the nature of bound CO at metal surfaces, a question of major importance in Fischer–Tropsch and synthesis gas chemistry (Chapter 8). Blyholder reasoned that, when CO is adsorbed in the undissociated state (the so-called α form), there is a transfer from the highest filled orbital of the CO molecule (5σ) to empty metal orbitals combined with considerable back-donation from the metal to the lowest unoccupied orbital of the adsorbate, i.e. the antibonding 2π orbital (see Fig. 1.24). UPS confirms the essential validity of this model, since the molecular orbitals can be 'seen' intact. (Both IR and HREELS, as well as thermal desorption spectroscopy (TDS; see Section 3.8), provide corroboration for this view.) But CO in the adsorbed state can also undergo dissociation:

$$CO_{ad} \rightarrow C_{ad} + O_{ad}$$

a reaction of decisive importance in the catalytic conversion of synthesis gas. UPS, coupled with XPS and AES, provides unambiguous proof of this dissociation, which takes place readily at modest temperatures on iron, molybdenum and tungsten. Palladium, however, at least at the temperatures where dissociation occurs on the other metals, shows little aptitude to fracture the adsorbed CO, a fact that helps us to understand why palladium is a good methanol synthesis catalyst and iron a poor one, bearing in mind that the essence of the surface reaction is to add hydrogen to an intact bound CO (see Section 8.2). The retention of the molecular integrity of CO on iron at low temperature (*ca* 80 K) is recognised (Fig. 3.24) by UPS from the peaks at 7.8 and 10.8 eV below the Fermi level. The first of these peaks is interpreted as overlapping emissions from orbitals derived from the CO 5σ lone pair and 1π levels, the 11.8 eV peak is due to the lone pair (the CO 4σ level). On raising the CO-covered

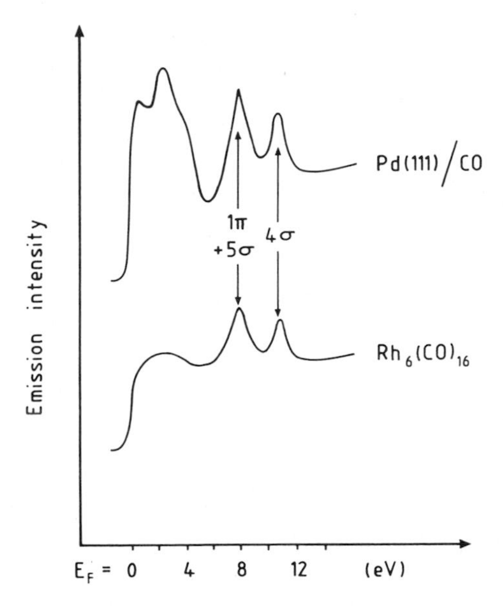

Figure 3.24 The retention of the molecular integrity of CO on platinum (and on iron at low temperature) is indicated by the UPS spectrum, which is similar to that of solid metal carbonyls like $Rh_6(CO)_{16}$. (Based on work of G. Ertl.)

iron to *ca* 430 K, a quite different UP spectrum results. The principal feature now is a peak at 5.6 eV below the Fermi level; this is attributable to adsorbed oxygen since this very peak is obtained when the iron absorbs oxygen dissociatively. XPS confirms this interpretation: it also suggests the existence of two types of surface carbon, that with a binding energy of 283.3 eV being ascribable to a carbidic species and that at 284.8 eV to free graphite.

It will be concluded that the ethos underlying the assessment from UPS spectra as to whether the CO, or any other adsorbate, remains intact on adsorption entails comparing the spectrum of the adsorbed entity with that of the free (gas-phase) molecules, making due allowance for the unavoidable loss of the vibrational modes in the UPS of the adsorbate. A striking example is contained in the work of Madix, who studied CH_3OD adsorbed on Cu(110) at 80 K. The difference spectrum produced by subtracting the spectrum of the clean surface from that of the adsorbate–adsorbent complex exhibits essentially all the features found in the UP spectrum of the gas-phase molecule, indicating that the CH_3OD remains intact. Spectral features which are associated with the lone pairs of electrons on the oxygen were shifted to higher binding energy on adsorption. Evidently the methanol is linked to the surface via the oxygen, an interpretation substantiated by the observed decrease in work function, there being a dipole layer with the positive end away from the surface (see also Eqs. (15) and (16)).

An added refinement to straightforward UPS is achieved in angle-resolved studies where photoelectron spectra, usually stimulated by unpolarized HeII light, are recorded as a function of the angle between the normal of the surface and the direction of the analyser of the emitted electrons. The direction of emission is related to the moments of the exciting photon and the excited electron. Angular distribution of photoemitted electrons can therefore provide information about the orbital quantum numbers of locally bound electrons or about the wave vector of delocalized electrons in a solid or its surface. By comparing observed with calculated intensities of the angularly resolved peaks a decision can be reached as to the orientation taken up by a surface-bound species, or by spatially localized energy bands in a solid. The angular plots of the 4σ and 5σ peaks in CO bound to {100} faces of platinum reveal, for example, that two-thirds of the CO molecules that are non-dissociatively adsorbed are in a tilted configuration, the angle of molecular tilt being less than *ca* 15°. On surfaces of ruthenium, however, the CO molecules are adsorbed in a vertical fashion on {100} faces; but at the step sites on terraced surfaces of this metal, the CO molecules are tilted at around 30° to the surface normal of the {100} planes (i.e. the molecular axis points along $[55\bar{1}]$. From Lambert's work, angle-resolved UPS (ARUPS) reveals that acetylene on palladium lies flat on the surface at low temperatures, with the triple bond parallel to the plane of the Pd atoms, whereas at higher temperatures the C–C axis is perpendicular to the surface.

3.5.4.5 Inverse Photoemission: A Means of Probing Unoccupied States

Photoemission spectroscopy reveals the density of occupied electronic levels, whether they are discrete or merged into bands. By contrast, inverse photoemission (or

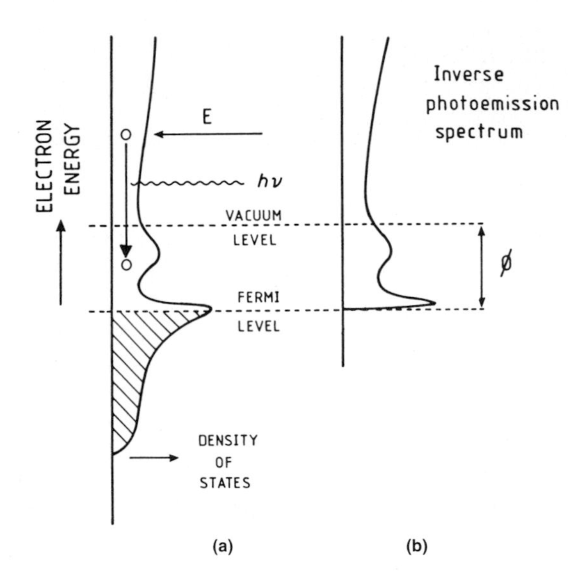

Figure 3.25 The essentials of inverse photoemission. In (a) we see how the unoccupied states are probed; and in (b) the ranges of energy explorable by inverse (IPES) and direct (PES) photoelectron spectroscopies are compared. (Based on work of P. Woodruff.)

'*bremsstrahlung* isochromat spectroscopy' as it is sometimes called) reveals features in the unoccupied levels. The basic principles are illustrated in Fig. 3.25. An electron of energy E is incident on the solid with density of states as indicated, and then decays radiatively. The spectrometer is set to detect photons of constant energy $h\nu$; and the spectrum is generated by sweeping E and monitoring the emitted photon flux, which replicates the unoccupied density of states.

In both conventional photoelectron spectroscopy (PES) and its inverse (IPES), strictly speaking it is the joint density of states between initial and final states which is detected rather than just the initial or final density of states. In PES the initial state is occupied and the final state unoccupied, so that the entire range of energy is involved in optical excitation. But photoelectrons excited into the region of energy space between the Fermi and vacuum levels, E_F and E_V, are unable to escape and be detected, thus rendering this region inaccessible in PES. In IPES, however, both initial and final states are unoccupied, with the occupied region being inaccessible. The entire unoccupied energy range is available for investigation in IPES. It is well suited for probing the region between E_F and E_V.

IPES has only recently been used to probe the electronic structure of species adsorbed at adsorbents and catalysts. Difference spectra, recorded for the clean and 'active' surface, reveal interesting spectral features. Thus an empty level derived from the 2π orbital of CO adsorbed on a copper surface was found at an energy of 0.9 eV below E_v and to have a halfwidth varying from 1.9 to 2.6 eV as a function of increasing coverage. Bearing in mind that it is easier to detect photons than electrons when they have to traverse through the atmosphere in which a catalyst is placed, IPES is inherently a better proposition than PES for future in-situ studies of catalysts.

3.6 The Structure and Crystallography of Surfaces: Measuring the Symmetries, Order and Disorder, and Deducing Bond Lengths and Bond Angles in the Adsorbed State

The technique that has so far proved most useful for the quantitative assessment of interatomic distances in surface layers is low-energy electron diffraction (LEED), which comes into its own when long-range order prevails. Scattering of neutrons or neutral atoms also yields quantitative information of a similar kind; but such experiments are far less readily carried out than LEED studies, which can be accomplished in a set-up capable also of monitoring Auger or photoelectron spectra and/or temperature-programmed desorption (TPD; see Section 3.9).

3.6.1 Two- and Three-Dimensional Surface Crystallography

In LEED, electrons of a well-defined energy are scattered from the outer layers of a surface which may be free from or covered with adsorbed species. If the electrons used have energy in the range 10–500 eV, their de Broglie wavelength falls between 3.9 and 0.64 Å, comparable with interatomic distances:

$$\lambda = \frac{h}{(2meV)^{1/2}} = \left\{\frac{150.4}{V}\right\}^{1/2} \tag{21}$$

where V is expressed in volts and λ in Å. The elastically scattered (i.e. diffracted) electrons can provide information about the periodicity of the adlayer, of the substratum (now almost universally termed the 'substrate') and of any coincidence structure set up by a combination of substrate and overlayer (see below). A schematic diagram of the apparatus used to record LEED patterns is shown in Fig. 3.26.

If the surface has long-range order and is free of microdomains that are not in registry with one another, a diffraction pattern consisting of sharply defined spots is displayed on the fluorescent screen. When the surface lacks long-range order, the diffraction spots broaden, are less intense, and some diffuse brightness appears between them. If misregistry of crystalline microdomains at the surface is pronounced, the spots give way to diffraction rings. A typical set of diffraction patterns is shown in Fig. 3.27.

To understand how the separation distance between the spots in a LEED pattern are reciprocally related to the real-space distance of atoms at the solid surface, we first use a one-dimensional analogy (Fig. 3.28). For constructive interference we have

$$n\lambda = a \sin\theta \tag{22}$$

Clearly the smaller the separation distance, a, in real space, the larger the value of $\sin\theta$, and hence of θ, and so the larger the distance between the incident beam and

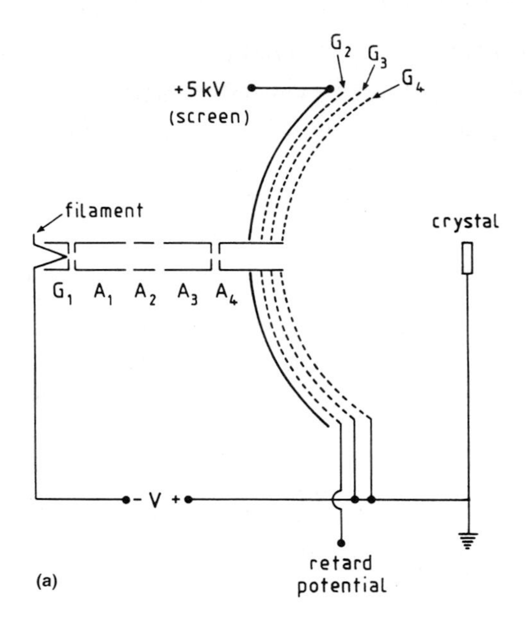

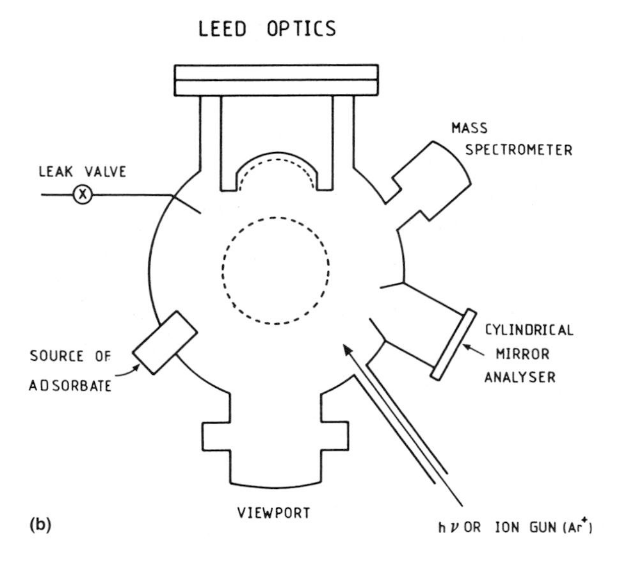

Figure 3.26 (a) Outline of system used to record LEED patterns. G_1, A_1, A_2, A_3 and A_4 are electron gun lenses; G_2, G_3 and G_4 are grids and V is the beam energy. (b) Horizontal section through main chamber showing schematically the various surface probes used to study crystal surfaces. The dotted line denotes the path of the crystal on rotation.

the diffraction spot. Extending this picture to the two-dimensional surface net or mesh, we may write, by analogy, for $n = 1$:

$$\lambda = d_{hk} \sin \theta_{hk} \tag{23}$$

where h and k are the two-dimensional Miller indices (see Chapter 5) for the mesh. A rigorous proof of Eq. (23) can be given in terms of the so-called 'Ewald con-

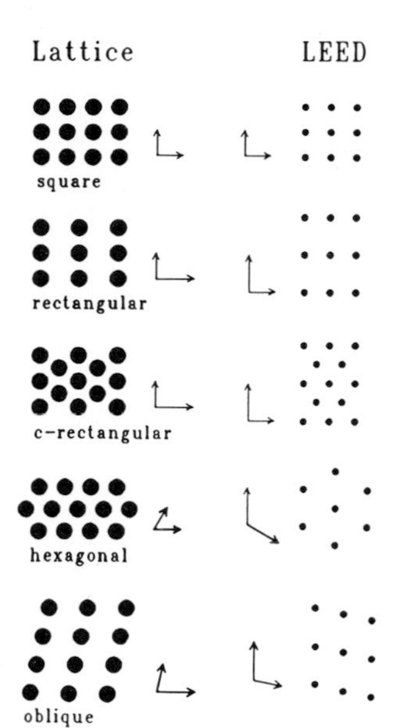

Figure 3.27 The five different surface lattices, base vectors of the real and reciprocal lattices, and the corresponding LEED patterns. With permission from J. W. Niemantsverdriet, *Spectroscopy in Catalysis*, VCH, Weinheim 1993.

struction', which entails drawing a circle of radius $1/\lambda$ 'through' the net of the reciprocal mesh.

In two-dimensional (2D) LEED studies, one focuses only on the symmetry and separation distances of the spot patterns: the intensities are ignored. In view of the reciprocal relationship between distances in LEED patterns and the real-space distances at the surface, the latter may be readily derived (Eq. (23)). From the diffraction pattern, we deduce the size and orientation of the unit cell (more correctly, the unit mesh) and, in particular, evidence for surface reconstruction in the outer layer of the clean solid, as well as for supermeshes (commonly, but erroneously, called superlattices) induced by the presence of overlayers (Fig. 3.29). We may also derive evidence for the occurrence of steps or kinks from the doubling of certain spots in the corresponding diffraction patterns. All this constitutes two-dimensional

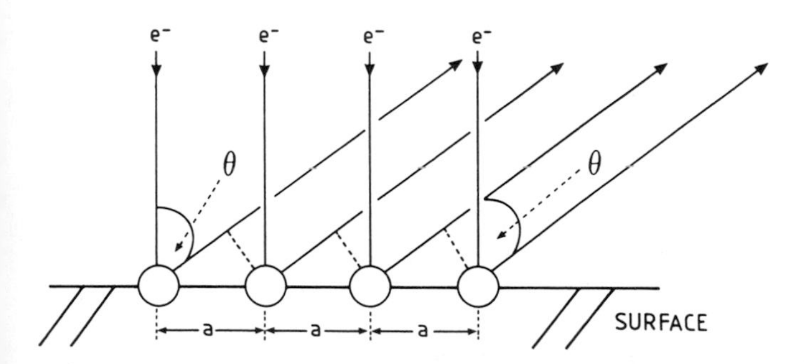

Figure 3.28 Diffraction of electrons from a one-dimensional array of atoms (see Eq. (22)).

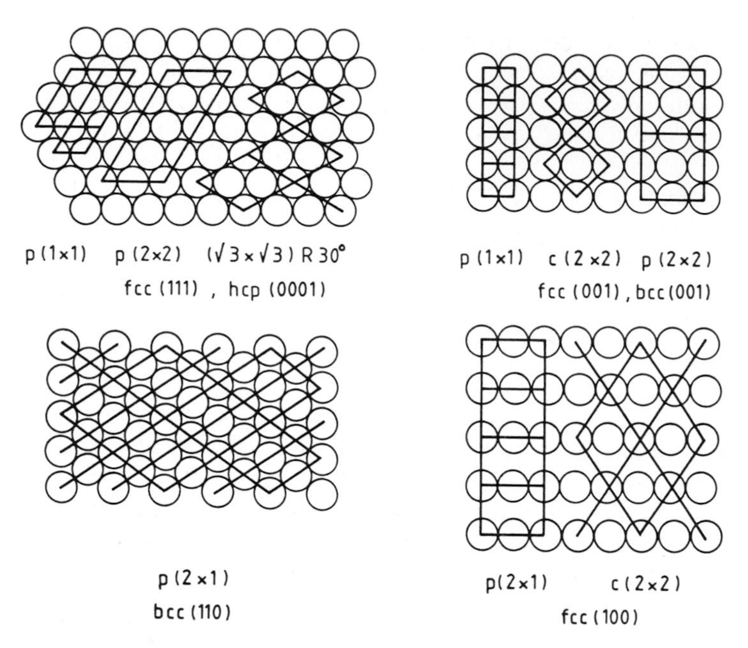

Figure 3.29 Common supermeshes on high-symmetry (low Miller index) crystal faces.

crystallography. A valuable aspect of LEED studies is that we may follow the variation of coverage, and of the ordered overlayer structures adopted at specific coverages, as a function of the exposure to the adsorbate.

In three-dimensional (3D) LEED studies attention is focused on the intensities of the diffraction spots, and the procedure is to record these intensities as a function of the energy of the electrons and of their direction of incidence. Just as modern four-circle X-ray, single-crystal diffractometry can, in most instances (with solids free from structural irregularities), yield from the measured, diffracted intensities the three-dimensional structure of the solid at the atomic scale, so, in principle, does three-dimensional LEED yield the total structure of the outer (two or three) layers of a solid surface. There are, however, many complications that obtrude in the methodology of three-dimensional LEED. First, because the interaction between electrons and solids (unlike that of X-rays and solids) is very strong, multiple scattering events are common; second, the problems of absorption (inelastic scattering) are acute. Until the late 1980s it was not possible to solve the 3D structures of surface layers from LEED intensities anything like as readily as solving 3D structures of bulk crystals from X-ray intensities. But remarkable progress has been made recently. First, data retrieval of intensities of diffraction spots as a function of electron energy and polar and azimuthal incidence angles has been revolutionized, thanks to the application of electronics and other devices at Erlangen, Germany. Second, Pendry and his collaborators (van Hove, Rous, Wander and Somorjai) have removed the laborious guesswork from the process of translating diffraction intensities to surface structure.

Until recently, the sequence of steps in LEED studies was to guess a plausible structure, to calculate the diffracted intensities, and to compare with experiment. The last of these steps was done by means of an R-factor measuring overall quality of agreement with experiment for the trial structure (when R is zero there is perfect agreement – never achieved in practice; when it is unity there is no agreement at all. An R-factor of 0.25 is perfectly acceptable). In tensor LEED, developed by Pendry et al., the first trial calculation is used to make a perturbation expansion with the original structure as a reference structure. This seminal development has yielded the automated determination of complex surface structures (now designated ATLEED). These procedures combine numerical search algorithms with efficient methods of determining the diffracted intensities for varying structures. In surface crystallography (encompassing the last three or four subsurface layers) one has all but reached the state of 3D X-ray crystallography of the mid-1980s whereby 'direct' methods of structure determination became routine practice. Typically 10 adjustable atoms (or 30 adjustable coordinates) can be readily determined on a laboratory workstation.

Theoretical developments, also instigated by Pendry (and Heinz), have recently made it possible to derive surface structures from diffuse LEED patterns. In other words it is no longer necessary to have a strictly ordered 2D structure for the positions of the individual atoms to be retrieved.

More exotic possibilities are currently under investigation; if realized, they would transform surface crystallography. One of these techniques is atomic resolution electron holography, an idea first mooted by Szökee (1986) and Barton (1988) following the conceptual breakthrough by Gabor (1948). Shirley et al. derived a 1D hologram in 1991 and King, almost simultaneously, successfully obtained images of iodine adatoms on an oriented silver surface from experimentally measured Auger intensity maps.

In LEED studies, the accuracy of the distances perpendicular to the surface is *ca* 0.02 Å (smaller than the atomic amplitudes at room temperature) and the accuracy of distances parallel to the surface *ca* 0.05 Å. It must be remembered, however, that the LEED technique requires well-defined single crystals. Moreover, the structures determined by LEED are only those stable in high vacuum. Also, the technique is inapplicable to the study of finely divided material: contrast solid-state NMR (Section 3.7.3).

3.6.2 Notations for Describing Ordered Structures at Surfaces

Some clean (bare) surfaces do not reconstruct, in the sense that they are found to possess the same structure, and essentially the same interatomic spacings parallel to the surface, as would be expected if, conceptually, the bulk solid were cut so as to expose a new surface. Some clean surfaces do reconstruct and are found to exhibit supercells. Many different types of supercell – some are rotated with respect to the underlying layers, others not – have been identified. A selection of the more common ones exhibited both by reconstructed clean surfaces and by ordered adlayers is shown in Fig. 3.29.

To understand the notation commonly used to describe these supercells, we first describe the unit mesh of the substrate by a pair of basis vectors $\boldsymbol{a}_1$ and $\boldsymbol{a}_2$, and the

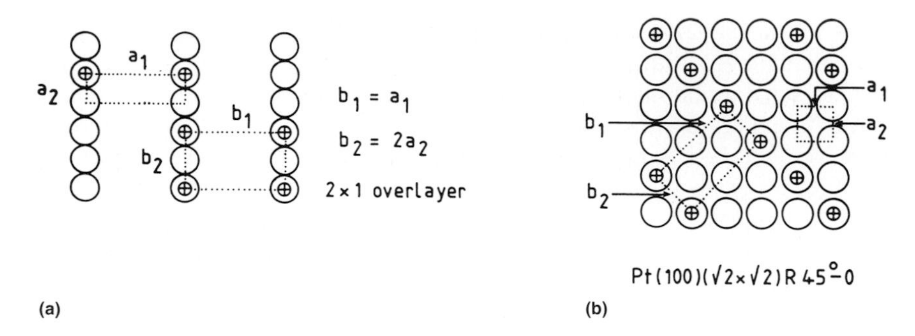

Figure 3.30 There is a direct relation, explained in the text, between reciprocal and real meshes at surfaces: (a) shows a (2 × 1) overlayer where the adsorbed species form a coincident supermesh; (b) shows a coincident supermesh overlayer on a Pt(100) surface.

unit mesh of the supercell by $\boldsymbol{b}_1$ and $\boldsymbol{b}_2$ (Fig. 3.30). The supercell is then described, in general, by:

c or p $(l \times m)\mathrm{R}\alpha$

where c and p refer, respectively, to a centred or primitive mesh; l and m are elongation factors ($l = |\boldsymbol{b}_1|/|\boldsymbol{a}_1|$ and $m = |\boldsymbol{b}_2|/|\boldsymbol{a}_2|$); and α is the angle through which the supercell is rotated (R) with respect to the cell of the substrate. The prefix p is often omitted, and the suffix Rα is always omitted when $\alpha = 0$. Sometimes the nature of the adsorbate is indicated using the elemental or compound symbol after the α value, and the element or compound formula (e.g., Pt or GaAs) is given, along with the Miller index of the surface in question, before the c or p (see Fig. 3.30). A c-mesh is generated from a p one by addition of an entity in the center of the latter. Note that the concept of unit mesh is not unique: the same two-dimensional lattice can be described by different unit meshes.

Another way of describing surface meshes is by matrix notation. The transformation matrix M is given by:

$$\begin{pmatrix} b_1 \\ b_2 \end{pmatrix} = \begin{pmatrix} m_{11} & m_{12} \\ m_{21} & m_{22} \end{pmatrix} \begin{pmatrix} a_1 \\ a_2 \end{pmatrix} = M \begin{pmatrix} a_1 \\ a_2 \end{pmatrix} \tag{24}$$

Clearly (see Fig. 3.30) a supercell described by the symbol p(2 × 1) is, in matrix notation, equivalent to

$$\begin{matrix} 2 & 0 \\ 0 & 1 \end{matrix}$$

Likewise a c(2 × 2) supercell (also described by $\sqrt{2} \times \sqrt{2}\,\mathrm{R}45°$) is equivalent to

$$\begin{matrix} 1 & -1 \\ 1 & 1 \end{matrix}$$

Such a supercell is formed on the (100) face of a methanation catalyst when a sulphur adlayer covers half the surface.

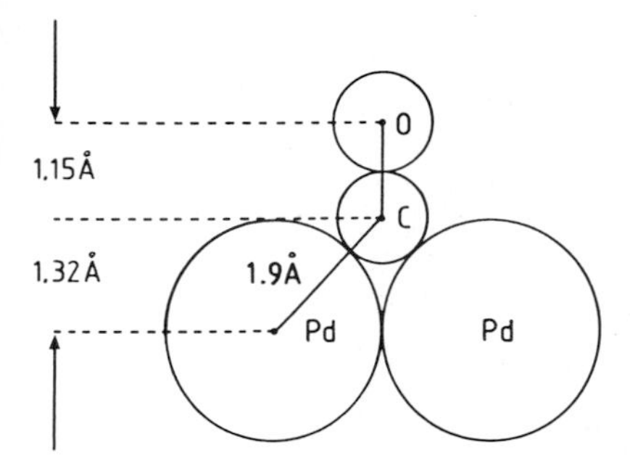

Figure 3.31 Bond lengths for species adsorbed at surfaces can be determined from LEED intensity data, as for these results with CO on palladium. (Based on work of G. Ertl.)

3.6.3 How do Bond Distances at Surfaces Compare with Those of Bulk Solids? What of Displacive Reconstructions?

Three-dimensional LEED yields bond distances for chemisorbed species that are entirely consistent with those we would expect from comparisons based on the structural properties of bulk materials. When CO is attached to a metal such as palladium we see that the bond lengths deduced from LEED intensities for the bridged species are as expected from the known structural properties of metal carbonyls (Fig. 3.31). Likewise, the ethylidyne moiety formed when ethylene is adsorbed at a Pt(111) surface has dimensions and symmetry entirely compatible with those exhibited by the organometallic analogues $Co_3(CO)_9CCH_3$ and $H_3Os_3(CO)_9CCH_3$. Tensor LEED reveals that an adsorbed molecule (e.g. ethylene on Rh(111) giving rise to bound ethylidyne, C_2H_3) can induce a displacive reconstruction in the substrate on which it is adsorbed (Fig. 3.32).

3.6.4 Atomic Scattering and Diffraction

Helium atoms possessing a thermal energy E, expressed in eV, have a de Broglie wavelength of $0.14/E^{1/2}$, expressed in Å. When the thermal energy is 0.02 eV, $\lambda = 1$ Å and such 'radiation' is readily diffracted from solid surfaces. The various processes that can occur when monoenergetic beams of neutral atoms are scattered from surfaces are quite complicated. But in view of the fact that scattered intensities contain information about the attractive potential, well depths, and the energetics of the quantized vibrations (phonons) at the surfaces, this is an area of growing importance, in that the technique provides a means of reaching some of the fundamental properties that govern the interaction of solids and gases.

3.6.5 EXAFS, SEXAFS, XANES and NEXAFS: Probing Bond Distances and Site Environments even when There is no Long-Range Order

The technique of *e*xtended *X*-ray *a*bsorption *f*ine *s*tructure (EXAFS) has its roots in the X-ray spectroscopic studies of Kossel and Kronig in the early 1930s. It is an immensely powerful tool since, unlike X-ray diffraction or LEED, it can be applied to structurally disordered as well as to ordered solids, to dispersed species on surfaces

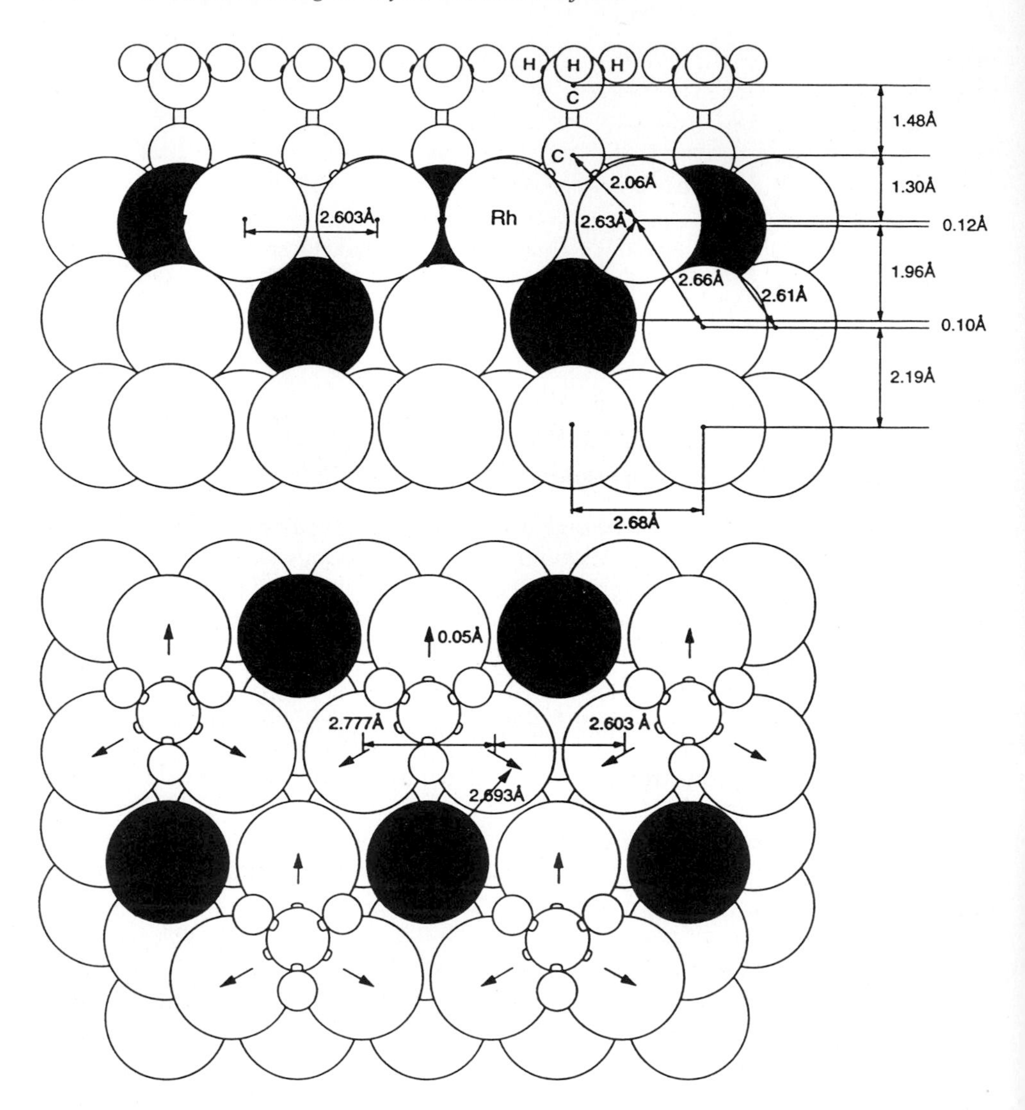

Figure 3.32 Side view (top) and top view (bottom) of Rh(111)(2 × 2)–C_2H_3 (ethylidyne, with guessed H positions). Dark atoms have relaxed perpendicular to the surface from bulk positions. Substrate relaxations are drawn to scale, indicated by arrows and labelled by their magnitudes. With permission from A. Wander, M. A. van Hove, G. A. Somorjai, *Phys. Rev. Lett.* **1991**, *67*, 626.

(as in the case of supported catalysts) and to materials in states of fluid suspension or to solid catalysts in contact with liquids. Apart from yielding distances that separate an arbitrarily selected central atom from other atoms in its first and successive neighboring shells, it also yields the coordination numbers and a measure of the atomic rigidity in those shells. EXAFS has enlarged our knowledge of the structural properties of proteins and organometallic clusters, and especially of solid and supported catalysts. An extension of the technique which focuses specifically on the environments of atoms at the exterior surfaces of solid (SEXAFS) is also an important new tool. XANES, which stands for *X*-ray *a*bsorption *n*ear-*e*dge *s*tructure, a related technique for which the underlying theory is more complicated because it involves multiple scattering, is also capable of yielding quantitative structural information when the material is not crystallographically ordered. Pre-edge features in the X-ray absorption spectrum of a solid or its surface are also structurally informative.

EXAFS is especially promising in that it can be used for in-situ studies of catalysts. It is also a very sensitive technique, being capable of probing the behaviour of less than 10^{12} atoms, which is equivalent to one-thousandth of a monolayer on a metal of 1 cm^2 surface area. In combination with X-ray diffraction, X-ray absorption offers a more or less complete structural picture of the local and extended structure of a solid catalyst under operating conditions.

3.6.5.1 The Origin of EXAFS and How it is Used

In the absorption of X-rays by matter (a process which primarily yields photoelectrons), a plot of absorption coefficient versus energy of the X-ray photon exhibits (except for a monatomic gas as absorber) an extended fine structure on the high-energy side of each absorption edge, K-, L-, M-, etc. Figure 3.33 shows three such extended structures when the photon energy of the X-ray exceeds 11.50 keV, the threshold value for the L_{III} ($=2p_{3/2}$) state of the core electron in platinum.

Fig. 3.33a shows three absorption edges due to the X-ray absorption of L-shell electrons. They differ in energy, their order being $L_{III} < L_{II} < L_I$. The electrons responsible for the absorptions are 2s (angular momentum quantum number $l = 0$) for L_I and 2p ($l = 1$) for L_{II} and L_{III}. The difference between L_{II} and L_{III} is caused by different couplings between spin and orbital angular momentum in each case. For L_{II}, both vectors are opposed, thus yielding the lowest possible total angular momentum quantum number, $j = l + s = 1/2$ (the $2p_{1/2}$ state). For L_{III}, spin and orbital angular momentum are aligned, producing the highest possible total angular momentum quantum number, $j - l + s = 3/2$ ($2p_{3/2}$).

The spectrum shown in Fig. 3.33a, consisting of distinct peaks superimposed on a step function, is a typical example for the spectra of transition metals with partially filled d states. The peaks (sometimes called 'white lines' because of their appearance in early recordings on film) are caused by resonances associated with the excitation of 2p electrons to unfilled d states. When the underlying step function is subtracted from the signal the peaks show a Lorentzian line shape.

For metals with completely filled d states (e.g. copper) the absorption edges take the form of simple step functions which are attributed to the excitation of electrons from 2p orbitals to unbound (continuum) states. It can be deduced theoretically that these steps should follow arctangent functions which are indeed reasonable approximations to measured spectra.

Fig. 3.33b compares the L_{III} absorption edges for the metals from rhenium through gold. For gold, the absorption edge is simply a step (no unfilled d orbitals which could lead to resonance) whereas for all other metals a more or less pronounced peak is superimposed thereupon. The intensity of the absorption threshold resonance increases with the number of unfilled d states, Re > Os > Ir > Pt > Au.

The fine structure that constitutes EXAFS consists of fluctuations in the absorp-

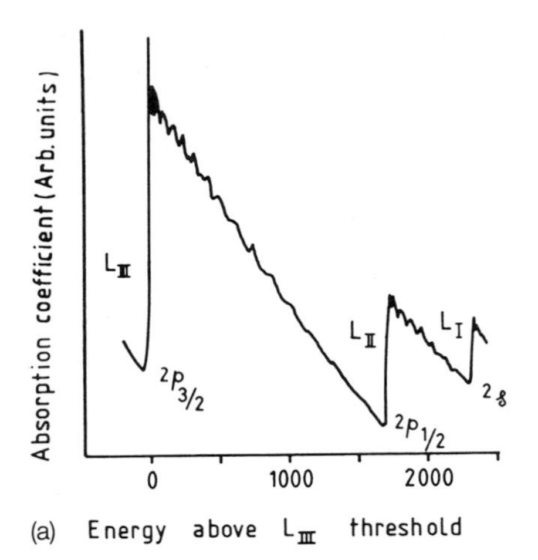

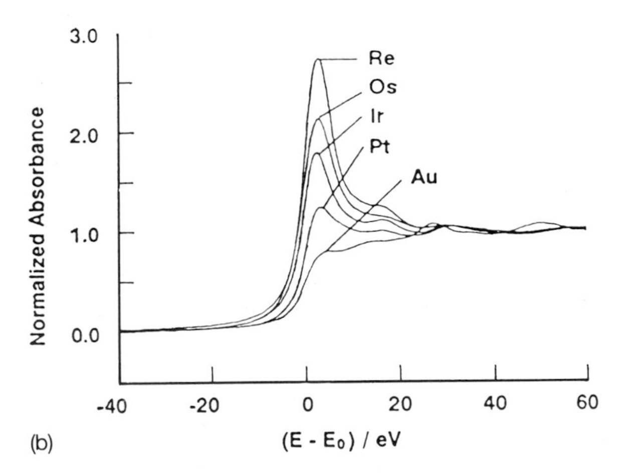

Figure 3.33 (a) Typical L-adsorption EXAFS spectrum of a thin foil of platinum. (b) X-ray spectra comparing the L_{III} absorption edges of the 5d metals (rhenium to gold). With permission from J. H. Sinfelt, G. Meitzner, *Acc. Chem. Res.* **1993**, *26*, 3.

tion coefficient beginning at an energy of *ca* 30 eV beyond the edge and extending over a range of some 1200 eV. This fine structure arises from back-scattering of the ejected photoelectron from the atoms near the absorbing atom. An indication of how a series of maxima and minima arises can be gleaned from Fig. 3.34. As the photon energy ω increases, so also does the energy of the internally emitted photoelectron, $\hbar\omega - E_b$. As the latter increases, the corresponding de Broglie wavelength decreases. Hence, for a fixed interatomic distance R_j, there will be a series of 'resonance' conditions when there will be constructive interference of the back-scattered and emitted waves, and this, in turn, is governed by the relative magnitudes of R_j and the wavelength. The smaller the value of R_j the greater will be the separation between the oscillations in the extended fine structure, and vice versa. Moreover, the greater the number of atoms located at a specific value of R_j (i.e., the greater the coordination number), the greater will be the amplitudes of the oscillations.

For photon energies less than E_b, excitation of photoelectrons is not possible – picture E_b to be E_L, the binding energy of the L-shell electrons in Pt – so that if one

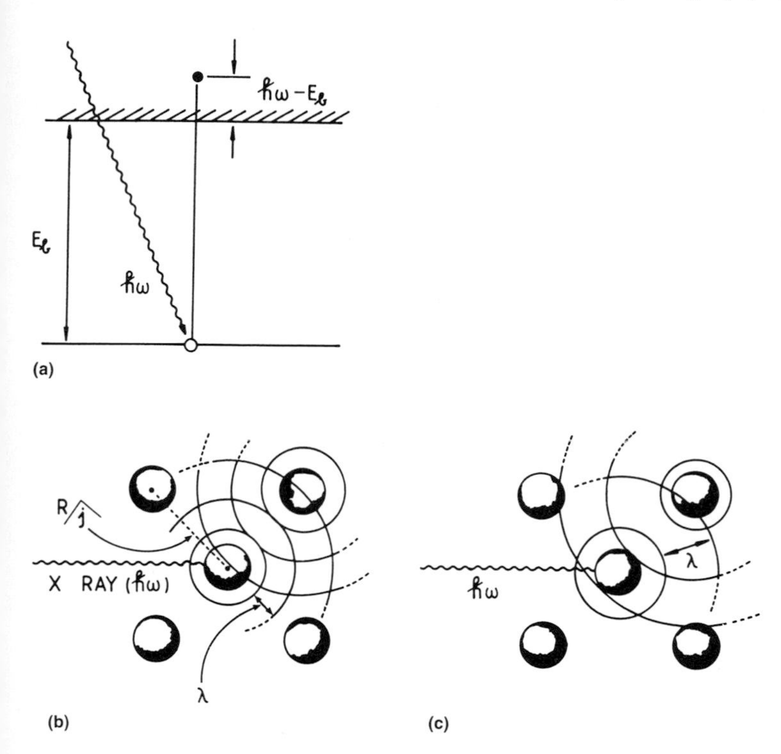

Figure 3.34 (a) A photoelectron of energy $\hbar\omega - E_b$ and wavelength λ_e is generated inside a sample by promotion from a core level. It is scattered by surrounding atoms. An oscillatory EXAFS pattern results from interference between the outgoing and scattered photoelectron wave. The wavelength varies with the incoming energy $\hbar\omega$. In (b) interference is constructive; in (c) it is destructive. With permission from P. Eisenberger, B. M. Kincaid, *Science* **1978**, *200*, 1441.

makes a transmission measurement with X-rays, and measures the absorption coefficient μ as a function of photon energy, a sharp rise is observed at the L-shell threshold energy. Since each element has its own unique set of core-electron binding energy levels (E_K, E_{L_I}, $E_{L_{II}}$, etc.), it is possible to study one kind of atom in the presence of many others simply by tuning the X-ray energy to the appropriate absorption edge.

EXAFS and related studies are now carried out (expeditiously) using synchrotron radiation sourecs. In recently designed 'storage rings' and other dedicated synchrotron facilities, photon fluxes of 10^{12}–10^{15} s^{-1} eV^{-1} (i.e. up to 1 W eV^{-1}) are available. Before long, with judicious use of Bragg–Fresnel lenses and X-ray mirrors, focused X-ray fluxes of 10^{19} s^{-1} will be available.

Sensitivities can be improved if, instead of measuring the flux of transmitted X-rays, one monitors the absorption by other means, e.g. by fluorescence yield (i.e. X-ray emission at a longer wavelength consequent upon the creation of a core-electron vacancy by the primary X-ray beam). Alternatively, the yield of secondary electrons or Auger electrons liberated from a solid is directly proportional to the extent of X-ray absorption. Methods based on secondary and Auger electron emission come into their own in SEXAFS (see Section 3.6.5.3), but are of no value for studying real catalysts under operating conditions.

Returning to Fig. 3.34, the wavelength λ_e of the electron emitted from the absorbing atom is related to the wave vector k by $\lambda_e = 2\pi/k$. In general, we have, for an electron of energy E, $\lambda = h/(2mE)^{1/2}$ where m is the mass of the electron. Clearly,

$$k = \frac{(2mE)^{1/2}}{\hbar} = \left\{\frac{2m(\hbar\omega - E_b)}{\hbar}\right\}^{1/2} \tag{25}$$

where E_b is the threshold (binding) energy and $\hbar = h/2\pi$. In the analysis of EXAFS data, it is useful to define a function $\chi(k)$ given by:

$$\chi(k) = (\mu - \mu_0)/\mu_0 \tag{26}$$

where μ and μ_0 are atomic absorption coefficients. The coefficient μ refers to absorption by an atom in the material in question, whereas μ_0 refers to absorption by an atom in the free state. Both are functions of k. Much theoretical work has gone into deriving a quantitative relationship between $\chi(k)$ and the structural features of the absorbing material. The generally accepted formula for EXAFS in a system with several distinct interatomic distances is

$$\chi(k) = \sum_j \frac{N_j}{kR^2_j} \cdot F_j(k) \cdot \exp(-2K^2\sigma_j{}^2) \sin[2kR_j + 2\delta_j(\boldsymbol{k})] \tag{27}$$

where the summation extends over j coordination shells. R_j is the distance from the central absorbing atom to atoms in the jth coordination shell; N_j is the number of atoms in the jth shell (coordination number); σ_j, the so-called Debye–Waller factor, is the root mean square deviation of the interatomic distance about R_j; $F_j(k)$ is a factor that accounts for electron back-scattering and inelastic scattering; and $\sin[2kR_j + 2\delta_j(k)]$ is the sinusoidal interference term, with $2\delta_j(k)$ being the phase-shift function.

Equation (27) obviously contains a great deal of information about the material being studied. In handling EXAFS data, well-tried methods have been evolved using this equation to retrieve the three key items of information,

interatomic distances, R_j,
coordination numbers, N_j,
Debye–Waller factors, σ_j,

from the observed EXAFS. Full details are given elsewhere. We simply note that Fourier transformation of the data yields a function $\phi_n(R)$:

$$\phi_n(R) = (1/2\pi)^{1/2} \int_{k_{min}}^{k_{max}} k^n \chi(k) \exp(-2ikR)\, dk \tag{28}$$

where R is the distance from the absorber atom, and k_{min} and k_{max} are the limits within which data are obtainable. The function $\chi(k)$ is multiplied by the factor k^n (where n is usually 3) so that there is a more or less uniform weighting of EXAFS oscillations over a range of k starting at about $k = 4\,Å^{-1}$, corresponding to photo-

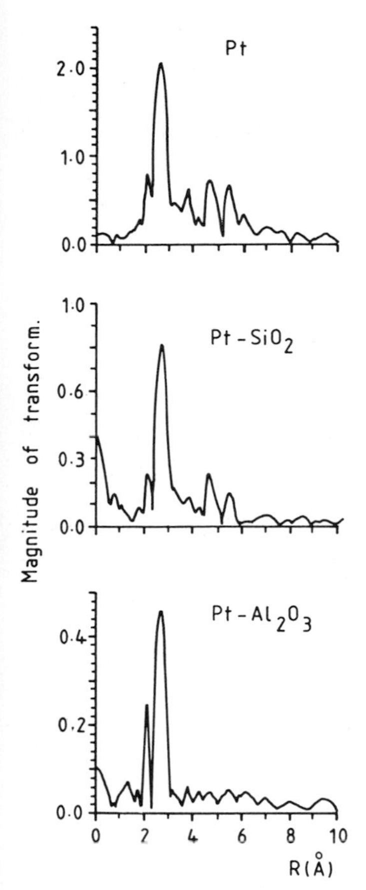

Figure 3.35 Transform plots derived from EXAFS data for platinum metal and for supported catalysts containing 1% Pt. (Based on work of J. H. Sinfelt.)

electron energy of about 60 eV above the edge. The transform function has real and imaginary parts, but only the magnitude of the transform which is everywhere positive is considered – see Fig. 3.35, taken from the work of Sinfelt and Meitzner.

The Fourier transform plot, typified by Fig. 3.35, exhibits a series of peaks at $R = R'_j = R_j - a_j$, the deviation a_j from a value of R_j corresponding to a particular coordination shell arising from the phase shift in Eq. (27). Clearly, with a material of known structure, we can evaluate the phase shifts $2\delta_j(k)$. This approach is often used in EXAFS studies, but phase shifts can also be theoretically computed.

By further extension of the kind of arguments presented above, bearing in mind that EXAFS data on model compounds of known structure can be made available, we proceed by means of iterative least-squares fitting procedures to arrive at values of R_1, N_1 and $\sigma_1{}^2$.

3.6.5.2 Applications of EXAFS to the Study of Catalysts

Many valuable advances in our knowledge of practical catalysts have been registered in recent years: most of these have emerged from a semiquantitative use of the technique.

Atomic Dispersion

Is there atomic dispersion in Rh/Al_2O_3 catalysts of low loading? IR results obtained from CO adsorbed on the supported metal have been interpreted to indicate that dispersion of the rhodium is on the atomic scale. This conclusion must, however, be invalid since Koningsberger et al. show EXAFS oscillations arising from Rh–Rh nearest neighbours. They also found that adsorption of CO at room temperature on the reduced catalyst significantly decreases the amplitude of the oscillations, implying that in very small crystallites there is disruption of Rh–Rh metallic bonds consequent upon chemisorption.

Bimetallic Cluster Catalysts

Are the conclusions concerning the nature of bimetallic cluster catalysts, drawn from the use of chemical probes, borne out by EXAFS measurements? One of the most striking facts pertaining to the behaviour of bimetallic catalysts comprising Group VII and Group IB metals is that pairs of metallic elements that are completely immiscible in the bulk (e.g. ruthenium and copper, or osmium and copper) may form bimetallic aggregates, the surface properties of which reveal extensive interaction between the two elements. Sinfelt et al., on the basis of hydrogen chemisorption studies, electron microscopy, catalyst selectivity and XPS, have shown that supported aggregates of Ru–Cu behave in such a manner as to indicate that copper tends to cover the ruthenium surface (just as if copper were chemisorbed on the Group VIII metal). Compelling evidence for this view comes from the progressive drop in extent of hydrogen chemisorption and of activity for ethane hydrogenolysis exhibited by ruthenium when increasing amounts of copper are incorporated into the bimetallic aggregates. These chemical probes yield results which are corroborated by LEED, Auger and thermal desorption spectroscopic experiments performed by Ertl et al. on model, single-crystal systems.

Until the advent of EXAFS it was very difficult to pinpoint any independent physical technique which could directly attack the problem of whether finely divided aggregates (e.g., 10–35 Å dia. particles supported on silica to the extent to *ca* 1 wt%) were bimetallic. X-ray diffraction shows no discernible peaks attributable to the constituent metals ruthenium and copper (or osmium and copper), owing to the degrees of catalyst dispersion (compare Fig. 3.4). EXAFS, however, offers strong proof for the validity of the conclusions based on the use of chemical probes. Specifically, the average composition of the first coordination shell surrounding Ru atoms is 90% Ru, while only 50% Ru surrounds a Cu atom in its first coordination shell.

Surface-Derivatized Catalysts

What can we tell about the structure of surface-derivatized catalysts? There is little doubt that, in the selective oxidation by O_2 of *o*-xylene to phthalic anhydride, neither V_2O_5 nor TiO_2 in their pure crystalline states exhibit much catalytic activity or selectivity. But when V_2O_5 is laid down so as to cover a monolayer of polycrystalline TiO_2 (anatase) there is a vast improvement in catalytic performance. This is an industrially important process which operates at *ca* 350 °C with 1 atm pressure. It has been argued that the root cause of this dramatic effect lies in the good epitaxial fit between the anatase and V_2O_5, the implication being that the resulting homogeneity of the surface confers high selectivity. EXAFS studies of titanium and vanadium are readily carried out as these elements, being adjacent in the Periodic Table, yield their respective X-ray absorption spectra in a single scan. In view of the rapid decay of EXAFS oscillations with increasing energy, the epitaxial model can be immediately discounted. There is considerable structural disorder in the active, monolayer phase. Moreover, by quantitatively fitting the EXAFS spectrum, it could be shown that the catalytically active vanadium oxide phase has two short and two longer V–O bonds of 1.65 and 1.90 Å. Detailed investigations by Wachs et al., using a range of other techniques including laser–Raman, in-situ, spectroscopic studies, further support and extend the general validity of this model. Polarized total-reflection X-ray absorption by Iwasawa et al. established that the V=O bonds of V_2O_5 on ZrO_2 are tilted to about 45° from the surface normal.

The structure of Wilkinson's catalyst for homogeneous selective hydrogenation, chlorotris(triphenylphosphine)rhodium [$(Ph_3P)_3RhCl$], is well known. But how is it linked when immobilized on a support made of polystyrene crosslinked with divinylbenzene? This answer cannot be provided by conventional X-ray crystallography owing to the scattering from the support and the high degree of non-crystallinity of the resulting catalyst preparation. EXAFS, however, as shown by Eisenberger et al., yields useful information. By first recording, for calibration purposes, the EXAFS spectrum of crystalline [$(Ph_3P)_3RhCl$], a check could be made on the reliability of the method: the EXAFS results of the 'free' catalyst agreed with the X-ray results with respect to coordination numbers and bond distances. The EXAFS spectrum for the polymer-bound catalyst showed that the ratio of phosphorus to chlorine nearest neighbours had changed from 3 : 1 to 2 : 2, and that two of the longer Rh–P bonds, initial lengths 2.35 Å, were replaced by two shorter bonds with Rh–P equal to 2.16 Å, the remaining Rh–P distances was unchanged at 2.23 Å. More significantly, the presence of two chlorine atoms in the rhodium(I) system clearly indicated that dimerization had occurred on immobilization, and that the chlorine atoms bridge two Rh atoms, which are probably in turn attached to the polymer by the short Rh–P bonds. This dimerization is implicated in the observed reduction in catalytic activity of the bound Wilkinson catalyst.

EXAFS has yielded quantitative data on the nature of Ti(IV) ions in mesoporous titano–silica epoxidation catalysts – see Thomas and Greaves (1994) and Maschmeyer *et al.* (1995).

3.6.5.3 SEXAFS

To carry out *surface* crystallographic studies with EXAFS (i.e. SEXAFS) we require a means of detection of the X-ray absorption signal which discriminates in favour of surface atoms relative to those atoms in the bulk. One way of doing this is to record the Auger electron yield consequent upon X-ray absorption – since electron escape depths are small, the Auger electron signal intensity will record the X-ray absorption of the atoms in the last two layers or so. Likewise we could use the total secondary-electron yield or even the yield of ions desorbed as a result of photon absorption. These days it is customary to record the X-ray fluorescence as a signal of absorption, this being readily detectable even under quite high gas pressures.

SEXAFS studies require a high ($> 10^{10}$ photons s^{-1}) photon flux incident upon the sample so as to give good signal-to-noise ratios in either the secondary-electron yield mode or the photon-stimulated desorption mode of detection. At present, such studies are feasible only by using monochromatized synchrotron radiation. The task of disentangling the SEXAFS spectra are much the same as those for ordinary EXAFS, as described in Section 3.6.5.1 (see Eq. (27)). Again, model compounds, in which there are known bond distances and atomic environments, are of value as standards. Often even a qualitative assessment of the SEXAFS structure affords meaningful insights. Thus, from the frequency of the oscillations (of the SEXAFS spectrum) we can conclude that the Al–O nearest-neighbour distance is longer in α-Al_2O_3 than in the surface oxide (a monolayer or so) formed when an Al(111) surface is exposed to O_2. It is to be noted that SEXAFS has almost zero prospects as an in-situ technique for operating catalysts.

3.6.5.4 XANES and Pre-edge Structure: Deducing Site Symmetry and Oxidation States

X-ray *a*bsorption *n*ear-*e*dge *s*tructure (XANES) gradually merges with the extended structure (EXAFS); and the active ingredient for the sensitivity latent in each of these spectroscopic tools is the internally photoemitted electron: the scattering of this electron modulates the absorption cross-section. XANES is distinguished in principle, however, from EXAFS because at low energies (near-edge) the absorption cross-sections are high and the modulation is correspondingly greater than in EXAFS. This means that the classical theory for EXAFS, so helpful in yielding structural parameters (via Eq. (27)) cannot be applied. A satisfactory theory of XANES needs to embrace the problem of multiple scattering, and unlike the situation that applies to EXAFS, analysis of data cannot proceed by Fourier transformation. A computational scheme has been developed by Durham and Pendry, based on a cluster method, which is flexible enough to be applicable to crystalline solids, free molecules, amorphous solids, organometallic complexes and solid surfaces. Briefly, the calculation proceeds by subdividing the cluster into shells of atoms (the scattering properties of each of which are described by a set of phase shifts), and the multiple scattering equations are solved within each shell in turn, the final step being calculation of multiple scattering between shells and the assembly of the whole cluster.

Pre-edge peaks in X-ray absorption spectroscopy are also very useful indicators of structural type and electronic state. For transition metals there are generally weak but distinct features in the spectra just before the onset of absorption (at K- or L-edges). Such features have been attributed to electronic transitions from 1s to some d-, s- (nondipole), and p-like (dipole-allowed) empty states. In their study of the compounds of vanadium, Wong et al. showed that the nature of the pre-edge peak (Fig. 3.36) is quite distinct for octahedrally, as compared with tetrahedrally, coordinated vanadium. With octahedral coordination, as in V_2O_3 or in the mineral roscoelite, there is a weak pre-edge absorption peak.

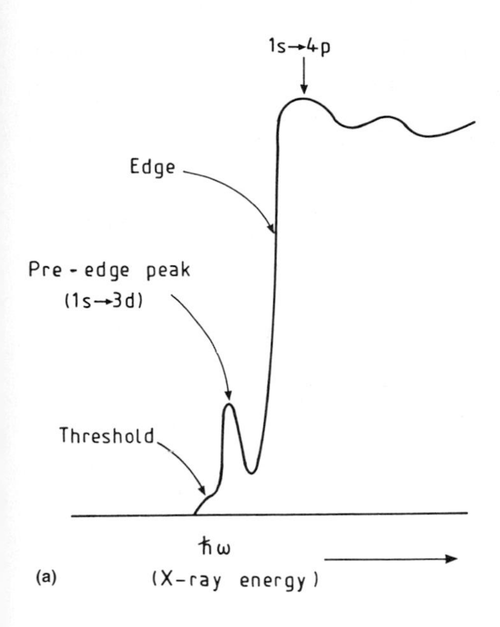

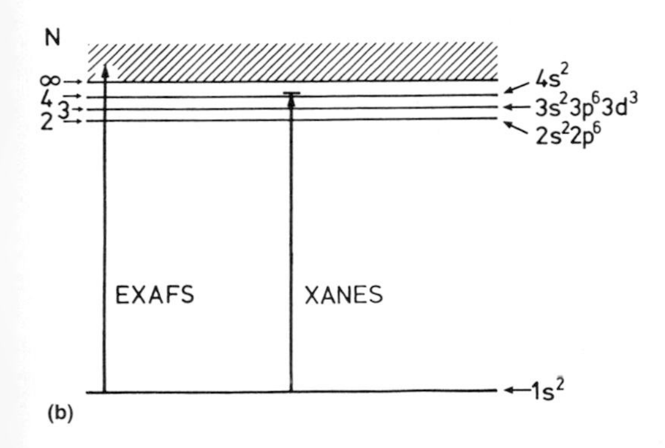

Figure 3.36 (a) Representation of the X-ray K-edge absorption spectra (XANES) obtained from solids containing vanadium. (b) Schematic energy level diagram for vanadium. The transitions are from the K-shell to empty valence states. XANES involves transitions to valence states and can thus be used to probe chemical bonding.

The transition of interest in the pre-edge region of the X-ray absorption spectrum is the V_{1s} to the empty manifold of the solid in question. For a transition metal atom in O_h symmetry, the 1s→3d transition is dipole-forbidden, but it is an allowed quadrupole transition. Although this is weak, it is still observable. For tetrahedral complexes, the lack of a centre of inversion permits a dipole transition 1s to the t_2 orbitals which can contain much metal character. Sankar and Rao et al. deduced the site symmetries of Co atoms in a range of model compounds and hydrodesulphurization catalysts in like fashion. Thus, in $CoAl_2O_4$, where Co is tetrahedrally coordinated, the 1s → 3d transition is more intense than in CoS or $CoMoO_4$ where, in each case, the Co is in octahedral coordination. Wong et al. found a strong correlation between oxidation state and the energy corresponding to a particular X-ray absorption feature (Fig 3.37).

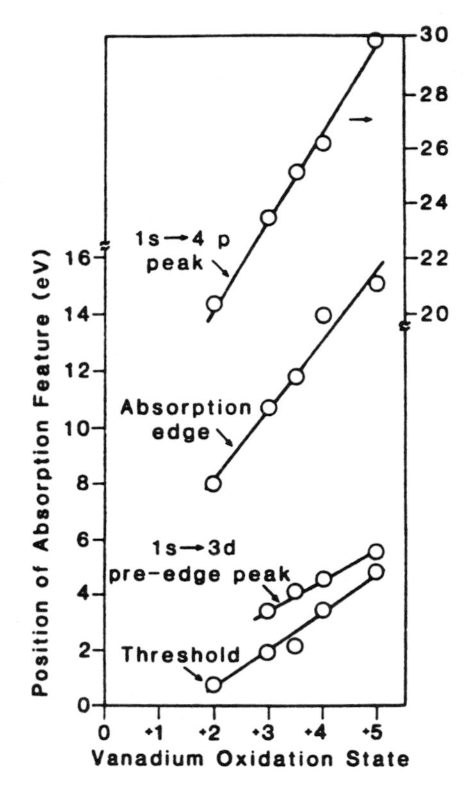

Figure 3.37 A strong correlation exists between the oxidation state of vanadium and the position of its absorption edge. (Courtesy J. Wong – see J. Wong, F. W. Lytle, R. P. Messmer, G. Maylott, *Phys. Rev. B* **1984**, *30*, 5596.)

3.6.5.5 NEXAFS

In addition to the above variants of X-ray absorption spectroscopy made possible by synchrotron radiation, near-edge X-ray absorption fine structure (NEXAFS) study is potentially as powerful as any for probing the nature of light elements ($Z < 10$) attached to solid surfaces. It has recently been put to good structural use by Stöhr, Sette, Madix and Gland in the USA. They arrived at intramolecular bond distances of simple species bound to metals.

In NEXAFS spectroscopy we focus on the energy region within 30 eV or so from the 1s excitation threshold (K-edge) of C, N and O. NEXAFS is dominated by intramolecular scattering resonances, there being little influence from scattering by atoms in the solid support. The observed resonances arise from transitions of the 1s core electron to bound states (e.g. empty or partially filled molecular orbitals) and/or to quasibound states in the continuum which have an enhanced amplitude on the molecule (so-called 'shape resonances'). The transitions are governed by dipole selection rules and analysis of their intensity as a function of the orientation of the electric vector, $\boldsymbol{E}$, enables the precise molecular orientation on the surface to be determined.

For molecules with π-bonding, NEXAFS exhibits a strong resonance which corresponds to a transition of a 1s electron into the antibonding π^* orbital. Hence, the presence or absence of the π^* resonance is a direct indication of the hybridization of the bond. In contrast to the π^* resonance, which is maximal when the $\boldsymbol{E}$ vector is

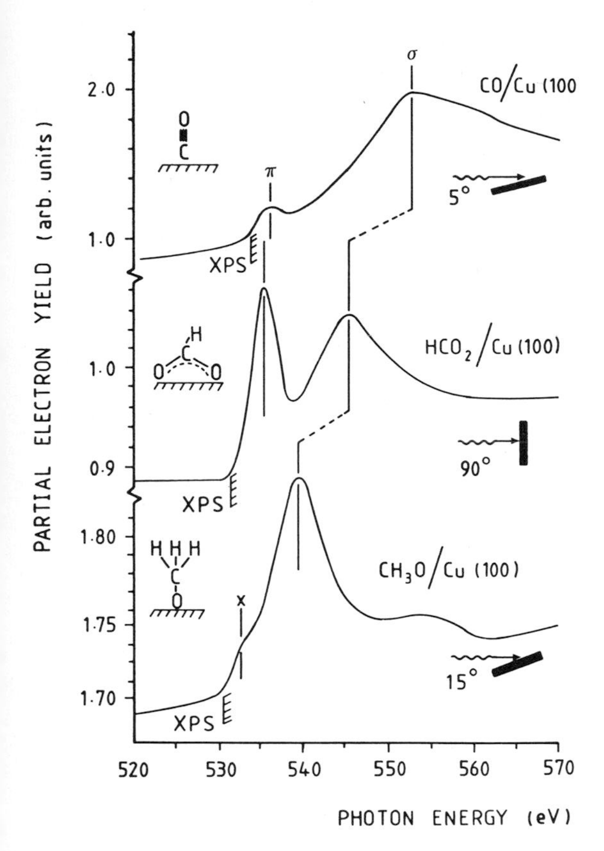

Figure 3.38 NEXAFS spectra above the oxygen K-edge for three molecules with C–O bonds attached to a Cu(100) surface. (Based on work of J. Stöhr and R. J. Madix.)

parallel to the π bond, the σ shape resonance is strongest if the $\boldsymbol{E}$ vector lies along the internuclear axis between the neighbouring atoms, and this resonance is present in all molecules. Since it arises from resonant scattering of the excited 1s photoelectron between atoms within the molecule, we see, by analogy with EXAFS, that its position in energy space bears a direct correspondence to the intramolecular bond length.

Figure 3.38 shows EXAFS spectra above the oxygen K-edge for three molecules with C–O bonds attached to a Cu(100) surface. The spectra are shown only for the X-ray incidence geometry which maximizes the intensity of the σ-shape resonance. Significantly, the position of the σ-shape resonance decreases in energy with increasing C–O bond length R. If we define Δ as the energy difference between the σ-shape resonance excitation energy (marked σ in Fig. 3.38) and the 1s binding energy, $E_B{}^{1s}$, relative to the Fermi level, we can write:

$$(\Delta - V_0)R^2 = C_0 \text{ (a constant)} \qquad (29)$$

Here V_0 is an 'inner potential'. The value of E_B^{1s} is obtainable from XPS data or directly from the absorption threshold (see label in Fig. 3.38). And by taking the known change of 0.3 Å (ΔR) in the C–O bond length between CH_3O and CO we can, for the two extreme cases CO/Cu(100) and CH_3O/Cu(100), determine V_0 and

C_0 in Eq. (29), knowing Δ. (This tacitly assumes that the respective C–O bond lengths of CO and CH_3O in gaseous and adsorbed states are the same, as indeed they appear to be.) The C–O bond lengths $R = (1.13 \pm 0.02)$ Å for CO/Cu(100) and $R = (1.43 \pm 0.02)$ Å for CH_3O/Cu(100), together with the measured values $\Delta = (18.8 \pm 0.5)$ eV and $\Delta = (8.3 \pm 0.3)$ eV lead to the values $V_0 = (-9.2 \pm 1.6)$ eV and $C_0 = (35.7 \pm 2.3)$ eV Å^2. We can now calculate that the value $\Delta = 13.8$ eV for HCO_2/Cu(100) corresponds to $R = (1.25 \pm 0.02)$ Å for the C–O bond length in formate ions coordinated to a variety of metals.

It has been found empirically that V_0 is surprisingly constant (*ca* 9 eV) for different, low-*Z* molecules with bonds between C, N and O atoms on various d-band metals. This is a great convenience, and already useful tracking of bond-length changes upon adsorption, e.g. the increase of 0.24 Å in the O–O bond when O_2 is bound to Pt(111), have been made. Again, this technique is invaluable for in-situ studies of the adsorbed phase on model catalysts, as may be gauged from the results of Sette et al., who probed the sodium-induced changes in bonding and bond lengths of CO on platinum by NEXAFS. With just 0.2 monolayers of sodium co-adsorbed, the C–O bond length increases by 0.12 Å. There is also a reduction and broadening of the 1s $\rightarrow 2\pi^*$ resonance. All this adds further support to the Blyholder of weakening of the C–O bond in the presence of electron-donating alkali-metal adsorbates by an increased back-donation of electrons into the $2\pi^*$ orbital (see Fig. 1.24).

We see therefore that the combined use of XANES (called NEXAFS if we deal with the K-edges of C, N and O) and pre-edge structure in X-ray absorption spectroscopy is an invaluable guide to site symmetry and local atomic environment. Complemented by the quantitative data that emerge from EXAFS studies, we see that X-ray absorption spectroscopy alone or in combination with X-ray diffraction (see Section 3.10) is of great importance in characterizing solid catalysts.

3.7 Other Structural Techniques for Characterizing Bulk and Surfaces of Catalysts

There are many other techniques that can be deployed for catalyst characterization. Some are related by the fact that they are based on resonance methods (ESR, NMR and Mössbauer spectroscopy), others are increasing in their popularity in view of more accessible instruments and sources (neutron scattering and electron microscopy). Yet others have a traditional flavour (magnetic susceptibility measurements) and some are sophisticated variants of older methods (temperature-programmed desorption).

3.7.1 Electron Spin Resonance (ESR): Probing the Nature of Catalytically Active Sites and the Concentration of Paramagnetic Intermediates on Surfaces and in the Gas Phase

ESR is applicable only in a relatively limited number of catalytically important situations, because it does not detect diamagnetic entities. Nevertheless, it is a valuable technique, both because of its high sensitivity – fractions of a monolayer can be quantitatively assayed – and its general capability to identify unambiguously trace amounts of paramagnetic species such as O^- or O_2^- in the presence of O^{2-}, and molybdenum(V) in the presence of molybdenum(VI). ESR can reveal the oxidation state, electronic configuration and coordination number of a paramagnetic ion. Under favourable circumstances the ground-state d-orbital configuration of the ion and any structural distortions (arising from the Jahn–Teller effect, for example) may be deduced; and the extent of covalency in the bonds attaching ligands to a central paramagnetic ion may be estimated. Recent developments, involving 'spin–echo' procedures, are particularly powerful in that they yield quantitative information relating to the environment of transition-metal species, such as copper(II) ions bound to organic reactants and located, for example, within intrazeolite cavities. In-situ experiments involving ESR were until recently uncommon, partly because low temperatures (77–300 K) are generally necessary for obtaining good spectroscopic data, but flow cells are now in common use. It is not difficult to study actual catalysts after transfer from a reactor section to an appropriate side-arm for ESR analysis. Che et al., as well as Lunsford et al., have demonstrated the advantages of attaching an ESR flow reactor to a matrix isolation system. With this technique, surface-generated gas-phase free radicals, such as methyl, peroxymethyl and π-allyl radicals formed at metal oxide surfaces, can be identified and their role in heterogeneous catalysis ascertained. It transpires that such radicals undergo simple coupling reactions (as in the dehydrodimerization of propylene to 1,5-hexadiene) and are often involved in the transport of intermediates from one region of a catalyst particle or catalytic reactor to another. Even though ESR is not exclusively a surface technique, one may readily distinguish between surface and bulk species by exposing the solid to O_2, which is itself paramagnetic and which will consequently, through magnetic dipole interactions, broaden the spectrum of those entities at the surface that contribute to the total signal.

Standard textbooks on physical chemistry give adequate detail pertaining to the theory and practice of ESR.

3.7.1.1 Examples of the Use of ESR in Heterogeneous Catalysis

Oxidation of Hydrocarbons

The oxidation of hydrocarbons in the presence of oxide catalysts has been much studied by ESR; Haber et al. have indicated that O^- ions favour non-selective conversions, whereas O^{2-} ions favour selective ones. Moreover, ESR measurements have led to the proposition that ions are formed sequentially thus:

$$O_2(g) \rightarrow O_2(ad) \rightarrow O_2^-(ad) \rightarrow 2\,O^-(ad) \rightarrow 2\,O^{2-}(bulk)$$

When simple alkanes interact with surface O^- ions, hydrogen atom abstraction processes tend to dominate, and the resulting alkyl radicals react rapidly with the metal oxide to form alkoxide ions. It is known that stoichiometric reactions between C_2H_6 and O^- ions on MgO give rise first to ethoxide ions, which subsequently decompose to C_2H_4 at elevated temperatures. Lunsford has shown that, with Mo/SiO_2

catalysts, it is possible to have as dominant surface species either O^- by using N_2O as the oxidant, or O_2^- by using O_2. In this way it is possible, in principle, to distinguish between the relative importance of O^-, O_2^-, or O^{2-}(bulk) ions.

With bismuth-rich molybdenum oxide catalysts, it was shown by Grasselli et al. (by a combination of ESR and an analysis of kinetic isotope effects) that the Bi–O bonds are capable of effecting the first hydrogen abstraction from propylene, a step which is of crucial importance in the catalytic production of acrolein from propylene.

Catalytic Polymerization of Ethylene

The nature of the active sites for the catalytic polymerization of ethylene using SiO_2-supported chromium has elicited much controversy: every oxidation state from chromium(II) to chromium(V) has been implicated as the active center. ESR shows Cr(III) ions to be effective catalytic centres when reaction proceeds under rather modest (laboratory) conditions at room temperatures; but the chromium(II) state seems the most dominantly active centre under typical industrial conditions *ca* 500 °C.

3.7.2 Electron Spin–Echo Modulation Spectroscopy (ESEMS): Probing the Environment of Paramagnetic Species

When a paramagnetic system is subjected to suitable sequences of microwave pulses, microwave echoes are generated owing to the re-formation of magnetization within the sample. Since the nearby nuclei are simultaneously excited, these echoes are modulated by the characteristic Larmor frequencies of the interacting nuclei. These modulations can be analysed to yield information about the distance between the paramagnetic ion and the interacting nuclei, the number of such nuclei and the isotropic hyperfine coupling.

In essence, ESEM spectroscopy utilizes the unpaired electron as a structural probe in such a way as to detect weak hyperfine interactions between the unpaired electron and surrounding nuclei in the first (and second) ‘solvation’ shells. Its particular merits as a structural tool are that it copes freely with powdered microcrystalline specimens, it is sensitive to paramagnetic species at quite low concentrations (*ca* 10^{-4} M), and it is well suited to providing quantitative values for the distances of surrounding protons and deuterons from a paramagnetic species.

Kevan and his co-workers, who have employed ESEMS for the examination of surfaces, especially the internal surfaces of zeolites, have estimated the reliability of the technique. They conclude that n, the number of equivalent nuclei at an average distance r, can be uniquely determined to the nearest integer up to about $n = 10$. The distance r can be uniquely determined to ± 0.01 nm.

Kevan’s combined use of ESR and ESEMS for the study of Cu^{2+} ions partially exchanged into Ca–X zeolite revealed that a trigonal bipyramidally coordinated species $Cu(O_3)_3(OH)_2$ (where O_3 refers to zeolite framework oxygens) was the dominant copper center in this zeolite, it being situated in a six-ring (hexagonal) window and coordinated to two hydroxyls as well as to the oxygens of the ring. ESEMS can be applied to many other catalytically significant ions, such as $(VO)^{2+}$ and Rh^{2+}, and to spin probes, such as hydroxymethyl radicals, which are also useful in characterizing the environment of a catalytically active site. The disadvantage of the technique for in-situ studies is that it is best used at low temperatures.

3.7.3 Nuclear Magnetic Resonance (NMR): A Technique Applicable, at High Resolution, to Solids and their Surfaces

Of all the spectroscopic methods used by chemists for structural elucidation, few have greater power than NMR. Besides NMR, no other single technique can so readily distinguish sp^2- from sp^3- hybridized carbon atoms in organic or organometallic compounds: no other can unambiguously identify ethers from acids, and alcohols from alkenes. Such discrimination works best when the species under investigation are in the dispersed state, usually in solution. The reason for this is that line-broadening effects caused by the magnetic influences of neighbouring atoms are absent when the species in question undergo rapid and random thermal motion. By the same token, species that are 'clamped' within a solid, or anchored firmly to a surface, yield NMR spectra that are so broadened as to be almost valueless so far as discrimination, based on chemical shifts, is concerned. However, various ways have been evolved to derive high-resolution NMR spectra even from solids or adsorbed and reactant entities at their surfaces and within their bulk. The future role of NMR for the study of catalysts, adsorbents and adsorbates is therefore likely to be considerable.

3.7.3.1 Basic Principles

The basic principles of NMR are broadly the same as for ESR, the essential difference being that now we monitor the reversal of nuclear magnetic moments rather than electron magnetic moments. For liquids, NMR principles are well known.

To appreciate the power of NMR, we need to recognize three important properties that can be extracted from the spectra: chemical shifts, fine structures, and relaxation times. Because of the particular electronic environment in which a certain nucleus in a molecule or continuous solid is immersed, the magnetic field B_{loc} experienced by that nucleus is different from the applied one: $B_{loc} = (1 - \sigma)B$, where, by definition, σ is called the shielding constant, and the extra field σB the chemical shift. Since protons in different chemical groups have different shielding constants, the resonance condition

$$h\nu = g_I \mu_N (1 - \sigma)B \tag{30}$$

is satisfied at different values of B for protons in different chemical environments. The well-known ^1H NMR spectrum of liquid ethanol shows three distinct peaks (see Fig. 3.39) in the intensity ratio 3:2:1, as expected from the formula, CH_3CH_2OH. The three distinct sets of protons stand out clearly from one another because their respective chemical shifts are very different.

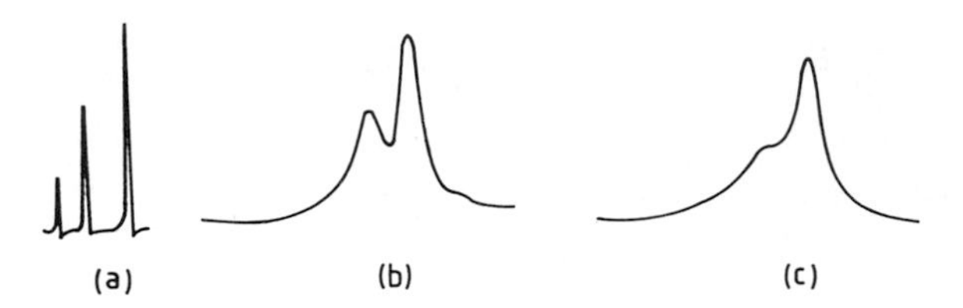

Figure 3.39 High-resolution ^1H NMR spectra of ethanol adsorbed on alumina at various degrees of coverage: (a) pure liquid; (b) 20 monolayers; and (c) three monolayers. (Based on work of E. G. Derouane.)

The splitting of NMR peaks into several lines, thereby generating fine structure, arises for much the same reasons as hyperfine structure does in ESR: because of magnetic interactions between the nuclei in the molecule. At high resolution, the ^1H methyl peak in the spectrum of ethanol consists of three lines with intensity ration 1 : 2 : 1. This arises because of the CH_2 next to the CH_3 group. The methyl line is first split into two by virtue of the two orientations of one of the methylene protons, and then each line is again split as a result of the interaction with the second proton. This phenomenon is termed *J*-splitting or spin–spin splitting. Because it is not directly evident in the NMR spectra of solids we shall not elaborate further on its features here, except to note that the magnitude of spin–spin splittings, because it reflects the coupling between two nuclei, is independent of the applied field and is an invaluable source of structural information. Unlike the situation that obtains for chemical shifts ($=\sigma B$), variation of the operating frequency of the spectrometer does not affect the magnitude of the spectral splittings due to coupling.

To appreciate what is meant by relaxation times, it is first useful to consider (Fig. 3.40) the way in which NMR spectra are recorded. After the radio-frequency pulse is applied to the sample, the spins in the sample will strive to re-establish the original equilibrium appropriate for the operating temperature and applied magnetic field. The decay of the longitudinal component (along z) may differ from the decay of the transverse components (in the xy plane). The time T_1, known as the spin–lattice (or longitudinal) relaxation time, is the time constant in the equation that describes the return of the magnetization, M_z, along the static field B_0 after a perturbation:

$$\frac{d\boldsymbol{M}_z}{dt} = -(\boldsymbol{M}_z - \boldsymbol{M}_0)/T_1 \tag{31}$$

T_2, known as the spin–spin (or transverse) relaxation time, is the time constant given by

$$\frac{d\boldsymbol{M}_x}{dt} = -\boldsymbol{M}_x/T_2 \tag{32a}$$

$$\frac{d\boldsymbol{M}_y}{dt} = -\boldsymbol{M}_y/T_2 \tag{32b}$$

In spin–lattice relaxation, the magnetic energy is transferred to the surrounding medium, appearing largely as translational and rotational energy. Spin–spin relaxation,

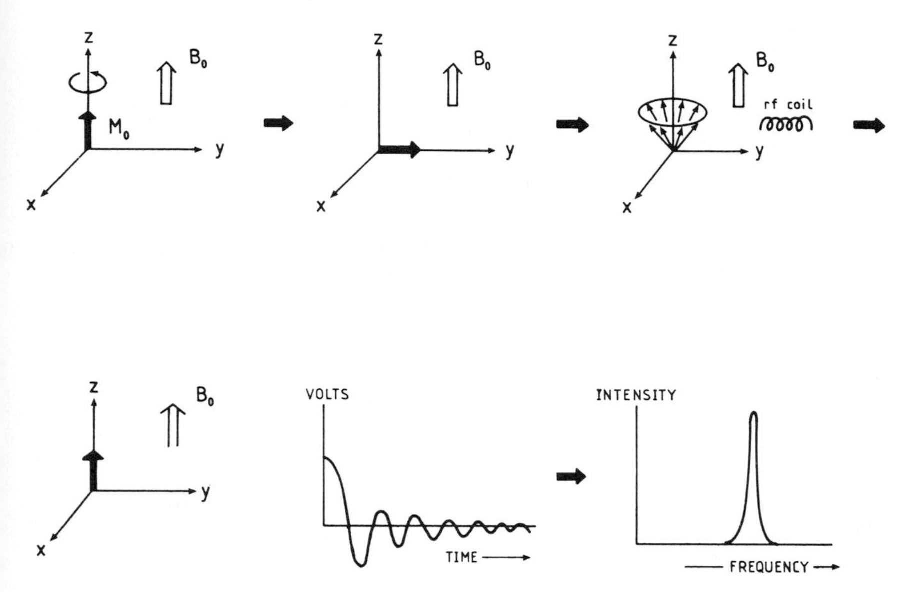

Figure 3.40 Diagrammatic outline of how NMR spectra are recorded. The original equilibrium between spins is disturbed by a burst of radio-frequency energy called a 90° pulse. The spins dephase according to their chemical shifts; and the spin system strives to re-establish its equilibrium via spin–spin and spin–lattice processes (see text). Equilibrium magnetization is re-established. The recorded free-induction decay is Fourier-transformed to yield the conventional spectrum.

on the other hand, involves mutual exchange of spins with neighboring nuclei. Large values of T_1 lead to broadened lines in the NMR spectrum, and this is one of the reasons why solids containing spin $\frac{1}{2}$ nuclei, which generally have large values of T_1, yield broad peaks. Nuclei with spin $> \frac{1}{2}$ (i.e., those that are quadrupolar) relax more efficiently so that solids containing such nuclei tend to yield narrower lines. In discussing relaxation, use is often made of correlation times, which are one-third of the relaxation times. It is also helpful to realize that, in general, relaxation time is the average time required for a population to change from its original condition to $1/e$ of the equilibrium condition.

Returning to Fig. 3.40, which summarizes the way in which NMR spectra are recorded, we note that the Fourier transform spectroscopic approach constitutes an important advance. In essence it entails the observation of the time-evolution of the nuclear spin states of the sample from a specially prepared initial condition followed by the transformation into the frequency domain of this time-resolved behaviour, from which we deduce information about the energy levels (chemical shifts, etc.) that are present. To be specific, the initial condition is prepared by exposing the sample to a short pulse (of controlled duration) of high-power radio-frequency radiation. This pulse distorts the equilibrium population of the nuclear energy levels; afterwards the system reverts towards the equilibrium population by emitting its excess energy. This emission takes the form of radiation at all the allowed transition frequencies, thereby yielding an oscillating decaying signal known as the free induction decay (FID), the oscillation resulting from the beats between all the contributing frequencies. The FID contains all the NMR spectral information associated with the sample; this information is retrieved by Fourier transformation using a computer. To enhance signals, the spectrometer can be programmed to accumulate the requisite number of FIDs prior to Fourier transformation. In this way, even spectra of nuclei of low abundance or low relative sensitivity can be obtained, which is one of the reasons why multinuclear NMR (involving ^{15}N, ^{17}O, ^{19}F, ^{31}P, ^{51}V, ^{103}Rh, ^{129}Xe and ^{195}Pt) can now be routinely obtained. Isotopic enrichment is also used to improve spectral intensity even further.

3.7.3.2 NMR Spectra of Solids

Whereas the spectral lines in the NMR of liquids are very narrow, those of solid samples are very broad. Thus, the ^{1}H NMR spectrum of water at room temperature is *ca* 0.1 Hz, whereas that of ice at low temperatures is 10^5 Hz. This enormous difference arises from the static anisotropic interactions to which the nuclei are subjected in the solid state. The spectral lines for liquids are very narrow because molecular tumbling averages the anisotropic interactions and effectively removes them from the spectrum. The predominant factors responsible for the anisotropic interactions are direct dipole–dipole internuclear interactions, quadrupole–field gradient interactions and chemical shift anisotropy.

The broadening of spectral lines in a solid arising from dipole–dipole interactions can be attributed to the fact that the z component of the local field of a nucleus 2 of magnetic moment μ at the site of a similar nucleus 1, a distance $r_{1,2}$ away, is given by the equation:

$$B_{\mathrm{loc}} \simeq \pm \frac{\mu}{r_{1,2}^3}(3\cos^2\theta_{1,2} - 1) \tag{33}$$

where $\theta_{1,2}$ is the angle between the line joining the two nuclei and the direction of the applied field. Indeed, the broadening of NMR lines for solids can be put to use in structural studies, since the linewidth is governed by the distance separating resonating nuclei within the solid. By measuring the mean-square width (known also as the second moment) of the absorption peak, internuclear distances can be obtained with good accuracy, since the second moment depends on the inverse sixth power of the internuclear distance. It can be shown that one of the major terms responsible for quadrupolar interaction also contains a factor $(3\cos^2\theta_{1,2} - 1)$, as in Eq. (33). Moreover, the broadening influence of the chemical shift anisotropy – which takes into consideration the fact that shielding constants depend on the orientation of the nuclear environment in the applied magnetic field – also contains a proportionality term $(3\cos^2\theta_{1,2} - 1)$.

We are now in a position to understand how broadened NMR lines in solids can be sharpened. To achieve this, the sample is spun rapidly about an axis at an angle θ_m to the magnetic field such that $\theta_m = \cos^{-1}(1/\sqrt{3})$; θ_m is 54°54′, the so-called 'magic angle'. Magic-angle-spinning NMR (MASNMR) is now a standard method of obtaining high-resolution spectra of solids.

MASNMR is often employed alongside two other additional means of improving the quality of solid-state NMR signals: decoupling and cross-polarization. The process of decoupling is carried out by a double-resonance technique, i.e. two radiofrequencies are needed. One is used to observe signals due to the nuclei under investigation; the other irradiates strongly the resonance of the nucleus to be decoupled. This procedure comes into its own in eliminating broadening influences arising from heteronuclear interactions (e.g., ^{13}C and ^{1}H in hydrocarbons, ^{29}Si and ^{1}H in silica gels and silicates). Cross-polarization (CP) is designed to overcome the problems associated with low sensitivity of rare spins and their long relaxation times (e.g. in improving ^{13}C signals). CP is based on the fact that the relatively strong spin polarization of abundant spins (e.g. ^{1}H) can be transferred to enhance the weak polarization of rare spins (e.g. ^{13}C). It results in higher signal intensity for ^{13}C, and the effective relaxation time of these rare spins is greatly reduced as a consequence of opening up for them a new relaxation path. This means that more FIDs can be accumulated in a given time, thereby accelerating the rate at which spectra can be recorded.

3.7.3.3 Applications of NMR to the Study of Catalysts, Adsorbents and Adsorbates

We subdivide our discussion into two parts. The first deals with the more traditional use of NMR, whereby information is obtained via chemical shifts, linewidths, and related features. The second illustrates the advantages of MASNMR and related refinements such as CP.

Probing Catalyst Structure via NMR Chemical Shifts and Linewidths

Knowing the chemical shift of a species in the pure (liquid) state, we may deduce the way in which a particular reactant is bound to a catalyst. When alkanols are adsorbed on to γ-alumina, the resulting ^{1}H NMR spectrum tells us that it is principally the OH group that is involved in the linkage to the surface because of the preferential broadening of its ^{1}H line.

The width of an NMR line is further reduced, the more mobile the species containing the nucleus in question. If the linewidth of the ^{1}H resonances in bound ethanol is recorded as it is progressively taken up by the high-area microcrystalline γ-Al_2O_3, a decrease in the linewidth is seen first, signifying greater mobility in the adsorbed state as surface coverage increases. But the linewidth remains constant as adsorption builds up in the micropores. With further uptake the linewidth increases, signifying that the alcohol is compressed in the filled micropores. Subsequent uptake leads, as expected, to a diminution in the linewidth as the adsorbate enters the macropores.

It has been shown from NMR linewidth studies that the temperature required to stop motion in the adsorbed phase is much lower than the freezing point of the corresponding liquid. This lowering can be as much as 100 K. From an analysis of line shapes of aromatic molecules in ZSM-5, it is found, not surprisingly in view of the relative shapes of the molecules and the channel dimensions, that *p*-xylene is capable of translational diffusion and rotation whereas *o*-xylene, in view of the 'rigid lattice' NMR anisotropy pattern that it exhibits, is capable only of very slow diffusion.

The identity and microstructure of zeolitic catalysts can be neatly probed by ^{129}Xe NMR. Fraissard has demonstrated that the ^{129}Xe chemical shift is the sum of several terms corresponding to the different interactions involving the adsorbed atom:

$$\delta = \delta_0 + \delta_S + \delta_E + \delta_C \tag{34}$$

where δ_0 is the reference chemical shift, δ_S corresponds to collisions between Xe and the walls of the zeolite cage, δ_E arises from the intrazeolite electric field and δ_C corresponds to collisions between the Xe atoms themselves. All the terms on the right-hand side of Eq. (34) can be evaluated separately. It has been shown that for a mixture of various microporous catalysts (e.g. two different zeolites) the ^{129}Xe spectrum will have as many components as there are different structures, and that the intensities of the peaks will be proportional to the number of cages of each type.

The work of Gates et al. is a good example of how conventional ^{1}H NMR has elucidated details of the performance of immobilized (anchored) homogeneous catalysts. A silica-supported organorhodium catalyst, such as Si–O–Rh(allyl)$_2$, has just one ^{1}H NMR signal at $\delta = 1.2$ ppm from tetramethylsilane (TMS). This arises from the protons of the allyl ligands bound to Rh. The width of the resonance suggests that all the allyl protons experience essentially the same average environments, and corroborates the view that all the Rh nuclei in the surface allyl complexes are present in discrete molecular entities.

Probing Catalyst Structure by Magic-Angle-Spinning And Other Specialized NMR Methods

Major progress had been achieved in characterizing catalysts by the sophisticated solid-state NMR techniques that have blossomed since 1980. These techniques have shed light on, or confirmed the existence of, new active sites, reaction intermediates and numerous other physicochemical properties of catalysts.

A good illustration of the way in which ^{1}H MASNMR can be used as a non-invasive method of determining the acidity of zeolite and amorphous silica–alumina catalysts is contained in the work of Pfeifer and Freude, of Klinowski and Lunsford and of Haw and Nicholas. These workers and others examined NH_4^+-exchanged Y-zeolite after various kinds of heat treatment, typical of those used to activate such catalysts. In the high-resolution ^{1}H MASNMR spectrum, recorded at 270 MHz, the resonance at the highest magnetic field (1.8 ppm with respect to TMS) arises from non-acidic OH groups attached to Si in the framework. These are the so-called terminal hydroxyl (silanol) groups, comparable with those found in silica gels. The MASNMR line at medium field (4.2 ppm) arises from the acidic structural OH, of the type studied earlier by broad-line NMR; and the signal at 7.2 ppm is in the main attributable to residual NH_4^+ ions. An important extra feature of this approach to acidity measurements is that it pinpoints the nature of the acid centre and it is strictly quantitative (contrast IR measurements, in which intensities of spectral lines are susceptible to variation with change of environment). ^{1}H MASNMR studies of the variation of acidity (as reflected in the intensity of the acidic OH peak) as a function of bulk composition of silica–alumina gels proved revealing. The intensity shows a maximum in the range 20–30 wt % of Al_2O_3, as does the rate coefficient for the catalytic cracking of cumene, thus quantitatively demonstrating the importance of Brønsted acidity in this kind of catalysis.

Veeman *et al.* used ^{1}H MASNMR to study to acidity of zeolite H-ZSM-5 and its borosilicate analogue known as H-boralite (where B replaces the Al in the framework). They were able to distinguish terminal and water hydroxyl groups from the acidic groups (**1**), and H-ZSM-5 was found to be more acidic than boralite.

H
O
T Si

T = Al or B

1

The MASNMR spectra of adsorbed trimethylphosphine (TMP) has been used by Lunsford to determine the concentration of Brønsted and Lewis acid sites on pure and chlorinated γ-Al_2O_3 samples. Chlorination with $CHCl_3$, CCl_4 or $AlCl_3$ promoted the formation of Brønsted acid centres, which are characterized by the protonated adduct of TMP. This adduct has a ^{31}P chemical shift of *ca* −3.8 ppm and a J_{P-H} scalar coupling of 517 Hz. Functional relationships were observed in these samples between the loss of Brønsted acid sites and the decrease in yields of both cracking and isomerization products of *n*-hexane at 150 °C.

One further example of the use of MASNMR illustrates a technique which Fraissard established with conventional ^{129}Xe NMR: the probing of different environments within the pores of a catalyst. Van Santen and Derouane, using ^{13}C CP MASNMR, have identified the siting of tetrapropylammonium (and other such ions used as 'templates' in zeolite synthesis) to be at channel intersections in ZSM-5. Extending this strategy, Melchior, Vaughan and Jacobson have demonstrated that the ^{13}C signal of the ttramethyl cation when trapped inside a zeolite is sensitive to whether it is enclathrated in α- or β-cages.

Solid-state NMR techniques have greatly extended our ability to characterize zeolites and clay catalysts. The formation of these solids in their embryonic states of crystallization and the details of the short-range Si,Al ordering and other relevant properties can be monitored. Other ways in which ^{29}Si and ^{27}Al MASNMR have elucidated the structure of zeolite catalysts include:

(a) distinguishing all five possible Si(*n*Al) building units Si(*n*Al) represents an SiO_4 tetrahedron linked to $nAlO_4$ tetrahedra and to $(4 - n)$ other SiO_4 tetrahedra – see Fig. 3.41;
(b) providing a non-destructive means of determining framework Si/Al ratios from the intensities of the Si(*n*Al) peaks;
(c) determining the number of crystallographically non-equivalent tetrahedral (Si) sites in the unit cell; and
(d) distinguishing unambiguously and quantitatively between tetrahedrally and octahedrally coordinated Al, even in noncrystalline material – see Fig. 3.41.

The recent development by Lippmaa and Pines of double-rotation solid-state NMR, which, in a sense, uses two magic angles – a rotation within a rotation at two differ-

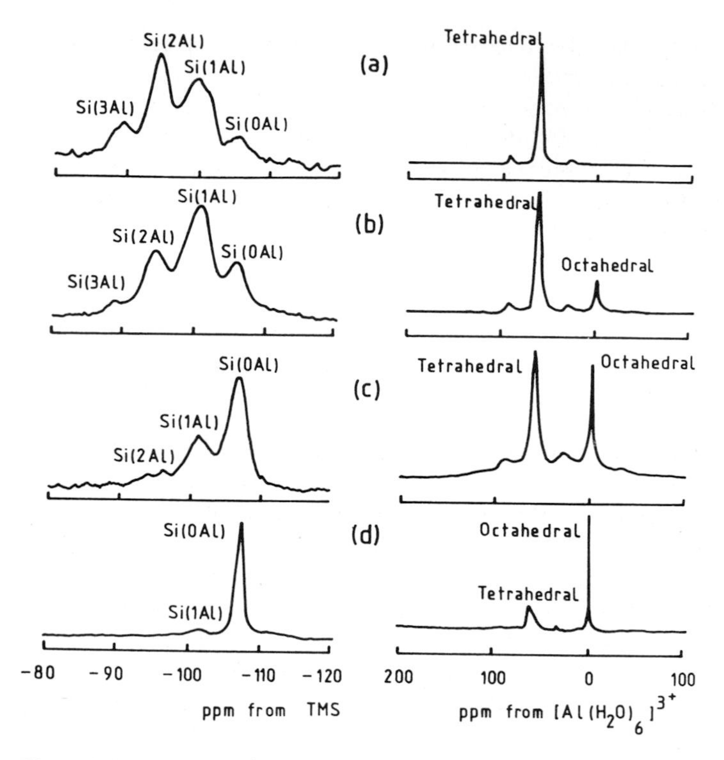

Figure 3.41 When NH_4^+-exchanged zeolite-Y is 'dealuminated', by hydrothermal treatment so as to produce a catalytically active and stable acid catalyst, the composition and Si/Al ordering of the framework is modified, as shown by the ^{29}Si MASNMR spectra (on the left). At the same time, the aluminium content of the framework, which has tetrahedral bonding, diminishes and the jettisoned aluminium, accommodated in the intrazeolite cavities, increases, as seen from the ^{27}Al spectra (right). (After J. Klinowski, J. M. Thomas, C. A. Fyfe, G. C. Gobbi, *Nature (London)* **1982**, *296*, 533.

ent angles with respect to the external magnetic field – narrows the broad lines of quadrupolar nuclei (spin $> \frac{1}{2}$). It facilitates the structural elucidation of noncrystalline oxides, especially by use of ^{17}O-enriched samples.

3.7.3.4 Future Prospects for the Study of Catalysts by Solid-State NMR

We can expect future in-situ utilization of MASNMR procedures, especially for the elucidation of catalytic processes (such as enzymatic reactions) that proceed readily at room temperature, to be widespread. This particular development, already underway, is discussed in Section 3.10. Fundamentally new NMR strategies are continually being developed. Two-dimensional NMR, already commonly used for solutions and liquids, promises to be of great importance in the characterization of solids via the use of both resolved and correlated spectra. NMR imaging, and possibly functional NMR, important techniques in medical diagnostics, could be put to a very good use in the characterization of commercial catalysts before and after use. NMR imaging utilizes a linear gradient of the magnetic field in addition to the homogeneours field B_0. Consequently, identical nuclei in different parts of the sample experience different external magnetic fields, and therefore resonate at different frequencies. This is how the spatial discrimination (which, at present, is not much better than a few microns) is achieved. But already one can see how the distribution of magnetically dilute nuclei (e.g. ^{13}C in the carbonaceous deposit) in a poisoned catalyst could be imaged by tomographic methods.

Zero-field NMR, pioneered by Pines, is a technique especially suitable for nuclei with low quadrupolar frequencies such as ^{2}H, ^{14}N and ^{27}Al, all of which figure often in catalyst preparation. It is also a refined method of measuring internuclear distances in powdered solids.

Great scope also exists for the deployment of double-resonance methods (apart from cross-polarization); and one important approach utilizes *s*pin–*e*cho *do*uble *r*esonance (SEDOR). This has the important advantage of being highly surface-selective: it can, for example, detect only those ^{195}Pt nuclei on the surface of platinum metal that are in close proximity to the ^{13}C nuclei in a chemisorbed layer of CO. It is instructive to outline the principle of the technique. Provided the two nuclei (of the isotopes under investigation) are near enough to be coupled, the magnetic resonance of one can affect the resonance of the other, and the strength of this coupling depends strongly on the internuclear distance. If, in general, a $\pi/2$ pulse is followed a time τ later by a π pulse, then a spin echo forms at time 2τ. During the first time interval τ the spins dephase, but during the second they rephase. However, when spin I is near to spin S, the latter gives rise to a local field which may increase or oppose the applied field. This has no effect on the resonance of spin I, as dephasing during the first interval τ is exactly matched by the rephasing during the second interval τ. If, however, spin S is also flipped with a π pulse when spin I is given its second π pulse, spin I dephases during the second time interval also, producing a smaller echo at 2τ.

Recent work by Slichter and Sinfelt using principally the SEDOR technique succeeded in retrieving quantitatively valuable information pertaining to acetylene chemisorbed on high-area, supported, platinum particles typical of those used in catalytic hydrocarbon reforming. They found that the surface of the platinum particles (10–30 Å average diameter) was covered by bound acetylene to the extent of 0.11–0.50 monolayer at 77 K (after outgassing the catalyst at 300 °C). Of this bound phase, 77% was present in a grouping CCH_2 and 23% was present as HCCH. The C–C bond length of the CCH_2 species was determined to be 1.44 Å, midway between the lengths of a single and double carbon–carbon bond (1.54 and 1.34 Å, respectively). The CCH_2 group is believed to be akin to that which exists in triosmium hydridocarbonyl, $H_2Os_3(CO)_9CCH_2$.

This work constitutes a significant achievement, if only because it illustrates how the nature of hydrocarbon groupings can be deduced de novo by NMR techniques for states of chemisorption comparable with those involved in practical heterogeneous catalysis. Neither LEED nor X-ray diffraction, nor probably EXAFS or SEXAFS, is capable of yielding such delicate information.

3.7.4 Mössbauer Spectroscopy: A Means of Determining Valence, Spin States, and Site Environments of Ions

Some elements possess isotopes which have states of nuclear excitation such that transitions between the ground and excited states result in the absorption or emission of γ-radiation. The archetypal case is ^{57}Fe, the nuclear excited state ($^{57}Fe^*$) being some 14.4. keV above the ground state. This excited state, which is itself a daughter product of the radioactive decay of ^{57}Co ($t_{1/2} = 270$ days), decays rapidly to the ground state (Fig. 3.42). The basis of Mössbauer spectroscopy is the resonant absorption of γ-photons by the sample under investigation. Conversely, in Mössbauer emission, the sample – containing ^{57}Co, for example, as a dopant in a well-defined electronic environment – serves as an emitting source. Resonant absorption will take place only when the $^{57}Fe^*$–Fe separations in emitter and absorber are precisely matched. Resonance is achieved, first, by 'clamping' the source atoms firmly in a solid so as to prevent the recoil of the $^{57}Fe^*$ atoms during their decay – an unbound $^{57}Fe^*$ has a recoil velocity of *ca* 100 m s^{-1}. Second, the absorber is moved steadily at an adjustable velocity so as to Doppler-modify the energy of the γ-photon. So sharp are the γ-lines (better than 10^{-7} eV) that a modulating velocity of a few millimetres per second suffices. A Mössbauer absorption spectrum therefore records absorption as a function of Doppler velocity of sample (or the velocity of the source in the case of Mössbauer emission spectroscopy).

The Mössbauer effect cannot be observed in gases or liquids and is feasible only in those solids in which there is a substantial recoil-free fraction of emitters – in other words, when the emitting atoms are firmly integrated into the structure. We immediately see that Mössbauer spectroscopy inevitably becomes less sensitive at high temperatures, and that signals from surface layers (where bonding is generally incomplete) are less than those emanating from layers in the bulk. The Mössbauer effect is recorded readily for relatively few isotopes: ^{57}Fe, ^{119}Sn, ^{151}Eu, ^{181}Ta, as well as isotopes of most of the rare-earth elements and of Ni, Ru, Sb, Kr, Se and I.

From measurement of Mössbauer isomer shifts, quadrupolar splitting and magnetic hyperfine interactions, deductions can be made regarding the valence (oxidation number), spin state and type of coordination an of ion. It is also possible to deduce transition temperatures for magnetic ordering and, occasionally, to determine particle sizes. The various factors governing splittings and interactions are discussed in physicochemical texts. We now focus on the use of this technique.

3.7.4.1 Specific Applications

Although, in principle, by appropriate use of enriched samples and by harnessing the conversion electrons liberated during de-excitation (see Section 3.7.4.2) especially of model catalysts, Mössbauer spectroscopy could be used to probe the chemical properties of the outer layers of solids, its chief use to date has been to characterize the state of certain key elements of bulk catalysts. Catalysts containing Fe, Co or Sn (be they supported metals, bimetallic clusters, mixed oxides, or metal chalcogenides) are in this category. Of late, however, the technique has proved very valuable in the study of zeolite, clay and intercalated catalysts.

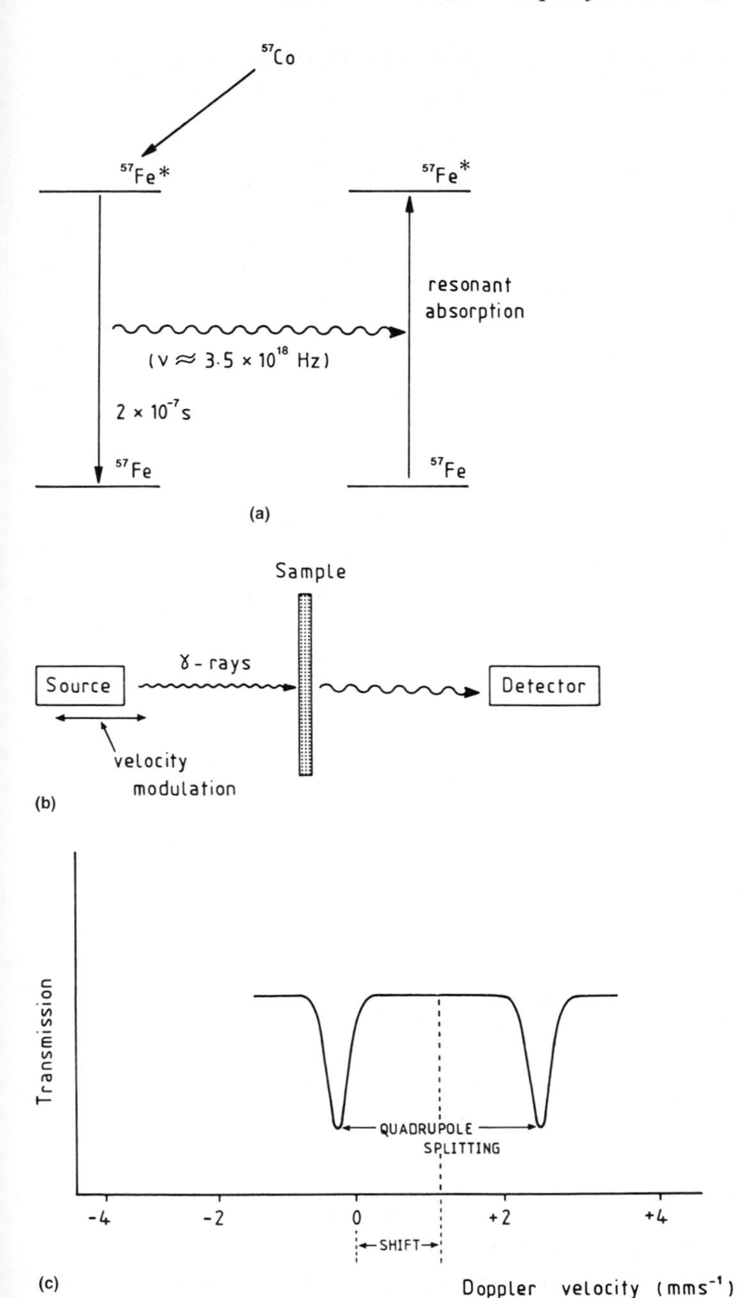

Figure 3.42 (a) The transitions involved in the Mössbauer effect with ^{57}Fe. (b) The set-up used experimentally. The energy of the γ-rays is modified when emitted from a moving source (Doppler effect). (c) Sketch of a typical ^{57}Fe Mössbauer spectrum showing splitting and isomer shift.

Isomer shift and quadrupole splitting generally leave little doubt as to the chemical state of the element of interest in the as-prepared or expired catalyst. By use of in-situ cells, insight can be gained into the status of the elements in functioning catlaysts also. Just as Mössbauer methods have shown that, in $CuFeS_2$ for example, the component cations are cuprous and ferric and not cupric and ferrous, so also have they revealed that:

(a) the iron atoms in scheelite-based, mixed, selective-oxidation catalysts such as $Bi_3(FeO_4)(MoO_4)_2$, are in the high-spin trivalent state;
(b) in ^{57}Fe-doped Co–Mo/Al_2O_3 catalysts the iron atoms are present as high-spin ferric ions in octahedral surroundings; and
(c) in montmorillonite catalysts for the conversion of some organic molecules, ferric ion sites in the clay function as Lewis acid sites.

Mössbauer spectroscopy has also been used to investigate the conditions required for the preparation of highly dispersed particles of iron on various supports. It was established that, on MgO supports, iron particles of diameter in the range 15–30 Å could be produced. There is likely to be increasing use of the Mössbauer effect in future studies of catalysis, especially as new analogues, rich in transition metals, of zeolites and clays are prepared. Mössbauer spectroscopy would be a routine tool for establishing whether Fe, CO, Ni, etc., are in tetrahedral framework sites or in octahedral sites in intrazeolite cavities, or are in interlamellar spaces.

3.7.4.2 Conversion-Electron Mössbauer Spectroscopy (CEMS): A Double-Resonance Technique of an Unusual Kind

Unlike transmission (i.e. absorption or emission) Mössbauer spectroscopy, which is best suited in the catalytic context for the study of relatively high-area solids, CEMS is applicable to samples possessing quite low surface-to-bulk ratios, such as foils and gauzes. This technique relies on the fact that when certain excited nuclei relax (^{57}Fe and ^{119}Sn are good examples) the de-excitation process occurs with a high yield of emitted conversion electrons: only a fraction of the de-excitation involves γ-emission. Then electrons emanate predominantly from the K-shell, and for ^{57}Fe they have a kinetic energy of *ca* 7 keV. For the same reasons that XPS is a surface and subsurface tool, these electrons have an escape depth of a few thousand Ångstrøms.

Hence, if, in place of recording the absorption of γ-rays by the sample, we monitor instead the electrons emitted on resonance, we thereby record the Mössbauer spectra of the surface region from which the electrons emerge. The cell assembly and detector system is so designed as to enable both the Doppler energy (hence the Mössbauer spectral parameters) and the γ-stimulated electron emission to be recorded simultaneously. A further refinement of this technique is energy analysis of the conversion electrons: this affords information about change of composition as a function of depth from the exterior surface. Figure. 3.43 shows both a CEM spectrum of the oxide formed on a steel surface heated to high temperature and a transmission Mössbauer spectrum of the removed oxide scale: peaks due to Fe, α-Fe_2O_3 and Fe_3O_4 are assigned. It is possible that this combination of two distinct kinds of Mössbauer spectroscopy could be of value in characterizing automobile exhaust catalysts such as those based on 'Fecralloy', platinum and Al_2O_3.

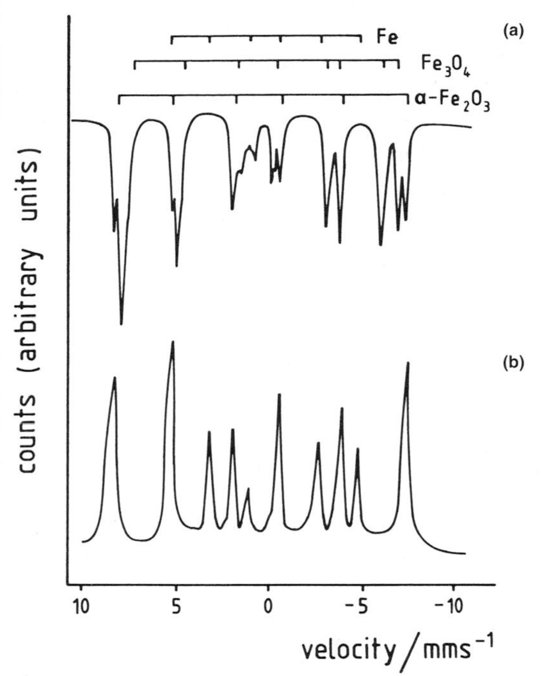

Figure 3.43 (a) Conversion-electron Mössbauer spectrum of a thin steel sample following heat treatment in air at 650 °C. (b) Transmission (conventional Mössbauer) spectrum of the oxide film after its removal from the steel.

3.7.5 Electron Microscopy

The view held by many physical scientists, that electron microscopy (which is often tacitly equated with scanning-electron microscopy, SEM) is capable only of producing images of the morphology of particles, is a mistaken one. Modern electron microscopy is capable of much more than merely producing vivid SEM photographs of crystal morphology. A rollcall of the accomplishments of transmission electron microscopy (TEM) merits emphasis (see Fig. 3.44):

1. Crystallographic phases can be identified from diffraction patterns which, at the same time, tell us whether the material is polycrystalline or in single-crystal form.
2. High-resolution electron microscopy (HREM) can, in a unique fashion, identify known phases and reveal the structure of previously unknown phases, as well as intergrowths in, or between, recognized phases.
3. The X-rays emitted during electron-microscopic analysis tell us the composition of the material under study. (Elements from $Z = 11$ (sodium) upwards are routinely detectable using thin beryllium windows for the lithium-drifted solid-state detectors used for energy-dispersive analysis. With 'windowless' detectors, elements from carbon ($Z = 6$) upwards can be qualitatively determined, and from oxygen upwards quantitatively so.)
4. A combination of electron-energy-loss (EEL) microscopy and EEL spectroscopy (EELS) is valuable not only for light-element analysis (lithium to sodium), but for recording plasmon spectra and Compton profile spectra that serve as bases for qualitative analysis and local bonding respectively.

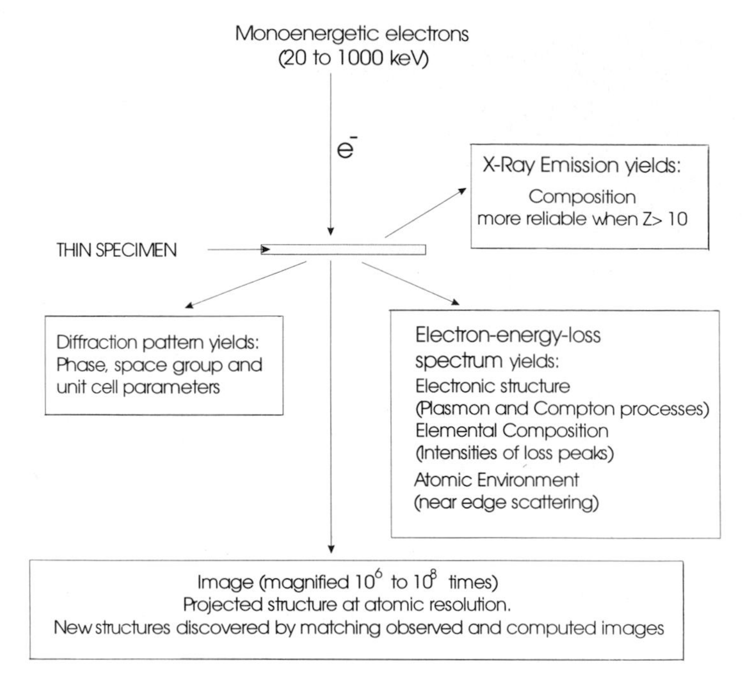

Figure 3.44 The atomic and electronic structure, as well as the elemental composition, of mintue and intergrown (20–10^4 Å thick) solids can be determined by transmission electron microscopy (TEM). The sketch summarizes the kind of information retrievable from an electron-microscopic study of catalysts.

5. EEL spectroscopy, carried out in an electron microscope, can, in ideal cases, determine oxidation states and site symmetries.
6. Surface topography, with monatomic step resolution, is recordable in the image mode.
7. Specialized modes of operation, including the ability to alternate between conventional transmission microscopy, scanning and scanning transmission modes, are a feature of many commercial microscopes. This renders them useful for the study of both dispersed particles and larger, uniform samples.
8. Controlled-atmosphere model catalytic experiments, up to modest pressures (*ca* 0.5 bar), are feasible.
9. One of the great triumphs of HREM has been the discovery by J. S. Anderson of infinitely adaptive structures. Certain oxides, which may be readily rendered grossly non-stoichiometric, display an 'infinite' capacity to adopt a new structure that is characteristic of the precise degree of non-stoichiometry.
10. Another major contribution made by HREM to catalysis is that it has provided the key to the determination of the structure, and hence to the understanding of the mode of action, of the following uniform heterogeneous catalysts: theta-1, zeolite ZSM-25, zeolite beta, zeolite ECR-1, ECR-35 and MCM-22. It has also helped to elucidate the subtle interrelationships between ZSM-3, CSZ-1, CSZ-3, ECR-4, ZSM-3, ZSM-20, ECR-30 and ECR-32.

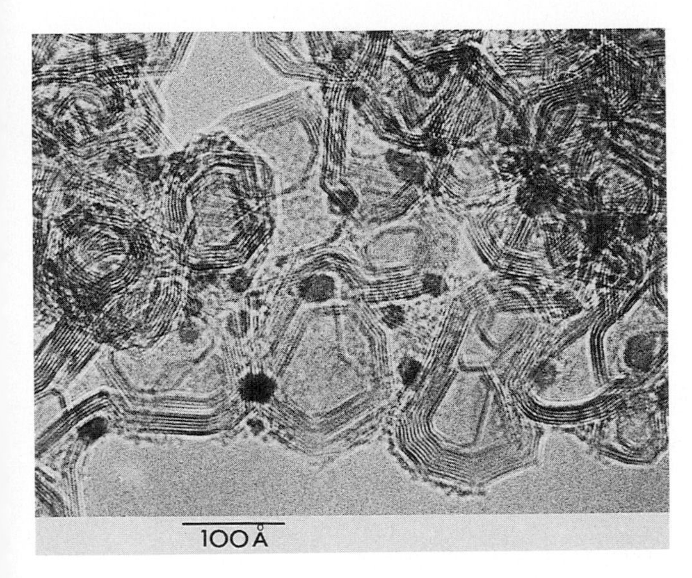

Figure 3.45 Typical high-resolution micrograph of British Petroleum ruthenium catalyst dispersed on a graphitic carbon support. This is a promising possibility for the synthesis of ammonia. The intraplanar distance of the graphitic sheets is 3.4 Å.

The sample is placed in a good vacuum (better than 10^{-6} Torr, and preferably 10^{-8} Torr), so as to avoid accumulation of contamination. The siting of lenses or other recording devices are not shown in Fig. 3.44: the position of these vary according to particular emphases, requirements and aims. In HREM, but not in STEM, an objective lens is used after the sample, which needs to be very thin (10–500 Å thickness) to permit ready penetration and to minimize multiple scattering events.

In HREM the strategy, ideally, is as follows. A series of images is recorded as a function of the objective lens focusing condition (with a fixed accelerating voltage), and also as a function of sample thickness. If the broad structural details of the solid under study are known (say from X-ray crystallography), it is possible to calculate what the image should look under different defocus settings and different sample thicknesses. This approach has been invaluable in the characterization of zeolites.

Figures 3.45 and 3.46 illustrate the value of HREM and Fig. 3.47 indicates the potential importance of EELS, allied to electron microscopy, in characterizing catalysts. The power of modern microscopy is emphasized when we recall that, as an analytical tool, it can detect less than 10^{-21} g of material and that it can image phases intergrown in others when less than a unit-cell width of the minority phase is present.

3.7.5.1 Scanning Probe Microscopy: STM and AFM

Scanning tunnelling microscopy (STM), introduced by Binnig and Rohner, and scanning force microscopy, better known as atomic force microscopy (AFM), introduced by Binnig and Quate, can provide real-space images at atomic resolution. In STM, a sharp metallic tip is scanned over the surface at a height of 5–10 Å (Fig.

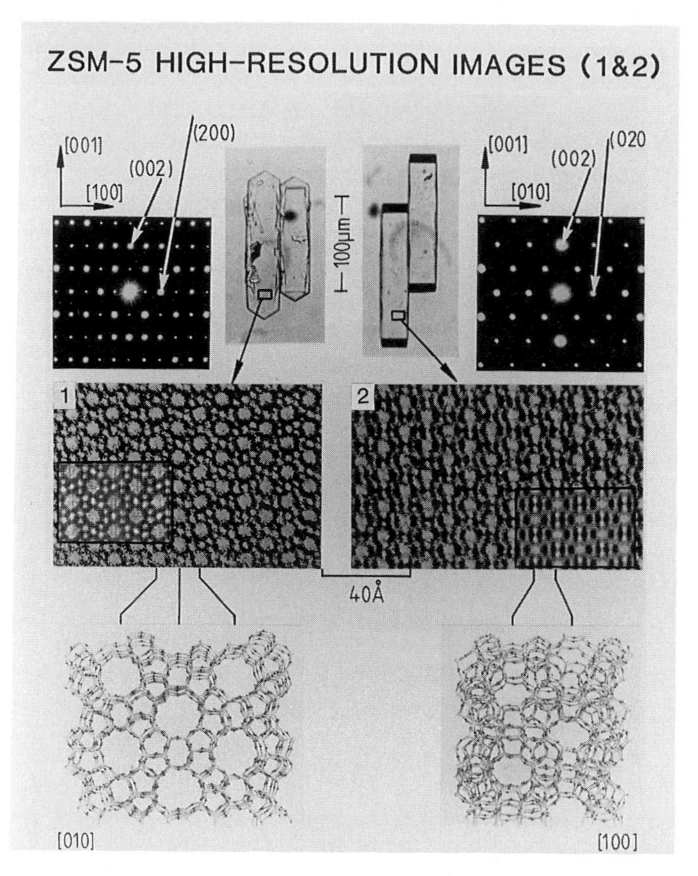

Figure 3.46 High-resolution micrographs (middle) together with selected-area diffraction patterns and optical micrographs of ZSM-5 looking down the [010] direction (top left) and the [100] direction (top right). Corresponding views of the structural model are shown at the bottom. (Based on work of G. R. Millward and J. M. Thomas).

3.48). When a voltage bias (typically $-2.5\,V \leq V_{bias} \leq 2.5\,V$) is applied between tip and sample, quantum mechanical electron tunnelling occurs from the tip to the sample (under positive sample bias) or from the sample to the tip (negative bias). The tunnelling current density (which is a sensitive measure of the tip–sample separation, the local density of states of the sample and the bias voltage) is measured as a function of tip position to yield a topograph of constant density of states. AFM employs a hard tip mounted on a soft cantilever spring. As the tip is scanned over the surface the deflection of the cantilever is measured as a function of tip position. The deflection reflects the atomic structure and is sensitive to the local compressibility of the sample. In general, STM is sensitive to the density of filled valence levels or empty low-energy levels lying between the sample Fermi energy and the applied V_{bias}. It is applicable only to conducting or semiconducting samples. AFM, on the other hand,

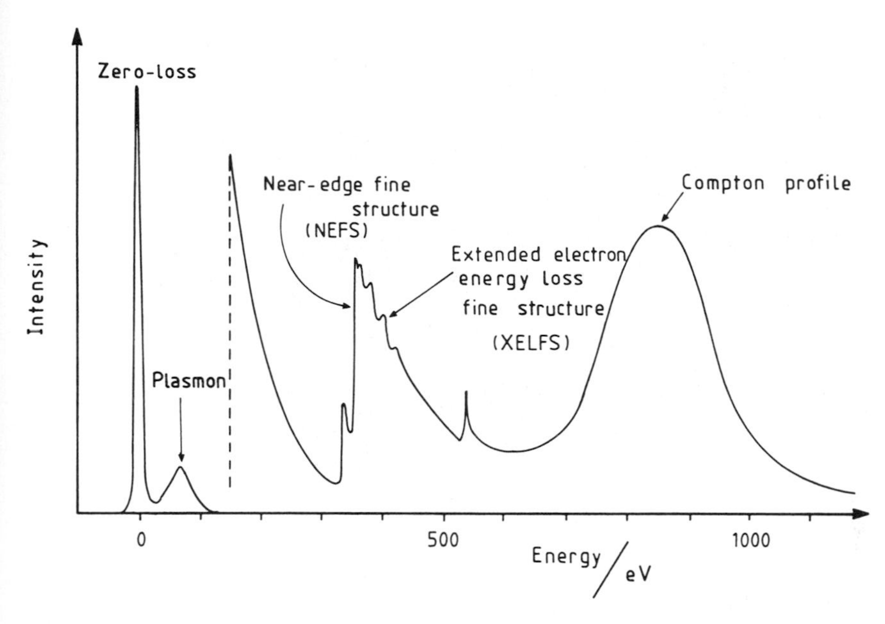

Figure 3.47 Electron-energy-loss spectroscopy (EELS) on microscopic regions of a catalyst specimen can identify and quantify the elements present (especially those with $Z < 10$). The near-edge fine structure (NEFS) and extended electron-energy-loss fine structure (EXELFS) as well as the Compton profile may yield information about the electronic structure of the solid and the bonding of the constituent atoms in optimum circumstances. (Based on work of B. G. Williams and J. M. Thomas).

may be used on insulators as well, and is sensitive to the total density of filled electronic states of the surface atoms of the sample: it thus provides information about the local atomic structure.

Scanning probe microscopes are normally operated at room temperature. The tunnelling current readily passes through an extremely thin, insulating layer such as vacuum, air or liquid. Instruments have recently been customized (by Behm, Somorjai, Sata and others) to operate at variable temperatures up to *ca* 725 K under atmospheric pressure; and there have been several studies of metal and transition-metal chalcogenide surfaces of catalytic interest. Recent work by Chianelli, Whangbo

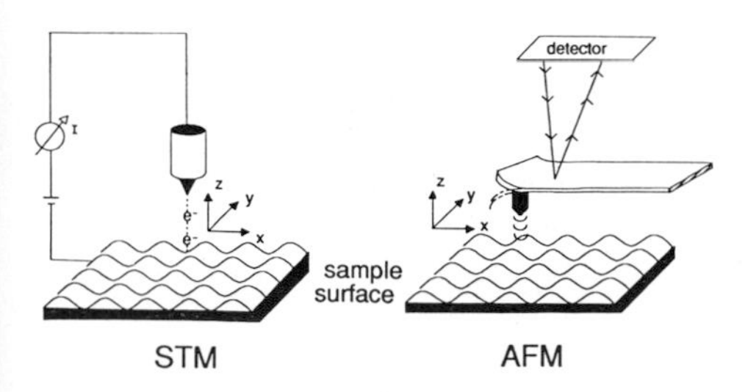

Figure 3.48 Schematic diagram of the scanning tunnelling microscope and the atom force microscope, highlighting their common feature of a localized probe that scans the surface and collects data on a nanometre scale. With permission from J. Frommer, *Angew. Chem., Int. Ed. Engl.*, **1992**, *31*, 1298.)

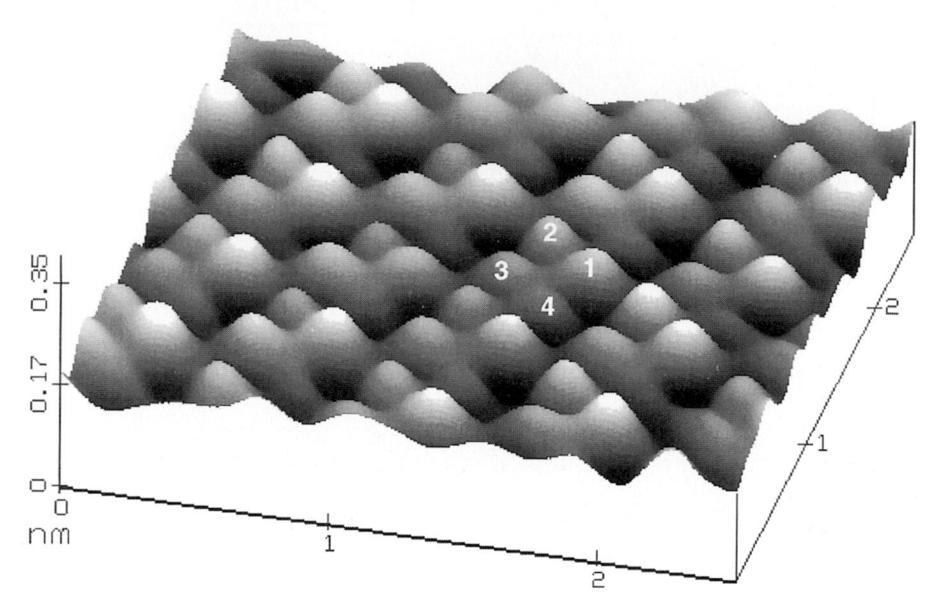

Figure 3.49 STM image of exterior surface of a model ReS_2 catalyst. Extended Hückel calculations (see Chapter 5) show that the high-electron-density protrusions demarcate positions of individual atoms of sulphur (labelled 1 to 4). With permission from S. P. Kelty, A. F. Ruppert, R. R. Chianelli, J. Ren, M. Whangbo, *J. Am. Chem. Soc.* **1994**, *116*, 7857.

and Kelty on the model hydrodesulpurization catalyst ReS_2 (Fig. 3.49), in which they identify the individual atoms of chalcogenide at the exterior surface, as well as that of Ertl on oxygen-induced restructuring of a copper surface, of Somorjai and co-workers on microfaceting of metals in the presence of reacting gases, and of Iwasawa on a video STM study of the catalytic decomposition of formic acid on TiO_2 surfaces, augurs well for the future.

3.7.5.2 Optical Microscopy

Whereas the attainable resolution of the electron microscope is very great, the optical microscope is limited in its resolving power by the wavelength of visible light, i.e. to a fraction of a micrometre. Nevertheless, quite useful information can be gleaned via the agency of optical microscopy, especially the recent confocal laser variant. Enhanced reactivity in solids generally, and enhanced catalytic activity at emergent dislocations in solids, have been profitably studied by optical microscopy. It is also a very convenient method by which to evaluate the relative catalytic effciencies of a range of metallic additives in the gasification of carbon. Controlled-atmosphere studies can be readily carried out using commercially available or otherwise-fashioned hot stages. The technique is well suited to monitoring the regeneration or fouling of metallic catalysts.

3.7.5.3 Ellipsometry: A Non-invasive Technique

In ellipsometry, plane-polarized light is specularly reflected from the surface of interest. Although most measurements use visible or near-infrared light for reasons of experimental convenience, any radiation across the whole spectrum from ultraviolet to infrared could be used. No interfaces between two media of different dielectric properties are ever ideal, and in general the reflected light is not plane-polarized but elliptically polarized. After reflection of light from its surface, the optical constants of any film are determined from the ellipticity of the reflected light. The parameters measured in practice are ψ and Δ, where $\tan \psi$ is the amplitude ratio of the resolved components of the electric vector of the incident light parallel to and perpendicular to the plane of incidence, and Δ is the phase difference of the two components. Light from a source is wavelength-selected by a monochromator, polarized and reflected from the specimen. The reflected light is passed through a compensator, when it reverts to plane polarization, and can be analysed in an analyser and then collected in a photomultiplier.

The power of ellipsometry lies in the ability to detect changes in optical properties corresponding to changes in film thickness of small fractions of a monolayer, and to do so in a completely non-destructive way – unlike, for example, AES or SIMS. For incident light of wavelength 5000 Å and a thin film of refractive index 1.5, the ultimate sensitivity in phase-shift measurement corresponds to a change in thickness of 0.05 Å. The technique is thus well suited to charting surface reactions from the very earliest stages, when initial adsorption on a clean surface occurs, through to production of reactant layers several monolayers in thickness. The technique has been used effectively to monitor growth of oxides on metals such as palladium used in the three-way auto-exhaust catalyst. It could well be adapted in future to more extensive use for the in-situ study of catalysts. Two powerful new variants, ellipsomicroscopy for surface imaging (EMSI) and reflection anisotropy microscopy (RAM), have recently been introduced by Rotermund *et al.* (1995) for recording pattern formation at surfaces under arbitrary pressures; and surface plasmon microscopy, developed by Krischer *et al.* (1995) is well suited for *in situ* studies of electrode surfaces.

3.7.6 Neutron Scattering: A Technique of Growing Importance in the Study of Catalysts

At first sight, neutron scattering may seem singularly ill-suited to probing the surfaces of heterogeneous catalysts. Compared with electrons, or even X-rays, neutrons are only weakly scattered by matter. For example, it takes some 10^7 molecular layers of water to halve the intensity of an incident beam of neutrons; and water is an unusually strong neutron scatterer! Clearly, neutron scattering studies come into their own with solids of very high areas and for investigating interlamellar species (e.g. in clay catalysts) or intracavity guest reactants or products in zeolitic catalysts. Neutrons are very useful in characterizing catalysts because they have the ability to penetrate container and/or support materials comprising elements of high atomic number. This enables catalysts to be studied under extreme conditions with the specimens housed inside high-temperature (600 °C) stainless steel reactors under very high pressure (500 bar). Moreover, neutrons are not subject to selection rules when they are inelastically scattered; they show elastic and scattering characteristics which are theoretically well understood; and they display wide-ranging differences in scattering cross-section among the nuclei. This enables us to examine selectively the presence and behaviour of one nucleus (hydrogen or deuterium, especially) among others.

Neutrons can therefore be used, when *elastically* scattered, to probe by diffraction the structure of adsorbents (or catalysts), adsorbates (or surface reactants) and reactant–catalyst complexes. When *inelastically* scattered, neutrons are a powerful source of spectroscopic information. An advantage of neutrons over X-rays is that wavelengths may be varied over a wide range, normally up to 20 Å, but even up to several hundred Ångstrøms.

At present, the most common source of neutrons for diffraction experiments is the nuclear reactor, but in recent years there has been a growth in the number of facilities producing neutrons by spallation from metal tragets bombarded with high-energy (*ca* 500 MeV) protons. The pulsed neutron beams that are emitted by such a source can conveniently be used to collect diffraction patterns by time-of-flight (TOF) methods. The neutron detector (typically a $^{10}BF_3$ counter) is placed at a fixed scattering angle, and the different scattering planes, which give rise to the peaks in the diffraction patterns, can be sampled by measuring the times that neutrons with different wavelengths (hence different speeds) take to reach the detector (Fig. 3.50): the neutron speed is sufficiently slow to permit this. This experiment can be described by the Bragg equation (Eq. (1)). One of the features of the TOF method is that it can readily provide powder diffraction patterns with exceptionally high resolution.

The energies of thermal neutrons are comparable with those of most molecular motions, so that, in the course of inelastic scattering, the range of energy transfer extends from 10^{-5} to 500 MeV (i.e. 0.0001–4000 cm^{-1}). This spans the energy ranges

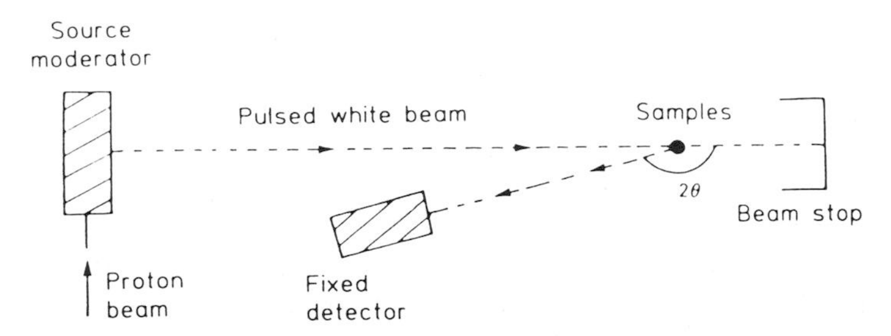

Figure 3.50 A schematic diagram of a time-of-flight (TOF) powder neutron diffractometer. A high-energy pulsed proton beam of energy *ca* 500 MeV is directed onto a metal target producing pulses of 'white' neutrons by spallation. In the simplest diffractometer arrangement, the diffraction pattern of a powder can be collected at a single, fixed detector by measuring the scattered intensity within each pulse as a function of the time-of-flight (see text). Long-wavelength (low-energy) neutrons within each pulse will arrive after the short-wavelength (high-energy) neutrons. Data from successive pulses (typically 10^6) are summed to give adequate statistics. With permission from A. K. Cheetham, A. P. Wilkinson, *Angew. Chem., Int. Ed. Engl.* **1992**, *31*, 1558.

covered by IR, Raman and HREEL spectroscopy. It therefore means that subtle changes in rotational, vibrational and diffusive motions of sorbed or surface species can be quantitatively identified.

The intensity of neutron scattering from a particular element is determined by the so-called scattering length (b) of its constituent isotopes, and although the magnitudes of these scattering lengths have a weak dependence upon atomic number ($\propto Z^{1/3}$), there are sizeable fluctuations superimposed upon this from nucleus to nucleus. If the element comprising the scattering sample has just one isotope, of zero nuclear spin, then the scattering lengths of all the nuclei of that element will be the same, and coherent scattering will be the result. If, however, the spin wave function of a neutron is changed in a scattering event, the latter is no longer coherent with the rest of the neutrons being scattered. Spin-incoherent scattering by protons is very pronounced; by deuterons it is much less so; and the ratio of coherent to incoherent scattering cross-sections for protons to deuterons is *ca* 90 : 1.

The energies E of scattered neutrons are determined either using diffraction to measure their wavelength λ ($E = h^2/2m_n\lambda^2$, where m_n is the mass of the neutron) or by TOF techniques. In a typical scattering experiment, the partial differential scattering cross section $d^2\sigma/d\Omega\, dE'$, is determined (by definition, $\sigma = 4\pi b^2$). The extent of scattering, $d^2\sigma$, into the solid angle between Ω and $\Omega + d\Omega$ with final neutron energies between E' and $E' + dE'$ is what one generally handles. In an 'elastic' experiment (i.e. in the purely diffractive mode), there is zero change of energy; in an 'inelastic' experiment, it is finite.

Greater knowledge of the structure of microcrystalline catalysts, of the atomic details of sorbed or intercalated reactants, of the nature of certain active sites, of the ordering of physically adsorbed phases and also of the texture catalysts has emerged from studies of elastic neutron scattering. Because of the advantageous cross-sectional characteristics of certain atomic nuclei, it is possible to obtain this knowledge when the catalysts (or model systems) are subjected to the extreme conditions that prevail in typical catalytic pressure vessels.

Studies of incoherent inelastic neutron scattering (IINS) have yielded powerful insights into the nature of bound reactants, especially of hydrogenous molecules, on realistic catalysts such as Raney nickel, palladium black, cation-exchanged zeolites and MoS_2. Here, advantage is taken of the absence of any spectroscopic selection rules ('surface', or otherwise) in assigning 'loss' peaks to various stretching and tilting modes. Although the resolution achievable in IINS spectroscopy is rather low compared with that of optical techniques (such as IR), it is comparable with that obtainable with HREELS (Section 3.5.3). Moreover, it is relatively easy with IINS, but not with IR or Raman studies, to focus on vibrational frequencies less than *ca* 300 cm^{-1}.

3.7.6.1 Determining the Atomic Structure and Texture of Microcrystalline Catalysts, the Nature of the Active Sites and the Disposition of Bound Reactants

Provided a catalyst is monophasic and well ordered, its atomic structure can be determined by neutron diffraction even though it exists as a microcrystalline powder. This situation obtains for many catalysts, especially zeolitic ones, and considerable progress has been made recently in arriving at the details of framework structure and location of exchangeable cations, and of model catalytic reactants sorbed within intrazeolite cavities. The method used for this purpose is the Rietveld neutron powder profile procedure. It was devised for analysing complex diffraction patterns by curve-fitting in which the least-squares refinement minimizes the difference between the observed and calculated profiles, rather than individual reflections. In order to do this with neutron diffraction patterns, it can normally be assumed that the reflections have a Gaussian distribution, and that the calculated intensity at each point on the profile is obtained by summing the contribution from the Gaussian functions that overlap at that point. Besides the conventional parameters in the least-squares refinement (atomic coordinates and temperature factors of the individual atoms), additional parameters are required: the lattice parameters, which determine the positions of the reflections, a correction factor for setting the zero point of the detector, and three parameters that describe the variation of the halfwidth of the Gaussian distribution with scattering angle. Rietveld refinement procedures have been used to great effect (especially by Cheetham, Wright and Cox) in the study of uniform heterogeneous catalysts.

Figure 3.51 illustrates the kind of information that can be derived by Rietveld analysis. Here, there is proof, established in quantitative detail, that La^{3+} ions in zeolite-Y polarize their hydration shells so much that $(LaOH)^{2+}$ ions together with ‘free’ protons, loosely attached to the framework, are formed. The $(LaOH)^{2+}\cdots H^{\delta+}{-}O^{\delta-}$ complex constitutes the active (acid) site in La–Y.

Even when the microcrystalline solid has only one-dimensional order, as with montmorillonite and other clay catalysts (Section 8.8), considerable headway can be made in pinpointing the disposition and location of intercalated species. This is accomplished by a one-dimensional Fourier analysis of the resolved $00\,l$ peaks.

The texture of catalysts (i.e. their mean size, surface area and particle size distribution) can be derived from small-angle neutron scattering, just as it can from small-angle X-ray scattering (SAXS) (Section 3.3.3.1). As with X-rays, the neutron method relies on diffraction by inhomogeneities such as pores or particles, the scattering length density of which differs from that of the surrounding matrix.

The information obtained can then be used, in much the same way as that derived from SAXS, to determine the sizes or the surface areas of these inhomogeneities and their size distribution. Of especial interest in catalysis are the prospects the technique offers of characterizing microporous materials in situ during sintering and gasification processes, and of exploring the structure of catalyst support precursors.

3.7.6.2 Determining the Structure of, and Identifying Functional Groups in, Chemisorbed Layers at Catalyst Surfaces

Here we illustrate how inelastic neutron scattering (INS), as a spectroscopic tool not limited by restrictive selection rules and despite its relatively low sensitivity, can reveal a great deal about species attached to, or submerged beneath, solid surfaces.

Vasudevan and others used inelastic neutron scattering to study the nature of the hydrogen contained within, and on, high-area MoS_2, a typical hydrodesulphurization catalyst. Inelastic neutron scattering spectra could be recorded with the catalyst

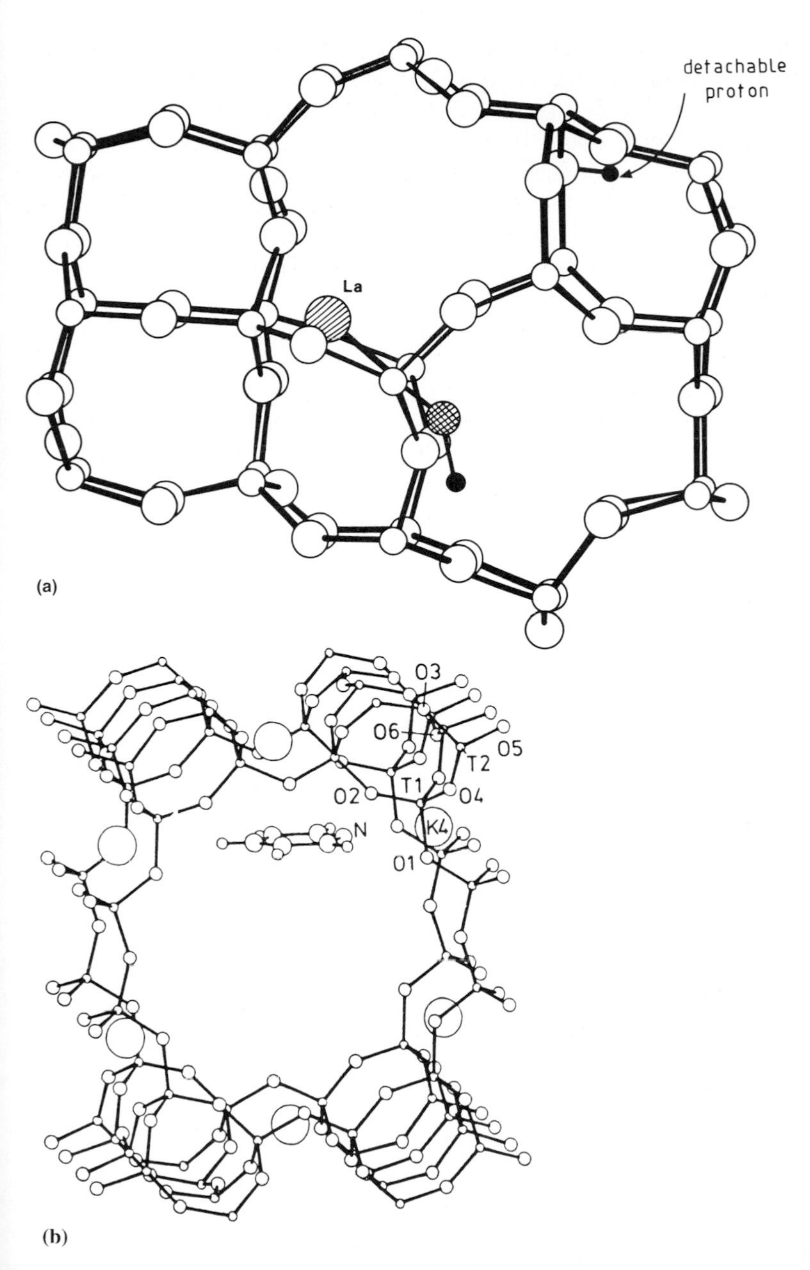

Figure 3.51 (a) This picture of the catalytically active site (known now in atomic detail) in La^{3+}-exchanged zeolite-Y was derived from Rietveld profile analysis of the neutron powder diffractogram (see text and Section 8.3). (Courtesy A. K. Cheetham, M. M. Eddy, J. M. Thomas *J. Chem. Soc. Chem. Comm.* **1984**, 1337.) (b) The location of pyridine in the catalyst potassium zeolite-L, determined by neutron diffraction. The nitrogen of the pyridine forms a Lewis acid–base complex with the potassium, whilst the aromatic ring enjoys short-range interactions with the aluminosilicate framework. (After P. A. Wright, J. M. Thomas, A. K. Cheetham, A. K. Nowak, *Nature (London)* **1985**, *318*, 611.)

exposed to H_2 pressures of up to 60 atm in a stainless steel vessel capable of being heated to 400 °C. The spectra of MoS_2 equilibrated with 1 atm of H_2 showed loss peaks at 622 and 872 cm^{-1}, the first of these being assignable to S–H deformation–vibrations. The 872 cm^{-1} peak could be due to either Mo–H or Mo–OH (owing to some surface contamination) deformation modes. A second peak at *ca* 400 cm^{-1} appears when the pressure of H_2 is increased to 60 atm. This represents a second site for bound hydrogen, also thought to be attached to sulphur but retained in the interlamellar regions. INS also readily picks up hydroxylated molybdate on goethite and hydrated molybdate on alumina catalysts. Moreover, it is capable of distinguishing *ab*sorbed from *ad*sorbed hydrogen in $PdH_{0.014}$. It has also established that very few of the H atoms attached to the surface of Raney nickel are linked to one Ni atom, and Renouprez has shown that, when water vapour is adsorbed on Raney nickel, the dissociation that ensues is best represented by $H_2O \rightarrow 2H_{ad} + O_{ad}$ rather than $H_2O \rightarrow H_{ad} + OH_{ad}$.

INS spectra of benzene adsorbed onto the surface of a commercial platinum black catalyst (of surface area 8 $m^2\,g^{-1}$) yield rich detail in the region 0–1000 cm^{-1}. The spectra indicate that the plane of the benzene lies parallel to the metal surface.

3.8 A Miscellany of Other Procedures

The list of possible techniques that may be invoked to probe surface character is ever-expanding; apart from the numerous procedures already outlined above, there are others of which we should be aware. For those catalysts (e.g. zeolites and clays as discussed in Chapter 8) that facilitate the production of carbenium ions, surface acidity is a prominent feature. The nature, strength and surface concentration of acid sites, typically those that can donate protons, are clearly fundamental for the ranking of catalysts for processes such as reforming, cracking, isomerisation and alkylation. A number of methods can be called upon, including solid-state NMR and IR spectroscopy. The latter is useful when pyridine is employed as a probe molecule. Characteristic of Brønsted activity is the presence IR-active peaks at 1490, 1540, 1620 and 1640 cm^{-1}; and characteristic of Lewis activity are peaks at 1450, 1580, 1600 and 1630 cm^{-1}. In addition, IR spectral shifts suffered by probe molecules such as CO and N_2 when they are bound (end-on) to the Brønsted OH site of a molecular sieve catalyst provide a quantitative guide to surface acidity, as shown by Knözinger, Kazansky, Marchese and Wakabayashi. H_2 and other small molecules may also be used as probes for surface acidity. The magnitudes of the red shift of the OH stretch frequency and of the fundamental vibration of the diatomic molecule are larger for CO than for N_2 because of the former's stronger ability for σ donation. ^{13}C NMR shifts of bound acetone furnish another method of probing surface acidity, as shown by Gorte and co-workers, who have also used calorimetry to record enthalpy changes upon uptake of ammonia and amines.

Many other assessments of surface acidity can be made, including ammonia titration and the use of amines allied to visual indications so as to employ the Hammett acidity function (H_0). The latter quantifies the extent to which reaction in solution occurs between a base B and a Brønsted acid to form its conjugate acid (BH^+):

$$H_0 = -\log a_{\mathrm{H}} + (f_{\mathrm{B}}/f_{\mathrm{BH^+}})$$

where $a_{\mathrm{H^+}}$, f_{B} and $f_{\mathrm{BH^+}}$ are, respectively, the activity coefficients in solution of the protons, the base and its conjugate acid. For practical purposes, a large number of organic amines ('Hammett indicators') are available, covering a wide range of pK_a. Sometimes difficulties arise because the organic amines are sterically too cumbersome to reach the acid sites in a microporous catalyst. If the dimensions of the bases used present no problem, it is sometimes convenient to use photo-acoustic spectroscopy, an alternative to IR spectroscopy, to estimate quantitatively the acid sites on an oxide catalyst. In addition, a number of progressively more demanding, proton-catalysed, organic reactions may be used to rank a particular series of acid catalysts, such as the isomerization of 2-methyl-2-pentene developed by Kramer and McVicker. In this way strongly acid, mixed oxides (e.g. SiO_2–Al_2O_3, P_2O_5–Al_2O_3 or zeolites) can be distinguished quantitatively from weakly acidic or non-acidic ones (SiO_2–MgO, MgO–Al_2O_3, ZnO_2–Y_2O_3 and B_2O_3–SiO_2).

Chromatography is not only a method of analysing reaction products: it is also capable of determining the amounts adsorbed and enthalpies of adsorption under actual reaction conditions. For a linear isotherm (Henry's law valid; see Chapter 2), the specific reaction time V_g is related to the net retention time t_R by the expression

$$V_g = (273 t_R F)/M T_f$$

where F is the flow rate, M the mass of the catalyst or adsorbent and T_f the temperature of the flowmeter. Since these three terms are constant, a plot of log t_R versus $1/T$ yields the heat of adsorption. Other variants of this technique, used to good effect by Waugh and co-workers, are frontal chromatography, stopped-flow gas chromatography, deuterium-exchange chromatography and vacancy chromatography.

Raman spectroscopy, either in its conventional form or, if feasible (as in the case of surface layers of PdO on palladium catalysts) in its resonant form, where a strong electronic resonance leads to enhancement of the Raman signal, is often a viable tool and in principle capable of adaptation for in-situ investigations. Using Raman spectroscopy, Chinese workers (Liu et al.) have identified the presence of the superoxide species O_2^- on a functioning Tl–La–O catalyst for the oxidative coupling of methane in the range 680–860 °C. Surface-enhanced Raman effects are also useful in this context.

*P*ositron *a*nnihilation *s*pectroscopy (PAS) has recently been developed by Lahtinen and Vehanen for surface studies of catalysts. The method consists of three sub-techniques: positron annihilation lifetime measurement; one- and two-dimensional angular correlation of the annihilation radiation; and Doppler-broadened annihilation radiation measurement. This approach, in the hads of Huang et al., has yielded surface acidities of microporous catalysts.

*P*ositron *e*mission computed *t*omography (PET) is a relatively new three-dimensional imaging technique, pioneered at Shell Amsterdam and by Bridgewater, capable of mapping quantitatively the concentration of positron-emitting tracers. The labelling of reactants with positron-emitting nuclides (e.g. ^{11}C, ^{13}N and ^{15}O) is used to study the oxidation of CO and the reduction of NO by CO, reactions of interest in auto-exhaust catalysis. PET permits in-situ transient experiments, just as with other labelled molecules (see Section 3.10.1 below).

The muon is an elementary particle that has spin $\frac{1}{2}$ and a magnetic moment about three times that of the proton. The frequencies of its resonance or precession signals provide a direct and accurate measurement of local magnetic or hyperfine fields. In a sense, the muon is a sensitive microscopic magnetometer. Implanted into any surface, its spin polarization may be monitored to define the sites it occupies in the condensed state and to yield information on local structure or dynamics. As yet, no significant application of muons has been made to catalysis.

3.9 Determining the Strength of Surface Bonds: Thermal and Other Temperature-Programmed Methods

The temperature at which species are desorbed from the surface of a heated solid obviously reflects the strength of the surface bond; the higher the temperature, the stronger the bond. This is why the pioneers of surface science, notably Irving Langmuir and J. K. Roberts, prepared atomically clean surfaces of the refractory solids tungsten and graphite by outgassing them at the highest attainable temperature (*ca* 3000 K). It is also the reason why temperature-programmed methods, irrespective of whether the desorbed entities are monitored by recording the composition and pressure of the gas phase or by probing the surface concentration directly (using XPS, AES or work-function changes), are now so popular in characterizing the surface properties of catalysts and adsorbents. The term 'flash desorption' describes the act of purging a surface, usually by heating the solid in an ultrahigh vacuum or in a stream of non-reactive gas. Flash desorption spectroscopy (FDS) simply records the desorption peaks as a function of temperature. Nowadays, FDS has become synonymous with TPD, temperature-programmed desorption, the principles of which we describe below. TPD is related to temperature-programmed reduction (TPR), to temperature-programmed reaction spectroscopy (TPRS) and, in a more general sense, to cyclic voltammetry, which is a valuable means of characterizing electrocatalysts.

3.9.1 Temperature-Programmed Desorption (TPD) or Flash Desorption Spectroscopy (FDS)

It was seen in Chapter 2 that measurements of the temperature coefficient of the rate of desorption of a species yield estimates of the heat of adsorption of that species or the surface grouping from which it formed. It follows from

$$E_\mathrm{d} = -\Delta H + E_\mathrm{a}$$

that, if the activation energy of adsorption E_a is zero, the activation energy of desorption E_d equals the heat of adsorption, $-\Delta H$. If the adsorption process is activated, however, then E_d sets an upper limit to the heat of adsorption. Irrespectively of whether we are concerned with the release of simple molecules such as CO, H_2 and O_2 from metal surfaces, or alkenes, alkanes and other products from oxide catalysts, the principles involved are the same.

If there are N_a species adsorbed on a given surface and m is the kinetic order of the desorption, then:

$$\frac{-dN_a}{dt} = vN_a^m \exp\left(\frac{-E_d}{RT}\right) \tag{35}$$

where v is the frequency factor for the desorption (Chapter 2). In a non-isothermal desorption, with a linear, programmed rate of temperature rise, at time t

$$T_t = T_0 + \beta t \tag{36}$$

where β is dT/dt, and T_0 the starting temperature, so that

$$\frac{-dN_a}{dT} \cdot \beta = vN_a^m \exp\left(\frac{-E_d}{RT}\right) \tag{37}$$

The value of m for which the plot of $\ln[(dnN_a/dt)\beta]$ versus $1/T$ is linear is the kinetic order, and the slope of this plot yields E_d, a quantitative measure of the strength of the bonding responsible for the peak in question. Figure 3.52 illustrates the second-order desorption of oxygen from rhodium. Note that the peak intensity in the TPD spectrum increases with increasing coverage; that the coverage can be monitored both from TPD and AES data; and that there is only one type of bound state, characterized by an E_d of $210 \pm 3\,\text{kJ mol}^{-1}$, for the oxygen. For other situations several distinct states of adsorption, each with its own strength of bonding, can be detected. Even single-crystal surfaces can exhibit a multiplicity of binding states. On the {111} faces of tungsten, for example, there are five discrete states with peak maxima ranging from 120 to 650 K, whereas on the {211} planes there are only two, with maxima at *ca* 400 and 650 K.

In using Eq. (37) we have tacitly assumed that v and E_d are independent of coverage. Pursuing these assumptions, we arrive at a rather more convenient method of extracting E_d. Thus the temperature T_p at which the desorption rate is maximal (i.e. at the peak), can be found by setting $-d/dT(dN_a/dT) = 0$. Hence, from Eq. (37), we have:

$$\frac{d}{dT}\left\{\frac{v}{\beta} N_a^m \exp\left(\frac{-E_d}{RT}\right)\right\} = 0 \tag{38}$$

or

$$\frac{v}{\beta} N_a^m \left(\frac{E_d}{RT^2}\right) \exp\left(\frac{-E_d}{RT}\right) = \frac{mv}{\beta} N_a^{(m-1)} \exp\left(\frac{-E_d}{RT}\right)\left(\frac{dN_a}{dT}\right)$$

It therefore follows that

$$\frac{E_d}{RT_p^2} = \frac{v}{\beta} \exp\left(\frac{-E_d}{RT_p}\right) \tag{39a}$$

for a first-order desorption ($m = 1$), and

$$\frac{E_d}{RT_p^2} = 2(N_a)_p \frac{v}{\beta} \exp\left(\frac{-E_d}{RT_p}\right) \tag{39b}$$

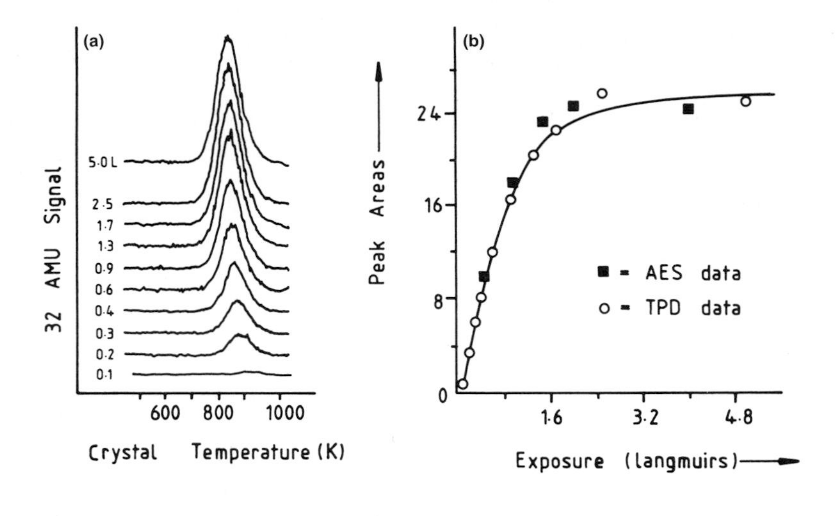

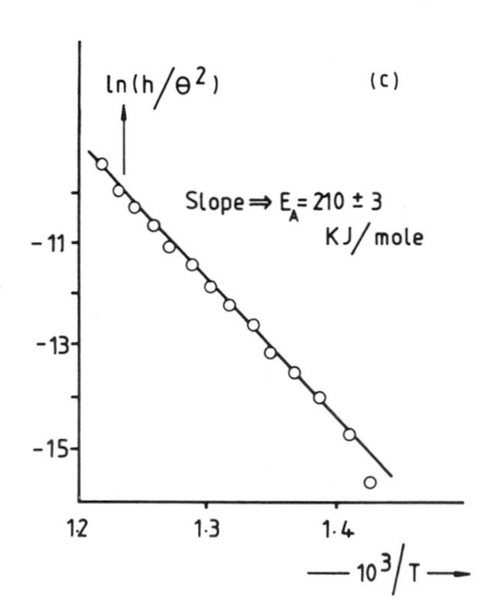

Figure 3.52 (a) Temperature-programmed desorption (TPD) spectra of O_2 from a rhodium surface (dosage in Langmuirs). (b) Oxygen uptake curve recorded both by AES and thermal desorption measurements. (c) Proof that the desorption is kinetically of second order (see text and Eq. (37)). (Courtesy R. M. Lambert).

for a second-order desorption ($m = 2$). We see that the value of T_p at a fixed value of β is independent of initial coverage, a characteristic feature of first-order desorption kinetics. By contrast, in a second-order desorption process the peak (T_p) shifts to lower temperatures as the coverage increases.

It can be shown that a plot of $\ln(T_p{}^2)$ versus $1/T_p$, obtained from data covering a range of values, yields a straight line, from the slope of which E_d may be extracted.

When thermally driven desorption occurs in a container of volume V to which is attached a pump that has an effective pumping rate of pressure built up from desorption of dp/dt, a simple equation relates the desorption rate to the observed pressure p. Thus, the rate of gas evolution is $-(dN_a/dt) \cdot kT/V$; and the rate of evacuation is Sp/V, where S is the effective pumping rate. Hence,

$$\frac{dp}{dt} = -\frac{dN_a}{dt} \cdot \frac{kT}{V} - \frac{Sp}{V} \tag{40}$$

But if

$$\frac{Sp}{V} \gg \frac{dp}{dt}$$

the rate of desorption is given by:

$$\frac{-dN_a}{dt} = \frac{Sp}{kT} \tag{41}$$

which means that the rate of desorption is simply proportional to the partial pressure. Note that integration of this equation yields the initial concentration of adsorbed species,

$$N_a = \frac{S}{kT} \int p\, dt \tag{42}$$

In other words, the area of the desorption peak is directly proportional to the surface coverage.

TPD studies are a valuable source of information on the mechanistic features of catalysed reactions. On nickel surfaces, the TPD peak for H_2 (at *ca* 360 K) is moved to lower temperatures if the surface is pre-exposed to CO. The indication here is that the chemisorbed CO displaces hydrogen from more to less strongly chemisorbed sites.

3.9.2 Temperature-Programmed Reaction Spectroscopy (TPRS)

If, instead of simply subjecting a solid rich in surface groups to programmed heating in vacuo, one does so in H_2, the spectrum of peaks detectable by mass spectrometry or gas chromatography is the result of either temperature-programmed reduction (TPR) or temperature-programmed hydrogenolysis (TPH). In general, bearing in mind that heating may take place in any appropriate reactive environment, we speak of temperature-programmed reaction spectroscopy (TPRS), which many workers have put to good use in clarifying the nature of the bonding and functional groups at catalyst surfaces.

McNicol at the Shell Laboratories, and many others, have harnessed TPR. They were able, by the hydrogen titration which is involved in the technique, to quantify the amounts of ions in given valence states. This cheap, simple and sensitive technique has been applied to supported monometallic (e.g. nickel on Al_2O_3), bimetallic (Pt/Ru, Co/Rh, transition-metal exchanged zeolites) and many other systems, including the ICI Cu/ZnO catalyst for the synthesis of methanol.

3.9.3 Magnitude of the Heat and Entropy of Adsorption

Firm conclusions can often be drawn from the magnitude of the heat of adsorption, which is measured either indirectly by application of the Clausius–Clapeyron equa-

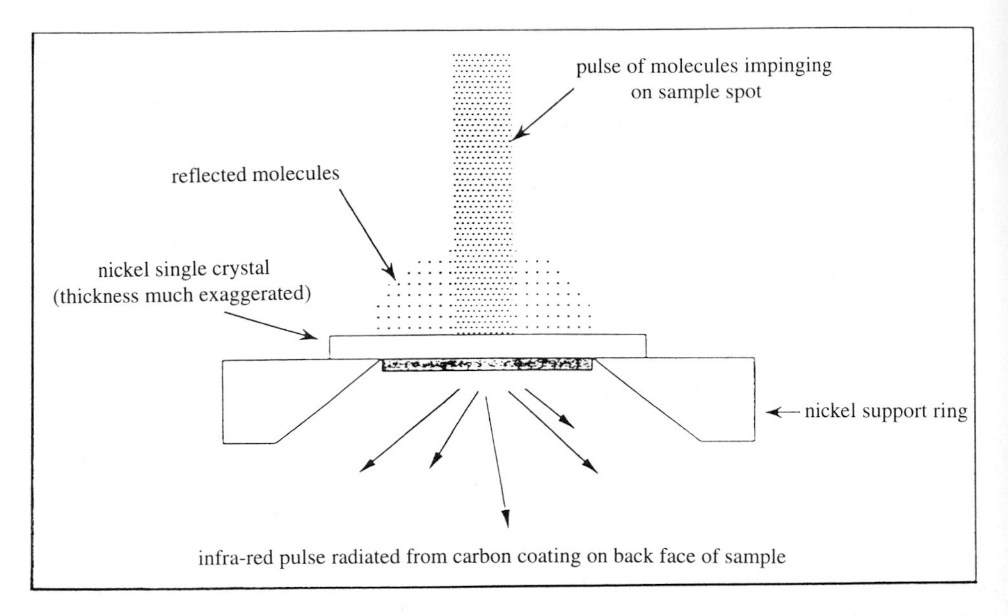

Figure 3.53 Design of ultrasensitive calorimeter for determination of heats of adsorption on single-crystal films. With permission from C. E. Borroni-Bird, N. Al-Sarraf, S. Andersson, D. A. King, *Chem. Phys. Lett.* **1991**, *183*, 516.

tion to appropriate, reversible isotherms (Section 2.4.1) or directly using calorimetry or modified differential scanning calorimetry. The greater the magnitude of the enthalpy change, the more likely it is that one is witnessing chemisorption rather than a weaker interaction between adsorbate and catalyst. The decline in enthalpy, discussed fully in Section 2.7.4, reveals whether mutual repulsion or modification in bond type with increasing coverage is operative. When bases such as ammonia or pyridine are used as surface probes, the heats of adsorption can sometimes reveal whether Brønsted or Lewis acid sites at the catalyst surface are dominant.

Although not suitable for investigations of typical industrial catalysts, an ultrasensitive and accurate method for measuring heats of adsorption on thin-film single-crystal substrates has been developed by King and Andersson. The principle of the method is summarized in Fig. 3.53. With this calorimeter, which has a heat capacity of *ca* $1 \times 10^{-6}\,\mathrm{J\,K^{-1}}$, heats of adsorption for 0.04 monolayers doses of diatomic gases on model catalyst surfaces may be routinely measured.

Since the mobility of an adsorbed species is directly related to the entropy of that species, knowledge of the entropy change accompanying adsorption is useful. It is evaluated as follows. From two adsorption isotherms at temperatures T_1 and T_2, the corresponding values for the equilibrium pressures p_1 and p_2 are selected for a given amount adsorbed. The difference in Gibbs free energy, ΔG_1, between the three-dimensional gas standard at temperature T_1 and the adsorbed species in equilibrium with gas at a pressure p_1 is given by

$$\Delta G_1 = -RT_1 \ln \frac{p_0}{p_1} \tag{43}$$

where p_0 is the standard pressure (760 Torr). Similarly, ΔG_2 for the same amount adsorbed at temperature T_2 is given by

$$\Delta G_2 = -RT_2 \ln \frac{p_0}{p_2} \tag{44}$$

But

$$\Delta G_1 = \Delta H - T_1 \Delta S \quad \text{and} \quad \Delta G_2 = \Delta H - T_2 \Delta S \tag{45}$$

where ΔH is the differential heat of adsorption and ΔS is the differential entropy of adsorption, both of which are calculable, at a variety of adsorbed amounts, from a series of equations such as (43)–(45).

The next step is to convert ΔS values corresponding to various amounts adsorbed to the respective $\Delta S^\ominus$ values, $\Delta S^\ominus$ being the difference in differential molar entropy between the three-dimensional gas in its standard state and in its adsorbed standard state. The standard state for the model of immobile adsorption is different from the standard state for that of mobile adsorption. For immobile adsorption the standard state is that corresponding to half-coverage of the surface, i.e. at $\theta = \frac{1}{2}$. For mobile adsorption it is convenient to use a standard state which is similar to the standard of the normal three-dimensional gases. One may define a standard state by dividing the standard volume of the three-dimensional gas (1 bar pressure) by an arbitrary chosen thickness of the adsorbed film, 6 Å. The resulting area per molecule, designated A_0 is 22.53 Å^2. After noting the difference in standard adsorbed states, we may proceed to calculate the values of $\Delta S^\ominus$. For the model of immobile or site adsorption we have the change in differential molar entropy $\Delta S_i^\ominus$ given by:

$$-\Delta S_i^\ominus = -\Delta S - R \ln \frac{\theta}{1-\theta} \tag{46}$$

And, for the mobile adsorption, the change $\Delta S_m^\ominus$ is given by

$$-\Delta S_m^\ominus = -\Delta S - R \ln \frac{A_0}{A} \tag{47}$$

where A refers to the area per molecule at the amount adsorbed that is under consideration. It is clear that, to evaluate $\Delta S^\ominus$, it is necessary to know not only the amount adsorbed but also the maximum amount which can be adsorbed (so as to yield θ values) and the actual surface area of the solid (so as to yield A values). Chapter 4 describes how to arrive at values of the monolayer capacity and the surface area, so that the experimental values of $\Delta S_i^\ominus$ and $\Delta S_m^\ominus$ can be obtained.

The final step in the method of assessing the mobility of the adsorbate from entropy data is to compare experimental values of ΔS^0 with those calculated theoretically from partition functions (see Section 2.5.3) on the basis of an assumed model. Thus, if a non-linear molecule is adsorbed in an immobile manner, there will be a loss of three degrees of translational freedom, three degrees of rotational freedom, and a small amount of vibrational freedom – often the vibrational freedom is retained on adsorption. From the partition functions, the entropy changes $\Delta S_i^\ominus$ associated with the loss of translational and rotational freedom can be calculated in terms of the atomic masses and moments of inertia. If the theoretical value of $\Delta S_i^\ominus$ agrees with the experimental value derived as explained above, it may be concluded that the adsorbate is immobile. Likewise, if the theoretical and experimental values of $\Delta S_m^\ominus$ agree well, the adsorbate may be regarded as mobile.

Although entropy computations of this type have yielded useful information about adsorbate freedom, this approach to the study of mobility has been largely supplanted by more reliable direct techniques, such as STM. It turns out that the entropy approach is, in the final analysis, insensitive. It cannot reliably be used to distinguish between the possibilities of immobile and mobile layers for certain adsorptions, simply because the effective areas of molecules on surfaces, or the surface areas of the solids themselves, are known to an inadequate degree of precision, thus affecting significantly the magnitude of $\Delta S_i^\ominus$ and $\Delta S_m^\ominus$ extracted from equations such as (46) and (47).

3.10 In-Situ Methods of Studying Catalysts: The Current Scene and Future Prospects

The reductionist approach to catalysis, whereby the key steps thought to be important in heterogeneous turnover – diffusion, chemisorption, surface migration and reorganization, desorption, etc. – are studied individually at a fundamental level, has much to commend it. It is surely enlightening that, as Zare and Madix have discovered, NO molecules desorbing from Pt(111) surfaces rotate like a helicopter blade in a plane parallel to the surface, but that those scattered from the surface bounce off in a cartwheel fashion. It is also illuminating that, as shown by Roberts' photoemission studies, chemisorption of O_2 proceeds via the sequence: $O_2(g) \rightarrow O_2{}^{\delta-}(s) \rightarrow O^{\delta-}(s) \rightarrow O^{2-}(ad)$. Likewise, the surface science approach to the study of methane synthesis from $CO + H_2$ mixtures has been elegantly substantiated by Goodman, who showed that the rate of formation of CH_4 over a Ni(100) or a Ni(111) single-crystal surface is identical to that observed over high-surface-area nickel catalysts supported on alumina (Fig. 3.54). (We note in passing that Goodman's work also demonstrates the reality of structure-insensitive catalytic reactions, as defined in Chapter 1.)

But the surface science approach, backed up by an impressive armoury of increasingly sophisticated techniques – the majority of which require high or ultrahigh vacuum and are fundamentally incapable of studying reactions at above a Torr or so of reactant pressure – cannot possibly address the enigma of catalysis under real-life operating conditions. The surface science approach frequently restricts study of model catalysts to such low pressures that they involve analysis of an unreal or artificial surface. If a metal M forms an oxide MO, the decomposition pressure P_{O_2} of which is higher than the pressure of the reactant O_2 in the model study, clearly only the metal and not the oxide surface can be investigated. Yet under real-life catalytic conditions it is the oxide and not the metal that covers the surface. It is for these

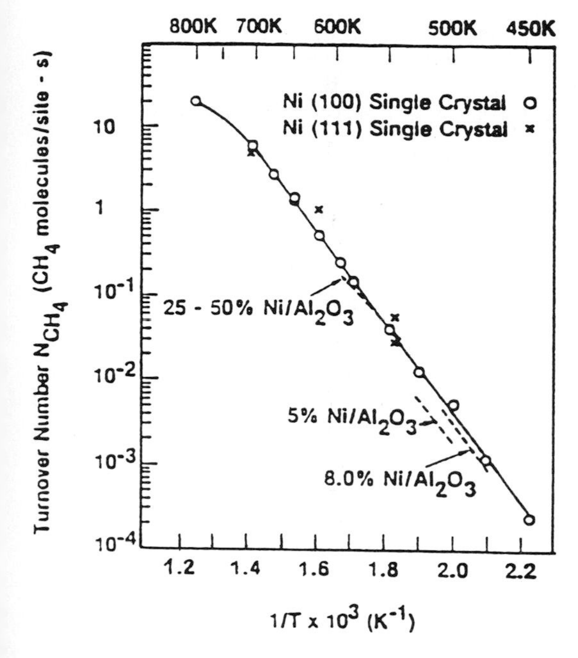

Figure 3.54 Comparison of rate of synthesis of methane over Ni(100) and Ni(111) single crystals and supported nickel catalysts for reaction at 120 Torr and a H_2/CO ratio of 4 : 1. With permission from R. D. Kelley, D. W. Goodman, *Surf. Sci.* **1982**, *123*, L743.

and other reasons that it is necessary to turn increasingly to those techniques that are capable of retrieving structural and other information about catalysts under operating conditions. There is, however, an intermediate stage, whereby certain 'surface science tools' may be adapted for studies at intermediate pressures: controlled-atmosphere electron microscopy (as in the work of Gai-Boyes and Baker), photoemission electron microscopy (as developed by Rotermund) and fluorescence yield X-ray emission from light elements (as developed by Stöhr, Madix and Gland) are illustrative examples.

Table 3.4 lists all the techniques that can be used under typical operating catalytic conditions. Some of these are not yet fully proved (e.g. positron annihilation spectroscopy, muon beams). Others are under-exploited (e.g. sum-frequency generation, ellipsometry, infrared thermography and acoustic emission) but are often quite well suited to monitoring changes in the catalyst during activation, use and regeneration.

Broadly speaking, in-situ techniques for studying catalysts under operating conditions fall into four categories: those that are kinetic and indirect, involving the use of isotopic labelling and chromatography or mass spectrometry; those that are spectroscopic, such as FTIR, X-ray absorption or inelastic neutron scattering; those that are based on diffraction and entail either X-ray or neutron beams; and those that are combinations of two or more of these.

3.10.1 Isotopic Labelling and Transient Response

Ever since H. S. Taylor used deuterium as a label to investigate the reactions of hydrocarbons at surfaces, isotope substitution has remained a valuable approach

Table 3.4 A selection of in-situ methods for characterising catalysts.

Spectroscopic and optical
Infrared: diffuse reflectance;[(a)] transmission (compressed discs); reflection–absorption (with or without polarisation modulation)
Raman, resonant Raman (laser-stimulated)[(a)]
X-ray absorption (XRA):[(a)] pre-edge; near-edge; extended-edge fine structure
Inelastic neutron scattering
Mössbauer[(a)]
Magnetic resonance:[(a)]multinuclear, MASNMR, 2D, etc.; ESR (EPR); spin–echo
Fluorescence:[(a)] lifetime and emission
Sum-frequency generation
Ellipsometry and Ellipsomicroscopy
Scanning tunnelling
Confocal laser microscopy
Conventional hot-stage microscopy
Diffraction
X-rays:[(a)] conventional; energy-dispersive; position-sensitive detection
Neutrons*
Scanning probe methods
STM
AFM
Tracer and other methods
Positron emission spectroscopy[(a)]
IR thermography[(a)]
Nuclear-chemical reactions
Acoustic emission
Kinetic and temporal
Transient response[(a)] (isotopic labelling)
Temperature-programmed desorption (TPD)[(a)]
Temporal analysis of products (TAP)
Temperature-programmed reaction spectroscopy (TPRS):[(a)]Reduction; desulphurization, etc.
Chromatography:[(a)]frontal; vacancy; stopped-flow, etc.
Microreactor studies[(a)]
combined approaches (examples)
XRD/XRA[(a)]
FTIR, XRD, multinuclear NMR
Mössbauer, TPR, XRD
FTIR/microreactor/TPD

(a) Readily adaptable for study of commercial catalysis.

for the catalyst scientist. Major discoveries have emerged as a consequence. One example is the proof, by using ^{18}O, that certain selective oxidation catalysts (e.g. Bi_2MoO_6) release structural oxygen to incoming reactants, and subsequently make up for the depletion in the solid by incorporating gaseous O_2. The establishment of the mechanisms of methanol oxidation (Chapter 8) and of hydrocarbon conversion, encompassing the hapticity and reorganization of bound ligands, are other examples. Radioactive isotopes, too, as demonstrated by the pioneering studies of Thomson, have been invaluable in clarifying the nature of poisoned surfaces. The use of ^{14}C is perhaps less widespread nowadays because NMR spectroscopy and mass spectrometry can generally cope with natural-abundance or slightly enriched ^{13}C-containing species.

Labelling has been particularly helpful in probing surfaces kinetically by the method of transient response, which has proved powerful in the hands of Tamaru (who pioneered the technique), Bell, Sachtler and Mims. We outline the principle and importance of this in-situ method by comparing the nickel- and platinum-catalysed methanation reaction:

$$CO + 3\,H_2 \rightarrow CH_4 + H_2O$$

Picture, for simplicity, the metal surface with just one type of surface intermediate implicated in the overall reaction. By definition, we have the rate R related to the average lifetime of the surface species, τ, and the number N of such intermediate species given by

$$R = N/\tau \tag{48}$$

Recalling the notion of turnover frequency (TOF; see Section 1.3.1) we have

$$\frac{R}{N_s} = \text{TOF} = \frac{1}{\tau}\cdot\frac{N}{N_s} = \frac{\theta}{\tau} \tag{49}$$

where N_s is the number of surface-exposed atoms of catalyst. Our task is to determine both N and τ. We do so by interrupting the supply of reactant and monitoring the gradually decaying rate (Fig. 3.55). The new rate R^* is related to the rate at steady state, R_{ss}, by

$$R^* = R_{ss}\exp(-t/\tau) \tag{50}$$

What, in effect, has happened is that the initial steady-state population of surface intermediates falls off from the point of interruption:

$$\frac{dN}{dt} = -\frac{N}{\tau} \tag{51}$$

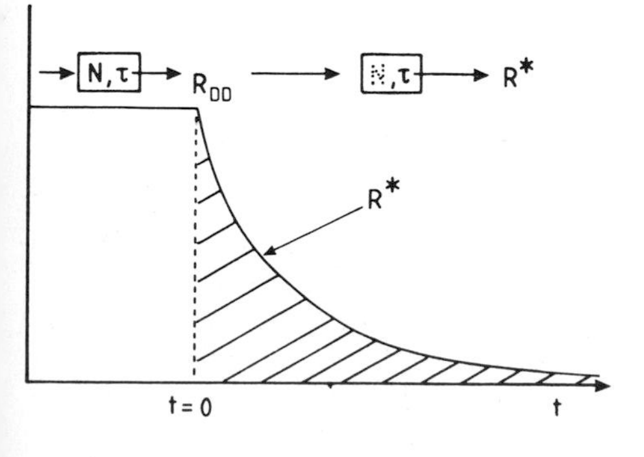

Figure 3.55 When there is an interruption at $t = 0$ to the supply of reactants reaching a catalyst surface, at which there are N intermediates each with a lifetime of τ, the steady-steady rate, R, falls off exponentially to a new rate R^*.

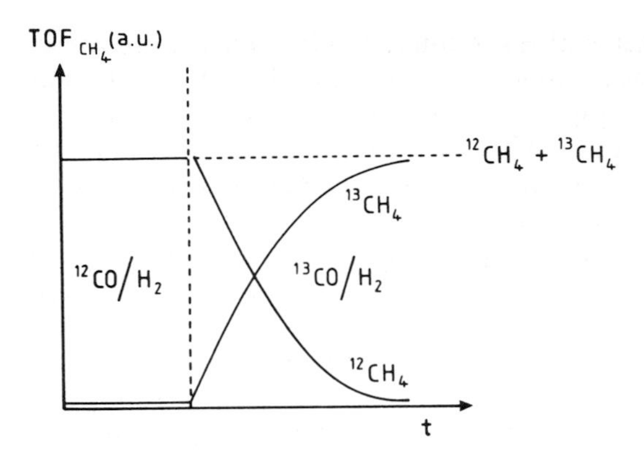

Figure 3.56 Isotopic labelling (see text) greatly assists in elucidating catalytic reactions such as the hydrogenation of CO schematized here (Courtesy P. Biloen).

so that

$$N(t) = N_{ss}\exp(-t/\tau) \tag{52}$$

A plot of $\ln(R^*)$ versus t therefore yields τ. The value of N is obtained if all the products corresponding to the shaded area in Fig. 3.55 are collected:

$$N = \int_0^\infty R^* \mathrm{d}t = \int_0^\infty \frac{N}{\tau}\exp\left(\frac{-t}{\tau}\right)\mathrm{d}t \tag{53}$$

This procedure for obtaining τ and N is reliable only if τ is independent of surface coverage. To avoid making this assumption, the stream of reactant, instead of being interrupted, can be replaced with an isotopic variant, i.e. $^{12}CO/H_2$ makes way for $^{13}CO/H_2$. Experimentally the situation is now as shown in Fig. 3.56. The surface is not perturbed under these circumstances, and τ may indeed be taken as constant. Biloen and co-workers, in comparing Pt/TiO_2 and Ni/SiO_2 as methanation catalysts at 225 °C ($H_2/CO = 3:1$, 1 bar pressure) found that, although the turnover frequencies for these catalysts were about equal (5.8×10^{-4} and 3.2×10^{-4} per metal surface atom per g), there was over a hundredfold difference in lifetime of surface intermediate (0.2 s and 91 s, respectively) and a corresponding difference in surface coverage. They further concluded that the intermediates involved in the C–O bond rupture step are different: CH_xO (on nickel) and CH_yO (on platinum) with $x < y$.

The above account deals with a simplified picture. Clearly, even for this reaction, greater insight can be gained by using CO/D_2 as well as labelled CO/H_2 mixtures. Furthermore, for other catalysed reactions many distinct types of surface intermediate could be implicated and, in principle, identified with the aid of the correct suite of transient response. Bell found that in an in-situ infrared study of Fischer–Tropsch synthesis (Section 8.2) on a RU/SiO_2 catalyst two distinctly different forms of carbon, C_α and C_s, are deposited on the surface.

Gas-chromatographic procedures can often complement transient-response studies. Both vacancy chromatography and stopped-flow gas chromatography are useful, as is deuterium-exchange chromatography introduced by Ozaki. In the last-named, the procedure is to pass a stream of H_2 continuously through the bed of catalyst. A small sample of D_2 is injected into the gas stream; exchange of deuterium with hydrogen takes place on the catalyst surface and HD emerges from the reactor (column) after a volume of gas has passed which is greater than that which passes between the injection and emergence of a sample of non-exchanging gas (e.g. helium). The difference between these two volumes, known as the deuterium retention volume, is the effective volume of exchangeable hydrogen on the surface. Changes in this retention volume as a function of catalyst pre-treatment, use or deactivation can help pinpoint the causes of catalytic activity.

*T*emporal *a*nalysis of *p*roducts (the TAP experiment) was introduced in the mid-1980s by workers at the Monsanto Company, St. Louis, USA, so as to improve greatly the resolution of the transient-response approach. Provided the individual kinetic steps in a heterogeneous reaction do not occur at a rate significantly faster than the transient, the approach outlined above, which is applicable to non-steady-state reactor systems with a minimum reactor residence time of about 1 s, is reliable. The TAP experiment, designed by Gleaves and co-workers, operates in the millisecond time regime and so offers an improvement in resolution of two or more orders of magnitude. This is accomplished by injecting an extremely narrow gas pulse into one end of a small cylindrical reactor and continuously evacuating the other end. After the pulse traverses the reactor it exits into the vacuum (of 10^{-10} Torr or so) and travels as a molecular beam through a differentially pumped system. A portion of the pulse is sampled by a mass spectrometer and its composition is determined as a function of time. Shortening the length of the reactor bed decreases the residence time and increases the resolution, the magnitude of which is ultimately governed by the width of the initial pulse, which currently may be reduced to little more than a microsecond. Unlike a TPD or TPRS experiment, a carrier gas is not used in TAP: the pulse moves as a result of the pressure gradient across the reactor.

At a typical pulse intensity of 10^{15} molecules per pulse with a catalyst possessing an active surface area of $10\,m^2\,g^{-1}$, a single pulse will address about 10^{-4} of the surface area per gram of catalyst. It is certainly feasible to operate with lower pulse intensities, thereby addressing as little as one-millionth of the active area of $10\,m^2\,g^{-1}$. Herein lies the key to TAP's ultrasensitivity.

The mathematical analysis required to handle the TAP data is straightforward – the partial differential equations are linear, and the variables are easily separated. Solutions have been derived for situations which include multiple sites of adsorption, gas-phase reaction, or reaction to form multiple products. A good example of the value of the TAP approach is contained in the work of Schlögl et al. (1994), who elucidated the mechanism of the selective oxidation of methanol to formaldehyde.

TAP also assists in the discovery of new catalytic reaction chemistry for several reasons. First, the time-resolved mass spectrometry inherent in the method enhances the detection of unexpected products; and second, because TAP is a (rapid) transient experiment, it reveals the nature of the individual steps occurring on the catalyst. Lastly, in view of its ultrasensitivity it can detect minute amounts of products and reaction intermediates which escape detection by conventional methods. Thus, Ebner et al. discovered that in the ammoxidation of methanol over a Mn–P–O catalyst, ethylenimine, which is thermodynamically favoured below 350 °C, is formed in addition to the expected HCN:

$$3\,CH_3OH + 2\,NH_3 + \tfrac{3}{2}O_2 \longrightarrow C_2H_4NH + HCN + 6\,H_2O$$

(Methylamine is a well-known intermediate in the overall ammoxidation of methanol.)

3.10.2 Infrared, Raman, NMR, Mössbauer and X-Ray Absorption Spectroscopy for In-Situ Studies

Whereas the IR reflection–absorption technique has value for in-situ studies of model systems composed of single-crystal surfaces, it is in general ill-suited for powdered catalyst operating under realistic conditions. Several ingenious cell designs, based on dispersive or Fourier-transform modes, have been tested, the one by Hegedus and co-workers being of fast response and rugged, with the cell itself behaving as a well-mixed CSTR reactor (see Chapter 7). Such reactors are eminently suitable for the study of automobile-exhaust catalysts, where, *inter alia*, the reactions of CO and NO are monitored. Results of the kind recorded in situ by Wolf (Fig. 3.57) show the growth and decline of Pt–CO and Pt–NO surface groupings in real time. Robust heatable–evacuable cells have been designed for diffuse reflectance IR Fourier transform spectroscopy (DRIFTs), which have the advantage of recording both surface-bound and gas-phase species. However, reflectance spectra recorded thus can be quite complicated, because they consist of a combination of transmission

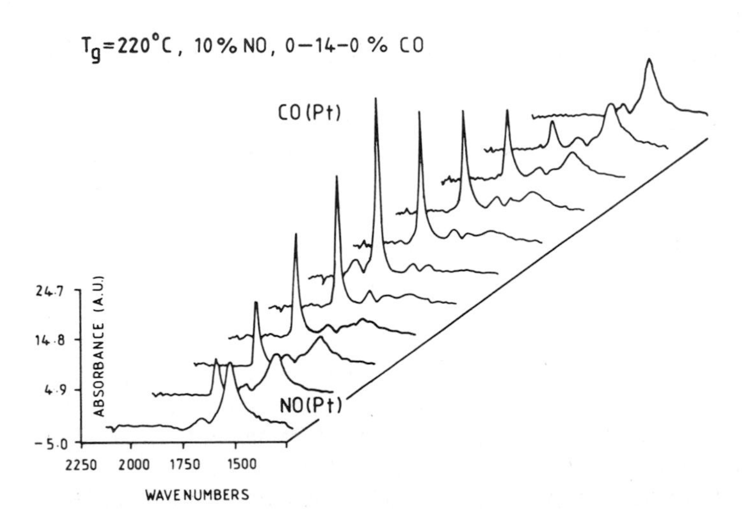

Figure 3.57 A platinum auto-exhaust catalyst converts CO and NO to N_2 and CO_2. In this experiment adsorbed NO is seen gradually to disappear and then reappear as the CO in the feed gas is first increased and then decreased. The infrared spectra also show the rise and fall of adsorbed CO (Courtesy E. E. Wolf).

and reflection effects, both internal and external. DRIFTS is also well suited to reveal dominant oxidation states and site symmetries of transition metal and lanthanide ions present in a catalyst, but without necessarily distinguishing surface from bulk residence. Using the DRIFTS technique, Baerns et al. disentangled the reaction steps in the oxidation of CH_4 to CO and H_2 over a Rh/γ-Al_2O_3 catalyst at 700 °C. Product distribution and the resulting IR band intensities of the respective adsorbates were strongly influenced by oxygen coverage and carbon deposits on the surface. CH_4 was dehydrogenated to carbonaceous deposits and H_2, and was simultaneously oxidized to CO_2 and H_2O. OH surface groups in the support were involved in the conversion of CH_x to CO via a reforming reaction. Corroborative evidence for this proposed mechanism came from TAP studies of the same system.

Joint work between the Topsøe Company, Denmark and the University of Wisconsin has combined in-situ FTIR and on-line catalytic activity measurements. The Topsøes have successfully investigated selective reduction of NO over vanadia–titania ('Denox') catalysts and also the presence of surface hydroxyl (and SH) groups and the acidic properties of Mo/Al_2O_3 hydrotreatment catalysts. They were led to the conclusion that SH groups may not only be determinants of the catalytic activity of reactions proceeding via Brønsted acid sites, but may also be the source of hydrogen during a catalytic cycle.

Laser Raman spectroscopy is well suited for in-situ work over wide ranges of temperature and pressure. It has the added advantage of being relatively insensitive to the type of support material (Al_2O_3, SiO_2, TiO_2) commonly used in heterogeneous catalysis. Extensive in-situ Raman studies of catalytically active transition metal oxides (nominally V_2O_5, MoO_3 and NiO) monolayers bound to 'inert' supports such as TiO_2 and Al_2O_3 have been performed by Wachs et al. V_2O_5 on anatase is of particular interest in that it is a highly selective catalyst for the conversion of *o*-xylene to phthalic anhydride.

Small crystallites of V_2O_5 (barely detectable by XRD) correspond, at 7% loading on the high-area anatase, to rather more than a monolayer. The sharp Raman band at 997 cm^{-1} is attributable to the symmetric stretch of the terminal V=O link in the V_2O_5 structure. This band is unique to V_2O_5. The peak intensity diminishes and there is broadening as the temperature is raised. Introduction of H_2 eliminates it; but it reappears when O_2 or air is admitted. In a separate study, Wachs et al. showed that, in the temperature range 350–575 °C, the optimal conditions for the catalytic conversion of *o*-xylene, a complete monolayer of surface vanadia remains intact on the anatase, but as the reaction proceeds it is partially reduced, with a drop in V=O intensity.

Raman spectroscopy not only reveals local, molecular modes, for the study of which it has so far been largely used in catalysis; it also reveals the phonon, i.e. collective, modes of the solid under investigation. Little advantage has been taken as yet of this other dimension. But one illustrative example deserves mention and that is in the field of graphite interlamellar compounds (graphite intercalates), which are selective catalysts for a number of processes. The Raman spectrum can simultaneously reveal the extent and nature of the intercalation – whether the guest species enters every interlamellar space of the graphite, or every alternate one – as well as the precise molecular identity of the intercalated species. Doubtless, similar information could be retrieved in the study of transition metal chalcogenides, which often have marked catalytic activity in hydrodesulphurization and related reactions.

In-situ NMR studies are readily carried out on catalysts (such as solid enzymes or clay intercalates) which function in the vicinity of room temperature. Often, all that needs to be done is to place the sample with reagents in an ordinary NMR cylindrical glass container; the mobility of the reactant is usually so high that sharp ^{1}H,^{13}C signals are obtained. In general, however, the solid catalyst needs to be spun rapidly at the 'magic angle' (for nuclei of spin $\frac{1}{2}$) and at variable angles (for quadrupolar nuclei). Details of the experimental arrangement are given in the book edited by two protagonists in this field, A. Pines and A. T. Bell (See Further Reading, Section 3.12). Anderson, Klinowski, Haw, Dobson and others have gained valuable insights into the mechanisms of the early stages of conversions such as the methanol-to-petrol (gasoline) process on zeolite H ZSM-5, by adsorbing one or more of the reagents methanol, ethanol, dimethyl ether and water and heating the sample to reaction temperature (typically 520 K) in the NMR probe while spectra were acquired with magic-angle spinning. Zamaraev and co-workers in Novosibirsk have extended such approaches in in-situ NMR studies of catalytic dehydration of alkanols over solid acids. Heaton and Iggo in Liverpool constructed a multinuclear NMR cell for the in-situ study of heterogeneous catalysts: it is capable of operating up to 510 K and in the pressure range 0.001–300 bar. Catalyst activation is carried out in an appropriate gas stream inside the NMR cell. This set-up proved successful in exploring the surface chemistry of methanol and H_2 on $Cu/ZnO/Al_2O_3$ industrial catalysts.

Berry et al., as well as Sinfelt and Dumesic, have shown the value of Mössbauer spectroscopy for in-situ characterisation of bimetallic catalysts. By focusing on the Mössbauer active nuclei ^{57}Fe and ^{193}Ir, they monitored the production of silica-supported iron iridium catalysts in H_2 and the formation of iron carbides following subsequent treatment in H_2 and CO. In-situ ^{151}Eu Mössbauer spectroscopy has shown that the reduction of Eu^{3+} in H_2 is more readily achieved in Eu^{3+}-exchanged zeolite Y than in Eu_2O_3.

X-ray absorption spectroscopy (see Section 3.10.3) is much more amenable for in-situ investigations of catalysts operating at high temperatures and pressures than well-nigh all other spectroscopic techniques, provided there is access to a synchrotron. Already good examples abound of the power of XANES and EXAFS, recorded in situ, on model and real catalysts. Abruña backed up his study of K-edge electrodeposited copper on Pt(111) surfaces with ab initio calculations of the XANES features and thereby identified the location of Cu atoms in the threefold hollow sites of the underlying platinum. Meitzner and Iglesia focused on the state of gallium in a working BP Cyclar catalyst (gallium-doped H ZSM-5) for the dehydrocyclodimerization of propane and butane to benzene and toluene. They showed that Ga^{3+} species present in the as-prepared, calcined catalyst undergo reduction during H_2 pretreatment or propane dehydrocyclodimerization. The reduced form of gallium (in the active catalyst), designated GaH_x, with the metal in a $\delta+$ oxidation state, resembles zero-valent gallium in the energy of the X-ray absorption edge but lacks the fine structure expected of gallium metal. These reduced gallium species appear to change from a bulk oxide-like structure into isolated $Ga^{\delta+}$ atoms (oxidation state < 1) that are coordinated to the equivalent of single basic oxygen within the zeolite channels.

X-ray absorption as an in-situ technique is even more powerful when it is used in conjunction with X-ray diffraction (XRD), as described in Section 3.10.4.

3.10.3 In-Situ X-Ray and Neutron Diffraction Studies

Ideally, students of catalysis would like to perform in-situ experiments with a temporal resolution that matches that associated with the rupture and formation of bonds, that is, on the picosecond (10^{-12} s $\equiv$ 1 ps) timescale. Are such experiments feasible? Yes, in principle – just about! The ideal experiment, therefore, is one in which a catalyst with well-defined active sites having reactants bound to it is subjected to a series of 'snapshot' pictures, using the probing technique of choice, at intervals of 10 ps or so. Such an experiment has been done on small molecules of *gas-phase* species – undergoing not catalytic but stoichiometric reaction – by sophisticated laser techniques such as direct femtosecond (real-time) mapping of the trajectories in the chemical reaction. Typical reactions probed in this way are the dissociation of I_2–Ar and Bi_2. Zewail has shown that femtosecond electron diffraction is also feasible for gas-phase species. But X-ray diffraction on *solid* catalysts at this level of temporal resolution (where transition states are directly probed) is not likely to be a feasible proposition. The best that can be achieved using the most advanced synchrotron sources is 10–100 ps time resolution; and the most appropriate recording procedure appears to be Laue-geometry diffraction with polychromatic X-rays (as in the study of crystalline enzymes with bound reactants). There are lessons in XRD to be learned from the molecular biologist.

The molecular biologists and X-ray crystallographers have been successful in probing the mechanisms of enzymatic conversion by time-resolved high-resolution X-ray diffraction. Take, for example, the crystallographic analysis of the catalytic mechanism of haloalkane dehalogenases, reported recently by Verschueren et al. Crystal structures of haloalkane dehalogenase were determined in the presence of the substrate (i.e. the reactant), 1,2-dichloroethane. At pH 5 and 4 °C, the reactant is bound in the active site without being converted. Warming to room temperature causes the reactant's carbon–chlorine bond to be broken, producing a chloride ion with concomitant alkylation of the active-site residue. At pH 6 and room temperature the alkylate enzyme is hydrolysed by a water molecule activated by the hisidine–aspartic acid pair in the active site. These results, taken with a time resolution on the minute-to-hour scale, yield deep insights into the mechanism of catalysts by the dehalogenase: it proceeds by a two-step process involving an ester intermediate covalently bound at the aspartic acid (designated Asp124). Likewise, Ravichandran et al., examining a haemoprotein domain (as a prototype of one of the numerous variants of the microsomal *P*-450s), by crystal structure analysis were able to shed great light on the general mechanism for proton transfer in *P*-450 enzymes.

These two successful examples represent the logical extension of the early work of Phillips on lysozyme and of Klug and co-workers (cf. Colour Plate 3) on the breakdown of transfer RNA by lead ions (see Chapter 1). This explanation of the mode of action of an enzyme, by determining its structure in the presence and absence of the bound reactant (substrate) or inhibitors, marked a turning point in enzymology. Phillips discovered, much as speculated by Pauling, that enzymes are molecules that are complementary in structure to the activated complexes of the reactions that they catalyse (that is, they accommodate the molecular configuration that is intermediate between the reacting substances and the products of the reaction). The work of Klug and co-workers was the direct forerunner of that of Verschueren et al. in that it entailed taking a difference Fourier analysis of the (Pb)RNA above and below a certain temperature where solid-state reaction rates became appreciable.

The crucial fact here is that enzymes are catalytically active at temperatures conveniently close to room temperature and that the reactant is bound, like a tortured victim, within the well-defined cleft that serves as the locus of the active site. The secure bonding of the reactant to the active site arises because of the shape-selectivity of the enzyme.

Molecular sieve catalysts are also characterized by their shape selectivity. And they too may accommodate organic reactant (or inhibitor) species in well-defined cavities that function as the active site. The great difference between zeolitic (molecular) sieve catalysts and their enzymatic analogues is in the temperature range at which catalysis proceeds. Because the zeolites generally become active catalysts at much higher temperatures, and because the reactants are generally less strongly bound in the catalytic cavity, where they execute rapid translational, vibrational and rotational motion, it has not yet proved feasible to perform the snapshot (before and after) sequence of structural studies by diffraction that is now almost a routine feature of the crystallographic study of enzymes. Even so, considerable progress has been achieved in monitoring the process of activation of zeolitic catalysts, as in the tracking (by Rietveld powder profile refinement analysis) of the movement of transition-metal ions (such as Ni^{2+}) from buried sites to more accessible and catalytically active sites.

When the entire bulk of a solid is implicated in catalytic turnover – as is the case for so-called 'uniform' heterogeneous catalysts such as zeolites and many mixed oxides which function via the redox mechanism (of Mars and van Krevelan), where there is sacrificial use of structural oxygen in the solid – in-situ XRD is particularly useful. Many appropriate cells have been designed for this purpose by Rayment, Boudart, Kuroda and Iwasawa.

Pickering et al., in a time-resolved in-situ XRD study of $Li_x{}^{+}Ni_{1-2x}{}^{2+}Ni_x{}^{3+}O$ ($x = 0.45$), which is active in the oxidative coupling of methane (CH_4), found that, in the presence of gaseous O_2 the bulk structure remains essentially unchanged for long periods. During this time, the catalyst is selective for C_2 production. By contrast, in the absence of gaseous O_2, the bulk structure immediately starts to decompose, yielding successively a total of four rock-salt-type lithium–nickel oxide phases. Here the initial selectivity for C_2 production is 100% and declines as the solid breaks down. The in-situ measurements yielded direct correlation between the appearance of different solid phases during the experiments, with variations in the rate of production of gaseous products.

There are several approaches to time-resolved X-ray diffraction.

(a) A conventional angle-scanning diffractometer can be used to perform short scans (in a narrow range of 2θ). This approach is suitable only for the examination of processes over extended times, owing to the time difference between the recording of the beginning and end of the diffraction.
(b) A *p*osition-*s*ensitive *d*etector (PSD) can be used simultaneously to record either the entire monochromatic diffraction pattern or a substantial section of it. The timescale on which structural processes can be studied, usually a few seconds, is limited by the counting electronics of the detectors.
(c) In the *e*nergy-*d*ispersive *X*-ray *d*iffraction (EDXD) technique, data are collected using 'white' synchrotron radiation in an experiment similar to TOF neutron diffraction (see Fig. 3.50). The detector is again placed at a fixed 2θ value, but the different wavelengths are resolved, not by time-of-flight, but by an energy-dispersive X-ray detector. This has the advantage that the whole diffraction pattern is accumulated simultaneously. EDXD permits rapid collection of data; the ultimate timescale of the experiment is limited by the counting electronics. With its fixed-angle arrangement, this approach is better suited for the study of catalysts under severe conditions of temperature and catalysts. It has been used by Cheetham, Thomas and Jones to monitor structural changes in pyrochlore catalysts, typically Eu_2IrO_7, for methane reforming by CO_2.

As mentioned earlier, the great advantage that neutrons offer as in-situ probes of operating catalysts is that they readily penetrate steel-walled reactors and so may be used to characterize catalysts held in high pressures of reacting gas. The potentially powerful technique of in-situ neutron diffraction has yet to be fully exploited, although there has been modest use of the Rietveld powder profile method on variable-temperature runs of uniform heterogeneous catalysts. If high neutron fluxes are available, it is possible to collect complete diffraction patterns in a few minutes, provided that efficient detection systems are used. This offers much scope for in-situ studies of catalysts under operating conditions. As Cheetham has shown in comparable contexts, neutron diffraction is well suited for in-situ studies because: (i) most materials do not significantly absorb neutrons (this simplifies the construction of the reaction cell); (ii) processes involving light atoms (hydrogenation, dehydrogenation, cracking, isomerisation, etc.) can be followed; and (iii) a fixed-angle detector can be used with time-of-flight techniques (see Fig. 3.50), again considerably simplifying the construction of reaction cells. However, even the most advanced neutron sources offer considerably lower fluxes than those available with modern X-ray sources, especially synchrotron storage rings. It is therefore not surprising that synchrotron X-ray diffractometers are able to follow dynamic processes on a much shorter timescale.

3.10.4 Combined X-Ray Absorption and X-Ray Diffraction for In-Situ Studies of Catalysts

Combined techniques always tend to yield more than the individual ones separately or than their simple addition. This is well illustrated for homogeneous nickel cata-

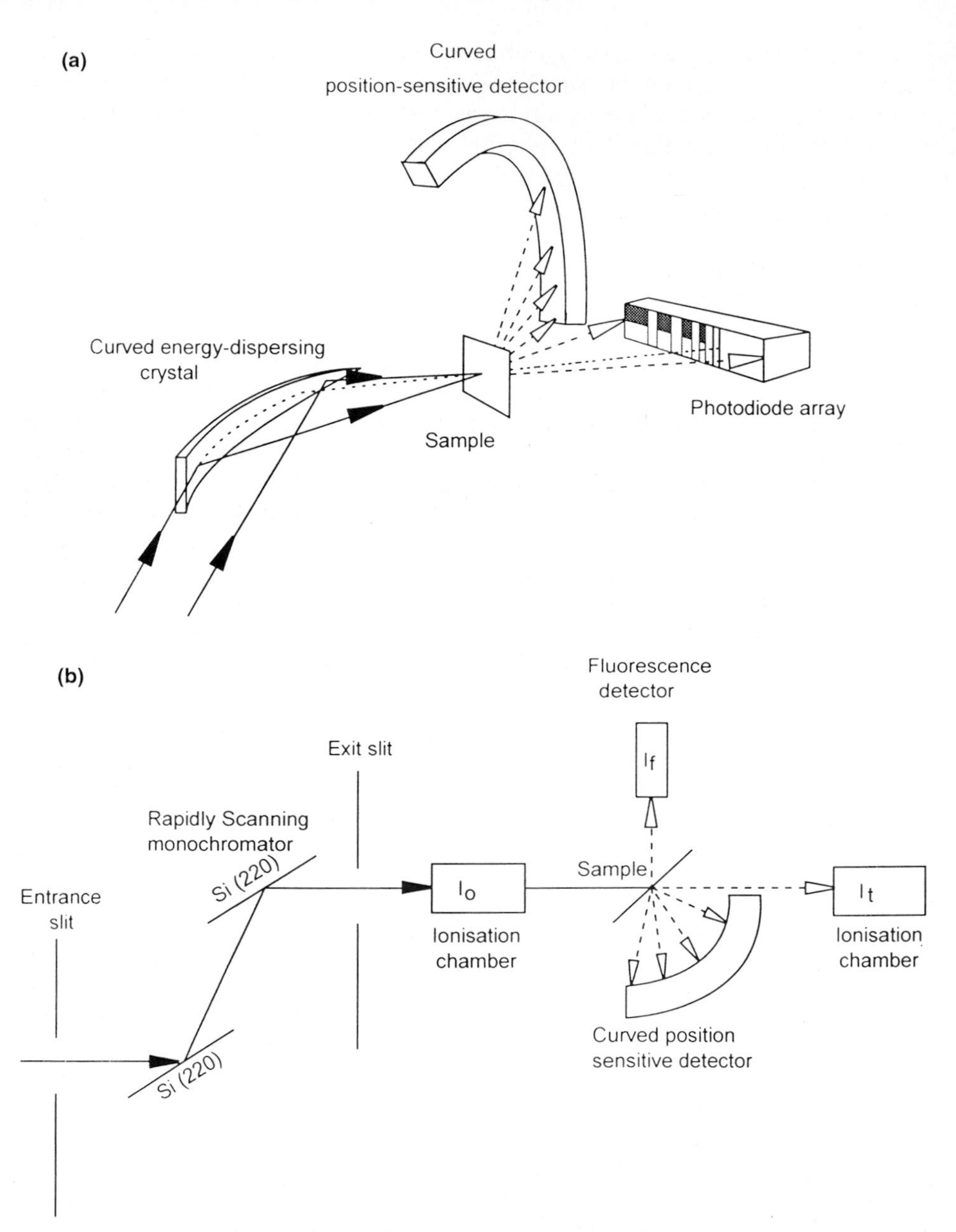

Figure 3.58 Two methods of employing combined EXAFS and XRD for in-situ studies. In (a) (top), a bent-tapered silicon crystal disperses the synchrotron radiation, and the X-ray absorption spectrum by the sample is recorded by a cooled photodiode array. The position-sensitive detector records the X-ray diffraction pattern. In (b) (bottom), a rapidly scanning silicon-based monochromator permits conventional (transmission) spectra as well as fluorescent yield X-ray absorption spectra to be recorded. (After J. M. Thomas, G. N. Greaves, C. R. A. Catlow, *Nucl. Instrum. Methods* **1995**, *397*, 1.)

lysts for propylene dimerisation in particular and alkene oligomerisation in general. Corker and Evans, combining in-situ NMR and EXAFS studies, were able to determine that the nickel coordination sphere of the highly active catalyst formed on addition of $Ph_2PCH_2C(CF_3)_2OH$ to $Ni(COD)_2$ (COD ≡ cyclo-octa-1,5-diene) is

$$Ni\{Ph_2PCH_2C(CF_3)_2O\}H\{Ph_2PCH_2C(CH_3)_2OH\}$$

Combined multinuclear solid-state NMR, FTIR and XRD have been used to good effect by van Bekkum et al. to establish the structure of supported heteropolyacid catalysts. For heterogeneous catalysts, however, where the aim is quantitatively to define the environment of the active site in the solid surface, the most attractive combination of in-situ techniques is X-ray absorption and X-ray diffraction, using one or other of the experimental arrangements shown in Fig. 3.58. In each case the sample is contained in a reaction vessel fitted with an appropriate X-ray transparent window, on a support capable of being heated up to 1000 °C in a controlled atmosphere. Gas chromatography is used to follow the changes in gas-phase composition.

For uniform heterogeneous catalysts like zeolites or the equally crystalline aluminium phosphate-based catalysts represented by MeALPOs, where Me is a divalent cation (Co^{2+}, Ni^{2+}, Zn^{2+}, Mn^{2+}, Mg^{2+}, etc.) that isomorphously substitutes for some of the Al^{III} in the framework, combined EXAFS–XRD is ideal for in-situ studies. Thus, framework-substitution of trivalent Al^{III} by divalent ions (such as Co^{II}) not only confers Brønsted acidity upon the resulting solid – a proton, loosely attached to one of the framework oxygens is required to maintain electroneutrality – it also introduces redox behaviour (see Fig. 3.59(a). When the cobalt is calcined in oxygen to Co^{III} the Brønsted acidity vanishes, only to reappear following reduction. In-situ studies of the special CoALPO-18 catalyst (for conversion of methanol to alkenes) by combined X-ray absorption and XRD yielded the information given in Fig. 3.59(b).

This is just one way in which, using synchrotron radiation, two distinct techniques are brought synergistically together. There are others. For example, advantage may be taken of *d*ifferential *a*nomalous *X*-ray *s*cattering (DAXS) where, by systematically changing (tuning) the X-ray wavelength used for diffraction through the X-ray absorption edge of a key constituent element in the catalyst, anomalously large changes in the scattering factor of that element are induced. Hence, the difference in diffraction intensities at slightly different photon energies close to the element's absorption edge will tend to stand out. In a DAXS in-situ study of $Pt/\zeta\text{-}Al_2O_3$, Liang et al. were able to discover that, at very low loadings of platinum, the metal formed microcrystalline clusters that had face-centred cubic packing.

Reflecting on the numerous techniques outlined in this chapter, one of the most important messages to note is that the scientist–technologist concerned with unravelling the intricacies of catalytic action now has available an impressive range of fundamental techniques capable of tracking the detailed structural and electronic changes that accompany the actual process of heterogeneous catalysis.

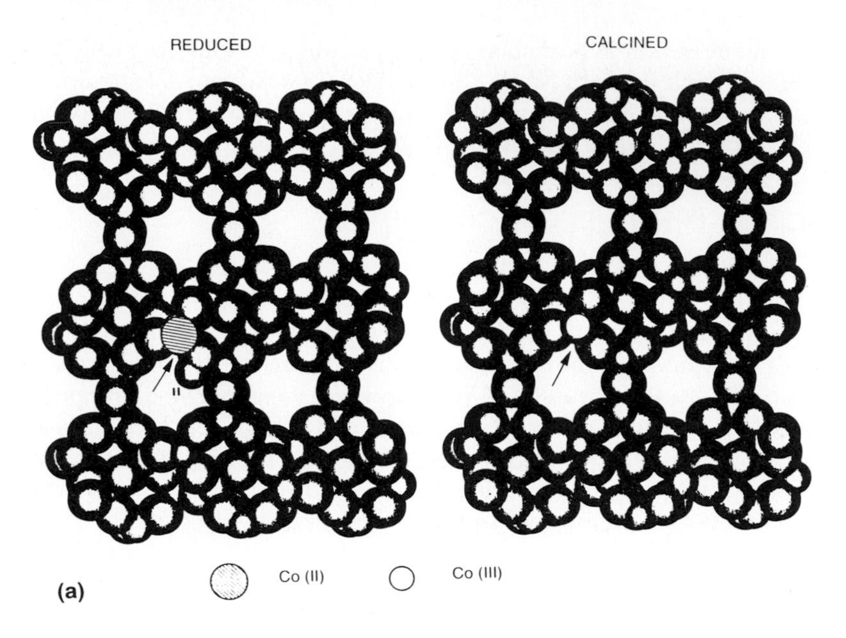
REDUCED
CALCINED
II
(a)
Co (II)
Co (III)

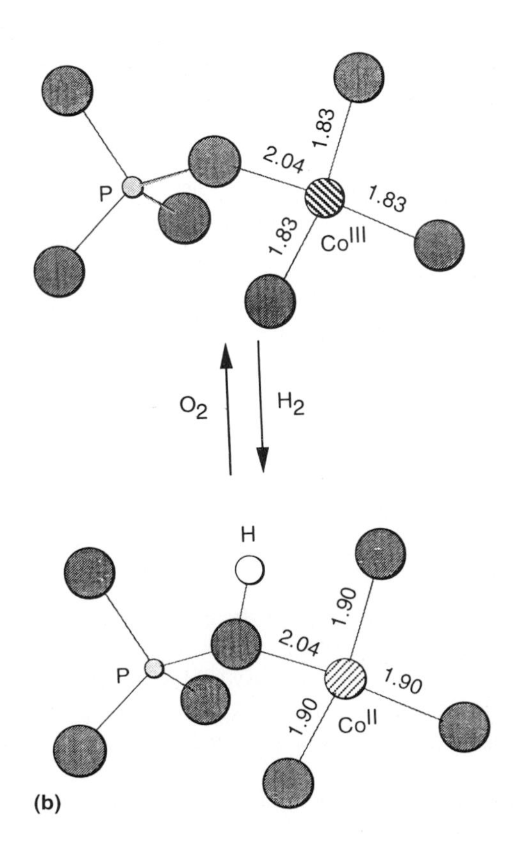
P
2.04
1.83
1.83
1.83
Co^{III}
O_2
H_2
H
P
2.04
1.90
1.90
1.90
Co^{II}
(b)

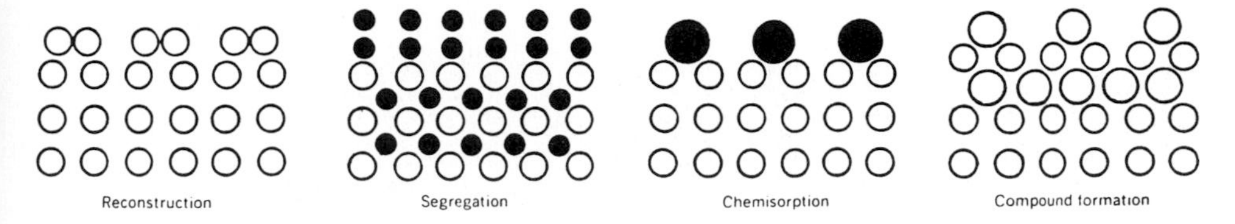

Figure 3.60

3.11 Problems

1 (a) From the accompanying list of binding energies for core electrons, compute the kinetic energies you would expect when soft X-rays (Al K_α) impinge upon a ruthenium catalyst previously exposed to carbon monoxide.

Core level	Binding energies [eV]		
	C	O	Al
K	284	532	1560
L_{II}	7	7	74

(b) It is surmised that several forms of 'elemental' carbon are present at the surface of an active supported ruthenium catalyst. What methods would you use to characterise the carbon?

2 With the aid of the schematic illustration shown in Fig. 3.60, amplify what is meant by surface reconstruction, surface compound formation, segregation and chemisorption.

Suppose that the solid in question is nickel, the segregated species carbon, the chemisorbed species sulphur and that the compound formed is an oxide, explain which experimental techniques you would use to identify the nature of the surface in each case. (P.J. Estrup, *Phys. Today* **1975**, *28*, 34.)

3 The dissociative chemisorption of H_2 on the (110) face of rhodium model catalyst leads to a sequence of lattice gas phases: p(1×3) at $\theta = \frac{1}{3}$, p(1×2) at $\theta = \frac{1}{2}$, (1×3)–2H at $\theta = \frac{2}{3}$ and a (1×2)–3H at $\theta = \frac{3}{2}$. Finally, at $\theta = 2$, a (1×1)–2H phase forms. Sketch the LEED patterns and the corresponding surface structural model for each of these phases.

Figure 3.59 (a) Representation of cobalt-containing ALPO-18 catalyst. The framework is closely similar to that of the aluminosilicate mineral chabazite. In ALPO-18 all the Si is replaced by P and a few per cent of the Al is replaced by Co^{II}. To conserve electroneutrality a proton is loosely bound to a framework oxygen, but is lost when the Co^{II} is oxidized to Co^{III}. The active site is arrowed. (b) Bond distances extracted from EXAFS analysis. (After J. M. Thomas, G. N. Greaves, G. Sankar, P. A. Wright, J. Chen, A. J. Dent, L. Marchese, *Angew. Chem., Int. Ed. Engl.* **1994**, *33*, 1871.)

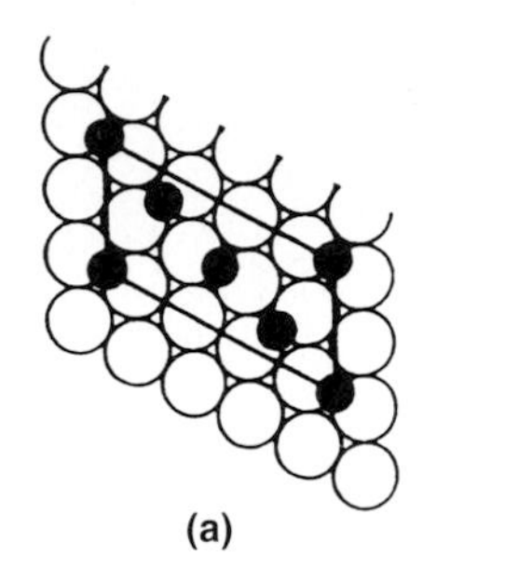
(a)

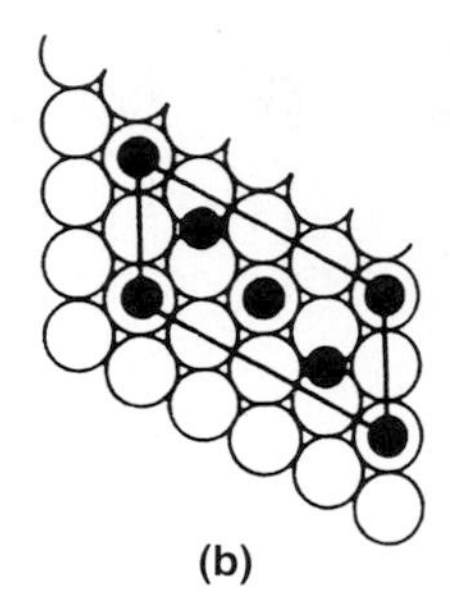
(b)

Figure 3.61

4 Two models (Fig. 3.61 a and b), each compatible with the observed LEED patterns, have been proposed for the (111) faces of platinum and nickel half-covered with non-dissociatively bound carbon monoxide. These models are represented as CO c(4 × 2).

Explain how, from additional spectroscopic evidence, you would distinguish between these two models. What light would SIMS (secondary ion mass spectrometry) studies shed on this problem?

5 At 300 K the highly electropositive metal yttrium chemisorbs chlorine dissociatively according to the equation

$$Cl_2(\text{gas}) + Y(\text{surface}) \rightarrow 2Cl(\text{adsorbed on Y}) \qquad \Delta H(\theta = 0) = -145\,\text{kJ mol}^{-1}$$

Estimate the radius of the adsorbed species, stating clearly any assumptions or approximations that you make. Explain

(a) how you would attempt to measure the extent of charge transfer between absorbate and substrate;

(b) how you would measure the strength of the Y–Cl chemisorption by a method other than calorimetry;

(c) how you expect the properties of the chemisorbed layer to change as surface coverage increases; and

(d) which crystal planes of yttrium should preferentially chemisorb chlorine.

Figure 3.62 shows the photoelectron spectra of the Y 3d core levels following chemisorption of chlorine at 300 K (A) and 900 K (B). What can you deduce from this?

Work function of yttrium = $310\,\text{kJ mol}^{-1}$
Electron affinity of the chlorine atom = $370\,\text{kJ mol}^{-1}$
Dissociation energy of Cl_2(gas) = $240\,\text{kJ mol}^{-1}$
$e = 1.602 \times 10^{-19}\,\text{C}$
$4\pi\varepsilon_0 = 1.11 \times 10^{-10}\,\text{F m}^{-1}$
$N_A = 6.022 \times 10^{23}\,\text{mol}^{-1}$

6 Outline briefly the principles which underlie the development of surface-sensitive electron spectroscopies.

The (100) plane of tungsten exhibits an electronically driven surface recon-

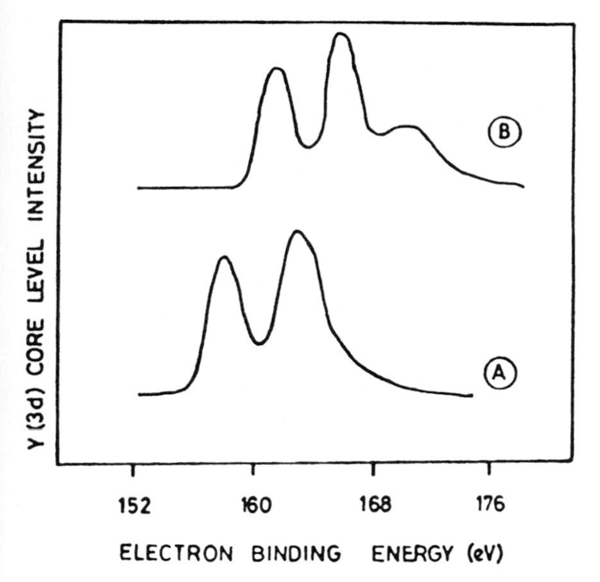

Figure 3.62

struction at low temperatures; the UPS ($h\nu = 40.8$ eV) shows a pronounced feature in the valence band, close to E_F, which is *not* characteristic of bulk tungsten. Explain how these phenomena arise.

At 300 K LEED shows that W(100) reverts to the normal (1 × 1) structure (Fig. 3.63 (a)); examination of the intensity/energy curves for the individual diffraction beams reveals the presence of many non-Bragg maxima. A *very small* dose of Cl_2 leads *immediately* to the formation of the LEED pattern in Fig. 3.63(b).

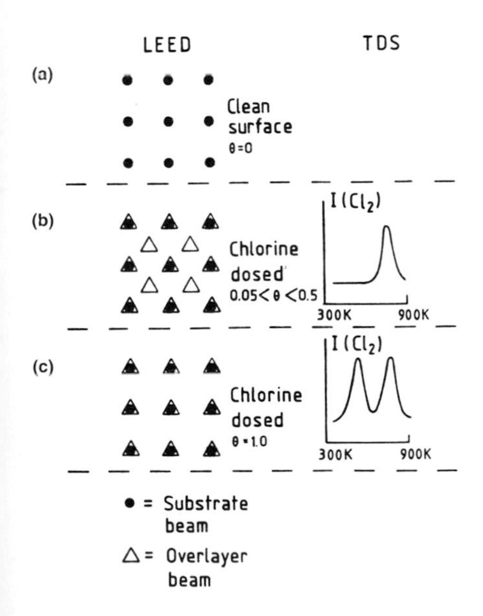

Figure 3.63

Further Cl_2 dosing up to a gross surface coverage of $\theta = 0.5$ increases the intensity of this pattern to a maximum value. Continued Cl_2 dosing causes some of the overlayer beams in Fig. 3.63 (b) to face until a (1×1) LEED pattern is eventually recovered at $\theta = 1.0$ (Fig. 3.36 (c)). Cl_2 thermal desorption spectra corresponding to $\theta = 0.5$ and 1.0 are also shown in Fig. 3.63 (b), (c).

At $\theta = 1.0$ the chlorine Auger signals at 180 eV and 2040 eV exhibit intensities of 3 and 1.2 units respectively; XPS ($h\nu = 1486$ eV) shows a single W(4d) core level binding energy of 910 eV. Very prolonged Cl_2 dosing leaves the 180 eV Cl Auger signal essentially unchanged, the 2040 eV signal increases to 2.6 units and a new XPS peak occurs which corresponds to a W(4d) core level binding energy of 916 eV.

Account for these observations in as much detail as you can.

7 The X-ray absorption spectrum (L-edges) at 100 K of a 2.5 μm platinum foil is shown in Fig. 3.64.

(a) Explain what the various features in this spectrum signify and, in particular, how Pt–Pt distances, along with other structural information, may be extracted from such a spectrum.

(b) A platinum–(10%)iridium cluster catalyst (Pt–Ir/Al_2O_3) used for hydrocarbon reforming is of such high dispersion as to yield very poor X-ray diffraction patterns. It can, however, be characterised by X-ray absorption studies using synchrotron radiation. What differences would you expect between the L-absorption spectrum of this dispersed catalyst and that of the platinum foil, and what kind of quantitative information can be gleaned from this spectrum?

	Energy [keV]		
	L_I	L_{II}	L_{III}
Ir	13.42	12.82	11.22
Pt	13.88	13.27	11.56

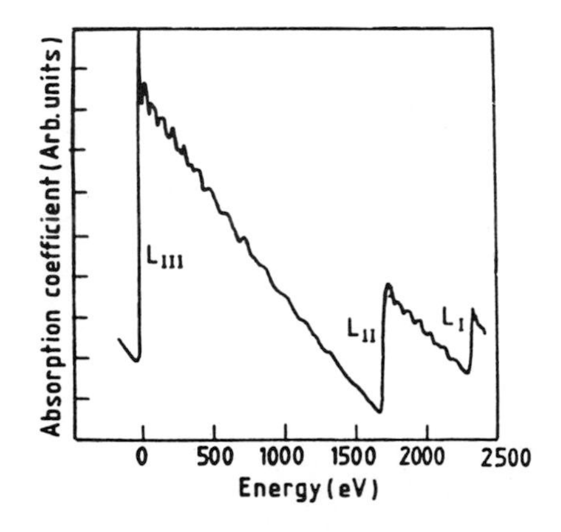

Figure 3.64

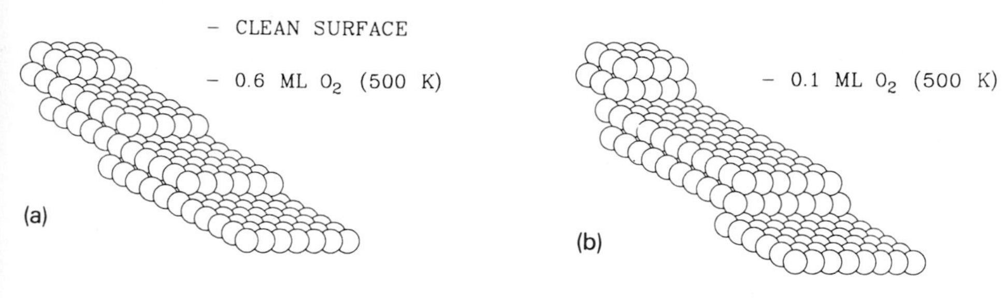

Figure 3.65

8 'Adsorbate-induced step-doubling and other reconstructions of metal surfaces are of great value in understanding the mode of operation of catalysts'. By reference to the paper by G. Hoogens and D. A. King (*Surf. Sci.* **1993**, *286*, 306), from which the strutures shown in Fig. 3.65 for the Rh{332} surfaces are taken, examine the meaning and the validity of this statement.

9 'The structure and bonding of transition-metal oxide catalysts have been clarified by studies of X-ray absorption near-edge structures (XANES)'.

Outline the principles of this method; and with the aid of the data given in Fig. 3.37 for a series of vanadium compounds elaborate on its scope and applicability.

What additional structural information is obtainable from studies of the extended X-ray absorption fine structure of these compounds?

10 The overall Si/Al ratio, determined by X-ray fluorescence, of a La–Y cracking catalyst is 2.58. The ^{29}Si MASNMR spectrum of this catalyst, along with the intensities of the various Si–nAl peaks ($n = 4, 3, 2, 1, 0$), is shown in Fig. 3.66. Determine whether the intrazeolite cavities of this catalyst are free of aluminium-containing species. [Derive the equation used for determining Si/Al ratios.]

11 Quasi-crystalline MoS_2 is an important hydrodesulphurization catalyst – it removes S atoms (as H_2S) from thio compounds such as thiophene in petroleum-

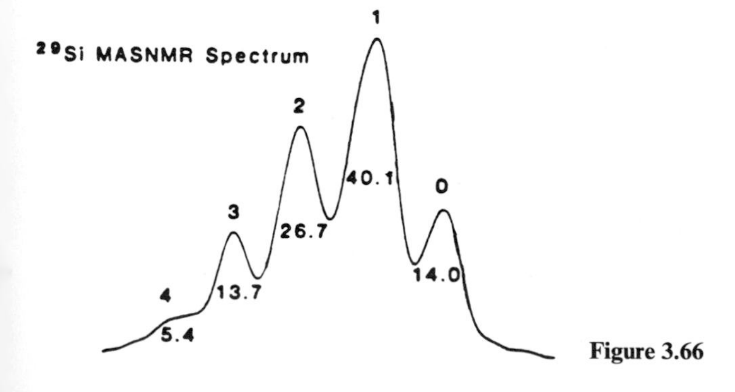

Figure 3.66

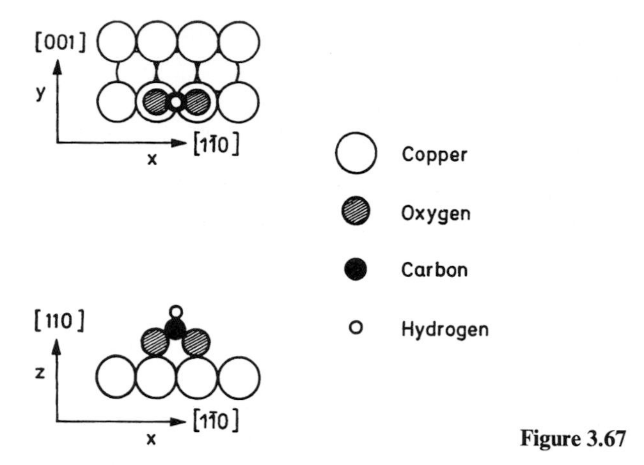

Figure 3.67

based products. A particular catalyst sample was shown, by N_2 BET studies, to have an area of 32 $m^2 g^{-1}$. Hydrogen uptake measurements revealed that a stoichiometric formula $MoS_2O.2H_2$ was established at 200 °C.

Is the H_2 present on the surface or in the bulk? How would you determine the chemical state of the bound hydrogen?

12 The structure shown in Fig. 3.67 has been proposed (by A. M. Bradshaw, D. P. Woodruff et al., *Surf. Sci.* **1988**, *201*, 228) for the formate species bound to a copper surface. By consulting the original paper, explain how photoelectron diffraction and SEXAFS data lead to this structure, in which the Cu–O bond distance is 1.98 Å.

13 Molecular oxygen when weakly chemisorbed onto a metal catalyst is suspected of acquiring a fractional charge *d* (less that one electronic charge). How would you go about estimating, by an experimental method, the magnitude of this surplus charge?

14 Briefly outline the essential principles of the techniques or procedures you would use to ascertain:
(a) whether hydrogen is dissociatively chemisorbed at a cobalt surface;
(b) whether carbon monoxide is associatively bound to a Ni(100) surface;
(c) the nature of the chemisorbed linkage when formic acid is bound to a copper surface;
(d) the extent of charge transfer between adsorbate and a metal adsorbent.

15 By reference to a short review (J. M. Thomas, G. N. Greaves, *Science* **1994**, *265*, 1675), summarize the key methods currently available for in-situ structural studies of heterogeneous catalysts. What are the advantages in pursuing combined studies (using synchrotron radiation) of X-ray absorption and X-ray diffraction?

16 Discuss the claim that NEXAFS is the best technique for determining the structure of adsorbed layers on solid catalysts.

17 Pt/TiO_2 and Ni/SiO_2 are useful methanation catalysts. Whereas the turnover frequencies for these catalysts at 225 °C are about equal (*ca* 4×10^{-4} per metal surface atom per g), the lifetimes of the surface intermediates differ by a factor of some 100. What do you make of this statement? Explain how the transient response method of kinetic analysis enables both the lifetime and coverage of the surface intermediates to be determined.

18 Explain the following spectroscopic observation. The HREEL spectrum of O_2 bound to a Pt(111) surface shows a peak at a lower frequency than that corresponding to the HREEL spectrum of the O_2^- ion.

19 After first consulting reviews by G. Ertl (*Angew. Chem., Int. Ed. Engl.* **1990**, *29*, 1219) and J. M. Thomas (*ibid.* **1994**, *33*, 913), trace how our understanding of real catalysts has progressed from studies of single-crystal surfaces. In particular, focus on the advances made through scanning electron microscopy and scanning tunnelling microscopy.

20 Summarize the experimental evidence for the following statements.
(a) H_2 is dissociatively chemisorbed on polycrystalline ZnO, but CO is non-dissociatively bound to Ni.
(b) –SiOH groups situated adjacent to substitutional Al are catalytically active sites in SiO_2–Al_2O_3 and zeolitic catalysts.
(c) There are several distinct chemisorbed states for oxygen on Rh(110).

21 It was once thought that surface ethylidyne groups played a key role in the hydrogenation of ethylene on Pd/Al_2O_3. Which techniques, and which particular strategies, would you use to test this belief?
(*Hint*: In-situ infrared spectroscopy and isotopic labelling are of great value in such investigations – see T. P. Beebe Jr., J. T. Yates Jr., *J. Am. Chem. Soc.* **1986**, *108*, 663.)

22 When *p*-xylene is intercalated by sheet silicate catalysts, the sharpness of both the ^1H and ^{13}C conventional NMR spectra show that the sorbed guest is quite mobile. Sketch the kind of spectra that you would expect for both the ^1H and ^{13}C nuclei of this xylene. How useful would such spectral measurements be in monitoring the catalytic addition of either water or methanol to intercalated 2-methyl propene (isobutene). (See C. A. Fyfe et al., *Angew. Chem., Int. Ed. Engl.* **1984**, *20*, 271.)

3.12 Further Reading

Many of the texts cited at the end of Chapter 1 (Section 1.6) should be consulted. Other worthy references, cited in this chapter, are:

M. Alario Franco, Zhou Wuzong, D. A. Jefferson, J. M. Thomas, *J. Phys. Chem.*, **1987**, *91*, 512.

A. Baiker, Proc. 11th Intl. Congress on Catalysis, Baltimore (Eds. J. W. Hightower, W. N. Delgass, E. Iglesia, A. T. Bell) Elsevier, Amsterdam, **1996**, p. 51.

J. Barton, *Phys. Rev. Lett.*, **1988**, *61*, 1356.

A. T. Bell, A. Pines (Eds). *NMR Techniques in Catalysis*, M. Dekker, New York, **1995**.

B. S. Clausen, H. Topsøe. *Catal. Lett.*, **1993**, *20*, 2.

M. L. Deviney, J. F. Gland (Eds.), *Catalyst Characterization Science: Surface and Solid State Chemistry*, ACS Symp. Series No. 288, American Chemical Society, Washington, DC, **1985**.

T. J. Dines, C. H. Rochester, J. Thomsen (Eds.) *Catalysis and Surface Characterisation*, Royal Society of Chemistry, Cambridge **1992**.

R. Doepper, A. Renken, *Chimia* **1996**, *50*, 61.

D. J. Dwyer, F. M. Hoffman (Eds.), *Surface Science of Catalysis: In Situ Probes and Reaction Kinetics*, ACS Symp. Series No. 482, American Chemical Society, Washington, DC, **1992**.

W. E. Farneth, R. J. Gorte, *Chem. Rev.*, **1995**, *95*, 615.

P. L. Gai-Boyes, *Catal Rev-Sci. Eng.* **1992**, *34*, 1.

L. F. Gladden, *Chem. Eng. Sci.* **1994**, *49*, 3339.

D. W. Goodman, *Chem. Rev.*, **1995**, *95*, 523.

J. F. Haw, J. B. Nicholas, T. Xu, L. W. Beck, D. B. Ferguson, *Acc. Chem. Res.* **1996**, *29*, 259.

M. A. van Hove, A. Wander, G. A. Somorjai, *Surf. Sci.* **1993**, *287/288*, 428.

B. Imelik, J. C. Vedrine (Eds.), *Catalyst Characterization: Physical Techniques for Solid Materials*, Plenum, New York, 1994.

Y. Iwasawa, *Proc. 11th Intl. Congress on Catalysis, Baltimore* (Eds. J. W. Hightower, W. N. Delgass, E. Iglesia, A. T. Bell), Elsevier, Amsterdam, **1996**, p. 21.

W. Jones, J. M. Thomas, M. J. Tricker, and R. K. Thorpe, *Appl. Surf. Sci.*, **1978**, *1*, 388.

R. W. Joyner, R. A. van Santen (Eds.), *Elementary Reaction Steps in Heterosgeneous Catalysis*, Kluwer, Dordrecht, **1993**. (Section 3 includes important articles by Knözinger, Iwasawa, Joyner, Salmeron and Ponec on in-situ methods).

D. A. King, P. Hu, *Appl. Surf. Sci.*, **1993**, *70/71*, 390.

D. C. Koningsberger, R. Prins (Eds.), *X-ray Absorption, Principles, Applications* (EXAFS, SEXAFS and XANES). Wiley, New York, **1988**.

K. Krischer, G. Flätgen, B. Pettinger, K. Doblhofer, H. Junkes, G. Ertl, *Science*, **1995**, *269*, 668.

J. H. Lunsford, H. Knözinger, *J. Phys. Chem.* **1993**, *97*, 13810.

B. J. McIntyre, M. Salmeron, G. A. Somorjai, *Catal. Lett.*, **1996**, *39*, 5.

T. Maschmeyer, F. Rey, G. Sankar, J. M. Thomas, *Nature*, **1995**, *378*, 159.

J. A. Niemantsverdriet, *Spectroscopy in Catalysis*, VCH, Weinheim, **1993**.

S. T. Oyama, W. Zhang, *J. Phys. Chem.*, **1996**, *100*, 10759.

J. B. Pendry, K. Heinz, W. Oed, *Phys. Rev. Lett.*, **1988**, *61*, 2953.

J. B. Pendry, in *New Methods of Modelling Processes within and on the Surfaces of Solids* (Eds. C. R. A. Catlow, A. M. Stoneham, J. M. Thomas) Oxford University Press, Oxford, **1993**, p. 101.

I. J. Pickering, J. M. Thomas, *J. Chem. Soc. Farad. Trans.*, **1991**, *87*, 3067.

A. Reller, D. A. Jefferson, J. M. Thomas, *Proc. Roy. Soc.*, **1984**, *A394*, 223.

M. R. Roberts, *Surf. Sci.*, **1994**, *299/300*, 769.

H. Rotermund, G. Haas, R. U. Franz, R. M. Tromp, G. Ertl, *Science*, **1995**, *270*, 608.

D. Sanfilippo (Ed). *The Catalytic Process from Laboratory to the Industrial Plant* (Proc. of the Third Seminar on Catalysis, Rimini, June 1994, published by the Italian Chemical Society: This is an invaluable source).

R. Schlögl, H. Schubert, U. Tegfmeyer, *Catal. Lett.*, **1994**, *28*, 383.

N. Sheppard, C. De La Cruz, *Adv. Catalysis,* **1996**, *41*, 1.

A. Szöke in *Short Wavelength Coherent Radiation: Generation and Applications*, D. T. Attwood and J. Baker (Eds.) Amer. Inst. Phys. Conf. Proc. No 147, New York **1986**.

O. Terasaki, G. R. Millward, J. M. Thomas, *Proc. Roy. Soc.*, **1984**, *A395*, 153.

J. M. Thomas, *Ultramicroscopy* **1982**, *8*, 13.

J. M. Thomas and G. N. Greaves, *Science* **1994**, *265*, 1675.

J. M. Thomas, J. Klinowski, *Adv. Catalysis*, **1985**, *33*, 199.

J. M. Thomas and K. I. Zamaraev, *Topics Catal.*, **1994**, *1*, 1.

I. E. Wachs, F. D. Hardcastle, *J. Phys. Chem.* **1991**, *95*, 5031.

R. Wiesendanger, H.-J. Guntherod (Eds.) *Scanning Tunneling Microscopy II* (Second Edn.) Springer, Berlin **1995**.

B. G. Williams, T. G. Sparrow, J. M. Thomas, *Acc. Chem. Res.* **1985**, *18*, 324.

E. E. Wolf, F. Qin, *Catal. Lett.* **1996**, *39*, 19.

4 The Significance of Pore Structure and Surface Area in Heterogeneous Catalysis

4.1 The Importance of Pore Structure and Surface Area

Gas reactions catalysed by solid materials occur at the exterior and interior surfaces of the porous catalyst. The rate of product formation is a function of the available surface area and so it follows that the greater the amount of surface area accessible to the reactants, the larger is the throughput (amount of reactant converted to product per unit time per unit catalyst mass). The only exceptions to this principle are those catalytic reactions in which the rate of conversion is limited by mass transport of reactants from the gas phase to the solid catalyst material (an example is the oxidation of ammonia by a platinum catalyst at relatively high temperatures, as discussed in Chapter 8). It is customary to disperse metal catalysts throughout the entire internal surface area of some suitable porous support, such as high-area silica or γ-alumina, in order to create a large specific surface area which is entirely accessible because of its open pore structure. On the other hand, depending on their mode of preparation, metal oxide catalysts often have a sufficiently high surface area and open pore structure for them to be employed directly, although chemical modification will often promote additional activity.

If the pores of the catalyst material or support are sufficiently wide not to impede the passage of reactants or products and the internal surface is also energetically homogeneous, then the rate of conversion of reactants to products is directly proportional to the specific surface area. We shall see later in this chapter that a narrow pore structure limits the reaction rate. Under these circumstances the reaction rate is either proportional to the square root of the specific surface area or independent of it, depending on the mode of diffusion within the pore structure. Furthermore, because of the *a-priori* heterogeneous energy distribution amongst catalyst sites inherent in most catalyst preparations, the effect of which is to render certain areas of the catalyst surface more active than others, the activity of the catalyst depends on the way in which this heterogeneity is distributed over the available surface area. Such *in-situ* heterogeneity, however, is usually only a small fraction of the total chemically active surface, and departure from a direct proportionality between rate of reaction and specific surface area cannot be accounted for solely by virtue of heterogeneity.

One of the earliest applications of surface area measurement was the prediction of catalyst poisoning. If, on continued use, the activity of a catalyst declines more rapidly than any decrease in surface area, then poisoning may be suspected. Conversely, if a decrease in surface area is concomitant with reduced activity, then thermal deactivation is indicated. Another application provides a method of assessing the efficacy of catalyst supports and promoters. A support or promoter may either increase the surface area available for adsorption and subsequent reaction or it may increase the catalyst activity per unit surface area. Hence, surface area measurement is an important expedient in predicting catalyst performance and determining the role which the catalyst surface plays in any heterogeneous gas reaction. It should be emphasised, however, that often only a small fraction of the surface area determined by physical techniques is chemically active.

Surface area is by no means the only physical property which determines the extent of adsorption and catalytic reaction. Equally important is the pore structure of the catalyst material or support, which, although contributing to the total surface area, must be regarded as a separate factor. This is because the distribution of pore sizes in a given catalyst preparation may be such that some of the internal surface area is completely inaccessible to large reactant molecules and, furthermore, may restrict the rate of conversion to products by impeding the diffusion of reactants and products throughout the porous medium. Accordingly, it is an advantage to know something about the pore structure of a catalyst. Commercial catalysts usually have a high internal surface area. If this were not so, the external surface, being quite small, would quickly become poisoned and the catalyst would rapidly lose activity. To be able to predict the correct pore size necessary to achieve a required activity and selectivity requires a pore model of the catalyst. Commercial catalysts do not have simple pore structures, so that one aspect of the general problem of reaction rates and selectivity in catalyst pores is the selection of an appropriate model which will reflect the experimentally measured surface area and pore volume. Once a suitable model has been identified, the remaining problem of how the chemical kinetics (and hence the rate of formation of product) are affected by diffusion in the pore structure (intraparticle transport) can be tackled. Furthermore, a realistic prediction can then be made of any likely improvement in catalytic activity and selectivity resulting from a judicious choice of catalyst particle and pore dimensions.

Having emphasised the utility of surface area and pore size determination, we shall now review the more important methods of studying surface area and pore volume by methods which are particularly pertinent to an estimation of the internal surface area of porous solids. Subsequently we turn to the problem of modelling porous structures, before concluding this chapter with a discussion of diffusion and reactivity in porous catalysts.

4.2 Experimental Methods of Estimating Surface Areas

There now exist a number of methods for determining the surface area of a porous solid. These have been reviewed in various texts but three methods persist which are relatively simple to apply and do not require extensive and sophisticated experimental equipment. The first method we shall discuss is the volumetric adsorption of an inert gas, the second is the monitoring of adsorption by gravimetric means, and finally a dynamic method employing the continuous flow of an inert gas and adsorbate through an adsorbent bed will be considered.

4.2.1 The Volumetric Method

In Chapter 2 (Section 2.4.2) it was shown that various isotherm shapes describe the adsorption of a gas by a solid at constant temperature. Provided the gas–solid system conforms to isotherm types I, II, or IV, it is possible to employ a volumetric method for the assessment of the surface area of a porous or non-porous solid. The crucial question is whether, by some means, the completion of an adsorbed monolayer can be detected from the isotherm shape, and this is possible only for isotherm types I, II and IV.

If the isotherm is of type I, a Langmuir or empirical Langmuir equation will often describe the adsorption adequately. We recall from Chapter 2 that the form of the Langmuir equation is

$$\frac{V}{V_{\rm m}} = \frac{bp}{1 + bp} \tag{1}$$

Even if the equation does not obey the theoretical constraints implicit in the Langmuir theory, an empirical equation of the same form can be constructed to fit a type I isotherm. Although the constant b in such an empirical equation would not have the same theoretical significance as in the Langmuir theory, the asymptotic value of V at high relative pressure will conform to the volume of gas adsorbed in a monolayer when saturation of the surface has been achieved. Recasting Eq. (1) as

$$\frac{p}{V} = \frac{p}{V_{\rm m}} + \frac{1}{bV_{\rm m}} \tag{2}$$

a plot of p/V against p will yield a straight line, the slope of which is the monolayer volume $V_{\rm m}$.

When the isotherm is of either type II or IV, the monolayer capacity may be identified either by noting the ordinate value of the volume (when V is plotted against p) as the isotherm bends over sharply or by applying the Brunauer–Emmett–Teller (BET) theory. The former method is euphemistically referred to as the 'point B' method and is only reliable when there is a well-defined and sharp change of curvature in the isotherm as shown in Fig. 4.2, below. When point B is well defined, then the volume read off on the ordinate gives the monolayer volume $V_{\rm m}$. Alternatively, the BET equation (see Section 2.6.2.4)

$$\frac{p}{V(p_0 - p)} = \frac{1}{V_{\rm m}c} + \frac{(c-1)}{V_{\rm m}c}\frac{p}{p_0} \tag{3}$$

may be applied when a plot of $\{p/V(p_0 - p)\}$ against p/p_0 (where p_0 is the vapour pressure of the adsorbate at the adsorption temperature) yields a straight line, the slope and intercept of which will provide the value of $V_{\rm m}$.

Once the monolayer volume has been ascertained, it is a simple matter to calculate the specific surface area of the adsorbent. If the specific monolayer volume is recorded in m^3/g (expressed at standard temperature and pressure), then the number of moles adsorbed in the monolayer is $V_{\rm m}/0.0224$. Multiplying by Avogadro's number (6.023×10^{23} molecules/mole) gives the number of molecules in the monolayer.

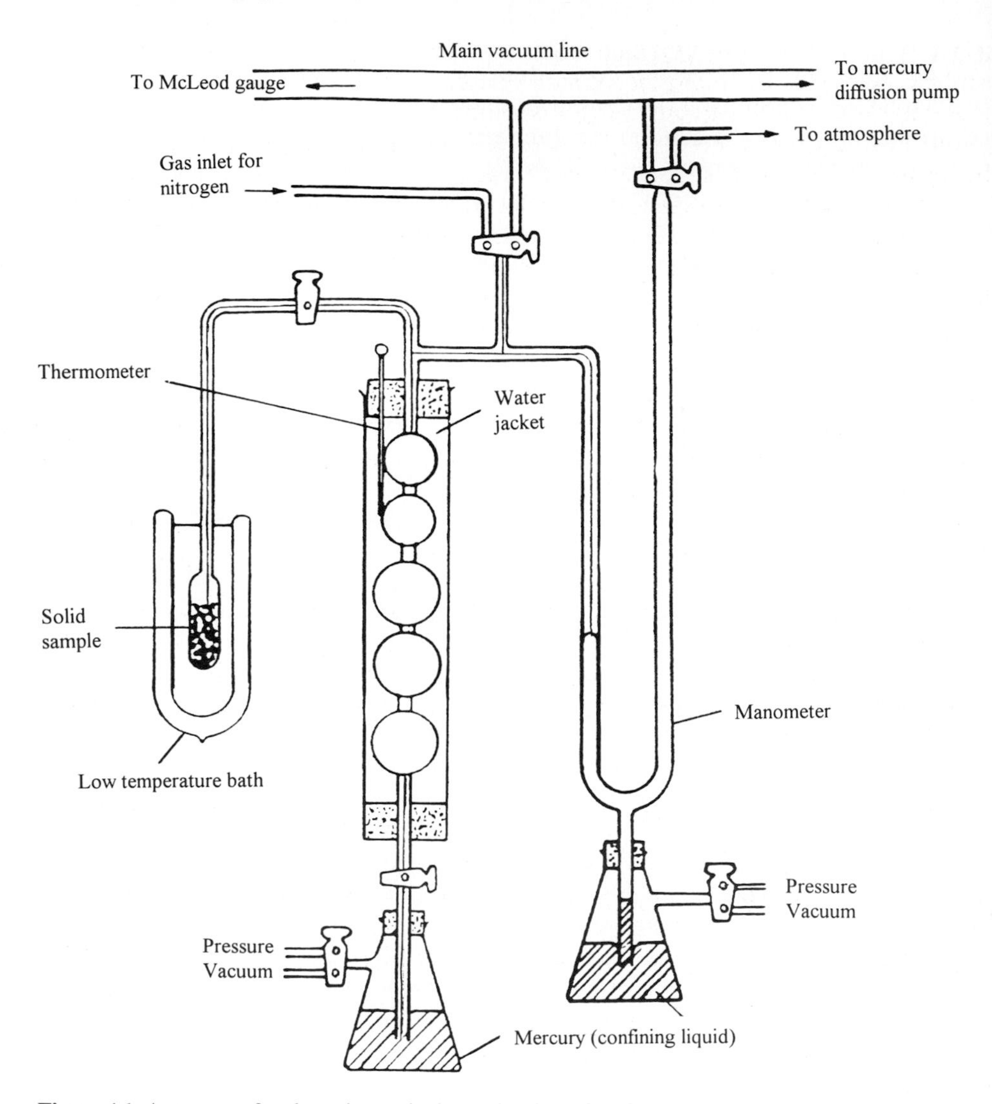

Figure 4.1 Apparatus for the volumetric determination of surface area.

If the area occupied by each adsorbate molecule is $A\,\text{m}^2$ (sometimes expressed as Ångstrøms squared, which is equivalent to $10^{-20}\,\text{m}^2$), then the specific surface area in m^2/g is clearly

$$S_g = \frac{V_m}{0.0224} \times 6.023 \times 10^{23} \times A = 2.69 \times 10^{25}\, V_m A\,\text{m}^2\,\text{g}^{-1} \tag{4}$$

Most adsorbents yield a type II or IV isotherm when nitrogen is used as adsorbate at 77 K (the boiling point of N_2 at 1.05 bar) and, provided the total specific surface area of the sample is not less than about $0.2\,\text{m}^2$, the use of nitrogen is quite satisfactory for the experimental determination of monolayer volume. A relatively unsophisticated form of apparatus for surface area determination for samples of more than about 20 m^2 is illustrated in Fig. 4.1. Nitrogen from a storage vessel is admitted to the previously evacuated and calibrated gas burette. The pressure is then recorded by means of the manometer when the mercury in the burette is at a particular graduation mark between two successive bulbs. This measured amount of nitrogen is then shared with the evacuated bulb containing the adsorbent sample at 77 K and the equilibrated pressure is measured. The apparent volume of the empty adsorbent bulb at 77 K will have been determined previously by volume-sharing experiments using nitrogen and by application of the gas laws. After a series of successive nitrogen doses delivered from the gas burette to the adsorbent, sufficient information will have been acquired to plot an adsorption isotherm or, alternatively, a linearised BET plot (see Eq. (3)) from which the value of V_m, and hence the specific surface area S_g, can be calculated. Alternative designs of apparatus, using nitrogen, have been described for measuring surface areas as low as 0.2 m^2.

Example

When nitrogen was employed to determine the surface area of a 0.50 g sample of silica gel, the following results were obtained:

Equilibrium pressure p [kPa]	0.8	3.3	5.0	6.3	7.5	9.0	11.2	18.7	30.7	38.0	42.7
Volume adsorbed (STP), $V \times 10^6$ [m^3]	3.1	6.4	6.7	7.0	7.2	7.4	7.7	8.5	9.9	10.7	11.5

The sample of silica gel was maintained at the normal boiling point of liquid nitrogen (*ca* 77 K). The area of plane surface which a single molecule of nitrogen would occupy is 16.2×10^{-20} m^2. Calculate the specific surface area of silica gel (a) using the point B method, (b) by linearising the Langmuir equation and (c) by the BET method.

Solution

(a) The plot of V against p is shown in Fig. 4.2, which yields an ordinate value of 7.7×10^{-6} m^3 at point B. Hence, $V_m = 15.4 \times 10^{-6}$ m^3 g^{-1}. From Eq. (4) the value of the specific surface area is 67.1 m^2 g^{-1}.

(b) The plot of p/V against p is shown in Fig. 4.3. It is valid only for adsorbed amounts less than or equal to the monolayer volume and so only the first seven points are extracted from the table above and plotted. The slope of the straight line is 11.6×10^4 m^{-3}, the reciprocal of which is 8.62×10^{-6} m^3. Hence $V_m = 17.2 \times 10^{-6}$ m^3 g^{-1}. From Eq. (4) then, $S_g = 75.1$ m^2 g^{-1}.

(c) To apply the BET Eq. (3), the saturated vapour pressure p_0 of nitrogen at 77 K is required. Now 77 K is the normal boiling point of nitrogen at 1.05 bar so $p_0 = 101.3$ kPa. Figure 4.4 shows the data plotted in the form $p/V(p_0 - p)$ against p/p_0. If the slope of such a plot is s and its intercept is i, then it is clear

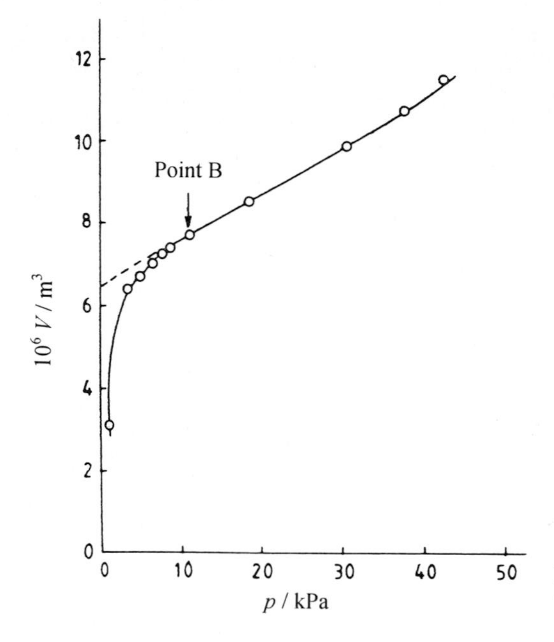

Figure 4.2 Location of 'point B' on isotherm.

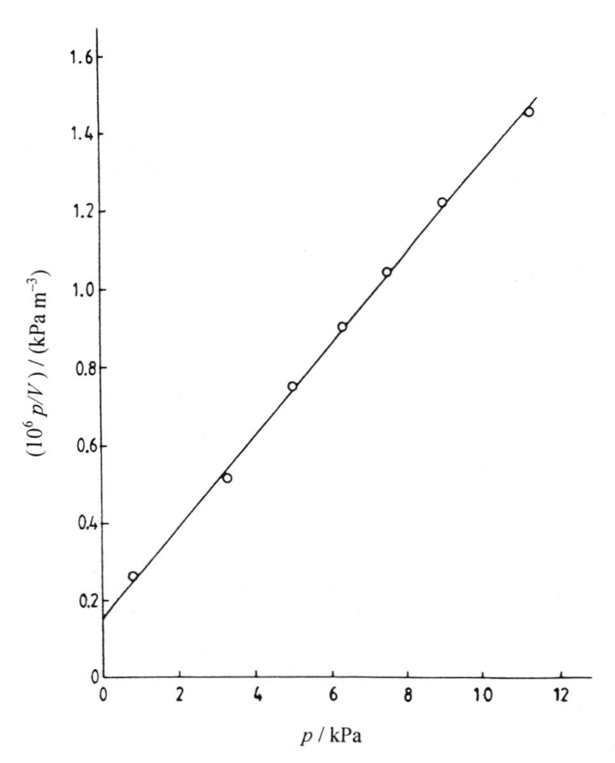

Figure 4.3 Plot of p/V against p to determine surface area.

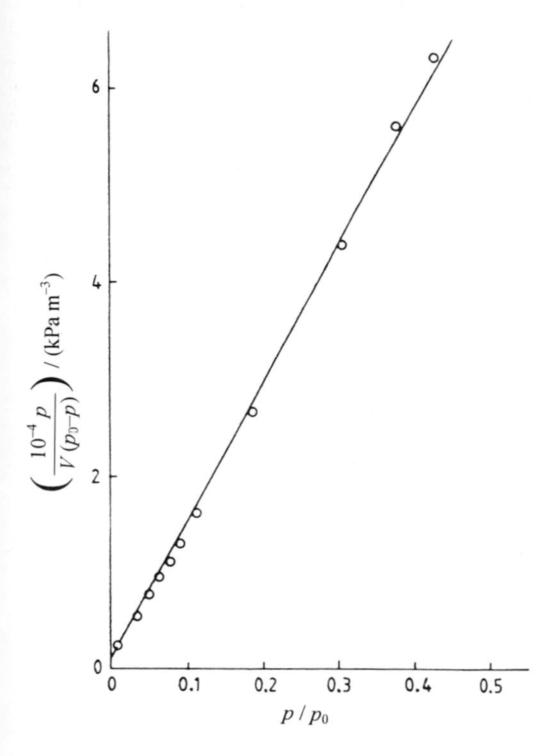

Figure 4.4 Linearized BET plot to determine surface area.

from Eq. (3) that $1/(s+i)$ is equal to V_m. From the graph displayed in Figure 4.4, $s = 13.8 \times 10^4\,m^{-3}$ and $i = 1.3 \times 10^3\,m^{-3}$. Recalling that these values are for 0.5 g of sample, then $V_m = 7.18 \times 10^{-6}\,m^3$ for 0.5 g or $14.4 \times 10^{-6}\,m^3\,g^{-1}$. Hence, from Eq. (4), $S_g = 62.6\,m^2\,g^{-1}$.

The discrepancy between the BET-derived value for the surface area and the values derived from the point B method and the linearised Langmuir equation illustrates the dangers in relying on the estimation of a single point either by inspection (point B method) or even by predicting point B from the Langmuir equation. In the example above, point B is not particularly well defined. Only if a sharp bend occurs in the isotherm at the monolayer volume is the point B method of value. On the other hand, the BET equation, even though it is now regarded as empirical, does take into account the possibility of multilayer adsorption during the process of adsorption and consequently it gives a more reliable estimate of monolayer completion.

When adsorbent samples of less than $0.2\,m^2$ are to be examined, it is advisable to employ an adsorbate with a much lower saturated vapour pressure than nitrogen, so that the volume remaining unadsorbed in the space between the calibrated dosing volume or burette and the sample bulk (often referred to as the dead space) is small. For this purpose krypton is suitable ($p_0 = 2.32 \times 10^{-2}$ bar at 77 K). Appropriate forms of apparatus have been described in standard texts for surface area measurement using krypton as adsorbate.

4.2.2 The Gravimetric Method

Very precise measurements of adsorbed amounts of gas can be made by recording the weight of the adsorbent during the course of adsorption. Great advances have been made in recent years in the construction of highly sensitive microbalances capable of recording mass changes of as little as 10^{-8} g. Two types of gravimetric balances are in common use, namely the spring balance and the beam balance, both capable of recording complete isotherms and hence of measuring surface areas of solids.

For physical adsorption measurement, which includes the estimation of surface areas using nitrogen or other inert gas as adsorbate, the spring balance is appropriate and relatively inexpensive to construct, though it does require skill in its assembly and care in operation. Its main advantage arises because the adsorbed amount is directly measurable and the amount of adsorbate remaining in the dead space volume is of no consequence. The uptake of adsorbate by adsorbent is followed by measuring the extension of a spring, on the end of which is suspended a small container holding the sample. Silica is a suitable material from which the spring can be fashioned; it has no elastic hysteresis so that calibration need not be repeated frequently. If l is the length of the quartz fibre and a its radius, the extension z for a mass m adsorbed is given by

$$z = \frac{lr^2}{\pi y a^4}(2m + l\mu) = km + \lambda \tag{5}$$

where r is the radius of the coil into which the spring is formed, y the torsional modulus of the quartz material and μ the mass per unit length of the silica. The sensitivity of the spring balance can be increased by a decrease in radius of the fibre material, by an increase in radius of the coil or by an increase in length (equivalent to increasing the number of turns in the helix), though there is a limit to the sensitivity which can be achieved because of the constraints imposed by mechanical strength and coil length. Sensitivities of 1 in 10^4 have been quoted in the literature for silica. Other material, such as a copper–beryllium alloy of a more robust nature than silica, has been used to form the helix. As Eq. (5) indicates, it is not necessary to know in advance the separate parameters l, r, a, y and μ. The spring balance can be calibrated *in situ* by adding known weights and noting the extension of the spring.

An alternative to the spring-type balance is the beam balance which may be pivotal, knifeedge or gravity-operated. Such balances are more sophisticated than the spring-type balance and require greater skill in operation. Figure 4.5 is an illustration of the arrangement used for the study of water vapour adsorption on polyurethane foam samples. The microbalance used is of the beam type, each balance arm constructed from fused quartz and balancing on pivots. Details of the design are sketched in Figure 4.5. A sensitivity of 10^{-6} g can be achieved with a microbalance system of this kind and is adequate for obtaining a precision of 1% in the weight of nitrogen covering an area of 1 m^2 (*ca* 2.8×10^{-6} g). Optical and electromagnetic arrangements for detecting the beam deflection when a weight change occurs are necessary; consequently, construction of a microbalance with high sensitivity requires great skill. Commercial instruments are, however, available.

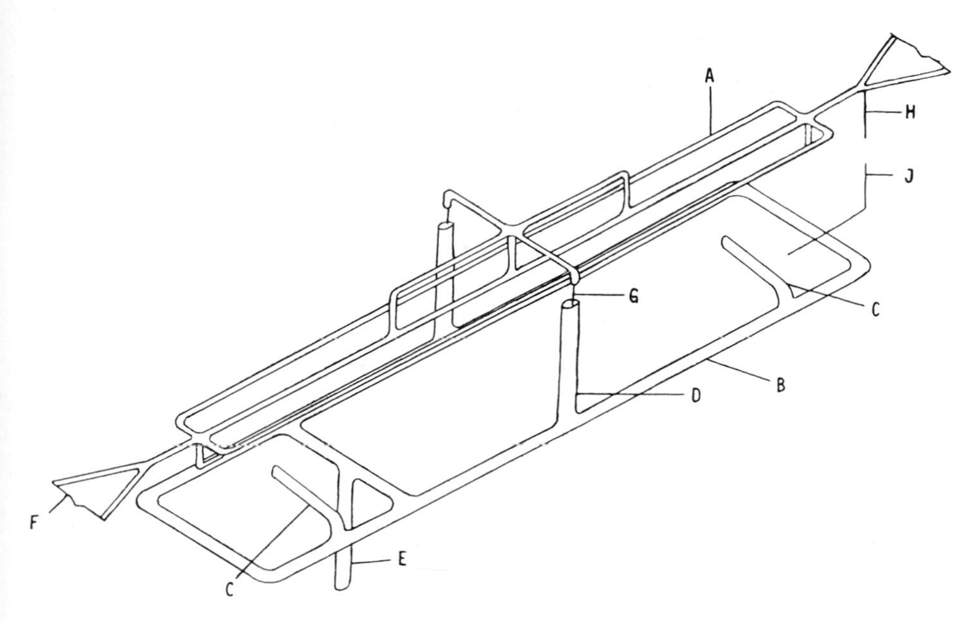

Figure 4.5 Sensitive beam-type balance used for gravimetric method of determining adsorption isotherm and hence surface area. A, balance beam; B, base frame; C, beam arrests; D, pivot supports; E, foot for securing beam assembly into balance envelope; F, thinned quartz supports for hangdown fibres; G, tungsten pivots 0.018 mm in diameter; H,J, reference wires. With permission from *Chem. Eng. Sci.* **1974**, *29*, 5499.

Two sources of error in measurement have to be taken into account when either a spring- or beam-type balance is used for recording gas adsorption. Unless rather elaborate precautions are taken to diminish or eliminate the temperature gradients between sample and counterweight in the limbs of the microbalance, serious disturbances arising from thermomolecular flow will be encountered. Quantitative estimates of the magnitude of such spurious effects and the means of eliminating them have been discussed (see Further Reading at the end of this chapter). Buoyancy forces are also often the source of errors, especially in the spring-type balance where a counterweight cannot be used to compensate for the effect. The forces due to buoyancy, however, can be calculated easily from a knowledge of the sample and holder volume and application of the gas laws.

4.2.3 The Dynamic Method

As changes in gas concentration can easily be monitored by gas chromatography techniques, the continuous flow of an adsorbable gas through a column of adsorbent affords a method of constructing an isotherm and hence of determining surface area. The equipment is easier to assemble than that for either of the techniques just described and does not require any means of evacuation. Hence it has become popular as a rapid method of surface area determination demanding little or no maintenance

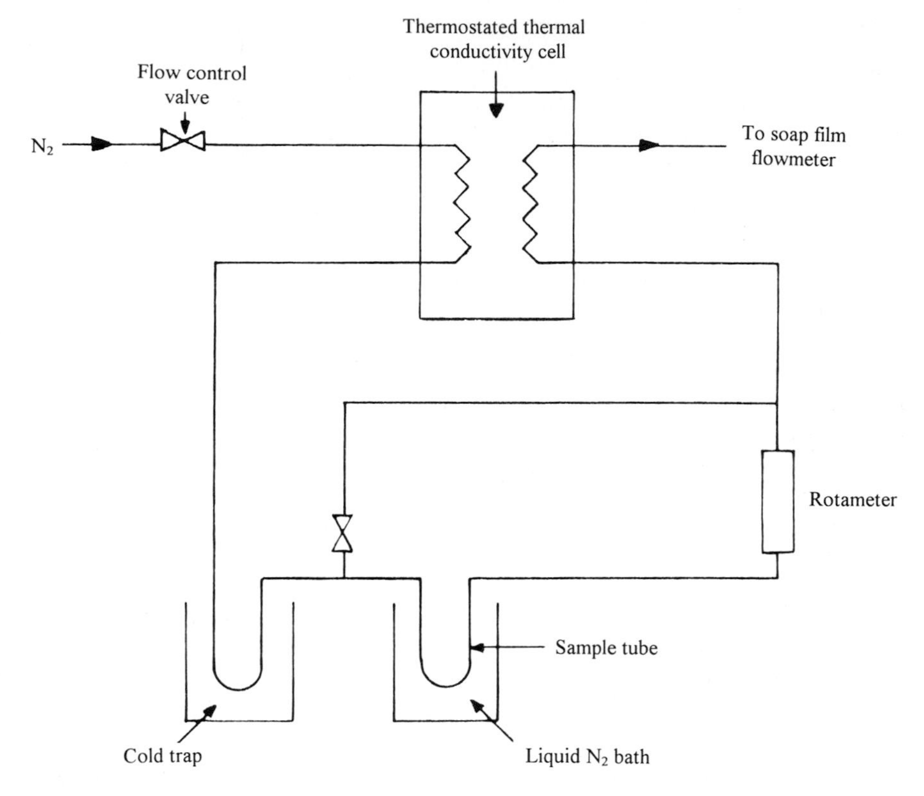

Figure 4.6 Equipment arrangement for dynamic method of estimating surface area.

as opposed to the volumetric and gravimetric techniques. Provided that the total sample area is only of the order of a few metres squared, accuracies in measurement of ±2% are attainable. Careful outgassing of the solid adsorbent sample is necessary, however, and the adsorbate must be thoroughly dried by passing it through a column containing a molecular sieve before it is allowed to flow through the bed of adsorbent.

Figure 4.6 is a schematic diagram of the equipment arrangement. Nitrogen is used as adsorbate and helium (a permanent gas not adsorbed significantly by most solids) is employed as carrier gas and diluent. A twin thermal conductivity cell, whose response is proportional to nitrogen concentration, is the means of detection and a millivolt recorder attached to the thermal conductivity cell indicates any difference in concentration between the reference and sample. A four-way sample valve of accurately known volume is incorporated to divert a sample of nitrogen and carrier gas through the adsorbent bed (brought to 77 K in a liquid nitrogen bath) and thence to the thermal conductivity cell, where it is compared with the gas which has plowed through the reference arm; this is the same nitrogen–helium mixture but it has not been passed through the adsorbent bed and therefore has not experienced adsorption. The volume of gas diverted through the sample will be adsorbed and equilibrium between gas phase and solid phase will be rapidly established. Consequently a

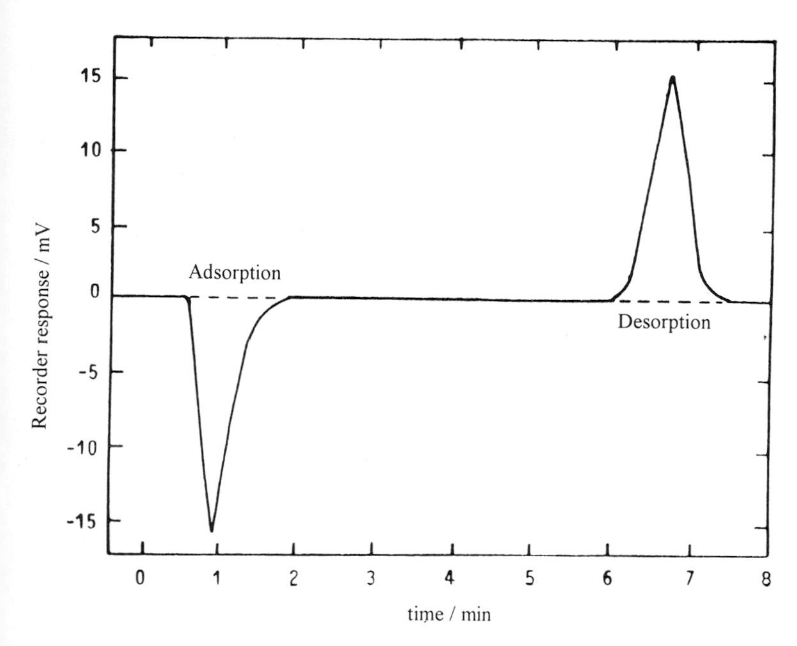

Figure 4.7 Record of gas adsorption and desorption by dynamic method.

downward peak (as shown in Figure 4.7) will appear on the recorder chart. The area contained between the peak and the steady horizontal recorder line is proportional to the amount of gas adsorbed at the partial pressure of nitrogen in the nitrogen–helium mixture. Calibration of peak areas can easily be accomplished by employing a number of different but known volumes of nitrogen. The nitrogen partial pressure can be altered by adjusting the flows of nitrogen and helium. Because the shape of the adsorption isotherm (concave or sometimes convex to the relative pressure axis) and also the dispersion of concentration by the adsorbent bed affect the sharpness or otherwise of the peak obtained, it is often better to record the peak area on desorption (immediately after removing the liquid nitrogen trap), when a sharper peak above the horizontal baseline is obtained. Repetition of adsorption–desorption sequences employing different nitrogen partial pressures quickly provides the data necessary to construct an isotherm and hence to determine the surface area of the sample.

4.3 Experimental Methods of Estimating Pore Volume and Diameter

One of the problems which is associated with catalysis by porous materials is the estimation of a mean pore diameter or, better still, a pore size distribution. Simple methods of determining pore volume are inadequate, since knowledge of, at least, a mean pore radius is essential if it is desired to predict the effect of pore size on reaction rates.

Two methods will be discussed: one is based on gas adsorption and is suitable for the estimation of pore sizes in the range 15–200 Å ($\triangleq 1.5 \times 10^{-9}$–$2 \times 10^{-8}$ m), and the other is based on the volume of mercury which can be forced under pressure into the pores of a solid and is suitable in the pore size range 100–10^5 Å ($\triangleq 10^{-8}$–10^{-5} m).

4.3.1 Gas Adsorption Method of Estimating Pore Volume and Diameter

The gas adsorption method of estimating pore volume and diameter is based upon the fact that gas condenses to liquid in narrow pores at pressures less than the saturated vapour pressure of the adsorbate. A simple equation relating the lowering of the vapour pressure above a cylindrical column of liquid contained in a capillary, to the radius r of the capillary, may be obtained by equating the work done in enlarging a spherical drop of liquid to the work done in adding molecules to the interior of the drop.

The work done in enlarging the surface area of a drop of liquid is $\gamma \Delta S$ where γ is the surface tension and ΔS the change in surface area (equivalent to $8\pi r \Delta r$ for a spherical drop of original radius r). Now the work done in adding molecules to the drop interior is $(\mu_0 - \mu)\,\Delta n$ where μ_0 is the chemical potential of the vapour over a plane liquid surface, μ the corresponding potential over the curved surface, and Δn the increase in the number of moles of liquid. Thus

$$8\pi r \Delta r \gamma = (\mu_0 - \mu)\Delta n \tag{6}$$

Now the increase in volume of the drop will be given by

$$V \Delta n = 4\pi r^2 \Delta r \tag{7}$$

where V is the molar volume of the liquid. From Eqs. (6) and (7), therefore,

$$(\mu_0 - \mu) = \frac{2\gamma V}{r} \tag{8}$$

Using the standard thermodynamic relations:

$$\mu_0 = \mu^{\ominus} + RT \ln p_0$$

$$\mu = \mu^{\ominus} + RT \ln p$$

where p_0 is the saturated vapour pressure and $\mu^{\ominus}$ is the standard chemical potential at unit pressure, substitution gives the Kelvin equation

$$\ln \frac{p_0}{p} = \frac{2\gamma V}{rRT} \tag{9}$$

which may be applied to the liquid inside a capillary of circular cross-section. If the angle of wetting between solid and liquid is α, then the component of the surface tension is $\gamma \cos \alpha$, and the right-hand side of Eq. (9) is modified by the factor $\cos \alpha$.

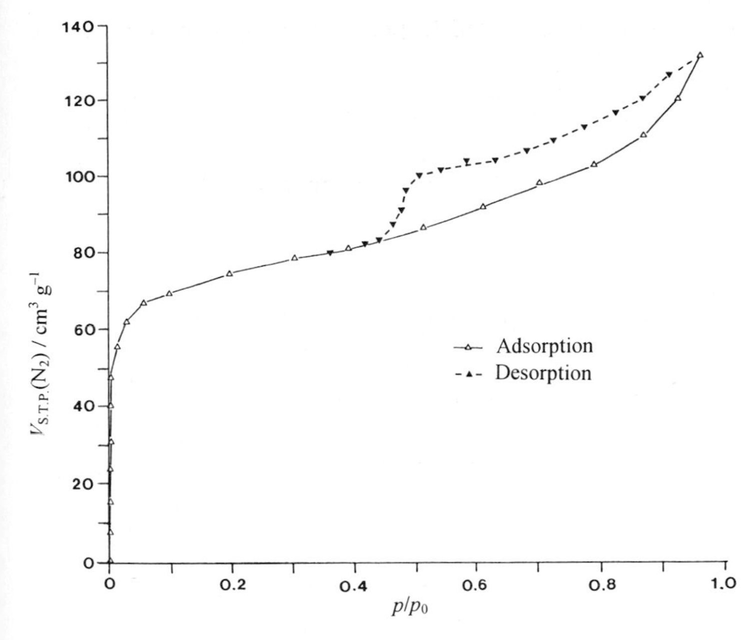

Figure 4.8 Adsorption–desorption isotherm of N_2 on 9% (w/w) NiO on Al_2O_3, determined at 77 K by the volumetric adsorption method.

In principle, then, Eq. (9) is a means of estimating pore radii from a gas adsorption isotherm. At any equilibrium pressure p, pores of radii less than r will be filled with the condensed vapour. Thus application of the Kelvin equation to all points on the isotherm at relative pressures greater than that corresponding to the monolayer volume (where capillary condensation begins to occur) will yield information concerning the volume of gas adsorbed in pores of different radii. In practice, the volume adsorbed at relative pressures up to about 0.998 can be measured and, for the purpose of numerical computation, it is necessary to assume, arbitrarily, that all pores are filled by condensate at the highest relative pressure recorded. We will also discover in Section 4.4 that, because of the various geometric forms which pores assume, the equilibrium pressure on desorption is greater than the equilibrium pressure when adsorbing. The relation between the shapes of these hysteresis loops (see Fig. 4.8, below) and the pore geometry will be revealed when discussing pore models, but if the Kelvin equation is applied to the isotherm, the equivalent *cylindrical* pore radius is deduced. It is also more sensible to apply the Kelvin equation to the desorption branch of the hysteresis loop rather than to the adsorption branch, because there is nothing arbitrary about assigning a wetting angle to a pore filled with liquid and possessing a well-defined meniscus. Thus, the Kelvin equation is more often than not applied to the desorption branch of the isotherm, in the form shown in Eq. (9) and assuming $\alpha = \pi/2$ radians.

Multilayer adsorption usually accompanies capillary condensation in the pores of solids. The Kelvin equation will therefore not give the correct radius. This is because the size of the pore entrance will have been effectively reduced by the thickness of the adsorbed layer. Thus the dimension of the cylindrical pore which will have been filled by capillary condensation at a pressure p is

$$r_\mathrm{p} = \frac{2\gamma V}{RT \ln p_0/p} + t \tag{10}$$

where t is the thickness of the adsorbed layer. This thickness may be calculated from the BET equation (Eq. (3)) by evaluating the number of molecules adsorbed at a given pressure p, dividing by the number of molecules adsorbed in a monolayer and multiplying by the effective height of a single layer of adsorbate (which is not necessarily equal to the molecular diameter and is dependent on the mode of molecular packing in the adsorbate layers). If σ is the effective height of a layer, then the adsorbed film thickness t is simply $(V/V_\mathrm{m})\sigma$. When the adsorbate packing is in the hexagonal mode, then for nitrogen $\sigma = 3.6$ Å but, if cubic packing prevails, $\sigma = 4.3$ Å. Other methods of calculating the film thickness have been discussed in the literature by Wheeler, who calculated t (in Angstrøm units) from the equation

$$t = 4.3\left(\frac{5}{\ln p_0/p}\right)^{1/3} \tag{11}$$

derived from and attributed to the form in which cooperative adsorption occurs.

The manner in which the total pore volume, and hence the pore size distribution, is computed from the desorption branch of an experimental isotherm (see Fig. 4.8) may be illustrated concisely by referring to each step of the following calculation procedure relating to the data provided in Table 4.1.

1. Compute the Kelvin radius r_k (in Ångstrøm units) from Eq. (9), which for nitrogen as adsorbate (surface tension $\gamma = 8.85 \times 10^{-5}\,\mathrm{N\,cm^{-1}}$ and molar volume $V = 34.6\,\mathrm{cm^3\,mol^{-1}}$) is given by

   $$r_\mathrm{k} = \frac{9.53}{\ln p_0/p}$$

   This is recorded in Column **3** of Table 4.1.
2. Compute the adsorbed layer thickness t from Eq. (4.11) (Column **4**).
3. The decrement ΔV in adsorbed volume is calculated by finding the difference between successive values in Column 2 (Column **5**).
4. Calculate the corresponding pore radius from

   $$r_\mathrm{p} = r_\mathrm{k} + t$$

   as given in Column **6**.
5. The amount Δt by which the adsorbed layer is thinned following a decrement in the amount adsorbed is likewise found from successive values in Column **4** (Column **7**).
6. Compute the decrement in gas volume ΔV_f by which the adsorbed film has diminished during each desorption step. This quantity is found as follows. The volume (at STP) of nitrogen which would occupy $1\,\mathrm{m^2}$ of surface is $0.23\,\mathrm{cm^3}$

(assuming a molecule of nitrogen occupies an area of 16.2 Å^2). As the height of a single layer of hexagonally packed nitrogen is 3.6 Å (calculated by finding the mass of nitrogen in hexagonal packing per unit area and multiplying by the density, 0.808×10^{-6} g m^{-3} of liquid nitrogen), then the required quantity is, in units of cm^3 g^{-1},

$$\Delta V_f = \frac{0.23}{3.6} \times \Delta t \times S_T = 0.064\ \Delta t S_T$$

where S_T is the total accumulated surface area (Column **12**). ΔV_f values are computed in Column **8** (no entries, however can be made in Column 8 until a finite value for the decrement in film thickness is recorded).

7. The volume ΔV_k attributed to desorption from pores containing liquid condensate rather than the adsorbed gas film is thus

$$\Delta V_k = \Delta V - \Delta V_f$$

given in Column **9**.

8. For a hypothetical cylindrical capillary, elementary geometry shows that the relation between volume decrements ΔV_p (corresponding to the complete pore radius) and ΔV_k (corresponding to the calculated Kelvin radius) is

$$\Delta V_p = \Delta V_k \left(\frac{r_p}{r_k}\right)^2$$

recorded in Column **10**.

9. The specific surface area ΔS corresponding to a volume ΔV_p contained by cylindrical capillaries is simply $2\Delta V_p/r_p$. However, because the quantity ΔV_p is recorded in cm^3 g^{-1} of gas at STP rather than as the actual volume of liquid condensate, one should correct by multiplying by the ratio of gas to liquid densities, ρ/ρ_l ($= 1.558 \times 10^{-3}$ for nitrogen). In units of m^2 g^{-1}, the surface area ΔS is thus

$$\Delta S = \frac{2\Delta V_p}{r_p} \times 10^4 \times 1.558 \times 10^{-3} = 31.16\,\frac{\Delta V_p}{r_p}$$

where the factor 10^4 accounts for the difference in units between ΔV_p (cm^3 g^{-1}) and r_p (Å).

10. The total accumulated surface area S_T in Column **12** is the summed values from Column **11** and is employed to calculate Column **8** (see step 6 of the computation).

11. The calculated values ΔV_p in Column **10** are (as mentioned in step 9) in cm^3 g^{-1} of nitrogen at STP. Correction to the actual volume of liquid filling the pore is given by

$$\Delta V_{pl} = \Delta V_p \times 1.558 \times 10^{-3}$$

where 1.558×10^{-3} is the ratio of gas to liquid densities (column **13**).

12. The final Column **14** gives the percentage of volume contained by pores of radius greater than the pore radius r_p from Column **6**.

Table 4.1 Computation of pore size distribution.[(a)]

1 p/p_0	**2** V [cm^3 g^{-1}]	**3** r_k [Å]	**4** t [Å]	**5** ΔV [cm^3 g^{-1}]	**6** r_p [Å] (**3**+**4**)	**7** Δt [Å]	**8** ΔV_f [cm^3 g^{-1}] 0.064 × (**7** × **12**)
0.960	131.20	–	–	–	–	–	–
0.914	125.50	106.0	16.4	5.70	122.4	–	–
0.871	119.90	69.0	14.2	5.60	83.2	2.2	–
0.827	115.00	50.2	12.8	4.90	63.0	1.4	0.70
0.777	111.50	37.8	11.6	3.50	49.4	1.2	0.76
0.728	108.5	30.0	10.8	3.00	40.8	0.9	0.83
0.680	105.8	24.7	10.1	2.70	34.8	0.7	0.78
0.633	103.6	20.8	9.5	2.20	30.4	0.6	0.77
0.579	102.9	17.4	9.0	0.70	26.4	0.5	0.74
0.539	101.1	15.4	8.6	1.80	24.1	0.4	0.74
0.505	99.0	13.9	8.3	2.10	22.3	0.3	0.56
0.485	95.45	13.2	8.2	3.55	21.4	0.1	0.54
0.474	90.95	12.8	8.1	4.50	20.9	0.1	0.42
0.459	87.1	12.2	8.0	3.85	20.2	0.1	0.31
0.442	83.7	11.7	7.9	3.40	19.5	0.1	0.52
0.417	81.6	10.9	7.7	2.10	18.6	0.2	0.68
0.389	80.2	10.1	7.5	1.40	17.6	0.3	1.05
0.361	79.1	9.4	7.3	1.10	16.7	0.2	1.16
0.336	78.0	8.7	7.1	1.10	15.9	0.2	1.14
0.312	77.0	8.2	7.0	1.00	15.2	0.1	0.99

(a) Bold Numbers at heads of column refer to column numbers; thus $\mathbf{6}^2/\mathbf{12}^2$ would indicate the square of the ratio of figures in columns **6** and **12**.

When the above computations are completed one can plot either a pore size distribution curve ($\Delta V_{pl}/\Delta r_p$ plotted against Δr_p where ΔV_{pl} entries are in Column **13** and Δr_p corresponds to decrements in r_p values in Column **6**) as illustrated in Fig 4.9, or a cumulative pore volume distribution curve (Column **14** plotted against Column **6**) as shown in Fig 4.10. The total surface area in Column **12** indicates the area contained by pores larger than the last recorded value in Column **6** and may well be less than the BET area if the solid contains micropores. Figures 4.9 and 4.10, respectively, show these distribution functions computed from the desorption branch of the isotherm in Fig 4.8.

4.3.2 Mercury Porosimeter Method of Estimating Pore Volume and Diameter

A more direct approach to pore size distribution is to measure the volume of liquid (one is chosen which does not wet the adsorbent) forced under pressure into the capillaries. The effect of interfacial surface tension is to oppose the entry of liquid into

9 ΔV_k [$cm^3\,g^{-1}$] (**5** − **8**)	10 ΔV_p [$cm^3\,g^{-1}$] (**9** × **6**^{2}/3^2)	11 ΔS [$m^2\,g^{-1}$] 31.16 × (**10**/**6**)	12 S_T [$m^2\,g^{-1}$]	13 ΔV_{pl} [$cm^3\,g^{-1}$] (1.55 × 10^{-3} × **10**)	14 Volume [%][b]
–	–	–	–	–	–
5.70	7.60	1.94	1.94	0.0118	91.22
5.60	8.15	3.05	4.99	0.0127	81.81
4.20	6.62	3.28	8.27	0.0103	74.17
2.74	4.69	2.96	11.23	0.0073	68.75
2.17	4.01	3.06	14.29	0.0062	64.13
1.92	3.80	3.40	17.69	0.0059	59.73
1.43	3.05	3.13	20.82	0.0047	56.21
0.04	0.10	0.12	20.94	0.0002	56.10
1.06	2.59	3.35	24.29	0.0040	53.12
1.54	3.93	5.49	29.78	0.0061	48.58
3.01	7.92	11.55	41.33	0.0123	39.44
4.08	10.91	16.29	57.62	0.0170	26.84
3.54	9.67	14.89	72.51	0.0151	15.68
2.88	8.06	12.85	85.36	0.0126	6.37
1.42	4.12	6.91	92.27	0.0064	1.61
0.35	1.05	1.86	94.13	0.0016	0.40
0.06	0.21	0.39	94.52	0.0003	0.16
0.04	0.12	0.23	94.75	0.0002	0.02
0.01	0.02	0.04	94.79	0.0000	0.00

[b] The values in column **14** are computed from $\left[\left(\sum \mathbf{13} - \sum_{\leqslant r_p} \mathbf{13}\right)/\sum \mathbf{13}\right] \times 100$ where $\sum \mathbf{13}$ is the sum of all the values in column **13** and $\sum_{\leqslant r_p} \mathbf{13}$ is the sum of the values in column **13** corresponding to values less than or equal to r_p.

the capillary. The force tending to impede the entry of a liquid into a narrow cylindrical channel of radius r is $2\pi r\gamma \cos\alpha$, where α is the contact angle between liquid and solid and γ is the surface tension. If a pressure p is imparted to the liquid (usually mercury, as it does not wet most solids), the force which tends to drive mercury into the pores is $\pi r^2 p$. Equating these two forces gives

$$r = -\frac{2\gamma \cos\alpha}{p} \tag{12}$$

as the radius of a pore which will accept the liquid driven in at a pressure p. For mercury, a contact angle of 140° and a surface tension of $4.8 \times 10^9\ \mathrm{N\,m^{-1}}$ are typical values. Equation (12) indicates that a pressure of $1.15 \times 10^{15}\ \mathrm{N\,m^{-2}}$ must be applied to fill pores of 100 Å radius. A pressure greater than this is, in ordinary circumstances, impracticable, so that the pressure porosimeter suffers from the disadvantage that capillaries of radius less than 100 Å remain unfilled and therefore escape detection. Nevertheless, since large pores are accounted for easily, the porosimeter is

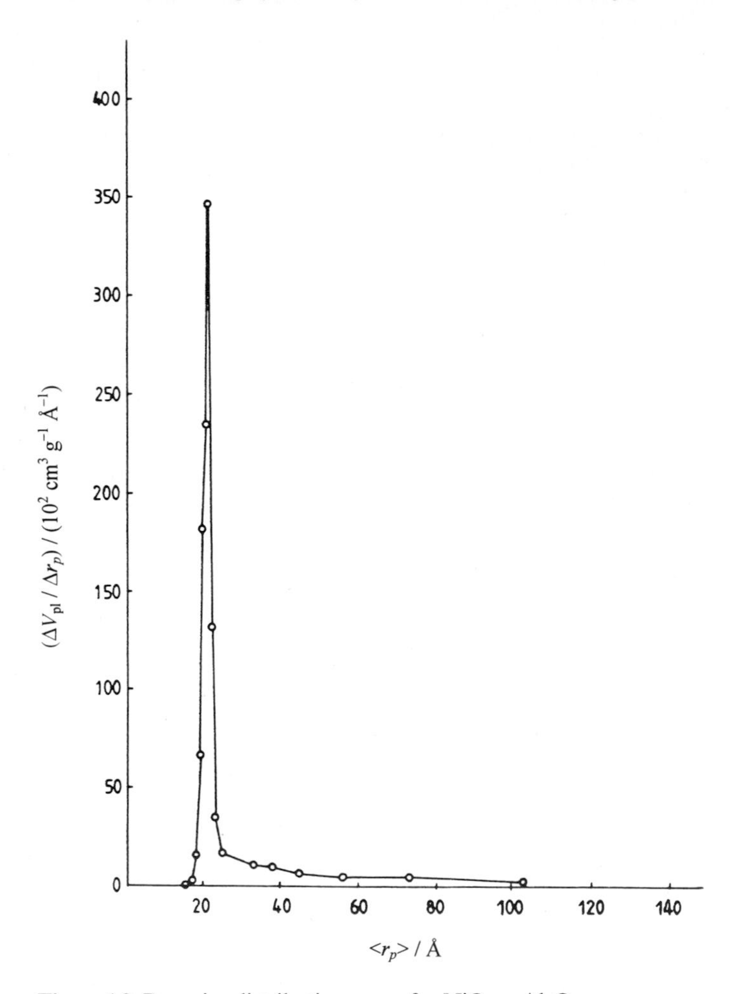

Figure 4.9 Pore size distribution curve for NiO on Al_2O_3.

especially useful in investigating the pore size distribution of porous materials containing pores up to about 10 μm radius.

The slope s to a plot of the volume V of liquid absorbed by the solid as a function of the applied pressure p will yield a value for dV/dp at a particular p and hence r. If the pore volume distribution function is $V(r)$, then

$$V(r) = \frac{dV}{dr} = \frac{dV}{dp} \times \frac{dp}{dr} = -\frac{sp}{r} \qquad (13)$$

where s is the slope of a curve of V as a function of p at a given value of p and hence r. When the right-hand side of Eq. (13) is plotted as a function of r, the resulting curve is the volume distribution function $V(r)$.

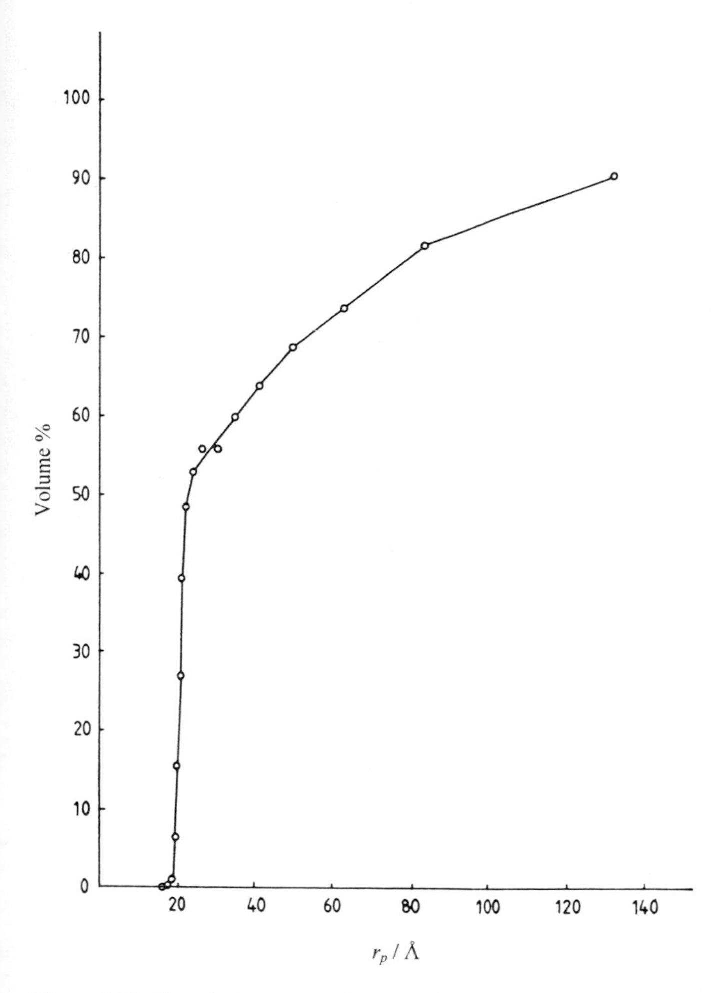

Figure 4.10 Cumulative pore volume distribution curve for NiO on Al_2O_3.

4.4 Models of the Pore Structure of Catalyst Materials

When interpreting the results of experiments in catalysis, it is always an advantage to have available a model of the porous structure of the catalytic material. Even a qualititative geometric description of the pore structure will give guidance concerning the ease or difficulty with which the reactants and products of the catalytic reaction can pervade the porous material. A quantitative mathematical description, on the other hand, is likely to prove useful in predicting reaction rates and whether, in particular, conversion would be impeded by virtue of the pore structure.

In this section we will first discuss what information about pore structure can be gleaned from the pore volume determinations outlined in Section 4.3, particularly when the phenomenon of hysteresis is evident. Elementary mathematical descriptions of porous structures conclude the discussion of pore modelling, bearing in mind that quantitative models are useful when deciding whether experimentally determined reaction rates have been influenced by intraparticle diffusion effects.

4.4.1 Hysteresis and the Shapes of Capillaries

Above relative pressures of about 0.2, porous adsorbents desorb a larger quantity of vapour at a given relative pressure than that corresponding to adsorption. Typical hysteresis loops are shown in Figs. 4.11–4.15 (types A–E), below. Several explanations have been proposed to account for such hysteresis. Zsigmondy assumed that during adsorption the vapour does not completely wet the walls of the adsorbent capillaries because an impurity, such as air, may be permanently adsorbed on the walls. Raising the pressure displaces any impurities until, at the saturated vapour pressure of the adsorbate, complete wetting takes place. On desorption, the angle of contact in the Kelvin equation is thus zero. Hence, for a given volume adsorbed, the pressure p_a on adsorption is greater than that on desorption p_d. Such an explanation is not acceptable for completely reversible hysteresis phenomena. McBain assumed that pores are shaped like old-fashioned ink bottles, with a narrow neck and a larger-diameter body. On adsorption, the neck will fill at relatively low pressures but the body will not fill until the pressure, as given by the Kelvin equation, is

$$p_a = p_0 \exp\left(\frac{-2\gamma V}{r_b RT}\right) \tag{14}$$

where r_b is the radius of the body of the capillary. The pore will not empty on desorption until the pressure is reduced to such an extent that the liquid in the neck is unstable, i.e. when

$$p_d = p_0 \exp\left(\frac{-2\gamma V}{r_n RT}\right) \tag{15}$$

where r_n is the radius of the neck of the pore. Since $r_n < r_b$, then $p_a > p_d$ and a given volume is desorbed at a pressure lower than that at which it is adsorbed. Now the bulk of the liquid in the body is in equilibrium with the vapour, so in this theory equilibrium corresponds to the adsorption branch of the hysteresis loop.

Experiments in which the adsorption of vapours on glass was studied by observing the change in colour of interference fringes indicated that the pore size distribution of the glass was discontinuous. The radii of pores deduced from the Kelvin equation showed consistent values on desorption, but on adsorption the calculated radii were twice as large. Prior to these experiments Cohan provided a satisfactory explanation for such an anomaly and deduced a relation between p_a and p_d for open-ended cylindrical capillaries. Cohan supposed that an annular ring of liquid forms in the capillary, but that a meniscus is not formed until the pore is full. Consequently, because the change in surface area on formation of an annular ring would be twice that when forming a spherical drop of the same radius, the relation between the pressure on adsorption and the pore radius is

$$p_a = p_0 \exp\left(\frac{-\gamma V}{rRT}\right) \tag{16}$$

Only when the pore is full is a meniscus formed, and then the pressure p_a on desorption will be given by the Kelvin equation (Eq. (9)). Hence, by combination of the Kelvin equation with Eq. (16),

$$\left(\frac{p_a}{p_0}\right)^2 = \frac{p_d}{p_0} \tag{17}$$

Therefore, in terms of the Kelvin equation, the radius corresponding to p_a is twice that corresponding to p_d.

Implicit in the above equations is thc cylindrical shape of the capillaries. The influence of various shapes of capillaries on the form of isotherms has been the subject of extensive investigation. The reverse problem has also been discussed and conclusions have been derived concerning the shapes of pores from the form of the adsorption–desorption isotherm.

Five types of hysteresis loops may be distinguished and examples are shown in Figs. 4.11–4.15. The types of capillary shape that could be responsible for each loop are also sketched.

4.4.1.1 Type A Hysteresis Loops

Those sorption isotherms (Fig. 4.11) in which either the adsorption or desorption branches are steep at intermediate relative pressures are characteristic of tubular-shaped capillaries open at both ends. The cross-section need not necessarily be circular, for a capillary with a polygonal cross-section will give rise to condensation along the inside edges until a cylindrically shaped meniscus is formed, after which, as shown by Eq. (16), the capillary will fill at a relative pressure corresponding to an effective radius $2r_c$ where r_c is the radius of the cylinder formed by the condensate in a continuum along the edges of the polygon. The relative pressure on desorption will correspond to the radius r_c according to the Kelvin equation. Thus the relation between p_a and p_d is given by Eq. (17).

If the open capillaries are composite, with slightly widened bodies of radius r_b, each of these cavities having a narrower neck of radius r_n, then the narrow parts will fill at a relative pressure corresponding to $2r_n$. Consequently, a spherical meniscus is formed at both ends of the wider spherical body. If $r_b \leqslant 2r_n$, then the vapour pressure is saturated or supersaturated with respect to the radius r_b and so the composite pore is completely filled at a relative pressure corresponding to $2r_n$. The pore is emptied at a relative pressure corresponding to the radius r_n at the open ends. If the pore has only one narrow neck and is accessible only through the wider body, then it will empty at a pressure corresponding to r_b. Hence, Eq. (17) is obeyed. If, on the other hand, the open tubular capacity has a rectangular cross-section of dimensions a and na (where $n > 1$)and a length $L \gg a$, then although a type A hysteresis loop is

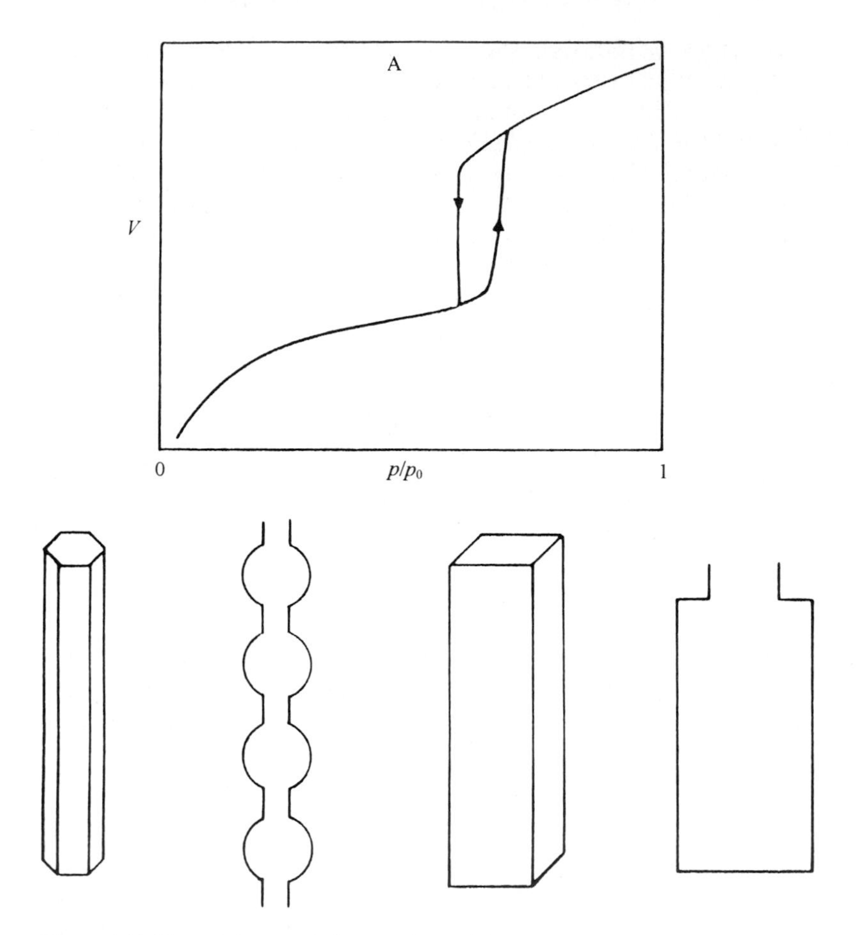

Figure 4.11 Type A hysteresis loop and possible pore structures.

formed, Eq. (17) is not obeyed. Capillary condensation starts along the inside edges until an inside meniscus with a radius $a/2$ is formed. The pores then fill at a pressure corresponding to an effective radius a. Emptying of such pores occurs when the effective radius of the inscribed cylinder is $na/(n+1)$. For square capillaries, Eq. (17) is obeyed, but in all cases for $n > 1$ we have

$$\left(\frac{p_a}{p_0}\right)^2 < \left(\frac{p_d}{p_0}\right) \tag{18}$$

If n is very large, then the adsorption and desorption branches coincide and no hysteresis is observed.

The ink-bottle-shaped capillaries, assumed by McBain to explain the phenomenon of hysteresis, also exhibit type A hysteresis. If one end of the capillary is closed and the radius of the body is r_b and that of the neck is r_n then, provided that $r_n < r_b < 2r_n$, the pore fills at a pressure corresponding to r_b, since, in this case, a meniscus forms only at the top of the body and is spheroidal in shape and not cylindrical. Consequently, Eqs. (6) and (7) apply to the change in surface energy and the number of moles transferred from vapour phase to liquid. The resulting relation is therefore identical with the Kelvin equation (Eq. (9)) and not Eq. (16). The pore will empty at a pressure corresponding to r_n as usual. The pressure on adsorption is therefore less than that on desorption and the inequality (16) is obeyed. If the radius of the body of the ink-bottle capillary is greater than the diameter of the neck, then the short necks are filled at a relative pressure corresponding to $2r_n$, but the whole pore will fill only at a relative pressure corresponding to r_b. In this latter case, the steep rise of the hysteresis loop is delayed until the whole pore is filled. Emptying takes place at a pressure corresponding to r_n. A wide hysteresis loop is produced and the inequality

$$\left(\frac{p_a}{p_0}\right)^2 > \left(\frac{p_d}{p_0}\right) \tag{19}$$

holds because, for this case, $r_b > 2r_n$.

4.4.1.2 Type B Hysteresis Loops

If a pore is formed by two parallel plates, a meniscus cannot form until the vapour pressure of the adsorbate is raised to the saturation vapour pressure. On desorption, the pore is emptied at a relative pressure corresponding to the width of the capillary, since this will be the effective radius of curvature of the meniscus. Thus the isotherm

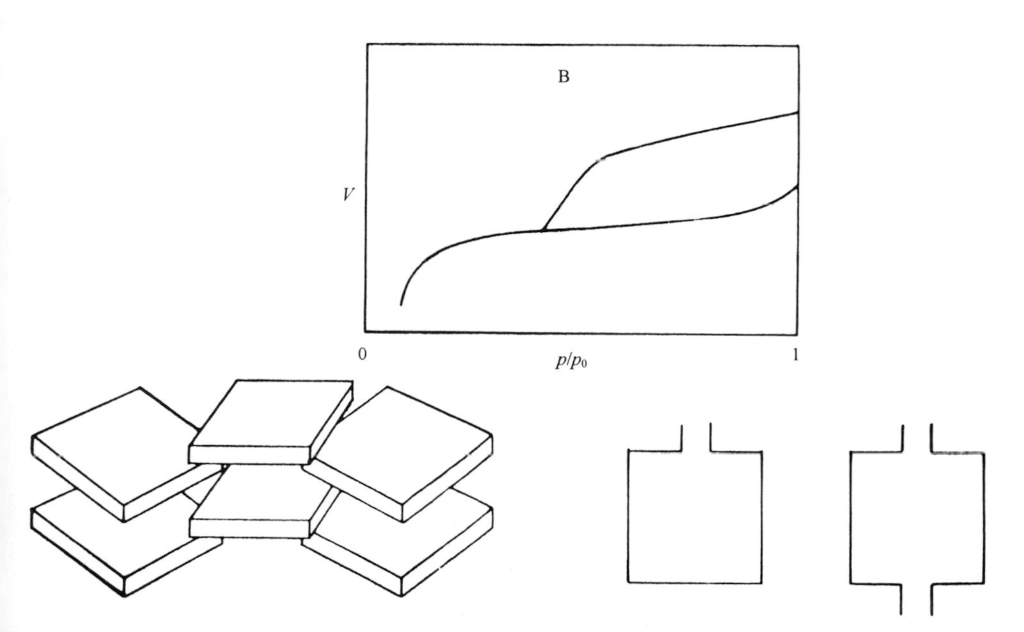

Figure 4.12 Type B hysteresis loop and possible pore structures.

has a steep adsorption branch at a relative pressure of unity and a sloping desorption branch at intermediate relative pressures. Such an isotherm (Fig. 4.12) is said to show type B hysteresis.

Graphite oxide, montmorillonites and aluminium hydroxides give rise to type B hysteresis loops. These materials are of a crystalline nature and the packing together of their particles in laminae leads to the formation of open slit-shaped capillaries with parallel walls. Such structures have been confirmed by examining the shapes of the capillaries using an optical birefringence technique.

Another type of capillary which may lead to type B hysteresis is an ink-bottle pore with a body which is so wide, *ca* 1000 Å, that the adsorption branch practically coincides with a relative pressure of unity.

4.4.1.3 Type C Hysteresis Loops

Type C hysteresis, on the other hand, may be characteristic of materials with spheroidal pores all with a circular cavity radius but with various-sized entrances. Open and closed ink-bottles with a heterogeneous distribution of neck radii would also cause such a hysteresis loop.

A heterogeneous distribution of capillary dimensions should exhibit the phenomenon of scanning, which is the ability of a system to follow a course across a hyste-

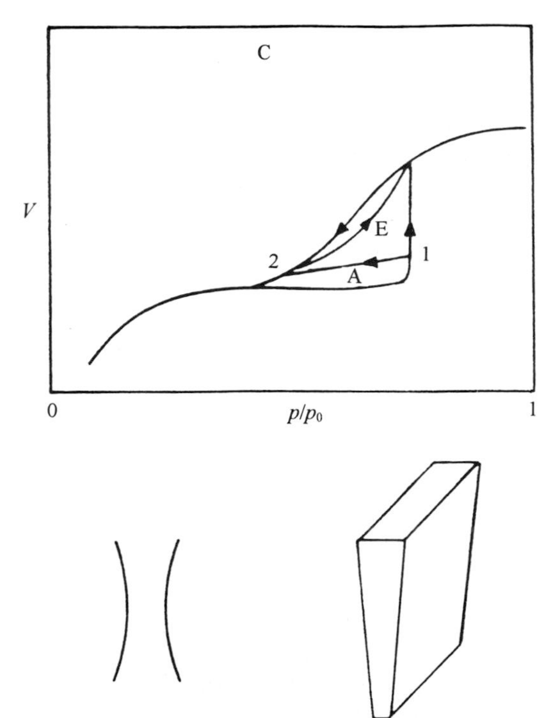

Figure 4.13 Type C hysteresis loop and possible pore structures.

resis loop. Consider point 1 on the adsorption branch of the type C hysteresis loop shown in Fig 4.13, and further suppose that this loop characterizes an assemblage of ink-bottle pores with various-sized necks. If the pressure is reduced, any capillaries which are partly filled will empty, but those which are completely full will retain the condensate until the pressure is lowered to p_2, the equilibrium pressure in the neck of the pore. Hence the hysteresis loop is crossed along path A. At no point on the desorption branch are there any partially filled pores. At point 2, a proportion of pores are empty and the remainder are full. If the pressure is now increased, those pores which are empty will fill again and the path B is followed. The adsorption branch is regained only at the saturation pressure because at point 2 some of the pores which had radii smaller than r_n have larger bodies and remain full. Therefore, there are fewer large pores to fill than under normal circumstances and, for a given volume adsorbed, the equilibrium pressure will be lower than on the original adsorption branch.

4.4.1.4 Type D Hysteresis Loops

A heterogeneous assembly of capillaries with large body radii and a sufficiently varying range of narrow, short necks will cause the adsorption branch of the type C loop to be displaced to higher relative pressures (Fig. 4.14). On account of the wide size range of narrow necks, the desorption branch will have a sloping character. Hysteresis loops of this type are rare, but have been observed when water is adsorbed on gibbsite partly converted to boehmite.

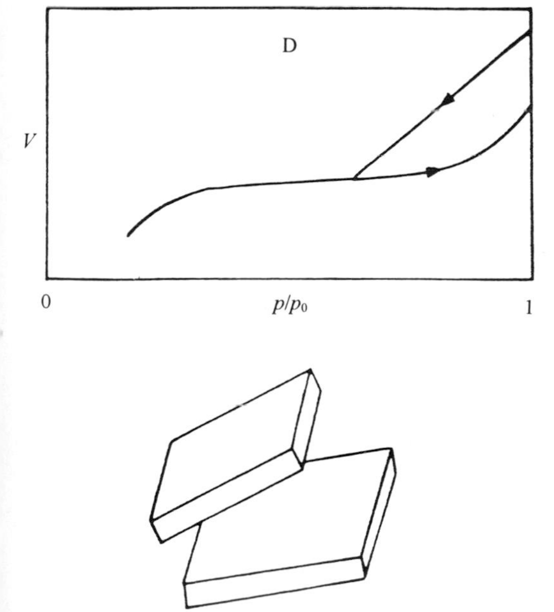

Figure 4.14 Type D hysteresis loop and possible pore structures.

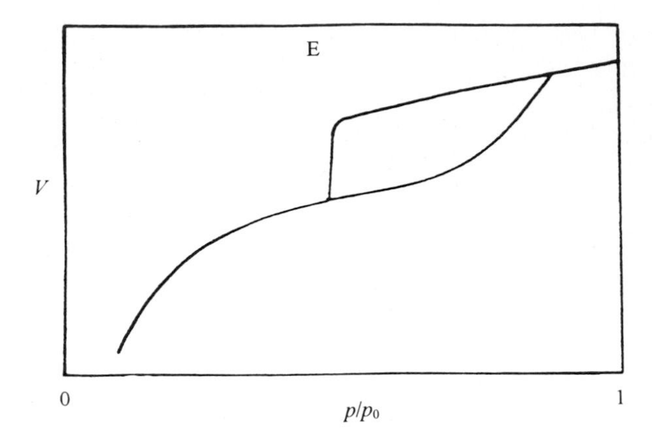

Figure 4.15 Type E hysteresis loop.

4.4.1.5 Type E Hysteresis Loops

This type of hysteresis loop, in contrast to type C, has a sloping adsorption branch and a steep desorption branch, both at intermediate relative pressures (Fig. 4.15). It may arise from the same types of open capillaries as are responsible for type A hysteresis when the effective radii of the bodies of the pores are heterogeneously distributed but the effective radii of the narrow entrances are all of equal size. Similarly, a distribution of various-sized spheroidal cavities with the same entrance diameter will give a type E hysteresis loop. Scanning behaviour can be expected from both types of pore.

Ink-bottle pores with large bodies of varying effective radii and small, narrow necks start filling at an effective cylindrical meniscus radius of $2r_n$ and continue filling until the whole body is full. They will empty at a relative pressure corresponding to r_n. Scanning behaviour will be absent for this type of pore, since on desorption the pores will empty spontaneously.

Type E hysteresis loops are well known and have been noted for the sorption of benzene by ferric oxide gel, for the sorption of water by silica gel, and also for nitrogen adsorption on many silica–magnesia and silica–alumina cracking catalysts.

4.4.1.6 Other Adsorption–Desorption Characteristics

The five types of hysteresis loop discussed above all have one vertical adsorption or desorption branch because the dimension of the cross-sections of either the wide or the narrow part of the pores are all of equal size and are predominantly of one shape. If such conditions are not fulfilled and there are pores whose bodies and necks have cross-sections covering a wide range of radii, hysteresis loops with sloping adsorption and desorption branches will result.

Tubular-type capillaries closed at one end will not exhibit hysteresis at all because, in this case, there will be no delay in the formation of a meniscus on adsorption. The Kelvin equation will therefore apply to both adsorption and desorption and p_a will equal p_d. The absence of hysteresis is also apparent for an assembly of open capillaries which are sufficiently narrow to preclude the adsorption of a layer more than four molecular diameters thick. For such narrow capillaries the thickness t of the adsorbed layers is not negligible in comparison with the radius; under these circumstances the radius r in the adsorption equation (Eq. (16)) should be replaced by $(r - t)$, and p_a is equal to p_d when the radius is two molecular thicknesses.

4.4.2 Geometric Models of Pores

Consideration has been given to geometric models of more complex pores such as non-intersecting capillaries and parallel-sided fissures, from which one derives an equivalent cylindrical pore size distribution as discussed in Section 4.3. For the simple case of non-intersecting capillaries the ratio of the volume V contained by pores to the surface area S created by these same pores is $2V/S$, which is the average pore radius $\bar{r}$. For other geometric configurations of pore assemblies one may write

$$\bar{r} = \frac{1}{F}\left(\frac{2V}{S}\right) \tag{20}$$

where F is a factor characteristic of the particular pore geometry ($F = 1$ for non-intersecting capillaries). Some indication of the pore geometry could therefore be deduced if an independent measurement of $\bar{r}$ is made (e.g. electron microscopy), thus leading to an evaluation of F. A Knowledge of F, however, does not always lead to a unique geometrical specification. Everett, in a classic paper (see Further Reading, Section 4.8.3), tabulated values of F for simple uniform pore structures in which only a single geometric parameter r defines the pore system. Table 4.2 gives the values of F for a number of simple uniform structures. More complicated uniform structures have to be characterised by two parameters and the factor F now depends on the ratio of the two parameters as well as the basic type of configuration. As Everett pointed out, the ratio of these two parameters determines the porosity of the structure, so that if one is able to obtain independent estimates of pore volume (e.g. by pyknometry and gas volume) and the geometric group to which the pore structure can be assigned (e.g. by electron microscopy), then a reliable value can be found for

Table 4.2 Relation[a] of volume contained by pores to surface area created by pores.

Basic porous structure	Proportionality factor F (Eq. (20))
Non-intersecting capillaries	1
Non-intersecting close-packed cylindrical rods	1
Cubic packing of spheres	0.613
Orthorhombic packing of spheres	0.433
Rhombohedral packing of spheres	0.229
Tetragonal-spheroidal packing of spheres	0.293

[a] $r = \frac{1}{F}\left(\frac{2V}{S}\right)$

Table 4.3 Relation between geometric radius and the experimental quantity $(2V/S)_{exp}$; $(2V/S)_{geom} = \alpha(2V/S)_{exp}$.

Basic structure	Proportionality factor α[a]
Non-intersecting cylindrical capillaries	$1 - \dfrac{d}{2r}$
Non-intersecting cylindrical rods	$1 + \dfrac{d}{2r}$
Intersecting cylindrical capillaries	$1 + \dfrac{d}{2R} + \dfrac{9d}{2R}\left(\dfrac{1 - d/2R}{1 + 10r/R}\right)$
Intersecting cylindrical rods	$1 - \dfrac{d}{2R} - \dfrac{9d}{2R}\left(\dfrac{1 + d/2R}{1 + 10r/R}\right)$

[a] Definitions:
d^3 = volume occupied by an adsorbed molecule
r = pore radius
$2R$ = spacing between intersecting capillaries or rods (pores)

F. The errors likely to arise in calculating a mean radius in this way from $2V/S$ for a non-uniform structure are unlikely to exceed 10%.

The ratio of pore volume to surface area will then, in general, give a mean pore radius, but at least one set of independent experiments is necessary to characterise the pore geometry. Although these measurements provide information about pore geometry, the following example suffices to show that the ratio $2V/S$, as deduced geometrically, is proportional to, but not equal to, the experimentally derived value. Everett calculated this proportionality factor for several geometric models and his results are shown in Table 4.3, which also includes the result of the example. Concave surfaces lead to an under-estimate of the geometrical surface while convex surfaces produce an over-estimate.

Example

Deduce the ratio $2V/S$ for a non-intersecting cylindrical capillary from purely geometric considerations.

Solution

Consider a cylindrical pore whose radius of curvature is large compared with the diameter of an adsorbed molecule. Suppose that the area which the single molecule occupies is d^2 and that the volume is d^3. Figure 4.16 illustrates the geometric problem of occupying the pore volume with cubes covering the surface and calculating the corresponding area.

The volume to be filled is $\pi r^2 d$, where r is the radius of the pore and d the linear dimension of the cube representing a molecule. Since the complete geometric surface

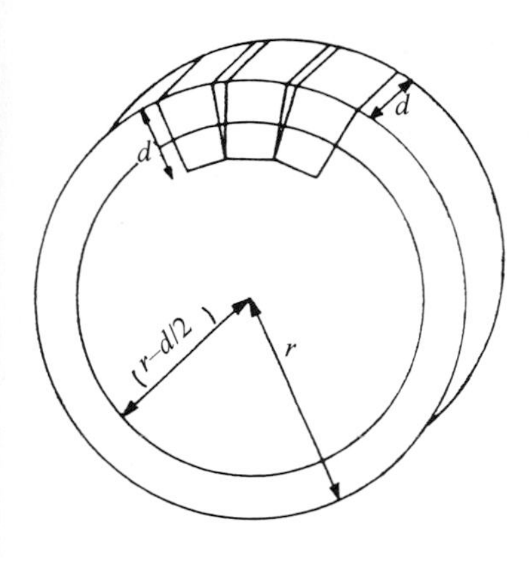

Figure 4.16 The problem of calculating a geometric value for $2V/S$.

of the pore is not covered by the squares, we can say that the locus of points representing the distance $(r - d/2)$ from the centre of the face of each cube to the pore wall describes a circle of which the radius will be the measured effective radius of the capillary.

The surface area of the imaginary cylinder formed by joining the centres of the faces of each cube is

$$2\pi(r - d/2)d$$

and, as defined by the problem, the volume occupied is $\pi r^2 d$. Hence the desired ratio is

$$\left(\frac{2V}{S}\right) = \frac{2\pi r^2 d}{2\pi(r - d/2)d} = \frac{r}{(1 - d/2r)}$$

4.4.3 Wheeler's Semi-empirical Pore Model

In order to provide a practical solution to the shortcomings arising from the somewhat arbitrary selection of points on an experimental isotherm, representing the completion of a monolayer and pore filling, and the ineluctable fact that the ratio of experimentally determined pore volume and surface area does not characterise a porous structure uniquely, Wheeler (1951), in a much-quoted paper (see Further Reading, Section 4.8.2), constructed a semi-empirical model of porous media which has been applied frequently to the prediction of chemical reaction rates occurring in porous catalysts.

Wheeler proposed that the mean radius $\bar{r}$ and length L of pores in a catalyst pellet are determined in such a way that the sum of the surface areas of all the pores constituting the honeycomb of pores is equal to the BET surface area, and that the sum of the pore volumes is equal to the experimental pore volume. If S_x represents the external surface area of a catalyst particle (e.g. determined by sedimentation) and there are n pores per unit internal area, the pore volume contained by nS_x cylindrically shaped pores is $nS_x \pi r^2 L$. If the total volume of the catalyst pellet is V_p and its porosity ψ, then the experimental pore volume is $V_p\psi$. Now the pellet porosity is the product of the specific pore volume V_g and the pellet density ρ_p, so equating the experimental pore volume to the volume of pores as described by the model,

$$V_p \rho_p V_g = n S_x \pi (\bar{r})^2 L \tag{21}$$

Similarly, we may equate the experimental surface area to the surface area of pores as given by the model. Now the mass of the catalyst pellet is the product of the pellet volume V_p and its density ρ_p, so the experimental surface area is $\rho_p V_p S_g$ where S_g is the BET specific surface area. The total area of nS_x straight cylindrical pores of average radius $\bar{r}$ and length L is $nS_x\, 2\pi\bar{r}L$. If the pore walls are rough and the pores intersect, then the latter quantity should be multiplied by a roughness factor τ and the factor $(1 - \psi)$, where ψ, the pellet porosity, is equivalent to the fraction of pore walls not interrupted by intersections. Thus we obtain

$$V_p \rho_p S_g = n S_x 2\pi \bar{r} L \tau (1 - \psi) \tag{22}$$

Dividing Eq. (21) by Eq. (22), the average pore radius is

$$\bar{r} = \frac{2V_g}{S_g} \tau (1 - \psi) \tag{23}$$

This model is obviously equivalent to the geometric non-intersecting pore model when the product $\tau\,(1 - \psi)$ is equal to unity. Wheeler also argued that as some pores in a practical pellet are parallel to the gas flow while others are at right angles, a reasonable assumption would be that the average orientation of pores to the direction of gas flow is 45°. In this case the number of pores in the pellet would be $\psi/\sqrt{2}\pi(\bar{r})^2$. Substituting this into Eq. (21) yields an average pore length

$$L = \sqrt{2}\,\frac{V_p}{S_x} \tag{24}$$

In terms of the pellet size d_p, the ratio V_p/S_x is $d_p/6$ for spheres and cubes. Thus, for many practical catalyst pellets, the average pore length is $\sqrt{2}d_p/6$ according to Wheeler's model. The roughness factor τ appearing in Eq. (23) was assumed to be 2 for practical purposes.

The Wheeler model thus provides an average pore radius $\bar{r}$ and pore length L in terms of the experimentally determinable parameters v, S_g and ψ. The only adjustable parameter is the roughness factor τ. The usefulness of this model was demonstrated by Wheeler by virtue of its successful application to the prediction of rates of catalytic reactions of industrial importance.

4.4.4 Mathematical Models of Porous Structures

4.4.4.1 The Dusty Gas Model

The most fundamental model of a porous structure is provided by the dusty gas model as critically reviewed and described by Jackson in a classic text (see Further Reading, Section 4.8.1). The basis of the theory is to suppose 'the action of the porous material to be similar to that of a number of particles, fixed in space and obstructing the motion of the particles of the moving system', as suggested by Maxwell as early as 1860. Thus if a number of gaseous species are diffusing through three-dimensional space, the imposition of one further species of much greater mass and fixed in space determines the foundations for writing flux equations for each species, describing diffusion and convective transport. Solution of the set of equations together with the appropriate boundary constraints and continuity conditions yields two effective diffusion coefficients. One of these relates to transport in narrow channels where the mean free path of the diffusing species is greater than the dimension of the free space within the porous medium, and is written

$$D_{eK} = K_0 D_K \tag{25}$$

where D_K is the Knudsen diffusion coefficient in a straight capillary (see Section 4.5.1.2). The coefficient D_{eK} is analogous to an effective Knudsen diffusion coefficient formulated to describe transport in narrow capillaries. K_0 depends on the pore structure geometry but can be estimated by means of carefully designed flux experiments (see Section 4.5). The other effective diffusion coefficient relates to transport in intraparticle free space, the dimensions of which are larger than the mean free path of the diffusing species: this coefficient is analogous to a bulk (or Maxwellian) diffusion coefficient and is written

$$D_{eM} = K_1 D_M \tag{26}$$

where D_M is the normal Maxwell diffusion coefficient for a bulk gaseous mixture which can be calculated (Section 4.5) from the kinetic theory of gases. K_1 is also best deduced from flux experiments.

In the dusty gas model then, effective diffusivities for species diffusing through the porous medium are related to diffusion coefficients for narrow capillaries (Knudsen diffusion) and bulk gases (Maxwellian diffusion) through the coefficients K_0 and K_1, both of which can be estimated experimentally. In addition to K_0 and K_1, however, it is necessary to consider a third parameter B_0, when forced flow due to total pressure gradients is prominent in the medium. Total pressure gradients might arise, for example as a result of a cracking reaction when the number of moles per unit volume of product within the catalyst pellet might far exceed (perhaps by a factor of 3 or more) the number of moles per unit volume of reactant. The flux due to forced flow (Poiseuille flow) is then added to the diffusive flux. The coefficient B_0 again depends on geometrical structure and can be determined by suitable experiments.

4.4.4.2 The Random Pore Model

The random pore model was originally developed to account for the behaviour of bidisperse systems which contain both micropores and macropores. Many industrial catalysts, for example, when prepared in pellet form, contain not only the smaller intraparticle pores but also larger pores consisting of the voids between compressed particles. Transport within the pellet is assumed to occur through void regions which are either narrow or more spacious and can be represented by average dimensions r_m (macro radius) or r_μ (micro radius). By employing statistical network theory, an overall effective diffusivity for a pellet may be deduced such that the effective diffusivity is

$$D_e = \varepsilon_m{}^2 \bar{D}_M + \frac{\varepsilon_\mu(1 + 3\varepsilon_\mu)}{1 - \varepsilon_\mu} \bar{D}_\mu \tag{27}$$

where $\bar{D}_M$ and $\bar{D}_\mu$ are given by

$$\frac{1}{\bar{D}_M} = \frac{1}{(D)_M} + \frac{1}{(D_K)_M} \quad \text{and} \quad \frac{1}{\bar{D}_\mu} = \frac{1}{(D)_\mu} + \frac{1}{(D_K)_\mu} \tag{28}$$

The parameters D and D_K, whether for macropore (denoted by subscript M) or for micropore (denoted by subscript μ) regions, are normal bulk and Knudsen diffusion coefficients respectively, and can be estimated from kinetic theory provided that the mean radii of the diffusion channels are known. Mean radii, of course, are obtainable from pore volume measurements as discussed in Section 4.3. For bidisperse systems two peaks (corresponding to macro and micro) would be expected in a differential pore size distribution curve, and provide the necessary information. Macro- and micro-voidages can also be determined experimentally.

4.4.4.3 Stochastic Pore Networks and Fractals

With the advent of high-speed computers the possibilities of extending the simple cylindrical pore model of Wheeler to computer simulations of stochastic pore networks, represented in two and even three dimensions, is now realistic. If a large number of cylindrical pore segments, the diameter of each pore being independent of adjacent pores, are arranged randomly into a two-dimensional network, then the assembled structure can be constructed to obey any preconceived experimental pore size distribution and can therefore be fitted to actual pore size distribution data. Figure 4.17 is an example of a simple two-dimensional stochastic pore network. Such stochastic pore networks also have the implicit property of representing hysteresis phenomena (discussed in Section 4.4.1) when pores fill at relative pressures greater than the relative pressures at which they empty. When a mercury porosimeter is used to study the size distribution of the relatively large pores in a porous

Figure 4.17 Simple two-dimensional stochastic pore network. With permission from R. Mann, *Trans. I. Chem. E.* **1993**, *71A*, 551.

Figure 4.18 Three-dimensional stochastic pore network. With permission from R. Mann, *Trans. I. Chem. E.* **1993**, *71A*, 551.

solid, it is well known that mercury becomes entrapped within the structure as the porosimeter pressure is reduced and retraction of mercury occurs: the stochastic pore network model predicts such mercury entrapment. It is clear therefore that computer simulation of pore structures using two-dimensional stochastic network assemblies is an important tool in understanding the behavioural nature of a porous solid. Figure 4.18 is a computer simulation of a three-dimensional stochastic pore network which illustrates that the concept of stochastic pore networks can be extended to represent pore size distribution in real three-dimensional space. Amongst the revelations which three-dimensional networks may depict is the possibility of the existence of planar pore spaces, clearly shown in Fig. 4.18; these planar pore spaces can be examined and compared with the polished sections of actual materials.

A further and more recent development in the simulation of pore structures is the application of the theory of fractals to the construction of pore geometries which can closely resemble naturally occurring porous structures; this technique may also replicate observed scanning electron micrographs of porous media. A one-dimensional fractal line generated from triangular forms of various dimensions can be constructed by repeated recursions of the basic form randomly selected. Figure 4.19 is a simple example of how such a fractal line may be produced. Such a procedure can be extended to two and three dimensions and visualised by the use of computer graphics and imaging techniques. In an important and imaginative paper Mann (see Further Reading, Section 4.8.3) has suggested that, by the application of fractals, realistic morphologies may be produced. Thus, combining such fractal surfaces with stochastic pore networks could be the basis for a computerised image reconstruction which, when compared with scanning electron micrographs of real porous surfaces, may lead to the construction of a mathematical model of the actual pore structure without even resorting to time-consuming experimental methods such as gas ad-

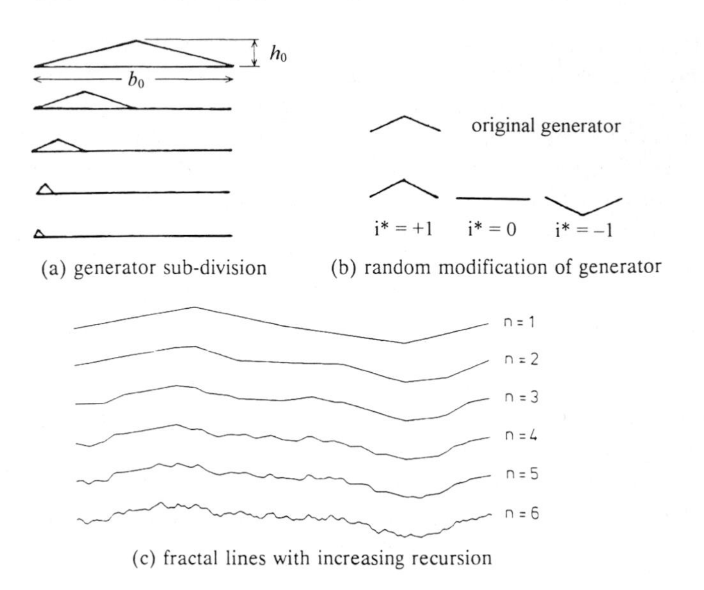

Figure 4.19 Generation of two-dimensional fractal lines. With permission from R. Mann, *Trans. I. Chem. E.* **1993**, *71A*, 551.

sorption and mercury porosimetry. When, therefore, randomly constructed pores are assembled into random networks and reproduced as a textured surface based upon fractal components, a three-dimensional view could be produced and manipulated to match a scanning electron micrograph.

4.5 Diffusion in Porous Catalysts

For gas–solid heterogeneous reactions, particle size and average pore diameter will influence the reaction rate per unit mass of solid if internal diffusion happens to be rate-limiting. The actual mode of transport within the porous structure will depend largely on the average pore radius and the conditions of pressure within the reactor. Before developing equations which will enable us to predict reaction reaction rates in porous solids, a brief consideration of transport in pores is pertinent.

4.5.1 The Effective Diffusivity

As implied in Section 4.4.4.1, the diffusion of gases through the tortuous narrow channels of a porous solid generally occurs by one or more of three mechanisms. When the mean free path of the gas molecules is considerably greater than the pore diameter, collisions between molecules in the gas are much less numerous than those between molecules and pore walls. Under these conditions the mode of transport is Knudsen diffusion. When the mean free path of the gas molecules is much smaller than the pore diameter, gaseous collisions will be more frequent than collisions of molecules with pore walls, and under these circumstances ordinary bulk diffusion occurs. A third mechanism of transport which is possible when a gas is adsorbed on the inner surface of a porous solid is surface diffusion. Transport then occurs by the movement of molecules over the surface in the direction of decreasing surface concentration. Although there is not much evidence on this point, it is unlikely that surface diffusion is of any importance in catalysis at elevated temperatures. Nevertheless, surface diffusion may contribute to the overall transport process in low-temperature reactions of some vapours. Finally, it should be borne in mind that when a total pressure difference is maintained across a pore, as is reputed to be the case for some catalytic cracking reactions, forced flow in pores is likely to occur, transport being due to a total concentration gradient.

Both Knudsen diffusion and bulk flow can be described adequately for homogeneous media. However, a porous mass of solid usually contains pores of non-uniform cross-section which pursue a very tortuous path through the particle and which may intersect with many other pores. Thus the flux predicted by an equation for normal bulk diffusion (or for Knudsen diffusion) should be multiplied by a geometric factor which takes into account the tortuosity and the fact that the flow will be impeded by that fraction of the total pellet volume which is solid. It is therefore expedient to define an effective diffusivity D_e in such a way that the flux of material may be thought of as flowing through an equivalent homogeneous medium. We may then write:

$$D_e = D\frac{\psi}{\tau} \tag{29}$$

where D is the normal diffusion coefficient (either Maxwellian or Knudsen as appropriate), ψ is the porosity of the particles and τ is a tortuosity factor.

We thus imply that the effective diffusion coefficient is calculated on the basis of a flux resulting from a concentration gradient in a homogeneous medium which has been made equivalent to the heterogeneous porous mass by invoking the geometric factor ψ/τ. Experimental techniques for estimating the effective diffusivity include diffusion and flow through pelletized particles. A common procedure is to expose the two faces of a compressed porous pellet of the material to two gas streams at the same pressure. One of the gas streams (passing over the upper face of the pellet) is composed of the gaseous component of interest diluted with an inert gas such as nitrogen or helium, while the second stream (passing over the lower face of the pellet) is pure diluent gas. Figure 4.20 is a diagrammatic sketch of the apparatus for measuring effective diffusivities. The catalyst pellet under investigation is held within a thermostated brass container in such a way that the only route for gas transport is by diffusion through the pellet. The procedure for operating the equipment is to introduce suddenly, by means of a gas valve, a steady concentration of the component of interest into the diluent inert stream passing over the upper face of the pellet and to monitor, as a function of time, the increase in concentration in the lower stream. Knowing the areas of the pellet faces and its length, the effective diffusion coefficient can be computed from the characteristics of the response curve obtained. An alternative to effecting a step change in concentration in the upper stream is the instantaneous injection of a pulse of the gaseous component, in which case the response

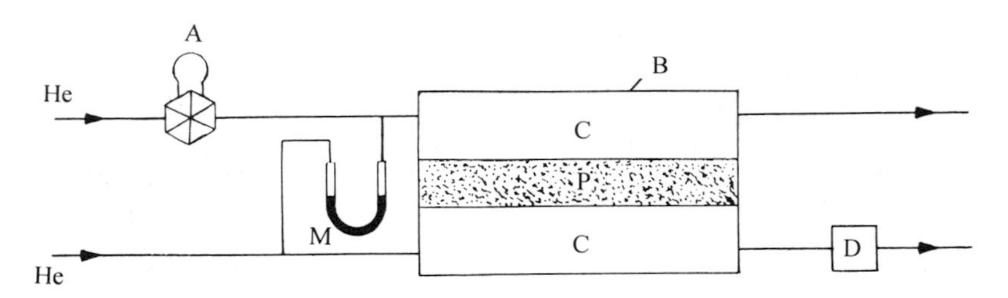

Figure 4.20 Apparatus for measuring effective diffusivity of a porous catalyst pellet. A, pulse or step input of sample gas; B, diffusion cell; C, upper and lower chamber of the cell; P, porous pellet; D, katharometer detector; M, manometer.

would be a skewed and diffuse peak, the moments of which yield information making it possible to calculate the effective diffusivity. Figure 4.21 shows the form of response obtained for a step change in concentration or, alternatively, a pulse injection. Details of the calculation procedure can be found in the original literature and other texts (see Further Reading at the end of this chapter).

Another interesting method relies on the measurement of the time lag required to reach a steady pressure gradient. Gas-chromatographic methods for evaluating effective diffusivities have also been employed.

Just as one considers bulk flow and Knudsen flow for homogeneous media, so one may have bulk (Maxwellian) or Knudsen transport in heterogeneous media.

4.5.1.1 Molecular (Maxwellian) Diffusion or Bulk Diffusion

A theoretical expression obtained for the molecular diffusion coefficient for two interdiffusing gases has been obtained by modifying the kinetic theory of gases and taking into account the nature of attraction and repulsion forces between gas molecules. The resulting expression for the diffusion coefficient has been successfully applied to many gaseous binary mixtures and represents one of the best methods for estimating unknown values. On the other hand, Maxwell's formula modified by Gilliland, which can be found in handbooks of chemistry or chemical engineering, also gives satisfactory results. Experimental methods for estimating diffusion coefficients rely on the measurement of flux per unit concentration gradient. Extensive tabulations of experimental diffusion coefficients for binary gas mixtures may be found in literature pertaining to gaseous diffusion.

To calculate the effective diffusivity in the region of molecular flow, the estimated value of D must be multiplied by the geometric factor ψ/τ, which is descriptive of the heterogeneous nature of the porous medium through which diffusion occurs.

The porosity ψ of the porous mass is included in the geometric factor to account for the fact that the flux per unit total cross-section is ψ times the flux if there were no solid present. The porosity may conveniently be measured by finding the particle density ρ_p in a pyknometer using an inert non-penetrating liquid. The true density ρ_s of the solid should also be found by observing the pressure of a gas (which is not adsorbed) before and after expansion into a vessel containing a known weight of the

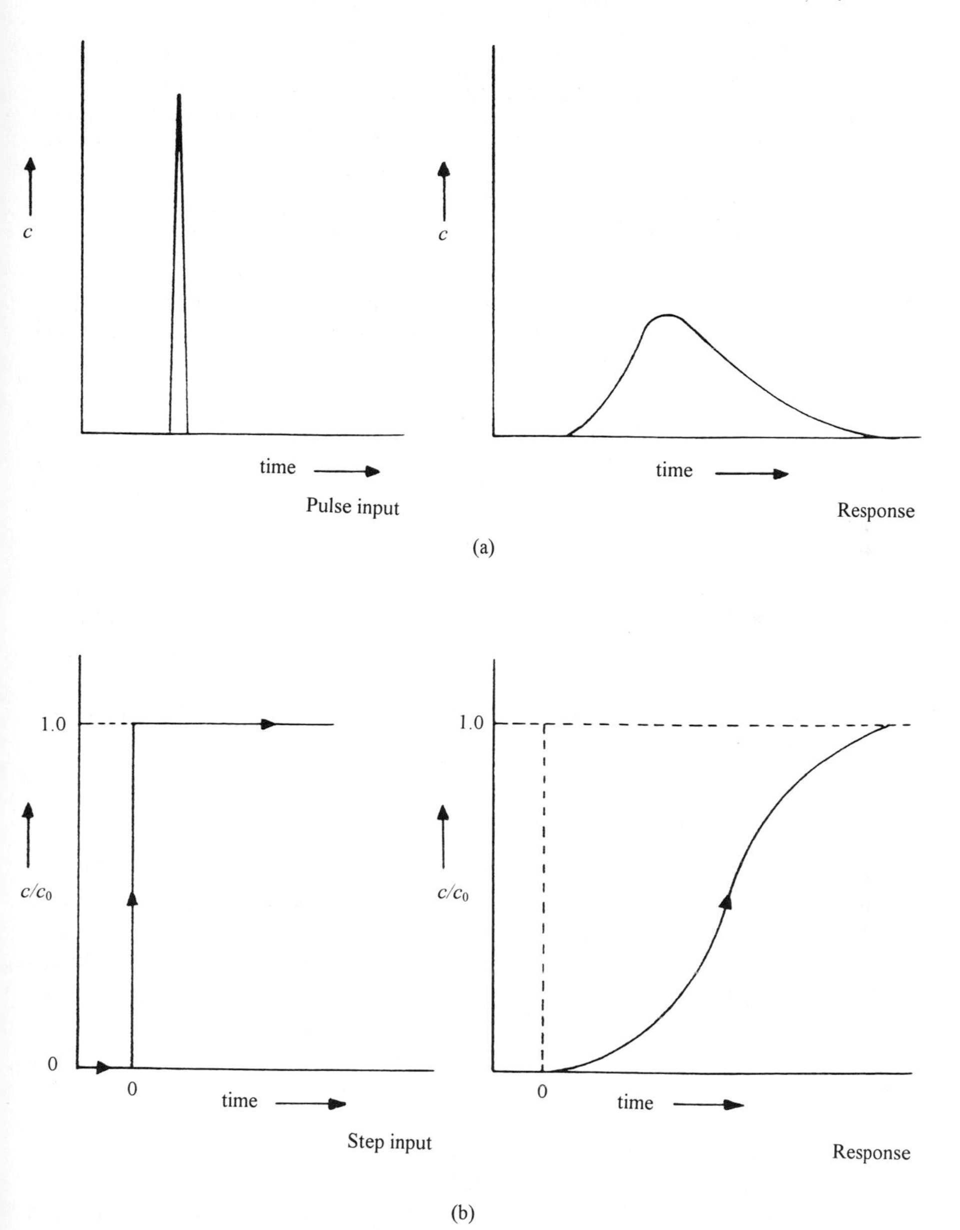

Figure 4.21 Response to a concentration change at the face of a porous pellet: (a) pulse input; (b) step input.

material. The ratio ρ_p/ρ_s then gives the fraction of solid present in the particles and $(1 - \rho_p/\rho_s)$ is the porosity.

The tortuosity τ is also included in the geometric factor to account for the tortuous nature of the pores. It is the ratio of the path length which must be traversed by molecules diffusing between two points within a pellet to the direct linear separation between those points. Theoretical predictions of τ rely on somewhat inadequate models of the porous structure, but experimental values may be obtained from measurements of D_e, D and ψ.

4.5.1.2 Knudsen Diffusion

In the region of flow where collisions of molecules with the container walls are more frequent than intermolecular gaseous collisions, Knudsen in a classic monograph published in 1934 demonstrated that the net flow of molecules in the direction of gas flow is proportional to the gradient of the molecular flux. From geometrical considerations it may be shown that, for the case of a capillary of circular cross-section and radius r, the proportionality factor is $8\pi r^3/3$. This results in a Knudsen diffusion coefficient:

$$D_K = \frac{2}{3} r \sqrt{\frac{8RT}{\pi M}} \tag{30}$$

where M is the molecular mass of the diffusing gas. This equation, however, cannot be directly applied to the majority of porous solids because they are usually not well represented by a collection of straight cylindrical capillaries. As discussed in Section 4.4.2, the pore radius is proportional to the ratio of the BET specific surface area to the specific pore volume, the proportionality factor depending on the pore geometry. Thus if, for the purposes of calculating a Knudsen diffusion coefficient, the pore model adopted consists of non-intersecting cylindrical capillaries and the radius computed from Eq. (20) is a radius equivalent to the radius of a cylinder having the same surface to volume ratio as the pore, then Eq. (30) may be applied. In terms of the porosity ψ, specific surface area S_g and particle density ρ_p (mass per unit total particle volume, including the volume occupied by pore space),

$$D_K = \frac{16}{3} \frac{\psi}{\rho_p S_g} \sqrt{\frac{RT}{\pi M}} \tag{31}$$

In the region of Knudsen flow the effective diffusivity D_{eK} for the porous solid may be computed in a similar way to the effective diffusivity in the region of molecular flow, i.e. D_K is simply multiplied by the geometric factor ψ/τ.

4.5.1.3 The Transition Region of Diffusion

Under conditions where Knudsen or molecular diffusion does not predominate, a relation for the effective diffusivity of a binary gas mixture is:

$$D_e = \frac{1}{\dfrac{1}{D_{eM}} + \dfrac{1}{D_{eK}} - \dfrac{x_A(1 + N_B/N_A)}{D_{eM}}} \tag{32}$$

where D_{eM} and D_{eK} are the effective diffusivities in the molecular and Knudsen regions of diffusion, N_A and N_B are the molar fluxes of the components A and B of the binary mixture and x_A is the mole fraction of A. This formula can be extended to represent a multicomponent mixture of gases.

4.5.1.4 Forced Flow in Pores

Many heterogeneous reactions give rise to an increase or decrease in the total number of moles present in the porous solid due to the reaction stoichiometry. In such cases there will be a pressure difference between the interior and exterior of the particle, and forced flow occurs. When the mean free path of the reacting molecules is large compared with the pore diameter, forced flow is indistinguishable from Knudsen flow and is not affected by pressure differentials. When, however, the mean free path is small compared with the pore diameter and a pressure difference exists across the pore, forced flow (Poiseuille flow) resulting from this pressure difference will be superimposed on molecular flow. The diffusion coefficient D_p for forced flow depends on the square of the pore radius and on the total pressure difference ΔP:

$$D_p = \frac{r^2 \Delta P}{8\mu} \tag{33}$$

The viscosity μ of most gases at about 1 bar pressure is of the order of 10^{-7} N s m^{-2}, so for pores of about 10^{-6} m radius D_p is approximately 10^{-5} m^2 s^{-1}. Molecular diffusion coefficients are of similar magnitude, so in small pores forced flow will compete with molecular diffusion. For fast reactions accompanied by an increase in the number of moles present, an excess pressure is developed in the interior recesses of the porous particle; this results in the forced flow of excess product and reactant molecules to the particle exterior. Conversely, for pores of radius greater than about 10^{-4} m, D_p is as high as 10^{-3} m^2 s^{-1} and the coefficient of diffusion which will determine the rate of intraparticle transport will be the coefficient of molecular diffusion.

Except in the case of reactions at high pressure, the pressure drop which must be maintained to cause flow through a packed bed of particles is usually insufficient to produce forced flow in the capillaries of the solid, and the gas flow is diverted around the exterior periphery of the pellets. Reactants then reach the interior of the porous solid by Knudsen or molecular diffusion.

4.6 Chemical Reaction in Porous Catalyst Pellets

When a chemical reaction occurs within a porous catalyst, the intrinsic rate of reaction may, depending on the pore structure and the ease or difficulty of intraparticle transport, be impeded. Thiele, in his much-quoted work (see Further Reading, Section 4.8.2) constructed mathematical models accounting for the effect of intraparticle diffusion on chemical reaction and was able to predict quantitatively how intraparticle diffusion effects reduced the reaction rate. For this purpose he employed a geometric model of a catalyst pellet with isotropic properties. He considered, in particular, that the effective diffusivity and effective thermal conductivity (analagous to the concept of molecular transport within a porous solid) are independent of position in the porous mass. Although idealised geometric shapes are used to depict the situation within a catalyst pellet, such models, as we shall see later, are quite good approximations to practical catalyst pellets.

The simplest case we shall consider is that of an isothermal first-order chemical reaction occurring within a rectangular slab of porous catalyst, the edges of which are sealed so that diffusion occurs in one dimension only. Figure 4.22 illustrates the geometry of the slab. Consider that the first order irreversible reaction

$$\mathrm{A} \rightarrow \mathrm{B}$$

occurs within the volume of the particle, and suppose its specific velocity constant on the basis of unit surface area is k_s. For heterogeneous reactions uninfluenced by mass transfer effects, experimental values for rate constants are usually based on unit surface area. The corresponding value in terms of unit total volume of particle would be $\rho_p S_g k_s$ where ρ_p is the apparent density of the catalyst pellet and S_g is the specific surface area per unit mass of the solid, including the internal pore surface area. We shall designate the specific rate constant based on unit volume of particle as k. The conservation of reactant A across the volume element of thickness Δx requires that

$$A_c D_e \left.\frac{\mathrm{d}c_A}{\mathrm{d}x}\right|_{x+\Delta x} = A_c D_e \left.\frac{\mathrm{d}c_A}{\mathrm{d}x}\right|_{x} - \mathrm{k} c_A A_c \Delta x \tag{34}$$

since, in the steady state, the flux of A into the element at $(x + \Delta x)$ must be balanced by the flux out of the element at x, minus the amount lost by reaction within the volume element $A_c \Delta x$. Note that for the co-ordinate system considered, because the concentration of the reacting component A decreases in the direction of decreasing x, the concentration gradient is positive, and hence the flux is negative. If the concentration gradient term at the point $(x + \Delta x)$ in Eq. (34) is expanded in a Taylor series about the point x and differential coefficients of order greater than two are ignored, the equation simplifies to:

$$\frac{\mathrm{d}^2 c_A}{\mathrm{d}x^2} - \frac{\mathrm{k} c_A}{D_e} = 0 \tag{35}$$

An analogous equation may be written for component B. By reference to Fig. 4.22 it will be seen that, because the product B diffuses outwards, its flux is positive. Reaction produces B within the slab of material and hence makes the term depicting the rate of formation of B in the material-balance equation positive, resulting in an equation similar in form to Eq. (35).

The boundary conditions for the problem may be written by referring to Fig. 4.22. At the exterior surface of the slab the concentration will be that corresponding to the conditions in the bulk gas phase. Then provided that there is no resistance to mass transfer in the gas phase,

$$c_A = c_{A_g} \quad \text{at}\, x = \pm L \tag{36}$$

At the centre of the slab considerations of symmetry demand that:

$$\frac{\mathrm{d}c_A}{\mathrm{d}k} = 0 \quad \text{at}\, x = 0 \tag{37}$$

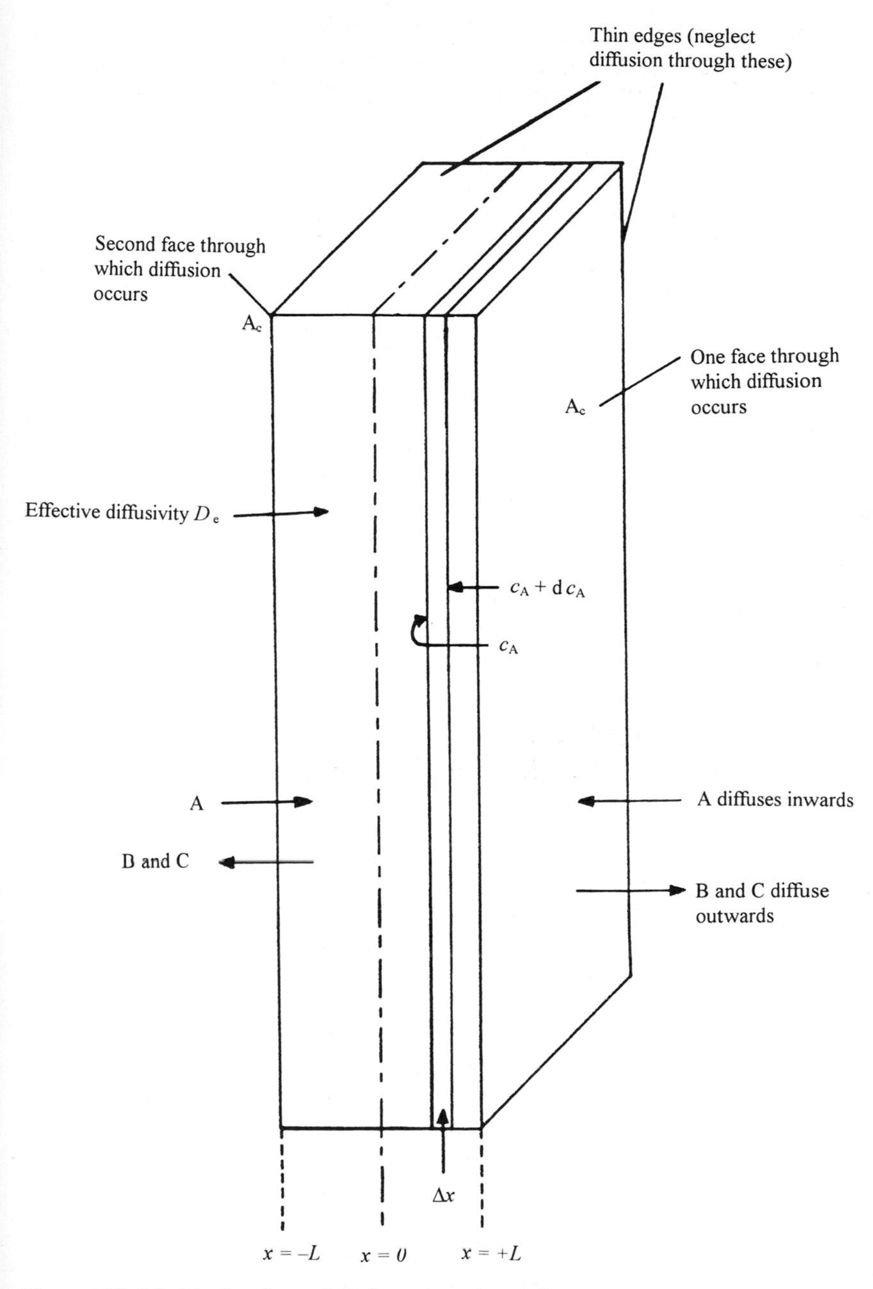

Figure 4.22 Model of wafer or slab-shaped catalyst pellet.

so that the net flux through the plane at $x = 0$ is zero, diffusion across this boundary being just as likely in the direction of increasing x as in the direction of decreasing x. The solution of Eq. (35) with the boundary conditions given by Eqs. (36) and (37) is

$$c_\mathrm{A} = c_{\mathrm{A_g}} \frac{\cosh \lambda x}{\cosh \lambda L} \tag{38}$$

where λ denotes the quantity $(k/D_\mathrm{e})^{1/2}$. Equation (38) describes the concentration profile of A within the catalyst slab. In the steady state the total rate of consumption of A must be equal to the total flux of A at the external surfaces. By reference to Fig. 4.22 this is seen to be $2A_\mathrm{c}D_\mathrm{e}(\mathrm{d}c_\mathrm{A}/\mathrm{d}x)_{x=L}$. Now in the absence of any resistance to diffusion within the pellet, the whole interior of the catalyst would be bathed in the gas phase concentration c_{Ag}, giving an intrinsic rate per unit volume of $2A_\mathrm{c}Lkc_{\mathrm{A_g}}$. The ratio of the rate of reaction when diffusion intrudes to the rate of reaction in the absence of diffusion is the effectiveness factor η. Thus

$$\eta = \frac{2A_\mathrm{c}D_\mathrm{e}(\mathrm{d}c_\mathrm{A}/\mathrm{d}x)_{x=L}}{2A_\mathrm{c}Lkc_{\mathrm{A_g}}} = \frac{1}{\phi}\frac{(\mathrm{d}c_\mathrm{A}/\mathrm{d}x)_{x=L}}{c_{\mathrm{A_g}}} \tag{39}$$

where $\phi(= \lambda L)$ is known as the Thiele modulus. Evaluating the concentration gradient at the exterior surface ($x = L$) from Eq. (38) we obtain

$$\eta = \frac{\tanh \lambda L}{\lambda L} = \frac{\tanh \phi}{\phi} \tag{40}$$

as the expression for the effectiveness factor for wafer or slab catalyst pellet geometry. If the function η is plotted from Eq. (40), corresponding to the case of an isothermal first-order irreversible reaction in a slab with sealed edges, it may be seen from Fig 4.25 (below) that when $\phi < 0.2$, η is close to unity. Under these conditions there would be no diffusional resistance, for the rate of chemical reaction is not limited by diffusion. On the other hand, when $\phi > 5.0$, $\eta = 1/\phi$ is a good approximation and for such conditions internal diffusion is the rate-determining process. Between these two limiting values of ϕ the effectiveness factor is calculated from Eq. (40) and the rate process is in a region where neither intraparticle diffusion nor chemical reaction is overwhelmingly rate-determining.

Only a very limited number of manufactured catalysts could be approximately described by the slab model but there appear to be many which conform to the shape of a cylinder or sphere. Utilising the same principles as for the slab, it may be shown (see the next example and Fig. 4.23) that, for a cylinder of radius r_0 sealed at the flat ends, the effectiveness factor is:

$$\eta = \frac{2I_1(\lambda r_0)}{\lambda r_0 I_0(\lambda r_0)} \tag{41}$$

where I_0 and I_1 denote zero- and first-order modified Bessel functions of the first kind.

For a sphere of radius r_0 (see Fig. 4.24 and the second example following, pertaining to a sphere):

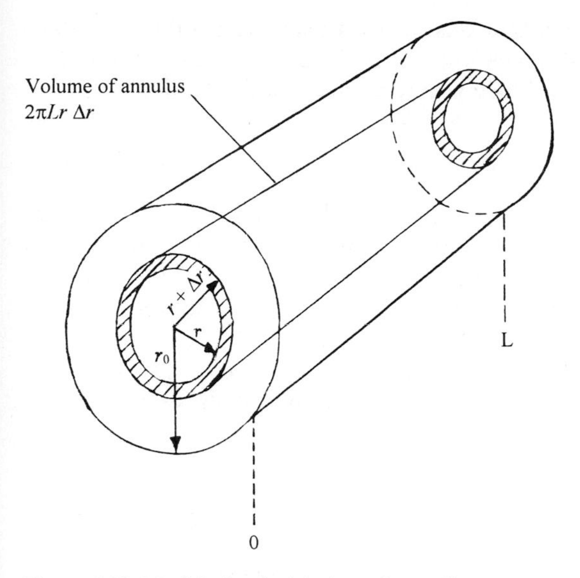

Figure 4.23 Model of cylindrical catalyst pellet.

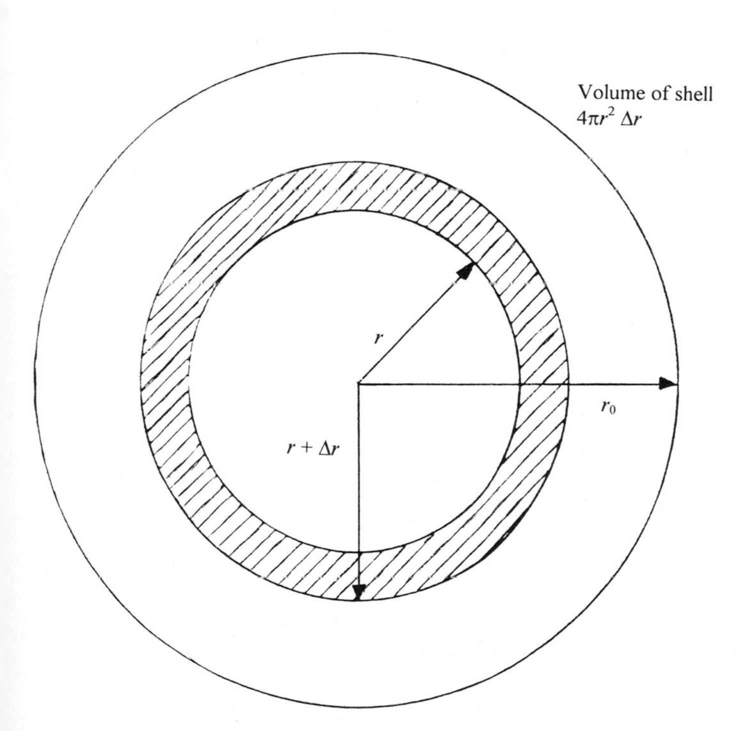

Figure 4.24 Model of spherical catalyst pellet.

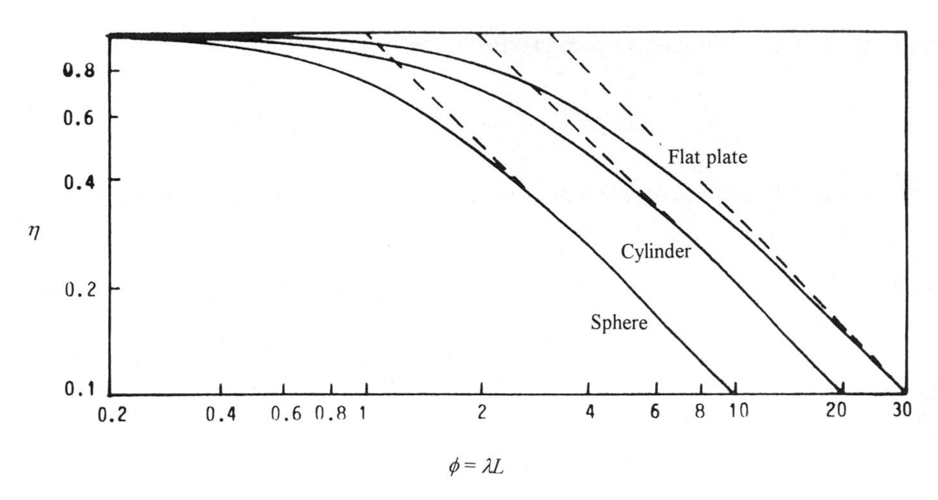

Figure 4.25 Effectiveness factor as a function of the Thiele modulus for isothermal catalyst pellets (slab, cylinder and sphere).

$$\eta = \frac{3}{\lambda r_0}\left\{\coth(\lambda r_0) - \frac{1}{\lambda r_0}\right\} \tag{42}$$

Hollow cylindrical catalyst pellets are sometimes employed, since excessive pressure drops in a packed bed of pellets can thus be avoided. A more complex expression for the effectiveness factor is obtained for such geometry. Figure 4.25 displays the curves $\eta(\phi)$ for the cylindrical and spherical catalyst pellets.

Example

Derive an expression for the effectiveness factor of a cylindrical catalyst pellet, sealed at both ends, in which a first-order chemical reaction occurs.

Solution

The pellet has cylindrical symmetry about its central axis. Construct an annulus with radii $(r + \Delta r)$ and r and consider the diffusive flux of material into and out of the cylindrical annulus, length L.

A material balance for the reactant gives (see Fig 4.23):

Diffusive flux in at $(r + \Delta r)$ − diffusive flux out at r
= amount reacted in volume $2\pi L r \Delta r$

i.e.

$$\left\{2\pi D_e L\left(r\frac{dc}{dr}\right)\Big|_{r+\Delta r}\right\} - \left\{2\pi D_e L\left(r\frac{dc}{dr}\right)\Big|_{r}\right\} = 2\pi L r \Delta r k c$$

Expanding the first term and ignoring terms higher than Δr^2:

$$\frac{d^2c}{dr^2} + \frac{1}{r}\frac{dc}{dr} - \lambda^2 c = 0 \quad \text{where } \lambda = \sqrt{\frac{k}{D_e}}$$

This is a standard modified Bessel equation of zero order whose solution is:

$$c = AI_0(\lambda r) + BK_0(\lambda r)$$

where I_0 and K_0 represent zero-order modified Bessel functions of the first and second kind respectively.

The boundary conditions for the problem are $r = r_0$, $c = c_g$; $r = 0$, c is finite. Since c remains finite at $r = 0$ and $K_0(0) = \infty$, then we must put $B = 0$ to satisfy the physical conditions. Substituting the boundary conditions therefore gives the solution:

$$\frac{c}{c_g} = \frac{I_0(\lambda r)}{I_0(\lambda r_0)}$$

For the cylinder:

$$\eta = \frac{2\pi r D_e L (dc/dr)_{r_0}}{\pi r_0^2 L k c_g}$$

From the relation between c and r:

$$\left(\frac{dc}{dr}\right)_{r_0} = c_g \frac{I_1(\lambda r_0)}{I_0(\lambda r_0)}$$

and since:

$$\frac{d}{dr}\{I_0(\lambda r)\} = I_1(\lambda r)$$

then:

$$\eta = \frac{2I_1(\lambda r_0)}{\lambda r_0 I_0(\lambda r_0)}$$

Example

Derive an expression for the effectiveness factor of a spherical catalyst pellet in which a first-order isothermal reaction occurs.

Solution

Take the origin of co-ordinates at the centre of the pellet, radius r_0, and construct an infinitesimally thin shell of radii $(r + \Delta r)$ and r (see Fig. 4.24). A material balance for the reactant across the shell gives:

Diffusive flux in at $(r+\Delta r)$– diffusive flux out at r
= amount reacted in volume $4\pi r^2 \Delta r$

i.e.

$$\left\{4\pi D_e L\left(r^2 \frac{dc}{dr}\right)\bigg|_{r+\Delta r}\right\} - \left\{4\pi D_e L\left(r^2 \frac{dc}{dr}\right)\bigg|_{r}\right\} = 4\pi L r^2 \Delta r k c$$

Expanding the first term and ignoring terms higher than Δr^2:

$$\frac{d^2c}{dr^2} + \frac{2}{r}\frac{dc}{dr} = \frac{kc}{D_e}$$

or:

$$\frac{1}{r^2}\frac{dc}{dr}\left\{r^2 \frac{dc}{dr}\right\} - \lambda^2 c = 0 \quad \text{where } \lambda = \sqrt{\frac{k}{D_e}}$$

Substituting $c = y/r$

$$\frac{d^2y}{dr^2} - \lambda^2 y = 0$$

Therefore

$$y = Ae^{\lambda r} + Be^{-\lambda r}$$

The boundary conditions for the problem are $r = r_0$, $c = c_g$; $r = 0$, c is finite. Now if, at $r = 0$, c is to remain finite then $y(0) = 0$. At r_0 we have $y(r_0) = c_g r_0$. Substituting these boundary conditions:

$$y = cr = \frac{c_g r_0 \sinh(\lambda r)}{\sinh(\lambda r_0)}$$

Now for a sphere:

$$\eta = \frac{(4\pi r_0{}^2 D_e (dc/dr)_{r_0}}{(4/3)\pi r_0{}^3 k c_g}$$

From the relation between c and r:

$$\left(\frac{dc}{dr}\right)_{r_0} = \frac{1}{r_0}\{\lambda r_0 \coth(\lambda r_0) - 1\}$$

Hence

$$\eta = \frac{3}{\lambda r_0}\left\{\coth(\lambda r_0) - \frac{1}{\lambda r_0}\right\}$$

The Thiele moduli for the cylinder and sphere differ from that for the slab. In the case of the slab we recall that $\phi = \lambda L$, whereas for the cylinder it is conveniently defined as $\phi = \lambda r_0/2$ and for the sphere as $\phi = \lambda r_0/3$. In each case the reciprocal of this corresponds to the respective asymptote for the curve representing the slab, cylinder or sphere. We may note here that the ratio of the geometric pellet volume V_p of each of the models to the external geometric surface area S_x is L for the slab, $r_0/2$ for the cylinder and $r_0/3$ for the sphere. Thus, if the Thiele modulus is defined as:

$$\phi = \lambda \frac{V_p}{S_x} = \frac{V_p}{S_x}\sqrt{\frac{k}{D_e}} \tag{43}$$

the asymptotes become coincident. The asymptotes for large ϕ correspond to $\eta = 1/\phi$ for any shape of particle because diffusion is rate-determining under these conditions and reaction occurs, therefore, in only a very thin region of the particle adjacent to the exterior surface. The curvature of the surface is thus unimportant when diffusion predominates.

The effectiveness factor for the slab model may also be calculated for reactions other than first-order ones. It turns out that when the Thiele modulus is large, the asymptotic value of η for all reactions is inversely proportional to the Thiele modulus, and when the latter approaches zero the effectiveness factor tends to unity. However, just as we found that the asymptotes for a first-order reaction in particles of different geometry do not coincide unless we choose a definition for the Thiele modulus which forces them to become superimposed, so we find that the asymptotes for reaction orders $n = 0$, 1 and 2 do not coincide unless we define a generalized Thiele modulus:

$$\bar{\phi} = \frac{V_p}{S_x}\left\{\frac{(n+1)}{2}\frac{kc_g^{n-1}}{D_e}\right\}^{1/2} \tag{44}$$

The modulus $\bar{\phi}$ defined by Eq. (44) has the advantage that the asymptotes to η are approximately coincident for all particle shapes and for all reaction orders except $n = 0$; for this latter case $\eta = 1$ for $\bar{\phi} < 2$ and $\eta = 1/\bar{\phi}$ for $\bar{\phi} > 2$. Thus η may be calculated from the simple slab model, using Eq. (44) to define the Thiele modulus. The curve of η as a function of $\bar{\phi}$ is therefore quite general for practical catalyst pellets. For $\bar{\phi} > 3$ it is found that $\eta = 1/\bar{\phi}$ to an accuracy within 0.5%, while the approximation is within 3.5% for $\bar{\phi} > 2$. It is best to use this generalised curve (i.e. η as a function of $\bar{\phi}$) because the asymptotes for different cases can then almost be made to coincide. The errors involved in using the generalised curve are probably no greater than errors perpetrated by estimating values of parameters in the Thiele modulus.

4.6.1 Effect of Intraparticle Diffusion on Experimental Parameters

When intraparticle diffusion is rate-determining, the kinetic behaviour of the system is different from that which prevails when chemical reaction is rate-determining. For conditions of diffusion control ϕ will be large and then the effectiveness factor η (=$(\tanh\phi)/\phi$, from Eq. (40)) becomes $1/\bar{\phi}$. From Eq. (44) it is seen therefore that η is proportional to $k^{-1/2}$. The chemical reaction rate, on the other hand, is directly proportional to k and so, because the chemical rate has to be multiplied by the effectiveness factor, the overall reaction rate is proportional to $k^{1/2}$. Since the specific rate constant is directly proportional to $\exp(-E/RT)$, where E is the activation energy for the chemical reaction in the absence of diffusion effects, we are led to the important result that for a diffusion-limited reaction the rate is proportional to $\exp(-E/2RT)$. Hence the apparent activation energy E_D, measured when reaction occurs in the diffusion-controlled region, is only half the true value:

$$E_D = E/2 \tag{45}$$

A further important result which arises because of the functional form of the Thiele modulus ϕ is that the apparent order of reaction in the diffusion-controlled region differs from that which is observed when chemical reaction is rate-determining. For nth-order chemical kinetics the chemical rate of reaction is proportional to c^n but the effectiveness factor, being equal to $1/\phi$ in the diffusion-controlled region, will, by Eq. (44), be proportional to $c^{-(n-1)/2}$. Hence the overall rate is proportional to $c^{(n+1)/2}$. The apparent order of reaction n_D therefore, as measured when reaction is dominated by intraparticle diffusion effects, is thus related to the true reaction order n by

$$n_D = (n+1)/2 \tag{46}$$

A zero-order reaction thus becomes a half-order reaction, a first-order reaction remains first-order, whereas a second-order reaction has an apparent order of 3/2 when strongly influenced by diffusional effects. Because k and n are modified in the diffusion-controlled region then, if the rate of the overall process is estimated by multiplying the chemical reaction rate by the effectiveness factor, it is imperative to know the true rate of chemical reaction uninfluenced by diffusion effects.

The functional dependence of other parameters on the reaction rate also becomes modified when diffusion determines the overall rate. Writing the overall rate for an nth-order reaction in terms of the catalyst bed volume, and substituting the general expression for the effectiveness factor at high values of $\bar{\phi}$ (where η is approximately equal to $1/\bar{\phi}$) and $\bar{\phi}$ is defined by Eq. (44), we obtain for the rate per unit bed volume

$$R_v = (1-e)kc_g{}^n\eta = (1-e)kc_g{}^n \frac{S_x}{V_p}\left\{\frac{2D_e}{(n+1)kc_g{}^{n-1}}\right\}^{1/2} \tag{47}$$

where $(1-e)$ is the fraction of reactor volume occupied by the particulate matter, and e is the bed voidage. Referring the specific rate constant to unit surface area, rather than unit reactor nolume, the term $(1-e)k$ is equivalent to $\rho_b S_g k_s$ where ρ_b is the bulk density of the catalyst and S_g is its surface area per unit mass. On the other hand, the rate constant k appearing as a square root in the numerator in Eq. (44) is based on unit particle volume and is therefore equal to $\rho_p S_g k_s$, where ρ_p is now the particle density. Thus, if bulk diffusion controls the reaction, the rate becomes dependent on the square root of the specific surface area, rather than being directly proportional to surface area in the absence of transport effects. We do not include the external surface area S_x in this reckoning since the ratio V_p/S_x, for a given particle shape, is an independent parameter characteristic of the particle size. On the other hand, if Knudsen diffusion determines the rate, then because the effective diffusivity for Knudsen flow is inversely proportional to the specific surface area (Eq. (31)) the reaction rate becomes independent of surface area.

Table 4.4 Effect of intraparticle diffusion on parameters involved in overall rate of reaction.

Rate-limiting process	Order	Activation energy	Surface area	Pore volume
Chemical reaction	n	E	S_g	Independent
Molecular diffusion	$(n+1)/2$	$E/2$	$\sqrt{S_g}$	$\sqrt{V_g}$
Knudsen diffusion	$(n+1)/2$	$E/2$	Independent	V_g

The pore volume V_g per unit mass (a measure of the particle porosity) is also a parameter which is important and is implicitly contained in Eq. (44) (because $k = \rho_p S_g k_s$). Since the product of the particle density ρ_p and specific pore volume V_g represents the porsity, then ρ_p is inversely proportional to V_g. Therefore, when the rate is controlled by bulk diffusion, it is proportional not simply to the square root of the specific surface area but to the product of $S_g{}^{1/2}$ and $V_g{}^{1/2}$. If Knudsen diffusion controls the reaction, then the overall rate is directly proportional to V_g since the effective Knudsen diffusivity contained in the quantity $(D_e/\rho_p)^{1/2}$ is, from Eq. (31), proportional to the ratio of the porosity ψ and the particle density ρ_p.

Table 4.4 summarises the effect which intraparticle mass transfer effects have on parameters involved explicitly or implicitly in the expression for the overall rate of reaction.

4.6.2 Non-isothermal Reactions in Porous Catalyst Pellets

So far, the effect of temperature gradients within the particle has been ignored. Strongly exothermic reactions generate a considerable amount of heat which, if conditions are to remain stable, must be transported through the particle to the exterior surface where it may then be dissipated. Similarly, an endothermic reaction requires a source of heat and in this case the heat must permeate the particle from the exterior to the interior. In any event it is possible for a temperature gradient to be established within the particle; the chemical reaction rate would then vary with position by virtue of temperature as well as concentration.

We may consider the problem by writing a material and heat balance for the slab of catalyst depicted in Fig. 4.23. For an irreversible first-order exothermic reaction the material balance, according to Eq. (35), is:

$$\frac{d^2 c_A}{dx^2} - \frac{k c_A}{D_e} = 0$$

A heat balance over the element Δx gives:

$$\frac{d^2 T}{dx^2} + \frac{(-\Delta H) k c_A}{k_e} = 0 \qquad (48)$$

where ΔH is the enthalpy change resulting from reaction, and k_e is the effective thermal conductivity of the particle defined by analogy with the discussion on effective diffusivity. In writing these two equations it should be remembered that the specific rate constant k is a function of temperature, usually of the Arrhenius form ($k = A\exp(-E/RT)$) where A is the frequency factor for reaction. These two simultaneous differential equations are to be solved together with the boundary conditions conforming to the absence of transport effects exterior to the particle:

$$c_{\mathrm{A}} = c_{\mathrm{A_g}} \quad \text{and} \quad T = T_{\mathrm{g}} \quad \text{at } x = \pm L \tag{49}$$

$$\frac{\mathrm{d}c_{\mathrm{A}}}{\mathrm{d}x} = \frac{\mathrm{d}T}{\mathrm{d}x} = 0 \quad \text{at } x = 0 \tag{50}$$

Because of the non-linearity of the equations the problem can only be solved in this form by numerical techniques. A solution to the problem may be presented in the form of families of curves for the effectiveness factor as a function of the Thiele modulus. Figure 4.26 shows these curves for the case of a first-order irreversible reaction occurring in spherical catalyst particles. Two additional independent dimensionless parameters are introduced into the problem and these are defined as:

$$\beta = \frac{(-\Delta H)D_{\mathrm{e}}c_{\mathrm{g}}}{k_{\mathrm{e}}T_{\mathrm{g}}} \tag{51}$$

$$\varepsilon = \frac{E}{RT_{\mathrm{g}}} \tag{52}$$

The parameter β represents the maximum temperature difference that could exist in the particle relative to the temperature at the exterior surface, for if we recognise that in the steady state the heat flux within an elementary thickness of the particle is balanced by the heat generated by chemical reaction, then:

$$k_{\mathrm{e}}\frac{\mathrm{d}T}{\mathrm{d}x} = -(-\Delta H)D_{\mathrm{e}}\frac{\mathrm{d}c}{\mathrm{d}x} \tag{53}$$

If Eq. (53) is then integrated from the exterior surface where $T = T_{\mathrm{g}}$ and $c = c_{\mathrm{g}}$ to the centre of the particle where (say) $T = T_{\mathrm{m}}$ and $c = c_{\mathrm{m}}$, we obtain

$$\frac{T_{\mathrm{m}} - T_{\mathrm{g}}}{T_{\mathrm{g}}} = \frac{(-\Delta H)D_{\mathrm{e}}}{k_{\mathrm{e}}T_{\mathrm{g}}}(c_{\mathrm{g}} - c_{\mathrm{m}}) \tag{54}$$

When the Thiele modulus is large, c_{m} is effectively zero and the maximum difference in temperature between the centre and exterior of the particle is $(-\Delta H)\,D_{\mathrm{e}}\,c_{\mathrm{g}}/k_{\mathrm{e}}$. Relative to the temperature outside the particle, this maximum temperature difference is therefore β. For exothermic reactions β is positive while for endothermic reactions it is negative. The curve in Fig. 4.26 for $\beta = 0$ represents isothermal conditions within the pellet. It is interesting to note that for a reaction in which $(-\Delta H) = 100\,\mathrm{kJ\,mol^{-1}}$, $k_{\mathrm{e}} = 10^{-5}\,\mathrm{kJ\,K^{-1}\,cm^{-1}\,s^{-1}}$, $D_{\mathrm{e}} = 10^{-1}\,\mathrm{cm^2\,s^{-1}}$ and $c_{\mathrm{g}} = 10^{-4}\,\mathrm{mol\,cm^{-3}}$, the value of $(T_{\mathrm{m}} - T_{\mathrm{g}})$ is 100 °C. In practice much lower values than this are observed, but it does serve to show that serious errors may be introduced into calculations if conditions within the pellet are arbitrarily assumed to be isothermal.

On the other hand, it has been argued that the resistance to heat transfer is effec-

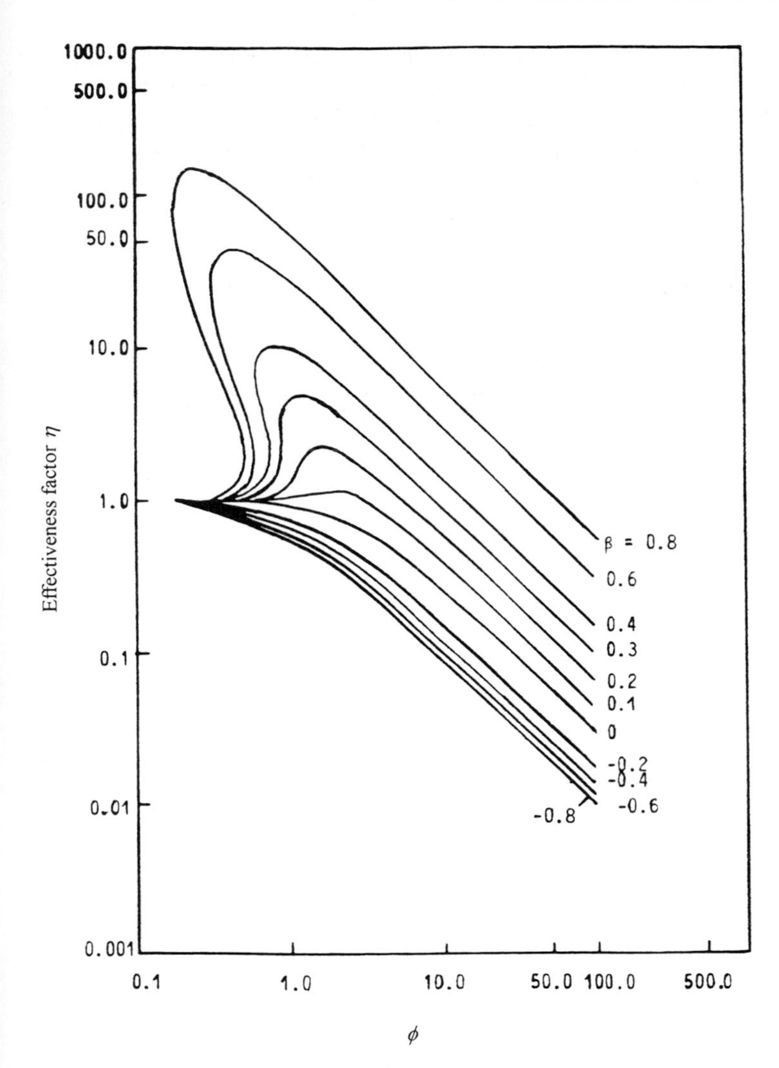

Figure 4.26 Effectiveness factor as a function of Thiele modulus for a (first-order) non-isothermal reaction in a spherical catalyst pellet.

tively within a thin gas film enveloping the catalyst particle. Thus, for the whole practical range of heat transfer coefficients and thermal conductivities, the catalyst particle may be considered to be at a uniform temperature. Any temperature increase arising from the exothermic nature of a reaction would therefore be across the exterior fluid film rather than in the pellet interior.

Figure 4.26 shows that, for exothermic reactions ($\beta > 0$), the effectiveness factor may exceed unity. This is because the increase in rate caused by the temperature rise inside the particle more than compensates for the decrease in rate caused by the negative concentration gradient which effects a decrease in concentration towards the centre of the particle. A further point of interest is that, for reactions which are highly exothermic and at low values of the Thiele modulus, the value of η is not uniquely defined by the Thiele modulus and the parameters β and ε. The shape of the curves in this region indicates that the effectiveness factor may correspond to any one of three values for a given value of the Thiele modulus. In effect there are three different conditions for which the rate of heat generation within the particle is equal to the rate of heat removal. One condition represents a metastable state and the remaining two conditions correspond to a region in which the rate is limited by chemical reaction (relatively low temperatures) and a region where there is diffusion limitation (relatively high temperatures). The region of multiple solutions in Fig. 4.26, however, corresponds to large values of β and ε seldom encountered in practice.

4.6.3 Criteria for Diffusion Control

In assessing whether a reactor is influenced by intraparticle mass transfer effects, a criterion has been developed for isothermal reactions based upon the observation that the effectiveness factor approaches unity when the generalised Thiele modulus is of the order of unity. It can be shown that the effectiveness factor for all catalyst geometries and reaction orders (except zero order) tends to unity when the generalised Thiele modulus falls below a value of 1. Since η is about unity when $\phi < \sqrt{2}$ for zero-order reactions, a quite general and safe criterion for diffusion control of simple isothermal reactions not affected by product inhibition is $(\bar{\phi})^2 < 1$. Since the Thiele modulus (see Eq. (44)) contains the specific rate constant for chemical reaction, which is often unknown, a more useful criterion is obtained by substituting $\mathrm{R_v}/c_\mathrm{g}$ (for a first-order reaction) for k to give:

$$\left(\frac{V_\mathrm{p}}{S_\mathrm{x}}\right)^2 \frac{\mathrm{R_v}}{D_\mathrm{e} c_\mathrm{g}} < 1 \tag{55}$$

where R_v is the measured rate of reaction per unit volume of catalyst particle.

It has been pointed out, however, that this criterion is invalid for more complex chemical reactions whose rate is retarded by products. In such cases the observed kinetic rate expression should be substituted into the material-balance equation for the particular geometry of particle concerned. An asymptotic solution to the material-balance equation then gives the correct form of the effectiveness factor. However, the results indicate that the inequality thus obtained is applicable only at high partial pressures of product. For low partial pressures of product (often the condition in an experimental differential tubular reactor) the reliability of the criterion will depend on the magnitude of the constants in the kinetic rate equation.

When reaction conditions within the particle are non-isothermal, a suitable criterion defining conditions under which a reaction is not controlled by mass and heat transfer effects in the solid is:

$$\left(\frac{V_\mathrm{p}}{S_\mathrm{x}}\right)^2 \frac{\mathrm{R_v}}{D_\mathrm{e} c_\mathrm{g}} \exp\left(\frac{\varepsilon\beta}{1+\beta}\right) < 1 \tag{56}$$

Example

Raw kinetic data provided by a research and development laboratory for a catalytic gas reaction in a packed tubular reactor are as follows:

(a) Experiments at constant temperature

Pressure $p \times 10^3$ [bar]	8.17	11.0	16.4	26.6	49.4
Rate $R \times 10^6$ [mol s^{-1}g^{-1}]	8.17	12.2	22.2	49.4	134

Surface area of catalyst = 324 m^2 g^{-1}

(b) Experiments at constant temperature

Temperature T [°C]	240	250	265	270	280
Rate $R \times 10^6$ [mol s^{-1}g^{-1}]	7.21	8.17	9.97	10.8	12.2

Surface area of catalyst = 324 m^2 g^{-1}

(c) Experiments at constant temperature and pressure

Surface area of catalyst [m^2 g^{-1}]	156	324	576	625
Rate $R \times 10^6$ [mol s^{-1}g^{-1}]	5.0	7.21	8.9	10.2

Determine whether the kinetics are influenced by intraparticle diffusion effects and hence find the true kinetic order and activation energy.

Solution

If intraparticle diffusion effects are prominent then, in the packed reactor, the rate per unit volume of bed is

$$\text{Rate} = (1 - e)kc^n\eta$$

where e is the bed voidage, n is the true order of reaction and η is the effectiveness factor. For a diffusion-controlled rate process

$$\eta = \frac{1}{\phi}$$

where ϕ is the Thiele modulus. Recalling that the rate constant k is on a volume basis and transposing to k_s on a surface area basis,

$$\text{Rate} = \frac{\rho_b S_g k_s c^n}{\phi}$$

Substituting for ϕ (from Eq. (44))

$$\text{Rate} = \rho_b S_g k_s c^n \left\{ \frac{2D}{(n+1)\rho_b S_g k_s c^{n-1}} \right\}^{1/2}$$

The above equation demonstrates the following features.

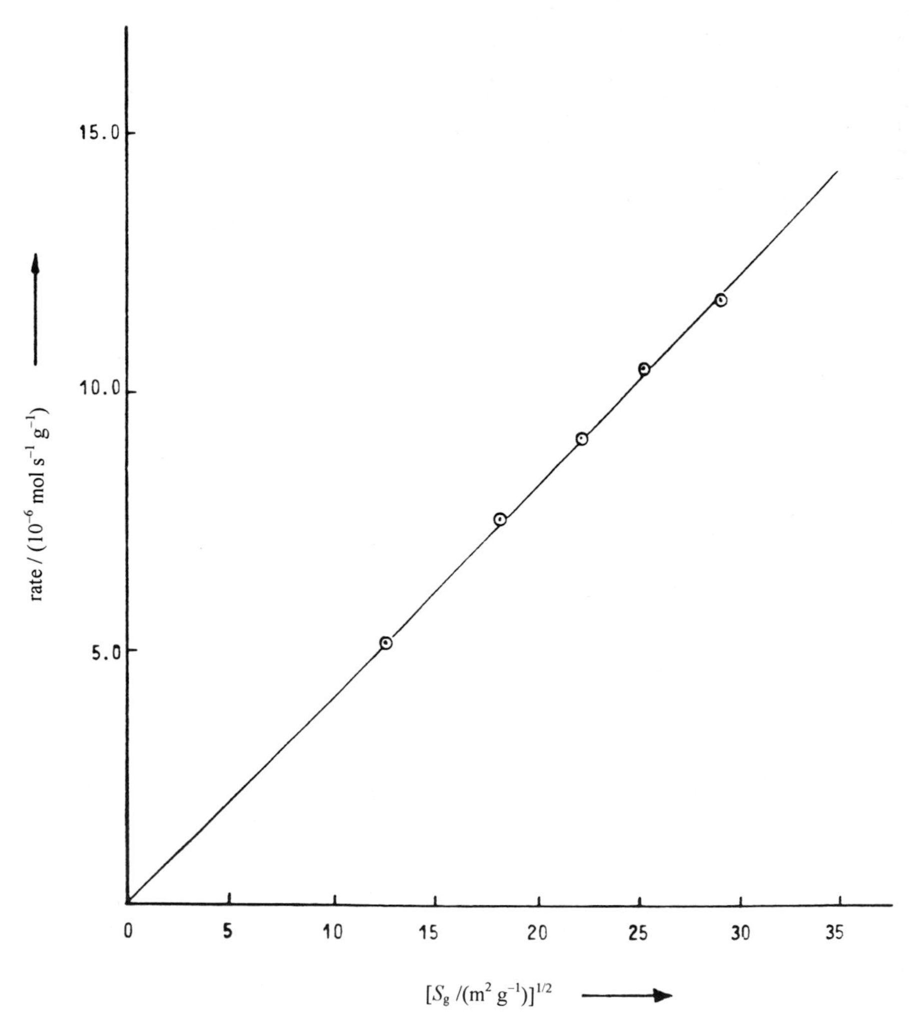

Figure 4.27 Rate of a diffusion-impeded reaction as a function of $S_g^{1/2}$.

(a) If intraparticle diffusion were predominant the rate would be proportional to $S_g^{1/2}$. Figure 4.27 shows a plot of the rate against $S_g^{1/2}$ and hence confirms that intraparticle diffusion is prevalent.

(b) The rate is proportional to $c^{(n+1)/2}$ and hence also to $p^{(n+1)/2}$. A plot of ln (rate) against ln p (Fig. 4.28) thus gives

$$\text{Slope} = \frac{n+1}{2} = 1.5$$

Hence the true order is $n = 2$

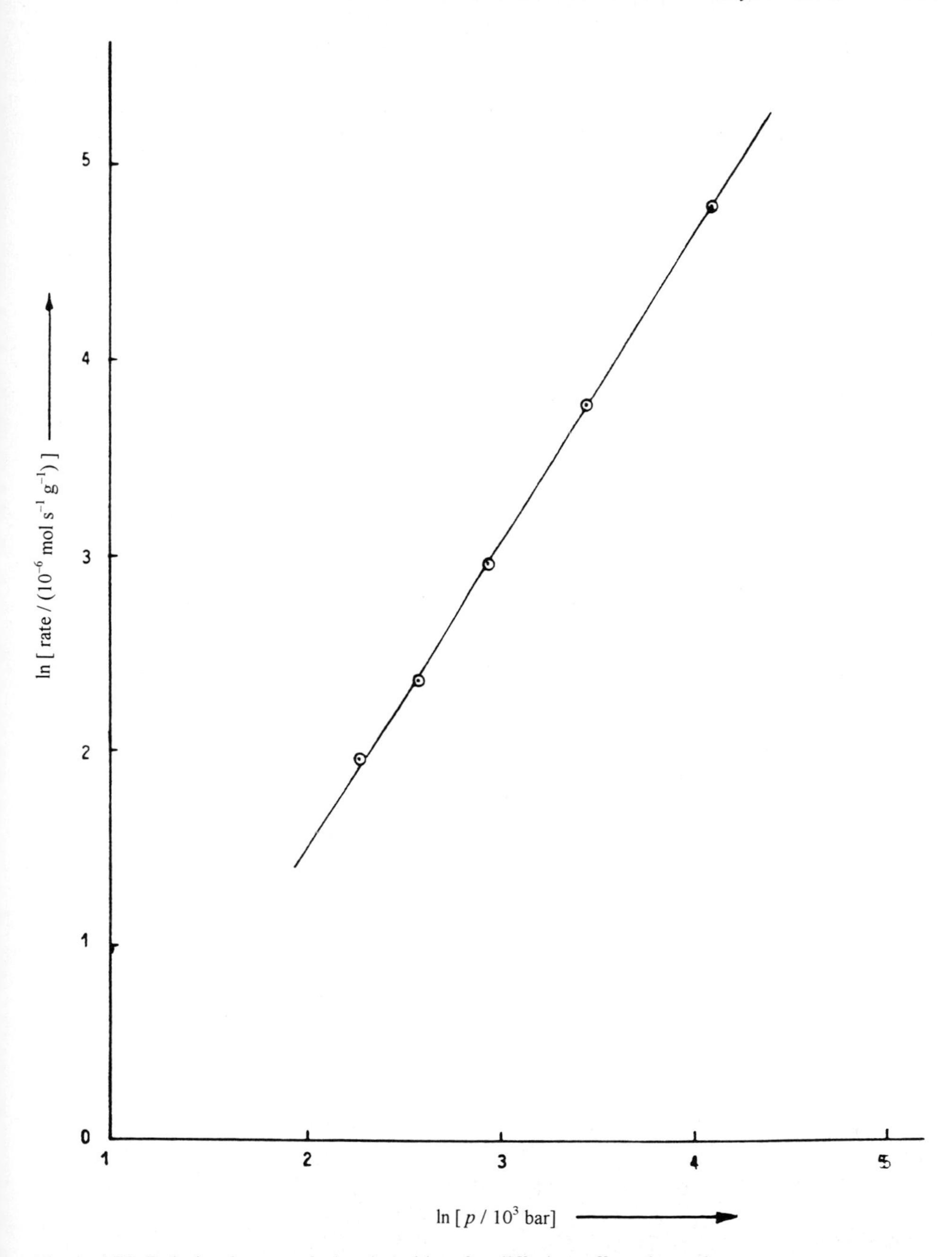

Figure 4.28 Relation between ln (rate) and ln p for diffusion-affected reaction.

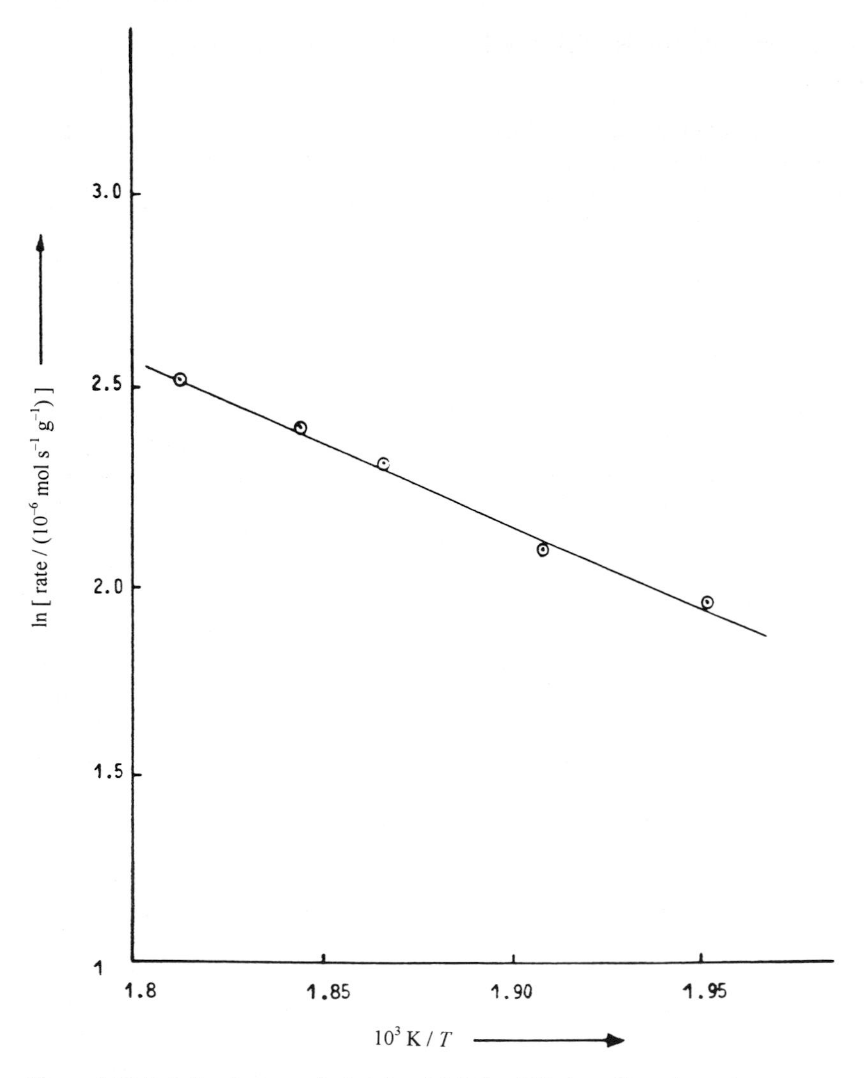

Figure 4.29 Relation between ln (rate) and $1/T$ for diffusion-affected reaction.

(c) The rate is proportional to $k^{1/2}$. Hence we can write

$$\text{Rate} = Kk^{1/2} = KA^{1/2} \exp(-E/2RT)$$

A plot of ln (rate) against $1/T$ (with T = absolute temperature [K]) thus yields

$$\text{Slope} = -E/2R$$

which, from Fig. 4.29, is 4.12×10^3 K. Thus the true activation energy is

$$E = 2 \times 8.31 \times 4.12 = 69.3\,\text{kJ mol}^{-1}$$

4.6.4 Experimental Methods of Assessing the Effect of Diffusion on Reaction

A common experimental criterion for diffusion control involves an evaluation of the rate of reaction as a function of particle size. At a sufficiently small particle size the measured rate of reaction will become independent of particle size and the rate of reaction can then be safely assumed to be independent of intraparticle mass transfer effects. At the other extreme, if the observed rate is inversely proportional to particle size, the reaction is strongly influenced by intraparticle diffusion. For a reaction inhibited by the presence of products, there is an attendant danger of misinterpreting experimental results obtained for different particle sizes when a differential reactor is used, for, under these conditions, the effectiveness factor is sensitive to changes in the partial pressure of product. However, provided the rate of reaction is not inhibited by products, the measured rate of reaction can be plotted as a function of catalyst particle size. When the specific rate R is no longer dependent on the diameter of the catalyst pellet (as at $\eta \rightarrow 1$), its value is assumed not to be influenced by intraparticle diffusion effects. Any other value of the rate $R(d_p)$ on the curve is one which is dependent on diffusion within the particle. The ratio of these two rates ($R(d_p)$ and R) is the effectiveness factor η for that particular particle size d_p.

An ingenious method developed by Hegedus and Petersen (see Further Reading, Section 4.8.1) (also applicable to an evaluation of catalyst poisoning) provides for the measurement, in one single experiment, of the reaction rate and the ratio of concentration at the catalyst exterior surface to the concentration at the pellet centre. Referring to the experimental system illustrated in Fig. 4.30, the catalyst pellet is sealed in a tube so that one face is exposed to the reactant, which is continually circulated by means of a small rotary displacement pump. The other face of the pellet is exposed to an infrared gas cell from which gas is continuously bled at a rate sufficient to ensure a steady-state concentration. This obverse face is equivalent to the centre-plane of a slab-shaped catalyst with its two faces exposed to reactant gas. Concomitantly with catalytic reaction within the pellet, reactant gas diffuses to the plane (labeled $x = 0$) exposed to the infrared gas cell. The concentration c_0 at the plane $x = 0$ is, in the steady state, the concentration of gas within the cell and is evaluated by the extent of infrared absorption. The concentration c_g at the other face ($x = L$) is the steady-state reactant concentration in the circulation loop (which behaves as a recycle reactor) and can be estimated by diverting a sample through a valve to a gas chromatograph.

Applying Eq. (38) describing the concentration profile in a slab-shaped catalyst pellet to the reacting system described (and illustrated in Fig. 4.30), one obtains

$$\frac{c_0}{c_g} = \frac{1}{\cosh \phi} \tag{57}$$

from which the effectiveness factor η may be found:

$$\eta = \frac{\tanh \phi}{\phi}$$

For the virtual half-slab catalyst pellet used,

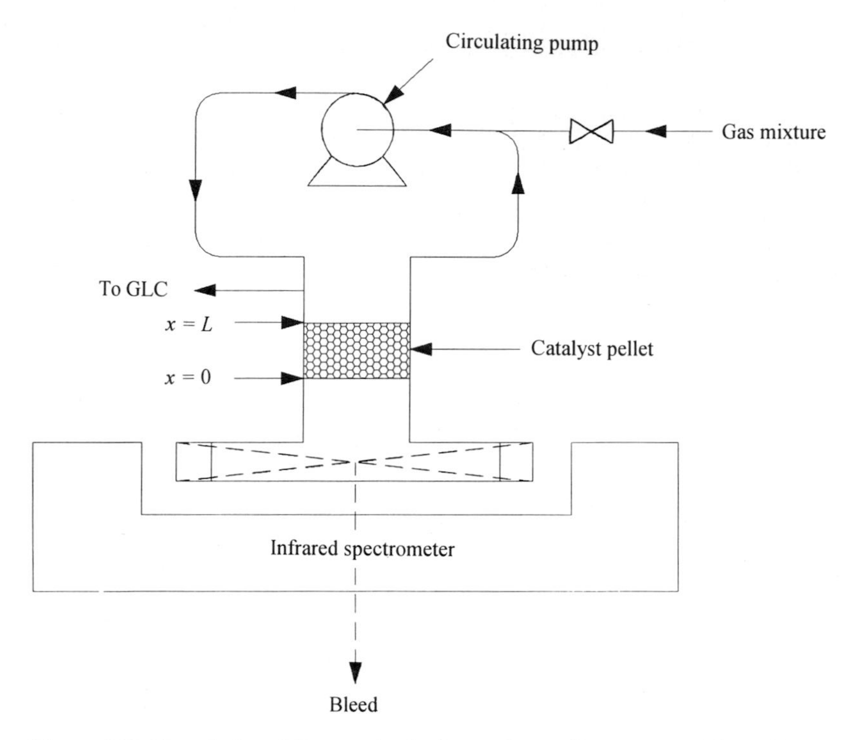

Figure 4.30 Hegedus' and Petersen's single catalyst pellet reactor. Adapted from E. E. Petersen, *Chemical Reaction Analysis*. Prentice Hall, Englewood Cliffs, **1965**.

$$\phi = \frac{L}{2}\left(\frac{\mathrm{R}}{c_g D_e}\right)^{1/2}$$

and so the ratio $\mathrm{R}/c_g(= k)$ emerges, provided that the pellet effective diffusivity D_e is known. Hence, by measurements of c_g (the steady concentration in the circulation loop) and c_0 (the concentration in the infrared cell), both the rate constant and the pellet effectiveness factor may be evaluated.

A similar analysis to the above yields information about reactions of other kinetic orders; it is, however, necessary to know the reaction order in advance and the functional form of the concentration profile analogous to Eq. (57) and corresponding to the appropriate kinetic order. Provided the pellet can be kept at a constant temperature (by surrounding the sealed tube containing the pellet with a temperature-controlled furnace), reaction rates and effectiveness factors can be estimated, in principle, at any reasonable temperature.

4.7 Problems

1 A sample of copper oxide powder was thought to have a specific surface area of less than $10\,\mathrm{m^2\,g^{-1}}$ and so it was decided to employ krypton as adsorbate for the volumetric method of determining surface area. The results in the Table were obtained for the volume V of krypton adsorbed on 5 g of sample at various pressures p.

p [mm Hg]	0.0208	0.1042	0.3125	0.5268	0.7292	1.0416	1.3542	1.8750	2.5000
V [cm^3]	2.375	2.708	3.333	3.708	3.958	4.250	4.375	4.583	4.750

The adsorption of krypton was at −183 °C, at which temperature the saturated vapour pressure p_0 of krypton is 20 mm Hg. If the area occupied by a single krypton molecule is 19.5×10^{-20} m^2, calculate the specific surface area of the sample (a) using the 'point B' method, (b) from a Langmuir plot and (c) employing the BET method.

Discuss reasons for the different values obtained. Which method is the most reliable and gives the more accurate result?

2 The accompanying data were recorded while investigating the pore size distribution of a porous activated carbon using a mercury porosimeter.

Pressure [bar]	Intruded volume [cm^3 g^{-1}]	Pressure [bar]	Intruded volume [cm^3 g^{-1}]
15.75	0.013	253.26	0.336
28.07	0.030	271.66	0.349
40.11	0.046	402.44	0.814
53.00	0.068	529.30	0.888
65.25	0.091	679.41	0.901
89.53	0.138	832.97	0.906
114.59	0.182	1059.87	0.910
140.95	0.222	1423.14	0.911
178.41	0.269	2093.53	0.915
234.50	0.321	4029.63	0.926

Assuming the contact angle for mercury is 140° and its surface tension is 4.8×10^9 N m^{-1}, plot a cumulative pore-volume curve as a function of pore radius, and a differential pore size distribution curve. From the results estimate the mean pore radius.

3 The adsorption–desorption data in the Table were obtained for a sample of porous material employing argon at −195 °C as the probe.

Volume retained on adsorption [cm^3]	11.0	14.1	16.5	24.2	33.0	35.0	38.2	38.3	38.5	39.5
Volume retained on desorption [cm^3]	11.1	14.0	17.0	18.0	19.3	22.1	24.5	32.3	36.0	39.2
Relative pressure	0.1	0.2	0.3	0.35	0.4	0.45	0.5	0.6	0.7	0.9

What conclusions can you draw from the above data regarding (a) the type of hysteresis and (b) possible pore shapes?

4 Compare and contrast the assumptions on which the pore models of (a) Wheeler, (b) Thiele and (c) Everett are based. Which of these models is most applicable to understanding the nature of porous structures and which model is the most satisfactory to apply when assessing the effect of physical parameters, such as specific surface area and specific pore volume, on chemical reaction rates?

5 Calculate the effective Knudsen diffusivity of hydrogen through a porous silica–alumina catalyst at a temperature of 20 °C and atmospheric pressure. The surface area found by the BET method is 150 $m^2\,g^{-1}$. Pyknometry yielded a value of 1.50 $g\,cm^{-3}$ for the pellet density and a solid density of 2.31 $g\,cm^{-3}$ was found by gas-sharing experiments. Assume a value of two for the tortuosity factor. What is the catalyst pore radius?

6 The first-order acid-catalysed hydration of sucrose to yield glucose was studied using different-size porous resin pellets, the effective diffusivity of which were $2.7 \times 10^{-7}\,cm^2\,s^{-1}$. From a series of kinetic experiments at 50 °C rate constants were determined. Results of the experiments were as follows:

Particle diameter [mm]	0.04	0.27	0.55	0.77
Observed rate constant [s^{-1}]	0.0193	0.0710	0.00664	0.00487

Using the above data, predict the overall reaction rate constant if catalyst pellets of 0.4 mm diameter were employed.

7 Pilot-plant investigations of a first-order heterogeneous catalytic reaction indicate that, at 1 bar total pressure and 430 °C, product is formed at a rate 0.24 $kmol\,s^{-1}$. The reactor volume is 100 m^3 and the voidage is estimated to be 0.4.

The rate of product formation varied with the catalyst particle size and so it was anticipated that the reaction is mass-transfer limited. Accordingly, a thorough laboratory investigation was initiated. The reaction was found to be first order in concentration, and measurements of the rate R as a function of specific surface area S_g, pore volume V_g and particle size d_p yielded the information in the Table.

d_p [mm]	7.00	5.00	3.50	3.16
S_g [$m^2\,g^{-1}$]	50	100	200	250
V_g [$mm^3\,g^{-1}$]	50	100	200	250
R [$kmol\,s^{-1}$]	0.17	0.24	0.34	0.38

(a) What mode of transport limitation controls the reaction?
(b) Estimate the value of the catalyst effectiveness factor assuming the effective particle diffusivity is $1.2 \times 10^{-6}\,m^2\,s^{-1}$.

8 The rate of a first-order heterogeneous catalytic reaction occurring at 500 °C and 1 bar absolute pressure in a continuously operated isothermal tubular reactor is $0.3\,\text{kmol}\,\text{m}^{-3}\,\text{s}^{-1}$. The reactant concentration may be assumed to be $1.58 \times 10^{-2}\,\text{kmol}\,\text{m}^{-3}$ and the reactor volume to be $1\,\text{m}^3$. Porous spherical catalyst pellets of $6 \times 10^{-3}\,\text{m}$ diameter are contained in the reactor and occupy 60% of the total reactor volume. If the effective diffusivity of the catalyst pellets is $1.5 \times 10^{-5}\,\text{m}^2\,\text{s}^{-1}$, calculate the effectiveness factor of the pellets.

Why, for severe diffusion limitation, can the effectiveness factor be calculated on the basis of simple slab geometry, despite the pellet shape?

9 Consider a catalyst pellet in a flowing gas stream operating under conditions such that there is a resistance to the transfer of reactant from fluid to solid as well as intraparticle diffusion effects. By carefully considering the boundary conditions applicable to an isothermal spherical pellet, show that the effectiveness factor for the pellet for a first-order reaction under these operating conditions is

$$\eta = \frac{3}{\lambda T}\left\{\frac{\coth(\lambda r - 1/\lambda r)}{1 + (2\lambda r/Sh)[\coth(\lambda r - 1/\lambda r)]}\right\}$$

where $\lambda = (k/D_e)^{1/2}$ and $Sh = h_D d_p/D_e$.

[Note: Sh is a dimensionless number (Sherwood number) which compares the external (interparticle) mass transfer coefficient h_D with the effective diffusivity D_e. The mass transfer coefficient is defined as the ratio of the molar flux to the concentration difference between bulk gas and the gas–particle interface.]

10 A catalytic reaction, with an exothermic heat of reaction of $210\,\text{kJ}\,\text{mol}^{-1}$, is effected using porous catalysts of effective diffusivity $5.2 \times 10^{-6}\,\text{m}^2\text{s}^{-1}$ and effective thermal conductivity $1.73 \times 10^{-4}\,\text{kJ}\,\text{m}^{-1}\,\text{s}^{-1}\,\text{k}^{-1}$. If the temperature and concentration at the exterior catalyst pellet surface are 70 °C and $5.43\,\text{mol}\,\text{m}^{-3}$, estimate the temperature difference between the centre of the catalyst pellet and the exterior surface.

The difference recorded experimentally is only about 2 °C. Account for the discrepancy between the experimental and estimated values.

4.8 Further Reading

4.8.1 General

S. J. Gregg, K. S. W. Sing, *Adsorption, Surface Area and Porosity*, Academic Press, London, **1967**.

J. M. Smith, *Chemical Engineering Kinetics*, 3rd Ed., McGraw-Hill, New York, **1981**.

R. Jackson, *Transport in Porous Catalysts*, Elsevier, Amsterdam, **1977**.

E. E. Petersen, *Chemical Reaction Analysis*, Prentice Hall, New York, **1965**.

J. J. Carberry, *Chemical and Catalytic Reaction Engineering*, McGraw-Hill, New York, **1976**.

R. Aris, *Mathematical Theory of Diffusion and Reaction in Permeable Catalysts*, Clarendon Press, Oxford, **1975**.

W. J. Thomas in *Perspectives in Catalysis*, (Eds: J. M. Thomas, and K. I. Zamaraev), Blackwell Scientific, London, **1992**, pp. 251–287.

4.8.2 Chemical Reaction in Porous Media

E. W. Thiele, *Ind. Eng. Chem.* **1939**, *31*, 916.

A. Wheeler, *Catalysis* **1955**, *2*, 118.

A. Wheeler, *Adv. Catal.* **1951**, *3*, 249.

4.8.3 Geometry of Pores and their Description

J. W. McBain, *J. Am. Chem. Soc.* **1935**, *57*, 699.

H. Cohan, *J. Am. Chem. Soc.* **1938**, *60*, 433.

D. H. Everett, in *The Structure and Properties of Porous Materials*, (Eds: D. H. Everett, F. S. Stone), Butterworths, London, **1958**, p. 95.

N. Wakao, J. M. Smith, *Chem. Eng. Sci.* **1962**, *17*, 825.

R. Mann, *Trans. IChemE.* **1993**, *71A*, 551.

5 The Solid-State and Surface Chemistry of Catalysts

Recent advances in the techniques of structural elucidation, both in computational chemistry and in the means of deriving the electronic and bonding properties of solids, have had a profound impact on our appreciation of the mode of operation of catalysts: they have even paved the way for more reliable methods of designing new catalysts and improving existing ones.

In this chapter, we present a synopsis of what is relevant in structural, computational and theoretical terms to understanding and predicting catalytic phenomena. Our approach, in the interests of brevity, is not rigorous, and so the reader will often be referred to other texts. What we convey is, *inter alia*, the relative importance of structure, symmetry, site environment, interatomic distance, degrees of flexibility and non-stoichiometry, variable valency, ease of atomic migration, and extent of electronic transfer to and from a surface. We also touch upon some of the interpretative frameworks for rationalizing the selectivity and activity of catalysts, issues that are also discussed in parts of Chapters 6 and 8.

5.1 Classification of Heterogeneous Catalysts

This is no easy task, for there is a wide variety of types of catalysts in common use; moreover, many of them are polyphasic and multicomponent. A metal may be dispersed minutely on a more or less inert oxide or carbonaceous support; a crystalline microporous aluminosilicate may have well-defined Brønsted acid sites located at the inner walls of its intra-particle cavities; transition-metal ions may be embedded in an active oxide support; a deintercalated graphite may yield finely divided, jettisoned, but at the same time supported, active metal crystallites, and one metal may be supported on the sulphide of another metal, which, in turn, is in contact with a high-area oxide of yet another.

Because of this range, classification according to structural type is somewhat arbitrary, although emphasis is naturally placed in any taxonomic system on the atomic structure of what is deemed to be the most active constituent. Classifications based on electronic properties or bond types are also convenient; and segregation of catalysts into metals or semiconductors or insulators sometimes serves a valuable purpose as, for example, in discussions of photo-catalytic phenomena. Division of catalysts according to their acid–base or redox properties is yet another basis of classification.

5.2 Structures

5.2.1 Metals and Alloys

Metals such as iron and ruthenium are used for the synthesis of ammonia, platinum for the hydrocracking of oil and silver for the oxidation of methanol to formaldehyde. Intermetallics such as Pt–Ir, Pt–Re and Pt–Sn are used extensively to convert petroleum naphtha into 'antiknock'-quality petrol (gasoline), i.e. in the reforming of hydrocarbons to high-octane products.

Most metals adopt a packing arrangement which maximizes the occupation of space. The constituent atoms take up either the hexagonal-close-packed (hcp) or the face-centred-cubic (fcc) structure, Figs 5.1(a) and 1(b), respectively. In either case, each atom has 12 nearest neighbours, six in the same plane as it, three above and three below. In the body-centred cubic (bcc) structure, Fig. 5.1(c), each atom has eight nearest neighbours and the structure is not as closely packed.

The closest packing of spheres (Fig. 5.2) is a feature of other solids also, the oxygens in certain metal oxides being a typical example. Two distinct kinds of interstices appear whenever spheres are in their closest-packed state: tetrahedral and octahedral. Whereas the number of octahedral interstices is equal to the number of spheres,

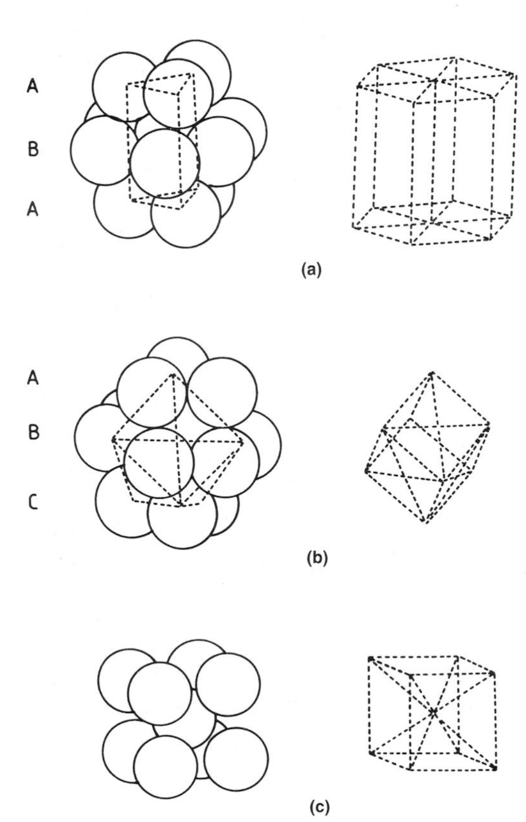

Figure 5.1 (a) Metallic catalysts of rhenium, ruthenuim and osmium are hexagonally close-packed, each atom being surrounded by 12 neighbours. The cluster compound $Rh.Rh_{12}(CO)_{24}H_3^{2-}$ has this arrangement of metal atoms. (b) In the face-centred-cubic structure (fcc), each atom also has 12 neighbours; this is the structure adopted by nickel, platinum and silver. (c) Body-centred cubic (bcc) structures are adopted by α-iron and vanadium. The unit cells are outlined on the right. The sequence of atomic layers in the hexagonally close-packed (hcp) structure is ABAB... and in fcc it is ABCABC....

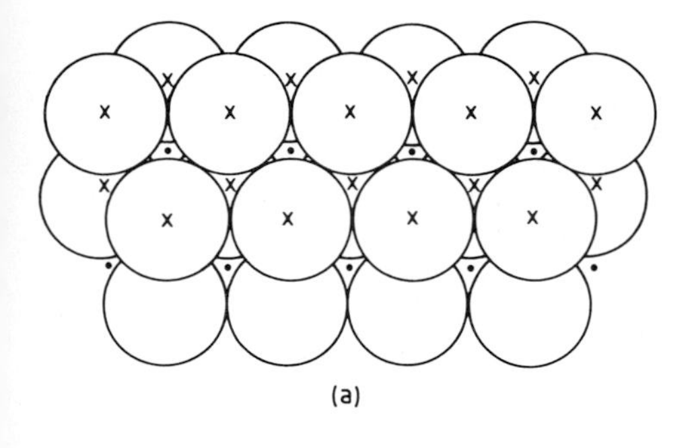

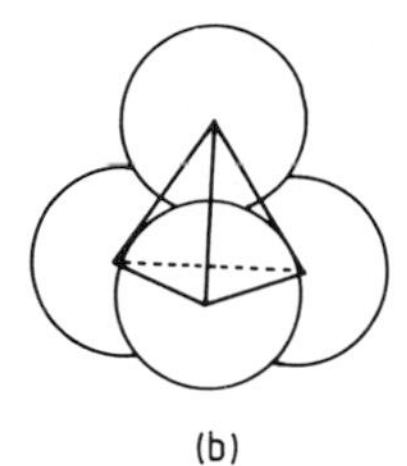

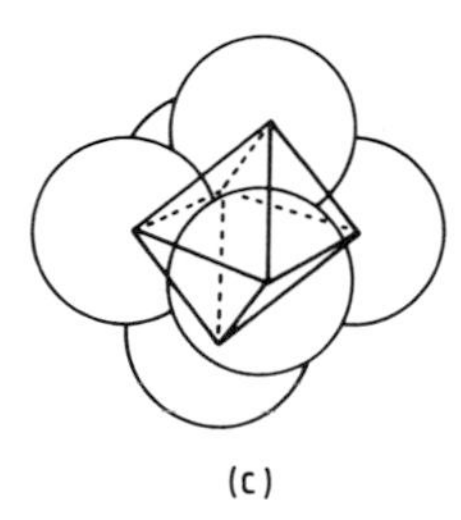

Figure 5.2 (a) Plan view of two layers of close-packed spheres. Tetrahedral (×) and octahedral (•) interstices, shown in elevation in (b) and (c), are present in close-packed layers, there being twice as many tetrahedral as octahedral ones.

the number of tetrahedral interstices is twice that of the spheres. (Note that the atomic packing in bulk metals is adopted in certain cluster compounds, the Rh atom arrangement in $Rh_{13}(CO)_{24}H_3^{2-}$ being the same as in a fragment of an hcp bulk metal.)

Large aggregates of metal atoms expose different kinds of crystallographic planes, the situation for a tungsten tip (bcc structure) being as shown in the hard-sphere model, Fig. 5.3. The number of neighbouring atoms (coordination number C_i) at the surface is smaller than that in the bulk to an extent that is governed by the precise location. Atoms in the flat terraces of the fcc metal shown in Fig. 5.4 have nine neighbours, those at a surface step seven, and those at a kink six.

5.2.1.1 Miller Indices and Miller–Bravais Indices

Planes are labelled using Miller indices, the significance of which is best appreciated by first considering a two-dimensional rectangular lattice. If the dimensions of the unit cell are a and b (Fig. 5.5) the sets of planes shown here can be distinguished by

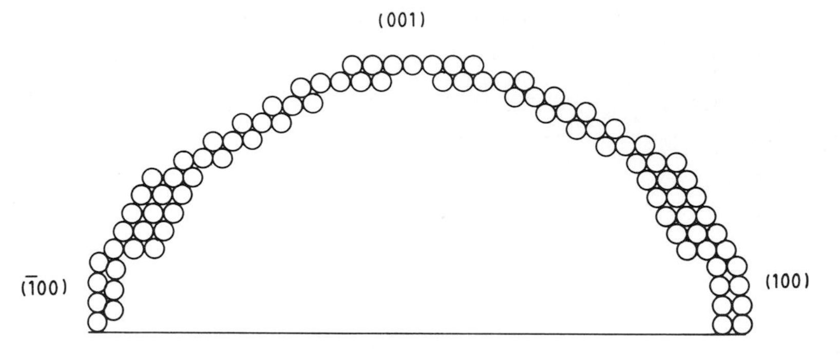

Figure 5.3 Hard-sphere model of a tungsten tip, showing cross-section of distinct crystallographic faces (see text).

the distances along the axes where one representative member of each set of planes intersects them. A sensible labelling scheme is to denote each set by the distances along the two axes to the points of intersection. If we select as the representative member the plane characterized by the least distances, the four sets shown here may be described as $(1a, \infty b)$, $(\infty a, 1b)$, $(1a, 1b)$ and $(-1a, 1b)$. But if, in addition, the distances along the unit cell axes are quoted in terms of the length of the cell in that direction, these planes can be denoted by $(1, \infty)$, $(\infty, 1)$, $(1, 1)$ and $(-1, 1)$ respectively. Now let us convert the two-dimensional lattice into a three-dimensional one. If the unit cell is of length c in the z direction; and, for simplicity, suppose all the planes intersect z at ∞c, then the full indices become $(1, \infty, \infty)$, $(\infty, 1, \infty)$, $(1, 1, \infty)$ and $(-1, 1, \infty)$. To eliminate the ∞, which is cumbersome, we take reciprocals of the indices. By definition, the Miller indices are the reciprocals of the numbers in round

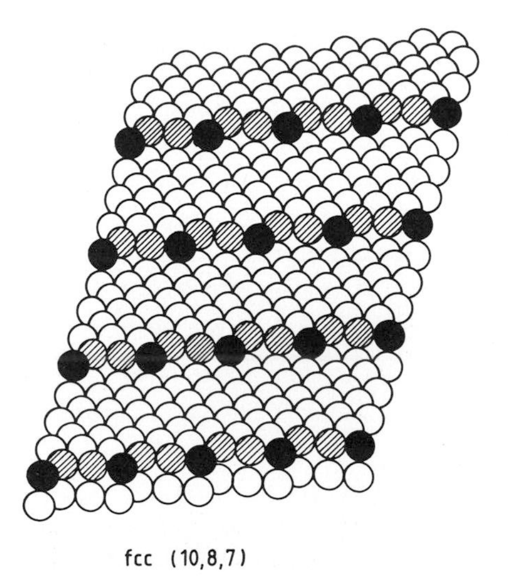

Figure 5.4 Atoms at terraces (empty circles) have nine nearest neighbours, those at steps (hatched) seven and those at kinks (dark) six in this surface of an fcc metal.

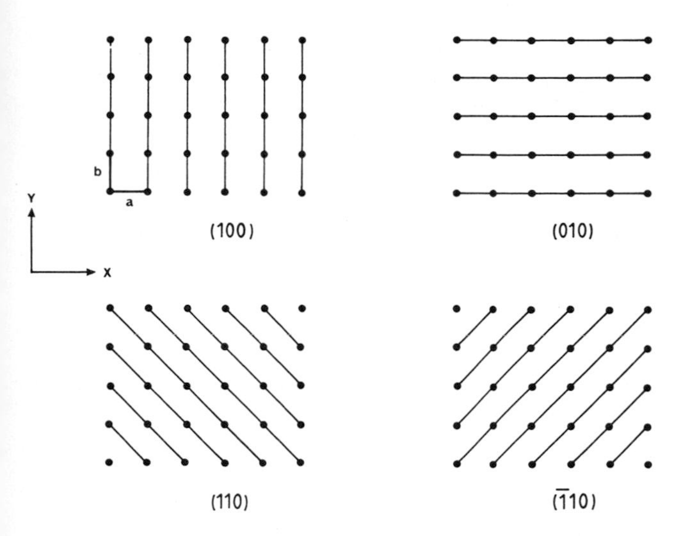

Figure 5.5 Miller indices of four planes in a two-dimensional rectangular lattice (see text).

brackets (parentheses), with fractions cleared. For the four planes considered here, the Miller indices are $(1, 0, 0)$, $(0, 1, 0)$ $(1, 1, 0)$ and $(1, 1, 0)$. For a plane with intercepts $(3a, 2b, \infty c)$ the Miller indices are $(2, 3, 0)$. Note that $(\frac{1}{3}, \frac{1}{2}, 0)$ on clearing the fractions needs to be multiplied by 6 thereby yielding $(2, 3, 0)$. Negative indices are written with a bar above the number, hence the $(\bar{1}\bar{1}0)$ plane in Fig. 5.5.

The Miller indices for a direction are calculated by the following procedure: (i) measure the coordinates of any point on that direction; (ii) divide the coordinates by

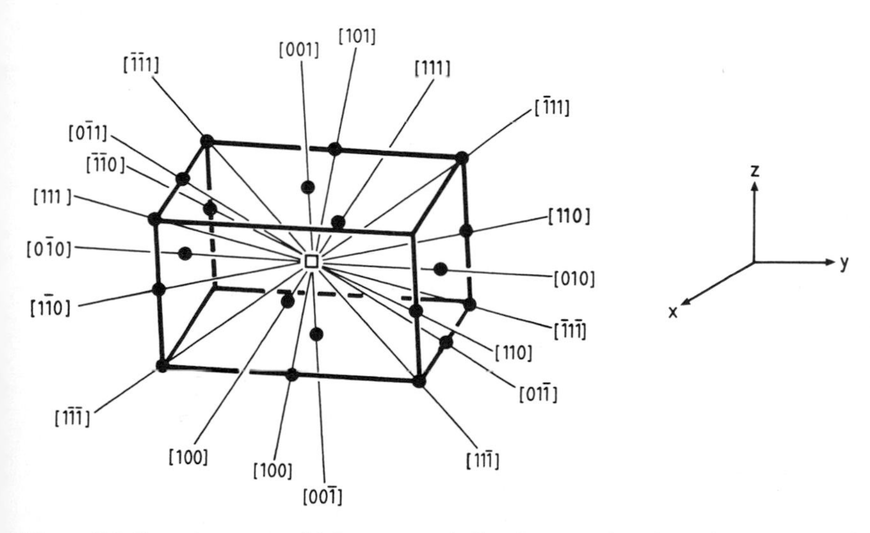

Figure 5.6 Some important (high-symmetry) directions passing through the centre of an arbitrary unit cell. The filled circles denote the points at which the directions emerge from the cell.

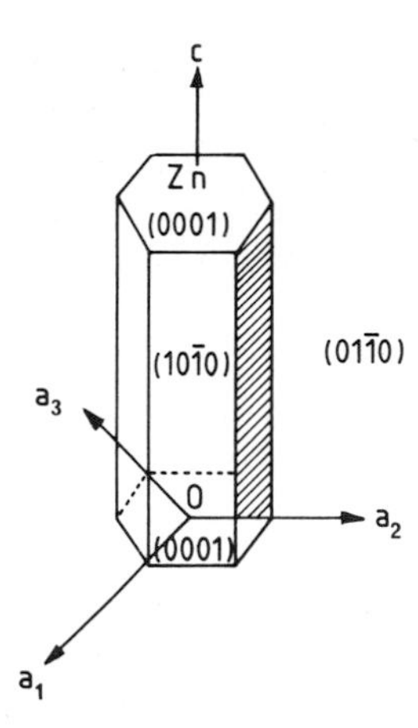

Figure 5.7 In the Miller–Bravais system for hexagonal structures, four axes (a_1, a_2, a_3 and c) are used. The (01$\bar{1}$0) plane is hatched.

the appropriate unit cell dimensions; (iii) rationalize the dividends; and (iv) place the rationalized dividends in square brackets. Some important directions are illustrated in Fig. 5.6.

Sometimes we wish to refer to a family of equivalent directions rather than a specific one. Thus, in the cubic system, the family denoted with the angle brackets $\langle 111 \rangle$ comprises eight specific directions [111], [11$\bar{1}$], [1$\bar{1}$1], [$\bar{1}$11], [1$\bar{1}\bar{1}$], [$\bar{1}\bar{1}$1], [$\bar{1}$1$\bar{1}$], and [$\bar{1}\bar{1}\bar{1}$]. Likewise a family of equivalent planes is denoted with curly brackets (braces). Taking again the cubic system we may represent the family of six equivalent planes, each having identical interplanar spacings (d) and being composed of an identical array of atoms: (100), ($\bar{1}$00), (010), (0$\bar{1}$0), (001) and (00$\bar{1}$), as {100}. In the literature on heterogeneous catalysis the round brackets are often wrongly used to describe families of crystal planes instead of the curly ones, and the square brackets wrongly used instead of the angled ones to describe families of directions.

To describe crystals that belong to the hexagonal system, three Miller indices are not enough because, with the Miller system, equivalent planes may turn out to have inequivalent indices. By using a fourth axis, this problem is avoided. The resulting Miller–Bravais indices (see Fig. 5.7) for a specific plane are written ($hkil$) and for a family of equivalent planes {$hkil$}. Miller indices {hkl} *can* be used for hexagonal crystals, but they can lead to ambiguity. Since $i = -(h + k)$ it is simple to convert from Miller indices to Miller–Bravais indices and vice versa. Thus (110) and (110) become, respectively, (11$\bar{2}$0) and (10$\bar{1}$0).

5.2.1.2 Transition-Metal Alloys and Bimetallic Clusters

It is well known that metals have the ability to form solid solutions with one another if their chemical identities are similar as, for example, with silver and palladium or with nickel and copper. The constituent atoms can be randomly distributed in the alloy, or ordered in such a way as to form superlattices. Some alloys which, when quenched from the molten state, are random solid solutions, will, on annealing, rearrange to the ordered state. Similar processes occur at surfaces, as we know from

LEED studies of alloys. Whereas most transition-metal alloys take up one or other of the hcp, fcc or bcc structures, some, notably the PtSn alloy catalysts used in petroleum refining, adopt a more open structure like that of NiAs, where each As atom is trigonally coordinated to six Ni atoms and each Ni atom is octahedrally coordinated to six As atoms, but there are two other Ni atoms which are sufficiently near to be regarded as bonded to the first Ni atom.

Remarkably, some transition-metal 'alloy' systems formed between two elements that are essentially immiscible in the bulk behave catalytically very much like other alloys formed from constituent metals that *are* completely miscible. As the copper content of Cu–Ni alloys increases, the unit-cell constant increases linearly (in conformity with Vegard's law), indicating that a true solid solution is formed within the bulk. But, from XPS and other data (Chapter 3), there is no doubt that copper tends to concentrate at the surface. An alloy consisting of only a few per cent of copper has a surface dominated by copper. This has a profound effect on the catalytic performance of the alloy.

The effect of copper on the activity of nickel for ethane hydrogenolysis differs greatly from its effect on the activity for cyclohexane dehydrogenation. There is a thousandfold drop in hydrogenolytic activity in a 5% Cu in Cu–Ni alloy. In contrast, the activity of the alloy remains essentially constant, at a value close to that of pure nickel, over a wide range of alloy composition. Sinfelt and the Dutch co-workers Ponec and Sachtler have rationalized this behaviour in terms of the differences in mechanisms of the two catalytic reactions, the key point being the preferential concentration of copper at the surface of a (miscible) Cu–Ni alloy.

Nowadays, several pairs of metallic elements that do not intermix in the bulk are used in the form of small aggregates ranging in size from about 10 to 60 Å in diameter as powerful catalysts in petroleum refining. The term 'alloy' would be a misnomer here; instead, we speak of bimetallic clusters, the term introduced by Sinfelt.

5.2.1.3 Highly-Dispersed Metals

The degree of dispersion, defined by the ratio of the number of surface atoms to the total number in the polyhedron under consideration, N_S/N_T, needs to be as close as possible to unity so as to maximize the utilization of a catalyst. We see that, for a crystallite of cubic shape (Fig.5.8(a)), N_S/N_T reaches unity when the size approaches 10 Å (assuming a unit cell spacing of 2.5 Å). Other polyhedra, such as the truncated octahedron shown in Fig. 5.8(b), approach dispersions of unity when the individual {100} faces indicated are *ca* 10 Å^2.

Many different types of polyhedra can be constructed from clusters consisting of between 10 and 100 atoms. Calculations of potential energy indicate that, for small values of N_T, the energy minimum does not correspond to the polyhedral fragments of an fcc structure, but, instead, to others formed by relaxation of some of the interatomic distances in the close-packed arrangement. For example, the binding energy per atom is higher and the surface energy smaller for an icosahedron compared with the cuboctahedron of the same number of atoms (Fig. 5.9). The icosahedron, with its characteristic fivefold symmetry, has 42 nearest-neighbour, atom–atom interactions, whereas the corresponding cuboctahedron has only 36. Moreover, only the triangu-

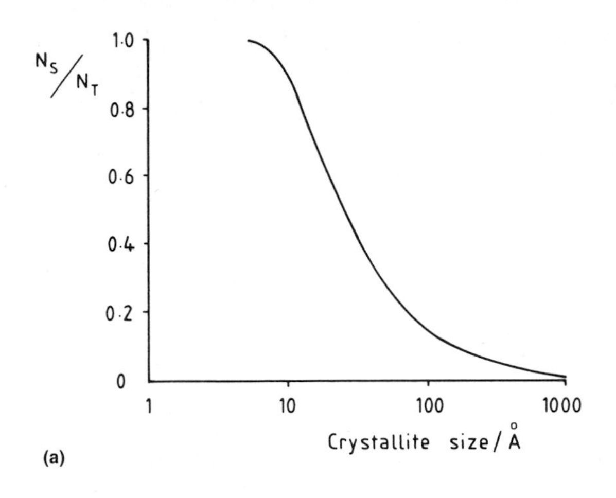

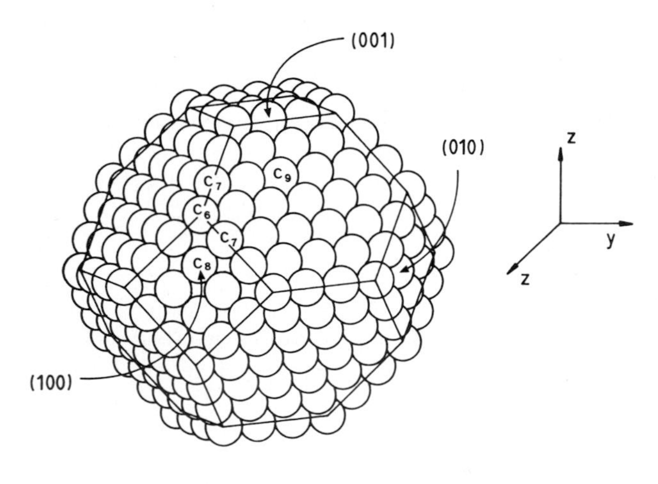

Figure 5.8 (a) The number of surface atoms N_S approaches the total number N_T when the size of the crystallites, assumed here to be cubes, approaches 10 Å. (b) When the {100} faces of truncated octahedra such as this one (with four-atom sides to the {100} faces) have three-atom or two-atom sides the values of N_S/N_T become 0.61 and 0.84 respectively.

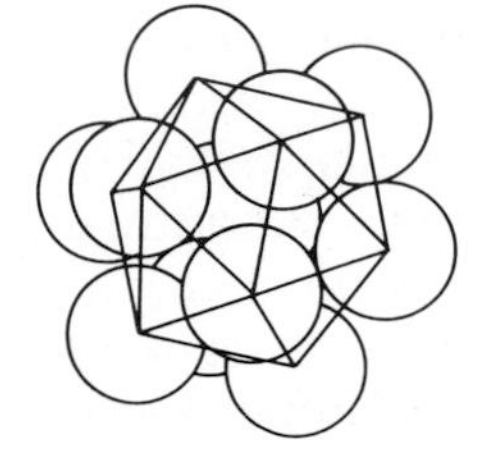

Figure 5.9 13-Atom clusters arranged in the form of a cubo-cotahedron (left) and icosahedron (right).

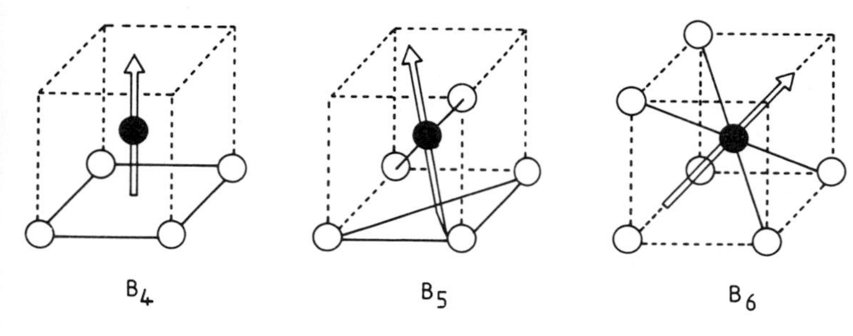

Figure 5.10 A surface site B_j has j nearest neighbours. Three distinct sites at the surface of a bcc metal are shown here.

lar fragments of {111} planes are exposed in a icosahedron, whereas there are fragments of both {111} and the higher-energy {100} planes in the cubo-octahedron, so that the overall surface energy of the icosahedron is smaller. These simple calculations do not take into consideration questions of elastic strain. High-resolution microscopy shows that highly dispersed metals can adopt unique structures involving microtwinning, planar faults and the incorporation of considerable internal strain.

For very small clusters it is not crystallographically sensible to talk of planes with Miller indices defined as for macrocrystals. It is more profitable to define surface struture with the notation of the Dutch workers Van Hardeveld and Hartog. As defined in Section 5.2.1, the coordination number C_i of a surface atom refers to the number of nearest neighbours to that atom. Examples of C_6, C_7 and C_9 surface atoms are shown in Fig. 5.4. A site is denoted by B_j when it has j nearest neighbours; B_4, B_5 and B_6 sites are schematized in Fig. 5.10.

5.1.3 Interstitial Phases

Close-packed arrays of transition-metal atoms can readily assimilate the smaller atoms of non-metals such as H, B, C, N, O and even Si, the resulting phases being interstitial solid solutions. The guest atoms can be accommodated either in the tetrahedral or octahedral sites (compare Fig. 5.2), of which there are many. The final composition is governed by the degree to which the sites available are occupied. If, for example, all the octahedral holes were occupied, the resulting structure is the same as that of rock salt, which is precisely the structure adopted by the interstitial phases TiC and ZrN. Likewise, TiH_2 may be regarded as an example of the well-known fluorite (CaF_2) structure but in this case the hydrogen enters all the available tetrahedral interstices of the parent titanium. A quarter occupancy of the tetrahedral sites in palladium yields Pd_2H.

Stoichiometric compounds such as Fe_4N and PdH_2 are often implicated as surface intermediates in catalysis, the former in the synthesis of ammonia and the latter in palladium-catalysed hydrogenations. Nitrided iron is also a good Fischer–Tropsch catalyst. There are also indications that non-crystalline phases, formed through partial occupancy of interstitial sites, are of importance catalytically. Indeed, the struc-

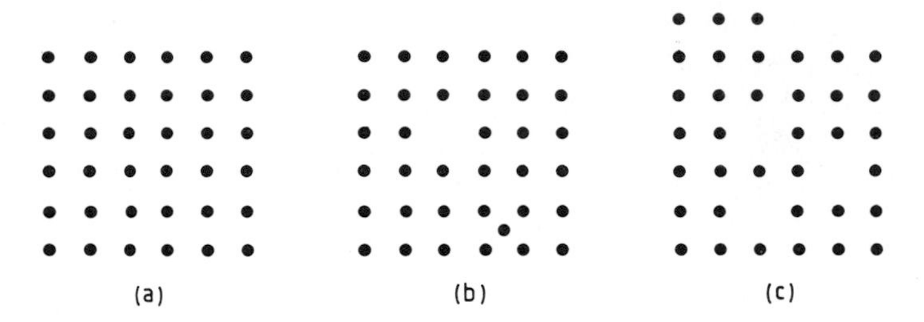

Figure 5.11 Two-dimensional representation of (a) an ideal lattice, (b) a lattice containing a Frenkel defect, and (c) a lattice containing Schottky defects.

tural chemistry of splat-cooled alloys, where molten multicomponent mixtures of metals, metalloids and non-metals are rapidly quenched to yield amorphous solids, hinges upon the formation of interstitial phases.

Occupation of interstitial sites occurs in many other parent structures besides those of metals. The Frenkel defect, which exists in many different structures, entails occupation of interstitial sites (Fig. 5.11). When, for example, partial solid solutions of the composition $(Sr_{1-x}La_x)F_{2+3x}$ are formed by the incorporation of LaF_3 into SrF_2, which has a fluorite structure, we find that three Sr^{2+} sites are not replaced by two La^{3+} (leaving a cation vacancy), but by three La^{3+} with supernumerary F^- ions that enter interstitial sites in the anion sublattice.

5.1.4 Simple Metallic Oxides and their Non-stoichiometric Variants

By 'simple' we mean oxides formed from a single metallic element. Such materials are rarely exactly stoichiometric and so are not accurately denoted by formulae such as MO, MO_2, MO_3, etc. In some solids, the degree of departure from exact stoichiometry may be quite small, involving less than 0.01% of regular sites being unoccupied or displaced from their normal positions, or more than a mere fraction of the available interstitial sites being tenanted. In others, however, the degree of non-stoichiometry may be gross. For example, ferrous oxide retains the same structural framework – rock salt – over the composition range $Fe_{0.82}O$ to $Fe_{0.96}O$. Again, the monoxides of titanium and vanadium, which also have the rock-salt struture, exhibit wide compositional ranges, $TiO_{0.90}$ to $TiO_{1.25}$ and $VO_{0.86}$ to $VO_{1.27}$. There are two extreme categories of structure that can accommodate gross non-stoichiometry: on the one hand, clusters of point defects form within a given structure – the fluorite framework is a good example – on the other, planar faults known as shear planes are created and ReO_3-based structures (see below) are prototypical of this feature. The occurrence in a solid of gross non-stoichiometry almost invariably means that there is scope for the insertion of guest ions of a kind likely to enhance catalytic activity of the parent solid. Gross non-stoichiometry also tends to facilitate solid-state migration within the cation or anion sublattices.

MgO is one of the best examples of a catalyst with the rock salt structure. MgO can accommodate substitutional impurities, such as Li^+ ions, which exert a dramatic

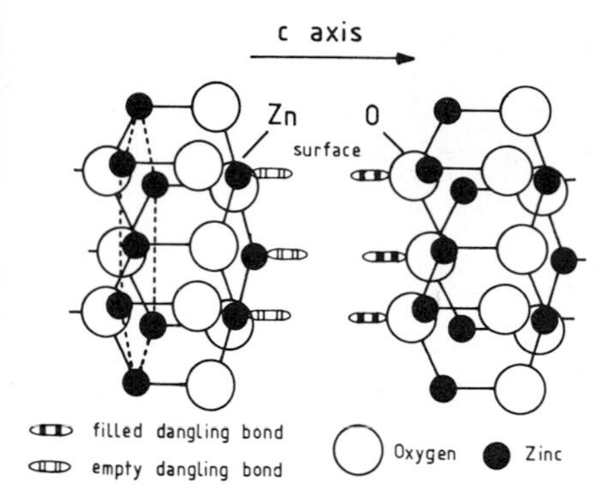

Figure 5.12 Representation of disposition of ions at the two unreconstructed distinct basal faces ((0001) and (000$\bar{1}$)) of ZnO. One of these exposes oxygen ions only, the other zinc ions only.

effect upon the catalytic performance of the resulting solid for the selective oxidation of methane.

ZnO, a versatile catalyst, crystallizes in the wurtzite structure mostly in the form of needles with the prism faces (i.e. $\{10\bar{1}0\}$ planes) very well developed (Fig. 5.12). Whereas the $\{10\bar{1}0\}$ faces are composed of an equal number of Zn^{2+} and O^{2-} ions, the two basal faces (0001) and (000$\bar{1}$) are different in that one exposes only Zn^{2+} the other only O^{2-} ions. Differences in general chemical, and especially adsorptive, behaviour between these two faces are expected and are indeed observed. It is probable that, as with many other oxide surfaces, OH groups abound at the surfaces of ZnO, these groups being formed as a result of reaction with water vapour. Non-stoichiometry in ZnO arises from native point defects like interstitial zinc (Zn_i) or oxygen vacancies (V_O) which can each be generated thermally:

$$ZnO(s) \longrightarrow Zn_i + \tfrac{1}{2}O_2(g)$$

$$ZnO(s) \longrightarrow Zn(s) + V_O + \tfrac{1}{2}O_2(g)$$

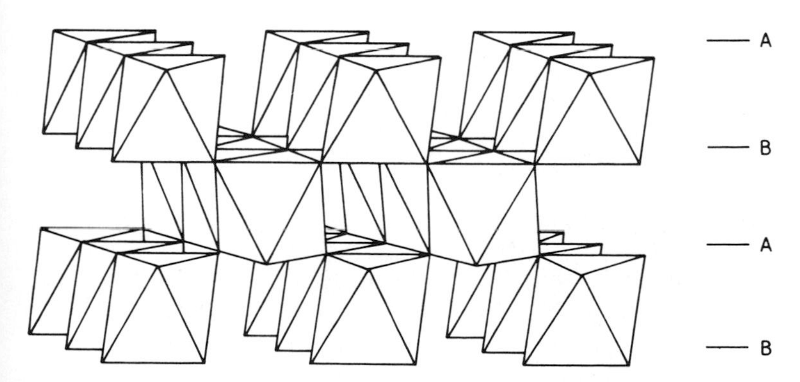

Figure 5.13 The rutile (TiO_2) structure consists of both edge-sharing and corner-sharing TiO_6 octahedra.

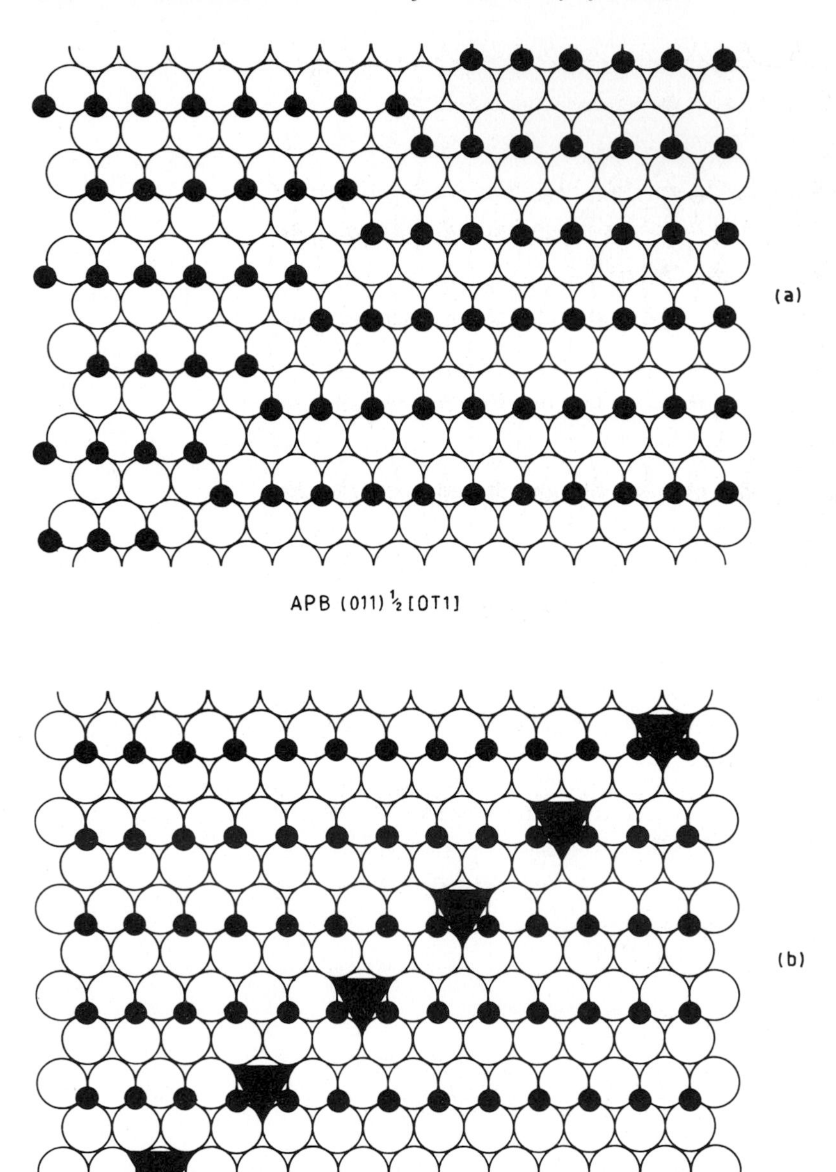

Figure 5.14 An antiphase boundary (APB) (a kind of stacking fault) is generated in the rutile structure when the Ti^{4+} ion sublattice is shifted as shown in (a). There is no change in stoichiometry when this disruption occurs. In (b), vacant oxygen sites have been created by reduction; and the planar fault introduced on the (121) plane (c) is a result of the collapse of this reduced structure. (d) This planar fault on (132) is, in effect, a 1 : 1 mixture of an (011) APB and a (121) CS plane.

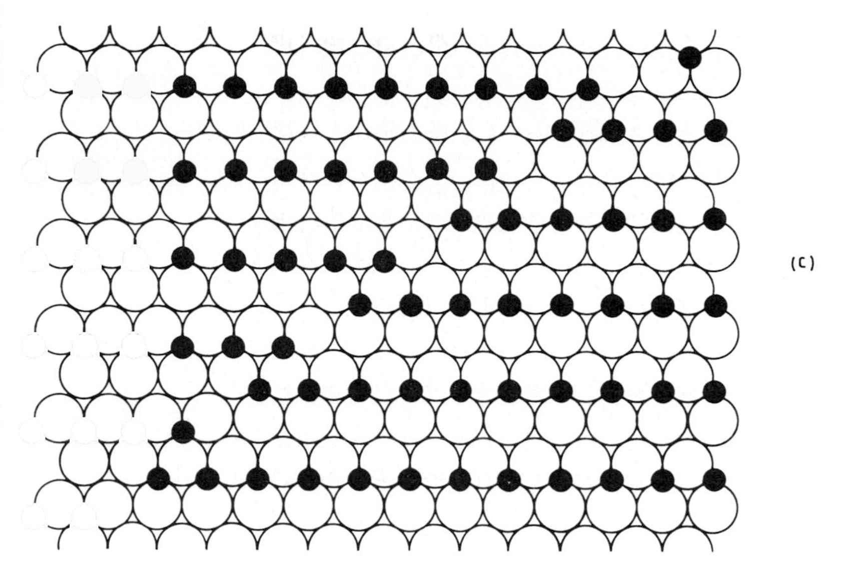

Trace of (121) CS plane

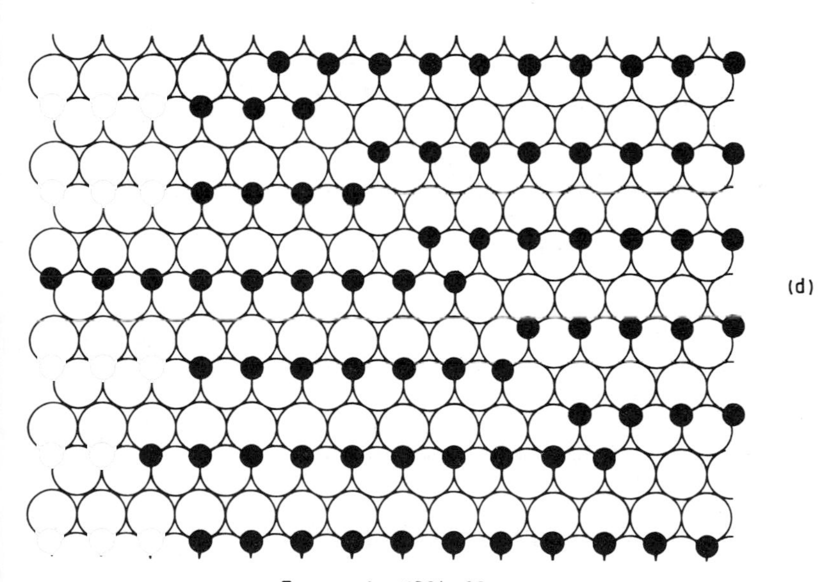

Trace of (132) CS plane

Non-stoichiometry, which can also be engendered by irradiation with the concomitant generation of 'free' electrons, not only influences the collective electronic properties of ZnO but also creates localized sites at the surface.

Defects may be accommodated in large concentration by the fluorite structure, which may be regarded as being derived from an fcc array of metal atoms with atoms of non-metal occupying all the tetrahedral sites, and also as an array of CaF_8 cubes that share all of their edges. Vacant cubes provide the interstitial sites into which supernumerary anions or other species may be inserted.

The fluorite structure is retained by the auto-exhaust catalyst component CeO_2 even when its non-stoichiometry is $CeO_{1.72}$. The high-temperature form of Bi_2O_3 is also a defect-fluorite structure, which is the archetype of many of the active bismuth molybdate catalysts for selective oxidation of hydrocarbons. It may also be regarded as the structural progenitor of the so-called A-type La_2O_3 structure.

In the rutile structure (Fig. 5.13), each MO_6 octahedron shares two opposite edges with neighbouring octahedra and is linked, via the remaining vertices, to four other octahedra, thereby forming connected chains that run through the solid. In an idealized form we may also regard this structure as a closest packing of O^{2-} anions with the M^{4+} cation occupying octahedral sites. We note that the α-PbO_2 structure, not as yet thought to be of great importance catalytically, is closely related to that of rutile.

Consideration of the rutile structure presents an opportunity to outline the principles that lead to the formation of homologous families of grossly non-stoichiometric oxides upon reduction of the parent TiO_2. Thus, in the range $1.750 \leq x \leq 1.972$ in TiO_x, we may represent the formula of the homologous series as Ti_nO_{2n-1}, $4 \leq n \leq 36$. What, in effect, happens is that, on reduction, some of the TiO_6 octahedra, lying on certain well-defined planes, are converted from their edge-sharing state (see Fig. 5.13) to a face-sharing state, and an appropriate number of Ti^{4+} ions is reduced to Ti^{3+}. The coordination number of the Ti ion remains at six. This conversion entails an operation called crystallographic shear (CS). In effect, looking down at the idealized picture, the situation is as schematized in Fig. 5.14 (compare Fig. 5.13). Careful thermodynamic measurements, in which Ti/O ratios are recorded as a function of oxygen pressure, leave no doubt about the existence of numerous non-stoichiometric materials of general formula Ti_nO_{2n-1} – see Fig. 5.15, which also shows the existence of a family of oxides of vanadium, V_nO_{2n-1}.

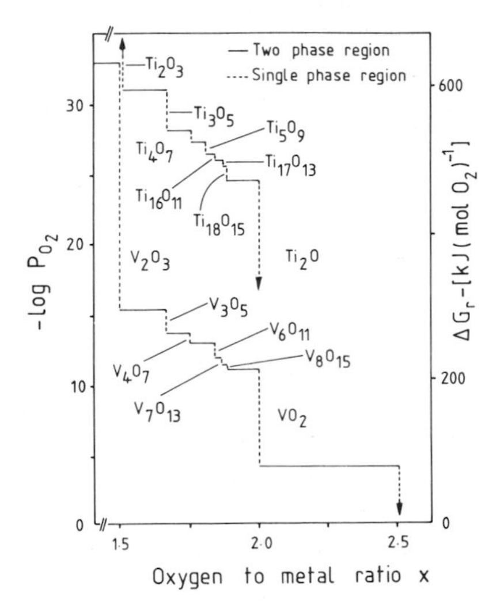

Figure 5.15 Equilibrium oxygen pressures in the Ti–O and V–O systems at 1000 K as a function of the metal/oxygen ratio.

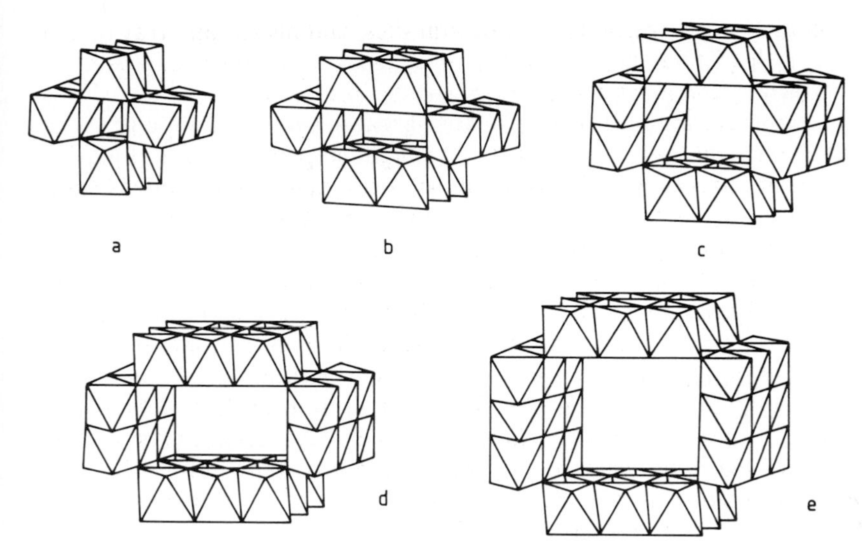

Figure 5.16 Mineral structures composed essentially of the oxides of manganese exhibit an interesting trend. Pyrolusite (β-MnO_2) has the rutile structure (a); ramsdellite (γ-MnO_2) has the structure of the mineral diaspore shown in (b); hollandite (α-MnO_2) has rectangular tunnels circumscribed by a 2×2 arrangement of edge-sharing octahedra (c). In romanechite (d) and todorkite (e), the tunnels are 2×3 and 3×3 octahedra respectively.

The way in which nominally simple structures, typified by α-MnO_2, give way to ostensibly more complex analogues foreshadows what we see frequently in ternary oxides. It is noteworthy that 'designed' catalysts, possessing desired diameters of tunnels and requisite compositions of their inner walls, can be fashioned by extending the principles that take us from rutile to another mineral type, hollandite (α-MnO_2). Nature produces minerals (Fig. 5.16) that exhibit the trend thought by some to be advantageous catalytically. The principles responsible for this trend could serve as a useful approach to the tailoring of catalysts for specific purposes. A Fischer–Tropsch catalyst, for example, made up of the appropriate elemental constituents and possessing well-defined tunnel dimensions could be capable of yielding 'non-Schultz–Flory' (see Section 8.2) products in the conversion of 'syn-gas' to alkanes or alkenes.

α-Al_2O_3 (corundum) is but one of the many known crystallographic forms of alumina, all of which have considerable prominence in heterogeneous catalysts. We need focus here on just three series, paying attention to the fact that they are all based on more or less close-packed O^{2-} ions with Al^{3+} in octahedral and tetrahedral sites and that differences arise when the O^{2-} layers are superposed on one another in different sequences. In the α-series, based on α-Al_2O_3 (corundum) just considered, the sequence of anions is ABAB.... In the β-series there are alternations of the close-packed layers, i.e. ABACABAC. And in the γ-species we have a strictly fcc packing sequence: ABCABC.... The γ-alumina structure is a defective version of the spinel structure (Section 5.2.6.3). It is best envisaged as a structure made up of O^{2-} ions in which one-third of the tetrahedral sites occupied in a normal spinel are vacant (denoted by V), i.e. as: $(Al^{0.67}\ V_{0.33})_{tet}\ (Al_2)_{oct}\ O_4$. The exterior surfaces of γ-Al_2O_3, like those of many other oxides, are rich in OH groups.

The β-series consists of χ- and κ-alumina, which are products of thermal decompositions of $Al(OH)_3$, the mineral form of which is known as gibbsite. (β-Alumina, a misnomer, is quite unrelated to the β-series of aluminas: it contains sodium and there are at least two known stoichiometries, $Na_2O \cdot 11Al_2O_3$ and $Na_2O \cdot 22Al_2O_3$. These are best pictured as regular layers of Na_2O and blocks of spinel, which is discussed later.) Members of the γ-species are obtained on decomposition of the hydroxide structures bayerite, norstrandite and boehmite, and may be conveniently divided into a low-temperature group, consisting of γ- and η-alumina, and a high-temperature group, δ- and θ-alumina. *In vacuo*, the three trihydroxides (gibbsite, bayerite and norstrandite) decompose at low temperatures into an essentially non-crystalline (amorphous) product, ρ-alumina, which at higher temperatures transforms into γ- or η-alumina and eventually into θ-alumina. The dehydration process in air, however, is best represented by Scheme 5.1.

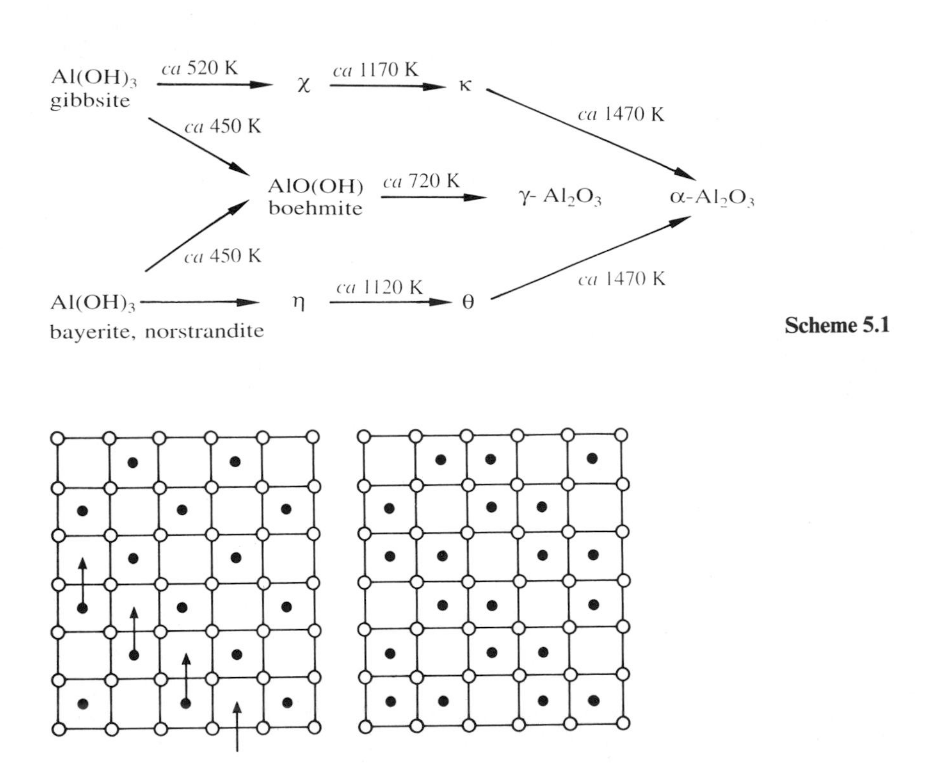

Fluorite (CaF_2) A – type La_2O_3

Figure 5.17 Idealized representation of how the A-La_2O_3 structural type is related to, and derived from, the fluorite structure. Each small square with a filled circle in its centre denotes a projection of MO_8 cubes.

Another structure, taken up especially by lanthanide oxides, corresponding to a stoichiometry M_2O_3, is the so-called A-type La_2O_3 structure. It can be shown that this structure results from a parent fluorite structure as a consequence of crystallographic shear along an edge of the anion cube following reduction. The net effect is shown in the idealized projection shown in Fig. 5.17. The effect of the CS operation is to change the MO_8 cubes from being all edge-sharing in the fluorite structure to edge- and face-sharing cubes in the A-type La_2O_3 structure.

Structures derived from or related to the framework of the ReO_3 structure are so numerous, varied and important that they are accorded a separate section.

5.2.4 Shear and Block Structures Based on ReO_3

We have seen that many solids such as rutile (Fig. 5.13) can eliminate an accumulation of individual vacancies by crystallographic shear, CS. But in no class of oxide is the dominance of shear planes more pronounced than in those derived from the ReO_3 structure. ReO_3 itself may be pictured as an infinite, three-dimensional array of corner sharing octahedra. Upon reduction, oxygen vacancies are created, and there is good electron-microscopic evidence for believing that these vacancies condense and form discs or planar arrays on well-defined low-index planes such as {120}

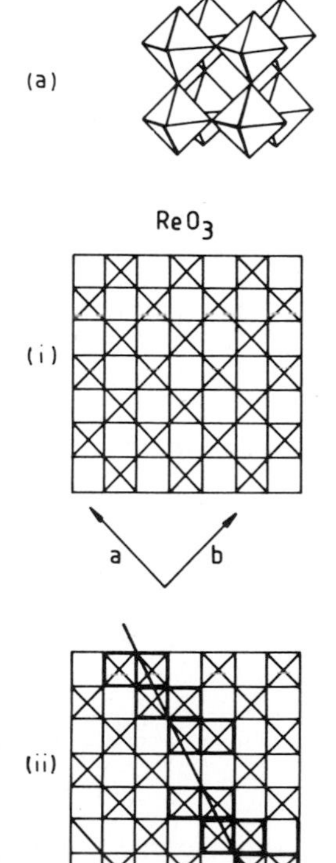

Figure 5.18 Formation of a CS plane in a crystal of ReO_3. (a) The ReO_3 structure. (b) The defect-free structure at (i), on reduction, develops a plane of vacant oxygen sites which collapses to (ii), a CS plane on (130) where each W atom is still in octahedral coordination.

or {130}. These planar arrays of vacancies, such as those shown on the (130) plane in Fig. 5.18 can be eliminated by a change in anion coordination along the (130) direction, the cation coordination remaining octahedral. The resulting planar fault is the CS plane. It is known that, as reduction of ReO_3 or WO_3 (which have the same structures) proceeds, more CS planes are introduced, and as equilibrium is reached they become regularly spaced. Sub-stoichiometric oxides (MO_x) which have within them regularly spaced CS planes of type {120} have a general formula MO_{3n-1}, n being the integer that defines the individual member of the homologous series. Likewise, oxides with regularly spaced {130} CS planes belong to the homologous series MO_{3n-1} (see Fig. 5.19). Examples of well-identified members of such homologous oxides abound: $(Mo, W)_{14}O_{41}$; $(Mo, W)_{12}O_{35}$; $(Mo, W)_{10}O_{29}$; Mo_9O_{26} and Mo_8O_{23}, all of which are members of the M_nO_{3n-1} series and $Mo_{18}O_{52}$ and $Mo_{19}O_{55}$, which are members of the M_nO_{3n-2} series.

The importance of block structures, which we consider next, in the context of catalysis and non-stoichiometry, is that they constitute extraordinarily flexible means of

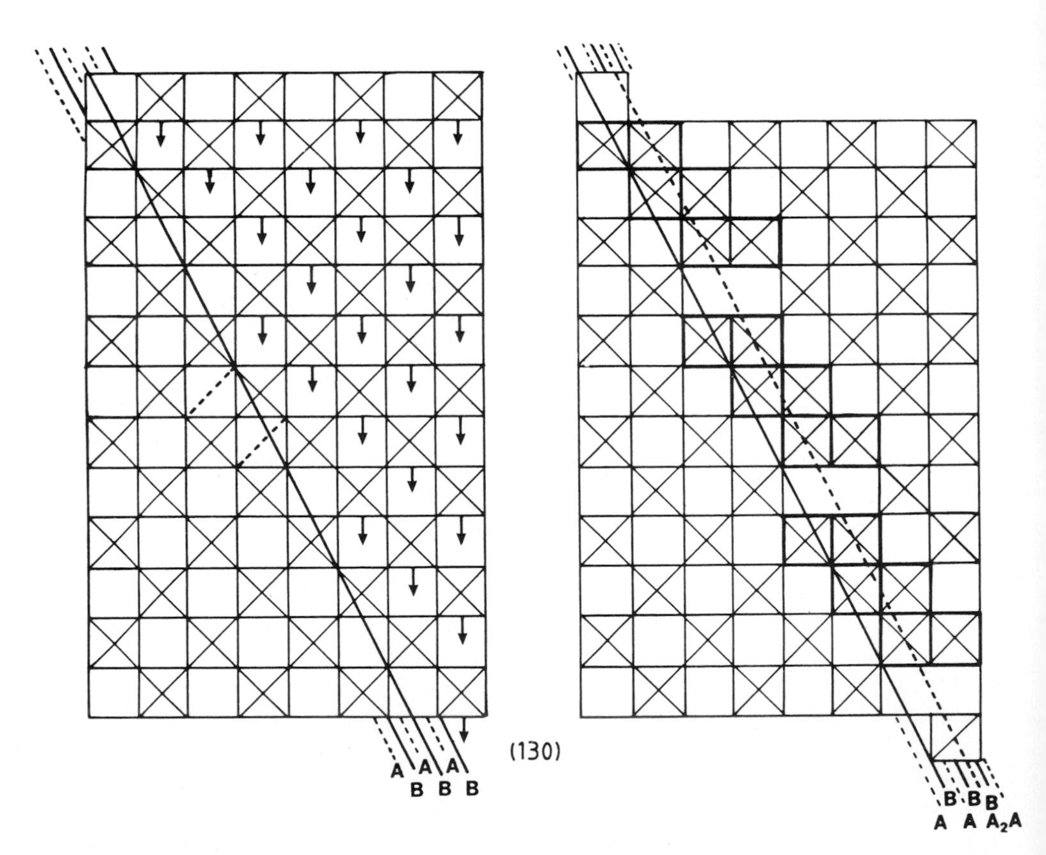

Figure 5.19 Non-stoichiometric oxides based on the ReO_3 structures with regularly spaced {130} CS planes have the general formula MO_{3n-1} (see text).

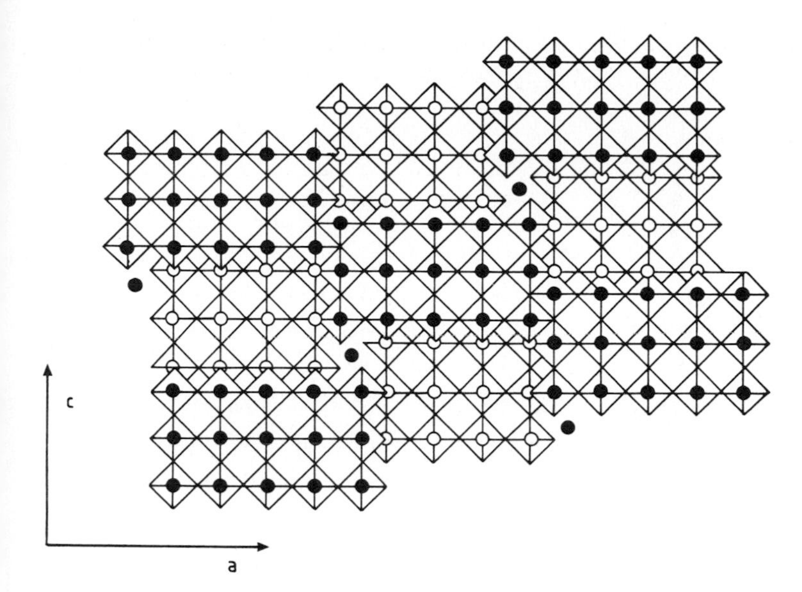

Figure 5.20 One of the many polymorphic forms of Nb_2O_5 has a structure which can be pictured as that formed by collapse of the basic ReO_3 structure along two CS planes, (106) and (209). The resulting block structure has units described as (3 × 4) and (3 × 5) octahedra, which are linked by edge-sharing with other such blocks at two different levels. Tetrahedral sites (shown as isolated filled circles) separate some of the blocks.

accommodating exceedingly small changes of M/O ratio whilst retaining a high degree of crystalline order. Block structures are found in the numerous polymorphic modifications of the oxides of niobium – of the 14 or so known modifications of Nb_2O_5, ten are block structures – but also in the ternary systems M–Nb–O where M is Ti, Mo, W or P. They are formed by the intersections of CS planes in two orthogonal directions in the parent oxide (Fig. 5.20). The resulting structure is divided into blocks of dimensions determined by the particular CS planes that intersect. In Fig. 5.20 the two distinct types of blocks that are formed are denoted (3 × 4) and (3 × 5), the integers referring to the number of octahedra sharing vertices. Because of the edge-sharing boundaries, there are two levels of blocks, those with the 'full circle'-centred octahedra and those, (empty circle-centred) half a unit cell below or above them in Fig. 5.20. In some block structures there is only one size of block (e.g. in $Ti_2Nb_2O_7$ the block dimension is (3 × 3)). It is to be noted that, at the block intersections, there are tunnels (often partially or totally occupied) running perpendicular to the plane of the blocks.

Recognizing the variety and subtlety of the changes that are permitted in block structures, the idea was mooted by J. S. Anderson in 1973, and has subsequently been vindicated, that for *any* proportion of the combining atoms in certain classes of solids an ordered periodic structure is possible. This is the notion of 'infinite adaptability'.

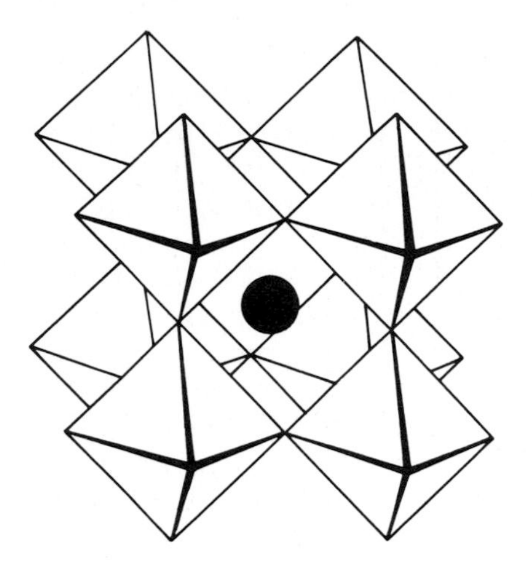

Figure 5.21 Perovskite ($CaTiO_3$) has a structure in which the Ca^{2+} ion is in 12-fold coordination at the centre of a cube of TiO_6 octahedra.

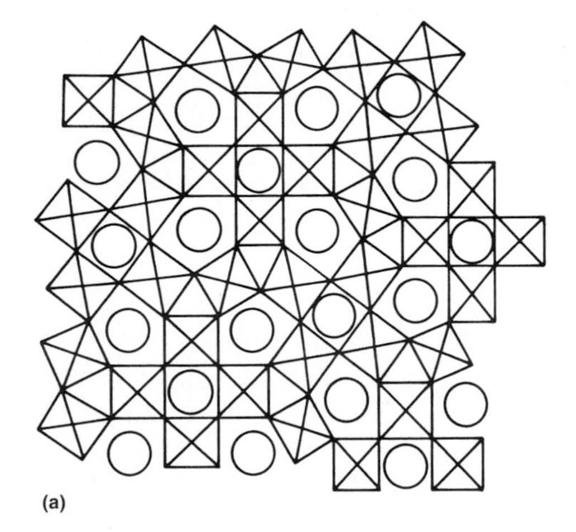

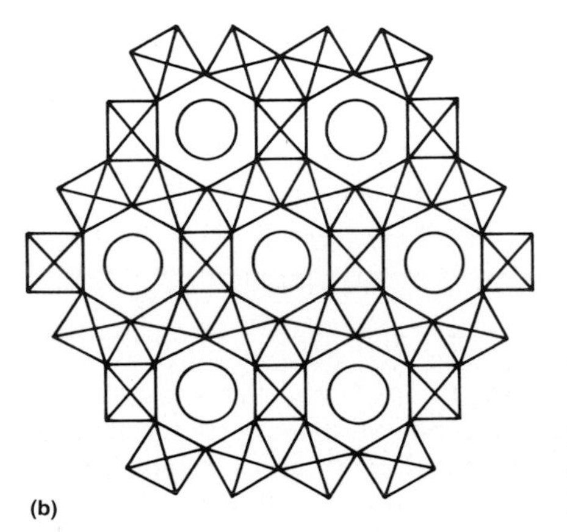

Figure 5.22 Projected drawings of tetragonal (a) and hexagonal (b) tungsten bronzes, TTB and HTB, respectively.

5.2.5 More Complicated Metallic Oxides

When extra cations A^{n+} are inserted into the interstices present in the ReO_3 structure (Fig. 5.21) a perovskite structure (general formula AMO_3 or ABO_3), to which we return in Section 5.2.6.1, is formed. The guest ion A^{n+} is in 12-fold coordination with the surrounding oxygens, and well-known examples of this structure are the bronzes $LiWO_3$ and $NaWO_3$. These perovskite structures are synonymous with the cubic tungsten bronzes. But when other bronzes are formed by the incorporation of alkali metals into WO_3, tetragonal and hexagonal tungsten bronzes are formed (Fig. 5.22).

The cubic (perovskite), tetragonal and hexagonal $A^{n+}MO_3$ bronzes differ from one another in both the number and size of the interstitial sites available to the A^{n+} ions. In the cubic structure, each A^{n+} ion is surrounded by eight MO_6 octahedra situated at the vertices of a cube. Since the number of cubic interstices equals the number of octahedra, this structure can be adopted by the alkali-metal bronzes of formula $A_x^+MO_3$, for values of x up to unity. The tetragonal structure is composed of three- four- and five-membered rings of octahedra (i.e. tunnels). Since the numbers of three-, four-, and five-membered rings are in the ratio 2 : 1 : 2 and since the trigonal prismatic (unlike the cubic and pentagonal) tunnels are too small to accommodate alkali cations, the TTB structure can be adopted only by A_xMO_3 bronzes in which $x \leq 0.60$. Likewise, the hexagonal tungsten bronze (HTB) structure, which contains trigonal prismatic and hexagonal prismatic tunnels in the ratio 2 : 1, can be adopted by bronzes only with $x \leq 0.33$.

More complicated bronzes, the structures of which are based on octahedra sharing both vertices and edges, are known. These include molybdenum bronzes A_xMoO_3,

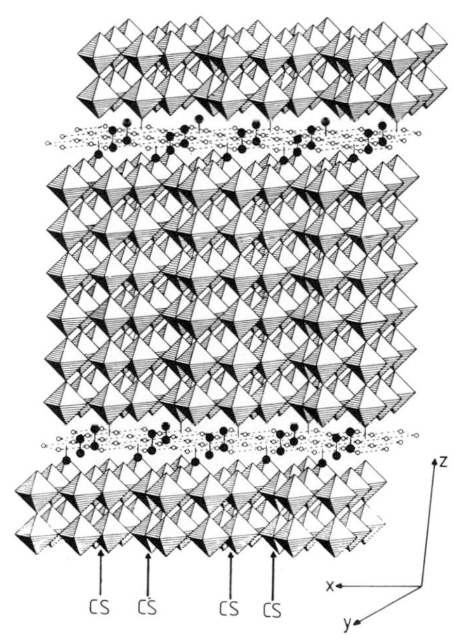

Figure 5.23 A new structure, discovered by HREM, in which layers of $Bi_2O_2^{2+}$ are separated by slabs of $(W, Mo)O_6$ octahedra. The catalyst $Bi_2(Mo, W)_{12}O_{35}$, effective in acrolein production from propylene, has this structure. (Based on work of D. A. Jefferson, J. M. Thomas and R. K. Grasselli, *J. Chem. Soc. Chem. Comm.*, **1983**, 594).

vanadium bronzes $A_xV_2O_5$, titanium bronzes A_xTiO_2 and some oxyfluoride bronzes such as $Na_xV_2O_{5-x}F_x$. And for the tungsten bronzes A_xWO_3 with $A = Ba^{2+}$, Bi^{3+}, Sn^{2+}, Pb^{2+} as well as the alkali-metal ions K^+, Rb^+ and Cs^+, so-called intergrowth structures, first discovered by the Swedish worker Kihlborg, are formed. These consist of narrow strips of the HTB intergrown with lamellae of WO_3 that are *n* octahedra wide. Electron micrographs leave no doubt as to the reality of these structures. A rather bizarre example of a bronze structure known to have catalytic activity in the conversion of propylene to acrolein is shown in Fig. 5.23, for the material of overall stoichiometry $Bi_2(Mo, W)_{12}O_{35}$. Here layers of $Bi_2O_2{}^{2+}$ are separated by slabs of $(W, Mo)O_6$ octahedra, of which some share vertices and others, as a result of crystallographic shear, share edges.

5.2.6 Even More Complicated Oxides

There are very many complex oxides which are either potentially capable of, or are already demonstrated to be of value in, heterogeneous catalysis. We shall focus on three structural archetypes; those based on the minerals perovskite ($CaTiO_3$), scheelite ($CaWO_4$) and the spinels (general formula AB_2O_4). We shall see that other structures, and other families of structures, can be derived from these important examples.

5.2.6.1 Perovskites and their Defective Variants

The properties of perovskites that make them so important in heterogeneous catalysis are:

1. their ability, invaluable for designed redox catalysts, to stabilise mixed-valence states of ions of useful transition metals, such as cobalt, iron, manganese and titanium;
2. the fact that unusual valence states are stabilized within its framework, e.g. Ni^{3+} and Mn^{4+} in $LaNiO_3$ and $CaMnO_3$ respectively;
3. the tendency of noble metals to be stabilized when highly dispersed on the surfaces of perovskites, e.g. palladium on $Ce_{0.1}Fe_{0.6}Co_{0.4}O_3$ is a good three-way auto-exhaust catalyst;
4. the relative ease with which their structural oxygen can be released (to oxidize a reactant hydrocarbon selectively); and
5. the advantageous electronic band gaps that they often possess (e.g. $SrTiO_3$, which is a popular photo-electrochemical material).

Perovskites find much scope as auto-exhaust catalysts and in other situations where redox processes hold sway, in dehydrocyclisation of hydrocarbons (Fig. 5.24), in fuel-cell components, in membrane catalysts for the conversion of natural gas to fuels, and in the photocatalytic breakdown of water.

The perovskite structure (Fig. 5.21) can accommodate numerous elements of the Periodic Table in one valence state or another. This, and its remarkable capacity to accommodate a multitude of different kinds of defects, give rise to numerous distinct structural patterns. These may be readily classified as follows.

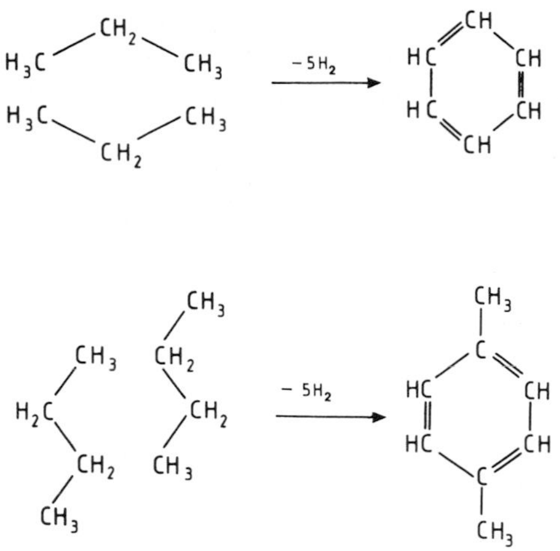

Figure 5.24 Dehydrogenation of alkanes to yield aromatics, which have higher octane numbers, occurs catalytically on certain perovskite oxides (*see* A. Reller, J. M. Thomas, D. A. Jefferson, *J. Phys. Chem.*, **1983**, *87*, 913).

(a) *Vacancies in the A sites.* When only some of the A sites are occupied the structure that results is identical to the cubic tungsten bronze, A_xWO_3. But A-site deficiency is also a feature of materials such as A_xBO_3 when B is Ti, Nb or Ta.

(b) *Vacancies in the B sites.* These are not so favoured energetically as their A-site analogues. An example is $Ba_2Ce^{4+}_{0.75}Sb^{5+}O_6$, there being seven B-cation vacancies, in ordered array, per unit cell.

(c) *Perovskites with anion vacancies.* These are much more common than perovskites with cation vacancies, and in grossly non-stoichiometric ABO_{3-x} there is a pronounced tendency for the vacancies to order. Whereas in the defective perovskites $SrTiO_{2.5}$ and $SrVO_{2.5}$ the anion vacancies are random, in the anion-deficient system $CaMnO_{3-x}$ over the range $0 \leq x \leq 0.5$ five distinct compositions ($CaMnO_{2.5}$, $CaMnO_{2.556}$, $CaMnO_{2.667}$, $CaMnO_{2.75}$ and $CaMnO_{2.80}$) with ordered vacancies have been identified. In $CaMnO_{2.5}$ anion vacancies convert the MnO_6 octahedra to MnO_5 square pyramids. In some other non-stoichiometric perovskites BO_6 octahedra are converted to BO_4 tetrahedra as x increases from 0 to 0.5. Thus in $Ca_2Fe_2O_5$ ($CaFeO_{2.5}$), which has the brownmillerite structure, there are alternate sheets of FeO_6 octahedra and FeO_4 tetrahedra.

(d) *Perovskites with anion excess.* These are somewhat less common than their anion deficient counterparts. A few systems do show 'apparent' oxygen excess: $LaMnO_{3+x}$ and $EuTiO_{3+x}$. But neutron diffraction studies of $LaMnO_{3.12}$ reveal that the oxygen excess is accommodated by vacancies at the A and B sites with partial elimination of La as La_2O_3. The composition of this perovskite is correctly represented by $La_{0.94}\square_{0.06}Mn_{0.98}\square_{0.02}O_3$, where $\square$ stands for cation vacancies.

5.2.6.2 Perovskites as the Prototypes of New Homologous Series

Solids of composition A_2BO_4 can be envisaged as made up of alternating layers of ABO_3 perovskite and AO rock-salt structures. If two rock-salt slabs were inserted

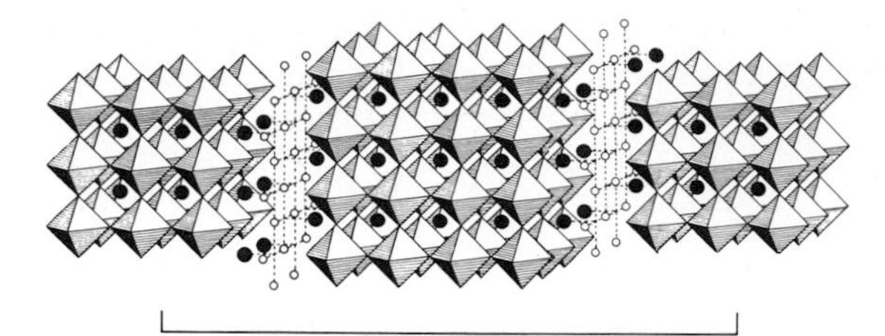

Figure 5.25 Ordered phases, discovered electron-microscopically by Jefferson, of two intergrown Aurivillius phases (see text). The composition of this solid is $Bi_9Ti_6CrO_{27}$.

between the perovskite layers the formulae would be $A_3B_2O_7$. Proceeding in this fashion we arrive at the homologous series $A_{n+1}B_nO_{3n+1}$, individual members of which (e.g. La_2NiO_4; $Sr_3Ti_2O_7$; $La_4Ni_3O_{10}$, etc.) are known as Ruddleston–Popper phases.

Another family of perovskite-derived complex oxides has been discovered by the French crystallographers Carpy and Galy. This has the general formula $A_nB_nO_{3n+2}$. The key structural feature here is the perovskite-like slab of thickness n octahedra, cut parallel to (110). The formula of each slab is $A_{n-1}B_nO_{3n+2}$, but when they are stacked we introduce, in effect, an extra sheet of A at the join, thereby producing $A_nB_nO_{3n+2}$.

A third family of complex oxides based on perovskites is associated with the name of the Swedish worker Aurivillius, who found that bismuth titanates formed a homologous series represented by $(Bi_2O_2)^{2+}$ $(A_{n-1}B_nO_{3n+1})^{2-}$. In general, A is a large

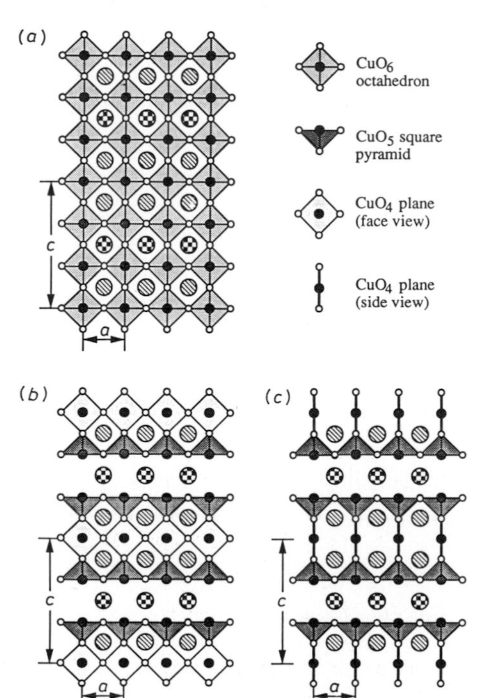

Figure 5.26 Derivatives of the ideal perovskite structure, ABO_3. (a) Hypothetical $YBa_2Cu_3O_9$, showing ordering of Y and Ba which gives a tripling of the c parameter. (b) and (c) $YBa_2Cu_3O_7$, derived from (a) by removal of oxygen atoms at $\frac{1}{2}0\frac{1}{2}$, $0\frac{1}{2}\frac{1}{2}$ and $\frac{1}{2}00$. (b) is viewed down [100], (c) down [010]. (Reprinted from I. J. Pickering and J. M. Thomas, *J. Chem. Soc., Faraday Trans.* **1991**, *87*, 3067.)

cation that can readily fit into the interstices in the perovskite layers, B is the octahedral cation and n the number of layers of corner-shared octahedra in each perovskite slab. Usually n runs from 1 to 5, and a well known example is $Bi_4Ti_3O_{12}$, with $n = 3$ and $A \equiv Bi$. These 'Aurivillius' phases are capable of even more subtle intergrowths: ordered intergrowths between two distinct phases (say of $n = 4$ and $n = 3$) are possible. The general formula is now $Bi_4A_{2n-1}B_{2n+1}O_{6n+9}$, and an example is shown in Fig. 5.25.

A fourth family, of great interest as warm superconductors but also as heterogeneous catalysts for the oxidation of CO, the partial ammoxidation of toluene to benzonitrile and the partial oxidation of CH_4, is the renowned system $YBa_2Cu_3O_{6+x}$ ($0 \leq x \leq 1$) where Y^{3+} and $2Ba^{2+}$ occupy the A positions. These atoms are ordered in the c direction as Ba–Y–Ba–Ba–Y–Ba, and Figs. 5.26 and 5.27 show schematically this tripled perovskite structure, having the formula $YBa_2Cu_3O_9$. In structural solutions for all compositions of x in $YBa_2Cu_3O_{6+x}$ it has been found that the oxygen site in the plane of the yttrium atoms (i.e. $00\frac{1}{2}$) is void, yielding the five-coordinate Cu(2) site. In addition there are vacancies in the basal plane at the $0\frac{1}{2}0$ and $\frac{1}{2}00$ sites, which vary depending on the precise oxygen composition and give rise to the variable occupancy and complicated structural relationships. For $YBa_2Cu_3O_7$, the $0\frac{1}{2}0$ site is fully occupied and the $\frac{1}{2}00$ site is empty. Figure 26 (b) and (c) shows two projections of this structure.

The $YBa_2Cu_3O_7$ structure consists of puckered Cu–O sheets in the ab plane, together with Cu–O chains in the b direction. These are linked through a common oxygen, giving rise to copper sites of five-coordinate, square-based pyramidal copper sheets and square-planar copper chains. The Y^{3+} cation lies between the square-based pyramidal sheets and the Ba^{2+} occupies a site between four square-based pyramids. The structure of the non-superconducting tetragonal end-member, $YBa_2Cu_3O_6$, is closely related (Fig. 5.27). There is a further oxygen vacancy at the $0\frac{1}{2}0$ position, so

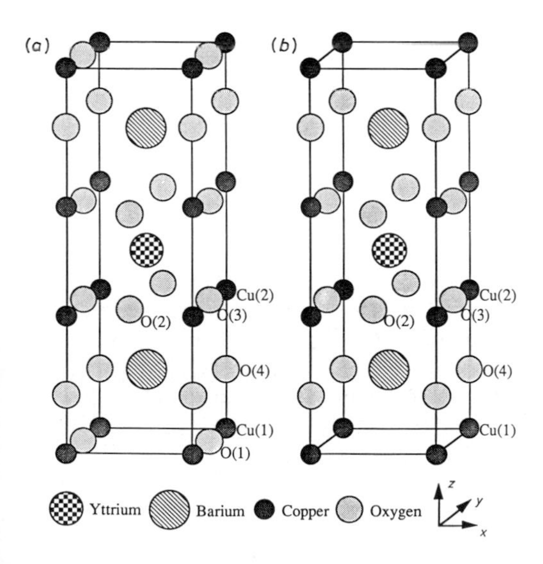

Figure 5.27 The structures of (a) $YBa_2Cu_3O_7$, and (b) $YBa_2Cu_3O_6$. (Reprinted from I. J. Pickering and J. M. Thomas, *J. Chem. Soc., Faraday Trans.* **1991**, *87*, 3067.)

that the Cu–O chains in the *b* direction disappear and Cu(1) becomes linear, twofold coordinated.

From the $YBa_2Cu_3O_7$ end-member, oxygen can be removed by reducing the occupancy of the O(1) site at ($0\frac{1}{2}0$). As this happens the Cu–O chains in the *b* direction are gradually destroyed, decreasing the difference between the *a* and *b* unit-cell axes. Further removal of the oxygen can occur until the $0\frac{1}{2}0$ site is vacant, as in the case of $YBa_2Cu_3O_6$.

The non-stoichiometric $YBa_2Cu_3O_{6+x}$ phases are good catalysts for the oxidation of CO to CO_2; and Bovin and his Swedish colleagues found that catalysis of the oxidation of toluene was much higher when $x \approx 0$ than when $x \approx 1$, presumably because the number of active sites is increased at low x. Cu(1) is two-coordinated when $x = 0$ but four-coordinated when $x = 1$; the coordinative unsaturation clearly helps. Over and above these considerations, the key feature of all the anion-defective perovskites is that they facilitate sacrificial loss of structural oxygen as well as solid-state migration of oxide ions.

5.2.6.3 Spinels, Scheelites and the Bismuth Molybdates

One of the most widely used of all solid catalyst precursors (magnetite, Fe_3O_4, for the synthesis of ammonia) is a spinel. Magnetite is also a good catalyst for the water-gas shift reaction. Some of the phases produced when this oxide is reduced to the catalytically active state are also spinels, as are some of the catalysts for the oxychlorination of ethylene to 1,2-dichloroethane. Spinels have a general formula AB_2O_4 and in the unit cell there are 32 oxygen ions with eight of the 64 tetrahedral and 16 of the 32 octahedral sites occupied (see Fig. 5.2). When the A ions are all housed by the tetrahedral sites and all the B ions are in the octahedral ones we talk of 'normal spinels'. The spinels MAl_2O_4 (M = Mg, Fe, Co, Ni, Mn or Zn) have this structure. But in certain other spinels, the eight tetrahedral positions are occupied, not by the eight A ions, but by one-half of the B ions, the rest of which, together with the A ions, are arranged at random in the 16 octahedral positions. These 'inverse spinels' are therefore formulated $B(AB)O_4$. Examples of inverse spinels include $Fe(MgFe)O_4$ and $Zn(SnZn)O_4$ as well as magnetite itself, $Fe^{III}(Fe^{II}Fe^{III})O_4$. In some spinels both M^{II} and M^{III} ions are randomly distributed in tetrahedral and octahedral sites. It is convenient to describe different modes of distribution by a parameter η, which gives the fraction of divalent metal ions in octahedral sites. Thus, for the normal spinel $\eta = 0$; for the inverse spinel $\eta = 1$; and for the randomly distributed situation $\eta = 0.67$.

There are several ways in which defective spinels can arise, the most common being those in which, by appropriate change of cation valence, vacancies occur in the cation sublattices, both tetrahedral and octahedral. Certain oxides are found to be soluble in a spinel host, with the result that the extra cations are housed in the available octahedral and/or tetrahedral sites. In addition to the formation of defects of this kind, replacement of the ions present in the original (normal or inverse) spinel by foreign ions can be effected, thereby offering abundant scope for the design of multicomponent, spinel-based catalysts.

The mineral scheelite, $CaWO_4$, has the crystal structure shown in Fig. 5.28. Like

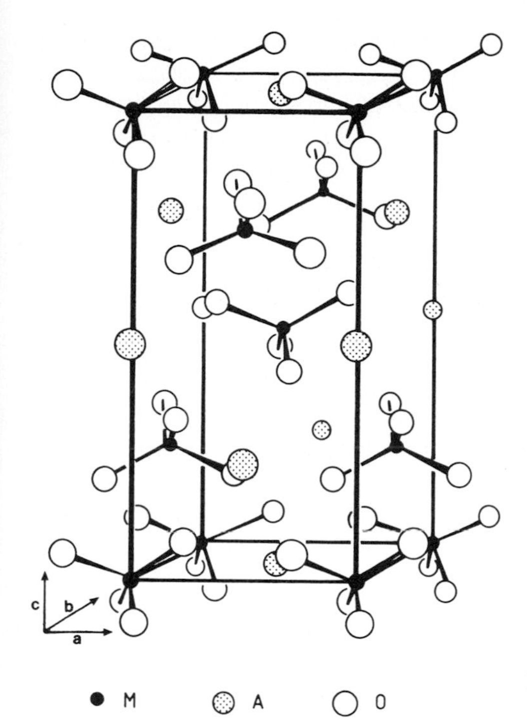

Figure 5.28 The unit cell of the scheelite ($CaWO_4$) structure.

perovskites, spinels and fluorite, the scheelite ABO_4 structure is capable of accommodating a large fraction of the elements of the Periodic Table. All the A cations are structurally equivalent to each other, as is the case for all the B cations and all the BO_4 tetrahedral units, respectively. The A category is coordinated to eight oxygens, each in a different tetrahedron. To date, the most prominent catalyst possessing a structure close to that of scheelite is α-bismuth molybdate, $Bi_2(MoO_4)_3$, which possesses a defect-scheelite structure $Bi_{2/3}\square_{1/3}MoO_4$ (just as δ-Bi_2O_3 has the defect-fluorite structure $Bi_2O_3\square$). This particular polymorph of bismuth molybdate is one of several (e.g. Bi_2MoO_6 and $Bi_2Mo_2O_9$) known to be efficient in selectively converting propylene with oxygen to acrolein, and, in the presence of ammonia, to acrylonitrile:

$$CH_3CH{=}CH_2 + O_2 \longrightarrow CH_2{=}CHCHO + H_2O$$

$$2\,CH_3CH{=}CH_2 + 3\,O_2 + 2\,NH_3 \longrightarrow 2\,CH{=}CHCN + 3\,H_2O$$

Two different cations can order on the A sites, as in $KEu(MoO_4)_2$, and also on the B sites, as in $Bi_3(FeO_4)(MoO_4)_2$. In general, long-range order between different cations on the A or M sites does not occur.

Very recently, as a result of combined high-resolution electron-microscopic, computer simulation and diffraction studies, it has come to light that all but one of the seven known bismuth molybdates can be pictured as being derived from the defect-fluorite structure. In so far as evolving strategies for designing better catalysts for the selective oxidation of hydrocarbons is concerned, it is useful to picture the fluorite or defect-fluorite structure as made up of bismuth oxide as the 'solvent', within which an exceptionally large family of other metal oxides (of molybdenum, niobium, vanadium, silicon, germanium, etc.) can be housed.

5.2.6.4 Heteropolyions (Keggin Structures) as Catalytically Significant Entities

In the mid-1820s Berzelius discovered that new heteropolyanions were formed by the acidification of solutions containing both phosphate and molybdate. The material prepared by him contains the phosphomolydbate ion $(PMo_{12}O_{40})^{3-}$. Since that time numerous other heteropolyanions have been prepared, chiefly with molybdenum and tungsten but with over 30 different heteroatoms, including many non-metals and most transition metals, frequently in more than one oxidation state. The free acids and the salts of small cations are water-soluble, but the salts of large cations (Cs^+ and Ba^{2+}) are usually insoluble. Heteropolyacids and their acid salts are of great interest catalytically: they are used commercially in petrochemical contexts for the hydration of alkenes, the formation of ethers and esters and for the isomerization of alkanes. They may be used (as intumescent solids – see below) or in supported form (on microporous or mesoporous supports).

Cationic equivalents of the Keggin ions – named after the crystallographer who determined their structure – such as $[Al_{13}O_4(OH)_{24}(H_2O)_{12}]^{7+}$ are used to pillar clays and other layered structures, where the intention is both to improve access to reactants by separating, to a fixed degree, the individual sheets and to enhance their thermal stability. Bulky heteropolyanions based on molybdenum or tungsten may be used to pillar hydrotalcite-based catalysts where the individual sheets are positively charged. Both molybdenum and tungsten are important elements for catalytic hydrodesulphurization.

In these ions, the heteroatoms are situated inside cavities formed by MO_6 octahedra (M = Mo, W or, in the case of the cationic species, Al) of the parent M atoms and are bonded to the oxygens of adjacent MO_6 octahedra. The inner cavity, within the 12 circumscribing octahedra, houses a tetrahedrally coordinated ion (e.g. P, As, Si, etc.) – see Fig. 5.29.

Significantly, the pentahydrate of the acid $H_3PW_{12}O_{40}$ is actually a hexahydrate containing $H_5O_2^+$ ions. A more appropriate formulation is therefore $(H_5O_2)_3{}^+$ $(PW_{12}O_{40})^{3-}$ and $(H_5O_2)_3{}^+$ $(HSiW_{12}O_{40})^{3-}$ and $(H_5O_2)_3{}^+$ $(H_2BW_{12}O_{40})^{3-}$ are the true formulae of the isomorphous analogues.

Just as clays swell when they take up water and organic molecules, so do solid heteropolyacids. They can assimilate ample quantities of water and a range of other polar molecules; and the resulting solid is a quasi-liquid in which hydrated protons and other incorporated molecules have great mobility, just as they do in the interlamellar spaces of clays. Reactants have access to many protons in such a swollen system because the surface area is large (as in three-dimensional porous catalysts, especially the zeolites, discussed later). The Japanese workers Misono and Ono, as well as Polish, French and Dutch scientists, have exploited this kind of catalytic chemistry, symbolized in Colour Plate 4.

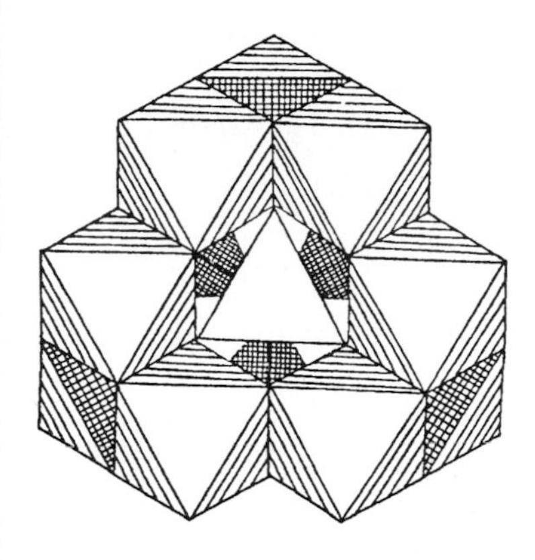

Figure 5.29 The Keggin ion shown here may be regarded as a central tetrahedron surrounded by 12 octahedra.

5.2.7 Clays, Zeolites and Related Structures

It is easier, though not straightforward, to define a zeolite than a clay. The Swedish nobleman Cronsted in 1756 used the term "zeolite" to describe the property of the mineral stilbite (idealized formula $Na_4Ca_8[Al_{20}Si_{52}O_{144}].56\,H_2O$) which, upon heating, behaves as if it were boiling (zeo-lite: the stone that boils). Any microporous material, possessing cages and/or channels of molecular dimension (3 to 10 Å diameter) and capable of accommodating copious quantities of water of other polar molecules in its intracrystallite spaces, is generally deemed to be a zeolite. Originally, the term referred exclusively to aluminosilicate minerals possessing the above properties, e.g. mordenite, offretite and clinoptilolite. Nowadays, since it is known that numerous elements in the Periodic Table (quite apart from the expected ones – aluminium, gallium, boron, germanium, etc.) are capable of being incorporated (by laboratory synthesis or post-synthesis isomorphous substitution) into the framework structure of crystalline microporous materials, the term 'zeolite' is used much more flexibly (and perhaps a shade confusingly). Whilst it is always true that a zeolite is also a molecular sieve solid, the corollary does not hold. Some 'artificial' molecular sieves cannot accommodate water and other polar species. In general, the traditional definition that a zeolite is an oxide framework of aluminium and silicon is honoured. The term 'molecular sieve' describes frameworks of any composition. More information about zeolites and related microporous materials based upon aluminium phosphate (ALPOs) are given in Section 5.2.7.2.

The term 'clay' generally refers to aluminosilicates of which the particle sizes fall in the micron (micrometre, μm) range and which also exhibit cation-exchange capacity. By this definition, clays encompass zeolites. Generally, however, the term refers chiefly to sheet silicates that have greater or less cation-exchange capacity. Most clays, but by no means all (e.g. some micas) swell upon uptake of water and of certain other, generally polar, solvents such as ethanol or glycerol.

The clays that are of premier catalytic interest are those known as montmorillonite and hectorite. Their structures are best discussed along with those of pyrophillite and talc on the one hand, and with beidellite and saponite on the other.

Table 5.1 Idealized formulae for some selected clays

Pyrophillite[(a)]	$(Al_4)^{oct}(Si_8)^{tet}O_{20}(OH)_4$
Montmorillonite[(a)]	$M^{n+}_{x/n} \cdot aH_2O(Al_{4-x}, Mg_x)^{oct}(Si_8)^{tet}O_{20}(OH)_4$
Beidellite[(a)]	$M^{n+}_{x/n} \cdot aH_2O(Al_4)^{oct}(Si_{8-x}Al_x)^{tet}O_{20}(OH)_4$
Muscovite[(a)]	$K_2(Al_4)^{oct}(Si_6Al_2)^{tet}O_{20}(OH)_4$
Talc[(b)]	$(Mg_6)^{oct}(Si_8)^{tet}O_{20}(OH)_4$
Hectorite[(b)]	$M^{n+}_{x/n} \cdot aH_2O(Mg_{6-x}Li_x)^{oct}(Si_8)^{tet}O_{20}(OH)_4$
Saponite[(b)]	$M^{n+}_{x/n} \cdot aH_2O(Mg_6)^{oct}(Si_{8-x}Al_x)^{tet}O_{20}(OH)_4$
Phlogopite[(b)]	$K_2(Mg_6)^{oct}(Si_6Al_2)^{tet}O_{20}(OH)_4$

[(a)] In these clays, *two*-thirds of the available octahedral sites (see Fig. 5.31) are occupied. These are termed *di*octahedral.
[(b)] Here, all (*three*-thirds) of the octahedral sites are occupied, and they are termed *tri*octahedral.

All these six so-called clay minerals, and many other members, are composed of two distinct types of connected sheets or layers, one consisting of corner-linked SiO_4 tetrahedra, the other edge-lined $A(O, OH)_6$ octahedra, where A, in the case of montmorillonite, pyrophillite and beidellite, is predominantly Al and, in the case of hectorite, talc and saponite, is predominantly Mg (see Table 5.1 and Fig. 5.30).

In both pyrophillite and talc the so-called TOT layers (i.e. tetrahedral–octahedral–tetrahedral) are neutral, so that no exchangeable cations (M^{n+}) can be accommodated in the interlamellar region. In pyrophillite, electrical neutrality of the layers is achieved by the presence of Si^{4+} in all eight tetrahedral sites in the unit repeat $A_4Si_8O_{20}(OH)_4$ of the oxygen framework and of Al^{3+} in two-thirds of the available octahedral (A) sites: in talc, all the tetrahedral and octahedral sites are occupied by Si^{4+} and Mg^{2+} respectively. Separate TOT layers are, therefore, rather loosely bound via the agency of weak dipolar and van der Waals forces. The six other sheet silicates (collectively known as smectites) listed in Table 5.1 bear a net negative charge on the TOT layers. This arises because of isomorphous substitution. In montmorillonite some of the Al^{3+} in the octahedral sublattice are replaced by Mg^{2+} ions, and in hectorite some of the Mg^{2+} in the octahedral sublattice are replaced by Li^+ ions. With beidellite and saponite, however, the isomorphous substitution takes place in the tetrahedral sublattice with Al^{3+} replacing some of the Si^{4+} ions. The residual negative charges on the layers of montmorillonite, hectorite, beidellite and saponite are counterbalanced in the natural state by coexisting interlamellar, hydrated cations, usually Na^+, Ca^{2+}, Mg^{2+}, etc. The precise degree of layer charge and

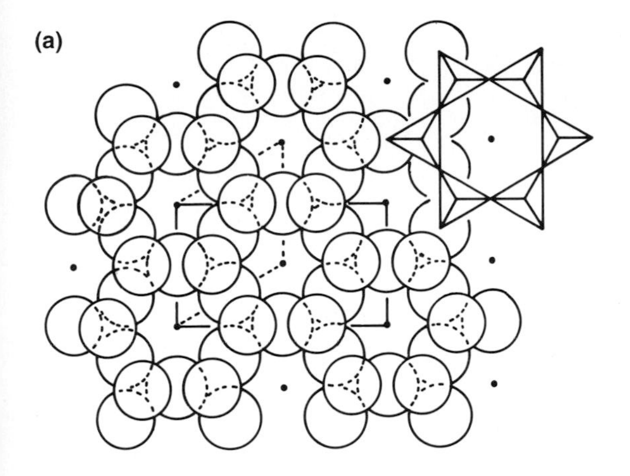

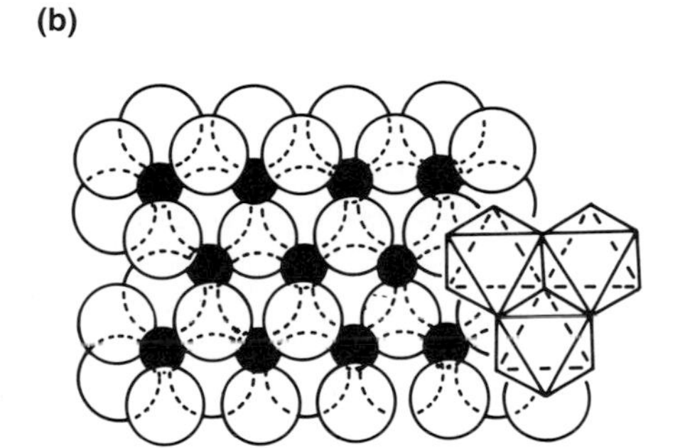

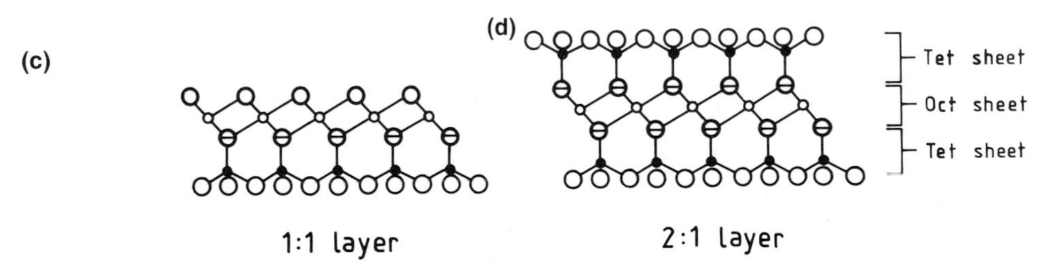

Figure 5.30 Plan views of (a) the corner-shared SiO_4 sheet and (b) the $A(O,OH)_6$ sheet to which it is attached in the so-called TOT (2 : 1) and TO (1 : 1) sheet silicates. In (c), SiO_4 'sheets' flank both sides of the central octahedral sheet; in (d) only one side.

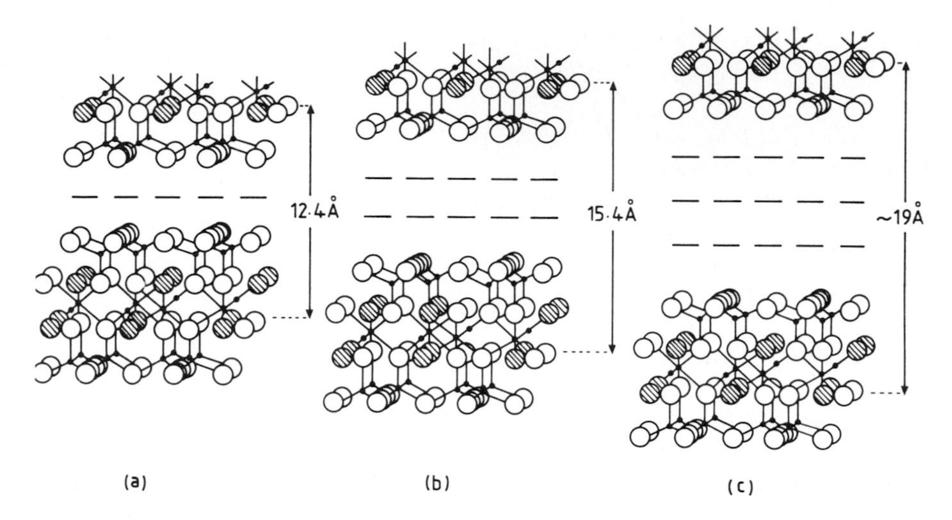

Figure 5.31 Expansion in basal spacing experienced by montmorillonite when (a) one-, (b) two- or (c) three-layer clays are formed on intercalation of γ-butyrolactone.

its distribution, as well as the particular nature of the interlamellar cation, are functions of the geological habitat and prior history of the clay in question. Nowadays clays can be readily synthesized: their cation-exchange capacities and degrees of crystallinity and catalytic performance are functions of their mode of preparation. Typically, the charge deficiency, which is the origin of the cation-exchange capacity, ranges from 0.4 to 1.2 units of electronic charge per Si_8O_{20}. These figures are to be compared with a value of 2.0 units for muscovite and phlogophite mica and zero for pyrophillite and talc. Put differently, the charge density in these smectites is such that unit charge occupies a basal area of between 45 to 100 $Å^2$.

The swelling of clays in water results from the extra hydration of the interlamellar cations. This is the best-known example of the important phenomenon of intercalation, which is simply the insertion of guest species into an accommodating host. The degree of swelling, however, is governed by the nature of the interlamellar cation, and the sorption isotherms often exhibit steps. The meaning of 'one-, two- and three-layer clays' is best illustrated diagrammatically; see Fig. 5.31. 'Layers' in this context refer to the interlamellar water. In some sheet silicates, the water is believed to take up an ice-like monolayer. NMR studies reveal that the interlamellar ion and associated water are rather mobile above room temperature, and capable of being displaced by various organic species (amines, lactones, acids, carbohydrates, etc.). As a consequence the interlamellar microenvironment is converted from a predominantly hydrophilic to a so-called oleophilic state, making it more conducive to the further insertion of organic species which would otherwise be difficult to intercalate into the original, water-rich smectite.

Catalyst scientists have realised that there are alternatives to the naturally occurring montmorillonites or hectorites as starting materials for novel conversions of organic species. The La Porte Company in England developed synthetic hectorites and fluorohectorites (where the OH^- ions attached to the octahedrally coordinated Mg^{2+} in the layer are replaced by F^-) and these are known as laponites. The fluoro form has higher thermal stability (because it cannot dehydroxylate) than the hydroxyl form. But neither can be produced, at least by hydrothermal synthesis, in a state of high crystalline order.

It is also possible, synthetically, to replace octahedrally coordinated Al^{3+} ions in montmorillonite by Ni^{2+}, Co^{2+} and other transition-metal ions resulting in what has been termed 'synthetic mica–montmorillonite', abbreviated SMM. Recently it has been shown that when Ni^{2+} or Co^{2+} is introduced into the octahedral sublattice of synthetic beidellite the resulting catalysts rival the performance of zeolites for the hydroisomerisation of paraffins.

Even further scope for designing and tuning clay catalysts to order results from the fact that, by judicious heat-treatment of Li^+ ion-exchange (synthetic or natural) montmorillonites, the charge density of the layers may be progressively diminished. This happens because the Li^+ ions diffuse through to, and become incorporated within, the octahedral network, thereby neutralising the charge that arose there from isomorphous replacement. These materials are known as charge-reduced clays.

In addition to the TOT sheet (or layered) silicate structures discussed above there are also a large number of TO structures, where a sheet of corner-linked SiO_4 tetrahedra abutts another one, thus edge-sharing octahedra. Kaolinite is one of the best-known examples (Fig. 5.32): its formula is $Al_2SiO_5(OH)_4$. This is a di-octahedral structure (in the sense that montmorillonite, its TOT analogue, is di-octahedral). There are other polymorphic forms of kaolinite, e.g. dickite and nacrite, which differ from it only in the sequence of the stacking of the TO layers. The serpentines (which embrace chrysotile, antigorite and lizardite) are tri-octahedral TO structures of ideal

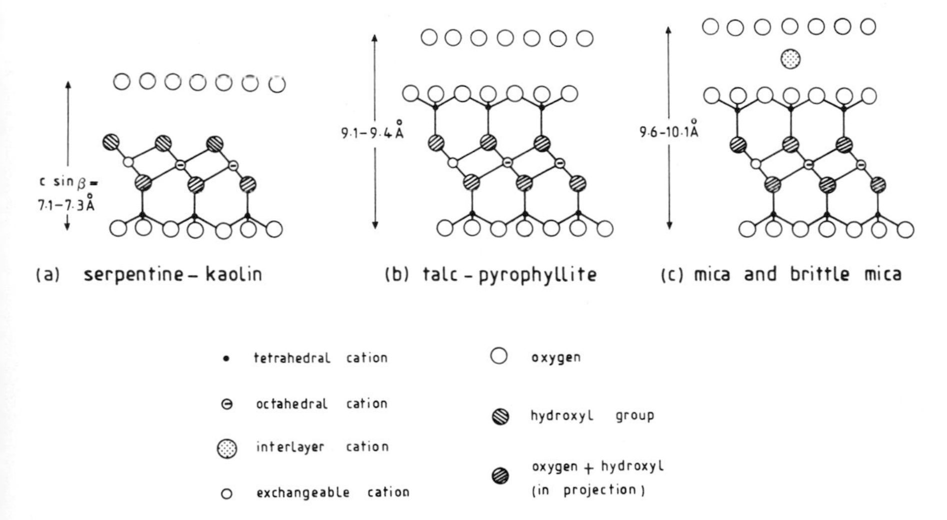

Figure 5.32 The kaolinite–serpentine structure, (a), is a T : O fusion (c.f. Fig. 5.31). Its relationship to talc–pyrophillite (b) and the micas (c) is illustrated here.

formula $Mg_3Si_2O_5(OH)_4$. When both aluminium and magnesium are present in the octahedra layers the resulting curvature of the TO structure depends on the precise amount of isomorphous substitution. Chrysatile is the material also known as asbestos.

Mineralogists nowadays refer to the TOT and TO structures (Figs 5.30 and 5.32) as smectites and kandites, respectively.

5.2.7.1 Pillared Clays are Effectively Two-Dimensional Zeolites

Even though naturally occurring and synthetic clays, which are usually rich in exchangeable Na^+, K^+, or Ca^{2+} cations, can be converted into viable acidic catalysts by directly or indirectly inserting protons into the interlamellar regions, such catalysts still suffer from the disadvantage of physical collapse at high temperatures. At *ca* 200 °C, the interlamellar solvent species tend to be expelled and the sheets cohere, with consequent loss of catalytic activity. This collapse may be prevented by inserting 'pillars', preferably of an inorganic character, which serve to keep the individual layers apart. But the generation of pillared clays has other advantages, not the east among them being the merit of incorporating extra 'pores' into the catalyst. Were it possible to space the pillars evenly, and in a controllable fashion, a new type of shape-selective catalyst would be produced. Indeed, the resulting high-area solid may be regarded as a two-dimensional zeolite, with acidic properties comparable with those of Y-type acidic zeolites (Section 5.2.7.2).

Pillaring can be achieved by contacting the clay mineral with solution rich in the cationic Keggin ion $[Al_{13}O_4(OH)_{24}(H_2O)_{12}]^{7+}$ (compare Fig. 5.29). Aluminium hydroxy polymers of this kind are readily prepared in the dispersed state from many solutions containing Al^{3+} ions by appropriate adjustment of pH so that, typically, OH/Al ratios are less than 2.3. Montmorillonoid, beidellitic and hectoritic clays can be effectively pillared in this way. Some brittle micas can also be pillared, as well as synthetic fluorophlogophite, by using other multinuclear ions, containing titanium or molybdenum. The resulting thermal stability of these pillared clays is impressive.

Pillared Clays and Fractal Dimension

Mandelbrot's work on the use of the widely applicable concept of fractals for the analysis of self-similar properties in general, and of tortuous, microporous or other seemingly ill-organized materials in particular, has prompted a recent analysis by Avnir and by van Damme of the nature of the geometrical constraints in porous solids of interest in catalysis such as carbons, silica-gels, zeolites and pillared clays in terms of fractals. The problem, in essence, is how quantitatively to describe irregular surfaces. Although irregular surfaces (or lines) cannot be generated by translational symmetry operations, they can often be characterized by so-called dilational symmetry: increasing or decreasing the resolution power of the technique used to probe the surface does not modify the general morphology of the surface itself. An example is shown in Fig. 5.33 for a line, which may be regarded as the elevation view of a surface. An N-fold magnification of the surface 'detail' reveals N^D smaller details morphologically similar to the previous one. D is, by definition, the fractal di-

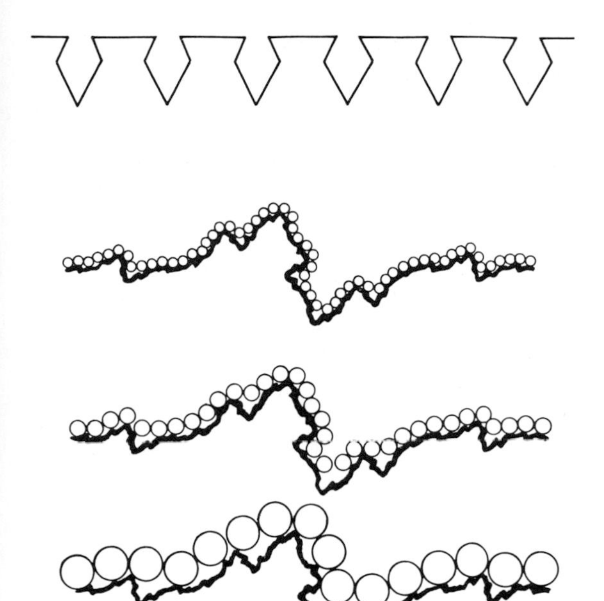

Figure 5.33 An illustration of the fractal dimension as applied to pillared clays (see text).

mension of the surface: it falls in the range $2 < D < 3$ (or $1 < D < 2$ for a fractal line). The higher the value of D, the greater is the tortuosity or space-filling character of the surface.

Avnir and Pfeiffer developed a method for measuring D based on the use of a range of differently sized adsorbate molecules to evaluate monolayer coverage. On a surface of fractal dimension D, the number of spherical molecules of radius r necessary to reach monolayer coverage scales down with increasing r as:

$$N \sim r^{-D} \text{ or } N \sim s^{-D/2} \qquad (1)$$

where s is the cross section of the probe molecule. The limit $D = 2$ corresponds to a regular or Euclidean surface; this limit is closely approached by faujasitic type zeolites (see below). For microporous silica gel, on the other hand, as well as microporous charcoal, D approaches 3. For pillared clays, irrespective of whether there was regular or irregular distribution of the pillars, the fractal dimension turned out to be close to 2.

5.2.7.2 A Synoptic Guide to the Structure of Zeolitic and Related Solid Catalysts

In the production of petrochemicals, as well as in the numerous reactions that they undergo – cracking, hydrocracking, hydration, dehydration, alkylation, isomerisation, oxidative addition and dehydrocyclisation – it is impossible to make progress that will lead to improvements in catalytic activity and selectivity without first understanding the structural principles of zeolitic solids. Ever since the mid-1960s, when Rabo and co-workers at the Union Carbide Laboratories and Plank and co-workers at the Mobil Laboratories announced dramatic examples of solid acid catalysis based on faujasitic zeolites, there has been continuing interest and growth in the number of zeolitic solids used for both laboratory-based and, more importantly, industrial-scale catalysis. Microporous aluminosilicate catalysts possessing the structures of well-known minerals (such as faujasite, ferrierite, mordenite and erionite) as well as many more that have no known naturally occurring analogues (for example, ZSM-5, theta-1, ZSM-23, zeolite Rho) are particularly well suited for the conversions of hydrocarbons, certain oxygenates and other species into useful products. They are, in effect, solids possessing three-dimensional (3D) surfaces, replete with cages and channels, which may or may not intersect. Linking the pores and distributed in a more or less spatially uniform fashion throughout their bulk are the active sites which are bridging hydroxyl groups, as shown in Scheme 5.2. These are the classic Brønsted acid centres, the intrinsic strength of which is a function both of the particular local environment of the aluminosilicate structure in which they occur and also the Si/Al ratio.

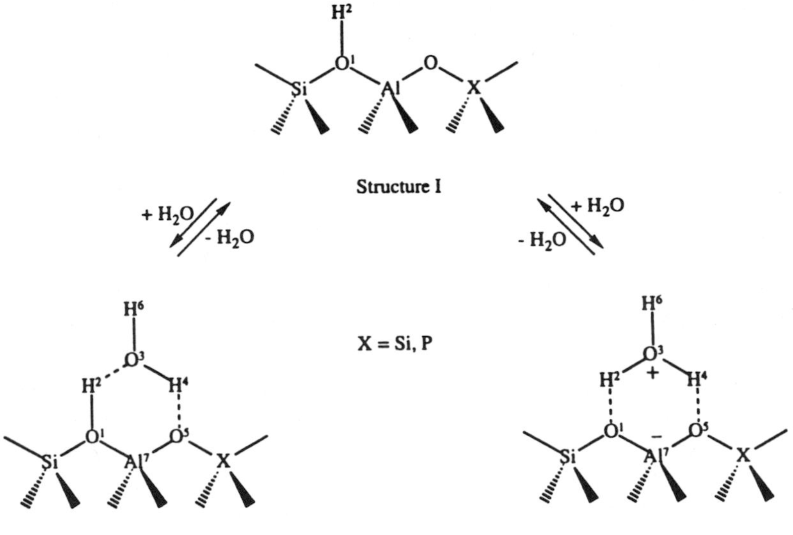

Structure I

$+ H_2O$ / $- H_2O$

X = Si, P

Structure II

Structure III

Scheme 5.2

Other 'heteroatoms', apart from Al, may be substitutionally accommodated inside an open-structure silica framework. For example, in Fe-ZSM5, trivalent iron ions occasionally take up the position of the tetravalent silicon. The local environment of the iron, readily determinable from extended X-ray absorption fine structure (EXAFS) studies, is described in Section 3.6.5.2. (This particular solid acid catalyst is reported to be effective in the catalytic conversion of methane directly to methanol using N_2O in place of dioxygen as an oxidant.) Other heteroatoms readily accommodated as trivalent ions into silicalite I (the parent siliceous extreme of ZSM-5) are gallium and boron.

Since the early 1980s a large family of related microporous solids, based on aluminum phosphate, $AlPO_4$, has been uncovered. As in aluminosilicates (where four-valent silicon is substitutionally replaced by trivalent Al^{3+}, Be^{3+}, Ga^{3+} or Fe^{3+}) in the ALPOs (aluminum phosphates), covering a wide range of structural types Brønsted acidity in the solid is engendered whenever a divalent ion (e.g. Co^{2+}, Zn^{2+}, Mg^{2+}, Mn^{2+}, etc.) replaces the trivalent aluminium, or whenever a tetravalent ion such as silicon replaces pentavalent phosphorus. The term MeALPO (metal–aluminium–phosphate) is given to the former and SAPO (silicon–aluminium–phosphate) the latter. It is often possible to have the two modes of substitution co-existing in one structure (e.g. as in CoAPSO-44, cobalt–aluminium–phosphorus–silicon type number 44 – see below).

Since 1992, when workers at the Mobil Research Laboratories announced the discovery of a large family of mesoporous solids (with aperture diameters ranging from some 20 to 100 Å), a further upsurge in interest in new crystalline (or partly crystalline) solid catalysts with open structures and vast, accessible internal areas (greater than 600 $m^2\,g^{-1}$) has occurred. It seems, from the work of Stucky and others, that mesoporous solids having compositions distinctly different from those siliceous ones first discovered by the Mobil group are relatively readily prepared. Examples of materials capable of being fashioned in a mesoporous state are the oxides of many metals including that of tungsten; doubtless many other compositions will be announced in due course.

Only a relatively small fraction of the numerous varieties of micro- and mesoporous inorganic materials that interest us here are preparable as single crystals, amenable to straightforward four-circle X-ray diffractometry. Very often, the synthesized new material, in the presence or absence of the organic template generally needed to induce the precursor mother liquor or gel to crystallise with an open structure – obtained after burning off the organic scaffolding – is in microcrystalline form. Powder (X-ray and neutron) diffraction methods are therefore indispensable; and when clues exist, from other measurements, about the nature of the framework of the new material, it is often possible to refine the structure by means of Rietveld profile analysis. When, however, good-quality crystals are obtainable – and a recent solid acid magnesium aluminium phosphate (MAPO), known as DAF-1 (Davy Faraday one), described below, is one such instance – there is a well-proven method (recently admirably summarised by Glusker et al.) of solving the structure.

Because only a rather small fraction of the many varieties of meso- and microporous inorganic materials of catalytic interest are preparable for single crystals, it has become increasingly necessary to solve crystal structures of new zeolites and MeALPOs using a multiple-pronged approach in which electron diffraction, high-resolution electron microscopy, solid-state NMR, powder X-ray diffraction, model building, as well as energy minimisation computational methods are all brought into play. The structure of MeALPO-36 was solved in this fashion. (Computational procedures involving lattice energy minimisation, together with simulated annealing and genetic algorithm techniques, as we shall see in Section 5.3 below, are playing an increasingly important role in solving and predicting complex crystal structures including zeolites and MeALPOs.)

Framework Density, Nomenclature and Secondary Building Units

As zeolitic materials do not constitute a readily definable family of crystalline solids, use is often made of the simple notion of framework density (FD), which is the number of tetrahedral (T) atoms per 1000 Å^3 in the structure. The well-known *Atlas of Zeolite Structures*, published by the International Zeolite Association, shows the distribution of these values for microporous and dense frameworks. Framework structures having FDs less than 21.0 T-atoms per 1000 Å^3 and based on three-dimensional extended tetrahedral networks are normally the only ones included in successive issues of the *Atlas*. Both the Linde type A and faujasite (zeolites X and Y) structures have FDs of close to 12.6.

Taking the rules adopted by the IUPAC Commission on Zeolite Nomenclature, all zeolitic and, ALPO (MeALPO) structure types consist of three capital letters. These 'codes' are not to be confused or equated with actual materials: structure types do not depend on composition, distribution of the various possible T-atoms (B, Be, Al, P, Si, Ge, Zn, Sn, Co, Fe, V, etc.), unit-cell dimensions or symmetry. In general, the codes (AEI, ATS, CHA, FAU etc.; see below) are derived from the names of the type species and do not include numbers and characters other than capital Roman letters. Recognizing that, in this category of solids, structural criteria alone do not provide an unambiguous numbering scheme, the indexing of existing and new members is facilitated by arranging the structure types in alphabetical order according to the code of the structure type, namely AEI (for $AlPO_4$-18, *A*LPO *ei*ghteen), AEL (for $AlPO_4$-11, *A*LPO *el*even), CHA (for *cha*bazite, $Ca_6[Al_{12}Si_{24}O_{72}]\cdot 4H_2O$), FER (for *fer*rierite, $Na_2Mg_2[Al_6Si_{30}O_{72}]\cdot 18\,H_2O$) and TON (for *T*heta-*one*, $Na_n[Al_nSi_{24-n}O_{48}]\cdot 4\,H_2O$).

In zeolitic and ALPO (MeALPO) structures the primary building units are almost invariably single TO_4 tetrahedra. The secondary building units (SBUs), which contain up to 16 T-atoms, are derived on the assumption that the entire extended framework may be envisaged as being composed of one type of SBU only. Invariably, SBUs are non-chiral; and it is to be noted that a unit cell always contains an integral number of SBUs. Figure 5.34 shows a selection of the more prominent SBUs and cages formed from them. The neutralizing, and on the whole exchangeable, cations are located at well-defined sites in the various cavities that exist within the structure, and the water molecules fill up the remaining voids. The water can be expelled upon heating and evacuation and may be replaced by a number of small organic or inorganic guests. By adjusting the valence or the size of the exchangeable cation, the molecular sieving and hence the shape-selecting property of a zeolite may be fine-tuned.

The microporosity of a zeolite can, in general, be further enhanced by increasing the Si/Al ratio of the macroanionic framework. This is known as dealumination:

$$\underset{\text{Hydrophilic}}{M_{x/n}[(AlO_2)_x(SiO_2)_y]\cdot mH_2O} \xrightarrow[+Si]{-Al,-M,-H_2O} \underset{\text{Hydrophobic}}{SiO_2}$$

Indeed, when the NH_4^+-exchanged form of zeolite Y is heated under hydrothermal conditions the process of dealumination is effected. The resulting structure is said to

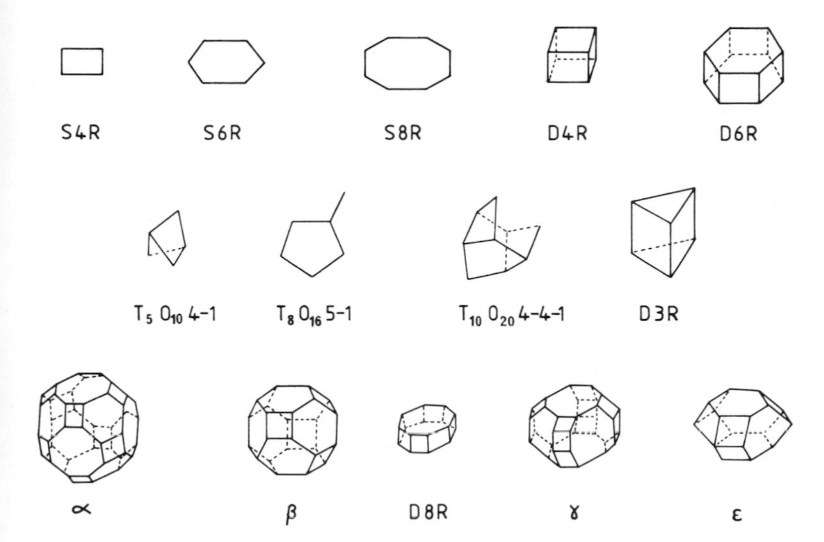

Figure 5.34 The top two rows show rings, or secondary building units (SBU), out of which zeolitic structures are composed. Cages (bottom row) are formed from these rings. The cages, in turn, are connected by various combinations of SBUs to yield full zeolitic structures such as those shown later. Each straight line denotes a T–O–T bond, where T now stands for either Al^{3+} or Si^{4+} and O for oxygen. (S4R, D8R etc stand for single four-ring, double eight-ring, etc.).

be ultrastabilized, in view of the face that it withstands high-temperature treatment at ~1000 °C without loss of structural integrity. During the course of stabilization the Si/Al ratio of the framework changes from an initial value of around 2.4:1 (depending upon the preparation of the original zeolite) to a final one of beyond 10 or 100:1. It has recently been found possible readily to dealuminate certain zeolites, e.g. those based on faujasite (zeolite Y), simply by exposure to the vapour of $SiCl_4$ at an elevated temperature and to achieve Si/Al rations of greater than 1000:1. Hydrothermal methods work well for the dealumination of other zeolites such as mordenite, offretite and ZSM-5 (to silicalite).

Zeolites X and Y are structurally analogous to the mineral faujasite. The building units are truncated octahedra, also known as sodalite cages, β-cages or tetrakaidecahedra – all three terms are synonymous. These cages, seen in Fig. 5.35a, are linked (in zeolites X and Y) to adjacent ones via hexagonal biprisms, thereby yielding larger supercages.

Some 50 distinct species of zeolitic minerals have been identified, as well as at least 140 synthetic species with a very wide range of aluminium contents.

There are several distinct ways of representing the structure of microporous solids such as zeolites. Two ways of portraying faujasite, the parent framework of zeolites X and Y, are shown in Fig. 5.35. Figure 5.36 shows the line drawings of the plan views of two catalytically important microporous solids: mordenite and zeolite L. Here, the vertices demarcate (Si, Al)O_4 tetrahedra and the lines joining the vertices oxygen bridges. In Fig. 5.37, the corresponding atomic representations of these same

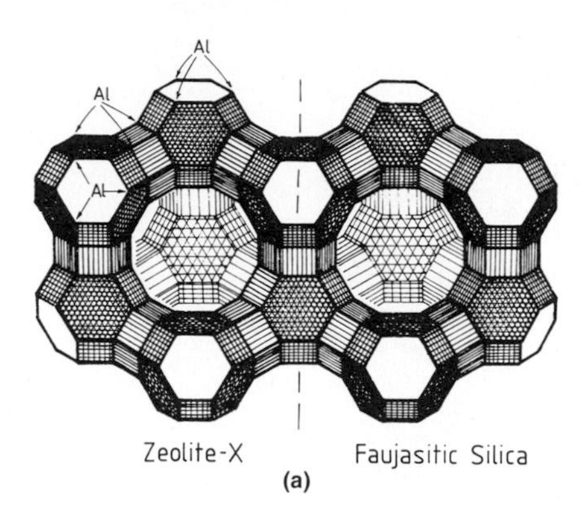

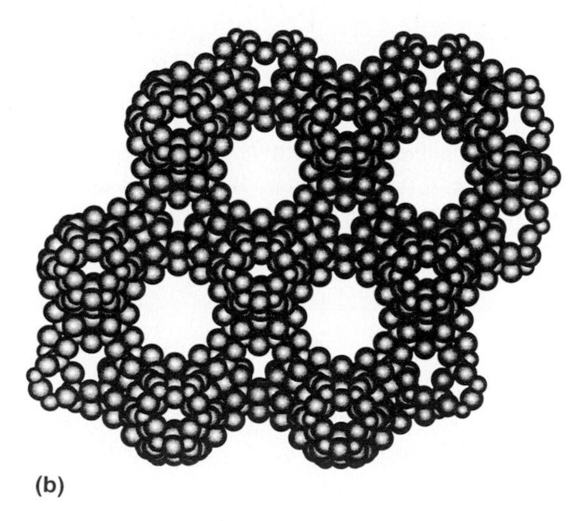

Figure 5.35 (a) Framework structure of the mineral faujasite, the archetype of zeolites X and Y. On dealumination (e.g. treatment with $SiCl_4$), a faujasitic zeolite can be converted to a faujasitic silica, where the Si/Al ratio of the framework is *ca* 100 : 1. (b) Another method of representing the faujasite structure. With permission from J. M. Thomas, R. G. Bell, C. R. A. Catlow, in *Handbook of Heterogeneous Catalysis* (Eds. G. Ertl, H. Knözinger, J. Weitkamp), VCH, Weinheim, **1997**.

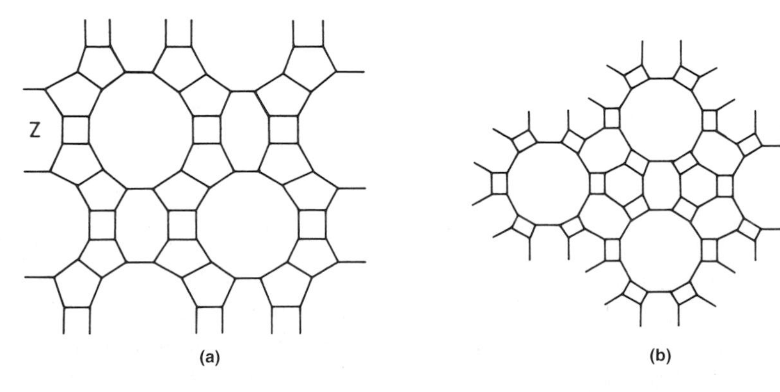

Figure 5.36 Plan view of framework structures of (a) mordenite and (b) zeolite L.

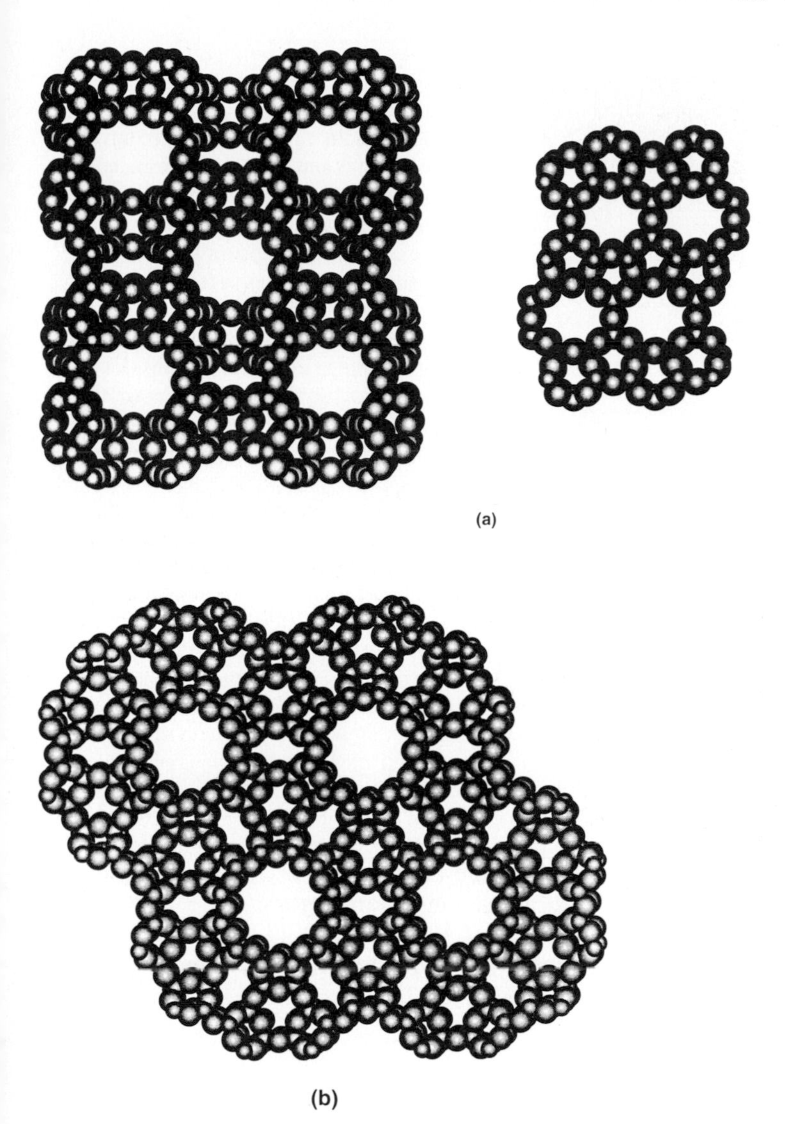

Figure 5.37 Atomic representations of the line analogues in Fig. 5.36: (a) mordenite, two mutually perpendicular views; (b) zeolite L. With permission from J. M. Thomas, R. G. Bell, C. R. A. Catlow, in *Handbook of Heterogeneous Catalysis* (Eds. G. Ertl, H. Knözinger, J. Weitkamp), VCH, Weinheim, **1997**.

structures are shown, the smaller spheres denoting Si(Al), the larger the oxygens. The closely related ZSM-5 (also known as MFI) and ZSM-11 (known as MEL) structures are depicted in Fig. 5.38. Colour Plate 5 shows the plan and elevation views of five zeolitic catalysts of great interest in the proton-induced isomerisation of but-1-ene to 2-methylpropylene (isobutene). ALPO-18 (known also as AEL) – see

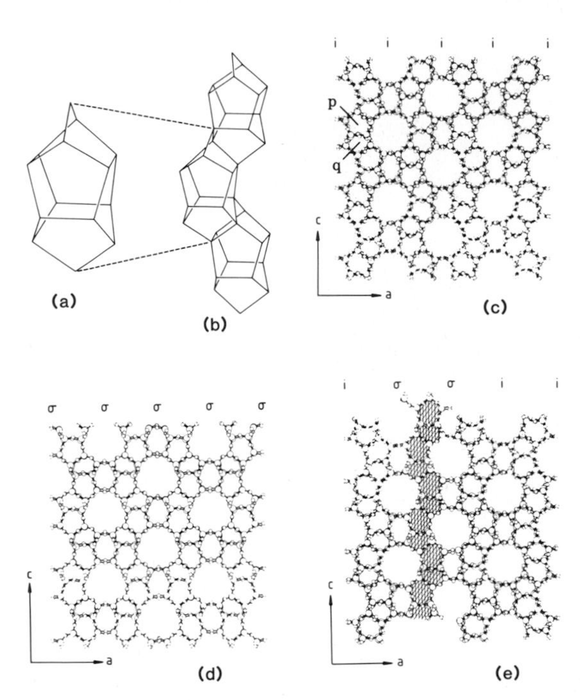

Figure 5.38 (a) Segment of the structure of ZSM-5 and ZSM-11 showing connected five-membered rings composed of linked tetrahedra (SiO_4 and AlO_4). Each connecting line represents an oxygen bridge. (b) The chains from which the ZSM-5 and ZSM-11 structures are built are themselves made up by linking the units shown in (a). (c) In ZSM-5, chains are linked so that {100} slabs are related by inversion (i). Here p and q refer to the larger and smaller five-membered rings, respectively. (The rings are in reality of equal size, but do not appear so in projection.) (d) In ZSM-11, chains are linked so that {100} slabs are mirror images (σ) of one another. (e) Representation of intergrowth of ZSM-5 and ZSM-11. (After J. M. Thomas and G. R. Millward, *J. Chem. Soc. Chem. Comm.* **1982**, 1380).

Colour Plate 6 – with cobalt ions substituted in the framework is a good catalyst for converting methanol to light alkenes (ethylene and propylene). This porous catalyst is a solid acid when the transition metal ion is divalent. ALPO-11-based catalysts (Fig. 5.39) are powerful in the so-called iso-dewaxing process, where straight-chain alkanes (the waxes) are isomerised to branched-chain analogues.

Even for the largest pore openings of *ca* 7.4 Å present in zeolitic (aluminosilicate) solids such as FAU, LTL and DFO, there are severe limitations for the size of molecules that can be subjected to shape-selective catalytic conversions. The recent pro-

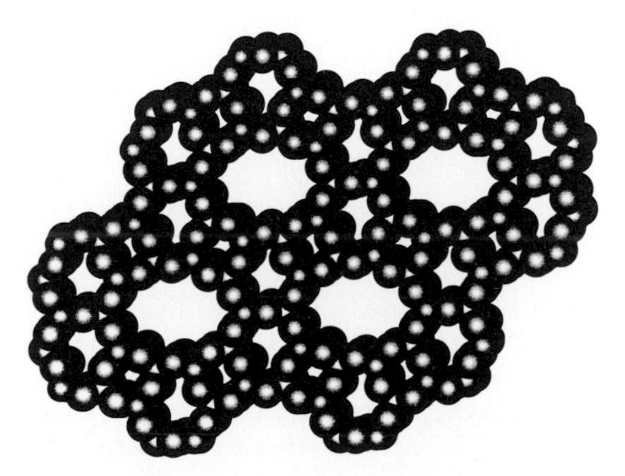

Figure 5.39 Atomic model of the ALPO-11 (AEL) structure, which has oval-shaped orifices.

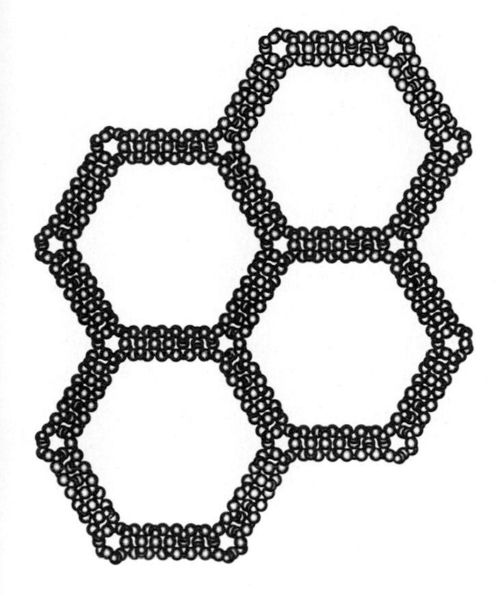

Figure 5.40 Model structure of an MCM-41 mesoporous silica.

duction of *meso*porous siliceous solids has, however, radically transformed prospects in this regard. With the so-called MCM-41 large-pore mesoporous structures reported by Kresge, Beck and co-workers at the Mobil Research Center, all this has changed. The dimensions of the pores (see Fig. 5.40) vary from 20 Å to 100 Å in regular fashion. They are, therefore, when suitably modified – as when tetrahedal Al or B replaces Si in the walls so as to confer Brønsted acidity – ideal as shape-selective catalysts for bulky molecules. Van Bekkum has demonstrated this elegantly for the Fries reaction, and Corma et al. have demonstrated that it is possible to alkylate 2,4-di(*t*-butyl)phenol with cinnamyl alcohol giving rise to 6,8,di(*t*-butyl)-2-phenyl-2,3-dihydro[4*H*]benzopyran (arising from intramolecular cyclization of the primary cinnamylphenol) together with 4-*t*-butylphenol and small amounts of 6-*t*-butyl-2-phenyl-2,3-dehydro[4*H*]benzopyran.

Given that many other solids, as recently demonstrated by Stucky et al. – among them the oxides of iron, vanadium, zirconium, and tungsten – can be produced as mesoporous solids, the prospects look bright for further advances in the (shape-) selective conversions of progressively bulkier organic reactants.

The structural drawings and models shown in Figs 5.35 to 5.40 (and Colour Plates 5 and 6) represent idealized states. It must bf remembered that subtle changes in pore dimensions and hence in accessibility to the interior surface of a zeolite occur simply by changing the nature of the exchangeable cation. We alluded earlier to the increase in effective aperture diameter that is caused by replacing two Na^+ ions with one Ca^{2+}. But quite major changes in the TOT angles of a zeolite can be effected simply by replacing one monovalent cation by another. Thus in going from Li^+ ion-exchanged to K^+ ion-exchanged zeolite X, the TOT angle changes from nearly 150° to close to 135°.

Acidity can be introduced into a zeolite in four ways (compare Fig. 1.18): (i) ion-exchange with NH_4^+ followed by thermal decomposition; (ii) hydrolysis of ion-exchanged polyvalent cations followed by partial dehydration; (iii) direct proton exchange; and (iv) reduction of exchanged metal ions to a lower valence state:

$$M^{n+}Z^- + \tfrac{1}{2}H_2 \rightarrow M^{(n-1)+}Z^- + H^+Z^-$$

The total acidity of a zeolitic catalyst depends on both the concentration of acidic sites and the strength of the individual sites. For highly siliceous zeolites the acidic sites are well separated and the ion-exchange capacity, MASNMR and IR absorption spectra as well as the catalytic activity for a test reaction (e.g. hexane cracking) scale with the aluminium content – see Fig. 1.11. The number of active sites in a zeolitic (Brønsted acid) catalyst can be determined in several ways, including ^{27}Al solid-state NMR, uptake of base and poisoning experiments using aliquots of amine. The results constitute eloquent proof both for the essential correctness of the identification of the source of catalytic activity and the reality of zeolites as *uniform* heterogeneous catalysts, in the sense that the high concentration of active sites is distributed throughout the bulk of the solid.

The influence of zeolite structure on zeolite acidity can best be understood if the zeolites are pictured as rigid ionizing solvents. Brønsted acid strength is a measure of the energy required to remove a proton from the site in question, but in practice the proton is added to a base to form a (carbo)cation (alkylcarbenium ion). Since the pores of the zeolite are of molecular dimension, the lining of the cavities (i.e. the oxide 'ions' of the framework) provide a more or less rigid solvation sheath for any cation formed within the zeolite. We see, therefore, how the pore system of a zeolite may influence a catalysed reaction both by affecting the ingress of reactant to, and egress of product from, the active site and by controlling what happens at the active site.

Fundamental studies, both experimental and theoretical, of the origins of Brønsted acidity in solid acid catalysts, and the nature of the reactive intermediates formed when alkanes are converted by them in processes such as isomerization or cracking, have intensified in recent years. It is becoming increasingly apparent from analyses on materials of identical stoichiometry but defferent framework structures (such as H^+ faujasite and H^+–ZSM-5) that the difference in value of the proton affinity for the bridging hydroxyl oxygen atom (Fig. 5.41) and the next oxygen atom is the key to determining the Brønsted acidity of these catalysts. Hydrogen–deuterium exchange reactions, along with dehydrogenation and cracking of *n*-alkanes on Brønsted acid zeolitic catalysts point to the dominance of pentacoordinated carbocations (carbonium ions) as the crucial intermediates for monomolecular conversions of alkanes.

The key points to note about zeolites in the context of their catalytic performance are that they have:

(a) sharply defined pore size distributions;
(b) high and adjustable acidity;
(c) very high surface areas (typically $600\,m^2\,g^{-1}$), the majority of which (~95%) is, depending upon crystallite size, internal and accessible through apertures of defined dimensions; and

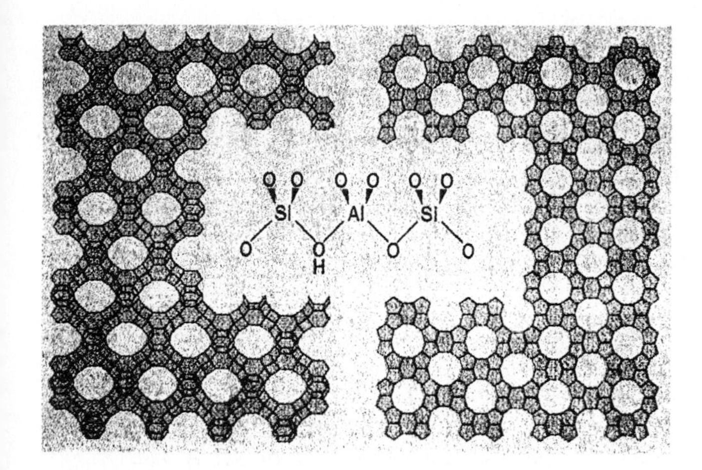

Figure 5.41 The acidity of the zeolite catalysts originate from bridging hydroxyl groups (see formula in the centre). Zeolite-Y (which has the same framework structure as the mineral faujasite, left) and ZSM-5 (right) are different structural forms, but can be synthesized with identical copositions. But even with exactly the same number of active sites they exhibit different catalytic activities. This is because of differences in the proton affinities of bridging hydroxyl oxygen atoms in their framework. (Reprinted from J. M. Thomas, *Angew. Chem., Int. Ed. Engl.* **1994**, *33*, 913.)

(d) good thermal stability (e.g. they are capable of surviving heat treatments in air up to 1000 °C, depending upon the composition of the zeolite framework).

Moreover, since the framework composition of zeolites can be changed from the extremes of low-silica to high-silica contents, the inner walls of the channels and cages can be more or less smoothly converted from the hydrophilic to hydrophobic extremes. Other attributes, apart from their relative ease of synthesis, that contribute to the attractive features of zeolitic catalysts are that:

1. the nature and siting of exchangeable cations can be adjusted and engineered;
2. the siting and energetics of potentially reactive organic species housed within the catalyst pores can also be engineered to some degree; and
3. the catalytically active sites are uniformly distributed throughout the solid, being accessible at the inner walls of the cavities.

Finally, in view of the fact that the active sites are situated predominantly inside the zeolite, and that all these sites, which are of very high concentration (far in excess of active sites on, for example, supported metal catalysts) are at the same time also in the bulk of the solid, zeolitic catalysts can be very well characterized by the powerful new tools that have recently become available for probing local environments within bulk solids. (Zeolite catalysts may in reality be viewed as three-dimensional surfaces). This is particularly true of high-resolution (solid-state) multinuclear NMR, of neutron and X-ray powder profile (Rietveld) methods, of high-resolution electron microscopy and of computer graphics techniques. Moreover, in-situ studies of zeolites by, for example, X-ray absorption and diffraction as well as neutron scattering are particularly apposite for zeolites, in contrast to most other heterogeneous catalysts.

5.3 Computational Approaches

5.3.1 A Resumé of Available Methodologies

A fundamental understanding of catalysis requires accurate knowledge firstly of the structures at the atomic level of the catalyst, and secondly of structural *changes* accompanying the key processes such as activation and deactivation. It also requires atomistic models for the active site and the ways in which reactant and product interact with them. Once accurate models of these structures have been developed, one proceeds to unravel the dynamic aspects involving diffusion and reaction which lie at the heart of catalysis. Currently, our knowledge of catalytic processes at the atomic level is being revolutionised by the application of computer modelling techniques which are able to predict with growing accuracy and reliability the structural and dynamic properties of matter at the atomic level. Here we shall focus on the modelling of the structures of crystals and surfaces and of sorbed molecules. We first outline the main methodologies used in contemporary computer modelling studies of crystal and surface structures.

There are two broad categories of technique: first, electronic structure methods in which one attempts to solve, at some level of approximation, the Schrödinger equation for the system under study; second, simulation techniques in which all electronic structure effects are subsumed into interatomic potential functions which describe the variations in the energy of the system as a function of nuclear coordinates. Both methods have important roles to play in modelling catalysts. We dwell in rather more detail on the former in Section 5.4, but we outline its essence here.

The electronic structure methodologies can themselves be divided into two general types of technique: the first, the Hartree–Fock methods solve the Hartree–Fock equations with explicit evaluation of the exchange energies; and in modern calculations, there is an increasing use of 'post'-Hartree–Fock techniques (configurational interaction or perturbation methods) to estimate correlation energies. A further subdivision of this class of technique is into ab-initio and semi-empirical approaches: the former calculates all integrals directly, while the latter parameterises many of the important integrals and sets the less important ones to zero. Although semi-empirical methods will have a continuing role, the constant growth in computer power means that ab-initio calculations can be performed on systems containing an increasingly large number of atoms.

An alternative approach, of growing power and popularity, is the so-called *l*ocal *d*ensity *f*unctional (LDF) method, the basis of which is the derivation by use of the variational principle of the *electron density* for the system, following a seminal study in 1964 by Hohenberg and Kohn, who established that the energy is a unique functional of the density of a system. With this approach, recently validated by Parr (1989) and Baker (1995), it is possible to calculate exactly the kinetic and 'Hartree' (electron–electron repulsion) energies. Exchange and correlation terms are evaluated in simpler approaches by using the local density approximation, i.e. these quantities are calculated as a function of the electron density using expressions appropriate for an electron gas of constant density. In practice, however, modern approaches add terms correcting for gradients in the density and for non-local effects. LDF methods have the advantage over Hartree–Fock methods of being more economical in computer time, thus allowing larger systems to be investigated. It is difficult, however, to quantify precisely the errors introduced by the local density approximation, although the accuracy of the results achieved by many LDF calculations suggests that these are small.

Calculations employing both approaches may be performed either on clusters or on periodic arrays of atoms. The latter is clearly more appropriate for modelling perfect crystal structures, while the former is well suited to describing defects and localised states, in particular active sites in zeolites or other catalysts. However, considerable care must be exercised in the choice of cluster size and in the approach used to 'terminate' the cluster and in particular to 'saturate' the dangling bonds of the peripheral atoms. If possible the cluster should be 'embedded' in an approximate representation of the surrounding lattice. The simplest procedure is to surround the cluster by point charges which at least describe the long-range Coulomb field in the surrounding lattice. More sophisticated procedures represent the embedding matrix by using a 'simulation' approach (i.e. employing interatomic potentials), although there are difficulties associated with the linking of the quantum mechanical cluster and the embedding region in an appropriate way.

The use of electronic structure methods is expanding rapidly with the growth in computer power. However, in many applications, simulations employing interatomic potentials are the most appropriate and economical methods for modelling structural, elastic, dielectric and lattice dynamical properties of crystals. They are, moreover, able to handle highly complex systems containing a large number of atoms.

As we saw in Chapter 2, simulation techniques comprise energy (and free-energy) minimisation procedures, Monte Carlo (MC) and molecular dynamics (MD) methods. All have been fruitfully used in crystal structure modelling. In the first (and simplest) method, we start by calculating the lattice energy (U), which is defined as the energy of the crystal with respect to component ions at infinity (ions are a sensible reference point for ionic and semi-ionic solids):

$$U = \frac{1}{2} \sum_{i} V_i \tag{2}$$

where the summation is taken over all ions in the unit cell and where V_i is the site potential of the ions which is evaluated by calculating its interaction energy with all other ions:

$$V_i = \sum_{ij} U_{ij} + \sum_{ijk} U_{ijk} + \cdots \tag{3}$$

where the first term represents the summation over all ion pairs, the second all ion triplets; higher terms may also be included, but this is relatively uncommon. Indeed, the majority of calculations include only the first term – the so-called pair potential approximation. This in turn may be usefully decomposed into Coulombic and non-Coulombic terms:

$$U_{ij}(r_i, r_j) = \frac{q_i q_j}{r_{ij}} + \phi_{ij}(r_{ij}) \tag{4}$$

where q_i is the charge on the atom or ion. The Coulombic term, which is purely two-body in nature, is handled by standard procedures (in particular the Ewald summation technique discussed along with other relevant procedures by Catlow (1995)). The non-Coulombic ϕ_{ij} is then usually approximated by analytical functions which generally include both attractive and repulsive components; the latter describes the Pauli repulsion due to overlap of closed-shell electron configurations, and the former attractive terms from dispersion (i.e. induced dipole–induced-dipole) and covalence effects. Several such functions are available, notably the Lennard-Jones potential:

$$V(r) = Ar^{-12} - Br^{-6} \tag{5}$$

the Morse potential:

$$V(r) = D\{1 - \exp[-\beta(r - r_e)]\}^2 \tag{6}$$

and the Buckingham potential:

$$V(r) = A\exp(-r/\rho) - Cr^{-6} \tag{7}$$

Morse potentials are more appropriate when covalent systems are being studied; D may then be interpreted as the covalent bond dissociation energy and r_e the equilibrium bond length. Buckingham potentials have been very widely used in the study of ionic and semi-ionic solids.

Inclusion of many-body effects may be achieved by the use of angle-dependent forces. A commonly used example is the simple 'bond-bending' terms of the type

$$E(\theta) = \frac{1}{2}k_B(\theta - \theta_0)^2 \tag{8}$$

where θ is the angle subtended by three atoms and θ_0 is an equilibrium value; k_B is the appropriate force constant. The use of such terms is most appropriate in systems with directional covalent bonding, e.g. silicates, where θ_0 will correspond to the tetrahedral angle subtended by O–Si–O bonds in the tetrahedral SiO_4 groups.

We note that atomic and ionic polarisation in solids is also essentially a many-body effect. Moreover, in ionic materials polarisation energies may be large, especially in the vicinity of charged defects and surfaces. The simplest way of modelling polarisability is to use the point-polarisable ion (PPI) model, which assigns a linear constant of proportionality, i.e. the polarisability, α, between the dipole moment μ and the field E acting on the atoms; that is,

$$\mu = \alpha E \tag{9}$$

However, as the dipole is a point entity, it cannot describe the physical basis of polarisability, which is the displacement of the valence-shell electron density in response to the applied field. These deficiencies are largely overcome by the shell model, which describes the development of a dipole moment in terms of the displacement of a massless shell of charge Y (representing the valence-shell electrons) from a core (representing the nucleus and core electrons), the core and the shell being connected by a harmonic restoring force for which the spring constant is k; the free-atom polarisability, α, is then given by $\alpha = Y^2/k$. Shell-model potentials have been very widely used in studies of lattice dynamic and defect properties of ionic catalysts.

Potentials of the type described above rest on the Born model description of ionic and semi-ionic solids, and as is clear from the above discussion, are based on the idea of ions (which may be polarisable) interacting via Coulomb forces and short-range pair potentials (which may be supplemented by three-body or even by higher-order terms). A contrasting approach is provided by molecular mechanics force fields, the conceptual basis of which is the covalently bonded network with energy terms associated with bond stretching and bending and with non-bonded interactions. For some systems, such as silica and silicates, both types of model may be sued, and, interestingly, as the work of Catlow, van Santen and Sauer shows, similar results are obtained despite the different approaches adopted.

Once the type of potential model to be used is established, it is next necessary to fix the variable parameters, i.e. those used in the description of the short-range potential $V(r)$, the shell model constants Y and k, and the effective atomic charges q (although we note that in many modelling studies of ionic oxides, and silicates, these have been fixed at integral fully ionic values). This is undertaken by two procedures: first, empirical methods, in which variable parameters are adjusted, generally via a least-squares fitting procedure, to observed crystal properties. The latter must include the crystal structure (and the procedure of 'fitting' to the structure has normally been achieved by minimizing the calculated forces acting on the atoms at their observed positions in the unit cell). Elastic constants should, where available, be included; and dielectric properties are required to parameterise the shell-model constants. Phonon dispersion curves also provide valuable information on interatomic forces; and force-constant models (in which the variable parameters are first and second derivatives of the potential) are commonly fitted to lattice dynamics data. (Empirically derived potential models are tested against phonon dispersion curves when the latter are available.) The second approach involves short-range potentials being calculated directly by use of ab-initio cluster or periodic boundary techniques, where the geometry is varied in a systematic manner with an effective potential being fitted to the resulting energy surface.

Having calculated the lattice energy using interatomic potentials, one then employs a minimisation algorithm, which adjusts all structural variables (cell dimensions and atomic coordinates) until a structure of minimum energy has been generated. This method is remarkably effective in simulating complex inorganic structures. It has, however, inherent weaknesses, first in its omission of explicit dynamic effects and second in the 'local minimum' problem, namely the fact that a minimisation procedure can only guarantee convergence to the nearest local minimum.

Fortunately an optimisation procedure which is much less susceptible to local convergence problems is available. This method, known as simulated annealing (introduced by Kirkpatrick et al. in 1983) introduces the possibility of accepting 'uphill' or energy-increasing moves in the minimisation of a given function. To control the fraction of energy-increasing modifications which are made, an analogy with the process of annealing a physical system may be drawn. Here the relative probabilities of existence of two states are readily represented at a given temperature by Boltzmann statistics. The temperature of the systems governs the population of high-energy states. On raising the temperature, states with widely differing energies will have increasingly similar probabilities, while at low temperatures, low-energy states will tend to dominate. By annealing a physical system in this fashion, the elevated temperature permits a range of states to exist and the subsequent gradual reduction in temperature shifts the equilibrium until low-energy states dominate. By equating the energy of the physical system to the object function to be minimised, with the temperature controlling the relative probabilities of two states, simulated annealing may be used as a general procedure for energy minimisation. (Because the system is permitted to sample a wide range of states in the high temperature of the procedure, simulated annealing is less susceptible to becoming trapped in local minima than traditional gradient minimisation methods).

The searching algorithm used in the majority of annealing studies is based on the so-called Metropolis algorithm of the Monte Carlo scheme, in which a particle is selected at random and its coordinates updated in a random fashion. The energy change associated with this perturbation in the structure is evaluated and the move either accepted or rejected on the basis of a comparison of the Boltzmann probability for the transition at the current temperature with a random number between 0 and 1. Application of this procedure over a sufficiently large number of moves leads to the generation of states which are representative of the equilibrium physical system at the specified temperature. Several schemes for governing the annealing of the system have been developed. For example, the effective heat capacity can be monitored to detect the onset of melting and freezing. Alternatively, the fraction of successful Metropolis moves may be monitored to determine the rate of cooling applied to the system. In addition to the use of the Monte Carlo procedure in crystal structure simulation, recent methods based on 'genetic algorithms' have been employed in which, as discussed later, structures 'evolve' towards configurations with predefined structural characteristics. In addition, molecular dynamics procedures can provide an effective means for exploring the physical structure of crystal systems. Such methods have proved successful in the refinement of data relating to sorbed molecules in zeolite pores. (In the molecular dynamics method, the actual equation of motion is solved numerically for each particle by calculating the force $\boldsymbol{F}$ on each particle due to all the other particles, and integrating the equation $\boldsymbol{F} = m\boldsymbol{a}$ (i.e. mass times acceleration). The trajectories of all the particles then yield a dynamic picture of how the assembly of particle changes with time.)

All the procedures described above are based on the notion of optimising the potential energy of the crystal calculated by an interatomic potential function. It is to be emphasised that the quality of the models used in describing the interatomic force is vitally important in determining the reliability of the resulting predictions.

But optimisation procedures may use target functions other than the potential energy of the crystal. Indeed, traditional crystal structure refinement methods minimise the discrepancy between calculated and experimental diffraction intensity. This procedure may be blended with some of the approaches described above, as in the powerful 'reverse Monte Carlo' technique, which refines structural data using an MC algorithm in which, however, the energy function is replaced by the deviation of calculated from experimental diffraction data.

Modelling techniques may also be used to probe more complex problems than those posed simply by the determination of crystal structure. Minimisation techniques have been used for many years in studies of both bulk and surface defects, and indeed methods based on the original approach of Mott and Littleton in the late 1930s in which relaxation of a region of structure surrounding the defect are coupled with a quasi-continuum model of the response of the remaining crystal. Such methods may be adapted to the study of surface defects and are likely to have a considerable role to play in modelling catalytic processes.

Another problem to which modelling techniques have been successfully applied concerns the determination of the energetics and location of sorbed molecules within microporous hosts.

5.3.1.1 Selected Applications

We consider three types of structural modelling: first, simulation of the three-dimensional structures of catalytic materials, including modelling of crystal structures of localised states (defects and impurities) within crystals; second, simulation studies of oxide surfaces, considering again both perfect and defective systems; and, third, the modelling of molecules within microporous solids and on surfaces.

Modelling of Three-Dimensional Structures

Since the mid-1970s, modelling of silicates and aluminosilicates has been a particularly active field owing both to their catalytic or adsorptive importance and to the complexity and variety of the structures involved. Of special interest are the microporous zeolites and aluminophosphates, where use of the Born model and molecular mechanics potentials has made it possible to model accurately the complex crystal structures of these solids.

An interesting example where atomistic modelling played an important role in structure solution was that of the microporous metal-containing aluminophosphate MeAPO-36, now used commercially for catalytic dewaxing. Like the zeolites, the MeALPO family of molecular sieves is composed of three-dimensional networks of corner-sharing TO_4 tetrahedra, where T represents mainly aluminium and phosphorus in the ratio 1 : 1 (although the aluminium can be substituted by minor amounts of metals (magnesium, zinc, cobalt, nickel, manganese, etc.) which, as described earlier, impart catalytic properties to the materials). A combination of 'molecular probe' gas adsorption and high-resolution electron microscopy on the microcrystalline MeAPO-36 indicated that the solid possesses a one-dimensional 12-ring channel system similar to that of ALPO-5, and consideration of the unit cell size and symmetry derived from electron diffraction suggested a model for the structure (Fig. 5.42). This model was then refined by a distance least-squares technique using experimentally derived unit-cell parameters and the maximum possible symmetry (*Cmcm*) which assumes disordered Al and P. The simulated X-ray diffraction pattern

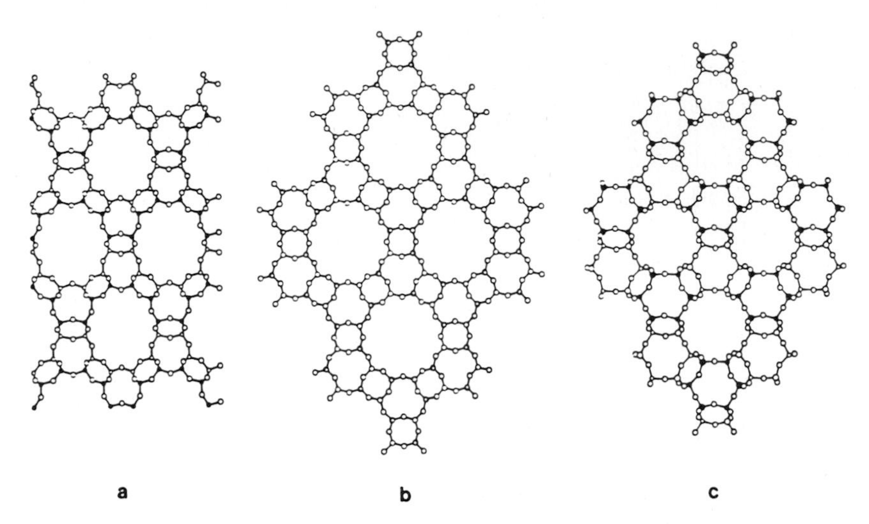

Figure 5.42 (a) and (b) Projections down [001] of the high-symmetry (Al and P disordered) models for MeALPO-36 (*Cmcm*) as a generalized structure and calcined ALPO-5, respectively; (c) projection of the structure of aluminosilicate cancrinite. For (a) and (c) the open circles representing tetrahedrally coordinated atoms are separated in the direction of the z axis by a distance of $c/2$ from those tetrahedrally coordinated atoms represented by closed circles. In (c) the smaller tetrahedrally surrounded atoms represent Al, the larger Si. Reprinted from P. A. Wright, S. Natarajan, J. M. Thomas, R. G. Bell, P. L. Gai-Boyan, R. H. Jones, J. Chen, *Angew. Chem., Int. Ed. Engl.* **1992**, *31*, 1473.

closely resembles experimental patterns collected at high temperature (greater than 350 °C) but does not explain peak splittings observed at lower temperatures and indicative of lower symmetry (Fig. 5.43).

Lattice energy minimisation calculations were then carried out using the so-called METAPOCS program in which no symmetry constraints were applied, and all atomic positions and cell parameters were allowed to vary. The T atoms were replaced by Si in a first simulation, and by ordered Al and P in a second calculation. The 'SiO_2' simulation used silicate potentials and included the effect of polarisability. The '$AlPO_4$' simulations used the partial charge potentials (of van Santen and co-workers).

While the 'SiO_2' simulation retained *Cmcm* crystallographic symmetry, the ordered '$AlPO_4$' simulation resulted in a reduction of space-group symmetry to C_1 and a distortion of the unit cell. The simulated XRD pattern closely resembles that observed at 150 °C, although the simulated unit cell ($a = 13.46$ Å, $b = 22.17$ Å, $c = 5.29$ Å, $\alpha = 92.2°$, $\beta = 92.0°$, $\gamma = 90.0°$) is about 2.5% larger than that observed experimentally ($a = 13.16$ Å, $b = 21.65$ Å, $c = 5.19$ Å, $\alpha = 90.2°$, $\beta = 92.0°$, $\gamma = 90.0°$). The 'solved' structure is shown in Colour Plate 7.

This computational approach, it should be noted, enables us not only to establish structural details in the absence of single crystal data, but also to predict them on the basis of a physically realistic model.

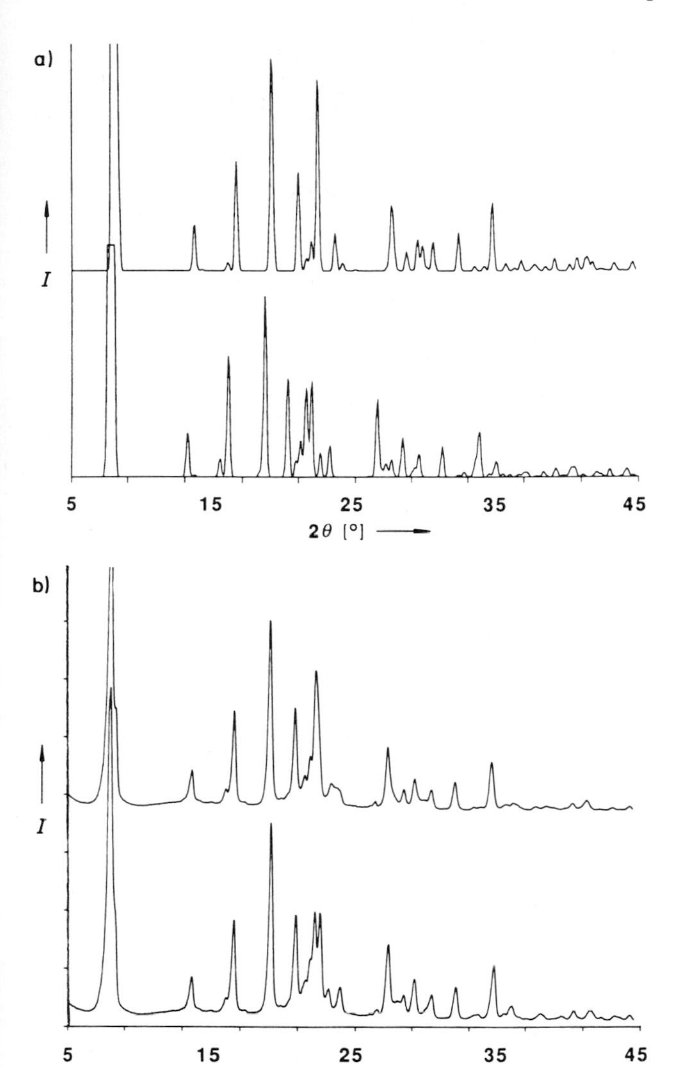

Figure 5.43 (a) Simulated X-ray diffraction patterns for the energy-minimized structures of SiO_2 (top) and $AlPO_4$ (bottom). (b) Experimental X-ray diffractograms for MgALPSO-36 at 375 °C (upper) and 150 °C (lower). *I* is the X-ray intensity in arbitrary units. Reprinted from P. A. Wright, S. Natarajan, J. M. Thomas, R. G. Bell, P. L. Gai-Boyan, R. H. Jones, J. Chen, *Angew. Chem., Int. Ed. Engl.* **1992**, *31*, 1473.

Extensive simulation (as well as quantum mechanical) studies have been made of the bridging hydroxyl groups – the root cause of Brønsted acid activity (see Fig. 5.41) of zeolitic catalysts – and precise details of the structure, infrared-active vibrational frequencies and proton affinities at such catalytic sites have been quantitatively determined using procedures developed by, *inter alia*, Catlow, Sauer and Van

Santen. It has been shown, for example, that when Fe replaces Al as the heteroatom in the framework of ZSM-5 (see Fig. 5.41), the bridging hydroxyl has a more strongly bound proton, a fact which rationalises the diminished acidity of (Fe)ZSM-5 in comparison with (Al)ZSM-5.

Modelling Surfaces

Following the pioneering work of Tasker, Colbourn and Mackrodt in the late 1970s, there have been several applications of minimisation methods to surface structural studies. The methods used entail cutting the crystal (computationally) so that a region within typically 5 to 15 layers of the surface is relaxed to equilibrium; the remainder of the crystal remains rigid – an entirely acceptable approximation, provided a sufficiently large relaxed region is used.

Early calculations on simple cubic oxides such as MgO demonstrated the occurrence of small but significant 'rumpling' effects in which there are differential displacements of the oppositely charged ions at the surface.

Recent calculations have shown that, in a large number of materials, impurities have lower energies at surface than bulk sites; this provides the driving force behind the widely observed phenomenon of surface segregation in which impurities are found preferentially to occupy surface layers. A recent example is the work of Sayle, Parker and Catlow which has shown that both Pd^{2+} and Pt^{2+} ions are extensively segregated to both the (110) and (111) surfaces of CeO_2, a result of relevance in the context of understanding the Al_2O_3/CeO_2/noble metal three-way auto-exhaust catalysts.

Several atomistic studies have demonstrated the importance of surface relaxations in determining the correct relative ordering of the surface energies of the various high-symmetry faces of corundum, Al_2O_3. The fully relaxed ordering for the five low-index surfaces being considered here (according to the electron-gas potential of Mackrodt) is:

$$(0001) < (10\bar{1}0) < (10\bar{1}2) < (11\bar{2}0) < (10\bar{1}1)$$

The basal plane (0001) is strongly stabilised by relaxation; in particular, the outermost aluminium is displaced towards the surface layer of oxygens by some 60% of the original spacing.

Computational chemistry, in the hands of Gay and Rohl, has been used to compute the equilibrium morphologies of many catalytically important solids, especially binary and ternary oxides. As yet no allowance has been made for the influence of solvent. Nevertheless, interesting insights have emerged.

Modelling the Structures and Energies of Sorbed Species

Direct experimental determination of the sites occupied by sorbed species is often difficult, although there have been a number of successes (as discussed in Chapter 3) employing high-resolution powder techniques with Rietveld refinement of both neutron and synchrotron data, NMR and EXAFS but, in general, unambiguous determination of the sorption site by these methods is not easy to achieve. There is therefore a strong incentive for the development of reliable theoretical tools for modelling sorption sites. An added bonus is that calculation of the associated energies may then be compared with thermodynamic data.

The whole range of simulation tools may, of course, be used in studying sorption; and great success has been achieved, especially by Bell and his school in Berkeley and by Smidt in Amsterdam, using sophisticated Monte Carlo and molecular dynamics simulations in modelling thermodynamic and transport properties of sorbed molecules. Here we focus on the use of simpler and more routinely applicable techniques for the identification of low-energy sites for sorbed molecules. This approach is complementary to the use of the regular MC and MD methodologies.

Minimisation methods have been used for several years in identifying low-energy sorption sites. (An early success was the work of Wright et al., who used simple minimisation procedures to identify sorption sites for pyridine in zeolite L. The site identified agreed well with those obtained from a Rietveld analysis of powder diffraction data, illustrating nicely the way in which simulation studies of this type may complement the analysis of experimental diffraction data.) Standard minimisation procedures have been used to study the sorption of linear alkanes in silicalite and purely siliceous faujasite. An important technical development in these studies was the inclusion of framework relaxation around the sorbed molecules, which was

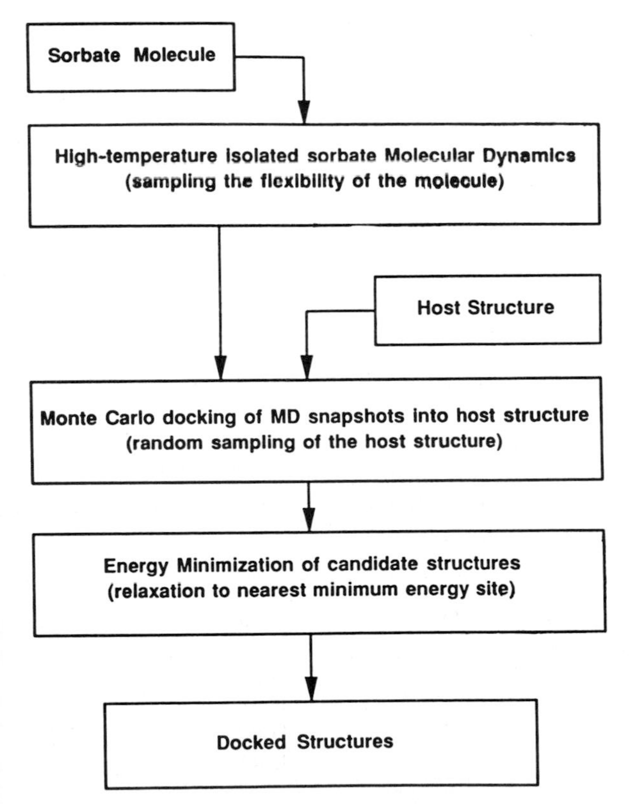

Figure 5.44 Summary of the 'docking' method for modelling the structures and energies of sorbed species. Reprinted from C. M. Freeman, C. R. A. Catlow, J. M. Thomas, S. Brode, *Chem. Phys. Lett.* **1991**, *186*, 137.

found to influence significantly the calculated sorption energy. Moreover, from these and related studies it has become clear that there are difficulties in using straightforward minimisation procedures to identify the lowest-energy sorption sites in these complex systems. The difficulties arise from the well-known 'multiple minimum' problem, i.e. the profusion of local minima in complex structures. This problem was overcome by a simple and ingenious procedure developed by Freeman et al., in which molecules are 'docked' in an automated manner into a zeolite host. The procedure involves three stages (see Fig. 5.44). Firstly, a molecular dynamics calculation is performed on the isolated molecule in order to generate a library of low-energy conformation states; secondly, a Monte Carlo method is used in which each configuration of the molecule is introduced into a succession of sites which are randomly generated within the zeolite. Their interaction energies with the zeolite cage are calculated and those which fall below a specified threshold value are accepted; this latter value is set to eliminate all conformations that are unacceptable due, for example, to being too close to the zeolite cage. Thirdly, a minimisation calculation is undertaken on each of the accepted configurations. In the initial work of Freeman et al., (1991), the zeolite cage was held rigid during the course of this minimization; but more recently, Shubin et al. have allowed the cage to relax during this final stage of the calculation. Provided the threshold energy is correctly set, the method is almost guaranteed to find the global energy minimum.

This approach has been applied to the problem of exploring the energetics (and sorption sites) of the four isomers of butene inside MFI (ZSM-5) which, for simplicity, was idealized in the computation to be the siliceous extreme (i.e. silicalite). It is immediately apparent from Table 5.2 that isobutene (2-methylpropylene) is less firmly bound to the MFI framework than any of the other three isomers. This immediately suggests that MFI is a good candidate catalyst for the isomerization of butene into 2-methylpropylene (the more desirable isomer since it is required for the production of methyl *t*-butyl ether (MTBE) by catalytic addition of methanol).

Similar computations show that other zeolitic frameworks, notably TON and FER, are equally good candidates for this isomerisation – again the binding energies of 2-methylpropylene are lower than for the other three isomers – thereby leading to its more facile diffusion through and out of the channels of the zeolite. (We note

Table 5.2 Calculated absolute and relative butene isomer binding energies in ZSM-5.

Isomer	Binding energy [kJ mol^{-1}]			
	Minimum		Average	
	Absolute	Relative	Absolute	Relative
2-Methylpropylene	−38.58	0.0	−28.09	0.0
But-1-ene	−54.30	−15.72	−44.41	−16.32
cis-But-2-ene	−45.45	−6.87	−31.64	−3.55
trans-But-2-ene	−48.76	−10.18	−38.44	−10.35

in passing that FER and TON are superior catalysts to ZSM-5 for this important isomerisation because neither of them has cross-channel intersections – like MFI – where unwanted dimerisation of the butenes could occur. Both FER and TON are now used as catalysts in the industrial production of 2-methylpropylene.)

The docking methodology has also been used, by Lewis and others, to shed light on the fundamental factors that govern the production of particular microporous structures from gel and other mother liquor precursors via selected templating organic molecules. Calculations in which the template molecule is 'docked' within a zeolite framework host suggest that, provided due account is taken of the interactions between the templating molecules inside the host, the maximum calculated binding energy for a given structure is the best criterion for the successful template for that structure. Although there is much experimental support for this important conclusion it is, as yet, premature to regard it as the crucial factor in the strategy of templating.

Cautionary Notes

Although great strides have been made through computer modelling in the solid-state and surface chemistry of catalysts, it would be naive to imagine that one will soon be able to 'enter' reactants, a hypothetical catalyst and desired products into a computer and to let the ensuing computation serve as a critical test and assessment for that catalyst. Computational studies of most chemical reactions still require a drastic reduction of the complexity of the system under study in order to render it amenable to present-day tools. This is mainly due to four causes. First of all, the unfavourable scaling behaviour of most modelling tools (in particular the quantum chemical ones); secondly the usually very large number of degrees of freedom possessed by reactant molecules for choosing a reaction coordinate; thirdly the very large number of 'trials' needed to be made by reactant molecules in order to surmount activation barriers; and last but not least, the very high degree of energetic accuracy or at least precision required to arrive at useful results.

With few exceptions, computational studies of reactions have been and are still done 'in vacuum at zero Kelvin', i.e. the influence of solvents and temperature for instance is usually not explicitly taken into account. Reaction coordinates are frequently predefined by chemical intuition or the search space is deliberately restricted beforehand. Overcoming these limitations continues to be a big challenge for computational chemistry. Often, classical force-field methods (i.e. molecular classical mechanics) and quantum chemical tools are combined to that end. Quantum molecular dynamics methods have been proved to possess potential for increasing the complexity of systems that can be studied (e.g. to include ionic surfaces, and solvent and temperature effects), although as yet they appear limited with respect to the activation barrier heights they can handle.

Significant effort is being expended on the development of so-called 'onion' or 'hybrid' models: these are models which tackle a system by means of quantum chemistry in the core (e.g. an active centre), then employ, for instance, semi-empirical methods followed by force-field calculations and/or finally a continuum model for the outermost shell. However, fundamental problems have still to be solved in order to impose the results of the 'outer shells' of these models as constraints on the 'inner' ones.

Some of the severe barriers to progress in modelling catalysis were recently identified by Kuipers and van Santen as follows.

(a) Modelling the chemistry of transition metal surfaces is extremely demanding.
(b) The size of molecular complexes that can be handled is still too limited for most homogeneous catalytic cycles.
(c) Solvent and medium effects in general cannot yet be included in catalytic cycles.
(d) (Competitive) physisorption effects, microscopic transport phenomena and shape-selectivity, as particularly relevant in zeolite catalysis, cannot yet be quantitatively integrated in the modelling of catalytic cycles.
(e) Searching the lowest equilibrium and transition states sometimes relies on wasteful brute-force methods.

Most of the quantum chemical, statistical chemical and dynamics methods that are currently extensively used in computational chemistry have been available since the early 1970s. Current successes stem largely from dramatically improved hardware facilities. Most critical is the accuracy of the electronic structure calculations (of the type adumbrated in Section 5.4) and the optimisation of geometry to determine the minimal potential-energy surfaces. Currently computations can be performed for transition metals on clusters of the order of 20 atoms. State-of-the-art calculations show accuracy in computed interaction energies of the order of 20 kJ/mol^{-1}. Therefore the primary factor that will determine future advances is the availability of computer resources and software that allow a size increase of the systems to be studied by a factor of 10 to 100. Clusters of this size are large enough to model several unit cells of complex systems.

Another fundamental consideration that has to be borne in mind is that a computer simulation of the overall kinetics of a catalytic conversion requires the integration of steps that have very different timescales. Approaches need to be developed that solve equations in the proper time domain using parameters obtained from studies on a much shorter timescale.

With molecular sieve catalysis it is often the case that an elementary step may be slow, but, nevertheless, diffusion within the pores may still be of importance when the dimensions of the catalyst particle are large. Moreover, the diffusivities may be strong functions of the concentration of reactant (product) species. In addition, the desorption rate of the product may also be a slow step. In such situations the overall rate of reaction has to be solved by addressing the differential equations that contain values for the diffusional constant which are obtained from separate simulations of molecular diffusion.

Massively parallel computers promise a significant increase in computational power, especially when broadly applicable parallelized algorithms are a reality. At that time molecular dynamics calculations of longer timescale on very large systems will become possible. However, their application will require accurate classical interaction potentials which will doubtless be developed from accurate quantum mechanical electronic structure calculations that focus on the precise evaluation of van der Waals interactions.

Notwithstanding the limitations that still face computational chemistry and the vast progress that must be made before new catalysts for a prescribed reaction are identified, tested and improved by modelling and simulation, it must not be forgotten that substantial advances have already been made. Atomistic simulations of solid-state structures and properties, the computation of idealized high-resolution electron-microscopic images of supposed structures, the determination of crystal structures (including the iterative analyses of X-ray and neutron diffractograms as well as NMR and X-ray absorption spectra) and the computational derivation of good-quality electronic properties are a reality. The fact that commercial companies providing modelling strategies and software packages are a prominent feature of the landscape of catalysis is itself significant.

5.4 A Chemist's Guide to the Electronic Structure of Solids and Their Surfaces

To understand why some catalysts facilitate electron transfer and others do not, and to appreciate why and how some molecular species can be readily bound to and transformed at a catalyst surface, we need to be acquainted first with the electronic structure of bulk solids and then with the principles which govern redistribution of electrons at a surface during chemisorption. Knowledge of the electronic structure of bulk solids assists in the classification of catalysts (Section 5.1); it also helps rationalise why certain metals are superior catalysts, why certain oxides are efficient photocatalysts, why certain materials are good catalyst supports, and how promotion and poisoning occurs. Knowledge of the electronic properties of surfaces is central to the whole phenomenon of catalysis, and progress in the design of better catalysts is dependent upon it.

Quantum mechanical techniques assisted by experimental procedure such as those discussed in Chapter 3 enable us to determine the energetics of electrons inside solids and to evaluate the widths and separation of energy bands as well as the distribution of energy levels in a given energy interval within these bands (i.e. the density of states). Quantum chemical techniques; especially LDF methods, also permit determination of the structure and energy of stable and reactive intermediates at catalyst surfaces. Indeed, the transitory nature of some of the intermediates implicated in catalysis are such that their structures can best be ascertained by computational procedures, rather than by direct, experimental ones.

It is not our intention in this chapter to delve thoroughly into the entire sweep of quantum mechanical calculations as they affect catalysis. The reader is referred to two very different and admirable monographs for further enlightenment in this regard: one by Hoffmann on *Solids and Surfaces* (1988), the other by van Santen on *Theoretical Heterogeneous Catalysis* (1991). Both of these, especially the former, highlight the importance of chemical intuition and the advantages of approximate methods. Specific references to ab-initio and other sophisticated approaches are cited in the Further Reading section that terminates this chapter.

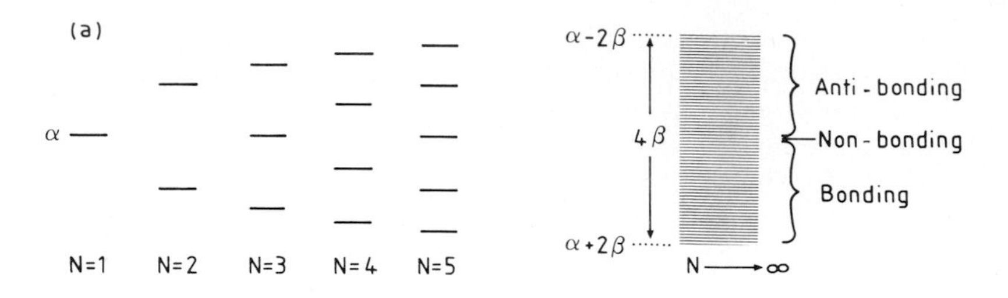

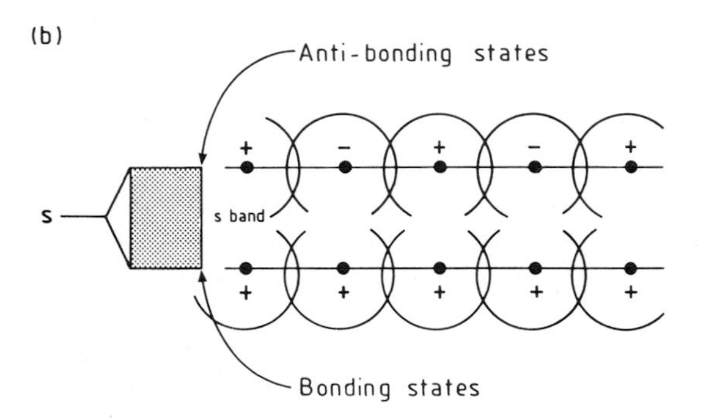

Figure 5.45 (a) The interaction of orbitals on neighbouring atoms in a solid gives rise to energy bands. The width of the band is designated 4β (see text). (b) The top and bottom of an *s* band consist of antibonding and bonding states, respectively.

5.4.1 Energy Bands

Whereas isolated atoms and small molecules have sharply defined energy levels corresponding to various orbital states, in solids – where there are myriad atoms – the interaction between orbitals on neighbouring atoms give rise to bands of energy, as schematized in Fig. 5.45(a). Here we picture, for simplicity, an infinite chain of N atoms (e.g. H or Li or Na) and the ensuing overlap of 1s, 2s or 3s orbitals. The mean energy of the band, α, is the energy of the s orbital on an isolated atom and is the so-called Coulomb integral:

$$\alpha = \int \phi_j H \phi_j \, \mathrm{d}\tau = \langle \phi_j | H | \phi_j \rangle \tag{10}$$

where H is the Hamiltonian and $\mathrm{d}\tau$ denotes integration over the whole three-dimensional space. (It is assumed that each lattice site j is represented by one orbital ϕ_j.)

By definition, the resonance integral β (otherwise known as the bond integral or transfer integral, sometimes designated t), is given by:

$$\beta = \int \phi_j H \phi_{j+1} \, \mathrm{d}\tau \tag{11}$$

It can be shown that the width of the energy band is 4β. What, in effect, governs the width is the degree of interaction between neighbouring atoms: the closer the atoms,

(a)

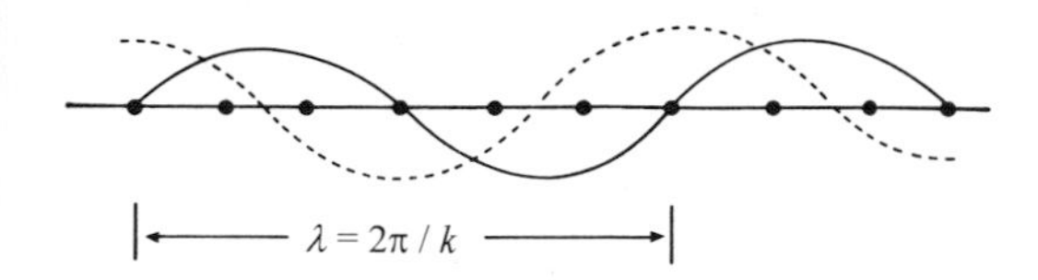

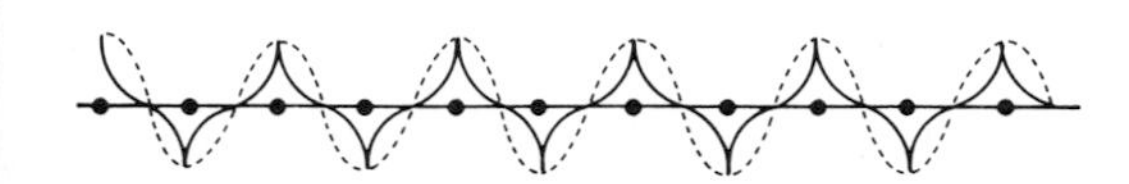

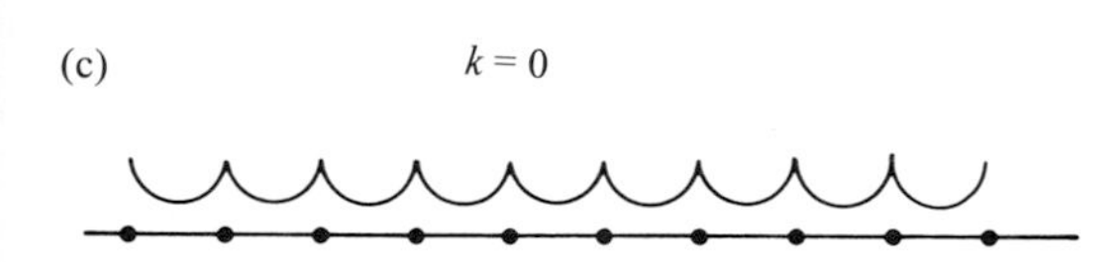

Figure 5.46 Wave functions for electrons along a chain. (a) Real (full line) and imaginary (broken) parts of a free-electron function. Bloch functions for different values of k formed by orbital overlap are shown in (b) and (c).

the greater the overlap and the wider the band. As a rough rule, the separation of the energy levels at the dimer ($N = 2$) stage corresponds to about half the width of the energy band ($N = \infty$). In effect, the energy band is composed of an infinite number of molecular orbitals extending over our N atoms, the separation of the energies of these orbitals being infinitesimal. We now have crystal (one-dimensional) orbitals or wave functions. All functions have a wavelike from, and the wavelength λ is determined by a quantum number k such that $\lambda = 2\pi/k$. When the wavelength has a value of π/a (i.e. two lattice spacings: see Fig. 5.46b) the atomic orbitals on adjacent atoms are out of phase with one another; when $k = 0$ the wavelength is infinite. The quantum number k can therefore be regarded as a label for the crystal molecular orbitals; and since it is the inverse of wavelength it is also called the wave vector. For the atomic array in Fig. 5.45(a) the crystal molecular orbitals ψ_k may be written:

$$\psi_k = \sum_N \exp(\mathrm{i}kNa)\phi_N \tag{12}$$

Clearly when $k = 0$,

$$\psi_0 = \phi_0 + \phi_1 + \phi_2 + \cdots = \sum_N \phi_N \tag{13}$$

and when $k = \pi/a$

$$\phi_{\pi/a} = \phi_0 - \phi_1 + \phi_2 - \phi_3 + \cdots = \sum_N (-1)^N \phi_N \tag{14}$$

The crystal wave function corresponding to $k = 0$ is the most bonding; the one for π/a the most antibonding. For a linear array of p orbitals the band that forms has its most antibonding combination at $k = 0$ and the most bonding combination at $k = \pi/a$:

$$\psi_0 = \phi_0 + \phi_1 + \phi_2 + \phi_3 + \cdots \quad (13a)$$
$$\ominus\oplus \quad \ominus\oplus \quad \ominus\oplus \quad \ominus\oplus$$

$$\psi_{\pi/a} = \phi_0 - \phi_1 + \phi_2 - \phi_3 + \cdots \quad (14a)$$
$$\ominus\oplus \quad \oplus\ominus \quad \ominus\oplus \quad \oplus\ominus$$

The number of allowed values of k is the number of translations (a) in the (1D) crystal. There are as many values of k as there are microscopic unit cells in the macroscopic crystal. There is an energy level, E, for each value of k – to be precise there are two for every value of k because we have degeneracy arising from the fact that $E(k) = E(-k)$. The space of k is obviously reciprocal space (cf. Section 5.2.1.1); and it is to be noted that the range of unique k given by $0 \leq k \leq \pi/a$ is called the Brillouin zone.

From a plot of E against k, known as the band structure, we may construct *d*ensities-*o*f-*s*tates plots (DOS) (Fig. 5.47) which tell us the relative bunching of energy levels within a given interval of energy. In general, the DOS plot is related to the inverse of the slope of the E/k plot: the flatter the band, the greater the number of levels at that energy.

For the linear array of either H, Li or Na, each atom contributes one electron. But since each energy level in the band may accommodate two electrons, the s band so formed is only half-filled. Pouring electrons into the near-continuum of energy levels in the band in conformity with the Pauli principle fills them up to an energy E_F, the Fermi energy. In the language of frontier orbitals, the Fermi energy represents the highest occupied molecular orbital (HOMO): it is also the energy of the lowest un-

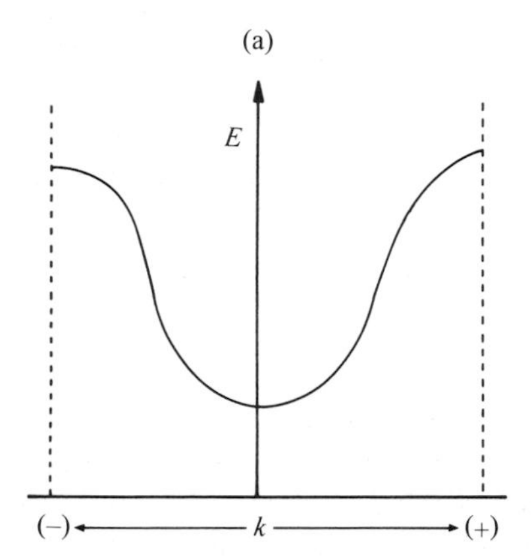

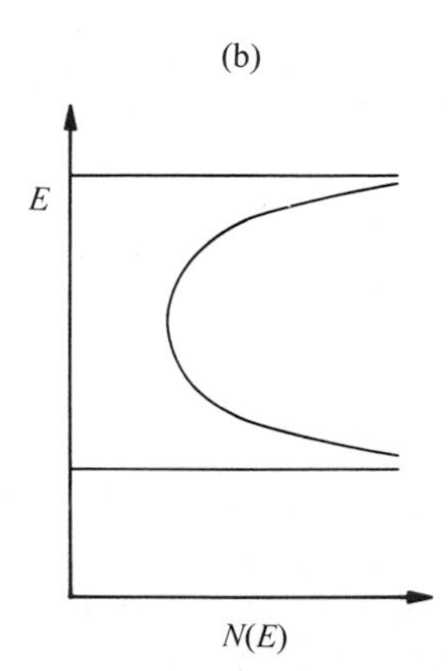

Figure 5.47 (i) $E(k)$ plot showing the k values for a chain of closely spaced atoms. (ii) The distribution of orbital energies, represented in a density of states plot, showing clustering at the top and bottom of the bond, for very large numbers of atoms.

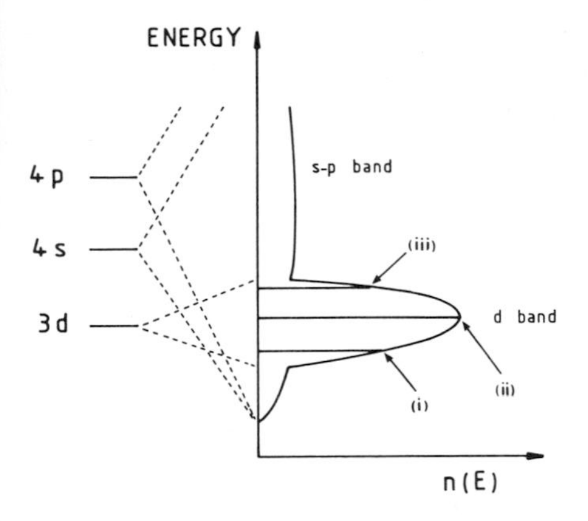

Figure 5.48 Typical density of states (DOS) plot for a transition metal in the first series. Horizontal lines show the position of the Fermi level early (i), in the middle (ii) and late (iii) in the series.

occupied MO (LUMO). And it follows that the work function of a metal is the energy required to take an electron from the Fermi level (i.e. E_F) to just outside the metal surface. We now understand what it is, electronically, that characterizes a metal: it is an incompletely filled energy band. When an electric potential difference is applied to a metallic solid, the electrons at the Fermi level (at the surface of the Fermi sea) can gain energy by promotion to empty higher levels. In an insulator, where the band is full, no such response is possible.

Superficially, it would appear that atomic arrays of Be, Ca or Ba, where each atom contributes two electrons to every energy level, should be insulating since their s bands would be precisely filled. Such arrays would still be metallic, however, because the s and p bands of these metals overlap, thereby leaving plenty of empty levels in the resulting s–p band.

Extending the above arguments to three dimensions, it is easy to see that, in principle, a transition metal in the first series will have a DOS plot such as that schematized in Fig. 5.48. The overlap of the d orbitals on the constituent atoms is naturally smaller than for the corresponding spatially more extended s and p orbitals; consequently the width of the d band is smaller than that of either the s or p band. Equally, because the number of d orbitals per atom is greater than the number of either s or p orbitals, these narrower bands attain higher values of $n(E)$ for a given E. It follows that early and late members of any series of transition metals have Fermi levels, E_F, at relatively low and high occupancies, respectively, of the d band.

5.4.1.1 Bands in 1D and 3D Crystals

To fix our ideas, let us take the $(PtH_4^{2-})_n$ polymer as an example where the monomer units are intact in the polymer. At a monomer separation of *ca.* 3 Å the major inter-unit-cell overlap is clearly between z^2 and z orbitals. Next highest in energy are xy, yz, π-type overlaps, the interactions all being rather small at this separation. Figure 5.49 is a sketch of what is to be expected in qualitative terms.

Following the logic of the preceding paragraph, we may now sketch the band structure to be expected of the well-known rutile structure (shown in Fig. 5.13), and

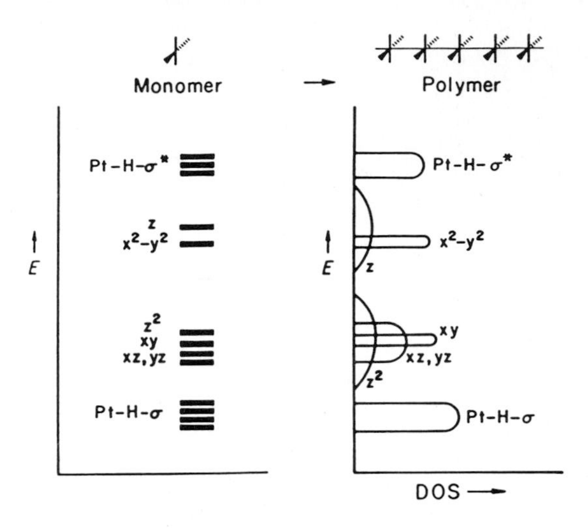

Figure 5.49 From discrete orbitals (for a monomer), left, to energy bands (for a polymer), right; see text. With permission from R. Hoffmann, *Solids and Surfaces: A Chemist's View of Bonding in Extended Structures*, VCH, Weinheim, **1988**, p. 29.

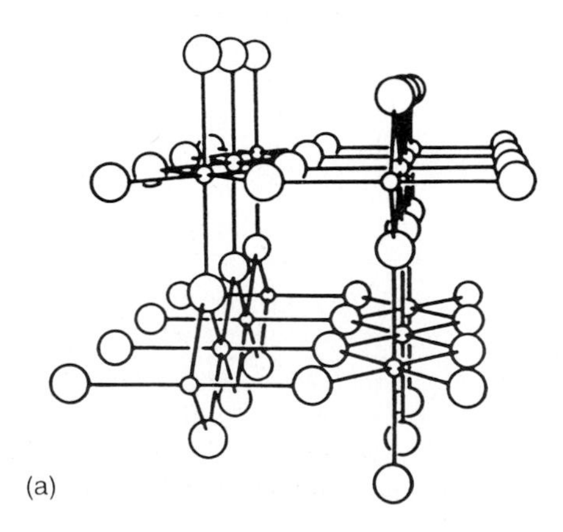

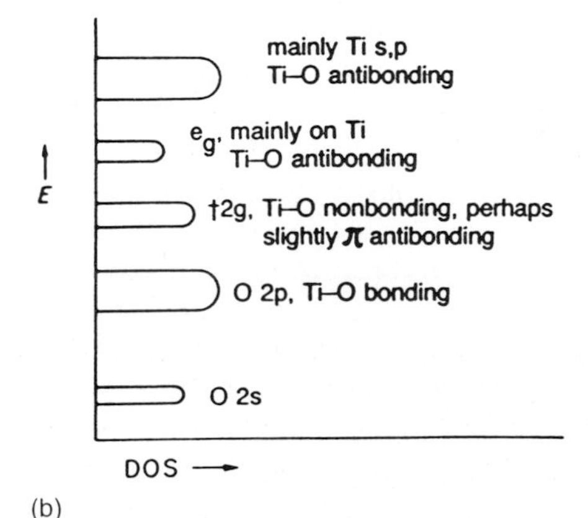

Figure 5.50 (a) Schematic of atomic structure of rutile, TiO_2 (cf. Fig. 5.3); (b) energy bands with qualitative description of DOS. With permission from R. Hoffmann, *Solids and Surfaces: A Chemist's View of Bonding in Extended Structures*, VCH, Weinheim, **1988**, p. 30.

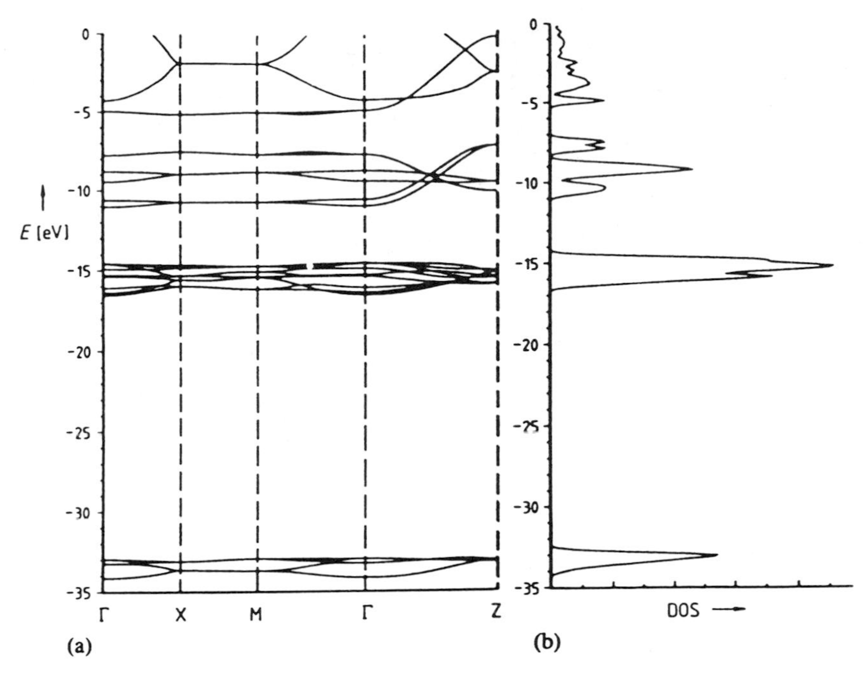

Figure 5.51 Band structure and density of states for rutile, TiO_2. With permission from R Hoffmann, *Solids and Surfaces: A Chemist's View of Bonding in Extended Structures*, VCH, Weinheim, **1988**, p. 31.

in its atomic equivalent; this is shown in Fig. 5.50(a) where each Ti^{4+} is octahedrally surrounded by O^{2-} ions. There are infinite chains of edge-sharing TiO_6 octahedra running in one direction in the crystal, but the Ti–Ti separation is always rather long. Even though there are no monomer units here, just an infinite assembly, there are, nevertheless, quite identifiable octahedral sites. At each, the metal d block must split into t_{2g} and e_g combinations, the classic three-below-two crystal field splitting. The only other feature to be recognized is that O has quite distinct 2s and 2p levels, and that there is no effective Ti···Ti or O···O interaction within the solid. We expect, in qualitative terms, a band sturcture – DOS schematized in Fig. 5.50(b).

However, in sketching the DOS as shown in Fig. 5.50(b) we are not addressing the more difficult task of computing the proper band structure of rutile, shown in Fig. 5.51. (This could be done rigorously as explained in texts listed under 'Further Reading' at the end of this chapter). In Fig. 5.51, part of the complexity stems from the fact that there are two formula units, $(TiO_2)_2$, per unit cell, so that the number of contributing orbitals is doubled, i.e. there are 12 O_{2ps} and six t_{2g} bands); moreover there are several zones (not just one as in Fig. 5.47), e.g. $\Gamma \rightarrow X$, $X \rightarrow M$, etc, where the arrows signify the directions in the 3D Brillouin zone. Hoffmann (1988) has explained how the portioning of the electron density between Ti and O in rutile may be worked out. We see from Fig. 5.52 that everything is more or less what we would have expected from Figs 5.49 and 5.50.

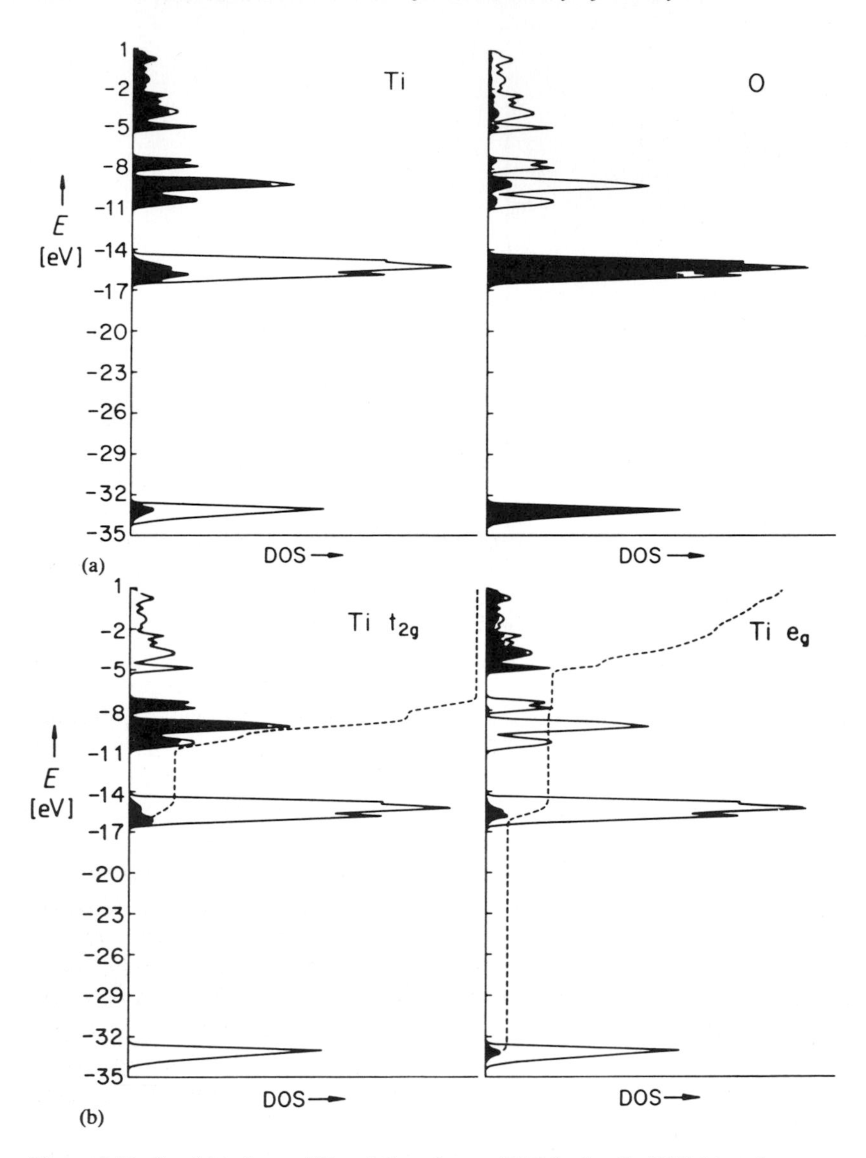

Figure 5.52 Combinations of Ti and O to the total DOS of rutile (TiO_2) are shown at top. At bottom the t_{2g} and e_g Ti contributions are shown; their integration (on a scale of 0–100%) is given by the broken line. With permission from R. Hoffmann, *Solids and Surfaces: A Chemist's View of Bonding in Extended Structures*, VCH, Weinheim, **1988**, p. 34.

5.4.1.2 Energy Bands in Ionic Solids

To illustrate the kind of situation that frequently obtains in ionic catalysts such as MgO, we focus on the highly ionic alkali halides where the s band of the metal ion and the p band of the halide are known, from spectroscopic and other measurements, to be widely separated.

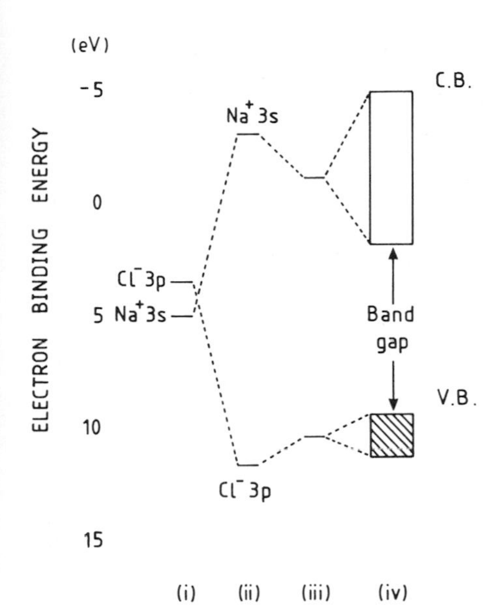

Figure 5.53 Energetics of formation of a crystal of Na^+Cl^- showing, *inter alia*, the importance of the Madelung potential (see text).

Consider the formation of an ionic solid M^+X^- from its constituent atoms M and X initially at infinite separation. Taking Na^+Cl^- as the specific ionic solid, we would expect the top (filled) level to be the chlorine 3p orbital into which an electron is placed on formation of the ion, and the (bottom) empty orbital, the sodium 3s (on the extreme left of Fig. 5.53) is, therefore, the difference between the electron affinity of a Cl atom and the ionization energy of a Na atom. When these gas-phase ions are brought into the solid they experience a strong electrostatic potential from the surrounding ions of opposite charge. To take exact account of this potential, successive shells of ions must be considered. This is effectively the same as doing a Madelung sum. The Madelung potential, V_M, experienced by an ion in a given structure which has a Madelung constant M, characteristic of the structure type, is given by

$$V_M = Mze/r$$

where r is the interionic distance. For NaCl the value of V_M is close to 9 eV. The Madelung potential is positive (attractive for an electron) at a Cl^- site: it contributes a repulsive energy for a Na^+ site. The situation is now as represented in (ii) of Fig. 5.53, and the separation in energy between the Cl^- 3p and Na^+ 3s states is *ca* 17 eV. In reality, the binding energy in the ionic solid is a good deal smaller than this because of the polarization of the surrounding crystal when an electron is removed. Polarization results in a lowering of energy when an electron is placed on a Na 3s orbital and likewise an increase when an electron is removed from the Cl 3p orbital (see Fig. 53(iii)). As the ions of the solid are placed sufficiently close together, the overlap of orbitals on neighbouring ions gives rise to the formation of one filled band (the valence band) and one empty band (the conduction band) separated by a band gap of *ca* 8 eV. This now gives us the band structure of a typical ionic solid. In MgO, for example, the band gap is between a valence band made up of O 2p orbitals and the Mg conduction band. We note that the Madelung term plays a major role in governing the magnitude of the band gap. Clearly we may expect band gaps to be largest where r values are smallest. It is not, therefore, surprising that LiF has one of the largest known band gaps, 12.5 eV.

5.4.1.3 Energy Bands in Transition-Metal Oxides: Understanding the Electronic Structure of the Monoxides of Titanium, Vanadium, Manganese and Nickel

All these oxides possess the NaCl structure. TiO and VO are metallic; the remaining monoxides are either insulating or semiconducting. Using the ideas given above we may now interpret these facts (compare Figs. 5.53 and 5.54).

Figure 5.54 shows how the ionic model is used to describe the electronic levels of a non-metallic oxide such as MnO. We note that:

(a) oxygen 2p forms a full valence band,
(b) metal 4s forms a broad but empty conduction band,
(c) metal 3d states can give rise to two processes, viz.

$$\left.\begin{array}{l}(3\mathrm{d})^n \rightarrow (3\mathrm{d})^{n+1} \\ \text{and} \\ (3\mathrm{d})^n \rightarrow (3\mathrm{d})^{n-1}\end{array}\right\}$$

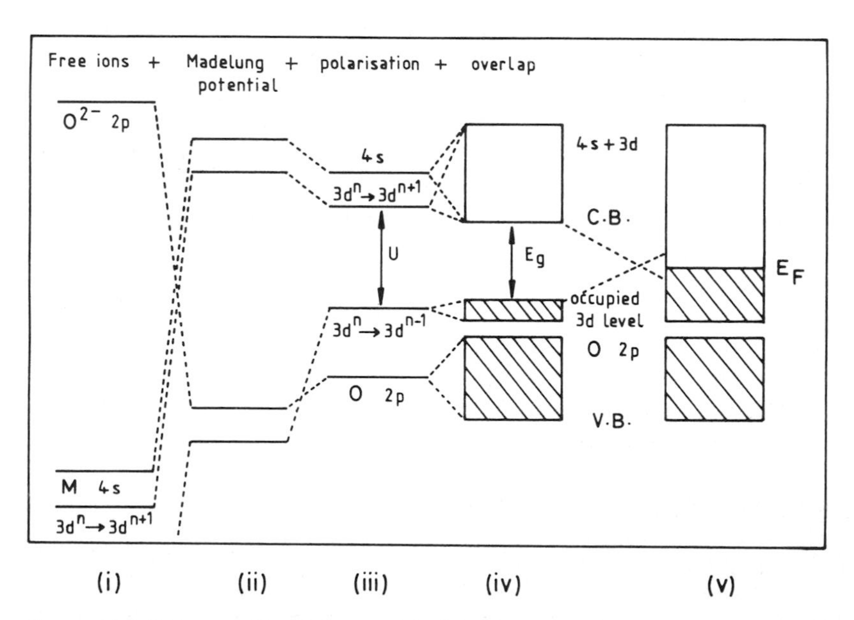

Figure 5.54 Conceptual breakdown of the influence of Madelung potential, polarization and Hubbard U (see text) in governing the nature of the electronic band structure of a transition-metal oxide such as MnO.

where n is the number of 3d electrons in the metal^{2+} ion. As in the case of NaCl (Fig. 5.53), the energies of the free ions (Fig. 5.54(i)) are influenced first by the Madelung potential and then by polarization. Finally, overlap between the orbitals on the constituent ions generates energy bands. The symbol in Fig. 5.54(iii), designated the Hubbard U, is the energy difference between the two 3d transitions described above. (In the gas phase this difference, being in this case the difference between the second and third ionization energies, is about 15 eV. In the solid state it is considerably less, owing to the influence of polarization. For many solids it falls in the range 3–5 eV.)

The electronic band structure given in Fig. 5.54 (iv) is that of a non-metallic oxide, where the d band width is insufficient to overcome the Hubbard U. (The Hubbard criterion for metallic behaviour is that the width of a band should be greater than the repulsion energy (U) for two electrons in the same crystal orbital.) Experiment (e.g. photoelectron spectroscopy) shows that the widths of the d bands in NiO and MnO are *ca* 1 eV. The empty band, at a higher energy E_g (equal to the band gap), is expected to be considerably broader since the energies of the processes:

$$\left.\begin{array}{l}(3d)^n \rightarrow (3d)^{n+1} \\ \\ (3d)^n \rightarrow (3d)^n(4s)^1\end{array}\right\}$$

and

are quite close, so that these levels can overlap in the solid into a single band with both 3d and 4s character.

Note that the band gap in these oxides is of quite a different character from that in MgO. Here the highest occupied level is of metal 3d character, not oxygen 2p as in MgO. In bridging the band gap with these transition metal oxides (by, say, the absorption of photons of requisite energy) the transition is from an occupied orbital on one cation into an empty orbital on a neighbouring one.

When the occupied 3d band and the unoccupied 4s, 3d bands (Fig. 5.54(iv)) overlap, the resulting band structure (Fig. 5.54(v)) corresponds to that of a metal. (Reverting to the Hubbard criterion, we may say that when band width wins out over repulsion, the band gap is eliminated). Such a band structure is applicable to TiO and VO. Why should these two oxides be so different from MnO and NiO (and also from FeO and CoO)? There are two reasons. First, the d orbitals are more diffuse earlier in the transition series, and give rise to greater overlap. Second, these oxides, unlike MnO and NiO, are *very* rich in structural defects (vacancies), with the result that the lattice is able to contract, thus further enhancing metal–metal interaction.

5.4.1.4 Energy Bands in Structures Related to ReO_3

We said earlier (Section 5.2.5) that a large family of solid structures, including the bronzes, shear and block ones, may be derived from the ReO_3 structure, which consists of an infinite, three-dimensional array of corner-sharing octahedra. Goodenough in 1976 showed how the band structure of this and many other oxides could be envisaged: we reproduce here the essence of his argument for ReO_3. One of the

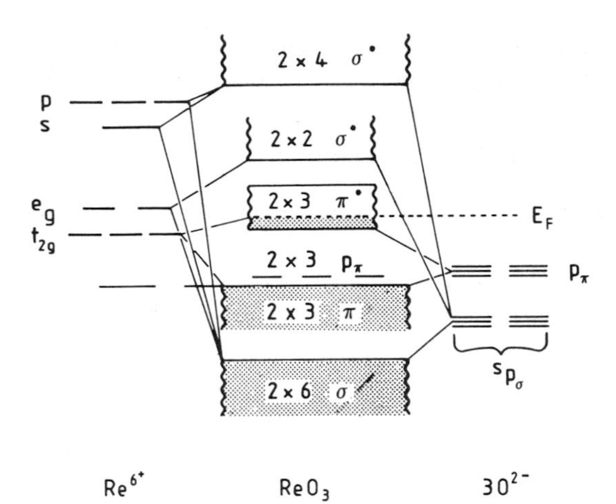

Figure 5.55 Energy-band diagram of ReO_3 (after Goodenough).

points to emphasize is that, just as the Madelung potential separates the energy levels of ions in simple ionic (non-transition-metal) solids such as NaCl and MgO, splitting of d orbitals in ligand fields plays an important role in determining the final nature of the band structure of transition metal oxides.

The energy diagram proposed by Goodenough is shown in Fig. 5.55. Each Re atom is surrounded by six O atoms, each of the latter being shared by two Re atoms. In octahedral coordination the d orbitals of an isolated ion are split into e_g and t_{2g} sub-levels which are, respectively, doubly and triply degenerate:

$$ML_6 \qquad \begin{array}{lll} e_g & - - & (d_{z^2}, d_{x^2-y^2}) \\ t_{2g} & - - - & (d_{yz}, d_{xz}, d_{xy}) \end{array}$$

Now the conduction and valence bands of ReO_3 arise from the covalent–ionic character of the bonding of cationic t_{2g} orbitals with p_π orbitals giving rise to a bonding (π) band and an antibonding (π^*) band, as shown in Fig. 5.55. Whereas in WO_3 (which has the same atomic structure as ReO_3) the π^* band is empty and the solid is insulating, in ReO_3 (which has the d^1, not the d^0, electronic configuration) the π^* band is partly populated, thus conferring metallic character on this oxide. This model of the band structure also explains why the tungsten bronze Na_xWO_3 ($x \geq 0.3$) is a metal: electrons fed into the solid by donation from the Na 3s orbitals enter the initially empty π^* conduction band, which then becomes partially occupied.

In perovskites (ABO_3 structure; see Section 5.2.6), the separate levels will, as in all solids, be broadened into bands. When the broadening is less than the ligand field splitting there may still be a gap between the lower t_{2g} band and the upper e_g. The point here is that a compound with six d electrons could have a full t_{2g} band and therefore be non-metallic. Stated chemically, we have a d^6 low-spin compound; and this is the situation likely to prevail in the first transition series for Co^{3+}, which has a d^6 configuration. Indeed the perovskite $LaCoO_3$ is non-metallic at low temperatures.

5.4.2 Fermi Levels in Insulators and Semiconductors

Earlier we noted that the concept of the density of states tells us that $N(E)\,dE$ is the number of allowed energy levels per unit volume of a solid in the energy interval E to $E + dE$. In the band gap, $N(E)$ is zero; and insulators and many semiconductors have band gaps ranging from *ca* 5 down to 0.1 eV. To understand the behaviour of electrons in such materials and to appreciate where the Fermi energy E_F is located in these cases, we use the Fermi–Dirac distribution function $f(E)$, defined by:

$$f(E) = \frac{1}{1 + \exp(E - E_F)} \tag{15}$$

This gives the fraction of the allowed energy levels which are occupied. For simplicity, consider a metal such as sodium. At 0 K, the Fermi–Dirac distribution corresponds to a sharp cut-off between filled levels up to E_F and empty ones above it (Fig. 5.56). At higher temperatures ($T_2 > T_1 > 0$ K), the distribution is smeared out, since some electrons are thermally excited to higher levels. Although at $T > 0$ K the boundary between filled and unfilled levels is no longer sharp, E_F still retains its fundamental meaning as it is the thermodynamic chemical potential for electrons in the solid. When two metals, or any group of solids, are brought into proper electrical contact, electrons will flow from one solid to another until equilibrium is reached when the Fermi levels are the same.

Consider now a typical semiconductor of which the DOS plot is simplified to that shown in Fig. 5.57(a). Since $f(E)$ represents the fractional occupancy, the actual concentration of electrons at an energy E is given by the product $f(E)N(E)$. In Figs. 5.57(a) and (b) we picture the electron distribution for a semiconductor or insulator in which the number of electrons excited into the conduction band is the same as the

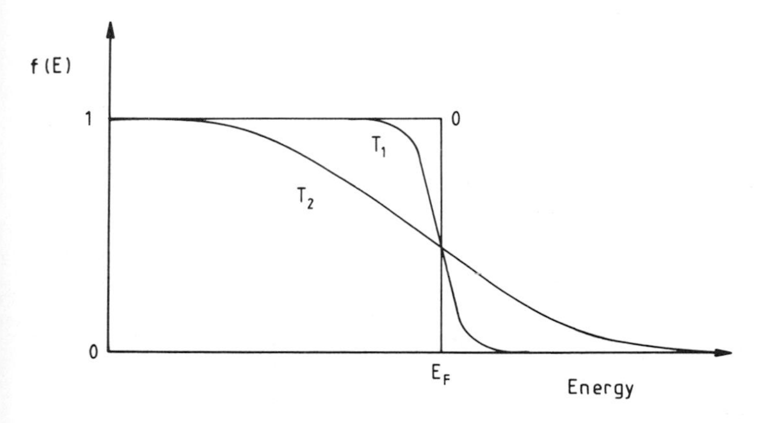

Figure 5.56 Theoretical curves of the Fermi–Dirac distribution function at absolute zero (0), and at higher temperatures ($T_2 > T_1$).

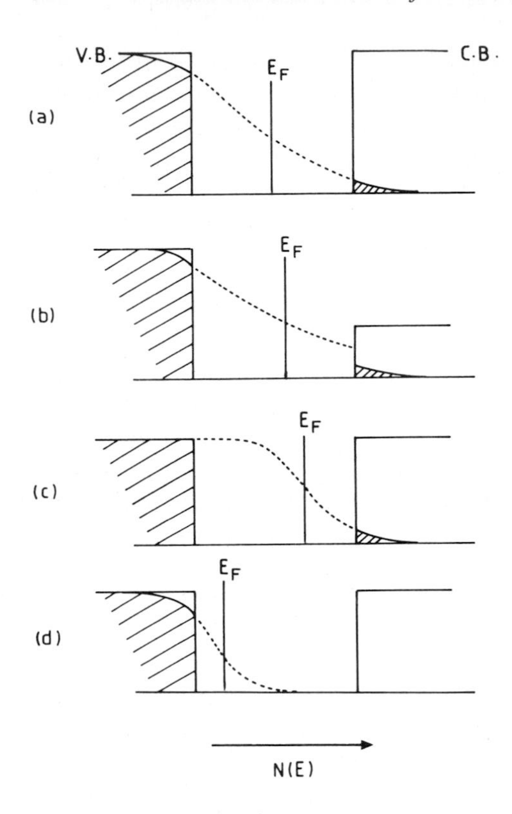

Figure 5.57 Fermi–Dirac distributions in (a) an intrinsic, pure semiconductor with the same density of states of electrons in the conduction band as of holes in the valence band; (b) a pure semiconductor with different densities of states in the conduction and valence bands; (c) *n*-type, and (d) *p*-type semiconductors.

number of holes (i.e. missing electrons) left in the valence band. Here, the fraction of occupied levels at an interval ΔE above E_F is the same as the fraction of unoccupied levels at ΔE below it. For Fig. 5.57(a) we note that the density of states in the conduction and valence bands are equal, so that E_F must fall in the middle of the band gap. The number of electrons excited into the conduction band (C. B.) equals the number of holes left in the valence band (V. B.); but when these densities of states are not equal, as in Fig. 5.57(b), the position of E_F shifts accordingly.

At room temperature, the magnitude of kT is only 0.024 eV, so that the fraction of electrons thermally excited will be rather low, unless the band gap is very small (as in the case of materials such as PbTe or InSb). Under these conditions the Fermi–Dirac distribution can be replaced by the Boltzmann equation ($n \propto \exp(-E_1/kT)$, where n is the number of species excited into an energy level E_1). Hence, for an energy E at the edge of the conduction band we have:

$$E - E_F = E_g/2$$

where E_g is the band gap. With $E_g \gg kT$ the Boltzmann equation yields for the number of excited electrons, n:

$$n \propto \exp(-E_g/2kT)$$

In other words, the electrical conductivity of a pure semiconductor will have an activation energy equal to half the band gap.

For semiconductors, where E_g normally exceeds 2 eV, the number of electrons excited into the C. B. is vanishingly small at room temperature. Indeed, impurity atoms can, when ionized (as in the case of substitutional P or As in silicon), make a dominating contribution to the electron population in the C. B. Likewise, impurities which can accept electrons from the top of the V. B. (such as substitutional B or Ga in silicon), can also exert dominating influences in the concentration of holes. Materials in the former category, which contribute *n*egatively charged carriers, are *n*-type semiconductors, whereas those in the latter, which contribute *p*ositively charged carriers, are *p*-type semiconductors. The effects of such impurities are shown in Fig. 5.57(c) and (d). In each case, E_F is shifted away from the centre of the band gap.

Interstitial zinc species in ZnO as well as oxygen vacancies in SnO_2 make these oxides *n*-type, whereas Ni^{3+} species in non-stoichiometric NiO make that oxide *p*-type. Impurities incorporated into a host structure in one form or another, and structural defects (e.g. unsaturated bonds, dislocations and planar faults such as tilt or twist boundaries), also result in the creation of extra energy levels situated within the band gap. Such manifestations are summarized in the next section.

5.4.3 Surface Electronic States and the Occurrence of Energy Levels Within the Band Gap

Both the American worker Schottky and the Russian worker Tamm long ago considered the electronic repercussions of the occurrence of a free surface. They concluded that extra states, so-called surface states, were introduced into the band gap merely as a result of the termination of the bulk structure at a surface. It is not difficult to appreciate why these states should occur. We saw in Fig. 5.53 that the Madelung potential is an important determinant of the magnitude of the band gap. Since atoms and ions at a surface have a lower coordination that their counterparts in the bulk, the surface Madelung potential is reduced from the bulk value. This effect should make the surface band gap less than that of the bulk, thereby generating both filled and empty levels (which might overlap and form bands) possessing energies within the band gap of the bulk solid (see Fig. 5.58, states (a) and (b)). We also saw (Figs. 5.53 and 5.54) that polarisation plays a part in determining the band gap: this works in the opposite way from the Madelung potential. We must also bear

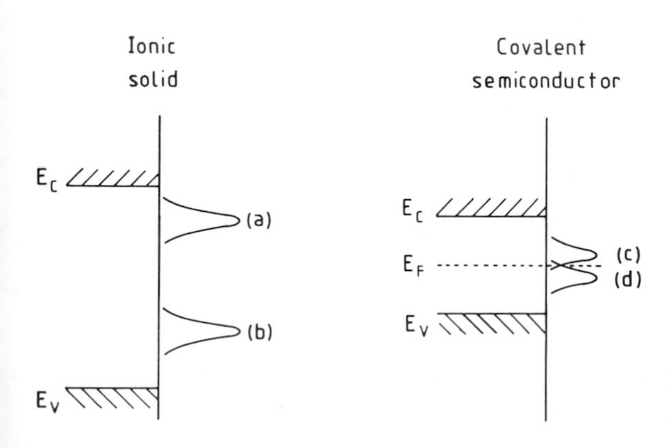

Figure 5.58 Intrinsic surface states. For ionic solids, left, the states (a) and (b) have conduction-band and valence-band character, respectively. For covalent solids the states (c) and (d) arise from dangling bonds and have antibonding and bonding character respectively.

in mind that substantial relaxation or reconstruction may occur at a free surface. In covalent semiconductors like silicon or germanium, dangling bonds at surfaces may pair off (see Fig. 5.58, states (c) and (d)): in ionic solids such as MgO the ions at certain surfaces move from their expected, idealized sites by up to about 10% of the interionic distances; and in a tetrahedral structure such as GaAs reorganization at the (110) surface is such as to lead to a marked protrusion of the As atoms.

It follows from the above that an array of adsorbed entities at an exterior surface may well give rise to surface states within the band gap. This is a well-recognized effect in the electrochemistry of semiconductors. A semiconducting or insulating solid that has been cleaned by reduction or ion-bombardment before examination or use as a catalyst, will tend to contain extra surface states situated in the band gap owing to the generation of 'abnormal' valence states and structural defects.

5.4.4 Band Bending and Metal–Semiconductor Junctions: Schottky Barriers

This topic, long recognized to be of central importance in photoelectrochemistry and photoelectrosynthesis (see Section 8.7.5), is also central to the energetics of adsorption on semiconductors, and in particular to the boundary-layer theory of chemisorption. Its likely importance in catalytic phenomena such as metal–support interaction is less widely appreciated.

We first consider an idealized semiconductor (typically ZnO) free of surface states. When contact is made between a metal and this *n*-type semiconductor, a potential difference is set up between the two (in a manner similar to that between two metals).

For the situation shown in Fig. 5.59, we see that electrons pass from the semiconductor to the metal until the Fermi levels (chemical potentials of the electrons) are equalized. The excess negative charge on the metal repels electrons near the surface of the semiconductor, thereby creating a layer which is depleted of conduction

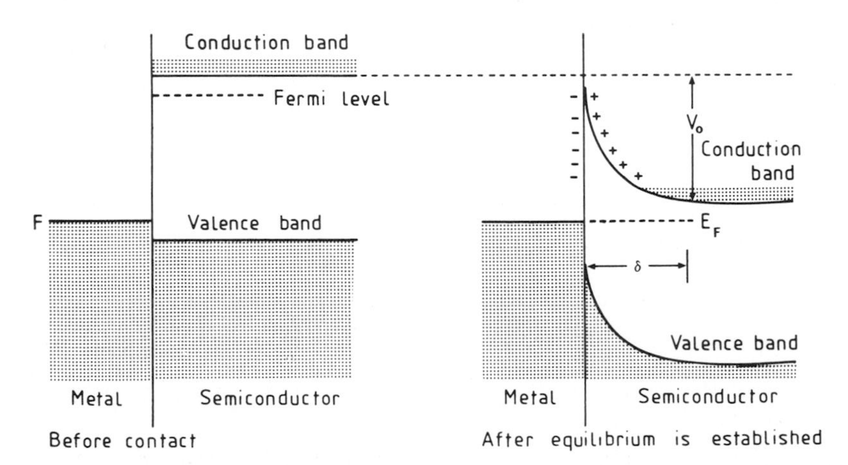

Figure 5.59 Energy bands at a metal–semiconductor contact before and after equilibrium is established.

electrons, the so-called depletion or barrier layer, which is a region of positive space charge because it contains the ionized donor impurities without the compensating charge of the conduction electrons. As a result of this potential difference V_0 between the conduction band at the surface and in the bulk semiconductor, the energy bands are distorted or bent. (It is because of this band bending that solids such as cuprous oxide can function as rectifiers.)

To estimate the magnitude of V_0 we use the Poisson equation in electrostatics. First assume that all conduction electrons are removed from the depletion layer, so that the charge density, ρ, is equal to en_d, where n_d is the concentration of donors (all assumed to be univalent) and e is the electronic charge. The Poisson equation is

$$\frac{d^2V}{dx^2} = -4\pi\rho/\varepsilon \tag{16}$$

where ε is the permittivity of the semiconductor and $x = 0$ at its surface. The two boundary conditions required to solve this equation are: $V = 0$ at $x = 0$ and $dV/dx = 0$ (no electric field) at $x \geq \delta$. We therefore have

$$V = -\left(\frac{(2\pi n_d e)}{\varepsilon}\right)\{(x-\delta)^2 - \delta^2\} \tag{17}$$

Hence the difference in potential at the surface $x = 0$ from that in the bulk of the semiconductor $x \geq \delta$ is:

$$V_0 = 2\pi n_d e\delta^2/\varepsilon \tag{18}$$

from which, knowing V_0, we may estimate δ. Semiconductors have ε typically about 10 (for Si, $\varepsilon = 12$), n_d is *ca* 10^{17} cm^{-3}, and $V_0 \simeq 1$ V. We see, therefore, that the depletion layer extends to *ca* 1000 Å.

This treatment is known as the Schottky barrier theory. The difficulty with it is that V_0 should be equal to the difference between the work functions of the metal and the semiconductor, i.e. it should depend very much upon the nature (or particular crystallographic face) of the metal used. Experimentally, more frequently than not, V_0 is independent of the metal. To account for this we revert, as Bardeen first realized, to surface states. We see, from Fig. 5.60, that there is band bending solely as a result of the transfer of electrons from within an n-type semiconductor to its empty surface states. Again we have a depletion layer, within which the bands are bent by the unbalanced positive charge of the ionized donors. This depletion layer acts as a barrier, just as in the case of a metal–semiconductor junction. If n_s is the concentration of surface states and n_d is again the bulk concentration of donors, then the thickness δ of the depletion layer is given by $n_s = n_d\delta$, and Eq. (18) becomes:

$$V_0 = \left(\frac{2\pi e}{\varepsilon}\right)\frac{n_s^2}{n_d} \tag{19}$$

an equation which shows V_0 to be independent of the metal. It is therefore clear that, even if only a fraction of a monolayer (of surface states) becomes occupied with electrons from the bulk semiconductor, band bending will occur. Similar arguments apply for a p-type semiconductor, but now a surplus layer forms where the bands bend and, in this case, the bands bend downwards.

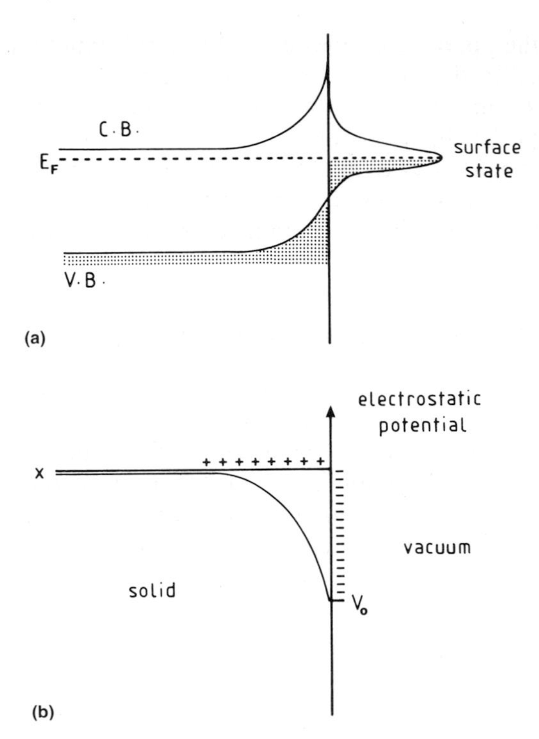

Figure 5.60 (a) Band bending occurs at the surface of an *n*-type semiconductor owing to the presence of surface states. (b) The corresponding charge distribution and variation of electrostatic potential.

5.4.4.1 Depletive Chemisorption on Semiconductors

Consider the chemisorption of a species capable of accepting electrons from the donors in an *n*-type semiconductor (Fig. 5.61). If A is the electron affinity of the adsorbed species and ϕ is the work function of the solid, the energy of chemisorption at zero coverage is clearly $(A - \phi)e$. As more species are adsorbed, a space charge

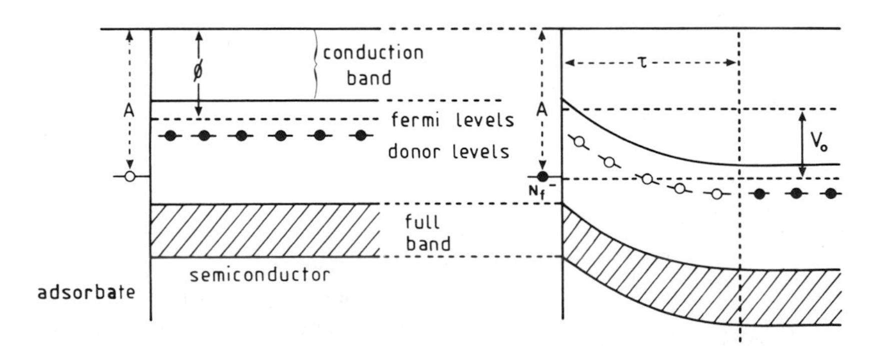

Figure 5.61 Diagrammatic representation of how a space charge builds up during anionic chemisorption on an *n*-type semiconductor: (a) before and (b) after chemisorption of the adsorbate (τ is the width of the band-bending region).

builds up in the boundary layer since the donor levels deeper in the semiconductor are called upon to yield their electrons. Transfer of electrons will continue until a potential barrier V_0 is formed (compare Fig. 5.60) when the potential energy of the electrons in the adsorbate equals that of the electrons in the solid. At this juncture, no further net adsorption can occur. It can be shown that the limiting concentration of adsorbed species on both *n*-type and *p*-type semiconductors is related to the magnitude to the band bending, conclusions first reached independently by Hauffe and Weisz.

5.4.4.2 The Bending of Bands when Semiconductors are Immersed in Electrolytes

Extending the above arguments, we would expect an *n*-type semiconductor to develop a depletion layer when it comes into contact with a solution containing a redox potential that is initially (see Fig. 5.62) below the Fermi level of the solid – compare Fig. 5.59. In this case, however, as well as the depletion layer that forms inside the semiconductor, a Helmholtz double layer forms immediately outside it. This thin layer consists of charge arising both from ions bound more or less to the solid surface

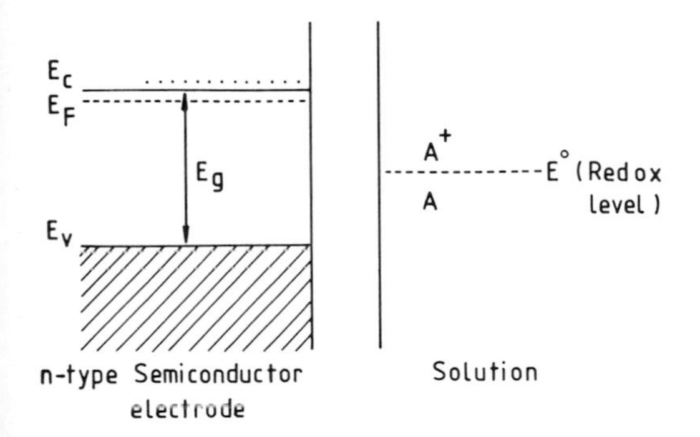

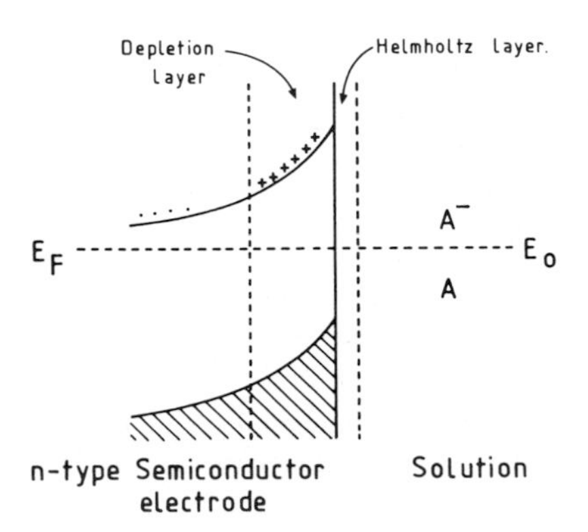

Figure 5.62 When an *n*-type semiconductor is brought into contact with an electrolyte solution, band bending occurs and a depletion layer as well as a Helmholtz layer is formed. (a) Band energies of semiconductor and solution without contact, (b) semiconductor and solution in contact.

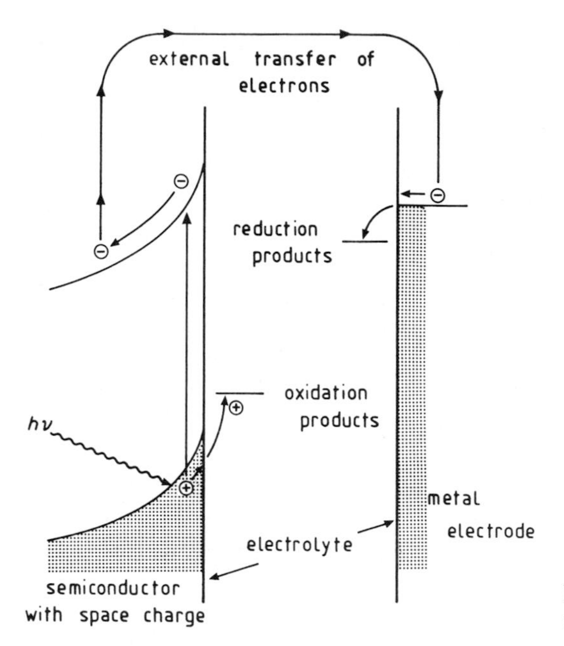

Figure 5.63 The essence of a photo-electrolytic cell (see also Section 1.2.2.2).

as a result of adsorption as well as from ions in solution attracted by the charged surface.

When a metal is also inserted into the solution and external connection made to the semiconductor, as indicated in Fig. 5.63, we see how, in principle, photo-electrolysis may take place by the production of electron–hole pairs, and their separation, within the depletion layer. We note that *n*-type TiO_2 has a band gap, E_g, of 3.2 eV, rather too large to be really useful in harnessing solar energy photo-electrochemically since only about 5% of the solar energy spectrum is greater than this value of E_g. By replacing some of the Ti atoms by V atoms and forming solid solutions $Ti_{1-x}V_xO_2$, the value of E_g can be reduced by *ca* 2.0 eV. with consequential improvement in the efficiency of solar energy conversion.

In considering the design of catalysts for more efficient harvesting of solar energy – a topic of growing importance – it is necessary to match the band gaps and Fermi levels of semiconductors and insulators with the redox properties of the electrolyte in which the solids are immersed. Fermi levels of solids are quoted by reference to that of an electron in vacuum infinitely separated from the surface. Electrode potentials of redox couples or metal electrodes, on the other hand, are referenced with respect to the normal hydrogen electrode (NHE). The difference between these two frames of reference, i.e. the energy required to remove an electron from the H^+/H_2 couple under standard conditions in solution to infinity, is 4.5 eV. The precise position of the H^+/H_2 couple under non-standard contact with a solid obviously depends upon the pH. Armed with this information we can label band-structure diagrams with energies referenced to the hydrogen electrode as shown in Fig. 5.64, based on the work of Gerischer, Nozik and Bolton.

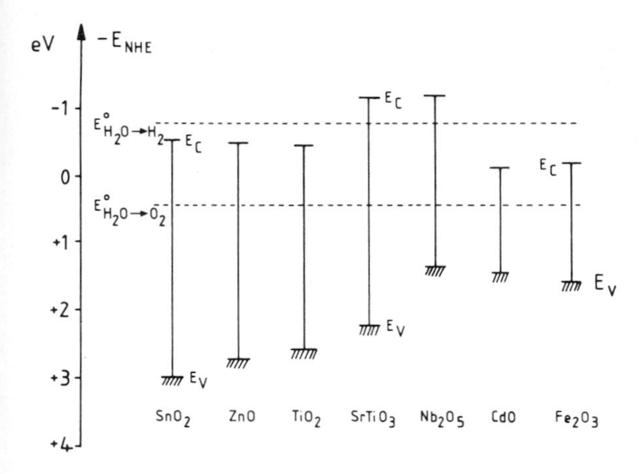

Figure 5.64 Position of valence-band and conduction-band edges in a variety of oxide semiconductors in contact with aqueous electrolytes (on the H_2 scale, NHE) at a pH of 13 (based on work of H. Gerischer).

5.4.5 Quantum Chemical Approaches to the Electronic Properties of Solids

'Independent particle' theories, by which we mean those in which it is assumed that, for an N-electron system, each electron moves in an average potential field produced by the remaining $N - 1$ electrons, can be utilized at either the 'first principles' or 'simple' level. By 'first principles' we mean using those quantum mechanical methods which are carried out self-consistently and which include the Coulomb interactions between electrons, between nuclei, and between electrons and nuclei, in addition to either a local or non-local exchange potential. By 'simple' we mean the utilization of procedures in which additional approximations beyond the notion of independent particles, e.g. semi-empirical parameters, are used in some methods, self-consistency is ignored in others, and electron–electron interactions are ignored in yet others. The former group encompasses ab-initio Hartree–Fock procedures and density functional approaches as well as pseudo-potential techniques; the latter encompass simple Hückel theory (and its equivalent in solid-state physics, the 'cubium' approach), extended Hückel and many other methods involving total or partial neglect of differential overlap (and endowed with acronyms such as CNDO, MINDO, etc.).

Quantum chemical techniques nowadays play an important part in the interpretation and prediction of catalytic properties (see, in particular, Section 6.3). Whilst more information can at present be gleaned about the electronic properties of the solid catalyst in the absence or presence of chemisorbed entities than about the fundamental features of catalysis itself, there are hopes that quantum chemical techniques will ultimately prove invaluable in pinpointing the characteristics of transition states at surfaces. Short-lived transition states are extremely difficult to study by direct experiment – even for the more favourable situations in homogeneous catalysis – and the hope is that reliable calculations of atomic configurations of atomic aggregates at energy maxima may reveal insights about reaction pathways that are not otherwise accessible. As outlined earlier, with the advent of powerful computers and the rapid recent progress in ab-initio and other quantum chemical techniques this is a burgeoning area of surface science. Conventional and theoretical lines between coordination, cluster and organometallic chemistry on the one hand, and the extended solids on the other, are nowadays much discussed. The reader is referred to some key papers and books (see Further Reading, at the end of this chapter) for guidance concerning the various theoretical procedures now in active use.

5.4.5.1 The Cluster and Thin-Slab Approach

It has been long been recognized that a 'fragment' of a metal consisting of some 20 or more atoms simulates the behaviour of a bulk metal, at least in certain properties. From the viewpoint of the theoretical catalyst scientist, clusters of metal containing N atoms (with N up to a few hundred) merit study not only because they are computationally quite manageable but also because they are particularly helpful in clarifying localized effects, such as the energetic or vibrational repercussions of 'placing' an adsorbed entity in a bridged or 'on-top' site.

Computationally, it is noteworthy that Messmer, using simple Hückel theory (see Section 5.4.6), showed that, for a simple cubic array of atoms, all the wave functions and energies (eigenvectors and eigenvalues) can be obtained in closed form for any size of cluster up to the infinite solid. Although the simple Hückel theory and the extended Hückel (EH) procedure, which has also been used for cluster calculations, both have their limitations (Section 5.4.6), they have been of considerable value in ascertaining what changes occur electronically as a result of adsorption on to a solid or in identifying trends in band structure along a related series of metals or solid compounds. Both methods lend themselves to the study of much larger clusters. Indeed, the solid may be regarded as semi-infinite or as a thin slab. A slab (thin film) is a solid of finite thickness (layers of, say, four or five atoms) but with infinite extent in the $\pm x$ and $\pm y$ directions, whereas a semi-infinite solid is one in which there is infinite extent in $\pm x$, $\pm y$ and $-z$. To take advantage of the translational symmetry of the solid, use is made of the Bloch theorem, which tells us that, in a translationally symmetric system, the wave functions in one dimension are of the form: $u_{nk}(x)\exp(ikx)$, where u_{nk} is periodic with the same periodicity as the 1D crystal potential $V(x)$, k is the wave vector, n the band index and a the lattice constant. Thus

$$u_{nk}(x+a) = u_{nk}(x) \tag{20}$$

The electronic states are solutions of the Schrödinger equation:

$$\underset{\sim}{H}\varphi_{nk} = E\varphi_{nk} \tag{21}$$

where $\underset{\sim}{H}$ is the one-electron Hamiltonian operator and includes the crystal potential $V(x)$ which itself has 1D translational periodicity $V(x+a) = V(x)$.

5.4.6 Hückel and Extended Hückel Calculations

Hückel theory, with all its faults and compensating conveniences, is a familiar route for calculating the eigenvalues and eigenfunctions of discrete molecules. Essentially the same procedures are employed for adsorbed species (e.g. 1,3-di-σ-adsorbed butane on platinum) and for the simplified model of a solid known as cubium.

For discrete species, the molecular orbitals φ are ϕ (linear combination of atomic orbitals; LCAO) and are assumed to be solutions of the Schrödinger equation: $\underset{\sim}{H}\varphi = E\varphi$. By making the substitution

$$\psi = \sum_{v} c_v \phi_v$$

we have:

$$\sum_{v} c_v(\underset{\sim}{H} - E)\phi_v = 0 \tag{22}$$

If this equation is multiplied by one of the atomic orbitals ϕ_μ and integration carried out over all three-dimensional space we arrive at the secular equations:

$$\sum_{v} c_v \int \phi_\mu(\underset{\sim}{H} - E)\phi_v = 0 \tag{23}$$

If we now define the quantities $H_{\mu v}$ and $S_{\mu v}$ by:

$$\int \phi_\mu \underset{\sim}{H}\phi_v \, \mathrm{d}\tau = H_{\mu v} \tag{24}$$

and

$$\int \phi_\mu E \phi_v \, \mathrm{d}\tau = E \int \phi_\mu \phi_v \, \mathrm{d}\tau = ES_{\mu v} \tag{25}$$

then the equations may be written:

$$\sum c_v(H_{\mu v} - ES_{\mu v}) = 0 \tag{26}$$

There is one equation of this type for each atomic orbital ϕ_μ in the set. By equating the secular determinant to zero:

$$|H_{\mu v} - ES_{\mu v}| = 0 \tag{27}$$

the allowed energies for each atomic orbital can be determined. Each energy is then substituted into this equation to arrive at the appropriate coefficients. In practice, Hückel theory involves several drastic assumptions:

$$S_{\mu v} = \begin{cases} 1 & \text{if } \mu = v \\ 0 & \text{if } \mu \neq v \end{cases}$$

$$H_{\mu v} = \begin{cases} \alpha_\mu & \text{if } \mu = v \\ \beta_{\mu v} & \text{if } \mu \neq v \text{ but } \mu, v \text{ are bonded neighbours} \\ 0 & \text{otherwise} \end{cases}$$

In general, values of the Hückel parameters α and β are determined empirically by fitting theoretical calculations to the observed (e.g. photoelectron spectroscopic data). Note, in particular, that the overlap integral $S_{\mu\nu}$ ($\mu \neq \nu$) is taken as zero, a chemically implausible assumption. In *e*xtended *H*ückel *t*heory (EHT), developed by Hoffmann, overlap integrals are not ignored: they are calculated from the respective atomic-orbital wave functions. Extended Hückel theory includes σ orbitals and, if appropriate, d and f orbitals also. The Coulomb integrals (i.e. the so-called diagonal matrix elements) are, as for simple Hückel theory, given by

$$\alpha_\mu = \int \phi_\mu \underset{\sim}{H} \phi_\mu \, \mathrm{d}\tau = H_{\mu\mu}$$

but the off-diagonal matrix elements or resonance integrals, $\beta_{\mu\nu}$, are expressed according to the Wolfsberg–Helmholtz approximation:

$$\beta_{\mu\nu} = -H_{\mu\nu} = \tfrac{1}{2} K (H_{\mu\mu} + H_{\nu\nu}) S_{\mu\nu} \tag{28}$$

where K is an empirical constant usually taken as 1.75. The semi-empirical nature of extended Hückel theory is seen by the fact that the energy parameters $H_{\mu\mu}$ and $H_{\nu\nu}$ are obtained from experimental ionisation data. In calculating $S_{\mu\nu}$ for all the considered orbitals, atomic orbitals are usually taken as being of *S*later *t*ype (STO), although linear combinations of Slater orbitals are sometimes used. To simplify the computations, STOs can be replaced by a linear combination of Gaussian functions ($\exp(-ar^2)$), the ingenious idea of the Cambridge theoretical chemist, S. F. Boys. Typically, descriptions of all-valence electron calculations use, say, STO-3G orbitals; this indicates that the STOs are expressed as a combination of three Gaussians.

EHT calculations for organic molecules have their limitations. For example, in optimising, the geometry of a molecule by EHT, the constituent atoms tend to collapse into one another – a consequence of the neglect of nuclear repulsion and the prescription that all off-diagonal matrix elements are attractive. The H_2 molecule is another, pathological, example: treatment according to EHT leads to a zero bond length and the final state of one superatom! EHT is not a self-consistent procedure; nevertheless, when an adroit choice of parameters is made, it is extremely useful both in exploring trends relating to the electronic properties of solids and in gaining fresh insights pertaining to chemisorption. Messmer, Hoffmann and co-workers, Baezold and Shustorovich in the USA have all made skilful use of EHT for the study of band structures and of surface bonding.

5.4.6.1 'Ab-Initio' Methods

Contrary to its literal meaning, 'ab initio' in quantum chemistry covers a multitude of methods, some of which do *not* start from the beginning. Ab-initio calculations range from the most sophisticated, which include all electrons (core and valence) and all correlation effects, to minimal or sub-minimal basis set *s*elf-*c*onsistent *f*ield (SCF) LCAO MO procedures.

The SCF molecular orbitals are defined by equations similar to Eqs. (26) and (27). They are taken to be eigenfunctions of an operator $\underset{\sim}{F}$.

$$\underset{\sim}{F}\varphi = E\varphi \tag{29}$$

so that, using the LCAO approximation

$$\varphi = \sum_{\nu} c_{\nu}\phi_{\nu}$$

the coefficients and energies are determined by the equations:

$$\sum_{\nu} c_{\nu}(F_{\mu\nu}ES_{\mu\nu}) = 0 \tag{30}$$

and

$$|F_{\mu\nu} - ES_{\mu\nu}| = 0 \tag{31}$$

where

$$F_{\mu\nu} = \int \phi_{\mu}\underset{\sim}{F}\phi_{\nu}\,\mathrm{d}\tau \tag{32}$$

The operator $\underset{\sim}{F}$ is defined by the elements of the full Hamiltonian. Detailed descriptions need not be given here. The key point, however, is that, in ab-initio SCF procedures of this kind, all atomic cores and all electrons are considered and there is an equation:

$$F_{\mu\nu} = H^{c}_{\mu\nu} + \sum_{\rho}\sum_{\sigma} P_{\rho\sigma}\left[(\mu\nu|\rho\sigma) - \tfrac{1}{2}(\mu\rho|\nu\sigma)\right] \tag{33}$$

H^{c} being the core Hamiltonian for an electron, and

$$(\mu\nu|\rho\sigma) = \iint \phi_{\mu}(1)\phi_{\nu}(1)\left(\frac{e^2}{r_{12}}\right)\phi_{\rho}(2)\phi_{\sigma}(2)\,\mathrm{d}\tau_1\,\mathrm{d}\tau_2 \tag{34}$$

which, interpreted physically, stands for the repulsion between an electron distributed in space according to the function $\phi_{\mu}\phi_{\nu}(1)$ and a second electron having the distribution $\phi_{\rho}\phi_{\sigma}(2)$. $P_{\rho\sigma}$ is known as the bond order and is written:

$$P_{\rho\sigma} = 2\sum_{i} c_{i\rho}c_{i\sigma} \tag{35}$$

It is summed over all occupied MOs, φ_i. In semi-empirical theories reference is often made to the Mulliken orbital population, N_{ρ}, which is the sum of the product of bond order and overlap, i.e.

$$N_{\rho} = \sum_{\sigma} P_{\rho\sigma}S_{\rho\sigma}$$

In the *H*artree–*F*ock (HF) method of ab-initio calculation, the self-consistent field procedure takes into account the indistinguishability of the electrons, but it neglects correlation effects. The *u*nrestricted *H*artree–*F*ock (UHF) method allows more freedom to the form of the orbitals by permitting the spatial form of the orbital to depend upon whether the electron has α or β spin. Both HF and UHF SCF methods can be improved by permitting so-called configuration interaction (CI) which, in effect, allows the wave function of the molecule (or solid or surface complex) to be described by a mixture of wave functions corresponding to different configurations. With the massive computers now available, one can undertake multi-configuration (MC) UHF SCF calculations.

In the *d*ensity *f*unctional *t*heory (DFT) approach pioneered by Parr and co-workers, what in effect has been done (viewed from the standpoint of understanding chemical reactivity) is that Fukui's concept of frontier orbitals, articulated originally within an approximate, semi-empirical theory, has been re-expressed within an exact (ab-initio) framework provided by the density functional formulation of Kohn and Sham, Parr and others, notably Morrel Cohen, in generalizing Fukui's theory, arrive at a reactivity index (or Fukui function) and the useful concepts of local softness and hardness. It is now possible to obtain commercially documents that present a detailed methodology of DFT and its implementation in such codes as DMol, produced by Newsam et al. at Biosym Technologies. In such an implementation, MOs are represented as linear combinations of a numerically generated basis set. A set of numerical grids is employed to construct the Hamiltonian matrix, overlap integral, charge density and exchange correlation potential. Such a numerical procedure permits flexibility in controlling both the accuracy and the computational expense of the ensuing calculation.

Density functional computations are an extremely popular, and on the whole successful, way of tackling problems of direct value in heterogeneous catalysis. An ambitious computation has been made by Catlow et al. to investigate the mechanism of the cyclotrimerization of acetylene catalysed by extra framework ions of Ni in zeolite-Y. Criteria based on the optimisation of orbital overlap allowed them to arrive at the most favourable orientation of the reactants for the reaction to proceed, as the key intermediates in the possible reaction pathway could be identified, and to elucidate the nature of the bonding between Ni^{2+} and acetylene during the course of catalysis.

It must not be thought that ab-initio methods are feasible only with molecular orbitals. There is an alternative approach for discussing the states of molecules and solids using wave functions generalized from the valence bond method, pioneered by Pauling, for discussing structure. This alternative approach, introduced by Coulson and Fisher and subsequently developed and elegantly applied by Simonetta, Goddard and others, is called the *g*eneralized *v*alence *b*ond (GVB) method. It differs from the HF method in that it has two orbitals, one for each electron, rather than one orbital per electron pair. The GVB wavefunction is the self-consistent generalization of the valence bond wavefunction, just as the HF wavefunction is the self-consistent generalisation of the MO wavefunction. Configuration interaction improves the quality of GVB calculations, just as it does with UHF analogues.

5.4.7 A Brief Selection of Quantum Chemical Studies

We first focus upon how semi-empirical methods clarify such properties and phenomena as band widths, Fermi levels and densities of state of metals; heats of adsorption on metals; and the adsorption of CO on nickel. We then show how ab-initio methods enlarge our understanding of the bonding of simple molecules to nickel and MgO, and of the mechanism of olefin metathesis. Examples of the use of quantum chemical interpretations are also given in Section 6.3, which deals with the poisoning of catalysts.

5.4.7.1 Band Widths, DOS and Fermi Levels of the Transition Metals

In handling metals in the first transition series, thin films possessing two-dimensional periodicity and typically three to five layers thick are subjected to an EH calculation of the type popularised by Hoffmann. Advantage is taken of the structural periodicity; and the determinant $|H - ES| = 0$ is solved by summing over several unit cells. The matrix elements take the form

$$H_{\mu\nu} = \sum_{r} \exp\,(i\boldsymbol{kr}) \int \phi_\mu \underset{\sim}{H} \phi_\nu \, \mathrm{d}\tau \tag{36}$$

where $\boldsymbol{r}$ is the position vector of a unit cell and $\boldsymbol{k}$ is the wave vector in the 2D Brillouin zone. The secular determinant is solved at typically 9 points in $\boldsymbol{k}$ space uniformly distributed within and enclosed by $\boldsymbol{k} = 0$ to $\boldsymbol{h} = \pi_a$ in two principal reciprocal space directions e.g. $(1, 0)$ and $(1, 1)$. The results are then averaged over the $\boldsymbol{k}$ points to compute energy or charge distribution by the Mulliken procedure (see Eq. (35)).

Figure 5.65 shows the resulting DOS plots for a four-layer slab of nickel, with the fractions of the DOS arising from the two inner and the two outer (i.e. the surface) layers distinguished. The absolute values of the energies are known to be erroneous. The calculated position of E_F, for example, exceeds the experimentally determined

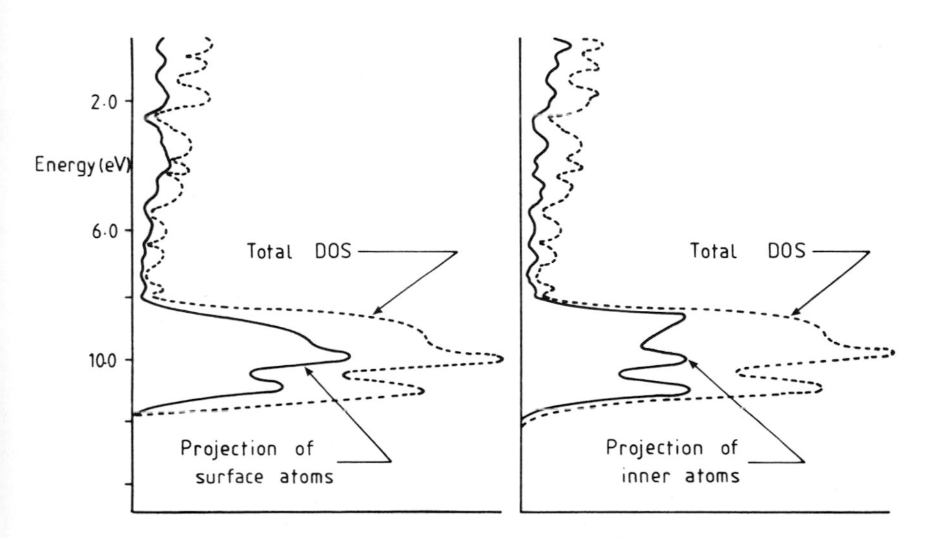

Figure 5.65 Projected density of states (DOS), calculated by EHT, on surface and inner layers of a Ni(100) four-layer slab (see text), with the fractions of the DOS arising from inner and surface layers distinguished (based on work of R. Hoffmann).

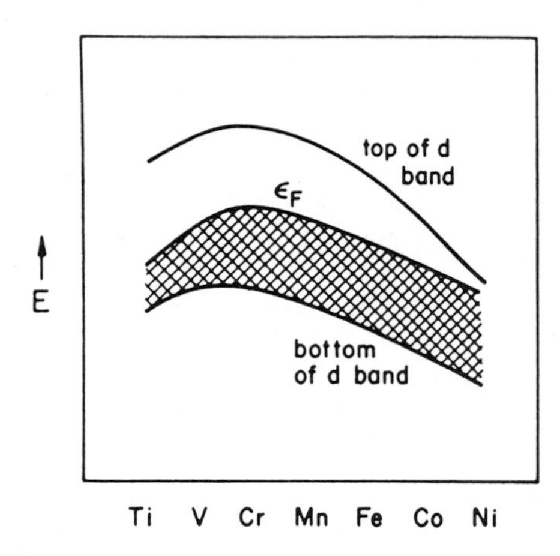

Figure 5.66 Qualitative indication of what happens to Fermi levels and d band energies along a transition-metal series. With permission from R. Hoffmann, *Solids and Surfaces: A Chemist's View of Bonding in Extended Structures*, VCH, Weinheim, **1388**.

work function of nickel by *ca* 3 eV. (This is typical of the error associated with EHT calculations.) Ignoring the deficiencies associated with the absolute values of the energy, and recognizing the merit of the EH method for coping with metal d orbitals, we note from Fig. 5.66 the trend in band widths (occupied and unoccupied) and in Fermi levels in the first transition series. E_F drops steadily in going from titanium to nickel. It is not surprising that the total band width decreases with increasing atomic number: this is a simple consequence of the increase in nuclear charge in proceeding from titanium to nickel. Note that the number of vacant sites in the d band is much smaller at the end than at the beginning of the transition series. The d-hole count, N_h, a useful concept (see Section 5.4.7.2), is proportional to W_{vac}/S_{occ}, the ratio of the vacant and occupied widths of the d band: it occurs in quantitative calculations of heats of adsorption.

5.4.7.2 Heats of Chemisorption from EHT Calculations

For simplicity, consider a metal of the first transition series interacting to from a monolayer with an adsorbate A, which may have lone-pair or singly occupied or vacant orbitals. If the total occupancy of the metal band is N_M (with, in general $N_M = N_d + N_s + N_p$, the occupancies of d, s and p bands respectively) then, after chemisorption to yield the monolayer A/M, the electron balance for surface metal atom is:

$$N_{AM} = N_M + \sum_{j=1}^{f} N_A^j = \text{constant} \tag{37}$$

where $N_A = 0$, 1 or 2 for each vacant, singly or doubly occupied adsorbate orbital, respectively, and f is the number of such orbitals. By definition, the heat of chemisorption (see Section 2.7), $-\Delta H_a = Q$, is given by:

$$-\Delta H_a = Q = E_0(A, M) - E(A + M) \tag{38}$$

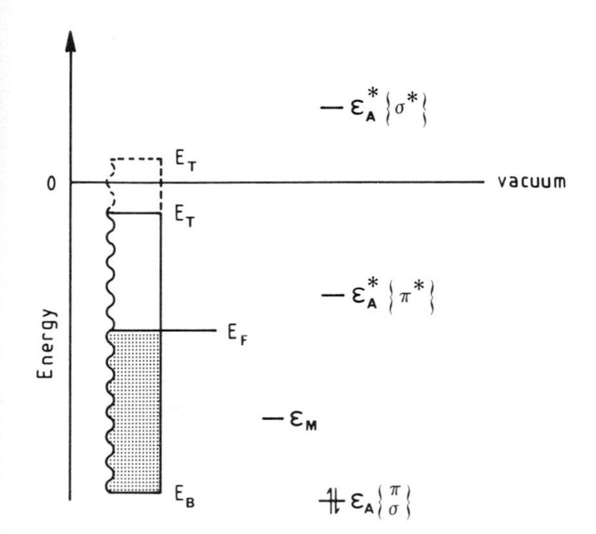

Figure 5.67 Energy diagram showing the interactions of the band structures of an adsorbate A and a metal M. The metal d band extends widely from the bottom (energy E_B) to the top (E_T), which may be below or above the vacuum level. The bands of the adsorbate, especially for low coverages, are very narrow. Typical positions of the adsorbate levels, occupied ε_A (σ or π) and vacant ε_A^* (π^* below but σ^* above vacuum level) as well as that of atomic metal orbital ε_M are shown.

where $E_0(A, M)$ is the total energy before chemsorption, and $E(A + M)$ the total electron energy after chemisorption, with the total electron count N_{AM} remaining constant (Eq. (37)). Now, for reasons of symmetry and energy, only some of the metal orbitals can interact with the orbitals of the adsorbate. (We assume that, within each irreducible representation for the relevant unit cell of the system A/M, an orbital in the adsorbate, say φ_A, interacts with one effective metal orbital φ_M which is some symmetry-adapted group orbital from all the available metal orbitals.) When we now picture the actual electronic interaction between A and M (Fig. 5.67), the orbitals of A, φ_A, arc taken to be quasi-degenerate and of initial energy ε_A, but the band of φ_M orbitals has a width $W(= W_{occ} + W_{unocc})$. Even if those metal orbitals not involved in the A–M interaction remain unchanged, their electron occupancies will change since they are implicated in the transfer of electrons or holes required to maintain a fixed Fermi level, E_F. (Tacit in this discussion is the constancy of E_F during chemisorption.) It follows, therefore, that the heat of chemisorption is made up of two main components, one from direct A–M interactions, the other from changes in orbital occupancies required to maintain the constancy of E_F.

Shustorovich and Baezold showed that, in order to evaluate Q, we should sum over all the occupied orbitals φ_{AM}, both bonding ($\varphi_{AM} \propto \varphi_A + \varphi_M$) and antibonding ($\varphi_{AM}^* \propto \varphi_A - \varphi_M$), the energies of which (ε_{AM}) lie below E_F. Proceeding in this way they showed that, for a donor (lone-pair) adsorbate, the heat of chemisorption, Q_D, is proportional to the d-hole count, N_h, and inversely proportional to $E_F - \varepsilon_A$, i.e.

$$Q_D \propto \beta^2 N_h/(E_F - \varepsilon_A) \tag{39}$$

where β is the resonance integral (off-diagonal matrix element).

Now we saw (Fig. 5.66) that E_F drops off rather slowly and N_h very sharply in going along the transition-metal series from titanium to nickel. We can therefore expect Q_D to decrease monotonically along this series.

For acceptor adsorbates (i.e. those containing vacant orbitals) these workers showed that the heat of chemisorption of the acceptor, Q_A, is proportional to the d occupancy, N_d:

$$Q_A \propto \beta^{*2} N_d/(\varepsilon_A^* - E_F) \tag{40}$$

where ε_A^* is the energy of antibonding orbital and β^* is again a resonance integral. This is opposite to the trend noted for donor adsorbates.

Experimental data for heats of chemisorption (see Section 2.7) are broadly in line with the predictions encapsulated in Eqs. (39) and (40). For strong acceptors, where $\beta^*/(\varepsilon_A^* - E_F) \gg 1$, the dependence of Q_A and N_d would be expected from Eq. (40) to be non-monotonic and may become parabolic. This is indeed what occurs with CO on transition metals.

Figure 5.67 gives us further insight into why metals often function as good catalysts. We first recall that the work functions of transition metals (i.e. the depth of E_F below the vacuum level) fall in the range 4.5–5.5 eV, i.e. they are much lower than the atomic d-orbital ionization energies, which range from 8 to 12 eV. In other words, $E_F - \varepsilon_M$ is *ca* 5 eV. We understand, therefore, why metal surfaces are better electron donors than metal complexes or small clusters, and can interact more effectively with σ^* and π^* vacant adsorbate orbitals. The second point to note is that these σ^* and π^* orbitals, because of their closer energetic proximity to E_F than the corresponding σ and π orbitals (Fig. 5.67), overlap more strongly with the metal d orbitals. Pursuing these points with characteristic clarity, Hoffman and Saillard have shown that what is principally responsible for the activation of small molecules such as CH_4 and H_2 when they approach transition-metal catalysts is the $M \rightarrow \sigma^*$ interaction.

5.4.7.3 The Adsorption of CO on Nickel

With the aid of EHT it is possible to calculate the density of states (DOS) curve for a thin slab (e.g. four (100) layers) of Ni before and after uptake of chemisorbed CO

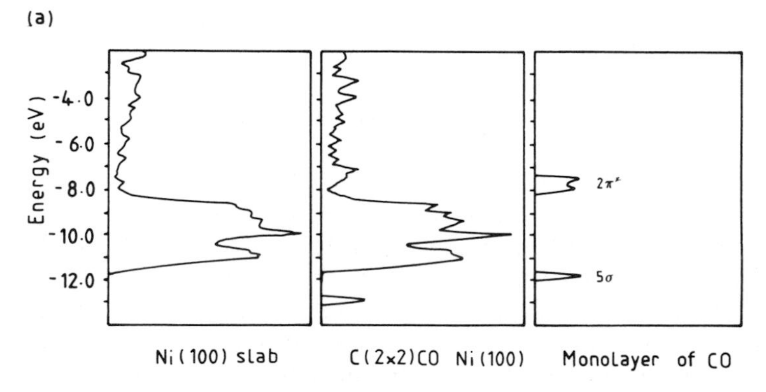

Figure 5.68 (a) Total density of states of a Ni(100) slab (left), the c(2 × 2) CO–Ni(100) system (middle) and a monolayer of CO (see text).

Table 5.3 Computed orbital populations in CO chemisorbed on metal surfaces of the first transition series (after Hoffmann)

	Electron densities in fragment orbitals					
	Ni(100)	Ni(111)	Ti(0001)	Cr(110)	Fe(110)	Co(0001)
5σ	1.60	1.59	1.73	1.67	1.62	1.60
$2\pi^*$	0.39	0.40	1.61	0.74	0.54	0.43

as a c(2 × 2) adsorbate. It is also possible to compute the band structure of the adsorbate itself, i.e. to ascertain the extent of overlap between 5σ and 2π orbitals on individual molecules when they are brought close together at the separation distances that they will attain in the chemisorbed state. The results are shown in Fig. 5.68. Table 5.3 enumerates orbital populations for the Ni(100) and five other metal surfaces.

The composite DOS (middle curve of Fig. 5.68) shows that the 5σ band has been depressed in energy. Further analysis shows that the net result so far as charge transfer is concerned is loss of electrons both from the CO 5σ (to the surface) and from a metal d_{z^2} (from the surface); compare Fig. 1.23. This figure includes the relevant orbital representations appropriate for the donation of electrons from the 5σ lone-pair orbital on the carbon end of the CO molecule to a solid surface, and back-donation from the metal d_{yz} (or d_{zx}) to the antibonding π orbitals (variously designated π^* or $2\pi^*$) of the CO. Whereas the former process, which could involve donation to partially filled or empty (i.e. virtual) s and d_{z^2} orbitals, strengthens the C–O link, the latter weakens it as it entails addition of electrons into the antibonding orbitals. These considerations are of central importance in the catalysed synthesis of methanol from syn-gas and in Fischer–Tropsch conversions (Sections 8.1 and 8.2).

5.4.7.4 Dissociative Chemisorption of CO

The work of Rhodin, Roberts, Ertl, Joyner and others established that, in general, early and middle transition metals break up CO; late ones simply bind it molecularly. Precisely how the CO is fractured is not known experimentally. Clearly, at some point the oxygen end of the molecule must make contact with the metal atoms, even though the common coordination mode on surfaces, as in molecular complexes, is through the carbon. In the context of pathways of dissociation, the discoveries of Bradshaw, Madey, Conrad and others, where CO is 'seen' lying down on some surfaces as depicted in Fig. 5.69, is illuminating. It is conceivable that such geometries prevail on the path to dissociation of CO to chemisorbed atoms. A. B. Anderson and Hoffmann have provided a useful theoretical model for CO bonding and dissociation. If we return to Table 5.3 we see a significant symptom pertaining to the bonding of CO on different metal surfaces, the population of CO 5σ and $2\pi^*$ orbitals.

Whereas the population of 5σ is almost constant, rising slowly as one moves from

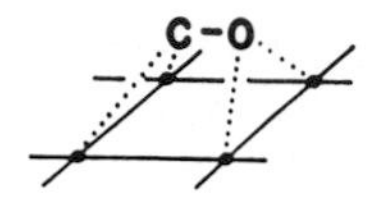

Figure 5.69 Schematic illustration, based on discoveries of Madey, Bradshaw and others, showing a possible intermediate state for bound CO about to dissociate on a metal surface. With permission from R. Hoffmann, *Solids and Surfaces: A Chemist's View of Bonding in Extended Structures*, VCH, Weinheim, **1988**.

the right (nickel) to the middle (titanium) of the first transition series, the population of $2\pi^*$, in contrast, rises sharply. Indeed, not much is left of the CO bond by the time we reach titanium. It is quite likely that, if one were to couple, dynamically, further geometric changes – allowing the CO to stretch and tilt towards the surface – one would presumably witness dissociation on the left side of the series (i.e. the left side of Fig. 5.66).

Figure 5.70 enables us to rationalize the trends described above. The 5σ level will interact more weakly as one moves to the left, but the dramatic effect is on $2\pi^*$. On the right, the $2\pi^*$ level lies above the metal d band. In the middle and on the left of the transition series, the Fermi level rises above $2\pi^*$, and so $2\pi^*$ interacts more, and is occupied to a greater extent. This holds the key to CO dissociation.

5.4.7.5 Insight from Ab-Initio Computations: Methanol Synthesis and Olefin Metathesis

In seeking to understand theoretically the bonding, energetics and mechanism of methanol synthesis on oxide catalysts, computational labour alone demands that oxides of low atomic number be tackled first. This philosophy motivated Hillier and co-workers to undertake an ab-initio (CI-SCF) study of the adsorption of CO, HCO, H_2CO and other small molecules on the (100) surface of MgO containing a guest Cu^+ ion. The surface itself was modelled by a small cluster of ions embedded in

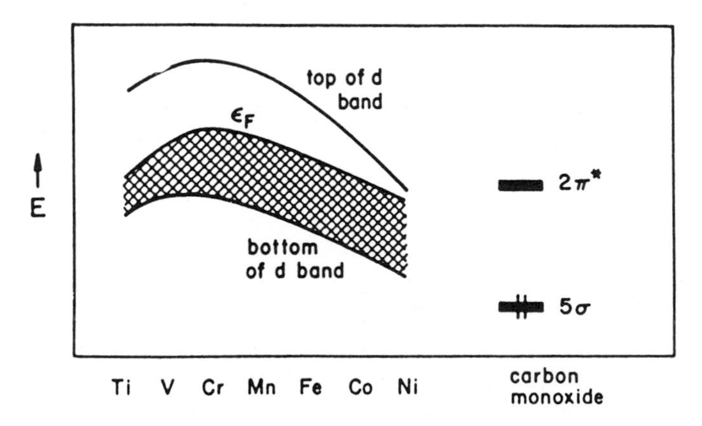

Figure 5.70 Superposition of the CO 5σ and $2\pi^*$ levels on the metal d band (see text). With permission from R. Hoffmann, *Solids and Surfaces: A Chemist's View of Bonding in Extended Surfaces*. VCH, Weinheim, **1988**.

a larger (5 × 5 × 2) point ion array to give the correct Madelung potential (Section 5.4.1) at the central ion. Their prior calculations had shown that lattice relaxation is of crucial importance. Indeed the guest Cu^+ ion relaxes some 1.0 Å out of the surface as a result of both electrostatic and elastic forces.

Reverting to Fig. 1.23 of Chapter 1, binding between the d^{10} Cu^+ ion and CO can involve both donation from the ligand 5σ orbital to the virtual 4s orbitals on the Cu^+ and back-donation from the $3d_{yz}$ and $3d_{yx}$ metal orbitals to the 2π virtual orbital. It transpires that this synergic donor–acceptor process is not viable for chemisorption at the non-defective (i.e. perfect) (100) MgO surface because only donation from the ligand σ orbital to the virtual 3s orbital of Mg^{2+} is possible. The $2p^6$ manifold in the closed shell of the Mg^{2+} ion is energetically inaccessible for any back-donation. At the sites of the guest Cu^+ ions, however, back-donation from the $3d^{10}$ manifold *is* feasible. This is a key factor in stabilizing bound CO at the catalyst surface. These calculations also revealed that it is important to take account of electron correlation in obtaining reliable estimates of the back-donation.

In order to cope satisfactorily with the problem of computing the contribution of electron correlation, the size of the cluster used must be small, especially for high-Z elements. This is the justification for modelling the interaction of CO and N_2 with transition-metal surfaces using clusters such as linear NiCO and NiN_2 and the CI GVB method. (It is salutary to recall that ab-initio HF SCF calculations of CO arrive at the wrong sign for the dipole moment, if electron correlation effects are not included!) It emerges that the $3d^{10}$ configuration is more important than $3d^9 4s^1$ insofar as bonding to N_2 is concerned.

The GVB ab-initio method in the hands of Goddard et al. has yielded useful insight into the mechanism of catalysis of olefin metathesis, which is central to the Phillips process of converting propylene to butadiene:

$$2\ \mathrm{(H)(H)C{=}C(CH_3)(H)} \longrightarrow \mathrm{(H)(H)C{=}C(H)(H)} + \mathrm{(H)(CH_3)C{=}C(CH_3)(H)}$$

Transition-metal complexes, such as $MoCl_6$, together with certain additives, catalyse olefin metathesis. It was once thought that the metal centre served to coordinate two olefins, thereby facilitating the overall reaction thus:

M ⇌ M ⇌ M ⇌ 2 ═ *

where the asterisks denote labelled carbons.

There is now little doubt that four-membered rings with a metal at one corner (a metallocycle) are implicated:

$$\mathrm{M{=}C} + \mathrm{{>}C^*{=}C^*{<}} \rightleftharpoons \mathrm{M{-}C{-}} / \mathrm{C^*{-}C^*} \rightleftharpoons \mathrm{M{=}C^*} + \mathrm{C{=}C^*}$$

With this background Goddard et al. examined, with GVB computation methods, how olefins react with metal oxo bonds:

$$\mathrm{H_2C{=}CH_2} + \mathrm{Cl_4Mo{=}O} \rightleftharpoons \mathrm{Cl_4Mo{-}O} / \mathrm{CH_2{-}CH_2}$$

$$\mathrm{H_2{=}CH_2} + \mathrm{Cl_2Mo({=}O)_2} \rightleftharpoons \mathrm{Cl_2Mo({=}O){-}O} / \mathrm{CH_2{-}CH_2}$$

5.5 Epilogue

Our discussions in this chapter should have convinced us that a quantum chemical approach to the interpretation of catalytic phenomena is indispensable if we are to gain a deep enough understanding to permit us to design better catalysts. Great studies have been taken since 1980 or so; and such is the pace of change in this domain of heterogeneous catalysis that it would be foolish to endeavour to single out any of the theoretical techniques that are likely, in the present perspective, to be of greatest value in the future. Rather, it is more important to emphasize anew some of the fundamental questions that concern us.

We should first recall that very many catalysts are multicomponent and multiphasic and that, consequently, the key to their better understanding is not likely to be found easily. A good specific example is the Pt–Re alloy catalyst supported on alumina, and pre-sulphided, which is so valuable in the reforming of hydrocarbons. We shall see in the next chapter (Section 6.3.1) that, although we know in broad terms *how* to modify such catalysts so as to boost a particular selectivity, we are still rather ignorant of *why* a change in catalyst performance takes place from a given chemical pretreatment. The multicomponent, multiphasic catalysts, of which there are many, are especially difficult to understand because it is no straightforward task to identify, let alone to characterize, their active sites.

We should also recall that an increasing family of catalysts (zeolites, clays and many mixed oxides typified by the bismuth molybdates) are monophasic and monocomponent, even though they may be, in operation, bi- or multi-functional. For this category of so-called uniform solid catalysts, identifying and characterizing the active sites is a more straightforward affair, as the techniques of solid-state chemistry and physics as well as those of surface science can be brought to bear to characterize these materials and to pinpoint their quintessential chemical features.

Reverting to the question of metallic or bimetallic catalysts either in supported or unsupported forms, it is clear that active sites in such systems can often be identified with ensembles or small clusters M_n (with n from 3 to 20). Theoretical studies of such clusters, apart from yielding desired properties such as the local or projected density of states or crystal orbital overlap populations from a 'static' model of a given structural entity, also reveal that small clusters can exist in a large number of geometric configurations very close or almost degenerate in energy. Moreover, interconversions between such configurations can freely occur by movement of one (not necessarily carrying the adsorbate) or more atoms in the cluster. Such processes can take place with little expenditure of energy. Simonetta and Gavezzotti, in a stimulating paper prompted by Falicov and Somorjai's correlation between catalytic activity and coordination number of atoms in a transition-metal surface, have argued that the change of relative position of one atom in a cluster may result in a significant change in its electron charge density, thereby producing the desired property of having the same atom to function as a good accepting and a good releasing centre. Apparently, configurational changes may occur rather readily in metallic clusters; they may take place several orders of magnitude more rapidly than the overall chemical reaction. More enquiries into the repercussions of such structural flexibility are required. But, at the very least, this picture offers an additional rationale for the occurrence of structure-sensitive catalytic reactions.

Lastly, we must never forget that although it is convenient to picture the skeletal atomic (or polyhedral) framework of a solid for some purposes, and also to emphasize the electron structure of the solid for other purposes, in the final analysis a unified approach, incorporating the reality of dynamical, atomic and electronic changes, is required. We shall be better placed to design superior catalysts when we can answer questions such as 'Why does this solid possess the bulk and surface structure that it has?' or 'How far can we modify the elemental components of this structural type without causing it to collapse?' and 'Why are there geometrical and stereochemical arrangements at surfaces that have no counterparts in other domains of chemistry?' These are also some of the key issues in inorganic chemistry of solids and in organometallic chemistry.

5.6 Problems

1 In common with many perovskite-based oxide catalysts, the superconducting compounds $YBa_2Cu_3O_{7-\delta}$ $(0 < \delta < 1)$ are capable of releasing or sorbing oxygen, depending upon the precise conditions of temperature and pressure. Explain how you would assess:

(a) the fraction of oxygen capable of reversible and irreversible uptake and loss,
(b) what structural changes occur in these solids during uptake or loss, and
(c) whether these 'warm' superconductors are also good selective oxidation catalysts.

(*Hint*: Consult the following reference for background information: I. J. Pickering, J. M. Thomas, *J. Chem. Soc., Faraday Trans.* **1991**, *87*, 3067).

2 Perovskites are good catalysts for a variety of reactions. Oxides crystallizing in this structure are, however, prone to exhibit defects of one kind or another. Summarize the principal types of structural defects that may occur in perovskites, and indicate which methods are best suited for the detection and characterization of each kind, or combination, of defect.

3 Alkaline-earth oxides are good catalysts for a variety of reactions involving the oxidation and photo-oxidation of hydrocarbons. Some of the mechanistic questions that arise devolve upon whether excitons and point defects are implicated. Elaborate upon ways in which such possibilities may be tested and suggest what other techniques (such as the sacrificial use of lattice oxygen) may have to be considered.
(*Hint*: First consult J. Cunningham et al., *J. Chem. Soc. Faraday Trans.* **1985**, *81*, 2027).

4 After first consulting P. Behrens (*Adv. Mater.* **1993**, *5*, 127), outline the ways in which microporous and mesoporous catalysts are characterized.

5 Tin dioxide (SnO_2) is the basis of many sensitive sensors for the detection of alcohol and other volatile liquids (see J. D. Wright, *Chem. Br.* **1995**, *31*, 374). Using the information given in this chapter sketch the band structure of p-type SnO_2, which is the active catalyst for the sensor, and offer a mechanism for its mode of action.

6 Direct, real-space solution of zeolite framework crystal structures by simulated annealing (see S. Kirkpatrick et al., *Science* **1983**, *220*, 671) is increasingly attempted as an alternative to conventional powder diffraction or model building methods. Explain the principles of simulated annealing. (Consult: M. W. Deem, J. M. Newsam, *J. Am. Chem. Soc.* **1992**, *114*, 7189).

7 By first consulting the review by J. M. Newsam and M. M. J. Treacy (*Zeolites* **1993**, *13*, 183) explain how the 'ZeoFile' anthology of data on known zeolitic structures may be used as in interactive facility for simulating powder, X-ray, neutron and electron diffraction patterns of all known and certain hypothetical zeolitic crystal structures.

8 Calculated and experimental crystal structures of the siliceous end-members of zeolites have been shown to agree quite closely. With the aid of the energy minimization computations developed by Catlow (see C. R. A. Catlow, *J. Chem. Soc., Chem. Comm.* **1990**, 782 and *Modelling of Structure and Reactivity in Zeolites* (Ed.: C. R. A. Catlow), Academic Press, London, **1992**), explain how the agreement for silicalite (Fig. 5.71) is achieved.

9 The energy diagram shown in Fig. 5.72 refers to the junction formed between a semiconductor and an electrolyte. (This is the basis upon which all photoelectrochemical cells operate – see R. J. D. Miller et al., *Surface Electron Transfer*

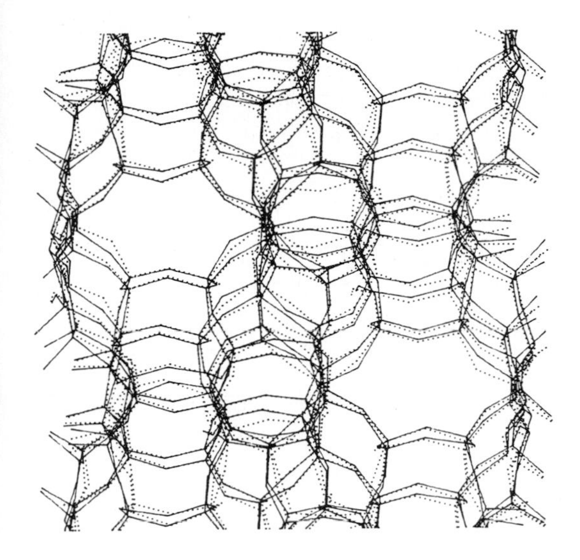

Figure 5.71 Crystal structure of silicalite: ···, calculatial; –, experimental.

Processes, VCH, Weinheim, **1995** and Section 8.7). Explain what all the symbols shown in this diagram mean, and outline how they are individually determined.

10 By anology with the work of Hoffmann described in this chapter on the interaction of CO with transition metals, consider qualitatively the theoretical principles that would favour dissociative as against non-dissociative adsorption of NO on rhodium and platinum.

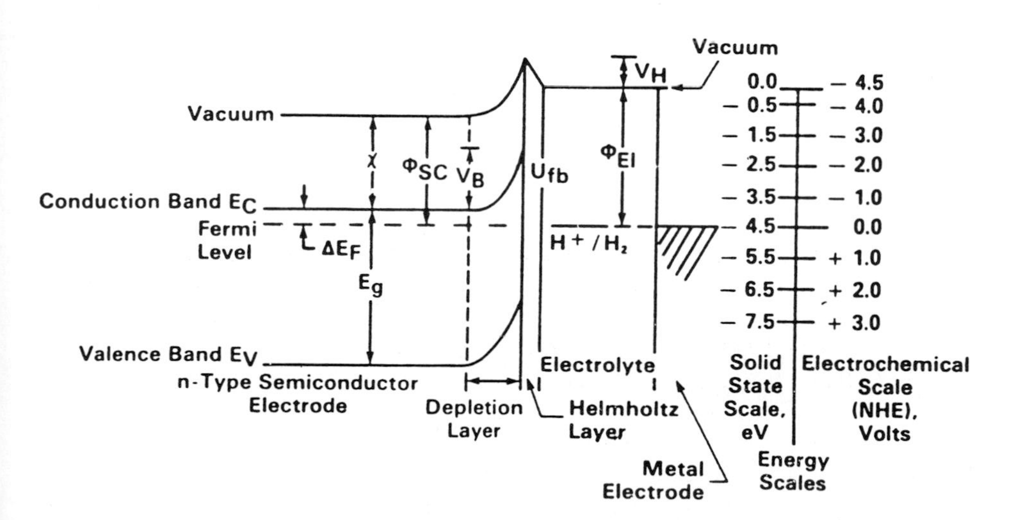

Figure 5.72 Energy diagram for the junction between a semiconductor and an electrolyte.

5.7 Further Reading

5.7.1 Structures

A. M. Bradshaw, *Faraday Discuss. Chem. Soc.*, **1990**, *89*, 1.

F. Cavani, F. Trifiro, A. Vaccari, Hydrotalcite-type anionic clays, *Catal. Today* **1991**, *11*, 173.

G. M. Clark, *The Structures of Non-molecular Solids*, Applied Science, London, **1972**.

M. E. Davis, New vistas in molecular sieve catalysis, *Acc. Chem. Res.* **1993**, *23*, 111.

J. P. Glusker, M. Lewis, M. Rossi, *Crystal Structure Analysis for Chemists and Biologists*, VCH, Weinheim, **1994**.

Y. Izumi, K. Unobe, M. Onaka, *Zeolite, Clay and Heteropoly Acids in Organic Reactions*, VCH, Weinheim, **1992**.

I. V. Kozhevnikov, Heteropoly acids and related compounds as catalysts for fine chemical synthesis, *Catal. Rev. – Sci. Eng.* **1995**, *37*, 311.

T. E. Madey, J. T. Yates, *Nuovo. Chim. Suppl.* **1967**, *5*, 483.

W. M. Meier, D. H. Olson (Eds.), *Atlas of Zeolite Structure Types*, Revised ed., Butterworths, London, **1993**.

M. O'Keeffe, A. Navrotsky (Eds.), *Structure and Bonding in Crystals*, Vols. I and II, Academic Press, New York, **1981**.

C. N. R. Rao, J. Gopalakrishnan, *New Directions in Solid State Chemistry*, Cambridge University Press, Cambridge, **1986**.

M. W. Roberts, *Chem. Soc. Rev.*, **1977**, *6*, 373.

A. F. Wells, *Structural Inorganic Chemistry*, 5th ed., Clarendon Press, Oxford, **1984**.

5.7.2 Computational Approaches

M. Allavena, E. Kassab, K. Seiti, *J. Phys. Chem.* **1991**, *95*, 9425.

M. P. Allen, D. J. Tildesley, *Computer Simulation of Liquids*, Oxford Science, Oxford, **1987**.

P. D. Battle, T. S. Bush, C. R. A. Catlow, Structures of Quaternary Ru and Sb Oxides by Computer Simulation, *J. Am. Chem. Soc.* **1995**, *117*, 6292.

A. T. Bell, Understanding catalysis at the molecular level, in *Science and Technology in Catalysis* (Eds.: Y. Izumi, H. Arai, M. Iwamoto), Kodansha, Tokyo, **1995**, p. 63.

C. R. A. Catlow (Ed.), *Modelling of Structure and Reactivity in Zeolites*, Academic Press, London, **1992**.

C. R. A. Catlow in *Handbook of Heterogeneous Catalysis*, (Eds. G. Ertl, H. Knözinger, J. Weitkamp), VCH, Weinheim **1997**.

H. Chuan Kang, W. H. Weinberg, Dynamic Monte Carlo simulation of surface-rate processes, *Acc. Chem. Res.* **1992**, *25*, 253.

H. Chuan Kang, W. H. Weinberg, Modelling the kinetics of heterogeneous catalysis, *Chem. Rev.* **1995**, *95*, 667.

C. M. Freeman, C. R. A. Catlow, J. M. Thomas, S. Brode, *Chem. Phys. Lett.*, **1991**, *186*, 137.

D. H. Gay, A. L. Rohl, MARVIN: a new computer code for studying surfaces, interfaces and morphologies, *J. Chem. Soc., Faraday Trans.* **1995**, *91*, 925.

A. R. George, C. R. A. Catlow, J. M. Thomas, *J. Chem. Soc. Farad. Trans.*, **1995**, *91*, 3975.

I. H. Hillier, S. P. Greatbanks, P. Sherwood, *J. Phys. Chem.* **1994**, *98*, 8134.

H. P. C. E. Kuipers, R. A. van Santen, *Peterhouse Catalysis Conf.*, March **1995**, (unpublished).

J. M. Newsam, Zeolite structural problems from a computational perspective, in *Proc. 9th Int. Zeolite Conf.*, Montreal, 1992 (Ed.: R. von Ballmoos), Butterworths, London, **1993**, p. 127.

D. S. Santilli, T. V. Harris, S. I. Zones, *Micropor. Mater.*, **1993**, *1*, 329.

J. Sauer, *Chem. Rev.* **1989**, *89*, 199.

R. A. van Santen, G. J. Kramer, *Chem. Rev.* **1995**, *95*, 637.

A. A. Shubin, C. R. A. Catlow, J. M. Thomas, K. I. Zamaracv, *Proc. Roy. Soc.*, **1994**, *A446*, 411.

5.7.3 Electronic Structures and Properties

M. Baerns, Z. Zhang, X. E. Verykios, Effect of electronic properties of catalysts for the oxidative coupling of methane, *Cat. Rev. – Sci. Eng.* **1994**, *36*, 507.

J. Baker, M. Muir, J. Andzelm, Density functional studies of proton transfer, *J. Chem. Phys.*, **1995**, *102*, 2063.

J. R. Bolton, *Science*, **1978**, *202*, 705.

J. K. Burdett, *Chemical Bonding in Solids*, Oxford University Press, Oxford, **1995**.

M. H. Cohen, M. V. Ganduglia-Pirovono, J. Kudrnovsky, Electronic and nuclear chemical reactivity, *J. Chem. Phys.* **1994**, *101*, 8988.

P. A. Cox, *The Electronic Structure and Chemistry of Solids*, Oxford Science, Oxford, **1987**.

B. Delley, *J. Chem. Phys.*, **1991**, *94*, 7245.

H. Gerischer, *Photoeffects at Semiconductor-Electrolyte Interfaces*, ACS Symp. Ser. No. 146, Washington, **1981**, p1.

B. Hammer, J. K. Norskov, Why gold is the noblest of all metals, *Nature* **1995**, *376*, 238.

R. Hoffmann, *Solids and Surfaces: A Chemist's View of Bonding in Extended Structures*, VCH, Weinheim, **1988**.

R. W. Joyner, R. A. van Santen (Eds.), *Elementary Reaction Steps in Heterogeneous Catalysts*, Kluwer Dordrecht, **1993**.

R. J. D. Miller, G. L. McLendon, A. J. Nozik, W. Schmickler, F. Willing, *Surface Electron Transfer Processes*, VCH, Weinheim, **1995**.

A. J. Nozik, *Farad. Disc. Chem. Soc.*, **1980**, *70*, 8.

R. G. Parr, W. Yang, *Density Functional Theory of Atoms and Molecules*, Oxford University Press, New York, **1989**.

R. A. van Santen, *Theoretical Heterogeneous Catalysis*, World Scientific, Singapore, **1991**.

S. Yoshida, S. Sakaki, H. Kobayashi, *Electronic Processes in Catalysis*, VCH, Weinheim, **1994**.

6 Poisoning, Promotion, Deactivation and Selectivity of Catalysts

6.1 Background

We saw in Section 1.3 and the following sections that the activity and selectivity of heterogeneous catalysts may change during the course of conversion. Seldom is there an increase in activity; it usually decreases owing to either chemical or physical reasons, or a combination thereof, which we shall discuss later. Sometimes the selectivity increases as a result of the deliberate or adventitious addition of poison. Poisoning, therefore, is not to be viewed as an entirely negative effect. On the contrary, especially with supported alloy catalysts, it may be strategically advantageous to introduce poisoning species prior to the commencement of the catalytic conversion.

From an analysis of earlier classic work on poisoning and deactivation of commercial catalysts and from recent elegant model studies of poisoning and its antithesis (promotion), a great body of related factual information has been established, and much theoretical understanding has emerged. For metals, well-recognised poisons are molecules or radicals containing elements in Group VB and VIB of the Periodic Table, notably nitrogen, phosphorus, arsenic, antimony, oxygen, sulphur, selenium and tellurium. But, as was recognised early in the 19th century (Section 1.2), molecules containing multiple bonds (e.g. CO) or those that are bulky and tenaciously adsorbed can also function as poisons. In the silver-catalysed conversions of ethylene, small quantities of chemisorbed chlorine improve the selectivity towards ethylene oxide and against complete oxidation to CO_2, but over-chlorination leads to poisoning. In zeolitic and other acid catalysts such as Al_2O_3, there is a loss in their activity for the cracking of hydrocarbons, but an improvement in their isomerisation facility, upon exposure to nitrogeneous bases such as pyridine or quinoline. The addition of sulphur to a platinum catalyst diminishes its activity for the hydrogenation of butadiene. Pt–Re alloy catalysts for conversions of *n*-alkanes are especially interesting. As expected, the initial activity of the sulphided catalyst is lower than that of the unsulphided catalyst, because of the poisoning effect of sulphur. But the sulphided catalyst displays a much smaller decline in activity, so that there is a higher total conversion.

Sulphur is a pervasive catalyst poison. It is present, in greater or lesser degree, in all transport fuels, so that efficient auto-exhaust catalysts need to be effective in the presence of this element. As base metals are very prone to poisoning by sulphur, all practical auto-exhaust catalysts (see Chapters 1 and 8) are based on the noble metals platinum, rhodium and palladium. Sulphur also destroys the performance of the Cu/ZnO catalyst for the synthesis of methanol (Chapter 8).

In this chapter we first consider how selectivity is governed by mass and heat transfer for both monofunctional and bifunctional catalysts. This is a question that affects unpoisoned and poisoned catalysts alike. It leads, in turn, to an analysis, largely from the viewpoint of the chemical engineer, of catalyst deactivation as a result of poisoning or other factors such as sintering or carbon laydown (fouling), and then to an assessment of the operational consequences of poisoning. Finally, we summarise modern views on our theoretical appreciation of the poisoning of catalysts, and draw appropriate lessons about its antithesis – promotion.

6.1.1 Effect of Mass Transfer on Catalytic Selectivity

6.1.1.1 Effect of Intraparticle Diffusion

Although the factors involving the chemical interaction between reactants and catalyst surface, the topography of the surface and – for zeolites and certain specially prepared graphites – the intracrystalline geometry are of central importance in determining the specificity with which a reaction occurs, it is nevertheless important to consider the effect which the porous nature of a commercially prepared catalyst has on reaction selectivity. First, it is preferable not to lose any of the prior advantages gained by research which has led to a particular catalyst choice for enhancing specificity. Secondly, the commercial operating conditions to be employed for the catalytic reaction should preserve product throughput or yield (whichever happens to be the choice on economic grounds) while sustaining maximum selectivity in favor of the desired product.

The effect which catalyst pore size has on the rate of heterogeneous catalytic reactions was considered by Thiele (see Further Reading, Chapter 4, Section 4.8.2) who originally laid the foundations for Wheeler's application of these principles to the question of selectivity (see Further Reading, Section 6.5.7). In Chapter 4 we discussed how chemical reaction rates are affected by the diffusion of reactants into, and products out of, the porous catalyst structure. There we showed that porous structures with narrow pores displaying low effective diffusivities, and thus providing only restricted access to the internal surface area (of which most heterogeneous catalysts are composed), limit the overall rate at which the reaction proceeds. If there is more than one chemical pathway by which the reactant can react, then, from considerations outlined in Chapter 4, it follows that the rate of reaction along the separate routes can be influenced to different extents by virtue of the relative abilities of the products (or reactants if more than one) to diffuse through the porous medium.

Although Wheeler's treatment of the problem invoked reaction in single pores, and could be applied to relatively simple porous structures (such as a straight non-intersecting cylindrical pore model) with moderate success, we will characterise the porous structure by means of the effective diffusivity D_e (discussed in Chapter 4), which can be measured for a given gaseous component. In order to develop the principles relating to the effects of diffusion on reaction selectivity, we will first discuss selectivity in isothermal catalyst pellets and confine our attention to three typical chemical reaction schemes.

For the first illustration we consider two parallel competing reactions. For instance, it may sometimes be necessary to convert into a desired product only one component in a mixture. The dehydrogenation of six-membered cycloparaffins in the presence of five-membered cycloparaffins without affecting the latter is one such example of a selectivity problem in petroleum reforming reactions. In this case it is desirable for the catalyst to favour a reaction depicted as

$$A \xrightarrow{k_1} B$$

when it might be possible for the reaction

$$X \xrightarrow{k_2} Y$$

to occur simultaneously. If both reactions were isothermal, first-order and unaffected by intraparticle diffusion, the ratio of the respective reaction rates, which would be a measure of the reaction selectivity, would be $k_1 c_{Ag}/k_2 c_{Xg}$ where the subscript g refers to conditions in the gas-phase exterior to the catalyst pellet. More specifically, the ratio of the rates of reaction of the two reacting components A and X is the intrinsic kinetic selectivity, designated S. Thus,

$$R_A/R_X = S = k_1/k_2 \tag{1}$$

If, however, both reactions were influenced by intraparticle diffusion effects, the rate of reaction of a particular component would (as shown in Chapter 4) be $D_e c_g \phi(\tanh \phi)$ where ϕ is the Thiele modulus for a first-order reaction.

As is often the case, the molecular weights of the diffusing reactants are similar and D_e can be regarded as effectively constant for the catalyst pellet. Under these circumstances

$$S = \frac{\phi_1 \tanh \phi_1}{\phi_2 \tanh \phi_2} \tag{2}$$

the asymptotic values of which are k_1/k_2 (for $\phi < 1$) and $(k_1/k_2)^{1/2}$ (for $\phi > 5$). We are therefore led to conclude, when parallel competing first-order reactions occur in isothermal pellets with large pores, that the intrinsic selectivity is unaffected. However, in large pellets with small pores the intrinsic selectivity is merely the square root of the value for the unimpeded reaction. Thus, for large Thiele modulus ϕ,

$$S = \phi_1/\phi_2 = (k_1/k_2)^{1/2} \tag{3}$$

When it is known that intraparticle diffusion impedes reaction, the corollary is that maximum selectivity is achieved by operating with small pellets and large-diameter pores.

Turning to the case of concurrent reactions of the type

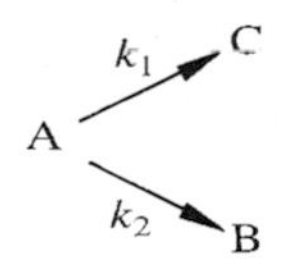

and exemplified by the catalytic decomposition of ethanol to yield either ethylene (dehydration) or acetaldehyde (dehydrogenation), if both reactions are of the same kinetic order, then the intrinsic selectivity will remain the same irrespective of whether or not diffusion influences the reaction rate. This is because both products are formed from the same reactant at a relative rate only dependent on the intrinsic kinetic constants. However, if the reactions are of different kinetic orders, then the reaction of the lowest kinetic order would be favored. If B were formed by a first-order reaction and C by a zero-order reaction, then, as the example below shows, if intraparticle diffusion effects were prominent, the rate of formation of B would be impeded with respect to the rate of formation of C.

Example

Two consecutive irreversible catalytic reactions

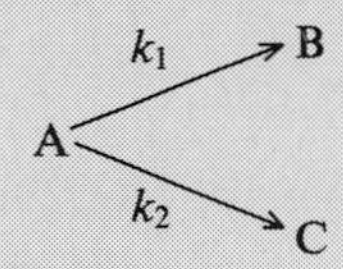

occur isothermally in a porous catalyst pellet. The desired product B is formed by a first-order chemical reaction, whereas the wasteful product C is formed by a zero-order reaction. It is known that diffusion influences both reactions. Deduce an expression for the selectivity.

Solution

For simplicity (as in Section 4.6), assume the catalyst pellet is a thin wafer. A steady-state material balance over an element of the pellet (see Fig. 4.22) of thickness Δx for component A yields

$$D_e \frac{d^2 c_A}{dx^2} = k_1 c_A + k_2$$

This is merely a statement asserting that, in the steady state, the change of diffusive flux (per unit pellet volume) of component A is balanced everywhere within the pellet by its total rate of reaction per unit volume. Rewriting this equation as

$$\frac{d^2 c_A}{dx^2} - f^2 c_A = g^2$$

where $f^2 = k_1/D_e$ and $g^2 = k_2/D_e$, we obtain for B and C, in a similar manner,

$$\frac{d^2 c_B}{dx^2} + f^2 c_A = 0$$

and

$$\frac{d^2 c_C}{dx^2} = -g^2$$

The boundary conditions are

$$x = \pm L, \quad c_A = c_{Ag}, \quad c_B = c_C = 0$$

at the gas–solid interface. At the pellet centre there is no net flux (pellet exposed to gas at both faces and unidirectional diffusion) so at

$$x = 0, \quad dc_A/dx = dc_B/dx = dc_C/dx = 0$$

The complementary solution to the first equation is

$$c_A = Ae^{fx} + Be^{-fx}$$

and the particular integral may be seen, by inspection, to be $(-g^2/f^2)$. Adding this to the complementary solution and determining A and B from the boundary conditions gives the concentration profile for A through the pellet:

$$c_A = \left(c_{Ag} + \frac{g^2}{f^2}\right)\frac{\cosh fx}{\cosh fL} - \frac{g^2}{f^2}$$

Defining the selectivity as the ratio of rates of formation of B and C and recalling that this is equivalent to the ratio of fluxes in the steady state,

$$S = \frac{(dc_B/dx)_{x=L}}{(dc_C/dx)_{x=L}}$$

To find both $(dc_B/dx)_{x=L}$ and $(dc_C/dx)_{x=L}$ we need only to integrate the differential equations for B and C once. Thus

$$\left(\frac{dc_B}{dx}\right)_{x=L} = -f^2\int_0^L c_A\,dx = -f^2\int_0^L\left\{\left(c_{Ag} + \frac{g^2}{f^2}\right)\frac{\cosh fx}{\cosh fL} - \frac{g^2}{f^2}\right\}dx$$

$$= -f\left(c_{Ag} + \frac{g^2}{f^2}\right)\tanh fL + g^2L$$

and

$$\left(\frac{dc_C}{dx}\right)_{x=L} = -g^2\int_0^L dx = -g^2L$$

Thus, finally,

$$S = \frac{f}{g^2L}\left(c_{Ag} + \frac{g^2}{f^2}\right)\tanh fL - 1$$

$$= \left\{\frac{k_1}{k_2}c_{Ag} + 1\right\}\frac{\tanh\phi}{\phi} - 1$$

The third important class of reactions, which is common in petroleum reforming reactions, may be represented by the scheme

$$A \xrightarrow{k_1} B \xrightarrow{k_2} C$$

and exemplified by the dehydrogenation of six-membered cycloparaffins to aromatics over metal and metal oxide catalysts. We suppose that B is the desired product and C the waste product. Assuming first-order kinetics and using the simple flat-plate model to account for diffusion effects (Section 4.6), the steady-state material balance equations for A and B, respectively, in the pellet are

$$D_e \frac{d^2 c_A}{dx^2} = k_1 c_A \tag{4}$$

$$D_e \frac{d^2 c_B}{dx^2} = k_2 c_B - k_1 c_A \tag{5}$$

Both of these equations are statements to the effect that, in the steady state, change of diffusive flux is balanced by chemical reaction. With the usual boundary conditions at the pellet periphery

$$c_A = c_{Ag} \quad \text{and} \quad c_B = c_{Bg} \quad \text{at } x = \pm L \tag{6}$$

and at the pellet centre where

$$dc_A/dx = dc_B/dx = 0 \quad \text{at } x = 0 \tag{7}$$

the equations may be solved to give the concentration profiles $c_A(x)$ and $c_B(x)$ within the pellet. To calculate the selectivity we require (as before) the ratio of the fluxes of A and B (equivalent, in the steady state, to the ratio of rates of reaction). It is convenient to calculate these at the pellet periphery where conditions are known. By this procedure we finally obtain for the selectivity the expression

$$S = -\frac{(dc_B/dx)_{x=L}}{(dc_A/dx)_{x=L}} = \left(\frac{k_1}{k_1 - k_2}\right)\left\{1 - \frac{\phi_2 \tanh \phi_2}{\phi_1 \tanh \phi_1}\right\} - \frac{c_{Bg}\phi_2 \tanh \phi_2}{c_{Ag}\phi_1 \tanh \phi_1} \tag{8}$$

For small values of the Thiele modulus (when diffusion effects are negligible), the above expression reduces to the selectivity we would calculate on chemical kinetic grounds (i.e. if chemical reaction were rate-controlling) when

$$S = 1 - c_{Bg}/kc_{Ag} \tag{9}$$

where $k = k_1/k_2$. Now although the ratio of fluxes, and hence the selectivity, are really point values, we may, without loss of generality, conveniently regard it as the rate of formation of B with respect to A, so that S becomes $(-dc_B/dc_A)$. Thus, integrating Eq. (9) from a value c_{Ai}, the value of c_A at the inlet to a fixed-bed catalytic reactor (see Sections 7.3.1.2 and 7.3.2.2), we obtain

$$\frac{c_B}{c_A} = \left(\frac{k}{k-1}\right)\left\{\left(\frac{c_A}{c_{Ai}}\right)^{(1-k)/k} - 1\right\} \tag{10}$$

which reflects the yield of B produced. If diffusion were rate-limiting, on the other hand, (large Thiele modulus such that $\tanh \phi$ is approximately unity) then Eq. (8) becomes

$$-\frac{\mathrm{d}c_{\mathrm{B}}}{\mathrm{d}c_{\mathrm{A}}} = \left(\frac{\sqrt{k}}{1+\sqrt{k}}\right) - \frac{1}{\sqrt{k}} \cdot \frac{c_{\mathrm{B}}}{c_{\mathrm{A}}} \tag{11}$$

which gives on intergration from a value c_{Ai}

$$\frac{c_{\mathrm{B}}}{c_{\mathrm{A}}} = \left(\frac{k}{k-1}\right)\left\{\left(\frac{1+\sqrt{k}}{\sqrt{k}}\right)\left(\frac{c_{\mathrm{A}}}{c_{\mathrm{Ai}}}\right)^{(1-\sqrt{k})/\sqrt{k}} - 1\right\} \tag{12}$$

Inspection of Eq. (12) and comparison with Eq. (10) shows that the yield of desired product under conditions of diffusion limitation is considerably reduced. The corollary which arises is that the yield of the desired product B would be increased if small catalyst particles with large pores (i.e. a small value of the Thiele modulus ϕ) were used. On the other hand, if the Thiele modulus is less than 0.3 (at about which value the yield of B would be in accord with Eq. (8), independent of diffusion effects) a large decrease in particle size would be necessary to achieve any further significant increase in selectivity. An increase in average pore size of the catalyst would also assist. When such a drastic reduction in pellet size is necessary to improve the yield, the use of a fluidized-bed reactor (see Section 7.3.2.3) would obviate the large pressure drop which occurs when employing small-size particles in a fixed bed.

6.1.1.2 Non-isothermal Conditions

In practice, of course, it is rare that the catalytic reactor operates isothermally. More often than not, heat is generated within the reactor by exothermic reaction and there is an attendant rise in temperature (for example, along the length of a packed tubular reactor) even though heat exchange equipment may be employed to remove some of the heat evolved. Consequently it is necessary to consider what effect non-isothermal conditions have on catalyst selectivity. The influence which the simultaneous transfer of heat and mass has on the selectivity of catalytic reactions can be assessed from a mathematical model in which diffusion and chemical reaction of each component within the porous catalyst are represented by a differential equation and heat released or absorbed by reaction is described by a heat-balance equation. The boundary conditions ascribed to the problem depend on whether interparticle heat and mass transfer are considered important. To illustrate how the model is constructed, we consider the case of two concurrent first-order reactions, both products B and C being formed from the same reactant. As was pointed out in Section 6.1.1.1, if conditions were isothermal, selectivity would not be affected by any diffusion effects within the catalyst pellet. However, we will see that non-isothermal conditions do affect selectivity, even when both competing reactions are of the same kinetic order. The conservation equations for each component in a flat slab-shaped porous pellet are

$$D_{\mathrm{e}} \frac{\mathrm{d}^2 c_{\mathrm{A}}}{\mathrm{d}x^2} - (k_1 + k_2) c_{\mathrm{A}} = 0 \tag{13}$$

$$D_{\mathrm{e}} \frac{\mathrm{d}^2 c_{\mathrm{B}}}{\mathrm{d}x^2} + k_1 c_{\mathrm{A}} = 0 \tag{14}$$

$$D_e \frac{d^2 c_C}{dx^2} + k_2 c_A = 0 \tag{15}$$

These merely assert that, for the steady state, the change in diffusive flux of a reacting component is exactly balanced by the rate at which that component is formed or disappears by chemical reaction. If it is assumed that the principal resistance to the transfer of heat generated by chemical reaction inside the pellet resides in a thin boundary layer of relatively stagnant gas bathing the pellet, then we may consider the temperature within the porous pellet to be substantially uniform, say at T_s. If the bulk gaseous temperature is T_g, the driving force for the transfer of heat from the gas–solid interface to the bulk gas is $(T_s - T_g)$ and the heat transferred can be written as $h(T_s - T_g)$ where h is the heat transfer coefficient from particulate to fluid phase. In the steady state this may be equated to the heat generated by reaction. Thus the appropriate form of heat balance is

$$h(T_s - T_g) = -(\Delta H_1 + \Delta H_2)(k_1 + k_2)c_A \tag{16}$$

Because both kinetic constants k_1 and k_2 are exponentially dependent on the temperature T_s within the pellet (according to an Arrhenius form of temperature dependence) and the reactant concentration c_A appears explicitly in the three mass-conservation equations and also in the heat-balance equation, the problem must be solved numerically rather than analytically. The boundary conditions at the pellet centre are, for reasons of symmetry,

$$x = 0; \quad \frac{dc_A}{dx} = \frac{dc_B}{dx} = \frac{dc_C}{dx} = 0 \tag{17}$$

At the catalyst pellet periphery

$$x = L; \quad c_A = c_{Ag}, \quad c_B = 0, \quad c_C = 0 \tag{18}$$

if we are concerned with an isolated pellet. Otherwise, for a pellet contained in a tubular reactor, the appropriate boundary conditions for concentration would be those which prevail at a given point along the catalyst bed and which must be sought iteratively when coupled to conservation equations for the catalyst bed. The constant pellet temperature T_s must also be sought iteratively. The technique for solution is thus to assume an initial value for T_s, solve the set of differential equations for the specified boundary conditions and then check that the heat balance equation is satisfied. If it is not, then a new value for T_s is assumed and the whole procedure repeated until convergence is obtained. It is a simple matter to calculate the selectivity once this has been accomplished by computing the ratio of fluxes of components B and C at the pellet boundary $x = L$ where conditions are known and apposite to the location of the pellet along the reactor length.

The effect of diffusion on catalyst selectivity in porous catalysts operating under non-isothermal conditions (see Section 4.6.2) has been examined by a number of workers. The mathematical problem has been comprehensively stated in a paper by Akhter et al. (1977) (see Further Reading, Section 6.5.3) which also takes into account the effect of surface diffusion on selectivity. For consecutive first-order exothermic reactions the selectivity increases with an increase in Thiele modulus when

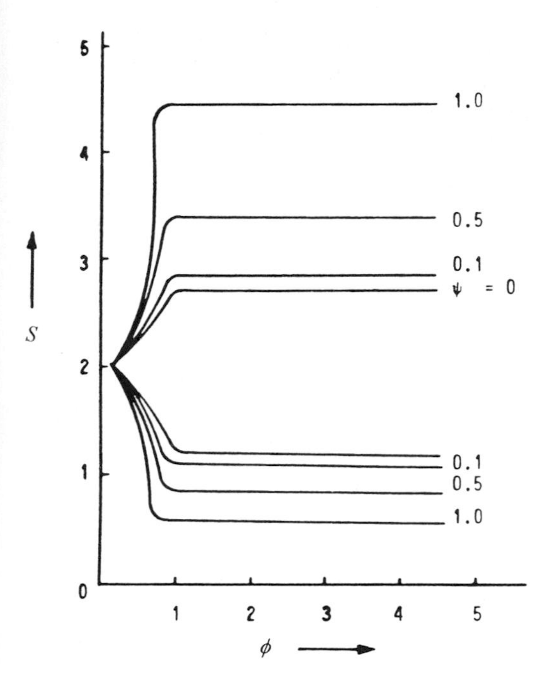

Figure 6.1 Effect of diffusion on catalytic selectivity for consecutive first-order reactions. The parameter ψ is a measure of the effect of surface diffusion. Curves (of selectivity S as a function of Thiele modulus ϕ) which ascend as ϕ increases relate to exothermic reactions, and those which descend relate to endothermic reactions. With permission from *J. Catalysis* **1977**, *50*, 205.

the parameter Δ (the difference between the activation energy for reaction and the heat of adsorption of the reactant) describing the desired reaction exceeds that for the undesired reaction. Figure 6.1 reveals that, whether or not surface diffusion is a mode of transport, the selectivity rises abruptly to an asymptotic value at Thiele moduli in the range 0.5–1.0. There would therefore be no point in increasing the Thiele modulus beyond 1 (increase in particle size and decrease in porosity) for such reactions in order to improve selectivity, because there would be an attendant decrease in product throughput as intraparticle diffusion becomes more difficult. Diametrically opposite effects to those just described are to be expected for consecutive endothermic reactions. A similar analysis concerning selectivity has also been applied to concurrent reactions influenced by diffusion and heat transfer.

6.1.1.3 Effect of Interparticle Mass and Heat Transfer

Mass and heat transfer between the bulk fluid phase and the external catalyst surface can have an effect on reaction rates, as Smith (1981) has described (see Section 6.5.1), and hence the selectivity, because of modified concentration and temperature driving forces. Consider the consecutive reaction

$$A \xrightarrow{k_1} B \xrightarrow{k_2} C$$

in which B is the desired product. The effect of a resistance to the mass transfer of reactant A from the bulk fluid to the catalyst surface (interparticle mass transfer) is to reduce the surface concentration below the bulk concentration values. The rate of formation of B from A is therefore decreased. Conversely, the surface concentration of B is greater than it is in the bulk fluid and the rate of disappearance of B by reaction to form C is increased beyond that which would have prevailed if there were no mass-transfer resistances and if surface concentrations were uniform and equal to those in the bulk fluid. Consequently the selectivity of formation of B with respect to C (which, as before, is the ratio of the net rates of formation of B and C) is reduced by the mass-transfer resistance. In terms of the unknown interfacial concentrations the selectivity is

$$S = \frac{\mathrm{R_B}}{\mathrm{R_C}} = \frac{k_1 c_{\mathrm{Ai}} - k_2 c_{\mathrm{Bi}}}{k_2 c_{\mathrm{Bi}}} \tag{19}$$

where the subscript i (in the present context) denotes interfacial conditions. These may be written in terms of known bulk gas-phase concentrations by invoking steady-state conditions between interface and gas phase. Thus, in the steady state, the rate of formation of B at the surface is balanced by its rate of mass transfer between fluid and solid:

$$k_1 c_{\mathrm{Ai}} = h_{\mathrm{DA}}(c_{\mathrm{Ag}} - c_{\mathrm{Ai}}) \tag{20}$$

and similarly for B:

$$k_1 c_{\mathrm{Ai}} - k_2 c_{\mathrm{Bi}} = h_{\mathrm{DB}}(c_{\mathrm{Bg}} - c_{\mathrm{Bi}}) \tag{21}$$

Eliminating the interfacial concentrations of A and B from Eq. (19) by employing Eqs. (20) and (21), the selectivity becomes

$$S = \frac{k_1}{k_2}\frac{c_{\mathrm{Ag}}}{c_{\mathrm{Bg}}}\left\{\frac{1 + (k_2/h_{\mathrm{DB}})}{(k_1/h_{\mathrm{DB}})(c_{\mathrm{Ag}}/c_{\mathrm{Bg}}) + (1/h_{\mathrm{DA}})(h_{\mathrm{DA}} + k_1)}\right\} - 1 \tag{22}$$

which is less than it would have been if there were no interparticle mass-transfer resistance and thus c_{g} were equal to c_{i} for both A and B.

Reactions described by other kinetic routes may be treated in a similar fashion. Although, for reasons already explained in Section 6.1.1.1, mass-transfer effects will not influence the selectivity of two concurrent reactions arising from the same reactant, heat transfer between fluid and solid does have an effect. Thus, when the temperature of the solid is uniform throughout (no intraparticle heat-transfer resistance) and equal to the interface temperature T_{i}, for the two first-order concurrent reactions

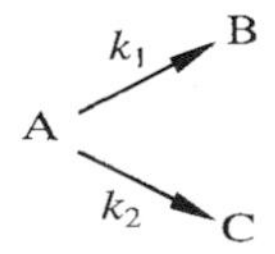

the selectivity is

$$S_{\mathrm{i}} = \frac{k_1}{k_2} = \frac{A_1 \exp\{-E_1/RT_{\mathrm{i}}\}}{A_2 \exp\{-E_2/RT_{\mathrm{i}}\}} \tag{23}$$

The selectivity under conditions when there is no external heat transfer resistance (when $T_i = T_g$) is

$$S = \frac{A_1 \exp\{-E_1/RT_g\}}{A_2 \exp\{-E_2/RT_g\}} \tag{24}$$

Comparing, then, the selectivity under conditions of interparticle heat-transfer resistance with that when there is no heat-transfer resistance:

$$\frac{S_i}{S} = \frac{\exp\left\{\frac{E_1}{R}\left(\frac{T_i - T_g}{T_i T_g}\right)\right\}}{\exp\left\{\frac{E_2}{R}\left(\frac{T_i - T_g}{T_i T_g}\right)\right\}} \tag{25}$$

For exothermic reactions $(T_i - T_g) > 0$, so if $E_1 > E_2$, the selectivity is greater when there is a heat-transfer resistance than when there is no resistance to the transfer of heat from solid to fluid, a result which is entirely consistent with the discussion in Section 6.1.1.2.

6.1.2 Bifunctional Catalysts

We saw in Section 1.2.1 that some heterogeneous catalytic processes require the presence of more than one catalyst type to achieve a significant yield of desired product. The conversion of *n*-heptane to isoheptane, for example, requires the presence of a dehydrogenation catalyst (such as platinum supported on alumina) together with an isomerisation catalyst (such as a silica–alumina or zeolite). In this particular case the *n*-heptane would be dehydrogenated by the platinum catalyst to *n*-heptene which, in turn, would be isomerised to isoheptene by the silica–alumina catalyst; the final step is the hydrogenation of isoheptene to isoheptane in the presence of platinum. In principle each of these reaction stages could be accomplished separately; alternatively, they could be effected within the same environment by mixing together the appropriate amounts of each catalyst, or even by dispersing platinum on the porous silica–alumina. Porous catalysts are normally employed to achieve this dual function because the large specific surface areas available enhance the product throughput. However, the porous nature of such bifunctional catalysts will also affect the conversion and yield of desired product so that it is quite natural to enquire whether there is an optimum amount of each catalyst to be employed in given circumstances.

Important pioneering work by Weisz (see Section 6.5.5) demonstrated that the simultaneous presence of a hydrogenation–dehydrogenation function (e.g. platinum) and an acidic function (e.g. silica–alumina) is necessary to achieve a satisfactory conversion of methylcyclopentane (a typical component present in naphtha feedstocks used in the petroleum industry) to benzene (suitable for increasing the aromaticity and anti-knock properties of a petroleum blend). Weisz further showed that intraparticle diffusion could affect the conversion. In 1964 Sinfelt (see Section 6.5.5) showed that hydrogen is also a necessary component of the reactant feed because during the isomerisation of *n*-heptane carbonaceous deposits (deleterious to the catalyst activity) occur unless the partial pressure of hydrogen is sufficiently high.

Theoretical considerations of optimum catalyst composition for bifunctional catalysts have led to the development of mathematical models which can, in principle, predict the optimum proportions of the two catalyst components which would lead to the maximum yield (and hence selectivity) or throughput of a desired product. Consider, for example, the conversion of reactant A into the product C and suppose that two catalyst components X and Y (for example, X could be platinum and Y a porous silica–slumina entity) are necessary for the reaction to proceed, via the intermediate product B, according to the kinetic scheme

$$A \xrightarrow{X} B \xrightarrow{Y} C \quad \textbf{Scheme 6.1}$$

In the absence of either the hydrogenation–dehydrogenation catalyst X or the isomerisation catalyst Y the product C cannot be formed from the reactant A. It follows that there is an optimum ratio of the amounts, or activities, of catalysts X and Y which will produce the maximum yield or throughput of C. Similar arguments can be advanced for more complex kinetic schemes. The problem was tackled originally by Gunn and Thomas (1965) (see Section 6.5.5), who assigned kinetic parameters to the simple sequence in Scheme 6.1 and also the more complex Schemes 6.2 and 6.3, and computed the steady-state flux requirements for each of the components A, B and C within a plug-flow tubular reactor (Sections 7.3.1.2 and 7.3.2.2) packed with a discrete mixture of catalysts X and Y, each active catalyst distributed within spherical particles of uniform size. The fraction of X necessary to maximise the yield of C was then calculated for given kinetic and diffusivity parameters. Figure 6.2 shows some of the results obtained and clearly demonstrates that, at least in principle, a maximum yield of the desired product can be obtained for a particular catalyst composition for each of the three schemes 6.1, 6.2 and 6.3. For similar values of kinetic parameters, Scheme 6.3, for example, requires a smaller fraction of catalyst X to be present in the reactor than Scheme 6.1 because, in the latter case, the reverse reaction of B to reform A tends to decrease the yield of C in accordance with the principle of Le Chetalier. For Scheme 6.3, however, in which an unwanted final product D is formed by the mere presence of X (which it is necessary to include in order to form the intermediate B which, in turn, is the precursor to C), an even smaller fraction of the X-type catalyst constitutes the optimum blend of discrete particles. Such calculations were performed with the constraint that the catalyst blend should be uniform along the reactor length. Similar computations in which both catalyst components are compounded in the same particle (curve 4) show that a higher yield of C can be obtained compared with that obtained from the optimum blends of discrete particles (curves 1, 2, 3). Once again, this result is intuitively ob-

$$A \underset{}{\overset{x}{\rightleftharpoons}} B \xrightarrow{y} C \quad \textbf{Scheme 6.2}$$

$$A \underset{}{\overset{x}{\rightleftharpoons}} B \quad \begin{matrix} \xrightarrow{y} C \\ \xrightarrow{x} D \end{matrix} \quad \textbf{Scheme 6.3}$$

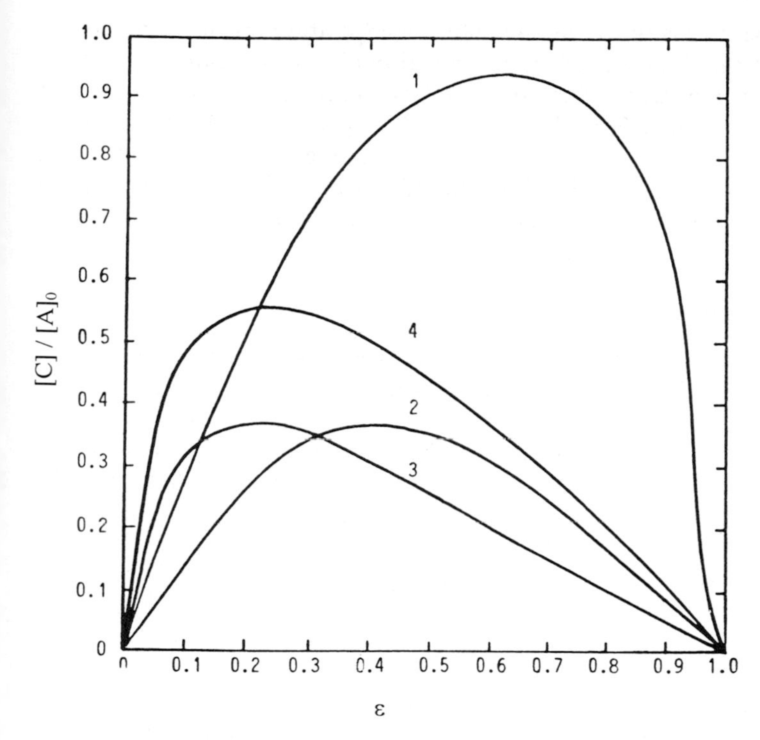

Figure 6.2 Optimum composition of a bifunctional catalyst: dimensionless concentration of C relative to A, c_C/c_A^0, plotted against relative catalyst composition ε (function of x/y). Curves 1, 2 and 3 refer to discrete catalyst particles containing X and Y in different pellets; curve 1 relates to Scheme 6.1, curve 2 to Scheme 6.2 and curve 3 to Scheme 6.3. Curve 4 describes the effect of compounding X and Y into single particles for Scheme 6.1. With permission from *Trans. I. Chem. E.* **1921**, *49*, 204.

vious for, in effect, by compounding the two catalysts into a single particle, the resistance to conversion is reduced by eliminating the need for the intermediate product to be transported through interparticle space from one catalyst type to the other. Many of these ideas were confirmed in principle by experimental studies which sought the optimum blend of catalyst functions for the formation of benzene from methylcyclopentane, a typical reforming reaction which aims to convert cyclic paraffins to aromatics. These interesting ideas concerning optimum catalyst formulations have been reviewed in the literature (see Section 6.5.5).

6.2 Catalyst Deactivation

The activity of a catalyst may decline during its operational life for several different reasons: the active sites of a catalyst may become poisoned by the adsorption of impurities in the feed stream; the fouling of the surface and blockages of pores by carbonaceous residues formed as a result of the breakdown or cracking of some hydrocarbons is sometimes a cause of activity loss; catalysts can also lose vital surface area by a process of sintering which occurs by crystal growth and agglomeration; actual loss of catalytic species may also contribute to the decline in activity of a catalyst if chemical transformations of the catalyst occur during reaction or where volatilisation is possible.

Quantitative approaches to the problems of poisoning and carbon laydown on catalysts have met with some success and an elementary description of these will be outlined here. When it is possible to predict quantitatively how the reaction rate is affected by either of these two deleterious processes, the modified reaction rate can be employed in the process design description subsequently required to devise strategies for reactor operations.

6.2.1 Deactivation Processes

The poisoning of catalysts by impurities present in the feedstock to a catalytic reactor results in a gradual decline of activity in the catalyst bed. The work of Maxted (1968) and of Rideal (1968) (see Section 6.5.1) in an earlier era has clearly demonstrated how a platinum catalyst is poisoned by sulphur compounds present in the feed to an experimental reactor in which carbonic acid is hydrogenated. The loss in activity is proportional to the amount of poison added to the feed (Fig. 6.3). Similarly, the dealkylation of cumene in the presence of an acidic-type catalyst such as

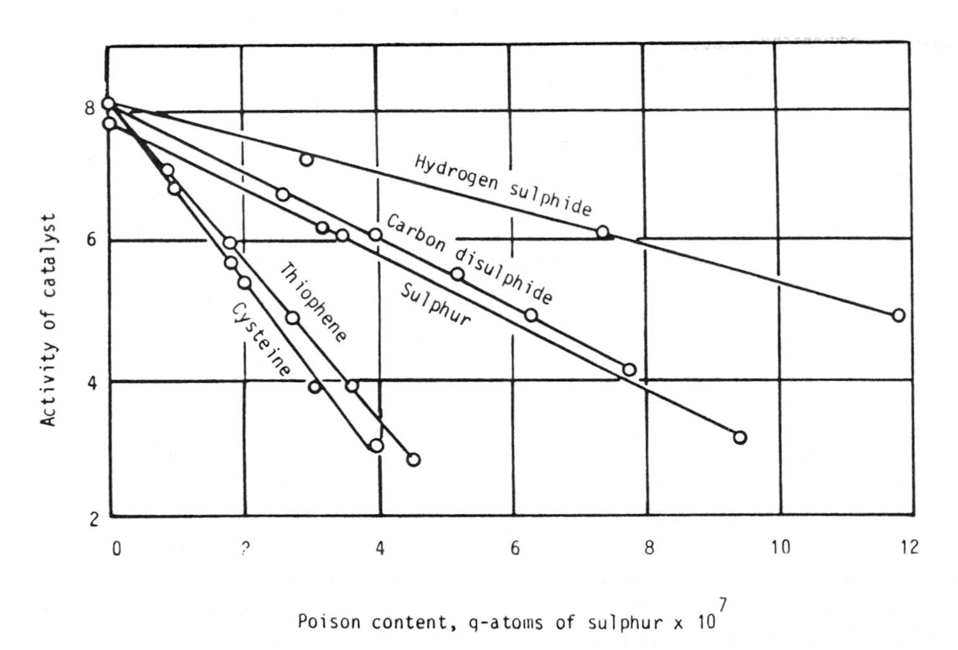

Figure 6.3 Loss in catalyst activity with the addition of poison. Loss in activity is directly proportional to the amount of sulphur-containing poisons added to the platinum-catalysed hydrogenation of crotonic acid. After E. B. Maxted, H. C. Evans, *J. Chem. Soc.* **1937**, *603*, 1004.

silica–alumina or a zeolite (as described in Chapter 8) is affected by basic compounds present in the feed which reduce the catalyst activity. In both instances the poisons are chemisorbed at the catalyst surface and therefore reduce the number of active catalyst sites available for chemisorption of the reactant. Strong poisons, as we saw earlier, are H_2S, NH_3, CO and organic heterocyclic compounds containing nitrogen or sulphur. They tend to be irreversibly adsorbed by the catalyst in competition with the reactant, thereby preventing the latter from utilizing those sites for the purpose of catalysis. Weak poisons are adsorbed reversibly and activity can often be restored. An example is the adsorption of oxides of sulphur present in motor-vehicle exhaust gases; these adsorbed impurities poison the catalyst monoliths installed in the exhaust manifold as a burner for destroying unburned toxic gases (see Chapter 8). When the engine operates under load the temperature within the monolith increases and the oxides of sulphur are desorbed.

Carbon laydown on catalyst surfaces occurs when unsaturated organic substances are formed during catalytic cracking of petroleum feedstocks and when the partial pressure of hydrogen is insufficient to prevent the inexorable build-up of carbonaceous residues. As a consequence, pores within the catalyst become blocked and this prevents access of reactant to active catalytic surfaces. Carbon may also be deposited on catalysts which are active in forming synthesis gas (syn-gas; a mixture of CO and H_2 formed by the catalytic reactions between steam and hydrocarbons) or during methanation and Fischer–Tropsch synthesis when mixtures of CO and H_2 react catalytically to form methane or other hydrocarbons and methanol. A necessary, but not sufficient, condition for a carbon-free catalyst surface during these reactions is that operating conditions should be selected so that carbon deposition cannot occur as a result of thermodynamic equilibria such as

$$2CO \rightleftharpoons C + CO_2$$

$$CH_4 \rightleftharpoons C + 2H_2$$

As the carbon formed by such equilibria is often thermodynamically more active than graphite, care must be taken when applying equilibrium data to predict operating conditions which would avoid graphite deposition. Kinetic manifestations of synthesis and reforming reactions can also be responsible for carbon deposition because, depending on catalyst composition and the nature of the hydrocarbon, reactions in which carbon is formed may be intrinsically faster than reactions in which it is consumed. This, of course, is a matter for careful catalyst selection and design. Nevertheless, occasional instability of operating conditions (for example, during plant start-up) leads to unexpected loss of activity by carbon deposition. Figure 6.4 shows carbon deposited on a reforming catalyst.

The phenomenon of sintering, in which there is loss of active surface area of a catalyst, occurs by several different mechanisms. A metal is usually supported on an inert material of high surface area to prevent excessive mobility of metal atoms. In spite of the dispersion of these metal atoms over the surface of the support, two-dimensional clusters of atoms can form as a result of surface diffusion. Larger clusters then grow and three-dimensional metal particles can form from such large clusters. The loss of surface area is evidently the result of the clustering of highly dispersed

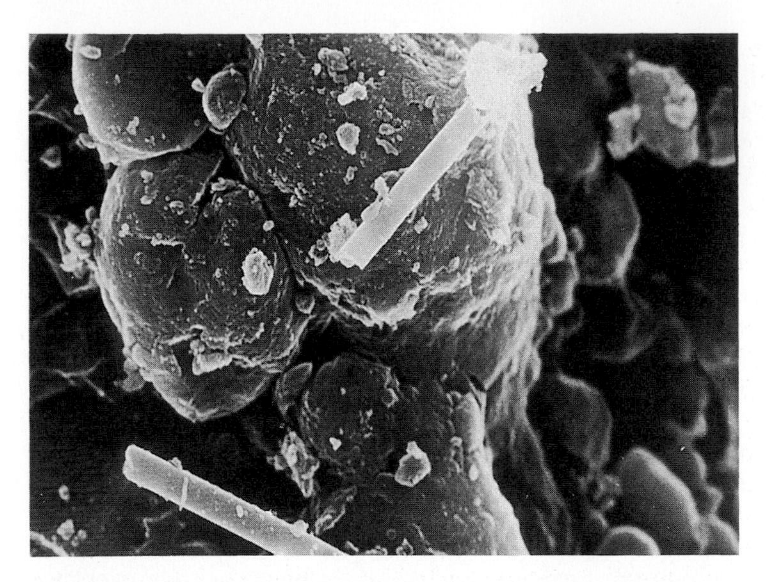

Figure 6.4 Carbon deposition on a 9% Ni, 1.5% Pt catalyst when CO was exposed to the catalyst at 300 °C. Micrographs at magnification 3000× are by scanning electron microscopy (SEM).

individual metal atoms. Crystallites with even lower surface areas can grow from three-dimensional particles by a process of migration and subsequent coalescence. Crystallites often grow at the sites of dislocations and steps in the support structure, for this is where they are most stable. Particle growth in an inert or reducing atmosphere is quite different from growth in an oxidising environment, where volatilisation of oxide (or chloride in an atmosphere of chlorine) can occur with subsequent deposition of the more stable metal at a crystallite centre. In an inert or reducing atmosphere particle growth is inversely related to the strength of cohesive forces in a metal crystallite. Experimental observations indeed demonstrate that the stability of metals, and hence their resistance to sintering, increases with increase in melting point of the metal. It has been shown (see Section 6.5), for example, that stability increases in the sequence nickel, palladium, platinum and rhodium, which concurs with the sequence of increasing melting points.

6.2.2 Deactivation Models

Four deactivation models will be outlined here. Each has its place in the architecture of catalysis and can be applied in various circumstances. Their usefulness and limitations are related to their application.

6.2.2.1 Steady-State Model

The manner in which catalytic activity is modified as a result of poisoning was first modelled quantitatively by Wheeler (1951) (see Section 6.5.1), who discerned two limiting types of poisoning which are termed 'uniform' poisoning and 'pore mouth' or 'selective' poisoning respectively. Wheeler's original models were expressed in terms of events occurring within a single pore and were subsequently extended to include porous particles of arbitrary shape by the simple expedient of equating measured specific surface areas and pore volumes to those for simple models of porous media composed of an assembly of single cylindrical pores. As in Section 4.6, and without loss of generality, we shall describe both models in terms of an effective diffusivity D_e of the reactant through the porous medium and the particle dimension rather than the gaseous diffusion coefficient and pore radius employed by Wheeler.

The uniform poisoning model assumes that the material responsible for poisoning pervades the entire porous particle uniformly but slowly. On the other hand, transport of the reactant by intraparticle diffusion is more rapid relative to the rate of poison deposition. Consequently the rate of the catalytic reaction within the particle is reduced in proportion to the fraction of catalytic sites poisoned. If ζ, then, is the fraction of sites uniformly poisoned, a fraction $(1 - \zeta)$ of the surface is unpoisoned and upon which reaction can proceed. The intrinsic reaction rate is thus reduced not only by virtue of the diffusional limitation caused by the porous structure and reflected in the effectiveness factor η, but also because only a fraction $(1 - \zeta)$ of surface is now active. It is a relatively straightforward matter to calculate the ratio of activity of the poisoned catalyst to the activity of an unpoisoned catalyst. As the example below shows, one compares the stationary flux of reactant to the particle surface with the flux in the absence of poisoning. For a first-order reaction occurring in a flat (slab)-shaped catalyst pellet poisoned homogeneously, this ratio is

$$F = \frac{(1-\zeta)^{1/2} \tanh\{\phi(1-\zeta)^{1/2}\}}{\tanh \phi} \tag{26}$$

where ϕ is the Thiele modulus for a first-order reaction occurring in a slab-shaped catalyst pellet. The two limiting cases of Eq. (26) correspond to extreme values of ϕ. When ϕ is sufficiently small for it to correspond with an effectiveness factor which is near unity ($\eta \rightarrow 1$), then the relative activity decreases linearly with the amount of poison added:

$$F = 1 - \zeta \quad \text{for } \eta \rightarrow 1 \tag{27}$$

The other extreme is when the reaction is controlled by diffusion limitations ($\eta = 1/\phi$). Then

$$F = (1-\zeta)^{1/2} \tag{28}$$

The relative activity now decreases less than linearly with the extent of poisoning.

Selective poisoning occurs with very active catalysts. Initially, the exterior surface is poisoned and subsequently, as more poison is added, an increasing depth of the interior surface becomes poisoned and inaccessible to reactant. If the reaction rate in the unpoisoned portion of the catalyst pellet happens to be chemically controlled without any diffusional resistance, then the reaction rate will fall off directly in proportion to the fraction of surface poisoned and the relative activity decreases linearly with the amount of poison added. However, if the reaction is diffusion-controlled, a different result is obtained, as shown in the second part of the example below. When diffusion limitation occurs in a selectively poisoned pellet, in the steady state the flux of reactant past the boundary between the poisoned and unpoisoned portions of the catalyst (see Fig. 6.6) will equal the reaction rate in the unpoisoned portion. Under such circumstances the relative activity becomes

$$F = \left(\frac{\tanh\{\phi(1-\zeta)\}}{\{1+\zeta\phi\tanh\{\phi(1-\zeta)\}\}}\right)\left(\frac{1}{\tanh\phi}\right) \tag{29}$$

For large values of the Thiele modulus, $\phi(1-\zeta)$ will be sufficiently large so that

$$F = \frac{1}{1+\zeta\phi} \tag{30}$$

The manner in which the relative activity falls off with increasing amount of added poison is thus clearly dependent on whether homogeneous or selective poisoning occurs. Figure 6.5 illustrates both of these extremes: curves 1 and 2 refer to homogeneous poisoning when the Thiele modulus is small and large respectively, while curve 3 illustrates the behaviour of the function represented by Eq. (29) for selective poisoning at moderate values of ϕ. Curve 4 shows that, for selective poisoning, when the fraction $\phi(1-\zeta)$ is sufficiently large (high catalyst activity), the relative activity falls off very drastically as poison is added.

This analysis of the problem of catalyst poisoning by Wheeler only considers the two extremes of homogeneous and selective poisoning. Furthermore, the models do not consider how the activity declines with increasing time of exposure of the catalyst to the poison.

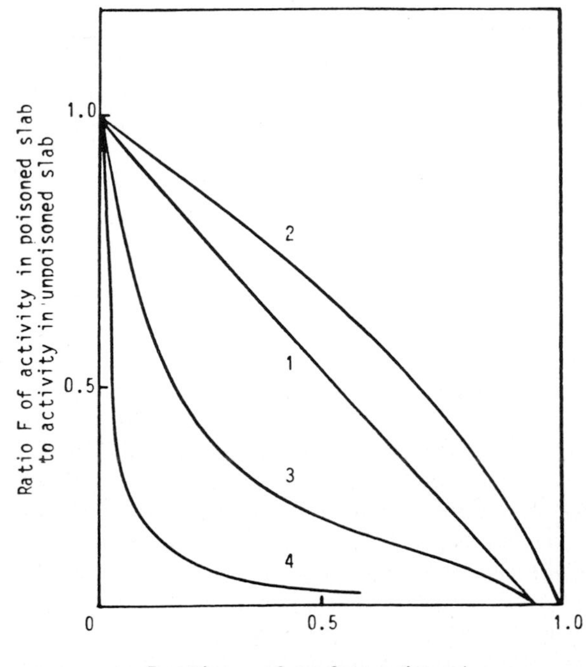

Figure 6.5 Wheeler's analysis of poisoning. Curves 1 and 2 represent homogeneous poisoning for small and large Thiele modulus respectively. Curve 3 is the function given in Eq. (29). Curve 4 displays behaviour for selective poisoning when the catalyst is highly active.

Example

A fraction ζ of the active surface of some porous slab-shaped catalyst pellets becomes poisoned. The pellets are used to catalyse a first-order isothermal chemical reaction. Find an expression for the ratio of the activity of the poisoned catalyst to the original activity of the unpoisoned catalyst when (a) homogeneous poisoning occurs, and (b) selective poisoning occurs.

Solution

(a) If homogeneous poisoning occurs, the activity decreases in proportion to the fraction $(1-\zeta)$ of surface remaining unpoisoned. In the steady state the rate of reaction is equal to the flux of reactant to the surface. The ratio of activity F of the poisoned slab to that of the unpoisoned slab will be equal to the ratio of the reactant fluxes under the respective conditions. Hence:

$$F = (\mathrm{d}c/\mathrm{d}x)'_{x=L}/(\mathrm{d}c/\mathrm{d}x)_{x=L}$$

where the prime denotes conditions in the poisoned slab. In Section 4.6 it was shown that the concentration of reactant diffusing into the pellet at a distance x from the exposed face is

$$c = c_\mathrm{g}\frac{\cosh(\lambda x)}{\cosh(\lambda L)} \quad \text{where } \lambda = (k/D_\mathrm{e})^{1/2}$$

If the slab were poisoned the activity would be $k(1-\zeta)$ rather than k and then:

$$c' = c_\mathrm{g}\frac{\cosh(\lambda' x)}{\cosh(\lambda' L)} \quad \text{where } \lambda' = (k(1-\zeta)/D_\mathrm{e})^{1/2}$$

Evaluating the respective fluxes at $x = L$:

$$F = \frac{(1-\zeta)^{1/2}\tanh\{\phi(1-\zeta)^{1/2}\}}{\tanh\phi} \quad \text{where } \phi = L(k/D_\mathrm{e})^{1/2}$$

(b) When selective poisoning occurs, the exterior surface of the porous pellet becomes poisoned initially and the reactants must then be transported to the unaffected interior of the catalyst before reaction may ensue. When the reaction rate in the unpoisoned portion is chemically controlled, the activity merely falls off in proportion to the fraction of surface poisoned. However, if the reaction is diffusion-limited in the steady state, the flux of reactant past the boundary between poisoned and unpoisoned surfaces is equal to the chemical reaction rate (see Fig. 6.6). Thus:

Flux of reactant at the boundary between poisoned and unpoisoned

$$\text{portion of slab} = D_\mathrm{e}\left(\frac{c_\mathrm{g}-c_\mathrm{L}}{\zeta L}\right)$$

$$\text{Reaction rate in unpoisoned length } (1-\zeta)L = D_\mathrm{e}\left(\frac{\mathrm{d}c}{\mathrm{d}x}\right)_{x=(1-\zeta)L}$$

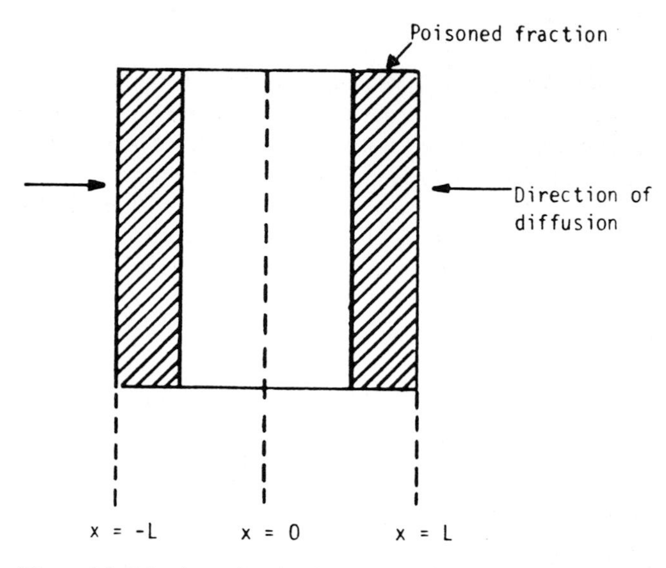

Figure 6.6 Selective poisoning in catalyst wafer. The shaded portions in the outermost regions of the wafer-shaped catalyst pellet are poisoned, whereas the clear innermost regions remain unpoisoned.

The concentration profile in the unpoisoned length is, by analogy with Eq. (38) in Section 4.6,

$$c = c_{\mathrm{L}} \frac{\cosh(\lambda x)}{\cosh\{\lambda(1-\zeta)L\}}$$

Therefore:

$$D_{\mathrm{e}}\left(\frac{\mathrm{d}c}{\mathrm{d}x}\right)_{x=(1-\zeta)L} = \frac{D_{\mathrm{e}}}{L} c_{\mathrm{L}}\phi \tanh\{\phi(1-\zeta)\} \quad \text{where } \phi = \lambda L = L(k/D_{\mathrm{e}})^{1/2}$$

In the steady state then:

$$\frac{D_{\mathrm{e}}(c_{\mathrm{g}} - c_{\mathrm{L}})}{\zeta L} = \frac{D_{\mathrm{e}}}{L} c_{\mathrm{L}}\phi \tanh\{\phi(1-\zeta)\}$$

Solving the above equation explicitly for c_{L}:

$$c_{\mathrm{L}} = \frac{c_{\mathrm{g}}}{1 + \phi\zeta \tanh\{\phi(1-\zeta)\}}$$

Hence the rate of reaction in the partially poisoned slab is:

$$\frac{D_{\mathrm{e}}(c_{\mathrm{g}} - c_{\mathrm{L}})}{\zeta L} = \frac{c_{\mathrm{g}} D_{\mathrm{e}}}{L} \left\{ \frac{\phi \tanh\{\phi(1-\zeta)\}}{1 + \phi\zeta \tanh\{\phi(1-\zeta)\}} \right\}$$

In an unpoisoned slab the reaction rate is $(c_{\mathrm{g}} D_{\mathrm{e}}/L)\ \phi \tanh \phi$ and so

$$F = \left(\frac{\tanh\{\phi(1-\zeta)\}}{1 + \phi\zeta \tanh\{\phi(1-\zeta)\}} \right) \left(\frac{1}{\tanh \phi} \right)$$

6.2.2.2 A Dynamic Model

Under many circumstances in practice, catalytic activity declines with the time for which the catalyst is on stream and as more and more poison enters a continuously operated catalytic reactor. Although the above analysis, of Wheeler, provides a framework of reference for assessing fractional activity following the addition of a known amount of poison during steady-state operation (when it is considered that the rate of addition of poison is negligible in comparison with the chemical reaction rate describing product formation), it does not give any quantitative indication of how the conversion of reactant to product may decline with time due to the continual addition of poison present in the process feed.

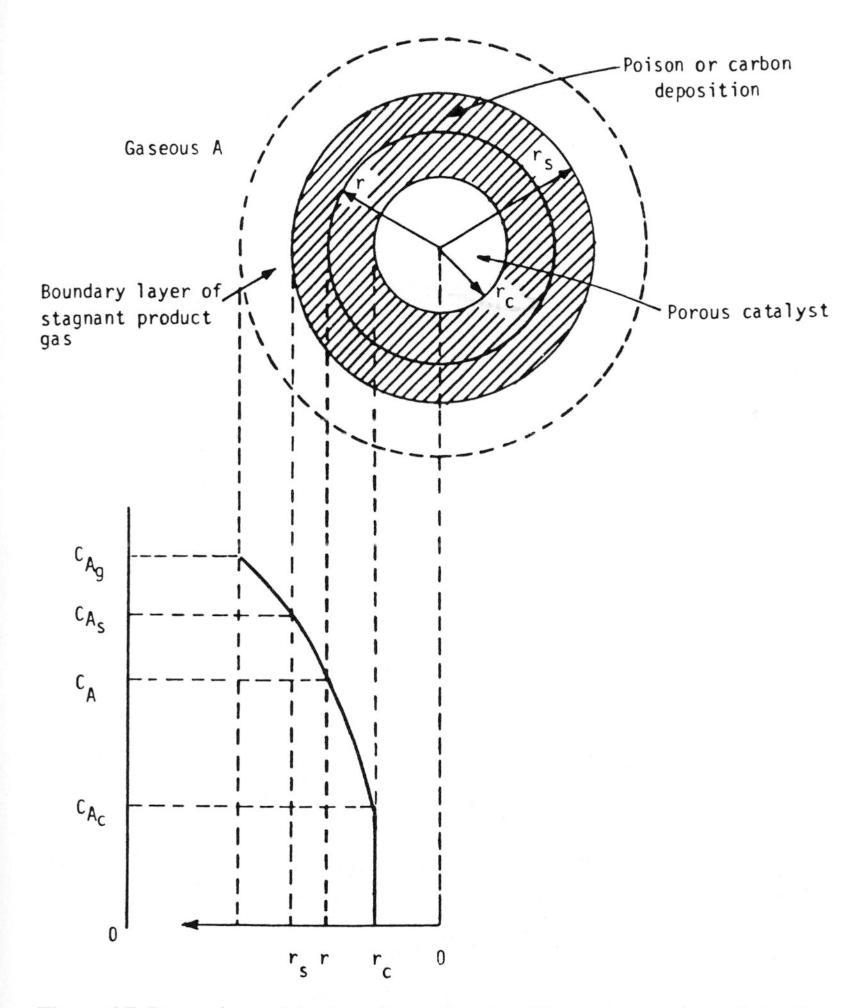

Figure 6.7 Dynamic model of catalyst poisoning. The upper portion of the diagram displays the partially poisoned (shaded area) spherical catalyst pellet and the lower part of the figure represents the expected concentration profile of gaseous reactant A within regions of the porous pellet.

We shall consider here a dynamic model based upon analogies with models of gas–solid reactions in which the solid reactant particle shrinks in size due to its consumption by chemical reaction to form product ash. Referring to Figure 6.7, we consider gaseous reactant A to be first transported across a gaseous, but relatively stagnant, boundary layer of gas surrounding the porous spherical pellet. At any particular moment, poison emanating from the gaseous feed stream will have diffused into the porous catalyst structure and, if diffusion of the poison is rapid in comparison with the rate at which poison is adsorbed (or coke deposited), there will be a shell of thickness $(r_s - r_c)$ which is poisoned and an unpoisoned core of radius r_c. As time elapses the core of radius r_c will diminish in size and the shell of poisoned (or coked) catalyst will thicken. If we consider both the rate of diffusion of reactant A through the poisoned porous shell of thickness $(r_s - r_c)$ and the reaction of A to form products within the porous unpoisoned core of radius r_c to be rapid in comparison with the rate of poisoning, then a pseudo-steady state can be invoked for reactant A in which the rate of diffusion N to the shell of external surface $4\pi r^2$ (calculated as the product of flux and surface area) is equated to the rate of reaction. Hence

$$N = D_e 4\pi r^2 \frac{dc_A}{dr} = \left(\frac{4}{3}\pi r^3\right)\eta k_A c_A \quad \text{at } r = r_c \tag{31}$$

where η is the pellet effectiveness for the first-order reaction considered. The gradient dc_A/dr in Eq. (31) at the boundary $r_c = r_s$ can be found by integrating the constant flux $(N = D_e 4\pi r^2 (dc_A/dr))$ between the limits r_c and r_s. The result is

$$N = \frac{4\pi D_e}{\left(\frac{1}{r_c} - \frac{1}{r_s}\right)}(c_{A_g} - c_{A_c}) \tag{32}$$

where c_{A_g} is the gaseous concentration of A at the pellet periphery (assuming little or no resistance to the transport of A from gas phase to the solid) and c_{A_c} is the interface concentration at $r = r_c$. Equating this flux to the right-hand side of Eq. (31) and dividing by $4\pi r_s^3/3$ to give the rate R_p for the whole pellet, one obtains

$$R_p = \frac{N}{4\pi r_s^3/3} = \frac{k_A c_{A_g}}{\frac{1}{\eta(1-\zeta)} + 3\phi^2\left[\frac{1-(1-\zeta)^{1/3}}{(1-\zeta)^{1/3}}\right]} \tag{33}$$

where $(1 - \zeta)$ represents the fraction of total surface unpoisoned (r_c^3/r_s^3) and ϕ is the Thiele modulus $(r_s/3)(k_A/D_e)^{1/2}$ (see Section 4.6). If there were no poisoning then $\zeta = 0$ and the rate is then $k_A \eta c_{A_g}$. Hence the ratio of reaction rates in the poisoned and unpoisoned pellets is

$$F = \frac{1}{\frac{1}{(1-\zeta)} + 3\eta\phi^2\left[\frac{1-(1-\zeta)^{1/3}}{(1-\zeta)^{1/3}}\right]} \tag{34}$$

If we now recall that the fraction of pellet poisoned is time-dependent, then a second relationship has to be developed which describes how the fraction of pellet volume remaining active varies as time elapses. The rate at which poison accumulates in the pellet and causes the boundary at r_c (see Fig. 6.7) to recede gradually to $r = 0$ (when the whole pellet is poisoned) is

$$(4\pi/3)d\{(r_s^3 - r_c^3)c_p^0\}/dt$$

where c_p^0 is the molar concentration of poison when the whole pellet is poisoned. Now, because we consider diffusion to be rapid in comparison with the deposition of poison, a pseudo-steady state can be invoked and the rates of the three processes

1. transport of poison across film from bulk gas to particle periphery,
2. diffusion of poison through the shell and
3. chemical deposition of poison

may be equated. We thus obtain

$$4\pi r_s^2 h_D(c_{P_g} - c_{P_s}) = 4\pi r_c^2 D_e \left(\frac{dc_P}{dr}\right)_{r=r_c} = 4\pi r_c^2 k_P c_{P_c} \tag{35}$$

The gradient (dc_p/dr) at $r = r_c$ can be evaluated (as for the reactant A) directly from the diffusion equation

$$\frac{d}{dr}\left(r^2 D_e \frac{dc_P}{dr}\right) = 0 \tag{36}$$

because diffusion is rapid. With the appropriate boundary conditions integration yields

$$r_c^2\left(\frac{dc_P}{dr}\right)_{r=r_c} = \frac{c_{P_s} - c_{P_c}}{\left(\frac{1}{r_c} - \frac{1}{r_s}\right)} \tag{37}$$

which can be substituted into Eq. (35). Eliminating the unknown concentrations c_{Ps} and c_{Pc} from the equations (6.35) and (6.37) and setting the three rate processes equal to the rate of accumulation of poison, one obtains an expression for the rate at which the core radius decreases. Integration of the resulting rate equation gives

$$t = \frac{r_s^2}{D_e c_{P_g}}\left\{\frac{(1-\sigma^3)}{3}\left(\frac{1}{\text{Sh}} - 1\right) + \frac{1-\sigma^2}{2} + \frac{1-\sigma}{\text{Da}}\right\} \tag{38}$$

where c_{P_g} is the gas-phase poison concentration (assumed constant for the purposes of describing this model), and σ is the ratio r_c/r_s (in turn equal to $(1-\zeta)^{1/3}$ as previously defined). Sh$(= h_D r_s/D_e)$ and Da $(= k_p r_s/D_e)$ are the dimensionless Sherwood and Damköhler numbers respectively.

Equations (34) and (38) are thus two coupled equations, the first describing how the ratio of reaction rates in poisoned and unpoisoned catalyst pellets depends on the fraction of pellet remaining poisoned (which will be denoted $F\{1-\zeta\}$) and the second implicitly expressing how the fraction poisoned is time-dependent (denoted $\zeta(t)$).

In effect, therefore, the information contained in these coupled equations will yield the desired relation $F(t)$ showing how the fractional activity of the catalyst pellet, exposed to a poison at concentration c_{P_g}, declines with time on stream and when the pellet is catalysing a particular reaction (in this example first-order).

6.2.3 Operational Consequences of Poisoning

When a strong poison to the main catalytic reaction is present in the feed stream to a continuously operated reactor, it is usually necessary to increase gradually the operating temperature in order to compensate for the loss in catalytic activity resulting from poisoning. If the reactor consists of a number of packed tubes contained within a shell and the reactor tubes are cooled (similarly to the operation of a shell-and-tube heat exchanger), either by an independent coolant or by the incoming feed stream, the strategy would be to decrease the mass flow rate so that the reaction can be sustained at a lower conversion. This is explained more graphically in Section 7.3.3, dealing with the thermal characteristics of catalytic reactors. If the appropriate operating parameters and characteristics were available, the dynamic model of poisoning could be applied and the reactor operating strategy deduced as a result of knowing how the catalytic activity $F(t)$ declines with time on stream.

6.3 Modern Theories of Poisoning and Promotion

As a result of the application of a range of sophisticated tools for studying catalysis at single-crystalline and polycrystalline surfaces, and especially because of the parallel use of techniques such as PES, IPES, LEED, AES, NEXAFS, isotopic labeling, TPD and work function measurements (Chapter 3), there has recently been a quickening in our awareness of the factors thought to be responsible for the dramatic or subtle effects exerted by even quite minor amounts (a few tenths of a monolayer) of certain additives on the catalytic properties of solids. Computational procedures, such as the Monte Carlo approach first used by Rideal for an investigation of poisoning, have also played their part, as we saw in Chapter 1. An admirable summary of model studies of the poisoning of catalysts by electronegative elements, with special reference to the role of nickel as a methanation catalyst ($CO + 3\,H_2 \rightarrow CH_4 + H_2O$), has been given by Goodman (1984) (see Section 6.5.2), and a graphic illustration of the reality of poisoning, taking us beyond the picture shown if Fig. 6.3, is to be found in Oudar's work on the systematic decline in hydrogenation rate of butadiene as a function of sulphur coverage of a Pt(110) surface (Fig. 6.8).

6.3.1 General Theoretical Considerations

That catalytic activity decreases in the presence of a poison is axiomatic. What we need to know are the reasons for this decrease, and for the concomitant changes of selectivity. The concept of ensembles, which figured eminently in the early Russian literature on catalysis, and later in the work of Martin et al., is based on the assumption that many adsorption complexes require more than just one surface atom

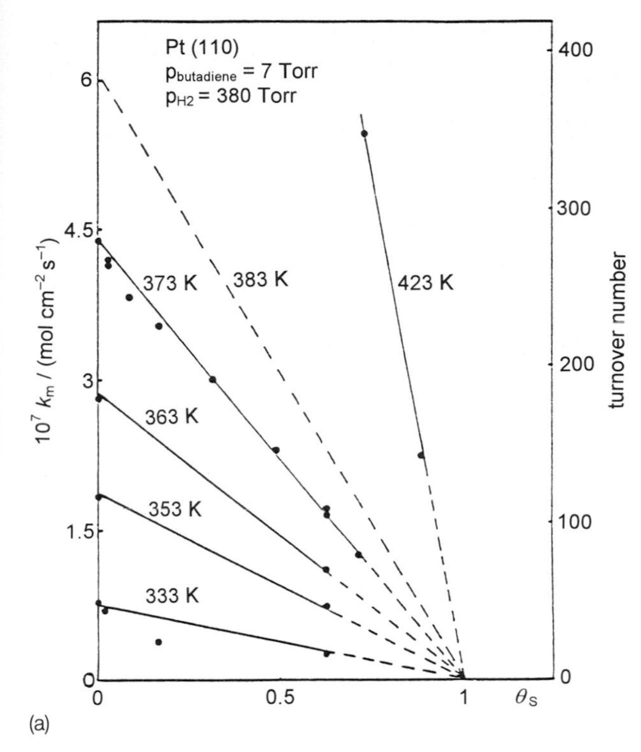

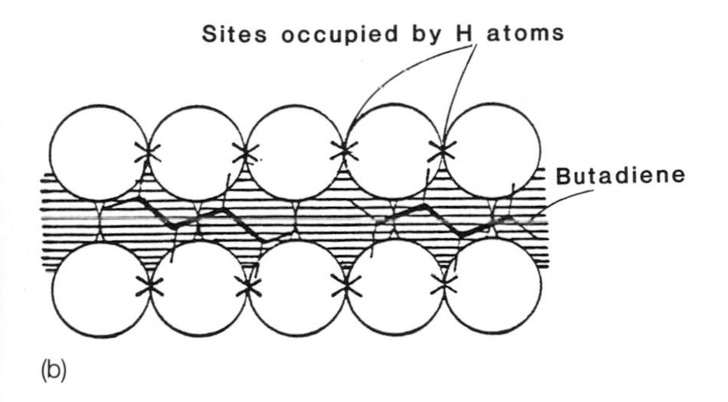

Figure 6.8 (a) The turnover frequency for the selective hydrogenation of butadiene on a platinum surface steadily falls to zero as the sulphur poison covers a monolayer at various temperatures. (b) Under steady-state conditions, it is found that the surface is covered with a layer of butadiene equivalent to two carbons per platinum atom. This model satisfies all the observed facts. With permission from J. Oudar, S. Pinol, C. M. Pradier, Y. Berthier, *J. Catalysis* **1987**, *107*, 445.

for the formation of chemical bonds. When adsorption is dissociative, this is self-evidently true. But even when monatomic species are adsorbed, they may be held at sites that are equidistant from three or more surface atoms with which bonds are formed. The sites occupied by bound sulphur atoms on the cluster model in Fig. 6.9 illustrate the point.

Since adsorption complexes can be formed only where ensembles of the required size are present, their abundance can control the selectivity of a catalyst. Thus, if the surface of an alloy AB is strongly enriched in B, the concentration of A_n ensembles with large n will be small on such a surface. Hence, when a given reaction has two or more reaction pathways available to it with similar activation energies, the pathway

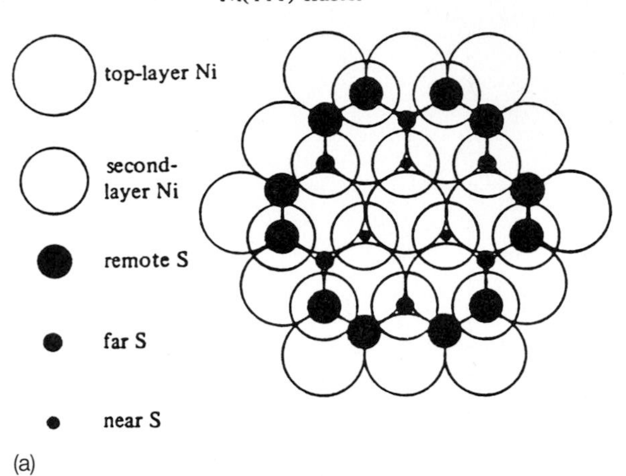
Ni(111) cluster
top-layer Ni
second-
layer Ni
remote S
far S
near S

(a)

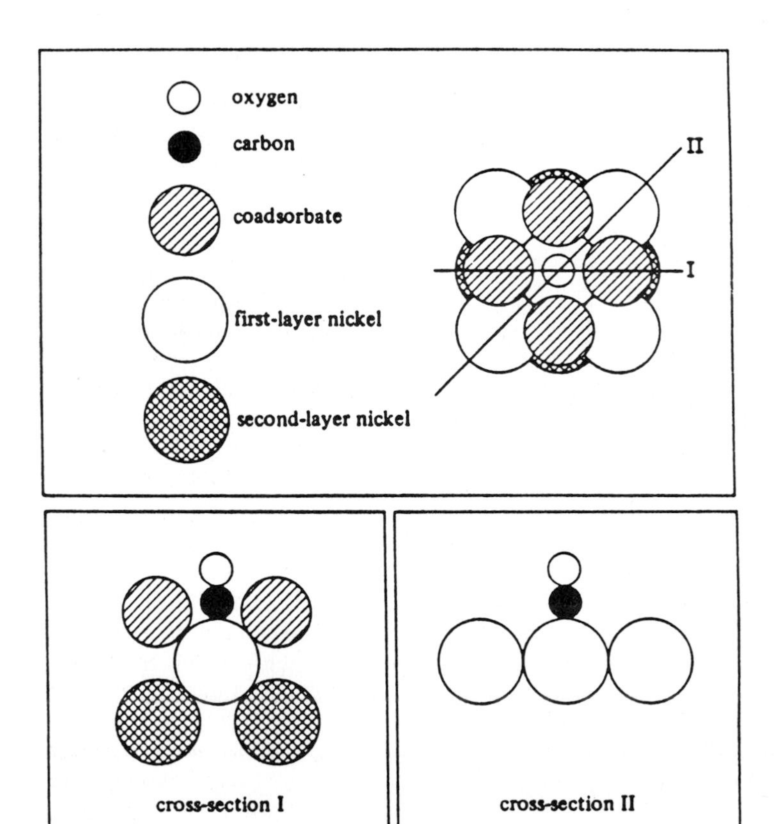
oxygen
carbon
coadsorbate
first-layer nickel
second-layer nickel
II
I
cross-section I
cross-section II

(b)

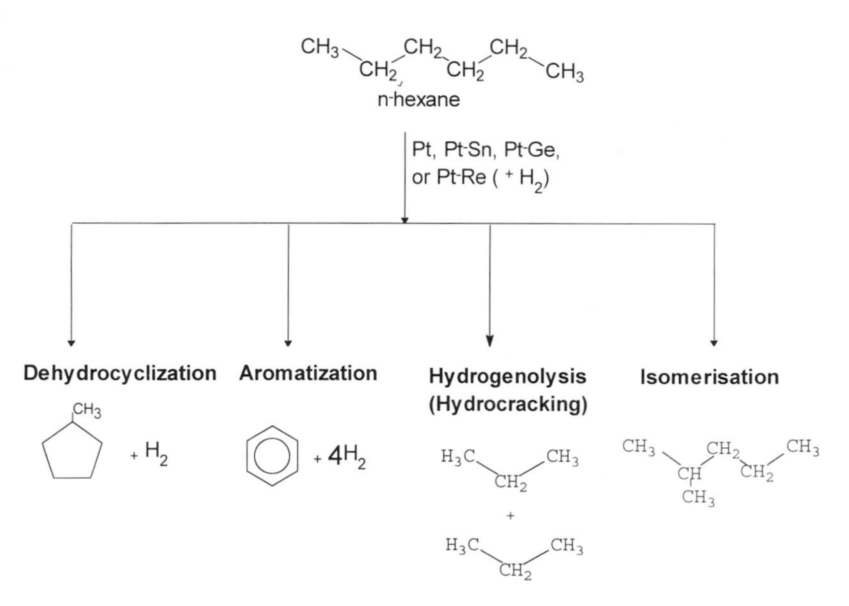

Figure 6.10 In the catalysed conversion of *n*-alkanes using platinum, Pt–Sn or Pt–Re catalysts, several possible processes can occur, each with their own ensemble requirements (see text). Poisoning is likely to affect each of these processes differently.

which demands the smallest A_n ensemble will be preferred. With this background, we can begin to appreciate qualitatively why, for example, when platinum (or Pt–Re) catalysts are modified by the addition of a poison such as sulphur, the original selectivity as well as the activity may be profoundly modified (Figs. 6.10 and 6.11).

That the initial activity of the sulphided catalyst is lower than that of the unsulphided one (Fig. 6.11) is not surprising, because the surface is, at the very least, physically poisoned by the sulphur. But what is remarkable (and extremely important commercially) is that the sulphided catalyst displays a smaller decline in activity, so that this catalyst shows a higher activity with continued use. Sulphidation also causes dramatic changes in selectivities; it reduces quite severely the selectivity for hydrocracking (hydrogenolysis) and reduces modestly that for dehydrocyclization. But the concomitant selectivity for isomerization and cyclization is increased significantly. It is noteworthy that these effects on selectivity are markedly different from those observed with the monometallic Pt/Al_2O_3 catalyst. In passing, therefore, we

---

Figure 6.9 (a) Plan view of cluster used in calculations of local density of states in the central adsorption site atop of the nickel atom on the (111) face. The near, far and remote poisons (S atoms) are shown. (b) Top view of another cluster used in a different kind of calculation. Here the Ni(100) surface with co-adsorbate (either poison or promoter) is simulated. With permission from J. M. McLaren, D. D. Veredensky, J. B. Pendry, R. W. Joyner, *Faraday Symp. Chem. Soc.* **1986**, *82*, paper 3.

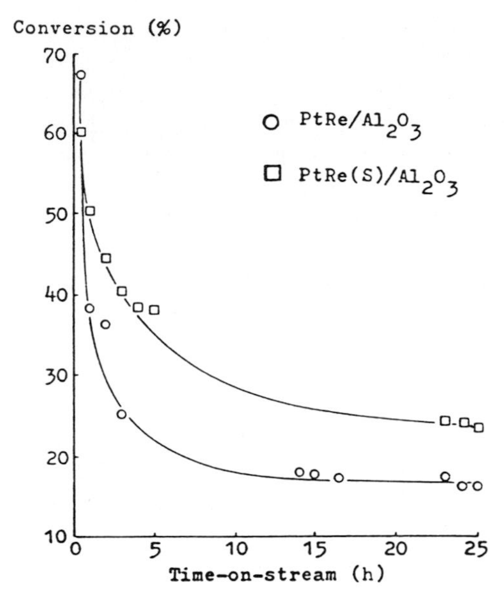

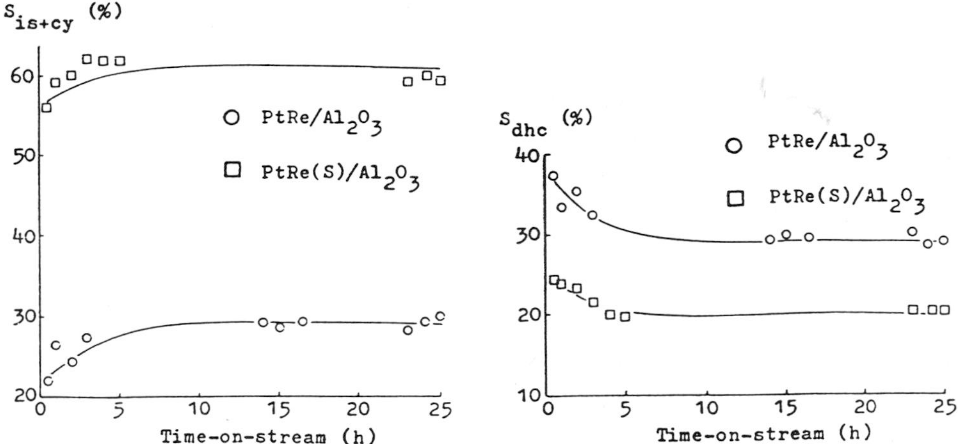

Figure 6.11 The performance of a supported Pt–Re alloy catalyst in the reforming of *n*-hexane is dramatically modified by prior sulphidation. Both the activity and various selectivities (see Fig. 6.10) are profoundly influenced. With permission from V. K. Shum, J. B. Butt, W. M. H. Sachtler, *J. Catalysis*, **1985**, *96*, 371.

note that, in so far as reforming of petrochemicals is concerned, the act of deliberate poisoning can lead to desired ends.

In addition to the ensemble effect, one recognizes that a ligand effect – to borrow a term from homogeneous catalysis – is also likely to operate. The ligand effect is based on the assumption that the nature and the strength of a chemical bond between the surface atom and of the adsorbate are influenced by the neighbours of that surface atom. Again, by analogy with homogeneous catalysis, and the localized picture there subsumed, it is thought that, in the ligand effect in heterogeneous catalysis, the influence of the first neighbours should dominate. Such arguments have often been invoked in explaining, for example, the role of subsurface oxygen species in silver catalysts, and of halogens or sulphur on various transition-metal catalysts. Two refinements have been made to the explanation of the ligand effect. It is recognized that, first, the electronegativity and, second, the effective size of the ligand (poison) are the essential factors. We shall see below, thanks to the cluster calculations of Pendry and Joyner, that what appears to be paramount – even more so than electronegativity or mere size – is the height of the poison above the surface.

6.3.2 Theoretical Interpretation of Poisoning and Promotion

To illustrate aspects of our present-day understanding of the role of additives in catalytic performance, we focus first on the methanation reaction and, in particular, the adverse role of electronegative elements on catalysis of this reaction by metals such as nickel. The facts have recently been highlighted by Goodman (see Fig. 6.12). Electronegative elements attached to a nickel surface dramatically decrease the rate

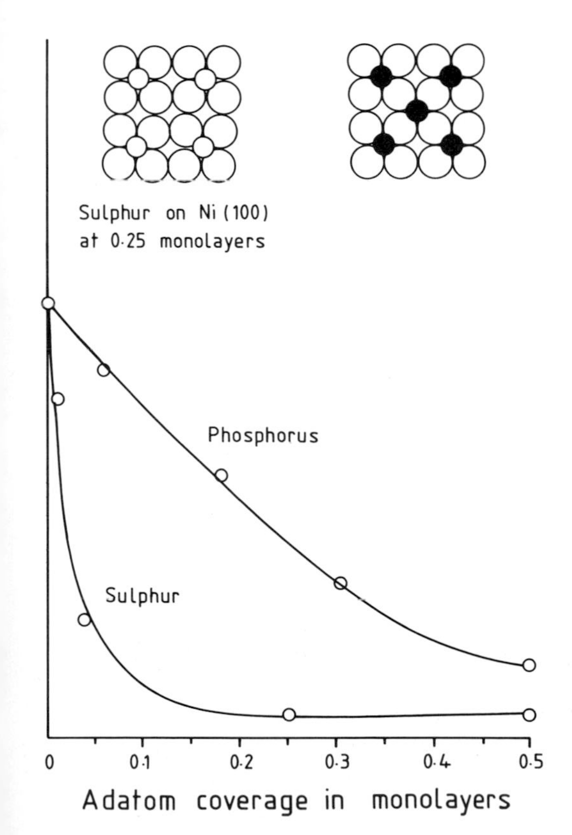

Figure 6.12 In the catalytic hydrogenation of CO, the rate of formation of CH_4 at a Ni(100) surface falls with increasing coverage of either sulphur or phosphorus. With permission from D. W. Goodman, *Acc. Chem. Res.* **1984,** *17*, 194.

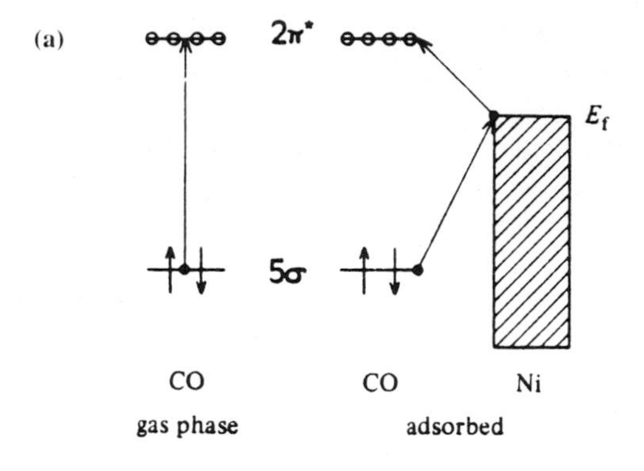

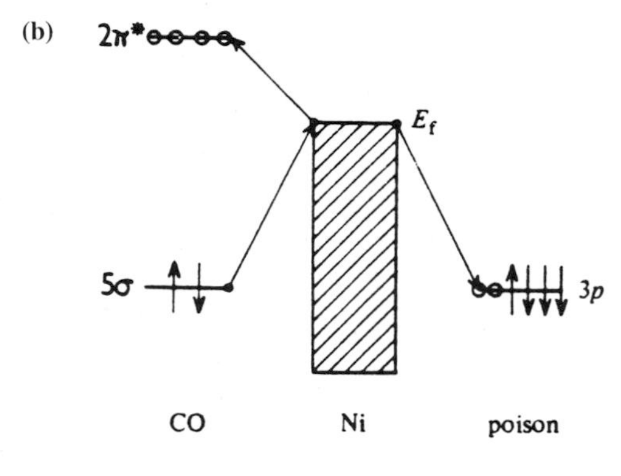

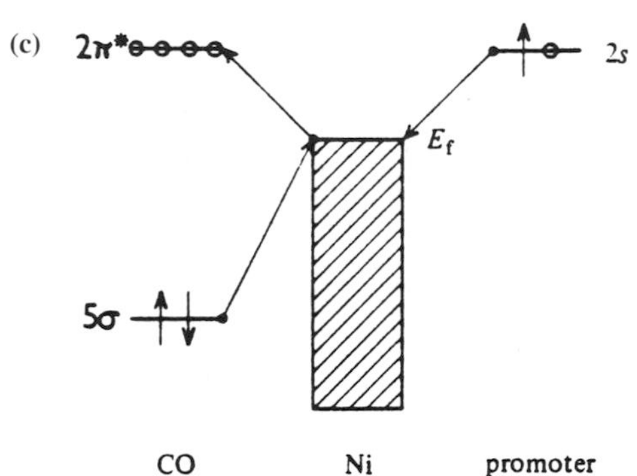

Figure 6.13 (a) When CO is adsorbed at a metal, the latter facilitates transitions of electrons from the 5σ to the $2\pi^*$ orbitals. Small arrows and open circles (at left) denote occupied and unoccupied spin orbitals, respectively. (b) and (c) Schematic illustration of the influence of poisons and promoters (cf. (a)). According to the indicated direction of flow of electrons, the poison decreases (b) and the promoter increases (c) the density of states of the metal at the Fermi level. With permission from J. M. McLaren, D. D. Veredensky, J. B. Pendry and R. W. Joyner, *Faraday Symp. Chem. Soc.* **1986**, *82*, paper 3.

of methanation and the ability of the surface to chemisorb CO. Moreover, there is a significant diminution in the binding energy of such CO as is adsorbed.

Several theoretical studies have indicated that there is a correlation between the electronic structure around the site of molecular adsorption and catalytic activity. Particular emphasis has been placed on the density of states at the Fermi level, which, for the adsorption of CO, can be readily illustrated, as in Fig. 1.23 (the so-called Blyholder model). In effect, the argument says that chemisorption and dissociation are related to charge transfer from the 5σ to the $2\pi^*$ orbital, and that the influence of the mediating metal catalyst, such as nickel, with a large density of states at the Fermi level E_F, is to provide an easier route via a two-step model, as represented in the simplified diagram in Fig. 6.13. Within this broad framework, we can immediately comprehend how traces of a CO-adsorbate will, through pumping or sucking away electron density at the site of adsorption (see Fig. 6.9), modify the strength of adsorption of CO to the surface.

Pendry, Joyner and co-workers have produced orbital contour plots (Fig. 6.14) for clusters that are representative of the Ni/CO, Ni/S/CO and Ni/Li/CO systems. Figures 6.14(a), (b), and (c) show the plots for the 5σ level of these three systems. Comparing the clean Ni/CO 5σ with that of the NI/S/CO shows that co-adsorbed sulphur atoms induce a rehybridization on the central nickel atom, from being

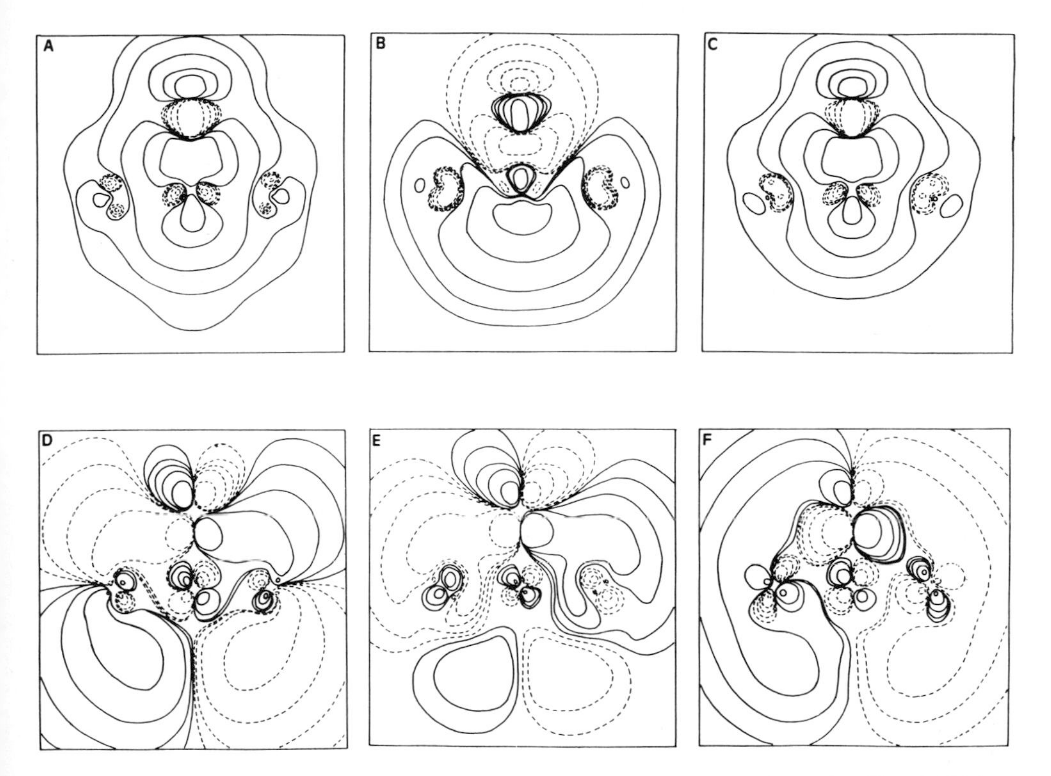

Figure 6.14 Orbital contours for the 5σ and $2\pi^*$ levels; the contours are ±0.001, ±0.0003, ±0.081 for CO adsorbed on Ni(100). Solid and broken lines represent positive and negative values, respectively. (a) The 5σ orbital for the system Ni/CO. (Note bonding nature of 5σ to the nickel.) (b) The 5σ orbital for the system Ni/S/CO. (Note the destabilization of the 5σ owing to the increased antibonding nature of this orbital. (c) The 5σ orbital for the system Ni/Li/CO (5σ orbital now similar to (a)). (d) The $2\pi^*$ orbital for the system Ni/CO, where we now see antibonding character between the carbon and nickel atoms. (e) The $2\pi^*$ orbital for the system Ni/S/CO. This level is essentially unaffected by the presence of sulphur. (f) The $2\pi^*$ orbital for Ni/Li/CO, where we see increased antibonding character, and increased back-donation. With permission from J. M. MacLaren et al., *Surf. Sci.* **1985**, *162*, 322.

mainly an s-like orbital to one whose main character is now p-type. The direct consequence of this is a reduction of the CO–metal interaction, since charge is drawn from this region into the metal, and surface charge is drawn towards the sulphur atom. In contrast, the broad similarity between Figs. 6.14(a) and (c) signifies that the influence of lithium in this energy region is small: there is still some bonding between the carbon and the central nickel.

From the corresponding $2\pi^*$ orbital contour plots (Figs. 6.14 (d), (e) and (f)), we see that, in each case, the $2\pi^*$ orbital is broadened by interaction with the central nickel, and that there is a similar Ni–C–O bonding character throughout. Combining the pictures that emerge from Figs. 6.13 and 6.14, we note that, in so far as the $2\pi^*$ orbital is concerned, the Ni/CO and Ni/S/CO systems show striking similarities. In both instances the $2\pi^*$ level lies well above the Fermi level, and although broadened by several electronvolts, remains mainly unoccupied. In contrast, the $2\pi^*$ orbital for Ni/Li/CO is pulled down to the Fermi level through an interaction with the lithium sp-orbital. There is strong broadening of this level, indicating that there is now some weight of the $2\pi^*$ orbital below E_F. In summary, therefore, we find that poisons exert a direct effect on the 5σ level, with a much weaker influence upon E_F and the $2\pi^*$ orbital, whereas promoters interact almost exclusively with the $2\pi^*$ near E_F, leading to a large shift towards E_F and a large broadening.

6.3.2.1 The Electronegativity of a Poison Seems to be of Secondary Importance

The influence of a range of electronegative poisons (carbon, sulphur, phosphorus, nitrogen and chlorine) on the catalytic activity of nickel in methanation and in facilitating dissociation of CO has been carefully assessed by Goodman, Ponec and van Santen. Experimentally, it emerged that the effectiveness of these elements as poisons was in the order Cl > S > N > C and P, i.e. they behave roughly in line with their Pauling electronegativities, which are 3.0, 2.5, 2.5 and 2.1 eV for chlorine, sulphur, carbon and phosphorus respectively. (Nitrogen is somewhat non-conformist, with an electronegativity of 3.0 eV.) Goodman and co-workers noted that carbon and nitrogen are less effective as poisons than their electronegativity would suggest.

Theoreticians have long recognized that the local density of states (Section 5.4.7), designated LDOS, is a general property of a catalyst surface that reflects the weight of electronic states, at a given energy and position, available for interacting with reactants. The LDOS may thus be used as a measure of (or to monitor) a catalyst's activity. A reduction in the LDOS near the Fermi level therefore signifies the influence of a poison, and the magnitude of this diminution should be a measure of the degree of poisoning.

Following a detailed theoretical assessment, Feibelman and Hamann concluded that the electronegativity of the adatom (CO-adsorbed with the reactant) determines its ability to poison. This, however, is not the conclusion reached by Pendry and Joyner. They too focused on changes in the LDOS at the site of CO adsorption on a Ni(100) cluster of the type shown in Fig. 6.9. They found that all the poisons carbon, sulphur, phosphorus and chlorine influence nearest-neighbour sites; and the effective range of sulphur and chlorine, the strongest poisons studied, was found to be 5 Å. Significantly, they discovered that the importance of electronegativity in determining poisoning strength is strongly modified by the size of the poison atom as reflected by its vertical distant (d) above the Ni(100) surface.

It is instructive to compare the influences of carbon and sulphur, the electronegativities of which are identical. The key point, however, so far as their role as poisons is concerned, is that they differ markedly in the distance at which they sit above the plane of the atoms in the catalyst surface. It is helpful that the precise locations of carbon and sulphur on the nickel surface are known from LEED data on the Ni(100)p(2×2)S and Ni(100)c(2×2)C overlayers. Carbon sits more or less in the surface, sulphur sits above it. The calculations for these locations show that sulphur is very much stronger than carbon as a poison. Carbon poisons the nearest-neighbour sites only weakly and its effect on the far sites is almost imperceptible. But when, computationally, the carbon is 'raised' to the same height as sulphur above the nickel surface; its poisoning is then equally strong. Even more important, the poisoning by this 'raised' carbon for far sites, 3.9 Å from the CO adsorption sites, is also strongly increased, showing that the effect is more than a simple overlapping of orbitals between CO and the poison. It seems that the critical ingredient is the presence of the nickel atom in the second layer immediately below the poison. If the poisoning atom gets too close to this second layer atom, its poisoning strength is considerably reduced.

Pendry and Joyner endeavoured to rationalise the experimental results of Goodman shown in Fig. 6.12. They were not able to achieve satisfactory agreement. Why? One reason is that, in the methanation reaction, poisoning of H_2 as well as of CO chemisorptior can occur. Another reason is that a more detailed analysis of the reaction kinetics is called for. CO dissociation may not even be the rate-limiting step in all circumstances. Furthermore, it may be quite inadequate to focus on just one nickel site as the locus of CO attachment. We must recall the reality of ensembles, discussed above. Calculations of ensembles, or at least of the changes in LDOS at a specified ensemble, are obviously desirable.

6.3.2.2 Other Factors Responsible for Promotion and Poisoning

It is instructive, following Ponec, to summarize the various ways, both physical and chemical, in which promoters can exert their influence on metal catalysts.

Physical effects
(a) Modifying the texture of the catalyst and/or the exposure of certain sites.
(b) Changing (stabilizing) the particle size and/or porosity.

Chemical effects
(a) Creation of parallel-working sites (as in bifunctional catalysts).
(b) Facilitating the formation of relevant intermediates.
(c) Exerting a direct (through-vacuum) or mediated (through-metal) intervention in the formation of essential intermediates; medium- or long-range interactions.

We have discussed rather fully how 'through-metal' promotion (and poisoning) operates, but hitherto we have not touched upon electrostatic ('through-vacuum') effects, nor on questions of charge mobility at metal surfaces. Changes in catalytic activity as a consequence of CO-adsorbed atoms can be correlated with changes in the electrostatic potential. In this model, the action of promoters is solely to reduce, and of poisons to increase, the electrostatic potential near the equilibrium site for adsorption. So far as the question of charge mobility goes, the essence of the argument is that the more mobile the charge at the Fermi level, the more easily is the surface able to respond to the presence of reactive species.

However, there are many other intriguing facts and pointers that are constantly emerging in relation to the key attributes of promotion and poisoning, so that it is premature to attempt the formulation of an all-embracing theory, even for metals and alloys alone. Thus, Roberts has demonstrated the crucial importance of transient oxygen species, designated O^-, in a wide range of promoter actions on metals. The activation of N–H and C–H bonds of impinging NH_3 or CH_4 on metal catalysts seems to proceed via the agency of these transient species, the lifetime of which is estimated to be *ca* 10^{-8} s at 295 K. These very species, which have a kinship with those present in doped MgO catalysts for methane dehydro-oxidations (Section 8.1) are also implicated in the promotion of methanol synthesis and water-gas shift reactions at copper catalysts. We shall see in Section 8.1 that dissociation of H_2 is an essential preliminary in methanol synthesis. It seems that the transient oxygen species, O^-, promotes the dissociative chemisorption of H_2, but may not be a vital part of the catalysis.

Promoter action and poisoning on non-metallic solids, including oxides and suplhides, are not amenable to so unified an interpretation as the catalytic behaviour of metals in syn-gas conversions, although there does seem to be some promise in a modification of the early notions of band-bending at semiconductor surfaces (see Section 5.4.1) and depletive and cumulative adsorption. In general, however, rather specific effects are invokes, as in the suggestion that promoter action, on the part of CO or nickel in MoS_2 catalysts, may be traced to the ability of the promoted material to facilitate the mobility, and hence the supply, of H atoms in the conducted phase. Schuit has argued that in CO-promoted MoS_2 hydrodesulphurization catalysts, the promoter activates the H_2, and that there is simultaneous transport of protons along the surface and electrons in the solid. In this sense, the explanation ties in with the promotional effects of tungsten and molybdenum bronzes in the catalysis of hydrogenation and ammonia synthesis, i.e. to the occurrence of facile spillover of hydrogen between metal and oxide support (see Section 1.3.3). There are, however, indications from the work of N. Y. and H. Topsøe, Chianelli and Daage that notions of surface acidity and basicity can also be of value in rationalizing promoter action on sulphides and oxides.

6.4 Problems

1 A mixed feed of *n*-butane and propane is dehydrogenated in the presence of a porous catalyst which remains at constant temperature. If the intrinsic chemical rate coefficient for the formation of butane is greater than that for propylene, how is the selectivity to the dehydrogenation of *n*-butane affected?

2 Methanol may be converted to formaldehyde in the presence of a silver catalyst supported on porous alumina. Carbon dioxide is also produced as a degradation product. If the reaction scheme is described by two first-order consecutive reactions, is the yield of formaldehyde greater of smaller for strong diffusion limitations than when diffusion limitation is negligible? Explain your reasons fully.

3 Write down the reaction–diffusion equations for two concurrent first-order reactions (such as the decomposition of ethanol to acetaldehyde on the one hand and ethylene on the other) occurring within an isothermal porous catalyst pellet (flat-plate geometry may be assumed – why?). From the differential mass-balance equations representing the model, demonstrate that selectivity would be unaffected by any intraphase diffusion effects. If non-isothermal conditions obtained, would the same conclusion be drawn? Write down the additional heat balance between bulk gas and particle, boundary conditions and any other relations necessary to solve the non-isothermal problem. Why is it that an interactive numerical solution to the problem is necessary?

4 Two first-order concurrent exothermic reactions,

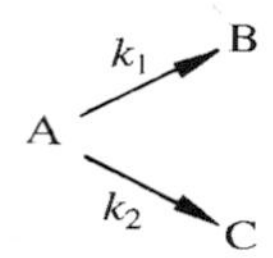

occur in the presence of catalyst particles sufficiently small to exclude the intrusion of intraphase mass-transfer effects. However, experimental measurements indicate that there is a temperature difference of 25 °C between the gas–particle interface and the bulk gas temperature of 500 °C. If the activation energy for the formation of B is 120 kJ mol^{-1} and that for the formation of C is 80 kJ mol^{-1}, how is the selective formation of B affetcted?

5 By how much is the reaction rate reduced when selective poisoning of a catalyst occurs, for which the Thiele modulus is 10 and the fraction poisoned, which is approximately independent of time, is 0.3? Compare the results you obtain from the steady-state model with that of the dynamic model.

6 It has been reported by H. Knözinger and P. Stolz (*Ber. Bunsenges. Phys, Chem.* **1971**, *75*, 1055) that 4-methylpyridine is more strongly bound to the Lewis acid sites of alumina than pyridine. But 2,4,6-trimethylpyridine is quantitatively displaced by pyridine. Why is this so? And what light do these facts shed on the feasibility of using substituted pyridines as probes for the acidic sites at the surfaces of solid catalysts?

7 The removal of nitrogen from organonitrogen compounds is a key process in the refining of petroleum (see B. C. Gates, J. R. Katzer, G. C. A. Schuit, *Chemistry of*

Catalytic Processes, McGraw-Hill, New York, **1979**). After first consulting T. C. Ho (*Catal, Rev. Sci. Eng.* **1988**, *30*, 117), describe the test reactions and other relevant features of this process.

8 The tarnishing that may occur when catalysts of metallic silver are handled in air prior to use can cause serious problems, since less than a monolayer of adsorbed sulphur profoundly affects the activity. Assuming that the sticking coefficient for H_2S on a surface of silver is 10^{-2} and independent of coverage, estimate how long it would take for a catalyst of area $10\,m^2$ to be tarnished (i.e. for a monolayer of Ag_2S to form) in an atmosphere consisting of 1 ppm of H_2S.

9 Discuss the proposition that activity and selectivity in hydrocarbon reactions catalysed by metals can be controlled by variations in the structure and composition of the solid catalyst. Explain the tabulated data, which refer to the hydrogenolysis (→*n*-hexane) and dehydrogenation (→cyclohexene) of cyclohexane by Cu/Ru alloy catalysts at 600 K.

	Copper content [atom %]			
	0	5	20	90
Hydrogenolysis activity [molecule $s^{-1}m^{-2}$]	10^7	10^4	10^3	10^2
Dehydrogenation activity [molecule $s^{-1}m^{-2}$]	10^8	10^8	10^8	5×10^7

How would you expect the behaviour of this system to depend on the size of the alloy particles?

10 (a) 'Changes in the X-ray absorption spectrum ($C_{1s} \rightarrow 2\pi^*$) of carbon monoxide adsorbed on Pt(111) before and after addition of sodium leave no doubt as to the mechanism of the promoter action of alkali metals.' Discuss.

(b) In terms of the band structure of graphite, what is meant by 'donor' and 'acceptor' graphite intercalates? For the graphite intercalation compounds of transition-metal hexafluorides, what correlation exists between the extent of charge transfer and layer spacing? How may the extent of charge transfer be monitored?

(c) In the selective oxidation of ethylene over $Ag–Al_2O_3$ catalysts it is found that electronegative additives have a pronounced favourable effect on reaction selectivity. However, it is found that the addition of alkalis to the catalyst also favourable affects the selectivity. Explain these observations with particular attention to mechanistic arguments.

6.5 Further Reading

6.5.1 General

J. B. Butt, *ACS Adv. Chem. Ser.* **1972**, *109*, 259.

L. L. Hegedus, *Catalyst Poisoning*, Dekker, New York, **1984**.

R. Hughes, *Deactivation of Catalysts*, Academic Press, London, **1984**.

E. B. Maxted, *Adv. Catal.* **1968**, *3*, 129.

J. Oudar, H. Wise, *Deactivation and Poisoning of Catalysts*, Dekker, New York, **1985**.

Z. Peterson, A. T. Bell (Eds.), *Catalyst Deactivation*, Dekker, New York, **1987**.

E. K. Rideal, *Concepts in Catalysis*, Academic Press, London, **1968**.

J. M. Smith, *Chemical Engineering Kinetics*, 3rd Ed., McGraw-Hill, New York, **1981**.

A. Wheeler, *Adv. Catal.* **1951**, 3, 249.

6.5.2 Studies of Model Surfaces

J. Benziger, R. J. Madix, *Surf. Sci.* **1981**, *109*, L155.

A. Corma, V. Fornes et al., *Prepr. ACS Symp. (Div. Petrol. Chem.)*, New York, **1986**, p. 184.

D. W. Goodman, *Acc. Chem. Res.* **1984**, *17*, 194.

R. M. Lambert, S. A. Tan, R. B. Grant, *Prepr. ACS Symp. (Div. Petrol. Chem.)*, New York, **1986**, p. 227.

G. A. Martin, C. F. Ng, *Prepr. ACS Symp. (Div. Petrol. Chem.)*, New York, **1986**, p. 208.

V. Ponec. *Adv. Catal.* **1993**, *32*, 149.

M. W. Roberts, *Int. Rev. Phys. Chem.* **1986**, 5, 57.

W. M. H. Sachtler, *J. Mol. Catal.* **1984**, *25*, 1.

W. M. H. Sachtler, Z. Zhang, *Adv. Catal.* **1933**, *39*, 129.

H. Topsøe, B. S. Clausen, et al., *Faraday Div. Symp.* No. 21, University of Bath, Sept. **1986**.

S. E. Wanke, P. C. Flynn, *Catal. Rev. Sci. Eng.* **1975**, *12*, 93.

6.5.3 Selectivity

S. M. M. Akhter, K. Blackmore, S. A. Shah, W. J. Thomas, *J. Catal.* **1977**, *50*, 205.

6.5.4 Theory of Poisoning and Promotion

C. T. Au, M .W. Roberts, *Int. Rev. Phys. Chem.* **1986**, *5*, 57.

G. C. Chinchin, M.S. Spencer, K. C. Waugh, D. H. Whan, *J. Chem. Soc., Faraday Trans. 1*, **1987**, *83*, 2193.

P. J. Feibelman, D. R. Hamann, *Surf. Sci.* **1985**, *149*, 48.

L. D. Marks, V. Heine, *J. Catal.* **1985**, *94*, 570.

J. M. McLaren, J. B. Pendry, R. W. Joyner, *Surf. Sci.* **1986**, *165*, 180.

J. M. McLaren, J. B. Pendry, D. D. Vvedersky, R. W. Joyner, *Surf. Sci.* **1985**, *162*, 322.

T. Okuhara, T. Enomoto, K. Tamaru, M. Misono, *Chem. Lett.*, **1984**, 1491.

J. B. Pendry, R. W. Joyner, J. M. MaLaren, *Surf. Sci.* **1986**, *165*, L80.

J. B. Pendry, R. W. Joyner, J. M. MaLaren, *Faraday Div. Symp.* No. 21, University of Bath, Sept. **1986**, Paper 21/3.

J. S. Rieck, A. T. Bell, *J. Catal.* **1986**, *99*, 262.

G. C. A. Schmit, *Int. J. Quantum Chem.*, **1977**, *XIL Suppl.*, 68.

R. A. van Santen, A. P. J. Jansen, in *Structure, Activity and Selectivity Relationships in Heterogeneous Catalysis* (Eds.: R. K. Grasselli, A. W. Sleight), Elsevier, Amsterdam, **1991**, p. 221.

6.5.5 The Bifunctional Effect of Catalysts

G. C. Bond, *Acc. Chem. Res.* **1993**, *26*, 490.

D. J. Gunn, and W. J. Thomas, *Chem. Eng. Sci.* **1965**, *20*, 89.

J. H. Sinfelt, *Adv. Chem. Eng.* **1964**, *5*, 37.

W. J. Thomas, *Trans. Inst. Chem. Eng.* **1971**, *49*, 204.

P. B. Weisz, *Proc. 2nd Int. Congr. Catalysis*, Paris, Vol. 1, **1961**, p. 937.

7 Catalytic Process Engineering

7.1 Statement of the Problem

Many important industrial processes may be classified as heterogeneous gas–solid catalysed reactions. The presence of the solid catalyst accelerates the conversion of reactants into products, but unless a considered choice is made concerning reaction conditions, catalyst geometry and reactor dimensions, the potential value of the process as an economic venture will be reduced. Problems such as the transfer of heat to or from a reactor will substantially influence its ultimate design. Similarly, the rate of transport of reactants and products, not only to and from the catalyst surface but also within the complex pore structure of the solid itself, will affect both the dimensions of the reactor unit required and the selection of catalyst particle size. The kinetics of the heterogeneous reaction will obviously determine the surface area and hence the mass of catalyst and reactor volume which is necessary to achieve a specified conversion within a given time in the case of a batch process, or within a given reactor length in the case of a continuous process.

It is the object of catalytic process engineering to assess the dimensions of reactor required and the amount of catalyst to use for an efficient conversion of a specified input of reactant into a desired product. This may be accomplished provided that certain operating conditions, such as the initial temperature, pressure and reactant concentrations, are chosen and that a decision is made concerning the type of reactor to be used. For example, a batch or continuous reactor may be used in which the conversion may be effected isothermally or adiabatically. Such operating variables constitute the design conditions. Different sets of design conditions will produce different estimates for the size of reactor. The optimum design will be that which is most economical in a pecuniary sense.

Specialised mathematical procedures, generally requiring the application of a numerical technique, can be employed to optimise the design. In practice, the final choice of operating conditions is often made on the basis of merely a few design calculations. Such calculations depend directly upon (a) the available kinetic data, (b) mass-transfer processes and (c) heat-transfer processes.

7.2 Kinetics of Heterogeneous Catalytic Reactions

7.2.1 The Overall Rate of Reaction

Chemical reactions, whether a catalyst is present or not, may be operated either batchwise or continuously. In the batchwise mode of operation the reactants, at some initial temperature and pressure, are introduced into the isolated reactor and the products of reaction are allowed to accumulate with the progress of time. The composition of the system changes monotonically with time until either chemical equilibrium is attained or the reaction is deliberately stopped. Examples of batch reactors are the laboratory flask equipped with a stopcock to introduce, and then isolate, reactants; and the autoclave used industrially in the pharmaceutical industry. On the other hand, continuous reactors operate by reactant feed flowing into the vessel inlet while the products of reaction continuously emerge from the outlet. Such a reactor, operated in the continuous mode, will tend towards a time-invariant state. Change of composition occurs spatially. Examples are the laboratory flow reactor, in which composition (independent of time for steady-state operation) gradually changes from inlet to outlet, and the laboratory continuous stirred-tank reactor in which the reactants enter the vessel (the contents of which are well mixed) and the products emerge from the outlet, a step change in concentration having occurred between the feed and reactor although the reactor operates in the steady state. The majority of industrial catalytic reactors are operated continuously; their various forms will be divulged later in this chapter.

An isothermal batch system, if left to run its own course, will eventually reach a time-invariant composition which, for such a closed system, is the state of chemical equilibrium. A continuously operating system, however, although it may tend towards a steady state condition, is not *per se* at chemical equilibrium. There are also many circumstances when a steady-state operating condition may not be achieved (e.g. in an autocatalytic reaction or for some reactions in which the heat generated by reaction cannot be removed sufficiently fast) and concentration oscillations and temperature fluctuations, or even gross perturbations, will occur (see Chapter 2). For the most part, however, we will concern ourselves with the operation of continuous reactors in the steady state.

Both chemical and physical rate processes occur simultaneously during a catalytic reaction in a continuous reactor operating in the steady state. In ammonia oxidation, for example, the reactants are transported by convection to the essentially non-porous platinum gauze surface where the ammonia and oxygen react chemically. Here we can distinguish two rate processes, those of convection and chemical reaction, and they will occur at identical rates if the steady state has been achieved. The convective flux N may be written as the product of a mass-transfer coefficient h_D and the driving force, which is the difference between the concentration in the bulk gas phase and at the interface.

$$N = h_D(c - c_i) \tag{1}$$

If air is in excess (usually the condition for ammonia oxidation) then, as a simplification, the kinetics of the chemical reaction may be treated as pseudo-first order. As the catalytic reaction occurs at the surface of the catalyst it is the interface concentration which determines the kinetics so

$$R = kc_i \tag{2}$$

where both the reaction rate R and the rate constant k are expressed on the basis of unit area.

Invoking the steady-state condition, the rates of both processes (measured in the same units) are identically equal and thus

$$h_D(c - c_i) = kc_i \tag{3}$$

We can now eliminate the unknown interface concentration c_i from these equations, and therefore rewrite the rate in terms of the known concentration c, thus:

$$R = \frac{h_D kc}{h_D + k} = \frac{c}{1/h_D + 1/k} \tag{4}$$

Two limiting cases are immediately apparent. If $h_D \gg k$ then $R = kc$ and the chemical reaction would be rate-limiting. If, however, $k \gg h_D$ then $R = h_D c$ and mass transfer would be rate-limiting. In fact, as we shall see in Chapter 8, the catalytic oxidation of ammonia is mass-transfer-limited. Little is known about the chemical kinetics, but it should now be evident that this is unimportant from the point of view of an engineer designing or operating the reactor: the overall kinetics are described by the product of the mass-transfer coefficient and the ammonia concentration.

The concept of resistance can usefully be applied when a process involving both fluid-to-solid transport and chemical reaction is analysed. Transport of reactant through the fluid phase to the solid must occur initially and is then followed sequentially by adsorption, surface reaction and desorption of product. Finally, the product of reaction is transported back into the bulk fluid phase. Thus, merely to consider mass transport and chemical reaction as the only significant steps is a gross oversimplification. Nevertheless, is serves to illustrate the point that, from Eq. (4), the overall rate coefficient $\bar{k}$ can be represented as

$$\frac{1}{\bar{k}} = \frac{1}{k} + \frac{1}{h_D} \tag{5}$$

In writing such an equation we imply that there is a resistance to chemical conversion and a resistance to mass transport. If the operating conditions were such that a particular process occurred in a regime intermediate between the chemically controlled region and the mass-transfer-controlled region, the overall rate would be a composite of the two rate processes and Eq. (5) would be employed rather than either one of the two limiting cases. Although the concept of resistance can also be applied in the case of a more complex process (where diffusion within the porous catalyst structure may also be significant) the overall rate coefficient cannot, in such a case, be written in the form of Eq. (5), for the problem would then involve the interface concentration c_i as a boundary condition as well as a variable dependent on penetration within the porous structure (see, for example, Chapter 4). Nevertheless, we may speak of intraparticle diffusion as rate-limiting because, for such circumstances, pore diffusion influences the overall kinetic behaviour of the system. Difficulties also arise if the rate of reaction is not a linear function of concentration. If the chemical reaction were anything but first order, the rate coefficients for chemical reaction and mass transfer would have different dimensions and one could then only refer to a step as rate-limiting over a restricted and specified concentration range.

Diffusion and mass transfer limited processes tend to occur in relatively high-temperature regions, whereas chemically controlled processes occur at lower temperatures. As we shall see in Chapter 8, the mass-transfer-controlled exothermic oxidations of methanol to formaldehyde, and of ammonia to nitric oxide, occur in the region of 700 K and 800 K respectively but the chemically controlled low-pressure methanol synthesis operates at a temperature just in excess of 500 K. An example of a typical pore diffusion-limited process is the reaction between carbon monoxide and hydrogen over a supported nickel catalyst (methanation) occurring at about 600 K. These differences is limiting effects at various temperatures occur because the temperature dependence of a chemical reaction is of an exponential form (the Arrhenius factor, exp $(-E/R_gT)$), while diffusion in particular, and mass transfer in general, are not so strongly influenced by temperature, the diffusion coefficient being dependent only on the square root of temperature. One way of illustrating this is to plot the functions $k(T)$, the temperature-dependent rate constant, and $D_e(T)$, the effective diffusivity (proportional to the bulk diffusion coefficient), against temperature.

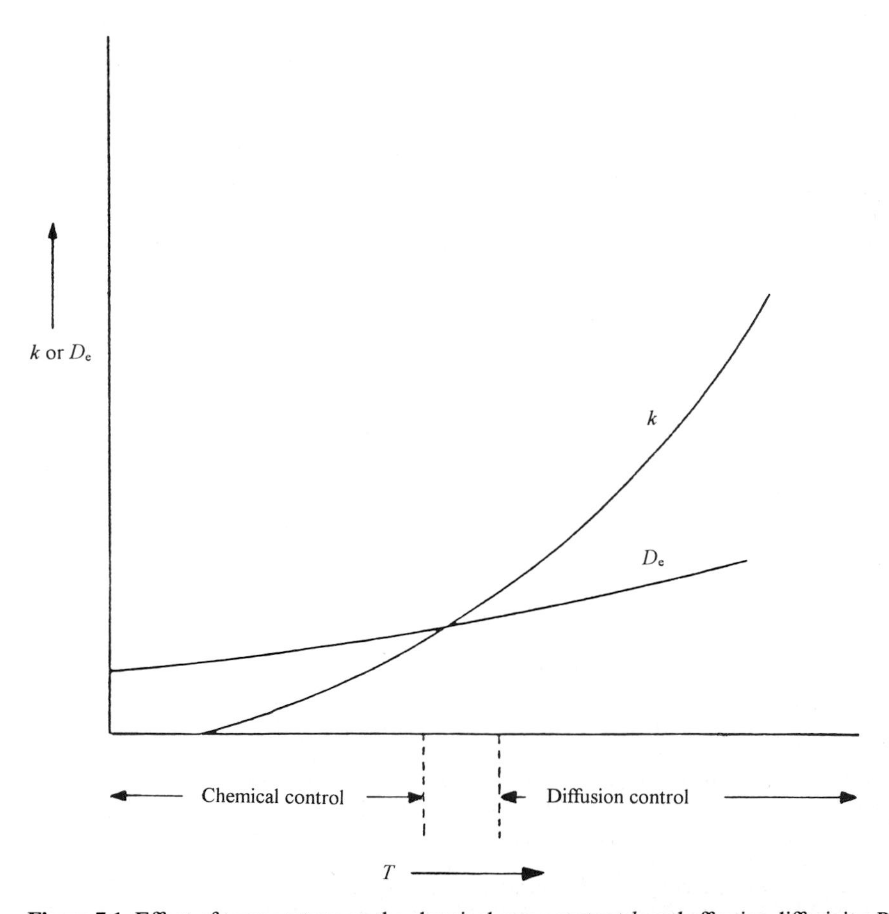

Figure 7.1 Effect of temperature on the chemical rate constant k and effective diffusivity D_e.

Such a plot is sketched in Fig. 7.1: we see how the chemical rate coefficient increases slowly at first and then rises sharply to overtake the diffusion coefficient.

Another way of demonstrating the same point is to plot the natural logarithm of the reaction rate against $1/T$. This is sketched in Fig. 7.2, where it is clearly seen that, in the higher regions of the $1/T$ scale (low-temperature regime), chemical reaction controls the rate and the slope of the straight line is $-E/R$. At relatively high temperatures (as discussed in Chapter 4) intraparticle diffusion will be rate-controlling, giving an apparent activation energy of one-half the true activation energy and consequently a straight line of slope $-E/2R$. At temperatures even higher than this, mass transfer through the gas phase from fluid to solid dominates the kinetics and, because the mass-transfer coefficient h_D is only weakly dependent on temperature (equivalent to only about 10 kJ mol^{-1} if it were to be assigned an activation energy), the slope of the curve of ln h_D as a function of $1/T$ would be less than half the slope in the chemically controlled region.

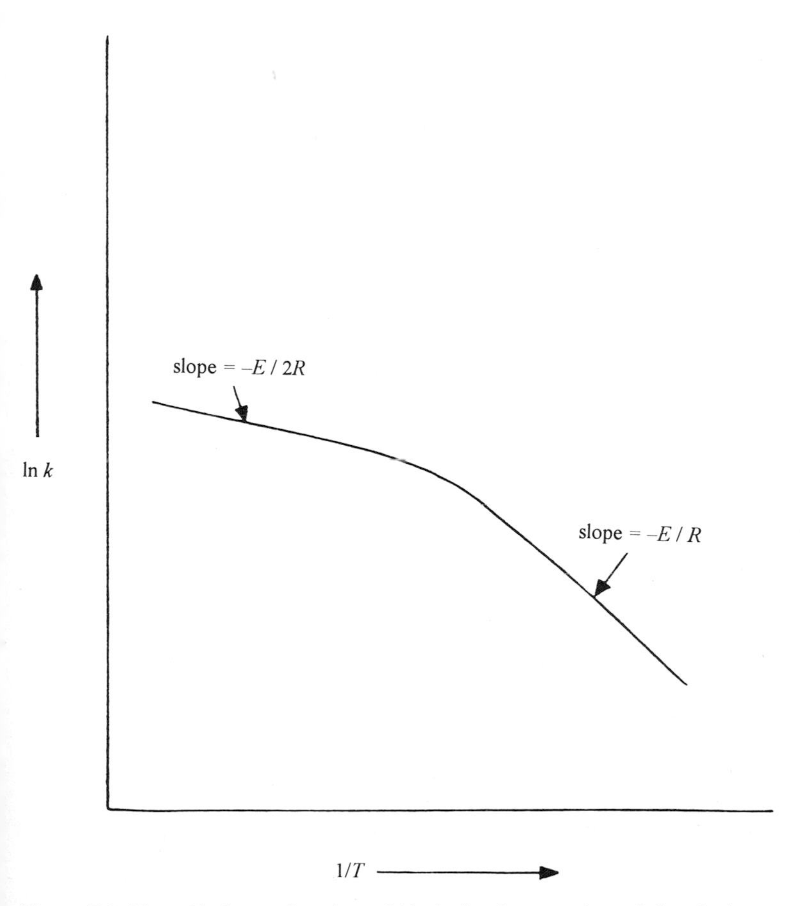

Figure 7.2 Plot of $\ln k$ as a function of $1/T$ indicating a region of chemical rate control (with activation energy E) and a region of diffusion control (with activation energy $E/2$).

7.2.2 The Rate of Chemical Reaction

The rate of a heterogeneous catalytic reaction uninfluenced by either mass- or heat-transfer effects would be, by definition, the rate at which the chemical transformation of reactants into products occurs when the catalyst (including the internal pore structure) is uniformly bathed with reacting fluid at the conditions of concentration and temperature prevailing in the bulk fluid. For purposes of reactor design and operation it is usual to express the overall rate and also the chemical reaction rate as the rate of formation of product (or alternatively the disappearance of reactant) per unit mass of catalyst. In SI units this would be $\mathrm{kmol\,kg^{-1}\,s^{-1}}$ though, of course, the literature abounds with alternative systems of units. For experimental purposes this manner of expressing reaction rates is also useful but it has become customary, when examining the fundamental character and activity of a catalyst surface, to speak of the turnover number (see Section 1.3): this is the number of reactant molecules transformed per second at a single active catalyst site. Many experimental results are also based on the rate of product formation per unit area of catalyst (usually, but not exclusively, the BET area).

7.2.3 Fundamental Kinetic Models

The earliest fundamental kinetic models of heterogeneous catalytic reactions arise from the researches of Langmuir, later extended and applied by Hinshelwood and further developed by Eley and Rideal (see Further Reading, Section 7.5).

The basis of the so-called Langmuir–Hinshelwood model is that the reactants are initially chemisorbed at the catalyst surface before rearrangement on the surface and subsequent desorption of the product. Consider, for example, reaction between A and B to form the product P. The chemisorption, surface reaction and desorption may be schematically represented as shown in Scheme 7.1, where the asterisks denote active catalyst sites and the species A–B bonded to one of these sites is the chemisorbed product P. The essence of model development is to assume that one or other of the kinetic steps chemisorption, desorption or surface reaction is rate-limiting (excluding any physical rate-limiting process such as intraparticle diffusion or fluid-to-solid mass transfer).

By way of illustration, we show how the overall chemical rate equation is developed. Suppose the adsorption of reactant A is the rate-determining step. The *net* rate of adsorption will be the difference between the rate of chemisorption of A and its rate of desorption. Applying the notions of Langmuir we can therefore write, for the *net* rate of adsorption of A,

$$\mathrm{R} = k_{\mathrm{dA}} p_{\mathrm{A}} \left(1 - \sum \theta\right) - k_{\mathrm{dA}} \theta_{\mathrm{A}} \tag{6}$$

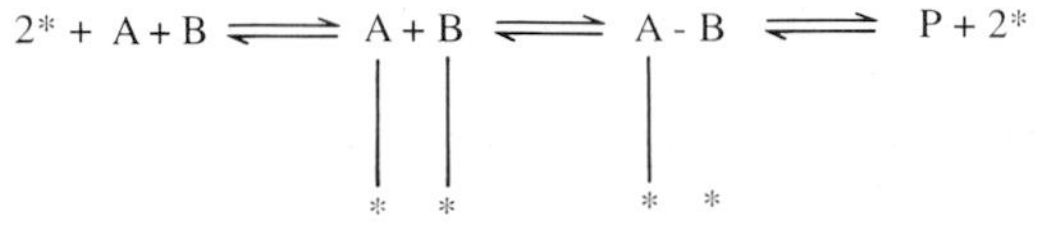

2* + A + B ⇌ A + B ⇌ A - B ⇌ P + 2*
(A and B each bonded to a site *; A - B bonded to a site *, with a free site *)

Scheme 7.1

where $\sum\theta$ represents the sum of fractions of active surface occupied by reactants A and B and product P. We next introduce an approximation assuming that, because the adsorption of A is rate-determining, the adsorption and desorption of B and P, and the surface reaction between adsorbed A and B to form adsorbed P, are each virtually at equilibrium. Such steady-state approximations have been discussed fully by Boudart, who concluded that they are justifiable for circumstances other than instantaneous phenomena. For the adsorption–desorption equilibria of B and P, then,

$$k_{aB}p_B\left(1-\sum\theta\right) = k_{dB}\theta_B \tag{7}$$

$$k_{aP}p_P\left(1-\sum\theta\right) = k_{dP}\theta_P \tag{8}$$

while the surface reaction, if it were at equilibrium, would be expressed by

$$k_s\theta_A\theta_B = k'_s\theta_P\left(1-\sum\theta\right) \tag{9}$$

Defining the adsorption equilibrium constant of A as

$$K_A = \frac{\theta_A}{p_A(1-\sum\theta)} \tag{10}$$

and recalling that the classic thermodynamic equilibrium constant for the overall equilibrium is

$$K = \frac{p_P}{p_A p_B} \tag{11}$$

then substitution of Eqs. (7)–(9) inclusive into Eq. (6) with subsequent utilisation of the equilibrium constants defined by Eqs. (10) and (11) gives, with minor algebraic manipulation,

$$\mathrm{R} = \frac{k_A\{p_A - p_P/p_B K\}}{1 + K_A p_A + K_B p_B + K_P p_P} \tag{12}$$

where k_B and k_P are the ratios of adsorption and desorption kinetic constants k_{dB}/k_{aB} and k_{dP}/k_{aP} respectively.

At this point we draw attention to the structure of Eq. (12). In the numerator it contains a driving-force term (contained within the braces) which describes the closeness of approach to true chemical equilibrium. The collection of terms in the denominator is sometimes referred to as the 'adsorption term' because the individual terms describe the retarding effects of adsorbed reactants and products. If, instead of the adsorption of A being rate-determining, one or other of the steps such as the desorption of product P, or the surface reaction between adsorbed A and B forming adsorbed P, were rate-determining, then a similar treatment (assuming steps other than that which is rate-determining to be at equilibrium) would yield rate equations similar in general structure to Eq. (12) but differing in detail. Table 7.1 provides a summary of some selected rate-controlling steps for a single reactant as well as for two reactants (both chemisorbed) forming product, and also incorporates the structure of the corresponding rate equation.

Table 7.1 General structure of Langmuir-type rate equations.

Reaction	Controlling step	Net rate	Kinetic constant	Driving force	Adsorption term
1 $A \rightleftharpoons P$	(a) Adsorption of A	$k_A p_A(1-\sum\theta) - k'_A\theta_A$	k_A	$p_A - \frac{p_P}{K}$	$1 + K_A p_A + K_P p_P$
	(b) Surface reaction, single-site mechanism	$k_S\theta_A - k'_S\theta_P$	$k_S K_A$	$p_A - \frac{p_P}{K}$	$1 + K_A p_A + K_P p_P$
	(c) Desorption of P	$k'_P\theta_P - k_P p_P(1-\sum\theta)$	$k'_P K_S K_A$	$p_A - \frac{p_P}{K}$	$1 + K_A p_A + K_P p_P$
2 $A \rightleftharpoons P + Q$	Surface reaction, A(ad) reacts with vacant site	$k_S\theta_A(1-\sum\theta) - k_S\theta_P$	$k_S K_A$	$p_A - \frac{p_P p_Q}{K}$	$(1 + K_A p_A + K_P p_P + K_Q p_Q)^2$
3 $A_2 \rightleftharpoons 2P$	(a) Dissociative adsorption of A_2	$k_A p_{A_2}(1-\sum\theta)^2 - k'_A\theta_A$	k_A	$p_{A_2} - \frac{{p_P}^2}{K}$	$(1 + K_A p_{A2} + K_P p_P)^2$
	(b) Surface reaction following dissociative adsorption of A_2	$k_S\theta_A - k'_S\theta_P$	$k_S K_A$	$p_{A_2} - \frac{p_P}{K}$	$1 + K_A p_{A2} + K_P p_P$
4 $A + B \rightleftharpoons P$	(a) Adsorption of A	$k_A p_A(1-\sum\theta) - k'_A\theta_A$	k_A	$p_A - \frac{p_P}{K p_B}$	$1 + K_A p_A + K_B p_B + K_P p_P$
	(b) Surface reaction	$k_S\theta_A\theta_B - k'_S\theta_P(1-\sum\theta)$	$k_S K_A K_B$	$p_A p_B - \frac{p_P}{K}$	$(1 + K_A p_A + K_B p_B + K_P p_P)^2$
	(c) Desorption of P	$k'_P\theta_P - k_P p_P(1-\sum\theta)$	$k'_P K_S K_A K_B$	$p_A p_B - \frac{p_P}{K}$	$1 + K_A p_A + K_B p_B + K_P p_P$

Note: If an Eley–Rideal mechanism prevails for a bimolecular–unimolecular reaction (such as 4 above) with chemisorbed A reacting with gaseous (or physically adsorbed) B, then (a) the term $K_B p_B$ in the denominator should be omitted, and (b) for surface reaction rate-controlling, the adsorption term is linear and not squared.

The Eley–Rideal model, developed subsequent to the Langmuir–Hinshelwood model, allows for the possibility that of two reactants only one might be chemisorbed, the second reactant being either physically adsorbed or reacting with the chemisorbed layer directly from the gas phase. We picture these events by the representation in Scheme 7.2.

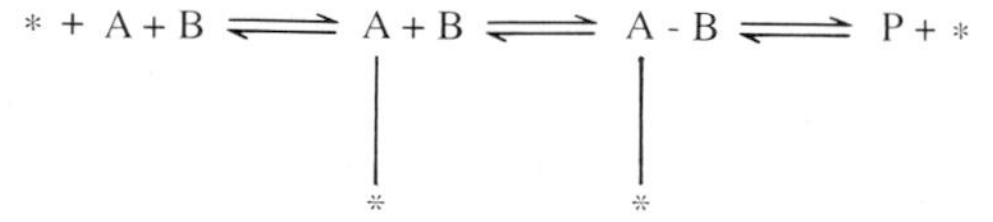

$$* + A + B \rightleftharpoons \underset{|\atop *}{A} + B \rightleftharpoons \underset{|\atop *}{A}\text{-}B \rightleftharpoons P + *$$

Scheme 7.2

To provide a further example of the nature of model development of this kind, we now suppose that the rate-determining step is the surface reaction between chemisorbed A and gaseous (or physically adsorbed) B to form the adsorbed product. The net rate of surface reaction will be the difference between the forward and reverse steps:

$$\mathrm{R} = k_s\theta_A p_B - k'_s\theta_P \tag{13}$$

We note here that the rate of the forward step is not proportional to the product of the surface coverage of both A and B but, because B is not chemisorbed, the rate is proportional to the product of the surface coverage of A and the partial pressure of B. Assuming, as before, that steps other than the surface reaction are at equilibrium, we may express the equilibrium adsorption of B and P by Eqs. (7) and (8) respectively, while the equilibrium adsorption of A is

$$k_{aA} p_A \left(1 - \sum \theta\right) = k_{dA}\theta_A \tag{14}$$

Making use, then, of Eqs. (7), (8) and (14) by first substituting into Eq. (13) and then defining the surface equilibrium state by

$$K_s = \frac{\theta_P}{\theta_A p_B} \tag{15}$$

and recalling the thermodynamic equilibrium relation (Eq. (11)), one finally obtains

$$\mathrm{R} = \frac{k_s K_A \{p_A p_B - p_P/K\}}{1 + K_A p_A + K_P p_P} \tag{16}$$

which may also be deduced from Table 7.1 by correctly applying the notes appended at the foot of the table.

The advantage of the Langmuir–Hinshelwood and Eley–Rideal approach to the deduction of a rate equation is that, given a prior insight into the reaction mechanism, the resulting kinetic equation will express the behaviour of the reacting system for a range of partial pressure conditions. However, the converse is by no means necessarily true and one should proceed with caution if the object is to confirm a mechanism by demonstrating that the kinetic rate equation is well obeyed. A number of differing mechanisms will give similarly structured rate equations, albeit with different kinetic and adsorption–desorption equilibrium constants. Furthermore, although similarly structured equations may be subjected to the rigours of statistical discrimination techniques, it is seldom found that the kinetic and equilibrium constants contained in the expression can be assigned satisfactory values which match with independently derived experimental values. Nevertheless, despite such difficulties and provided there is some independent evidence concerning the mechanisms of reaction, the Langmuir–Hinshelwood and Eley–Rideal models are helpful in predicting reaction rates and for use in process engineering calculations. The constants in the equations are then perhaps best looked upon as semi-empirical adjustable parameters. As long as the rate equation correctly reflects the rate of reaction over a modest range of conditions, then this is all that is required for process engineering design.

The fundamental notions of these earlier workers have been modified by Boudart (1968) (see Section 7.5.2). He asserted that, because of the inherent complexity of many catalytic reactions, it is often uncertain what surface species are involved. Furthermore, there may conceivably be a number of surface intermediates which are precursors to the final product, and consequently many separate elementary kinetic steps contribute to the overall reaction. Boudart therefore proposed some simplifying assumptions. It was assumed, first, that only the most prevalent surface species is important and, secondly, that only one chemical kinetic step is rate-determining. All other chemical steps are regarded as being in a state of quasi-equilibrium. Such assumptions, of course, bring about an enormous reduction in the complexity of the rate expression, for now such steps as adsorption or desorption, unless they happen unambiguously to be the rate-determining step, are disregarded. Two-step kinetic models (as referred to by Boudart) of this kind are certainly justified (as far as process engineering is concerned) if the model is capable of predicting rates of reaction over an acceptable range, for the resulting equation is usually much simpler than any listed in Table 7.1.

Arising from the structure of the Langmuir–Hinshelwood and other rate expressions is the frequently overlooked consequence of the behaviour of the reaction rate as a function of temperature. In the laboratory, initial rate data are often acquired for constant partial pressures of reactants, and the normal practice is to plot the logarithm of the initial rate as a function of reciprocal temperature, thereby deducing an activation energy. Departures from straight-line plots have already been referred to in Chapter 4, where it was noted that the intrusion of intraparticle diffusion effects causes such behaviour. There are, however, other possible reasons for observing non-linear effects based upon the fact that adsorption–desorption equilibria are temperature-dependent – as, of course, the celebrated van't Hoff equation describing the effect of temperature on chemical equilibria informs us. Hinshelwood, in his classic text on chemical kinetics, points out that a lowering of the activation energy would occur if the heat of adsorption of reactant were significant. Thus, suppose that independent evidence suggests that the surface reaction is rate-determining. Provided initial-rate data are used (when the partial pressure of product is negligibly small), the reaction rate (deduced from case 1b in Table 7.1) would be

$$\mathrm{R} = \frac{k_s K_A p_A}{1 + K_A p_A} \tag{17}$$

Now the term k_s is a kinetic constant and so it would conform with the Arrhenius temperature dependence $A\,\exp(-E_s/RT)$ where A is the pre-exponential factor and E_s is the activation energy of the surface reaction. K_A, on the other hand, is an adsorption–desorption equilibrium constant. Following the classic van't Hoff isochore with q as the *net* heat of adsorption of A, then K_s would be $\exp\,(q/R_gT)$. Substituting into Eq. (17),

$$\mathrm{R} = \frac{A e^{-E_s/RT} e^{q/RT} p_A}{1 + e^{q/RT} p_A} \tag{18}$$

Taking logarithms and differentiating,

$$-\frac{d(\ln \mathrm{R})}{d(1/RT)} = E_s - q - \frac{qK_s p_A}{1 + K_s p_A} \tag{19}$$

and because the third term on the right-hand side of Eq. (19) involves K_s and is therefore temperature-dependent, a plot of $\ln \mathrm{R}$ against $1/T$ would not be linear. Indeed, Satterfield 1970 (see Section 7.5) points out that for some circumstances a maximum could appear in such a plot.

Notwithstanding the fundamental basis of the Langmuir–Hinshelwood and Eley–Rideal ideas, a process engineer would just as soon use an empirical rate expression of the form

$$\mathrm{R} = kc_A^a c_B^b \tag{20}$$

which expresses the rate as powers of the concentration of reactants (and products if appropriate). It is an entirely empirical formulation of the rate of reaction and the indices would be deduced by standard curve-fitting procedures. Its use is entirely justified for process calculations if it is capable of describing and predicting rates over the desired range of temperature and pressure of operation. The kinetic constant k would, as usual, be of Arrhenius form. The virtue of Eq. (20) is its simplicity.

7.2.4 The Effect of Intraparticle Diffusion

As intimated in Section 7.2.1, intraparticle diffusion can be rate-limiting under certain circumstances. It is found that, for the majority of catalytic reactions occurring at process conditions, the reaction is, to a greater or lesser extent, influenced by intraparticle diffusion. Satterfield (1970) quotes a number of commercially important reactions and gives the process conditions and type of catalyst used. Even from such a summarised compilation it can be seen that in almost every case intraparticle diffusion has an effect on the reaction rate: in some cases the restriction is severe but for other reactions diffusion within the catalyst particle does not impede the reaction so seriously. Limitation of chemical reaction rate within porous catalysts is caused principally by the narrowness of pores within the catalyst structure. It is not difficult to imagine that reactants and products will be unable to penetrate the porous structure with ease unless the molecular dimensions of the reactants are sufficiently small for their transport along the pore channels to be unimpeded. It then the pores are narrow, chemical reaction will occur and products will be formed before much penetration into the catalyst interior has occurred. The reaction would effectively be taking place on the exterior catalyst surface, or perhaps only within a thin outer shell of the catalyst pellet. Of course, the temperature at which reaction occurs will also have an influence on whether or not diffusion is rate-limiting because, as shown in Section 7.2.1, the higher the temperature the more likely it is that diffusion in the catalyst pores will dominate and control the reaction.

In Chapter 4 the nature of intraparticle diffusion was discussed and it was shown how, within a porous catalyst pellet, the chemical reaction and diffusion processes are coupled. Solution of the differential equations describing diffusion concomitant with chemical reaction in the interior of the porous catalyst gave rise to the formulation of the reactant and product concentration profiles in the pellet, from which it was demonstrated that, for the steady state, it is possible to derive an effectiveness factor. As this effectiveness factor is a measure of the observed rate of reaction compared with the rate of reaction if resistance to intraparticle diffusion (or heat conduction) were completely absent, we have at our disposal a means of correcting the true chemical rate (if indeed the experimental data reflected the true kinetics) to allow for the intrusion of intraparticle diffusion effects.

Thus, for example, if we represent the true chemical kinetics as some (known) function of concentration $f(c)$, then the rate of reaction which would have been observed had there been diffusion limitation would be

$$\mathrm{R} = kf(c)\eta \tag{21}$$

where η (as in Chapter 4) is the effectiveness factor, which can be either calculated using the appropriate formula or, better still, measured experimentally.

7.2.5 The Effect of Interparticle (Fluid-to-Solid) Transport

Two well-known commercial catalytic processes may be cited as examples where mass transport of reactant through the gas phase to the solid catalyst (in the form of metal gauze or crystals) is rate-determining. Both catalytic reactions occur at relatively high temperatures, and as the catalyst in each case is non-porous, intraparticle diffusion effects are absent. The processes have already been referred to (Section 7.2.1) and are treated more fully in Chapter 8: they are the catalytic oxidation of methanol and of ammonia.

If, then, such a resistance to mass transfer occurs in a particular process, the concentration c of a reactant in the bulk fluid will differ from its concentration c_i at the solid–gas interface. The driving force for transport will be $(c - c_\mathrm{i})$ but because c_i is unknown it is necessary to eliminate it from the rate equation describing the gas-to-solid transfer process. We have already invoked the concept of the steady state in a continuous process (Section 7.2.1) and the elimination of the interface concentration was illustrated by a simple example, equating the rate of mass transport (Eq. (1)) to the chemical rate (Eq. (2)).

Analyses of this kind invoking a driving force for mass transport between bulk phase and gas–solid interface are based upon a film model in which the resistance to mass transfer is supposed to be confined to a film of finite thickness adjacent to the exterior surface of the solid. Figure 7.3 illustrates this elementary model. In the gas

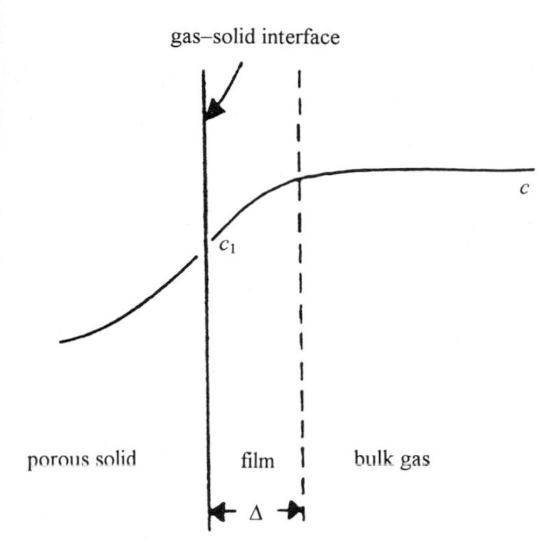

Figure 7.3 Reactant concentration profile from bulk gas through a relatively stagnant gas film to the gas–solid interface and thence within the interior porous solid.

film, of thickness Δ, ordinary bulk diffusion is supposed to occur so that the gaseous flux to the surface may be written

$$N = \frac{D(c - c_i)}{\Delta} \tag{22}$$

Now, although reasonable estimates for Δ can be arrived at by the application of boundary layer theory, it is more straightforward and pragmatic to replace D/Δ by a mass-transfer coefficient h_D which can be determined separately by experiment and correlation. In terms of either the pressure or concentration, the flux (which, in the steady state, is equal to the reaction rate) is

$$N = h_D(c - c_i) = k_g(p - p_i) \tag{23}$$

where k_g is the gas-to-film mass-transfer coefficient in terms of unit external surface area and unit pressure driving force.

In order to observe the effect of such interphase transport on the catalytic reaction, as before we equate the flux (Eq. (1)) to the chemical reaction rate (Eq. (2)) to obtain

$$\frac{c_i}{c} = \frac{1}{1 + k/h_D} \tag{24}$$

which is only another way of writing Eq. (3). The overall reaction rate is thus

$$\mathrm{R} = kc_i = \frac{kc}{1 + k/h_D} \tag{25}$$

(also given in a slightly different form by Eq. (4)). Comparing this overall rate with the rate kc if chemical reaction were rate-controlling, we can write a kind of effectiveness factor for interphase transport:

$$\eta_i = \frac{kc_i}{kc} = \frac{1}{1 + k/h_D} \tag{26}$$

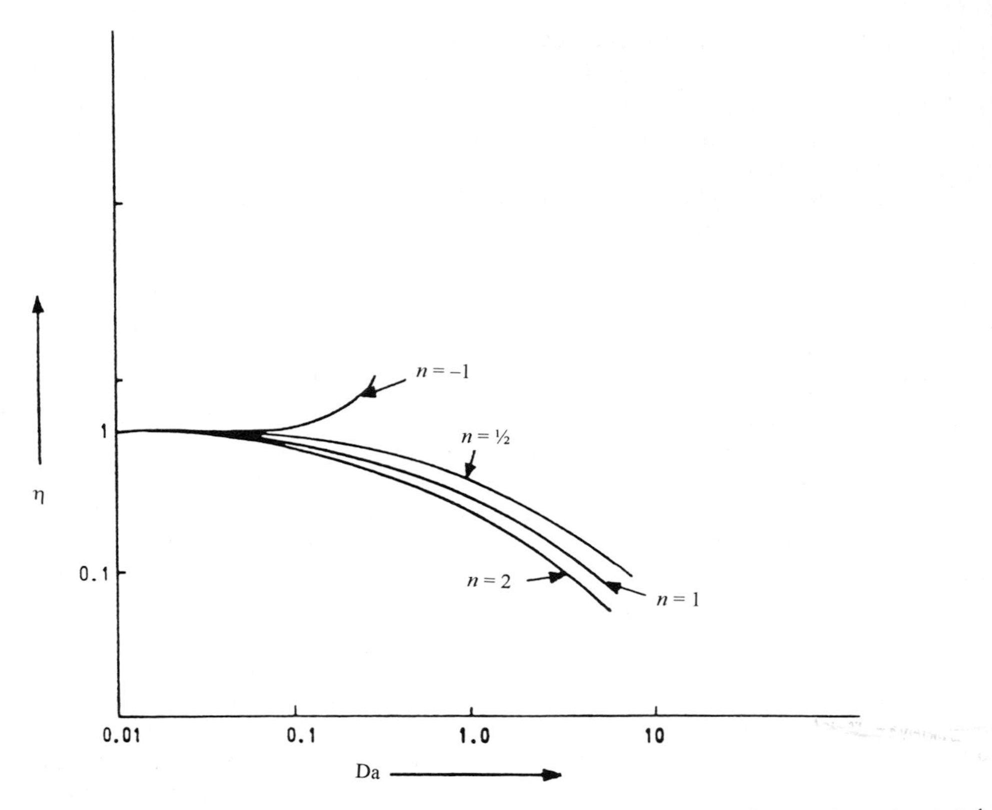

Figure 7.4 Effectiveness factor as a function of the Damköhler number Da for reaction orders 2, 1, $\frac{1}{2}$ and −1. With permission from G. Cassiere, J. J. Carberry, *Chem. Eng. Educ.*, Winter **1973**, p. 22.

This interphase effectiveness factor (referred to as the transport factor Tr by Petersen) has been expressed as a function of the Damköhler number, Da, which is a comparison of chemical reaction rate with the rate of mass transport for identical conditions. If, then, the Damköhler number is given by kc^{n-1}/h_D, where n is the true reaction order and in which the rate constant k is expressed on a unit area basis, it is obvious (see Eq. (26)) that for a first-order reaction the interphase effectiveness factor decreases monotonically with increase in Da.

The relation between η_i and Da has also been deduced for other reaction orders and is reproduced in Fig. 7.4. Intuitively we would expect that, with increased convection, the resistance to transport of gas through the interfacial gas film would become less. Indeed, the value of h_D always increases with increase in rate of gas flow and so, for a given chemical reaction, the Damköhler number diminishes; in the limit, η_i therefore tends to reflect the chemical reaction rate.

7.2.6 The Effect of Catalyst Deactivation

As outlined in Chapter 6, the chemical reactivity of many catalysts diminishes during use. It is recognised that deterioration of catalyst performance may be the result of poisoning by the presence of strongly adsorbed foreign material in the process feed-stock. Catalyst deactivation may also be ascribed to the formation (by concomitant side reactions) of coke precursors, resulting in eventual coke deposition and consequent blockage of surface area and pore structure. Catalyst decay may also occur because sintering causes loss of surface area.

The importance of understanding and being able to predict loss of activity during catalyst usage must not be under-estimated. Models of varying degrees of complexity were presented and described in Section 6.2.2. Their usefulness centred on their ability to predict catalyst effectiveness factors which, because of the inherent unsteady-state nature of deactivation, are functions of process time. For purposes of reactor design and the prediction of reactor operation, such computed effectiveness factors may be coupled with the chemical kinetics in an equation fashioned to describe the course of reactor performance. A problem which is equally, if not more, important is the uncoupling of the chemical rate from the overall reaction rate when it is affected by catalyst deactivation. In an illuminating paper, Levenspiel (1972a) (see Section 7.5.2) describes how this can be achieved, employing simple experimental techniques. Because the strategies employed in this search for the true chemical rate, unimpeded by deactivation effects, are not only instructive but also pragmatic, we outline their application here.

First we accept that the overall reaction rate is a product of two functions – the chemical rate (expressed empirically as in Eq. (21)) dependent on the concentration of reactant in the fluid, and the present reactivity of the catalyst pellet. Such a formulation is merely a modification of the concept that the effectiveness factor is a ratio of a mass-transport-limited rate to an uninhibited chemical rate. By transcribing this basic definition into the realm of catalyst activity one obtains the form for activity employed by Levenspiel. He defines the catalyst activity at a given moment as the ratio of the present rate of conversion of reactant to the rate of conversion on a fresh catalyst pellet. Although such a definition has some obvious deficiencies (e.g. it is seldom that fresh catalysts composed of the same material have identical activities and, furthermore, the activities of two fresh pellets taken from the same batch may differ) it is nevertheless useful for the development of an experimental method for arriving at the true chemical rate.

Parallel deactivation, in which the reactant itself is the cause of deterioration, is described by the routes

$$\mathrm{A(g)} \rightarrow \mathrm{B(g)}$$

$$\mathrm{A(g)} \rightarrow \mathrm{C(s)}$$

The uninhibited rate of chemical reaction is represented thus:

$$\mathrm{R} = k c_{\mathrm{A}}^{n} a \tag{27}$$

where n is the reaction order and a the activity as just defined. The decline in activity of the catalyst is expressed by

$$-\frac{\mathrm{d}a}{\mathrm{d}t} = k_{\mathrm{d}} c_{\mathrm{A}}^{m} a^{d} \tag{28}$$

where the indices m and d are exponents expressing the kinetics of the deactivation process. Both k and k_d are assumed to obey an Arrhenius form of temperature dependence. Provided m is not zero, the equations are coupled and the activity cannot be expressed as a function of time only and then substituted into the rate equation.

Series deactivation, in which the product of reaction decomposes and results in catalyst deactivation, is represented by

$$A(g) \rightarrow B(g)$$

$$B(g) \rightarrow C(s)$$

The kinetics of the main reaction forming the desired product B are, of course, still expressed by Eq. (27) but the deactivation rate is

$$-\frac{da}{dt} = k_d c_B^m a^d \tag{29}$$

Finally, independent deactivation in which the catalyst suffers structural modification during reaction (sintering) and whose rate is

$$-\frac{da}{dt} = k_d a^d \tag{30}$$

is independent of reactant or product concentration. Equation (30) may therefore be integrated directly and a expressed as a function of time only. Some isomerisation and also cracking reactions, in which the decline in activity depends on the sum of the concentration of reactants and products, also fall into this latter category whenever the resultant concentration is a constant independent of process time.

The merit of representing reaction and deactivation rates in this way is that the coupled equations can be inserted into an appropriate form of reactor equation (Section 7.3, q.v.) and integrated (either in closed form when reaction orders are simple, or otherwise numerically), thus producing a relation of the form $c_A(\tau, t)$ where τ is the fluid residence time within the reactor and t the (real) time during which the catalyst has been functioning in the process. One example will suffice to illustrate the method of analysis. Suppose we are concerned with a parallel deactivation process. Then provided we can choose some experimental means of controlling reactor inlet and outlet concentrations, Eqs. (27) and (28) can be manipulated to reveal both concentration and deactivation dependence.

Section 7.3 deals with the advantages of different types of experimental reactor for studying catalytic reactions. It will be evident from this subsequent description that the spinning-basket reactor, which maintains a small charge of catalyst as a batch within a perforated receptacle rotating at high speed in the reactant gas environment, fulfils our requirement for controlling concentrations. The reactant inlet concentration c_{A_0} can be adjusted at will by means of a combination of pressure regulation and flow control valves. The outlet concentration c_A would have to be continually monitored by means of chromatographic equipment and any out-of-balance signal from a predetermined set value used to control the inlet volumetric flow rate u. As shown in Section 7.3.1.3, the steady-state mass balance for the spinning-basket reactor is

$$uc_{A_0} = uc_A + WR \tag{31}$$

where the reaction rate is measured on the basis of unit mass of catalyst and the concentration in moles per unit volume of fluid. W is the mass of catalyst loaded in the reactor. If, then, both c_{A_0} and c_A are maintained at constant values, Eqs. (27) and (28) can be rewritten:

$$\mathrm{R} = k'a \tag{32}$$

$$-\frac{\mathrm{d}a}{\mathrm{d}t} = k'_{\mathrm{d}}a^d \tag{33}$$

Integrating Eq. (33) (remembering that, because of the definition of activity, the initial activity is unity) and substituting the result into Eqs. (31) and (32) yields (provided d is not unity)

$$(\tau')^{d-1} = \left(\frac{c_{A_0} - c_A}{k'}\right)^{d-1} \{1 + (d-1)k'_{\mathrm{d}}t\} \tag{34}$$

where τ' $(= W/u)$ is proportional to the fluid residence time τ $(= \mathrm{e}W/\rho_{\mathrm{b}}u)$ in the reactor. When first-order deactivation occurs ($d = 1$) one obtains

$$\ln \tau' = \ln\left(\frac{c_{A_0} - c_A}{k'}\right) + k'_{\mathrm{d}}t \tag{35}$$

To maintain the outlet concentration c_A constant, the inlet flow is continually adjusted and hence changes. By observing how τ' varies with elapsed time t, the relationships given by Eqs. (34) and (35) can be tested and thus the exponent d is determined. Its value may be anything between -1 and 3 and, as demonstrated by Levenspiel, varies with the Thiele modulus (Section 4.5). Once the deactivation kinetics have been elucidated, it is then possible to unravel the concentration and temperature dependence. Thus, for example, the concentration dependence could be found by repeating similar sets of experiments at differing levels of inlet concentration. The temperature dependence could be determined by monitoring, at constant volumetric flow rate, the change in outlet concentration with temperature. Care must be exercised concerning this latter point for, if the temperature coefficient associated with catalyst deactivation is large, then the catalyst activity may alter significantly during an experiment and thus invalidate the results.

The above analysis rests on the assumption of a quasi-steady state. Such an approximation is acceptable when the rate of deactivation is significantly less than the rate of the chemical reaction being studied. If it is not, resort to a different reactor type is necessary. A typical case of rapid catalyst deactivation is a petroleum cracking reaction where carbon laydown can occur in a matter of seconds. Catalyst must be continually regenerated during the process if product quality is to be maintained. Figure 7.5(f) (Section 7.3.1) is a sketch of a suitable reactor arrangement. The catalyst bed is fluidised by means of the inlet reactant flow and the cracking reaction occurs in the vessel labelled 1. The reactant flow rate is arranged to be sufficiently rapid for the catalyst to be carried over into a second vessel (2) where the catalyst is regenerated by burning off the carbon deposits with air or steam. The upward flow of air (or steam) in the regenerator must also be fast enough for the catalyst particles not only to be fluidised but to be carried over and returned to the reaction vessel, where reaction occurs in the presence of continually reactivated particles of catalyst. Levenspiel has shown that this type of reactor arrangement can also be subjected to analysis, although assumptions must be made about the nature of mixing in the particulate phase.

Before dismissing the subject of catalyst deactivation, it is pertinent to enquire what policy should be adopted if a high catalyst activity is to be sustained during industrial operation. The catalytic synthesis of ammonia may be cited as an example. As pointed out in Chapter 8, this reaction is thermodynamically reversible and exothermic. The bell-shaped curves labelled Γ_1 and Γ_2, referred to in Section 7.3.3 and illustrated by Fig. 7.18, correspond to rates of heat generation during reaction for different levels of catalyst activity. The straight lines, on the other hand, represent the rate of heat loss (by heat exchange with the cooler incoming feed). The point (locations such as S_1, S_2) at which the heat loss line is tangential to the highest point of the heat generation curve represents the optimum stable operating condition for a given catalyst activity. As the catalyst activity declines, so the rate of heat generation due to reaction becomes less and a lower bell-shaped curve would now represent the rate of heat generation for this rather less active catalyst. Consequently the point at which the heat loss line is tangential to the heat generation curve is displaced to a position of higher temperature. This means that the inlet temperature of the catalytic reactor has to be gradually increased to compensate for the decreasing activity of the catalyst. Such a procedure is indeed in accord with industrial practice.

7.3 Catalytic Reactors

7.3.1 Experimental Laboratory Reactors

Before embarking upon an exposition of industrial catalytic reactors we will describe the efficacy of various types of laboratory-scale catalytic reactors and outline the reasons for choice of a particular laboratory reactor type to accomplish specific measurements in different circumstances.

Laboratory reactors may be conveniently divided into two broad categories according to their mode of operation. Batch reactors are closed systems in which measured quantities of reactant interact with a charge of catalyst which is usually maintained at a fixed temperature. The concentration of reactant diminishes with time as the reaction proceeds and this is utilised as the basis of experimental observation. Continuous reactors, on the other hand, are open systems into which the reactants are continuously introduced and products extracted. The solid catalyst is often contained within the system (as for example in a fixed bed of solids), but continuous reactors in which solids flow either cocurrent or countercurrent to the flowing fluid are also used for special purposes. Continuous catalytic reactors can be further subdivided into reactors of different configuration whose performance is dependent on the mode of gas flow through the reactor. We shall outline here the basic features of the principal categories of catalytic reactors normally employed in the laboratory.

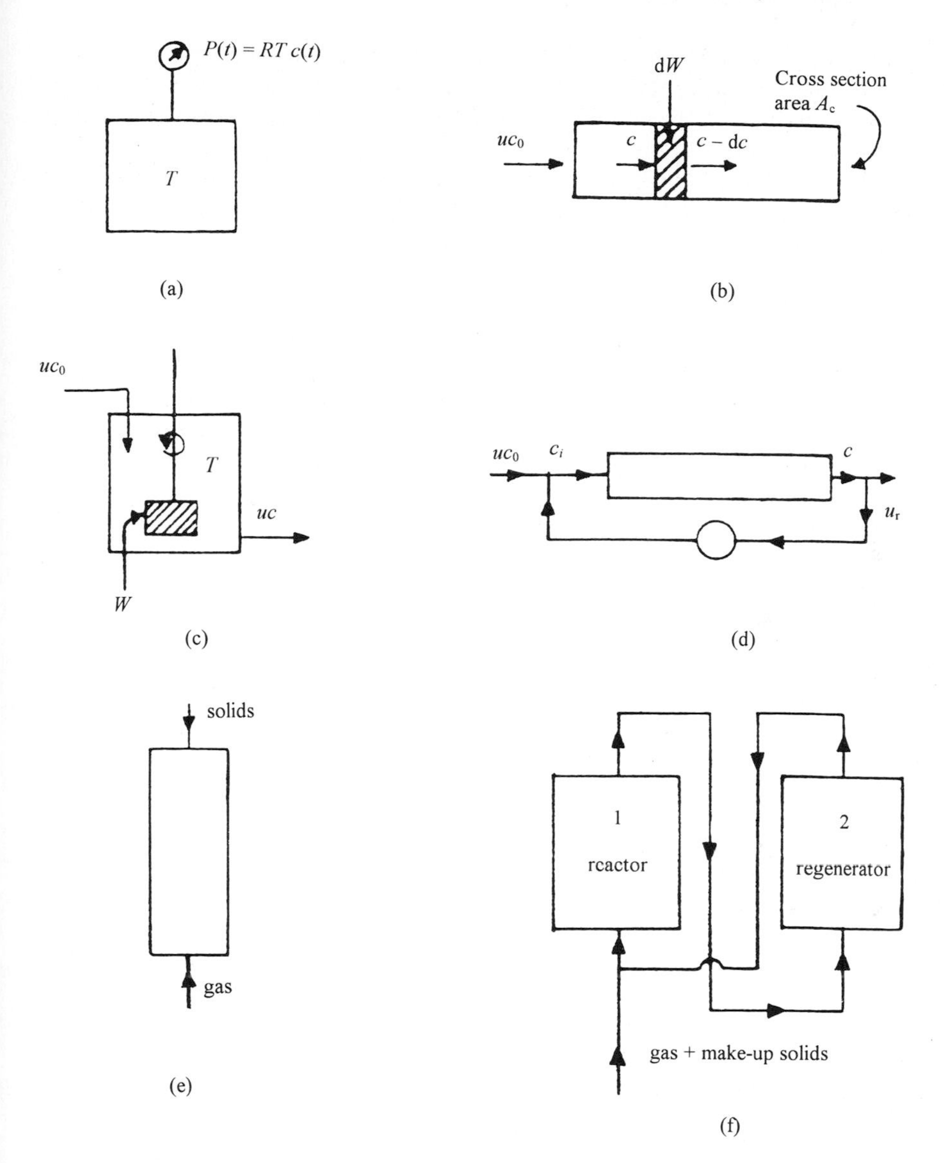

Figure 7.5 Catalytic reactor configurations: (a) batch; (b) continuous tubular; (c) continuous stirred-tank; (d) recycle; (e) raining-solids; (f) circulating-solids catalytic reactor–regenerator.

7.3.1.1 Batch Reactors

Some of the early classic experiments on the kinetics of catalysed gas reactions reported by Hinshelwood and by Rideal and Taylor were studied using batch-type reactors. The usual arrangement was to contain a specific mass (often just a few grams) of catalyst in a closed glass reaction vessel maintained at a constant temperature (Fig.7.5(a)). The gaseous reactants were then introduced into the vessel at a given instant and the progress of the reaction followed by recording the change of total pressure with time or, alternatively, by sampling the reaction products at incremental times and chemically analysing the product mixture. The rate of reaction could thus be represented as the number of moles of reactant reacted per unit time per unit area of catalyst (the specific surface area of the catalyst having been determined by one of the usual methods – see, for example, Chapter 4). For a first-order reaction the rate of reaction would be written

$$\mathrm{R} = -\mathrm{d}c/\mathrm{d}t = kc \tag{36}$$

the dimensions of the reaction velocity constant being LT^{-1}. Any other reaction order could be expressed equally well and the equation integrated to provide a concentration–time function which characterises the kinetics. The methods employed for determining reaction orders and temperature dependence have been adequately described in classic texts on kinetics and will not be repeated here.

The early kinetic experiments with catalysed gas reactions were often restricted to pressures well below 1 atm, for convenience of measurement. Provided that the reaction rate under these conditions was not too rapid, transport limitations were not likely to be too serious. For more rapid exothermic catalysed reactions, the true kinetics are prone to be affected by both inter- and intra-particle transport effects (see Chapter 4), particularly if the reaction is to be studied under more realistic conditions apposite to the pressure and temperature which one might select for industrial operation.

Thus the batch reactor has its limitations. Although pressure and temperature of operation are not serious constraints (reactors can be made of steel rather than glass), the batch reactor is unsuitable for fast exothermic reactions because the changes in concentration and pressure are difficult to follow and it is far from easy to maintain a constant temperature. Furthermore, one is often in the realm of speculation when estimating the effect of transport of heat and mass between fluid and solid, unless the gas is deliberately circulated at a rapid rate of flow through the solids.

7.3.1.2 Tubular Reactors

The laboratory-scale tubular reactor is simply a straight tube packed with catalyst and through which reactant gas continually flows. The dimensions of the tube often depend on experimental circumstances. A microreactor may be no more than 0.5 cm in diameter and 10 cm in length and would contain only a few grams of catalyst. Larger tubes are also used for experimental purposes (e.g. when a high conversion is required) but the larger the tube diameter, the more difficult is the radial transport of heat and hence it is less easy to achieve a uniform temperature. Isothermal conditions within the reactor are greatly assisted by restricting the conversion to only a few per cent. Provided that the analytical equipment at the reactor exit is commensurate with the small changes in concentration of reactants and products (which is the result of operating the reactor in this way), then steady isothermal conditions can be achieved for quite fast exothermic reactions. When a continuous reactor produces product at a low conversion level, the reactor is said to operate in a 'differential' mode.

The differential tubular reactor is a popular device because it is easy and inexpensive to construct, and moreover can be used for studying quite fast exothermic reactions. Tracer experiments show that the gas passes through the catalyst bed in plug flow; that is to say, the radial velocity profile is uniform. Figure 7.5(b) shows that, in passing through a differential volume element containing packed catalyst, the reactant and product concentrations (expressed as moles per unit volume of gas) change as a result of reaction within the element. If the volumetric flow rate of gas through the tube of cross-sectional area A_c is u, and the reactant concentration diminishes by an amount dc in traversing the length dz packed with catalyst of bulk density ρ_b, then a steady-state material balance gives

$$-u\,\mathrm{d}c = \mathrm{R}\rho_b A_c\,\mathrm{d}z \tag{37}$$

where R is the reaction rate expressed as moles reacted per unit time per unit catalyst mass. Expressing these differential quantities in terms of recorded concentrations at inlet and exit, it is obvious that the rate of reaction is directly measurable. Within the small volume $A_c\,\mathrm{d}z$ of tube, there will be a mass $\mathrm{d}W(= \rho_b A_c\,\mathrm{d}z)$ of catalyst. If the conversion is low (less than *ca* 5%), we can rewrite Eq. (37) in the more useful algebraic form

$$\mathrm{R} = \frac{u(c_0 - c)}{W} = \frac{e(c_0 - c)}{\rho_b \tau} \tag{38}$$

where $(c_0 - c)$ is the small difference in concentration between inlet and exit, and τ is the residence time of gas in the reactor under reaction conditions. As τ is defined as the ratio of reactor void volume to the volumetric flow rate at reaction conditions, it is equal to $eW/\rho_b u$. For low conversions, therefore, the reaction rate R may be measured directly.

Lest it be thought that the differential reactor is the panacea for all conditions of experimental operation, we should point out that Weekman (1974) and Sunderland (1976) (see Section 7.5.3) have both commented on its limitations. The problem of accurate chemical analysis of multicomponent mixtures becomes a major difficulty when only small conversions are allowed but this, of course, is strongly dependent on the nature of the component mixture. The differential reactor should be mounted vertically with downward flow of gases; such a disposition avoids loss of solid material from the reactor but extreme care must be taken when starting an experiment. Gas flow should be very gradually increased to the desired value to avoid channelling within the solids bed. If channelling were to occur, then some of the gas would by-pass the catalyst and the performance would not be represented by Eq. (38), which implicitly assumes that all fluid elements pass through the solid bed in plug flow. Another limitation of the differential reactor is that, for vapour–liquid feeds, good distribution of vapour and liquid within the bed is generally poor. Finally, if catalyst activity declines rapidly, then an arrangement in which the solid material (as well as the fluid) is in continuous flow is to be preferred, thus ruling out use of the differential reactor for such circumstances.

The tubular reactor may, of course, be operated at higher conversions if desired. It is then referred to as an 'integral' tubular reactor. Provided that the diameter of the tube is kept small, then there is still the possibility of maintaining isothermal conditions. However, any reaction but the mildest of exothermic reactions is likely to violate the desired condition of constant bed temperature. If isothermal conditions are maintained and plug flow is an adequate description of the fluid mechanics, then Eq. (37) describes the reactor performance. Because the conversion is no longer small and the concentration of reactant will decrease monotonically along the bed length, Eq. (37) has to be integrated and subsequently manipulated to extract a reaction rate which is an average for the concentration conditions in the bed. As an illustration of the procedure, suppose we are concerned with a first-order chemical reaction

$$A \rightarrow B + C$$

The molar flow rate F_A of reactant A at any cross-section of the tube is uc. We may thus rewrite Eq. (37) in the form

$$-dF_A = F_{A_0} dx = R\, dW \tag{39}$$

where F_{A_0} is the molar flow rate of A at the tube inlet and dx the incremental conversion for the element of catalyst mass dW. The quantity of catalyst which generates a conversion x at the reactor exit is thus

$$W = F_{A_0} \int_0^x \frac{dx}{R(x)} \tag{40}$$

The reaction rate is frequently expressed in terms of concentrations or partial pressures so that to integrate Eq. (40) it is necessary to reformulate the rate in terms of conversion. This is a relatively simple matter because the partial pressure may be rewritten in terms of a mole fraction and total pressure. Thus one may substitute

$$R = kp_A = ky_A P \tag{41}$$

in Eq. (40) where the mole fraction of A from the stoichiometry is clearly

$$y_A = \left(\frac{1-x}{1+x}\right) P \tag{42}$$

and P is the total pressure. Expressing the rate in such a manner therefore also accounts for any change in volume on reaction. Substitution of Eqs. (41) and (42) in Eq. (40) yields

$$W = \frac{F_{A_0}}{kP} \int_0^x \left(\frac{1+x}{1-x}\right) dx = \frac{F_{A_0}}{kP} \left\{ 2 \ln\left(\frac{1}{1-x}\right) - x \right\} \tag{43}$$

Should a diluent inert gas be present in the feed, the volume change on reaction will be affected. Such a change can easily be accounted for by incorporating the appropriate molar ratio of inert to reactant feed in the denominator of the reactant mole fraction. If this ratio is α, for example, Eq. (42) becomes

$$y_A = \left(\frac{1-x}{\alpha+1+x}\right)P \tag{44}$$

and Eq. (43) is modified to

$$W = \frac{F_{A_0}}{kP}\left\{(\alpha+2)\ln\left(\frac{1}{1-x}\right) - x\right\} \tag{45}$$

Reactions obeying more complex kinetic laws (including those where there is a volume change on reaction) would yield correspondingly more complicated expressions. The problem of elucidating the kinetics then becomes one of fitting an equation to the observed experimental data. This can be exceedingly time-consuming and is fraught with difficulties (e.g. the possibility that two or more integrated kinetic equations fit the experimental data equally well). Statistical significance tests then have to be introduced to analyse the data and it is evident that one would be committed to a very large experimental programme with no certainty of obtaining reliable kinetic data. The great variety of kinetic expressions corresponding to different kinetic laws that are obtained by integrating the basic reactor equation (Eq. (37)), and the procedures for extracting data, are well documented in some standard texts. Diffusion effects within the catalyst pellets contained in the bed are best avoided by using a small particle size. However, if the particle size is reduced too much the pressure drop over the bed may be large and will not only invalidate the calculation of rate data but may also cause experimental operating difficulties. Fluid-to-solid transport effects, on the other hand, are only eliminated at sufficiently high flow rates which may restrict the range of residence times available.

In summary, the use of differential tubular reactors for the analysis of rate data is satisfactory when endothermic or only moderately exothermic gas–solid catalytic reactions are to be studied, and provided that rapid catalyst deactivation does not occur. Reaction rates are then obtained directly by experiment and consequently the functional concentration and temperature dependence can easily be obtained (although their validity at higher conversions must be in some doubt). When data are required at higher conversions, the reactor can be operated as an integral reactor but then only reactions which are mildly exothermic or endothermic can be analysed, but with less accuracy and a great deal of numerical complexity and concomitant uncertainty.

7.3.1.3 Continuous Stirred-Tank Reactor

The continuous stirred-tank reactor (CSTR) is a well-known device for studying, at constant temperature, liquid-phase reactions. Modifications to accommodate gas–liquid reactions and gas–solid catalysed reactions have been described. We will outline here the adaptation of the CSTR to the experimental study of heterogeneous gas–solid catalytic reactions. First, we describe the principle upon which the equipment operates. A mass W of catalyst is contained in a perforated container which rotates at high speed in the environment of the reactant gas (Fig. 7.5(c)). As the gas is well mixed and in intimate contact with the catalyst particles, the concentration (expressed as moles per unit mass of gas) and temperature of the contents of the reactor will be virtually constant. The concentration c_0 of the gaseous reactant feed therefore drops instantaneously to the value c (as a consequence of chemical reaction and good mixing) which is the constant concentration of reactant at the exit of the reactor. If steady-state conditions obtain in the vessel, then the conservation of mass (material balance) demands that, as in Eq. (38),

$$\rho u c_0 = \rho u c + W\mathrm{R} \tag{46}$$

To put it another way, the difference in molar flow, $\rho u(c_0 - c)$, between the inlet and outlet of the reactor must be balanced by the total rate $W\mathrm{R}$ at which reactant disappears by virtue of chemical reaction. The performance of the CSTR can be assessed by rearranging Eq. (38). For example, for first-order kinetics one obtains (with $\mathrm{R} = kc$)

$$\frac{c}{c_0} = \frac{1}{1 + k\tau'} \tag{47}$$

where τ' is W/u $(= \tau\rho_b/e)$, i.e. it is proportional to τ, the fluid residence time in the reactor. Just as in the differential reactor, reaction rates are, most conveniently, directly measurable.

Although Eq. (46) is formally identical with the equations in Section 7.3.1.2 for the differential reactor, it should be appreciated that the fluid mechanical behaviour of the two gas–solid systems is entirely different. In the differential tubular reactor the solid is stationary and the gas is in plug flow. Provided that the particles of catalyst are small and the gas flow through the differential reactor is sufficiently high, then neither intra- nor inter-particle transport effects should intrude. However, interparticle (fluid-to-solid) transfer resistance can easily become apparent when gas flows are too low. In the continuous stirred-tank reactor, on the other hand, the gas is well mixed with the solid and one is more certain, particularly at high speeds of revolution of the catalyst container, that interparticle resistances are reduced to a minimum. Intraparticle diffusion effects can be minimized in both cases by selecting sufficiently small particles.

The usual procedure for both the differential tubular and CSTR reactors, when acquiring experimental data, is to ensure that interparticle transport and intraparticle diffusion are reduced to a minimum or preferably eliminated. To eliminate intraparticle diffusion, rates are measured for a number of particle sizes. When the particle is sufficiently small, diffusion effects within the particle will be virtually absent and the measured rate will remain constant even for further reduction of particle size. Fluid-to-solid transport resistance can be eliminated by increasing the gas flow to a sufficiently high value so that the rate of reaction does not change with variation of gas flow for the same residence time (and therefore for constant W/u). In the latter case, therefore, when studying the effect of gas flow, if u is increased then W must also be increased by the same proportion to maintain a constant residence time of gas in contact with solid.

Figure 7.6 is an illustration of a spinning-basket catalytic reactor often used in industrial and academic laboratories for studying gas–solid catalysed reactions and

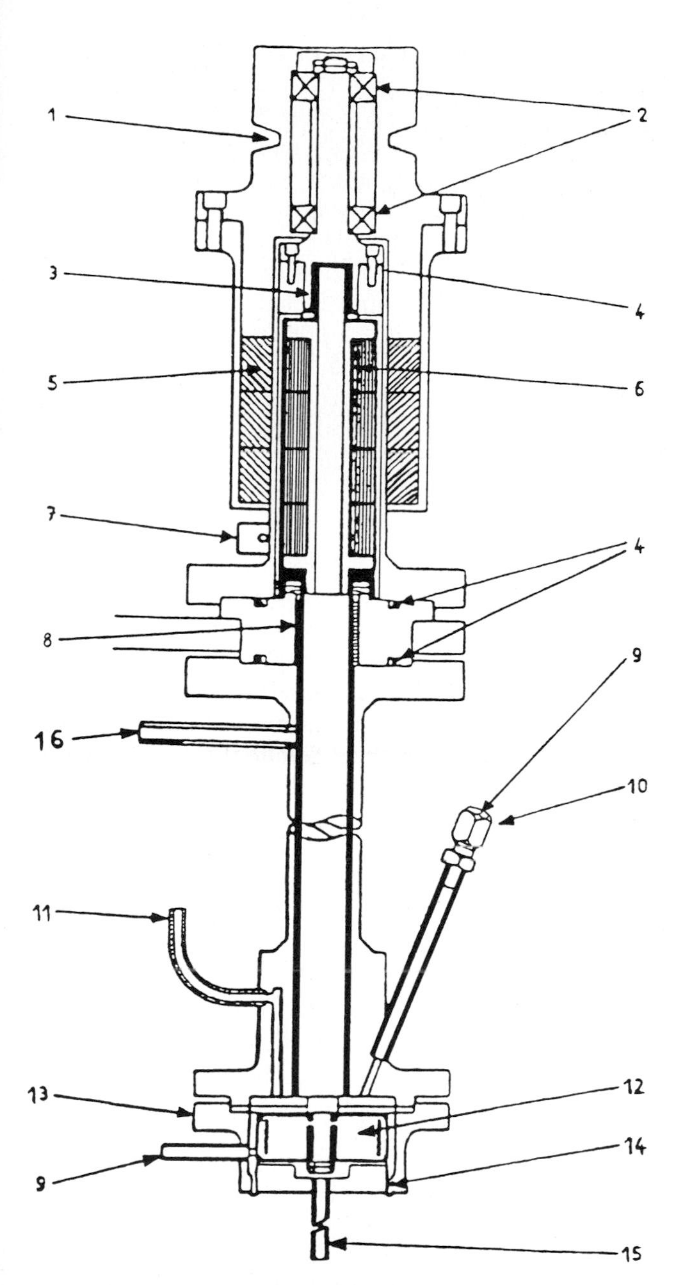

Figure 7.6 Details of the spinning-basket experimental reactor. 1, belt drive groove; 2, outer magnet bearings; 3, upper bearing; 4, O-ring seals; 5, outer magnet; 6, inner magnet; 7, reed relay; 8, lower bearing; 9, thermocouple entry; 10, pressure tapping; 11, gas inlet port; 12, catalyst basket; 13, reaction vessel; 14, baffles; 15, gas outlet port; 16, purge inlet port.

for the testing of catalyst performance. A small charge of catalyst pellets (a few grams) is contained in the perforated basket, which is rotated at high speed in a chamber through which the gases pass. The feed gas is introduced directly to the catalyst basket through the inlet pipe. Good fluid–solid mixing is thus achieved. The shaft to which the basket is attached is rotated by means of a magnetically coupled variable-speed motor. The basket and chamber are isolated from the environment by suitable seals and the shaft bearings are hardened steel balls. The basket and chamber can be surrounded by a furnace and heated if desired. Materials of construction will depend on the temperature and pressure conditions selected for operation. Such experimental reactors have been operated successfully at pressures of *ca* 2 MPa and 600 °C. Other rotating reactor devices have been reported in the literature. Carberry (1969) designed a rotating basket quite similar to the one described here, whereas Berty (1974) relied on rotation of an impeller to effect good gas circulation through a stationary chamber containing the catalyst (see Section 7.5.3).

7.3.1.4 Recycle Reactor

It is very useful to have an experimental device which allows one to vary the fluid mechanical behaviour of the system between the extremes of plug flow (as in the differential tubular reactor) and perfect mixing (as in the ideal CSTR). In this way it is possible to determine at what point interparticle fluid-to-solid transport effects cease to be important. The experimental arrangement for a recycle reactor is illustrated in Fig. 7.5(d). The equipment consists of a small tubular reactor and a circulating pump so that a proportion of the product gases from the exit of the catalyst bed can be recirculated and added to the continuous feed. It is essential that the circulating pump is capable of recirculating the gases at variable but steady flow rates.

The performance of the recycle reactor may easily be deduced by referring to Fig. 7.5(d), noting that the volume flow through the tubular reactor is $(u + u_r)$. Thus, applying Eq. (37), considering the case of a first-order reaction without volume change and integrating the result,

$$\tau_r{}' = -\int_{c_i}^{c} \frac{dc}{kc} = -\frac{1}{k} \ln \frac{c}{c_i} \tag{48}$$

where τ_r' $(= W/(u + u_r))$ is proportional to the residence time in the recycle reactor. At the mixing point where the recycle stream meets the incoming feed a material balance gives

$$(u + u_r)c_i = uc_0 + u_r c \tag{49}$$

Rearranging and setting the ratio of the recycle flow u_r to the inlet flow u equal to r,

$$\frac{c}{c_i} = \frac{c(1 + r)}{c_0 + cr} \tag{50}$$

Substituting this result directly into Eq. (48) yields

$$\tau_r' = -\frac{1}{k}\ln\left\{\frac{c(1+r)}{c_0+cr}\right\} \tag{51}$$

which describes the reactor performance for any recycle ratio r. Kinetic expressions other than first order could be substituted into the reactor equation (Eq. (48)) if desired. When integrated and coupled with a material balance at the point where the recycle and inlet streams mix (Eq. (49)), a result expressing the reactor performance would be obtained.

The two extremes of behaviour may be deduced by considering separately the cases for no recycle and for high recycle ratio respectively. Thus putting $r = 0$ in Eq. (51) directly gives an equation analogous to Eq. (48) describing a first-order reaction in a plug-flow tubular reactor but with a volume flow u rather than u_r. When, however, we put $r = \infty$ in Eq. (50), the ratio c_i/c tends to unity and a trivial result is obtained. What we really wish to know is how the natural logarithm of the argument behaves in the vicinity of unity. Let the argument be y. Expanding $\ln y$ in a Taylor series about unity, it can easily be shown that $\ln y$ is approximately $(y-1)/y$. Employing this result to expand the logarithmic argument in Eq. (51) we obtain

$$\ln\left\{\frac{c(1+r)}{c_0+cr}\right\} \approx \frac{c-c_0}{c(1+r)} \tag{52}$$

Substituting into Eq. (51) and rearranging,

$$\frac{c}{c_0} = \frac{1}{1+(1+r)k\tau} \tag{53}$$

Now the ratio of residence times when the reactor is operating in the plug-flow mode compared with the well-mixed mode will be τ/τ_r $(= \tau'/\tau'_r)$ and this is

$$\frac{\tau}{\tau_r} = \frac{eW}{\rho_b u}\,\frac{\rho_b(u+u_r)}{eW} = 1+r \tag{54}$$

Thus Eq. (53) reduces to Eq. (47) for large values of the recycle ratio r and the reactor then performs as a CSTR.

7.3.1.5 Flowing-Solids Reactors

It was pointed out in Section 7.2.6 that when a catalyst is subject to rapid deactivation, then it is necessary to resort to a reactor system in which the catalyst particles are in motion so that they have little time in the reaction environment to deactivate by more than a small amount. Figure 7.5 shows a sketch of two possible experimental reactor arrangements with flowing solids. In the first configuration (Fig. 7.5(e)) the catalyst particles are either conveyed pneumatically by the gaseous feed in an upward direction or alternatively flow downwards under the action of gravity and countercurrent to the gas. In the second arrangement shown (Fig.7.5(f)) the particles of catalyst are circulated from the reaction vessel (1) to a second vessel (2) where they are regenerated (usually by means of hot air). Depending on the size and configuration of the vessels in each case, the solids can be considered to be either in plug flow (raining-solids and transport line reactors with a tubular configuration) or mixed flow (larger-diameter vessels where the particles move in swirling eddies). Indeed, the principal difficulty in making use of these experimental reactors is defining the fluid mechanics of the solid and fluid phases.

If the rate of deactivation of the catalyst is linearly dependent on activity (first-order process given by Eq. (27) in Section 7.2.6 with $n = 1$) and the solids are in mixed flow, then the mean activity depends only on the residence time (events in first-order processes depend only on time). As far as the solid phase is concerned, the catalyst activity can be treated as though the particles are well mixed in a CSTR. Thus for a contact time τ $(= eW/\rho_b u)$, by analogy with Eq. (47) the average particle activity is

$$\bar{a} = \frac{1}{1 + k'_d \tau'} \tag{55}$$

where k'_d is the first-order deactivation constant. For a deactivation process which is anything other than first order, events will also depend on local concentration and so the activity of the particles has to be averaged over all possible residence times. For well-mixed flow the distribution function (termed the exit age distribution function – see Section 7.5) expressing the fraction of particles in the reactor outflow which have residence times between t and $(t + \mathrm{d}t)$ is

$$E(t)\,\mathrm{d}t = \frac{\mathrm{e}^{-t/\tau'}}{\tau'}\,\mathrm{dt} \tag{56}$$

where τ' is proportional to the average residence time. Hence the average activity of particles is

$$\bar{a} = \int_0^\infty aE(t)\,\mathrm{d}t \tag{57}$$

For a second-order deactivation process (or, for that matter, any deactivation which is not first order) events depend on initial particle activity as well as on time. For second-order deactivation kinetics, for example, the activity at any time t is $(1 + a_0 kt)^{-1}$. Thus the result of substituting into both the appropriate residence time distribution function and the activity–time relationship given by Eq. (57) is to give the time-averaged activity. For anything other than simple deactivation kinetics the integral has to be evaluated numerically. The correct result cannot be obtained by a simple material-balance formulation as is the case for first-order deactivation kinetics where the relative activity is independent of the initial activity (the same result as given in Eq. (55) can be obtained for a first-order reaction by substitution of Eq. (56) into Eq. (57) for the residence time distribution function and putting $\bar{a} = \exp(-k_d t)$ for first-order deactivation kinetics).

Now suppose the gas is also in mixed flow. Applying the CSTR equation (Eq. (47)) for the gas and recalling the reaction rate is $kc\bar{a}$, we obtain

$$\frac{c}{c_0} = \frac{1}{1 + \tau' k\bar{a}} \tag{58}$$

If the gas were in plug flow rather than mixed flow, the corresponding expression would be

$$\ln \frac{c}{c_0} = -k\tau'\bar{a} \tag{59}$$

analogously to Eq. (48), which applies to performance in a tubular reactor.

Combining the result of integrating Eq. (57) with either of the above equations for mixed flow or plug flow of gas respectively, one obtains an expression for the reactor performance. In many cases quite complex expressions are obtained and for fractional-order kinetics the result must be deduced numerically. For first-order catalyst deactivation coupled with mixed flow of both solids and gas it is a straightforward matter to obtain the result,

$$\frac{c}{c_0} = \frac{1 + \tau' k_d}{1 + \tau'(k + k_d)} \tag{60}$$

For plug flow of gas, on the other hand,

$$\ln \frac{c}{c_0} = -\frac{k\tau'}{1 + k'_d \tau'} \tag{61}$$

7.3.1.6 Slurry Reactors

Many heterogeneous gas–solid catalysed reactions necessarily occur in a liquid-phase environment. Thus, in some catalytic hydrogenation reactions, the substrate is a liquid and the catalyst is held in suspension in the liquid phase. The catalytic hydrogenation of high-molecular-weight olefins is a good example and will serve to illustrate the experimental procedures involved in a laboratory kinetic investigation.

We refer then to a study of the hydrogenation of α-methylstyrene employing a copper magnesium silicate catalyst (see Section 7.5.3.7). The rate of catalytic hydrogenation is determined by the net rate at which the various physical and chemical steps occur. We must thus consider: (i) mass transfer of hydrogen from the bulk gas to the gas–liquid interface; (ii) mass transfer of hydrogen from the interface to the bulk liquid phase; (iii) mixing and diffusion in the liquid phase; (iv) mass transfer from the bulk liquid to the surface of the catalyst particles suspended in the liquid; (v) adsorption and chemical reaction at the catalyst internal surface (which occurs in parallel with intraparticle diffusion); and (vi) the reverse sequence of steps involving the product cumene. The desorption of product may be dismissed as a rate-limiting step on the grounds that the product, being more saturated with hydrogen than the reactant, is likely to be less strongly adsorbed on the catalyst than the reactant and will be more readily desorbed. Provided the mixing in the bulk liquid is moderately intense, then mixing and diffusion in the liquid is efficient and should not be the cause of any resistance to conversion. As undiluted hydrogen was used in the experiments, the gas–liquid interface may reasonably be considered saturated with hydrogen, so it is most unlikely that there is any resistance to the transfer of hydrogen from the bulk gas to the gas–liquid interface.

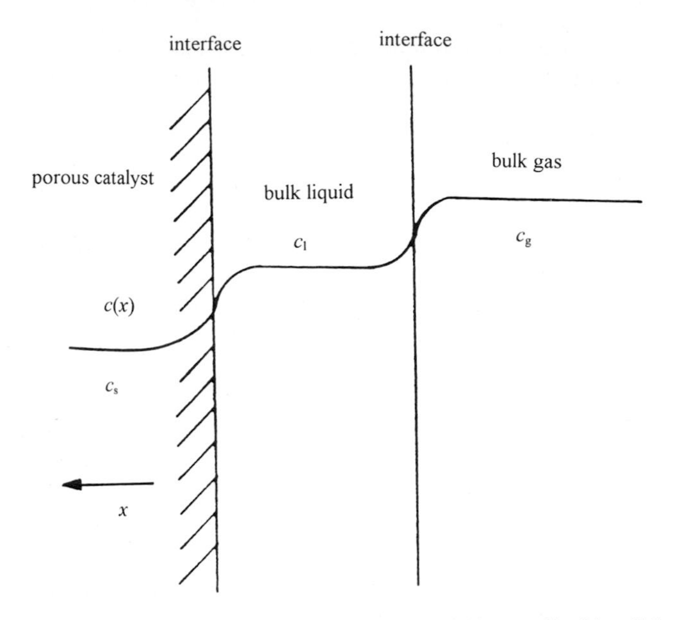

Figure 7.7 Reactant concentration profile within gas–liquid–solid catalyst system.

We take account, then, only of those transfer processes which have not been eliminated and which are considered to affect the overall rate of conversion. Figure 7.7 is a sketch of the concentration profile corresponding to such resistances. If we can assume that a steady state is developed during a particular phase of the reaction, these rate processes will occur at the same net rate R (measured in these experiments as moles of styrene converted per unit time per unit expanded volume of slurry). If A_b is the total gas bubble area and A_c the total catalyst external area, both on the basis of unit volume of slurry, then we may write

$$\mathrm{R} = k_L A_b (c_g - c_L) = k_c A_c (c_L - c_s) = k_s A_c c_s \tag{62}$$

in view of the fact that the chemical reaction rate was demonstrated to be pseudo-first order in hydrogen concentration and the intraparticle effectiveness factor approximately unity. Eliminating the unknown quantities c_L and c_s from these equations results in

$$\frac{c_g}{\mathrm{R}} = \frac{1}{k_L A_b} + \frac{1}{k_c A_c} + \frac{1}{k_s A_c} \tag{63}$$

If the gas bubbles forming the gas–liquid interface are spheres of diameter d_b and the gas hold-up in the liquid is H volumes of gas per unit volume of slurry (easily measured by observing the liquid expansion when hydrogen is bubbled through the liquid at a steady rate), then the total bubble area per unit volume assisting hydrogen transfer is $6H/d_b$. Similarly, for spherical catalyst particles of diameter d_p and pellet density ρ_p, the catalyst external area per unit volume of slurry is equal to $6m/\rho_p d_p$, where m is the mass of catalyst per unit volume loaded into the liquid. Equation (63) therefore becomes

$$\frac{c_g}{R} = \frac{d_b}{6k_L H} + \frac{\rho_p d_p}{6m}\left(\frac{1}{k_c} + \frac{1}{k_s}\right) \tag{64}$$

Clearly, then, by devising a suitable range of experiments it should be possible to evaluate the coefficients in the above equations.

The equipment used as reactor was a 250 cm^3 capacity impeller-agitated, glandless, stainless steel autoclave to which a condenser was attached to prevent escape of reactant or product vapour. The purified reactant was siphoned into the vessel containing a known catalyst weight in an atmosphere of nitrogen, the reactor was heated to the desired temperature and then hydrogen was bubbled through the liquid. No measurable amount of reaction occurred during such pretreatment, so zero time was taken as the moment when the impeller was started. Progress of the reaction was assessed by withdrawing, at various intervals of time, small samples of the liquid contents for chromatographic analysis.

A slurry reactor such as that described operates in a semi-batch mode. The gaseous reactant A is bubbled into well-mixed liquid B, where it reacts, and the unreacted fraction is continuously allowed to escape through a valve controlling the steady pressure. Thus reactant A flows continuously but steadily and a material balance gives

$$uc_{A_0} = uc_A + Vkc_A c_B \tag{65}$$

assuming the reaction is first order in both A and B. On the other hand, the reactant B is gradually depleted so a material balance for B gives

$$-\frac{dc_B}{dt} = kc_A c_B \tag{66}$$

Solving these two simultaneous equations,

$$t = \frac{1}{kc_{A_0}}\{\tau k c_{B_0} x - \ln(1-x)\} \tag{67}$$

where x is the fractional conversion of the reactant B, τ is the residence time (in this case V/u) of A in the liquid and subscript zero refers to initial conditions. For low conversions $\ln(1-x)$ may be approximated to $-x$ (MacLaurin series expansion near zero). Thus the conversion is directly proportional to time at low conversions (provided one is operating in the chemically controlled regime and the kinetics are as assumed). The slope of such a plot is the rate of conversion

$$\frac{dx}{dt} = \frac{kc_{A_0}}{(\tau k c_{B_0} - 1)} \tag{68}$$

At much higher conversions $\ln(1-x)$ is approximately $\{x/(1-x)\}$ (Taylor series expansion near unity) so now

$$t = \frac{1}{kc_{A_0}} \left\{ \tau kc_{B_0} x - \frac{x}{1-x} \right\} \tag{69}$$

and

$$\frac{dx}{dt} = \frac{kc_{A_0}}{\left\{ \tau kc_{B_0} - 1/(1-x)^2 \right\}} \tag{70}$$

the rate obviously falling off to zero as the reactant becomes depleted.

According to Eq. (64), at low catalyst loadings the reaction should not be dominated by gas-to-liquid interphase transfer. Measurement of the conversion rate at low catalyst loadings and low conversions should thus yield k_s which, because B is in excess, will be equivalent to kc_B. In this way the kinetic constant is extracted. Equation (64) also predicts that the reciprocal rate is related to the reciprocal catalyst loading. From the slope and intercept of such a plot, $1/k_L$ and the lumped parameter $(1/k_c + 1/k_s)$ can be obtained. On the other hand, at very high catalyst loadings the reaction will be dominated by resistance at the gas–liquid interface and so k_l can be separately assessed. Finally, it was shown, using a correlation embracing the liquid-to-solid transfer coefficient k_c, that the resistance to transfer of hydrogen from liquid to solid could, at most, be only slight.

When these experiments were repeated in a 2500 cm^3 capacity vessel there was, on a scale factor basis, considerably less agitation (the impeller Reynolds number was much smaller) and the reaction was controlled by hydrogen transfer through the liquid. These latter conditions quite probably approximate to those which would obtain in pilot-scale equipment.

7.3.2 Industrial Chemical Reactors

Chemical reactors employed in industrial catalytic processes may also be divided into two main categories – batch and continuous. Batch reactors are used when it is desired to make small quantities of materials such as pharmaceuticals, or to manufacture unusual products or catalyst preparations and other substances for commercial trials. Continuous reators are much the better economic prospect once it has been decided that large amounts of product are to be produced. In addition to many of the principles discussed in relation to experimental catalytic reactors in Section 7.3.1, other considerations must now be introduced, for once the economics of operation becomes of paramount importance it is necessary to develop strategies for assessing reactor size and performance. Compromises often have to be reached concerning the selection of reactor type and, for exothermic reactions, the method of heat removal is of prime importance. Finally, because of the difference in scale between the experimental and commercial catalytic reactors, questions concerning the dispersion of mass and heat within the reactor cannot be neglected.

In this section we will deal primarily with the principles of design and operation of catalytic reactors in a general context (citing some specific examples during discussion) and will defer the description of process details (including descriptions of reactors) until Chapter 8.

7.3.2.1 Batch Reactors

Some of the advantages of a batch reactor for the industrial production of small quantities of materials should not be overlooked when deciding whether a small-scale process should be batch or continuous. The batch reactor above all else is a versatile piece of equipment. It can be lined with corrosion-resistant materials such as enamel or rubber, or made from stainless steel, and it may be used on a large number of different occasions to execute a wide variety of reactions. Small batch reactors are quite easy to control and require much less auxiliary equipment such as pumps, control valves and pressure gauges than their counterparts, continuous reactors. Large batch reactors, however, do need rather more than simple control systems to ensure steady temperature and pressure operation. As it is difficult to effect good gas–solid mixing in a batch reactor, commercial application is normally restricted to catalytic reactions involving at least one reactant liquid. The catalyst particles used should be sufficiently small to ensure that when the impeller is rotated at a high speed the catalyst is in a state of suspension within the liquid, and well distributed. The impeller may be of the anchor type, or alternatively it may take the form of bladed paddles. In addition, it is not unusual to fit baffles to the tank interior to assist mixing. Cooling or heating the liquid contents may be effected by immersed coils through which cold water or steam flows. Figure 7.8 indicates the manner of operation of batch reactors.

When designing a batch process it is most important to assess the length of time required for a specified conversion of reactant to product. It is then only a matter of

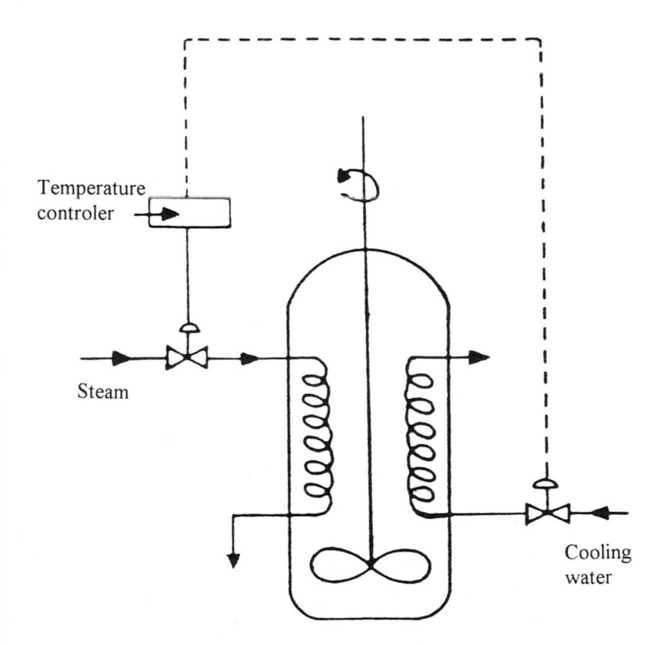

Figure 7.8 Batch reactor containing heating and cooling coils. Adapted from J. J. Coulson, J. F. Richardson, *Chemical Engineering* Vol. 3, Pergamon Press, Oxford, **1971**.

direct scale-up to find the volume of reactor required for a given production rate. The basic equation employed for batch reactor design is the integrated form of the rate equation, Eq. (35). Thus, for an isothermal reaction, the time required for a fractional conversion x is

$$t = -\int_{c_0}^{c} \frac{\mathrm{d}c}{\mathrm{R}(c)} = c_0 \int_{0}^{x} \frac{\mathrm{d}x}{\mathrm{R}(x)} \tag{71}$$

provided that the molar density of the reacting mixture remains constant, which is a reasonable approximation for most liquid-phase reactions. The rate is normally expressed on the basis of unit volume of the expanded slurry, and corresponding to this the concentration is on a volume basis. For more than a single reactant or a reversible reaction, it is more convenient to retain concentration rather than fractional conversion as the variable.

If several batches of product have to be made to meet a commercial requirement, then it is judicious to estimate the maximum rate of production and devise an operational strategy accordingly. Usually the reaction rate will decrease with increasing reaction time (one exception is an autocatalytic biochemical process) and it would therefore seem best to operate the reactor for a succession of short intervals so that a high average reaction rate can be sustained. However, two problems arise. First, it would be necessary to separate small amounts of product from a large amount of reactant and this is likely to be uneconomical. Secondly, the reactor has to be shut down, emptied, recharged with reactant and reassembled. This takes a finite time and must be accounted for when calculating the maximum production rate and hence the duration of each batch. Figure 7.9 is a typical concentration–time curve for isothermal conditions. If the shutdown time per batch is t_s then the maximum rate can be found by drawing a tangent to the curve from the point $(-t_s, 0)$. The maximum production rate is then $x_m/(t_s + t_m)$ where (t_m, x_m) is the point at which the line is tangential to the curve. The reaction time t_m is not necessarily the optimum for the process as a whole, however, for other factors such as product separation costs should also be considered.

If there are appreciable heat effects during reaction, the temperature may not remain constant despite the employment of immersed heating or cooling coils. The temperature T_c of the coolant is normally taken as a constant as it is usually either condensing steam (for heating) or a high flow rate of cold fluid (for cooling). Equating the difference between the heat released (or absorbed) in the tank volume V and the heat transferred to the cold fluid (or from condensed steam) with the change in enthalpy of the tank contents:

$$(-\Delta H)V\mathrm{R} - UA_c(T - T_c) = \frac{\mathrm{d}H}{\mathrm{d}t} = \left(\sum_i m_i c_{pi}\right)\frac{\mathrm{d}T}{\mathrm{d}t} \tag{72}$$

where $(-\Delta H)$ is the heat of reaction and U the overall heat-transfer coefficient for the exchange of heat through the coils of total heat-transfer surface A_c. The summation of the product of mass and heat capacity on the right-hand side of Eq. (72) is for all the components of the reacting fluid and must include the equivalent heat capacity of the equipment. To find the time required for a specified conversion entails

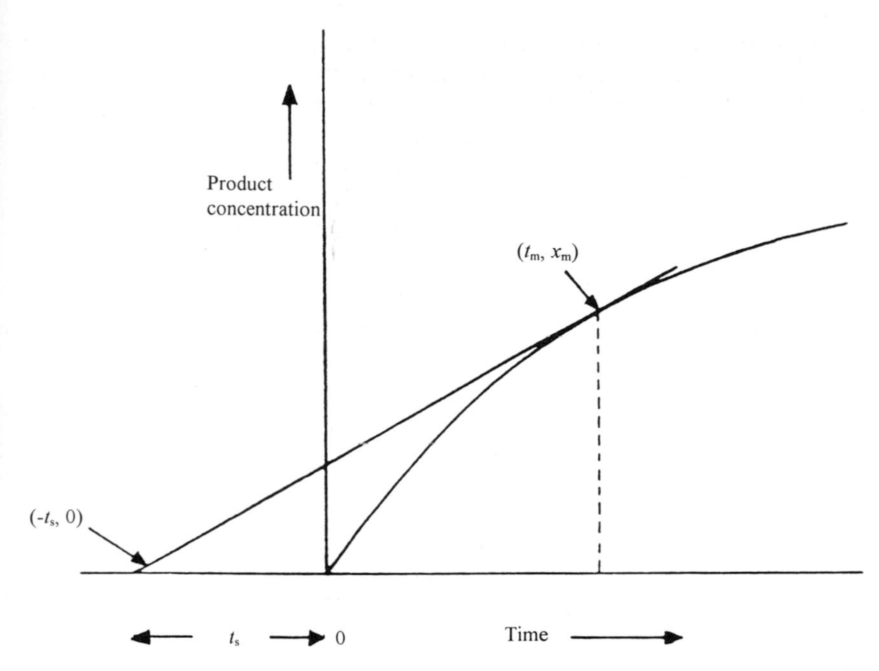

Figure 7.9 Form of product concentration–time plot in batch reactor. Adapted from J. J. Coulson, J. F. Richardson, *Chemical Engineering* Vol. 3, Pergamon Press, Oxford, **1971**.

the solution of the coupled reactor equation (Eq. (71)) and the heat-balance equation (Eq. (72)). Because the reaction rate is an exponential function of temperature, solution is usually accomplished numerically.

7.3.2.2 Continuous Tubular Reactors

The problems which a process engineer has to solve when contemplating the design of a chemical reactor packed with a catalyst or reacting solid are, in principle, similar to those encountered during the design of an empty reactor, except that the presence of the solid somewhat complicates the fluid dynamics and consequently the material- and heat-balance equations. The situation is further exacerbated by the designer having to predict and avoid those conditions which might lead to instability within the reactor.

The effect which the solid packing has on the flow pattern within a tubular reactor can sometimes be of sufficient magnitude to cause significant departures from plug flow conditions. The presence of solid particles in a tube causes elements of flowing gas to become displaced randomly and therefore produces a mixing effect. An eddy diffusion coefficient can be ascribed to this mixing effect and becomes superimposed on the transport process which normally occurs in unpacked tubes – either a molecular diffusion process at fairly low Reynolds numbers Re ($= \rho u_l d_p / \mu$ where d_p is the particle diameter, u_l the linear velocity of the fluid and μ the fluid viscosity) or eddy motion due to turbulence at high Reynolds numbers. Both transverse and longitudinal components of the flux attributed to this dispersion effect are of importance but operate in opposite ways. Transverse dispersion tends to bring the performance of the reactor closer to that which would be predicted by a simple design equation based upon plug flow. On the other hand, longitudinal dispersion is inclined to invalidate the plug-flow assumption such that the conversion would be less than would be expected if plug-flow conditions obtained. The reason for this is that transverse mixing of the fluid elements helps to smooth out the parabolic velocity profile which normally develops in an unpacked tube, whereas longitudinal dispersion in the direction of flow causes some fluid elements to spend less time in the reactor than they would if this additional component of flux due to eddy motion were not superimposed.

The magnitude of the dispersion effect due to transverse or radial mixing can be assessed by relying on theoretical predictions and experimental observations which assert that the value of the Peclet number Pe ($= u_l d_p / D$) for transverse dispersion in packed tubes is approximately 10 (see Section 7.5.4.2). At Reynolds numbers of around 100 the diffusion coefficient to be ascribed to radial dispersion effects is about four times greater than the value for molecular diffusion. At higher Reynolds numbers the radial dispersion effect is correspondingly larger.

Longitudinal dispersion in packed reactors is thought to be caused by interstices between particles acting as mixing chambers. Theoretical analysis of a model based on this assumption shows that the Peclet number for longitudinal dispersion is about 2, and this has been confirmed by experiment. Thus the diffusion coefficient for longitudinal dispersion is approximately five times that for transverse dispersion for the same flow conditions. The flux which results from the longitudinal dispersion effect is, however, usually much smaller than the flux resulting from transverse dispersion, because axial concentration gradients are very much less steep than concentration gradients if the ratio of the tube length to diameter is large.

Isothermal Conditions

The isothermal fixed-bed tubular reactor with no longitudinal dispersion effects represents the simplest form of reactor to analyse. No net exchange of mass or energy occurs in the radial direction, so transverse dispersion effects can be neglected. If we also suppose that the ratio of the tube length to particle size is large, then we can safely ignore longitudinal dispersion effects compared with the effect of bulk flow. Hence, in writing the conservation equation over an element dz of the length of the reactor (Fig. 7.10) we may consider that the fluid velocity u is independent of radial position; this implies a flat velocity profile (plug-flow conditions) and ignores dispersion effects in the direction of flow.

We suppose that a mass of catalyst W ($= \rho_b A_c \mathrm{d}z$ where ρ_b is the catalyst bulk density) is contained within the element dz. The difference between the quantity of reactant leaving and entering the element is $u\mathrm{d}c$ (where u is the volumetric velocity of the fluid and c is the concentration of product, again expressed as moles per unit volume of fluid). Equating this to the number of moles of reactant converted per unit catalyst mass contained within the volume element $A_c \mathrm{d}z$, we obtain (compare Eq. (37))

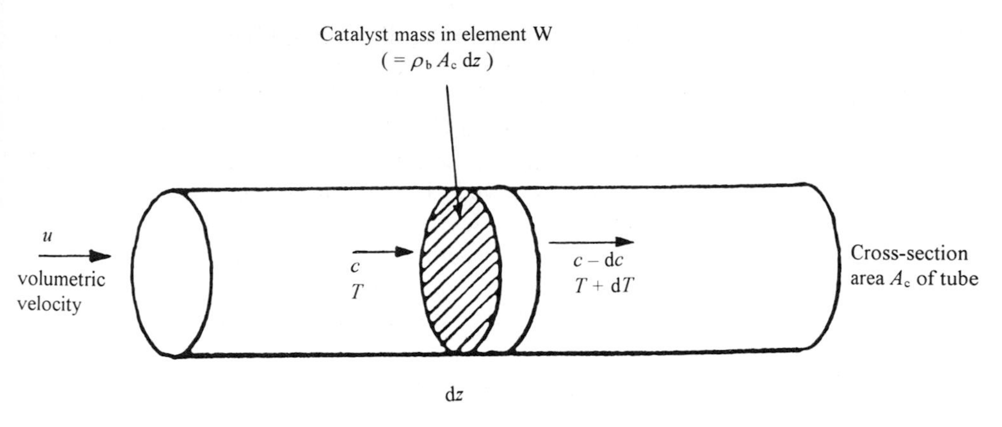

Figure 7.10 Element of catalyst packing within tubular reactor.

$$-\frac{\mathrm{d}(uc)}{\mathrm{d}z} = F_0 \frac{\mathrm{d}x}{\mathrm{d}z} = \rho_b \mathrm{R} \eta A_c \tag{73}$$

where F_0 is the inlet molar flow rate of reactant and dx/dz is the change in conversion per unit reactor length. The effectiveness factor η is now included to account for intraparticle diffusion effects and the rate of conversion R is on the basis of unit catalyst mass.

The usual procedure for integrating this equation is first to write the particle conservation equation (see Chapter 4) and solve analytically or numerically in order to calculate η for the reactor entrance. The reactor is then (notionally) divided up into incremental lengths (say 100) and Eq. (73) is solved for the first increment by recasting the equation into difference form. A new value of x at the next increment along the reactor is thus provided. The whole procedure may then be repeated at successive increments along the reactor until the specified conversion of reactant is attained.

For rapid estimation of reactor size the value of η is often taken either as unity (no intraparticle resistance) or as a constant numerical value for the whole of the reactor length. The length L of tube required to achieve a specified outlet concentration is then obtained by a direct integration of Eq. (73), with η set at unity (or other appropriate value):

$$L = \frac{F_0}{A_c \rho_b \eta} \int_0^x \frac{\mathrm{d}x}{\mathrm{R}(x)} \tag{74}$$

If the reaction rate is a function of total pressure as well as concentration, and there is a pressure drop along the reactor due to the presence of solid catalyst pellets, the conversion within the reactor will be affected by the drop in total pressure along the tube. As the pressure drop in a packed tube is normally linear (see Section 7.5.4.2), it is a simple matter to express the pressure in terms of position along the reactor tube and substitute in the rate equation (Eq. (73)) prior to integration (the reaction rate R is a function of partial pressures and hence, in many instances, total pressure).

In the event that axial dispersion is important, the reactor performance tends to fall below that of a plug-flow reactor. If dispersion can be looked upon as a diffusive flux, then a term $(A_c D_L d^2c/dz^2)$ can be added to the left-hand side of Eq. (73) (it will always have a sign opposite to the convective term $-d(uc)/dz$) and the equation can be solved with the appropriate boundary conditions to give the axial concentration profile $c(z)$. D_L, the dispersion coefficient, would assume units of L^2T^{-1}, similar to a diffusivity. However, Bischoff and Levenspiel (1963) (see Section 7.5.4.2), using a dispersion model, compared reactor performance for different extents of axial dispersion (measured by the parameter $D_L/u_l L$ where u_l is the linear gas velocity). As the Peclet number $Pe_L(= u_l d_p/D_L)$ is about 2, $D_L/u_l L$ may be regarded as $d_p/2L$, the particle diameter divided by twice the reactor length. For most industrial reactors this is a very small quantity indeed (*ca* 10^{-2}) and it is evident from the computations of Bischoff and Levenspiel that for such conditions the reactor performance deviates by no more than a few per cent from that obtained in a plug-flow reactor. In most circumstances, therefore, axial dispersion is seldom of sufficient magnitude to be of importance.

Adiabatic Conditions

Adiabatic reactors are more frequently encountered in practice than isothermal reactors. Because there is no exchange of heat with the surroundings, radial temperature gradients are absent. All of the heat generated or absorbed by the chemical reaction manifests itself by a change in enthalpy of the fluid stream. It is therefore necessary to write a heat-balance equation for the reaction in addition to the material-balance (Eq. (73)). Generally, heat transfer between solid and fluid is sufficiently rapid for it to be justifiable to assume that all the heat generated or absorbed at any point in the reactor is transmitted instantaneously to or from the solid. It is therefore only necessary to take a heat balance for the fluid entering and emerging from an elementary section dz. Referring to Fig. 7.10 and neglecting the effect of longitudinal heat conduction:

$$\rho u \bar{c}_p \frac{dT}{dz} = \rho_b R(-\Delta H)\eta A_c \tag{75}$$

where ρ is the density of the gaseous (or liquid) fluid, $\bar{c}_p$ is the mean heat capacity of the fluid and $(-\Delta H)$ is the heat of reaction. The reaction rate is now a function of temperature. Simultaneous solution of the mass-balance (Eq. (73)) and the heat-balance (Eq. (75)) with the appropriate boundary conditions gives both c and T as a function of z.

A simplified procedure for design is to taken both η and $(-\Delta H)/\rho\bar{c}_p$ as constant. If then Eq. (75) (the heat-balance equation) is divided by Eq. (73) (the mass-balance equation) and integrated, we immediately obtain:

$$T = T_0 + \frac{(-\Delta H)}{\rho \bar{c}_p} c \tag{76}$$

where T_0 is the inlet temperature. This relation implies that the adiabatic reaction path is linear. If Eq. (76) is substituted into the mass-balance (Eq. (73)),

$$F_0 \frac{dx}{dz} = \rho_b A_c \eta R(x, T_0) \tag{77}$$

where the reaction rate $R(x, T_0)$ along the adiabatic reaction path is now expressed as a function of the conversion x and the (constant) inlet temperature T_0. Integration then gives x as a function of z directly, and use of Eq. (76) gives the temperature profile. Because the adiabatic reaction path is linear, a graphical solution, also applicable to multi-bed reactors, is particularly apposite. If the design data are available in the form of rate data for various temperatures and conversions, they may be displayed as contours of equal reaction rate in the $T(c)$ or $T(x)$ plane. Figure 7.11 shows such contours, upon which is superimposed an adiabatic reaction path of slope $\rho\bar{c}_p/(-\Delta H)$ and intercept T_0 on the abscissa. The catalyst mass necessary to achieve a specified conversion x may be evaluated by computing

$$W = \frac{F_0}{\rho_b \eta} \int_0^x \frac{dx}{R(x, T_0)} \tag{78}$$

from a plot of $1/R(x, T_0)$ as a function of x. The various values of $R(x, T_0)$ are simply those points at which the adiabatic reaction path intersects the contours. This procedure is readily adaptable for numerical solution by computer if the kinetic data are carefully organised.

It is often necessary to employ more than one adiabatic reactor to achieve a desired conversion. The catalytic oxidation of SO_2 to SO_3 (see Chapter 8) is a case in point. In the first place chemical equilibrium may have been established in the first reactor and it would then be necessary to cool and remove the product before it enters the second reactor. This is one good reason for choosing a catalyst which will function at the lowest possible temperature. Secondly, for an exothermic reaction, the temperature may rise to a point at which it is deleterious to the catalyst activity. At this point the products from the first reactor are cooled prior to entering a second adiabatic reactor. To design such a system it is only necessary to superimpose on the rate contours the adiabatic temperature paths for each of the reactors. The volume requirements for each reactor can then be computed from the rate contours in the same way as for a single reactor. It is necessary, however, to consider carefully how many reactors in series it is economic to operate.

Should we wish to minimise the size of the system it would be important to ensure that, for all conversions along the reactor length, the rate is at its maximum (R_m). Since the rate is a function of conversion and temperature, setting the partial differential $(\partial R/\partial T)_c$ equal to zero will yield, for an exothermic reaction, a relation $T_{mx}(c)$ which is the locus of temperatures at which the reaction rate is a maximum for a given conversion. The locus T_{mx} of these points passes through the maxima of curves of c as a function of T shown in Fig. 7.11 as contours of constant rate. Thus, to operate a series of adiabatic reactors along an optimum temperature path, hence minimising the reactor size, the feed is heated to some point A (Fig. 7.12) and the

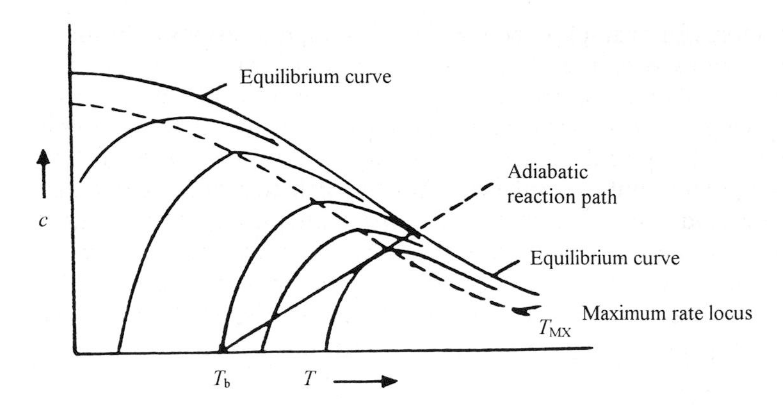

Figure 7.11 Product concentration–temperature plot for an exothermic equilibrium reaction with reaction rate as a parameter and adiabatic reaction path superimposed. After R. Aris, *Introduction to the Analysis of Chemical Reactors*, Prentice Hall, Englewood Cliffs, **1965**.

reaction allowed to continue along an adiabatic reaction path until a point such as B, in the vicinity of the optimum temperature curve T_{mx}, is reached. The products are then cooled to C before entering a second adiabatic reactor in which reaction proceeds to an extent indicated by D, again in the vicinity of the curve T_{mx}. The greater the number of adiabatic reactors in series, the more closely the optimum path is fol-

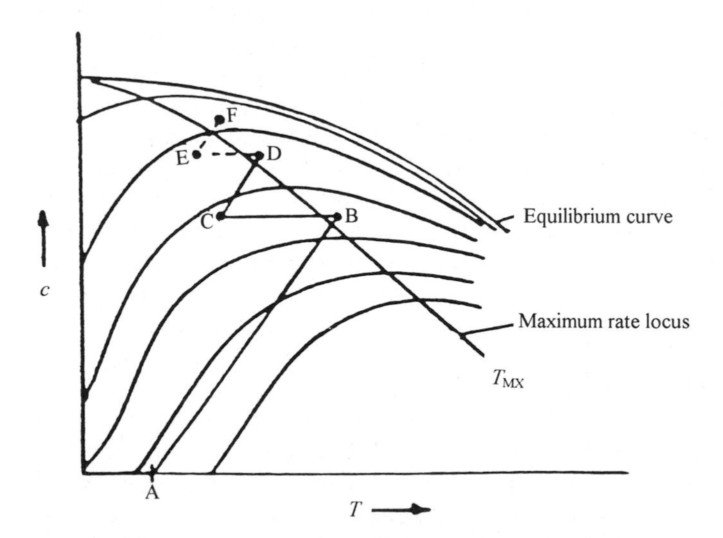

Figure 7.12 Product concentration–temperature plots (reaction rate as parameter) showing superimposed adiabatic reaction paths for three successive reactors (AB, reactor 1; CD, reactor 2; and, if necessary, the path EF, reactor 3). Intermediate cooling occurs between each reactor along the paths BC and DE. After R. Aris, *Introduction to the Analysis of Chemical Reactors*, Prentice Hall, Englewood Cliffs, **1965**.

lowed. For three reactors (for example), there will be six design decisions to be made corresponding to the points A to F inclusive. These six decisions may be made in such a way as to minimise the capital and running costs of the system of reactors and heat exchangers. However, such an optimisation problem is outside the scope of this chapter and the interested reader is referred to a book by Aris (1961) (see Section 7.5.4.2). It should be pointed out, nevertheless, that the high cost of installing and operating heat-transfer and control equipment so as to maintain the optimum temperature profile militates against its use. If the reaction is not highly exothermic, an optimal isothermal reactor system may be a sufficiently economic proposition and it may not be much larger than the adiabatic system of reactors. Each case has to be examined on its own merits and compared with other alternatives.

Non-Isothermal Conditions

When the reactor exchanges heat with the surroundings, radial temperature gradients exist and this causes transverse diffusion of the reactant. For an exothermic reaction, the reaction rate will be highest along the tube axis because the temperature there will be greater than at any other radial position. Reactants will therefore be rapidly consumed at the tube centre, resulting in a steep transverse concentration gradient causing an inward flux of reactant and a corresponding outward flux of products.

The existence of radial temperature and concentration gradients renders the simple plug-flow approach to design inadequate. In principle one could insert into the one-dimensional mass-balance (Eq. (73)) and heat-balance (Eq. (75)) additional terms to account for the dispersion of mass and heat in both the radial and longitudinal directions. Solution of two such simultaneous ordinary differential equations, by numerical means, would yield concentration and temperature profiles for the reactor. Even with such a degree of sophistication, however, the model is not fully representative of events. For example, such a (so-called) homogeneous model assumes that there is no real distinction between gas and solid and that the gas temperature at a point along the bed is the same as the catalyst temperature. We know from experience, however, that catalyst hot-spots sometimes develop. To take account of the difference between gas and catalyst temperatures, continuity equations (similar in form to the reactor equations) for the catalyst particles must be written, and the reactor and particle equations must be coupled by boundary condition statements to the effect that the mass and heat fluxes at the particle periphery are balanced by mass- and heat-transfer between catalyst particle and gas.

Although further advances in the theory of fixed-bed reactor design have been made, it is unusual for experimental data to be of sufficient precision and extent to justify the application of sophisticated methods of calculation. Uncertainties in the knowledge of effective thermal conductivities and heat transfer between gas and solid make the calculation of temperature distribution in the bed susceptible to inaccuracies, particularly in view of the pronounced effect of temperature on the reaction rate. A useful approach to the preliminary design of a non-isothermal fixed-bed reactor is to assume that all the resistance to heat transfer is in a thin layer near the tube wall. This is a fair approximation because radial temperature profiles in packed beds are parabolic with most of the resistance to heat transfer near the tube wall. With this assumption a one-dimensional model, which becomes quite accurate for small-diameter tubes, is satisfactory for the approximate design of reactors. Neglecting diffusion and conduction in the direction of flow, the mass and energy balances for a single component of the reacting mixture are, from Eq. (73),

$$-\frac{\mathrm{d}(uc)}{\mathrm{d}z} = \rho_\mathrm{b} \mathrm{R} A_\mathrm{c} \tag{79}$$

$$\rho u \bar{c}_\mathrm{p} \frac{\mathrm{d}T}{\mathrm{d}z} = \rho_\mathrm{b} A_\mathrm{c} (-\Delta H) \mathrm{R} - U(T - T_\mathrm{w}) \tag{80}$$

where the effectiveness factor has been taken as unity, T_w is the wall temperature and U is a wall heat-transfer coefficient representing the rate of heat transfer per unit temperature gradient per unit length of tube.

If the wall temperature is constant, inspection of the above equations shows that, for a given inlet temperature, a maximum temperature is attained somewhere along the reactor length if the reaction is exothermic. It is desirable that this should not exceed the temperature at which the catalyst activity declines. In Fig. 7.13 the curve ABC shows a non-isothermal reaction path for an inlet temperature T_0 corresponding to A. Provided that $T_0 > T_\mathrm{w}$, it is obvious that $\mathrm{d}T/\mathrm{d}c < (-\Delta H)/\rho \bar{c}_\mathrm{p}$ and the rate of temperature increase will be less than in the adiabatic case. The point B, in fact, corresponds to the temperature at which the reaction rate is at a maximum and the locus of such points is the curve T_mx described previously. The maximum temperature attained from any given inlet temperature may be calculated by solving,

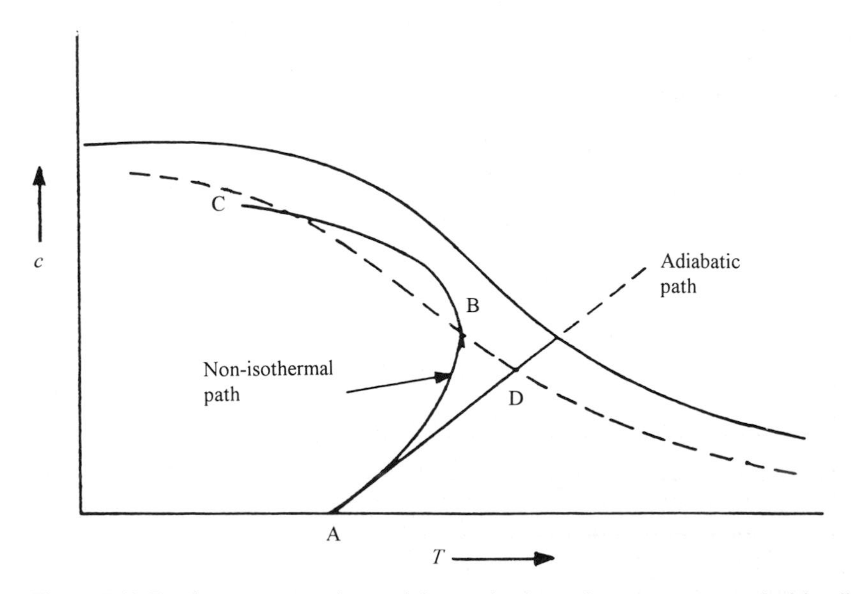

Figure 7.13 Product concentration path in non-isothermal reactor compared with adiabatic reactor. After R. Aris, *Introduction to the Analysis of Chemical Reactors*, Prentice Hall, Englewood Cliffs, **1965**.

using an iterative method, the pair of simultaneous equations (Eqs. (73) and (75)), and finding the temperature at which $\mathrm{d}R/\mathrm{d}T = 0$ or equivalently, $\mathrm{d}c/\mathrm{d}T = 0$ (see Section 7.3.3). We will see later that a packed tubular reactor is very sensitive to change in wall temperature (Section 7.3.3). It is therefore important to estimate the maximum attainable temperature, for a given inlet temperature, from the point of view of maintaining both catalyst activity and reactor stability.

An important class of reactors is that for which the wall temperature is not constant but varies along the reactor length. Such would be the case when the cooling tubes and reactor tubes form an integral part of a composite heat exchanger. The ammonia synthesis reactor (see Chapter 8) is a particular case. Figures 7.14 and 7.15 show, respectively, cocurrent and countercurrent flow of coolant and reactant mixture, the coolant fluid being entirely independent and separate from the reactants and products. However, the reactant feed itself may be used as coolant prior to entering the reactor tubes and again may flow cocurrent of countercurrent to the reactant mixture. In each case heat is exchanged between the reaction mixture and the cooling fluid. A heat balance for a component of the reaction mixture leads to

$$\rho u \bar{c}_{\mathrm{p}} \frac{\mathrm{d}T}{\mathrm{d}z} = \rho_{\mathrm{b}} A_{\mathrm{c}}(-\Delta H)\mathrm{R} - U(T - T_{\mathrm{c}}) \tag{81}$$

an equation analogous to Eq. (80) but in which T_{c} is now a function of z. The variation in T_{c} may be described by taking a heat balance for an infinitesimal section of the cooling tube:

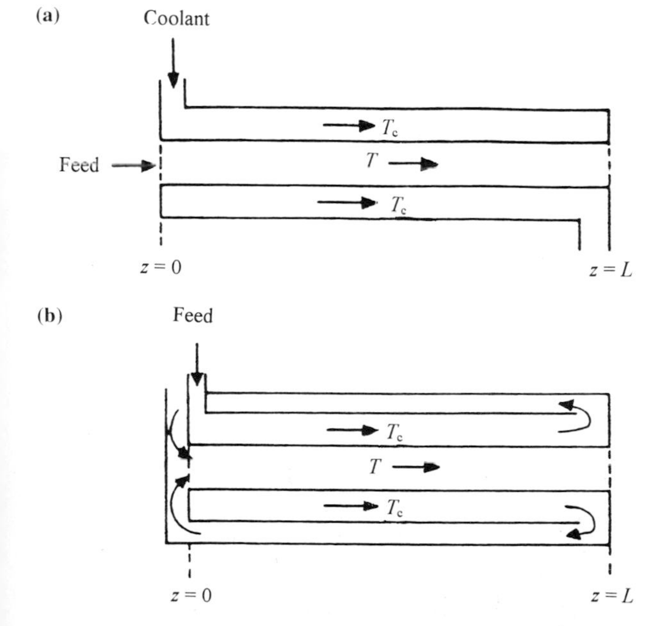

Figure 7.14 (a) Independent and (b) self-cooled tubular reactors operating in cocurrent mode. After R. Aris, *Introduction to the Analysis of Chemical Reactors*, Prentice Hall, Englewood Cliffs, **1965**.

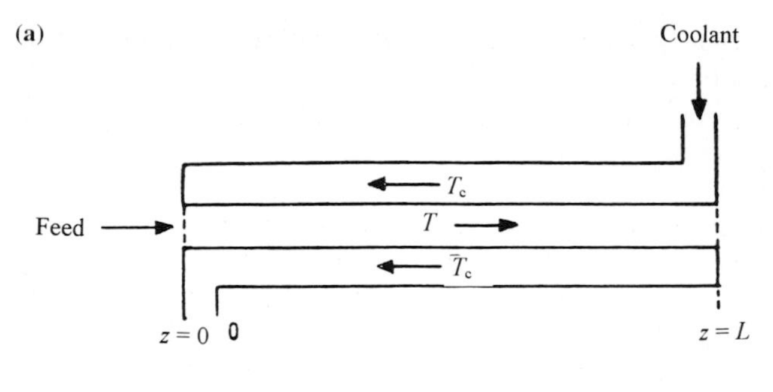

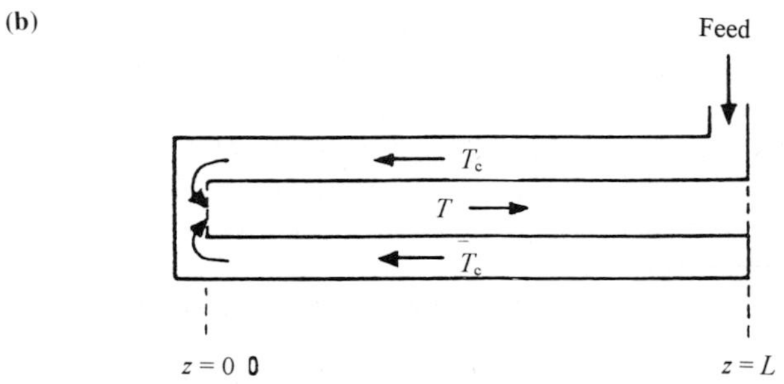

Figure 7.15 (a) Independent and (b) self-cooled tubular reactors operating in countercurrent mode. After R. Aris, *Introduction to the Analysis of Chemical Reactors*, Prentice Hall, Englewood Cliffs, **1965**.

$$(\rho u)_c \bar{c}_{pc} \frac{dT_c}{dz} \pm U(T - T_c) = 0 \tag{82}$$

where $(\rho u)_c$ is the mass flow rate of coolant and $\bar{c}_{pc}$ is the mean heat capacity of the coolant. If flow is cocurrent the lower sign is used; if countercurrent the upper sign is used. Since the mass flow rate of the cooling fluid is based upon the cross-sectional area of the reactor tube, the ratio $\rho u \bar{c}_p / (\rho u)_c \bar{c}_{pc}$ is a measure of the capacities of the two streams to exchange heat. In terms of the limitations imposed by the one-dimensional model, the system is fully described by Eqs. (81) and (82) together with the mass-balance equation derived from Eq. (73):

$$u\frac{dc}{dz} = \rho_b \mathrm{R}(c, T)\eta A_c \tag{83}$$

The boundary conditions will depend on whether the flow is cocurrent or countercurrent, and whether or not the coolant is independent of the reactant mixture.

The reaction path in the (T, c) plane could be plotted by solving the above set of equations with the appropriate boundary conditions. A reaction path similar to the curve ABC in Fig. 7.13 would be obtained. The size of reactor necessary to achieve a specified conversion could be assessed by tabulating points at which the non-isothermal reaction path crosses the constant-rate contours, hence giving values of $\mathrm{R}(c, T)$ which could be used to integrate the mass-balance equation, (Eq. 73). The reaction path would be suitable, provided that the maximum temperature attained is not deleterious to the catalyst activity.

7.3.2.3 Fluidised-Bed Reactor

The use of fluidised-bed reactors for gas–solid reactions has certain advantages over batch and tubular-type reactors. Apart from the mechanical advantage gained by the ease with which solids may be conveyed, the high wall-to-bed heat-transfer coefficient enables heat to be abstracted or absorbed by the reactor with little difficulty. Furthermore, because of the movement of solid particles, the whole of the gas in the reactor is substantially at the same temperature. Another advantage is that the external catalyst surface area offered to the gas is greater than that for a fixed bed and so reactions limited by diffusion in pores will yield higher conversions in a fluidised bed. It is outside the scope of this book to enter into a discussion of the mechanics of fluidisation, for which reference should be made to standard works (see Section 7.5.4.3). It will be sufficient to say that when a gas is passed upwards through a bed of solid particles, there is a pressure drop across the bed which increases steadily with the gas flow. A point is eventually reached when the upward drag on the solid particles by the gas is equal to the weight of the particles. If the gas flow is increased further, the upward drag increases and lifts the particles, thereby increasing the voidage of the bed. The fixed bed then continues to expand until it attains the loosest packing arrangement. Any further increase in gas velocity causes particles to separate from one another and to be freely suspended. The whole bed is now in the fluidised state. Any increase in gas flow is no longer matched by a corresponding increase in pressure drop, since the velocity of gas flow through the interstices between the particles is decreased as a result of bed expansion. Increasing the gas flow beyond the point of incipient fluidisation results in an increase in the voidage of the bed. A point is eventually reached when gas bubbles form within the bed. The fluidised bed then appears to be like a boiling liquid. The gas bubbles which form move upwards through the solid particles, which are in a state of continuous motion. Many industrial-scale fluidised beds are operated in this mode which is characterised by high gas flow maintaining a vigorous bubbling bed with large rising gas bubbles.

Various mathematical models of the fluidised bed exist (see Section 7.5.4.3). The Davidson–Harrison (1963) model of a fluidised bed in which a catalytic reaction occurs is described in terms of two phases – the gas bubble phase containing gas with only small amounts of associated fine particles, and the emulsion phase containing the greater part of the suspended solid matter. The description they provide is one of gas circulation within the bubbles and only a small amount of penetration of the gas into the cloud of fine particles surrounding each bubble. The velocity of bubble rise through the fluidised bed, cloud thickness and gas recirculation rate are all simple functions of the rising bubble size. Rowe (1964) further asserted that each bubble of gas draws along behind its path a wake of solids causing circulation of solids within the bed. Kunii and Levenspiel (1969) further extended these fundamental experimental observations and applied them to a model of a chemically reacting fluidised bed. They ascribed parameters to the intertransfer of reactant between bubble, cloud and wake, and also between cloud and wake and the emulsion phase. By writing

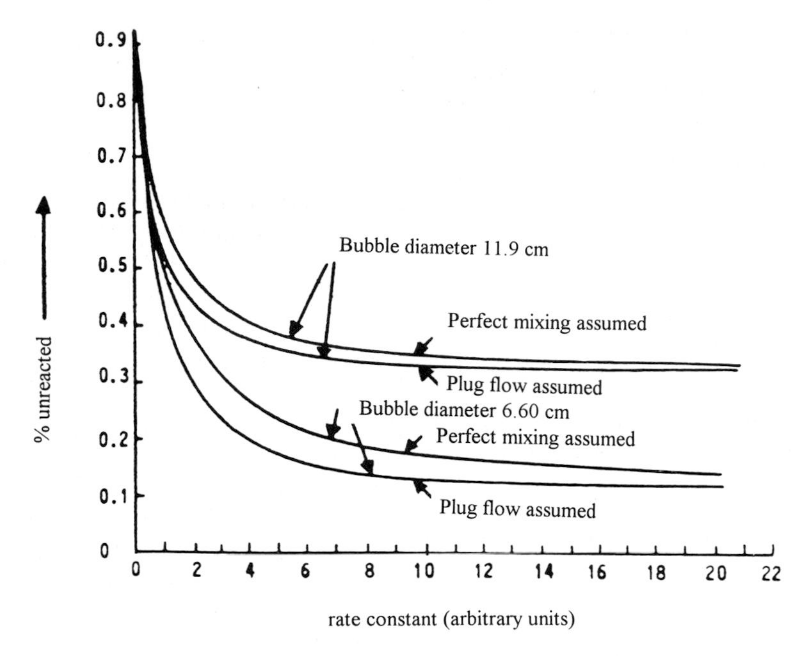

Figure 7.16 Effect of gas bubble size on performance of fluidised-bed reactors. With permission from J. F. Davidson, D. Harrison, *Fluidised Solids*, Cambridge University Press, **1963**.

simple material conservation equations for the bubble and emulsion phases and eliminating algebraically reactant concentrations in the cloud and emulsion phase, they expressed the reaction rate as the product of an overall reaction-rate coefficient and the reactant concentration.

All of the parameters in the above model of Kunii and Levenspiel, except the chemical rate constant, are governed by the bubble size and hinge upon the Davidson–Harrison description of the fluid mechanics of a bubbling fluidised bed. Hence the performance of a fluidised bed can be represented as a function of bubble size, assuming that the bubbles are in plug flow through the vessel. Figure 7.16 shows that, for a given chemical rate constant, large bubbles result in relatively poor performance. This is not at all surprising, because there must be a large proportion of gas passing through the bed which does not effectively contact the fine catalyst particles. On the other hand, small gas bubbles result in a performance intermediate between a well-mixed stirred-tank and a plug-flow reactor. In industrial reactors, large bed diameters and high gas velocities are employed which lead to vigorously bubbling beds with large-diameter bubbles. To avoid the problems associated with large gas bubbles it is often customary to insert baffles at regular spatial positions within the bed and thereby reduce the size of the bubbles. Alternatively, a sufficiently high gas velocity could be employed to carry the particles upwards and out of the bed to be returned again to the bottom of the bed via a loop. The bed then behaves as if it were a lean emulsion with little gas bubbling, and consequently gives a better per-

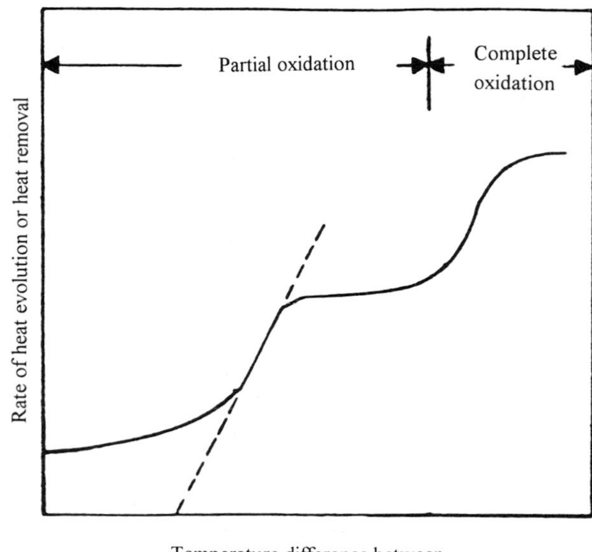

Figure 7.17 Choice of operating temperature for a catalytic partial oxidation reaction in a fluidised bed.

formance. Good gas distributor design is also of prime importance in maintaining high conversions.

The use of fluidised-bed reactors for some industrial partial oxidation reactions has certain advantages over tubular-type reactors. In particular, the whole of the gas in the fluidised bed is substantially at the same temperature and this mode of operation therefore leads to effective temperature control of the reaction environment. If partial and extensive catalytic oxidations are regarded as consecutive chemical reactions occurring in two steps, then the heat generated by chemical reaction as a function of temperature difference between the fluidised bed and its surroundings is a double sigmoid shaped curve, as shown in Fig. 7.17. To restrict conversion in the bed to the desired partial oxidation product, it is necessary to employ not only a highly selective oxidation catalyst, but also to ensure that the bed operates under such conditions that heat is removed from it at a rate which is commensurate with the rate at which heat is generated by the partial oxidation step. This, in turn, means choosing the appropriate temperature difference between bed and surroundings and, by careful design procedures, the bed size and materials of construction. Figure 7.17 illustrates schematically a correct choice of heat removal rate by the intersection of the heat-removal line with the heat-generation curve.

Three well-known examples of processes employing fluidised-bed operations are the oxidations of naphthalene and xylene to phthalic anhydride using a supported V_2O_5 catalyst, and the ammoxidation of propylene utilising a mixed oxide composition containing bismuth molybdate. Typically, this latter reaction is executed by passing a mixture of ammonia, air and propylene to a fluidised bed operating at about 0.2 MPa pressure and 400–500 °C, and with just a few seconds' contact time between the gas and fluidised catalyst particles.

7.3.2.4 The Trickle-Bed Reactor

The trickle-bed reactor evolved as a commercial unit because of the necessity of removing sulphur and nitrogen from low-boiling liquid petroleum feedstock fractions. While vapour fractions can be dealt with in the usual way in fixed-bed tubular reactors, for low-boiling liquids it is necessary to ensure that the liquid reactant is in intimate contact with hydrogen (the gaseous coreactant) in the presence of a porous solid catalyst. Both sulphur and nitrogen are usually present in the form of organic compounds – sulphur as alkyl sulphides, disulphides, mercaptans and thiophenes, and nitrogen as a variety of heterocyclic compounds. Removal of sulphur and nitrogen from the liquid petroleum feedstock is accomplished by reacting with hydrogen (at pressures up to 7 MPa and temperatures in the range 300–1000 °C) in the presence of a mixture of cobalt and molybdenum oxides supported on γ-alumina. Alternatively a mixture of nickel and molybdenum oxides supported on alumina is used, especially if hydrodenitrogenation reactions are involved in the processing. The catalysts employed are always presulphided before use (see Section 7.5.1, Topsoe 1996).

Contacting between gas, liquid and porous solid catalyst phases is usually accomplished by passing hydrogen downwards through a packed catalyst bed and through which the liquid petroleum trickles under the action of gravity cocurrently with the gaseous hydrogen flow. Alternatively, liquid and gas phases may flow cocurrently upwards and in this mode of operation the unit is said to be in flooded flow. Fine catalyst particles are often transported with the fluid with the latter arrangement, which therefore occasions catalyst loss, attrition and solid–liquid separation problems downstream. Countercurrent flow is generally avoided because the liquid flow pattern becomes irregular and unpredictable. There is much evidence to show that the catalyst particles are only partially wetted when the reactor operates in the trickle-flow mode, the most common arrangement in practice. Research with pilot-scale units has shown that the major resistances to mass transfer of reactant to catalyst are within the liquid film surrounding the wetted catalyst particles and also those associated with interparticle diffusion. The situation is not unlike that described for the experimental slurry reactor (Section 7.3.1.6). Equating the rate of mass transfer across the liquid film to the reaction rate (see Fig. 7.7), first order in hydrogen concentration,

$$\mathrm{R} = k_\mathrm{L} a_\mathrm{L}(c_\mathrm{g} - c_\mathrm{i}) = k c_\mathrm{i} \eta \tag{84}$$

where a_L is the solid–liquid interfacial area per unit volume of bed and η is the intraparticle effectiveness factor. Eliminating the unknown liquid film interfacial concentration c_i,

$$\frac{c_\mathrm{g}}{\mathrm{R}} = \frac{1}{k_\mathrm{L} a_\mathrm{L}} + \frac{1}{k\eta} \tag{85}$$

It is probable that the pores of wetted catalyst particles are filled with liquid. Hence, by virtue of the low values of liquid diffusivities (*ca* $10^{-5}\,\mathrm{cm^2\,s^{-1}}$) the effectiveness factor will almost certainly be less than unity. A criterion for assessing the importance of mass transfer in the trickling liquid film has been suggested by Satterfield (1980) (see Section 7.5.1), who argues that, if liquid-film mass transport were important, the rate of reaction could be equated to the rate of mass transfer across the liquid film. For a spherical catalyst particle with diameter d_p the volume of the enveloping liquid is $\pi d_p{}^3/6$ and the corresponding interfacial area for mass transfer is $\pi d_p{}^2$. Hence

$$\frac{\pi d_p{}^3}{6}\mathrm{R} = k_L \pi d_p{}^2 (c - c_i) \tag{86}$$

If the criterion adopted is that when $c_i < 0.95c$ liquid film mass transport is rate-limiting, then Eq. (86) reduces to the inequality

$$\frac{10 d_p \mathrm{R}}{3c} > k_L \tag{87}$$

which is readily applicable in practice as a relatively simple assessment of the importance of mass transfer through the trickling liquid. In the absence of any independent experimental data for k_L, as a first approximation it may be assumed to be D/δ, where D is the liquid diffusivity and δ the film thickness, which may be independently estimated from hydrodynamical considerations.

7.3.2.5 Metal Gauze Reactors

A reaction in which mass transfer predominates is the catalytic oxidation of ammonia. The heat of reaction is high (*ca* $226\,\mathrm{kJ\,mol^{-1}}$) and the reaction conditions, accordingly, are extreme. In the region of 600–700 °C and in the presence of platinum, conversion to nitric oxide is limited by mass transfer from the gas phase to the catalyst surface. Consequently it is not necessary to employ a large extent of metal catalyst surface area impregnated within an inert porous support; such a strategy would waste precious platinum and also lead to unnecessary pressure drops in a packed reactor. The configuration employed in practice is a pad consisting of many layers of finely interwoven platinum gauze. Although conversion to nitric oxide is complete after two or three layers, additional layers are used to sustain the length of time for which the reactor remains in continuous use; the gauzes at the top of the pad where the ammonia–air mixture enters the reactor in downward flow gradually disintegrate and hence become ineffective. To obtain good gas distribution through the pad of gauzes (of *ca* 3 m radius and 5 mm thickness) the inlet gases are expanded through a cone on to the gauze pad.

As the rate of ammonia oxidation is mass-transfer-limited, the overall rate of reaction will be represented by

$$\mathrm{R_v} = h_D a_v c \tag{88}$$

where $\mathrm{R_v}$ is the reaction rate per unit volume, h_D is the mass transfer coefficient for the transport of reactant between gas phase and catalyst surface, and a_v is the interfacial surface area per unit pad volume. If the gas can be regarded as in plug flow through the pad of gauzes, then Eq. (82) should be coupled with the reactor equation for plug flow,

$$-u_l \frac{dc}{dz} = R_v = h_D a_v c \tag{89}$$

where u_l is the linear superficial gas velocity, and the concentration and reaction rate are measured on a volume basis (compare Eq. (73)). Thus the exit concentration of ammonia from the pad is given by integrating Eq. (89), yielding

$$\frac{c}{c_0} = \exp\left(-\frac{h_D a L}{u_l}\right) \tag{90}$$

where c_0 is the concentration of ammonia in the feed. The mass-transfer coefficient is regarded as the average of its respective values at the inlet and exit gas temperatures. The increase in gas temperature between inlet and exit of reactor is calculated from a heat balance along an infinitesimal section of the reactor. If, in traversing such an infinitesimal length of bed, the gas temperature increases by dT and the corresponding decrease in concentration of reactant as a result of reaction is dc, then

$$\rho u_l \bar{c}_p dT = -(-\Delta H) u_l dc \tag{91}$$

Integrating forwards from the reactor inlet where the gas temperature and reactant concentration are T_0 and c_0 respectively,

$$T = T_0 + \frac{(-\Delta H)}{\rho \bar{c}_p}(c_0 - c) \tag{92}$$

Equation (92) signifies that the gas temperature increases linearly with extent of reaction (assuming the ratio $(-\Delta H)/\rho \bar{c}_p$ is approximately independent of temperature). The gauze temperature, on the other hand, is calculated by means of a heat balance between catalyst surface and bulk gas. Conservation of energy demands that, at a given point along the reactor where the gauze temperature is (say) T_s corresponding with a gas temperature T,

$$h(T_s - T) = (-\Delta H) h_D (c - c_s) \tag{93}$$

where h is the coefficient of heat transfer between gas and gauze.

Recalling that, because the reaction is mass-transfer-controlled, the interface concentration c_s at the gauze surface will be negligibly small, then the gauze temperature at the point in the reactor where the gas temperature is T will be

$$T_s = T + \frac{(-\Delta H) h_D c}{h} \tag{94}$$

7.3.3 Thermal Characteristics of a Catalytic Reactor

Many industrial catalytic processes are of an exothermic nature. Steam reforming of carbon monoxide, sulphur dioxide oxidation and ammonia synthesis are three common examples. Unless care is taken in the design and operation of these reactors, much of the heat generated by reaction can be wasted and the temperature within the reactor may rise to such an extent that the catalyst activity is destroyed by sintering.

Clearly, then, it is desirable to utilise economically the heat generated by the catalytic reaction. If the heat dissipated by reaction can be used to heat the cold incoming feed to a temperature sufficient to initiate and sustain a fast reaction, then by judicious choice of operating conditions the process may be rendered thermally self-sustaining. This may be accomplished by exchanging heat between the hot exit gases from the reactor and the colder incoming feed.

To illustrate the essential characteristics of reactor thermal stability and autothermal operation, consider the self-cooled countercurrent reactor, which is sketched in Fig 7.15(b) and exemplified by the ammonia synthesis reactor (Chapter 8). Equation (73) can be regarded as the mass-balance equation (provided longitudinal and radial dispersion effects are neglected and the effectiveness factor is unity), giving

$$-u\frac{dc}{dz} = \rho_b R(c, T) A_c \tag{95}$$

and Eq. (75) as the heat balance for the reactor tube, giving

$$\rho u \bar{c}_p \frac{dT}{dz} = \rho_b A_c(-\Delta H) R - U(T - T_c) \tag{81}$$

The extent of reaction and therefore the amount of heat exchanged with the cold incoming feed will be influenced by the temperature difference between the hot reacting gases and the cold feed at any point along the reactor–exchanger assembly. The heat-balance equation (Eq. (82)) for the cooling tube is thus coupled with the above equations

$$(\rho u_c)\bar{c}_{pc}\frac{dT_c}{dz} + U(T - T_c) = 0 \tag{82}$$

Combining these three simultaneous equations and recalling that for a self-cooled reactor $(\rho u)_c = \rho u$, we obtain

$$\frac{d}{dz}\left\{\frac{(-\Delta H)c}{\rho \bar{c}_p} + T - \gamma T_c\right\} = 0 \tag{96}$$

where $\gamma = \bar{c}_p/\bar{c}_{pc}$. Integration forward from $z = 0$ (where $T = T_0 = T_{c0}$, yet to be determined, and $c = c_0$) gives (assuming γ is approximately unity)

$$T - T_c = \frac{(-\Delta H)}{\rho \bar{c}_p}(c_0 - c) \tag{97}$$

where c_0 is the composition of the cold feed, assuming no reaction occurs in the unpacked, catalyst-free, cooling tubes. Substitution of this latter equation into the heat balance equation (Eq. (81)) eliminates T_c and produces

$$\frac{dT}{dz} = \frac{(-\Delta H)}{\rho u \bar{c}_p}\left\{\rho_b A_c R - \frac{\Gamma}{L}(c_0 - c)\right\} \tag{98}$$

where $\Gamma(= UL/\rho\bar{c}_p)$ is the heat-exchanging capacity of the system with a tube length L. Simultaneous solution of Eq.(98) with the mass-balance equation (Eq. (73)) yields c and T as functions of length z. Because the reaction rate is a function of both concentration and temperature it is usually necessary to integrate Eq. (98) numerically.

Forward integration of Eq. (98) provides a value for T_L, the exit temperature from the reactor. The exit concentration c_L is usually specified in terms of the conversion required. One is therefore in a position to calculate the heat generated by reaction

$$Q_g = (-\Delta H)(c_L - c_0) \tag{99}$$

The net heat loss from the reactor to the cooling tubes is

$$Q_l = \rho\bar{c}_p(T_L - T_{cL}) \tag{100}$$

where T_{cL} is the known coolant temperature at $z = L$ where the coolant enters the reactor assembly (see Fig 7.15(b)). To achieve a thermally self-supporting system the condition $Q_g = Q_l$ must be met. It follows that the temperature rise of the coolant may be expressed as

$$\Delta T = T_{c0} - T_{cL} = T_0 - T_L + \frac{(-\Delta H)}{\rho\bar{c}_p}(c_0 - c_L) \tag{101}$$

Clearly $(c_0 - c_L)$ is a function of T_0 (the value of T at $z = 0$) so the right-hand side of Eq. (101) is a function of T_0 only. Figure 7.18 illustrates plots of ΔT versus T_0 with the heat-exchanging capacity Γ as parameter. Each curve represents the temperature rise of the coolant stream as a result of heat generated by reaction. The left-hand side of the same equation can be represented as a straight line of unit slope through the

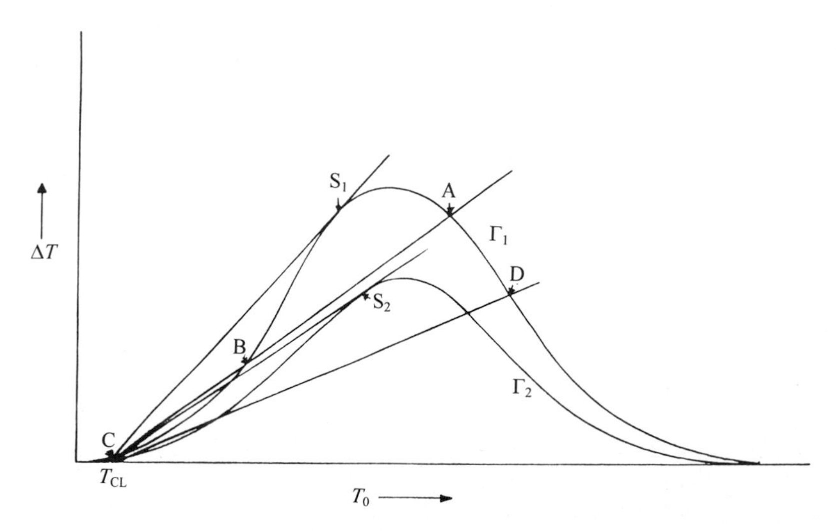

Figure 7.18 Choice of operating conditions in continuous tube catalytic reactors. A and C represent stable operating points while B is an unstable point. Γ_1 and Γ_2 are the bell-shaped reaction curves for two systems with differing heat-exchange capacities. The straight lines are the cooling lines.

point (T_{cL}, 0). At points of intersection of the straight line with any of the curves, conditions for thermal self-sufficiency will be met because it was with such a condition that Eq. (101) was obtained.

Examination of Fig. 7.18 reveals that there may be as many as three points of intersection of the straight line with the bell-shaped curve. Cases where there are no intersections are of no interest in the circumstances under discussion. Focusing attention on the curve labelled Γ_1 (a particular heat-exchange capacity of the system) we see that there are intersections located at A, B and C. Points A and C are stable operating points, while B represents an unstable condition. To appreciate that A, for example, is a stable operating point, suppose that there were a sudden perturbation in the feed temperature T_0 (in this case located at A) so that it increased slightly. The heat generated by the reaction (curve Γ_1) would be insufficient to heat the feed stream to above T_0 (the heat generation curve lies below the heat loss line to the right of A) and so the system would adjust itself and revert to the original reactor inlet temperature. Were the feed temperature to decrease suddenly, there would be enough heat generated by reaction for the feed stream to revert to T_0 at the reactor inlet (the heat generation curve lies above the heat loss line to the left of point C). Thus A represents a stable operating point at high conversion where the heat generated by reaction is sufficient to heat the feed stream to a satisfactory inlet temperature. The stable point C, however, is quite unsatisfactory as an operating point, for the heat generated by reaction only warms the feed stream to a relatively low temperature which is only sufficient to sustain a low conversion. A more desirable choice of operating conditions would be at point D on the curve where the heat-exchange capacity is still Γ_1 and only a single point of intersection with the heat loss line occurs. At such conditions a high conversion can be maintained and there is no danger that any violent perturbation in feed conditions would extinguish the reaction.

As the catalyst activity declines over long operating periods, heat generation curves such as those displayed in Fig. 7.18 (labelled Γ_1 and Γ_2, with Γ_1 representing the higher catalyst activity) are depressed downwards so that the reaction would eventually be quenched. The optimum stable operating condition for a given catalyst activity Γ_1 and coolant inlet temperature T_{cL} is when the heat loss line is tangential at S_1 to the heat generation curve at a high value of ΔT. With decrease in catalyst activity, therefore, S_1 is displaced to regions of higher temperature such as S_2 (corresponding to the heat generation curve for lower catalyst activity Γ_2). For continuing operation, therefore, more heat exchange capacity must be added to the system. This may be achieved by decreasing the mass flow rate. The temperature level along the reactor length therefore increases and the rate is sustained, albeit at a lower conversion concomitant with the decreased catalytic activity. During the operation of non-isothermal catalytic reactors, certain operating problems can arise because of the extreme exothermicity of some catalytic reactions. One of these operational difficulties is the sharp temperature increase of the reactant fluid along the reactor length. Figure 7.19 is a sketch of a typical fluid temperature profile for a catalytic oxidation reaction in a cooled fixed-bed tube. Near the tube inlet the temperature increases slowly at first, but as more heat is released by reaction, the temperature rises more steeply. At the tube exit where conversion is near completion, much less heat is evolved and the temperature is accordingly lower. The axial temperature

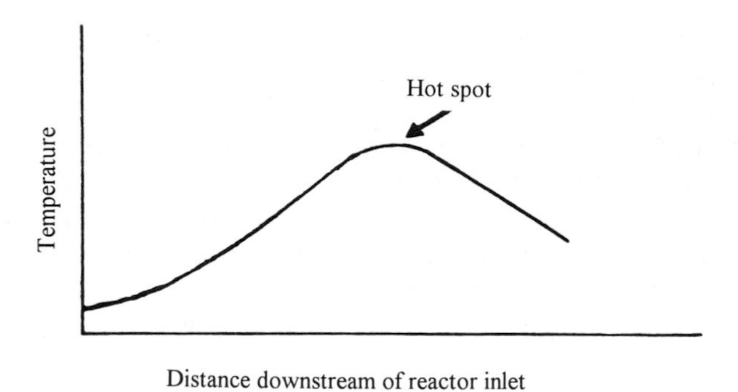

Figure 7.19 Hot spot along the length of a continuous catalytic tubular reactor.

profile thus reaches a maximum (known as the 'hot spot' in the parlance of process technology) somewhere along the reactor length. The more effective the cooling, the lower is the temperature maximum, and a satisfactory reactor design would take account of the heat transfer between reactant and coolant streams and ensure that the temperature maximum is well below any temperature which is deleterious to the catalyst activity. Random hot spots may also occur in packed catalyst tubes, due to uneven catalyst distribution causing inefficient heat transfer; these are impossible to predict and are best avoided by ensuring a uniform catalyst distribution throughout the reactor.

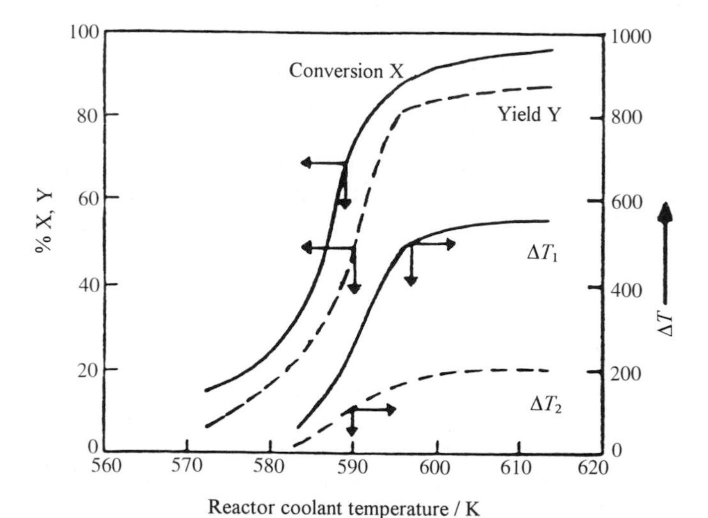

Figure 7.20 Sensitivity of reactor to change temperature difference ΔT between reactor and coolant. Conversion is represented as $\%x$ while the yield is described by y, both plotted as ordinate. The curves labelled ΔT_1 and ΔT_2 are differences between the temperatures of catalyst particles and reactant gas and between gas phase and coolant, respectively. With permission from J. J. Carberry, D. White, *Ind. Eng. Chem.* **1969**, *61*, 27.

The second operating difficulty is the extreme sensitivity of some reactor systems to a slight change in coolant temperature. This phenomenon is known as 'parametric sensitivity' and its effect is illustrated in Fig. 7.20 for the catalytic oxidation of naphthalene on a V_2O_5 catalyst contained in a cooled tubular reactor. The predicted sensitivity is due to the manner in which the heat-transfer coefficients (from catalyst particle to reactant gas and from the gas phase through the tube wall to the coolant) change with temperature and flow conditions. Unwelcome responses to small changes in coolant temperature are best avoided by a careful analysis of the system and a suitable choice of operating system which circumvents the sensitive operating region.

7.4 Problems

1 The catalysed synthesis of carbon disulphide from sulphur and pentane in the presence of vanadium pentoxide supported on alumina was studied in an experimental laboratory tubular reactor. Experimental conditions were arranged so that inter- and intraparticle transport effects were negligible.

It was observed that the reaction was less than first order with respect to both S_2 and C_5H_{12} and that the rate of CS_2 formation was retarded by H_2S. From first principles, show that a Langmuir–Hinshelwood, Hougen–Watson, type of kinetic expression in which either (a) surface reaction between chemisorbed S_2 and C_5H_{12}, or (b) the desorption of the product H_2S, is rate-determining satisfies the experimental observations. What additional experiments might help to discriminate between these two models?

2 A catalyst used to catalyse a first-order reaction is deactivated by poisoning according to a first-order deactivation process. A continuous experimental packed tubular reactor operated at constant temperature is used to test both the reaction kinetics and the validity of the assumption concerning the kinetic order of the catalyst deactivation. Develop a suitable expression which will enable the kinetics to be tested and describe how the rate constants (k for the reaction and k_d for the deactivation) may be evaluated experimentally. How may activation energies for the reaction and deactivation be determined?

3 An experimental study of the catalytic oxidation of sulphur dioxide revealed the information given in the Table.

SO_2 converted [%]	5.0	20.0	35.0	58.0	68.0	71.0	75.0
R [kmol s^{-1} kg^{-1} $\times 10^4$]	200	300	400	400	300	200	75

Calculate (a) the mass of catalyst required to achieve a 78% conversion of sulphur dioxide in a single pass through an adiabatically operated packed tubular reactor if the feed temperature is 420 °C and the volumetric flow rate is 3.8 m^3 s^{-1} (measured at STP); and (b) the exit temperature of the reacting gases. The composition of the feed is 7.0% SO_2, 83% N_2 and 11% O_2 by volume.

The following data may be used for your calculation: heat of reaction at 425 °C is 101 kJ mol^{-1}; mean heat capacity of reaction mixture is 19.4 kJ kmol^{-1} K^{-1}.

4 The rate of the catalytic dehydrogenation of ethylbenzene (molecular weight 106 g mol^{-1}) has been reported in terms of the partial pressures of reactant and products:

$$\mathrm{R} = k\left(P_{\mathrm{E}} - \frac{P_{\mathrm{S}}P_{\mathrm{H}}}{K}\right)$$

where the subscripts E, S and H refer to ethylbenzene, styrene and hydrogen, respectively. The specific rate constant was given as

$$\log_{10} k = -\frac{4770}{T} + 1.54$$

with k in units of kmol s^{-1} atm^{-1} (kg catalyst)$^{-1}$ and temperature T in Kelvin units. The equilibrium constant K is tabulated as a function of temperature below.

T/K	673	773	873	973
K/bar	1.7×10^{-3}	2.5×10^{-2}	0.23	1.4

Using a one-dimensional reactor model with no longitudinal dispersion, write down all the numerical steps necessary to organise the data for the purpose of calculating the amount of catalyst necessary to product 10^3 kg of styrene per day. The reactor is to operate adiabatically and consists of vertical tubes of 1.2 m diameter, packed with catalyst pellets. Specifications and data are as follows:

Feed temperature = 625 °C
Bulk density of catalyst bed = 1440 kg m^{-3}
Average pressure in reactor tubes = 121 kPa
Heat of reaction = 1.4×10^5 kJ kmol^{-1}
Feed rate of ethyl benzene = 1.70×10^{-3} kmol s^{-1}
Feed rate of steam used to supply heat for reaction = 34.0×10^{-3} kmol s^{-1}
Average heat capacity of reactant and product vapours = 2.8 kJ kg^{-1}

In your answer, it will be sufficient to:

(a) produce a numerical relation between the rate of conversion per unit reactor length and reaction rate;
(b) express the reaction rate in terms of conversion and the parameters k and K;
(c) find a numerical relation which gives temperature as a function of conversion at any point along the reactor length; and
(d) outline briefly how these relations can be organised to yield the information required.

5 Provide a preliminary estimate of the diameter of tubes to be installed in a fixed-bed catalytic reactor for the production of vinyl chloride from acetylene and hydrogen chloride. Plug flow of material through the tubes may be assumed and dispersion of heat, other than by convection, in the direction of flow may be neglected. The heat generated by reaction ($-\Delta H = 107.7\,\mathrm{kJ\,kmol^{-1}}$) is to be used to raise steam at 120 °C. Allow for a temperature drop of 10 °C through the tube wall and a maximum permitted catalyst temperature of 252 °C.

The overall rate of reaction may be expressed as:

$$\mathrm{R} = 0.12(1 + 0.024T)\,\mathrm{kmol\,kg^{-1}}$$

where T is the temperature in degrees centigrade in excess of 118 °C. The bulk density of the catalyst to be employed is $288\,\mathrm{kg\,m^{-1}}$ and the effective thermal conductivity is $6.92 \times 10^{-3}\,\mathrm{kJ\,m^{-1}\,s^{-1}\,K^{-1}}$.

[*Note*: For the preliminary estimate required, the temperature of the bed may be assumed to be constant at any radial position at all distances from the bed entrance. A zero-order Bessel equation is obtained when the heat conservation for the tube is written. This equation is of the form:

$$\frac{1}{x}\frac{\mathrm{d}}{\mathrm{d}x}\left(x\frac{\mathrm{d}y}{\mathrm{d}x}\right) + ay = 0$$

and the solution is: $y = AJ_0(x\sqrt{a}) + BY_0(x\sqrt{a})$ where J_0 is a zero-order Bessel function of the first kind and Y_0 a zero-order Bessel function of the second kind, and A and B are arbitrary constants determined by the boundary conditions. You should consult tables of Bessel functions when solving this problem derived from earlier, ca. pre-1960, technology.]

6 In a laboratory experiment a hydrocarbon, together with excess air, was passed at a total superficial mass velocity of $0.2\,\mathrm{kg\,m^{-2}\,s^{-1}}$ through a packed-bed tubular reactor. The catalyst employed was supported on inert non-porous spheres of 0.004 m diameter and $0.22\,\mathrm{m^2\,kg^{-1}}$ external surface area. The bed voidage was 0.43.

The observed reaction rate at 10% conversion was $24\,\mathrm{mol\,h^{-1}\,kg^{-1}}$ and the recorded gas temperature 445 °C. Calculate the temperature at the catalyst surface assuming that interparticle mass and heat transport are the dominant resistances. State any assumptions you make in any steps leading to your answer. Data required for your calculation are as follows: exothermic heat of reaction 105 kJ $\mathrm{mol^{-1}}$, mean heat capacity of reaction mixture $1.09\,\mathrm{kJ\,kg^{-1}\,K^{-1}}$ and viscosity $3.72 \times 10^{-5}\,\mathrm{N\,s\,m^{-2}}$. The Prandtl number may be assumed to have a value of 0.7.

Appropriate correlations for interparticle mass and heat transfer are:

$$j_\mathrm{D} = \frac{0.458}{e}(\mathrm{Re})^{-0.4}$$

$$j_\mathrm{H} = \frac{h}{Gc_\mathrm{p}}\left(\frac{a}{a_\mathrm{T}}\right)(\mathrm{Pr})^{0.67}$$

where e, G and c_p are symbols representing the bed void fraction, mass flow per unit area and mean heat capacity respectively; a and a_T are the areas available for mass and heat transfer respectively.

7 An irreversible exothermic catalytic reaction is carried out in a plug-flow tubular reactor equipped with an annular jacket in which a coolant is vaporized at a temperature T_j. The temperature at any point along the reactor may be assumed to be uniform in a radial direction.
Show:
(a) that the maximum in the longitudinal temperature profile, T_m, should not exceed a critical value, T_m', in order to avoid thermal instability; and
(b) that this critical value is given by:

$$T_m' = \left(\frac{E}{R(T_m - T_j)}\right)^{1/2}$$

were E/R is the ratio of the activation energy for the reaction to the universal gas constant.

8 Estimate the number of catalyst gauze layers required to obtain a 99.8% conversion to nitric oxide of an 11% (by volume) mixture of ammonia in air at 1 bar total pressure fed upwards through the gauzes at 0.4 kg m^{-2} s^{-1} and at an initial temperature of 60 °C. The platinum gauzes are available from manufacturers in the form of finely woven wire mesh with an available surface area of 1.19 m^2 per square metre of a single gauze layer. It is known that the reaction is mass-transfer-controlled, the value of the mass-transfer coefficient being 1.37 m s^{-1}. In practice, the layers of gauze form a catalyst pad and are contained in a reactor shell of about 4 m diameter.

What temperature will the gauze have reached at the location where the conversion is 90%?

9 Develop from first principles, assuming a pseudo-homogeneous reactor model, equations sufficient to describe catalytic reaction in a packed tubular reactor under non-isothermal operating conditions. State any assumptions made and discuss the significance of each term in the mass- and heat-balance equations for the reactor. You are not expected to derive any particle equations, but appropriate use should be made of the particle effectiveness factor to obtain the pseudo-homogeneous reaction rate.

By making suitable approximations, show that the above partial differential equations can lead to a simple set of ordinary differential equations which may be readily applied to the design of an industrial reactor. Discuss thoroughly but briefly *one* such application.

10 A non-isothermally operated packed tubular reactor in which an exothermic catalytic reaction occurs can be unstable by virtue of (a) perturbation in the feed temperature and (b) heat-transfer limitation between fluid and solid.

Discuss the principles involved in both of these types of instability.

7.5 Further Reading

7.5.1 General

J. J. Carberry, *Chemical and Catalytic Reaction Engineering*, McGraw-Hill, New York, **1976**.

K. G. Denbigh, J. C. R. Turner, *Chemical Reactor Theory*, 2nd Ed., Cambridge University Press, Cambridge, **1971**.

G. F. Froment, K. B. Bischoff, *Chemical Reactor Analysis and Design*, Wiley, New York, **1979**.

C. N. Satterfield, *Mass Transfer in Heterogeneous Catalysis*, MIT Press, Cambridge, MA, **1970**.

C. N. Satterfield, *Heterogeneous Catalysis in Practice*, McGraw-Hill, New York, **1980**.

C. N. Satterfield, T. K. Sherwood, *The Role of Diffusion in Catalysis*, Addison-Wesley, Reading, MA, **1963**.

J. M. Smith, *Chemical Engineering Kinetics*, 3rd Ed., McGraw-Hill, New York, **1981**.

H. Topsøe, B. S. Clausen, F. E. Massoth, *Hydrotreating Catalysis*, Reprint from *Catalysis* (Eds. J. R. Anderson, M. Boudart), Vol. 11. Springer, Berlin **1996**.

7.5.2 Kinetic Models

M. Boudart, *Kinetics of Chemical Processes*, Prentice-Hall, Englewood Cliffs, NJ, **1968**.

D. D. Eley, K. E. Rideal, *Proc. R. Soc.* **A 1941**, *178*, 429, 455.

C. N. Hinshelwood, *The Kinetics of Chemical Change*, 4th Ed., Oxford University Press, Oxford, **1940**.

I. Langmuir, *J. Am. Chem. Soc.* **1918**, *40*, 1361.

O. Levenspiel, *J. Catal.* **1972a**, *25*, 265.

O. Levenspiel, *Chemical Reaction Engineering*, 2nd Ed., Wiley, New York, **1972b**.

7.5.3 Experimental Chemical Reactor Configurations

J. M. Berty, *Chem. Eng. Prog.* **1974**, *70*, 78.

J. J. Carberry, *Catal. Rev.* **1969**, *3*, 61.

B. C. Gates, J. R. Katzer, G. C. A. Schuit, *Chemistry of Catalytic Processes*, McGraw-Hill, New York, **1979**.

P. Sunderland, *Trans. IChE* **1976**, *54*, 135.

V. W. Weekman, *AIChE J.*, **1974**, *20*, 833.

7.5.3.1 Slurry Reactors

P. H. Calderbank, *Trans. IChE* **1958**, *36*, 443.

A. Clarke, J. Lloyd Langston, W. J. Thomas, *Trans. IChE* **1977**, *5*, 93.

7.5.4 Industrial Reactors

7.5.4.1 Batch

J. C. Lee, in *Chemical Engineering*, 2nd Ed., Vol. 3, (Eds.: J. Coulson, J. F. Richardson, D. G. Peacock), Pergamon Press, Oxford, **1979**, Chapter 1.

7.5.4.2 Continuous Tubular Reactors

R. Aris, *The Optimal Design of Chemical Reactors*, Academic Press, New York, **1961**.

O. Levenspiel, K. B. Bischoff, *Adv. Chem. Eng.* **1963**, *4*, 95.

E. Singer, R. H. Wilhelm, *Chem. Eng. Prog.* **1950**, *46*, 343.

J. F. Wehner, R. H. Wilhelm, *Chem. Eng. Sci.* **1956**, *6*, 89.

7.5.4.3 Fluidised-Bed Reactors

J. F. Davidson, D. Harrison, *Fluidised Particles*, Cambrdige University Press, Cambridge, **1963**.

J. F. Davidson, D. Harrison (Eds.), *Fluidisation*, Academic Press, New York, **1971**.

D. Kunii, O. Levenspiel, *Fluidisation Engineering*, Krieger, New York, **1969**.

P. N. Rowe, *Chem. Eng. Prog.* **1964**, *60*, 73.

7.5.5 Thermal Characteristics of Reactors

H. Kramers, K. R. Westerterp, *Elements of Chemical Reactor Design and Operation*, Chapman and Hall, London, **1963**.

C. van Heerden, *Ind. Eng. Chem.* **1953**, *45*, 242.

8 Heterogeneous Catalysis: Examples and Case Histories

Our aims in this chapter are several. First, we bring to a focus the knowledge assembled earlier in the course of discussing our appreciation of certain commercially important reactions, such as the synthesis of ammonia and methanol. Second, we illustrate how, at a fundamental level, the key features of particular types of catalysis – which may, as yet, be far from commercial exploitation, e.g. photocatalysis and the harnessing of solar energy – hang together so as to offer a strategy for future research and study. Third, we set out as much to educate and stimulate the student of catalysis as we do to summarize accepted wisdom of particular processes for the benefit of the seasoned practitioners. Occasionally we emphasize laboratory-based studies more than real-life, scaled-up operations; and in the main we concern ourselves with those reactions that have either already assumed major technological significance or those that are potentially capable of doing so.

8.1 The Synthesis of Methanol

The basis of almost all present-day commercial units for the production of methanol is the ICI process, which converts a high-pressure gas mixture of CO, CO_2 and H_2, with greater than 99% efficiency, into the alcohol, using a catalyst containing copper, zinc oxide (ZnO) and alumina (Al_2O_3) at temperatures between 250 and 300 °C. This synthesis is of enormous industrial importance, since it is an effective method of converting 'syn-gas' (see Section 1.1.1) into a product which is pivotal as a precursor for other useful chemicals (e.g. formaldehyde and acetic acid), for high-octane fuels (e.g. petrol, by oligomerization and elimination of water), and for blending agents (e.g. by addition to olefins). In 1987, the world demand for methanol was *ca* 13×10^6 tons. There have been numerous studies of the mode of action of the catalyst; and since it is difficult to pin down the mechanistic details valid under reactor conditions, several model studies have been carried out, using chiefly CO/H_2 mixtures and supported metals.

Bearing in mind that syn-gas is conveniently made from steam-reformed natural gas:

$$CH_4 + H_2O \rightleftharpoons CO + 3\,H_2 \tag{1}$$

it is recognized that the following reactions can occur simultaneously during industrial synthesis:

$$H_2O + CO \rightleftharpoons CO_2 + H_2 \tag{2}$$

$$2\,H_2 + CO \rightleftharpoons CH_3OH \tag{3}$$

$$3\,H_2 + CO_2 \rightleftharpoons CH_3OH + H_2O \qquad (4)$$

One of the first questions that has to be answered, therefore, is whether methanol is synthesized primarily from CO or from CO_2? Other pertinent questions devolve upon:

(a) the state of the copper in the working catalyst;
(b) the roles played by ZnO and Al_2O_3;
(c) which reaction step is rate-determining;
(d) the nature of the active sites.

Recent studies by ICI scientists have led to plausible answers to these questions. But, as is seen below, they do not totally concur with those of Klier et al., whose efforts have involved slightly different catalyst preparations and gas ratios (70 : 30 H_2/CO) and intermediate (75 atm) gas pressure. The ICI process operates in the range 50–100 bar with a composition CO/CO_2/H_2 of 10 : 10 : 80. The model studies of Takeuchi and Katzer, who used rhodium-supported TiO_2 in a laboratory study, yield convincing proof that CO remains intact during the course of catalysed conversion.

8.1.1 Proof that CO is not Dissociated During Methanol Synthesis

This proof entails the use of isotopically labelled CO in the syn-gas mixture. To be specific, $^{13}C^{16}O$ and $^{12}C^{18}O$ (50 : 50 mixtures were used; other conditions were: $p_{CO} = 25$ Torr and $p_{H_2} = 610$ Torr; temperature 150 °C; and the model catalyst consisted of 3% rhodium of TiO_2. After 18% conversion the observed compositions were as shown in Table 8.1. Since the amount of cross-products (i.e. $^{12}CH_3^{16}OH$ and $^{13}CH_3^{18}OH$) is small, scrambling processes, which could operate via the sequences given below, are negligible.

A mechanism based on CO dissociation

$$H_2(g) \rightleftharpoons 2\,H(ad) \qquad (5)$$

$$CO(g) \rightleftharpoons CO(ad) \rightleftharpoons C(ad) + O(ad) \qquad (6)$$

$$CH_n(ad) + H(ad) \rightleftharpoons CH_{n+1}(ad) \quad (n = 0, 1, 2) \qquad (7)$$

$$O(ad) + H(ad) \rightleftharpoons OH(ad) \qquad (8)$$

Table 8.1 Composition of methanol produced from syn-gas after 18% conversion.

Mass number	Species	Percentage
32	$^{12}CH_3^{16}OH$	1
33	$^{13}CH_3^{16}OH$	54
34	$^{12}CH_3^{18}OH$	44
35	$^{13}CH_3^{18}OH$	2

$$\mathrm{CH}_m(\mathrm{ad}) + \mathrm{OH}(\mathrm{ad}) \rightleftharpoons \mathrm{CH}_m\mathrm{OH}(\mathrm{ad}) \quad (m = 0, 1, 2) \tag{9}$$

$$\mathrm{CH}_m\mathrm{OH}(\mathrm{ad}) + \mathrm{H}_{3-m}(\mathrm{ad}) \rightleftharpoons \mathrm{CH_3OH}(\mathrm{g}) \tag{10}$$

A mechanism based on non-dissociative adsorption of CO

$$\mathrm{H_2}(\mathrm{g}) \rightleftharpoons 2\,\mathrm{H}(\mathrm{ad})$$

$$\mathrm{CO}(\mathrm{g}) \rightleftharpoons \mathrm{CO}(\mathrm{ad}) \tag{11}$$

$$\mathrm{COH}_j(\mathrm{ad}) + \mathrm{H}(\mathrm{ad}) \rightleftharpoons \mathrm{COH}_{(j+1)}(\mathrm{ad}) \quad (j = 0, 1, 2) \tag{12}$$

$$\mathrm{COH_3}(\mathrm{ad}) + \mathrm{H}(\mathrm{ad}) \rightleftharpoons \mathrm{CH_3OH}(\mathrm{g}) \tag{13}$$

Clearly, the mass-spectrometric results favour the second mechanism.

8.1.2 The Role of CO_2: Evidence that it is the Main Source of CH_3OH

Working also with labelled oxides of carbon, but at higher pressures, Russian scientists concluded that methanol is formed from CO_2 rather than CO, i.e. the synthesis could, in essence, be formulated:

$$\mathrm{CO} + \mathrm{H_2O} \rightarrow \mathrm{CO_2} + \mathrm{H_2} \rightarrow \mathrm{CH_3OH} + \mathrm{H_2O} \tag{14}$$

At 250 °C, 50 atm total pressure and a gas composition that contained $^{14}CO_2$ in a mixture of equal numbers of moles of CO and CO_2 ($CO/CO_2/H_2$ is 10 : 10 : 80), along with a standard, commercial Cu–ZnO–Al_2O_3 catalyst, ICI scientists confirmed the conclusion of Kagan and Temkin et al. that CH_3OH is indeed formed predominantly from CO_2, the radioactivity being transferred to the methanol. Moreover, scrambling via the reaction

$$^{14}\mathrm{CO_2} + \mathrm{H_2} \rightleftharpoons {}^{14}\mathrm{CO} + \mathrm{H_2O} \tag{15}$$

is negligible, and the conclusion here, therefore, is that the reverse water-gas shift reaction (left to right in Eq. (15)) is slower than the rate of synthesis of methanol. Under somewhat different conditions (see below), Klier et al. have demonstrated that the primary source of carbon for methanol is CO and not CO_2, a point to which we shall return later.

8.1.3 The State of Copper in the Working Catalyst

Using a multiplicity of modern techniques for characterizing catalysts, it has been concluded that some copper is incorporated into the ZnO structure, there being evidence from Auger electron spectroscopy, XPS of copper core levels and ultraviolet–visible (UV–VS) absorption spectra (of ZnO) that the guest is in its Cu^{I} electronic state and in solid solution within the ZnO matrix. If this is the actual condition of the 'active' copper in the working catalyst, a plausible scheme for the catalytic action, which embraces the role played by ZnO and which specifies the active site, can be formulated as in Fig. 8.1.

Here we see that CO is non-dissociatively adsorbed on the Cu^{I} centre, and that H_2 is dissociatively bound to juxtaposed sites on the ZnO support. The mechanism also

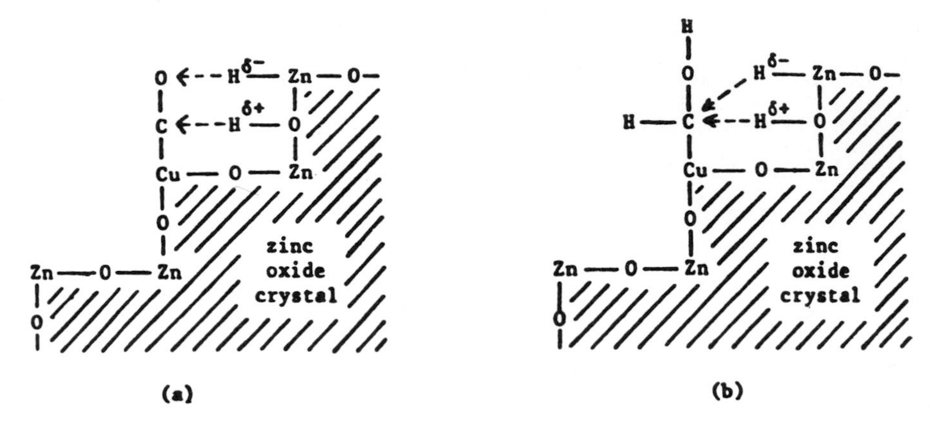

Figure 8.1 Proposed catalytic centre for methanol synthesis on copper–ZnO catalyst. (After F. S. Stone, private communication to JMT).

entails heterolytic splitting of H_2 (on Zn^{2+}–O^{2-} pairs). The electrophilic carbon end of the CO attached to an adjacent Cu^{I} in the surface is then attached by the positively charged hydrogen, and the nucleophilic oxygen end of the negative hydrogen yields CHOH(ad). A second dissociatively adsorbed H_2 attacks the CHOH, causing hydrogenolysis of the Cu–C bond, thereby releasing CH_3OH.

In this picture it is tacitly assumed that the function of the Al_2O_3 is, presumably, to create and maintain a large area for the ZnO, and, therefore, a high degree of accessibility to the Cu^{I} active site. Also tacit is the view that the function of the CO_2 is to keep the oxidation state as copper(I). It must also be remembered that the characterizing techniques (particularly AES, XPS and UV–VIS) which have led to this picture, are necessarily employed under conditions far removed from those experienced by the working catalyst.

Scientists at ICI laboratories set about determining the state of the copper by adsorbing nitrous oxide (N_2O) on a series of copper–ZnO–Al_2O_3 catalysts that had been subjected to standard reduction treatments (as in industrial practice); the resulting catalysts were felt to be representative of the initial state of the real catalyst surface. Even after pre-reduction with either H_2 or CO, some of the copper still remains in an oxidized state, probably as Cu^{I}, since XPS has shown Cu^{II} to be absent. But the number of free (i.e. Cu^0) surface copper atoms in the catalyst preparation can be evaluated from the amount of nitrogen (N_2) evolved after exposure of the catalyst to a pulsed N_2O at temperature and total pressure comparable with those used in synthesis. The number of accessible Cu atoms can be computed using the following equation if the liberated N_2 is monitored:

$$Cu + N_2O \rightarrow CuO + N_2\uparrow \tag{16}$$

It transpires that there is a linear dependence of synthesis activity on the surface area of the metallic copper determined by N_2O titration (Fig. 8.2). It further emerges that about 30% of the initial copper surface of a typical industrial catalyst is unavailable for reaction with N_2O under working conditions. Doubtless, in the real-life catalyst, the surface is in a dynamic state; and it is likely that the precise degree of dynamism

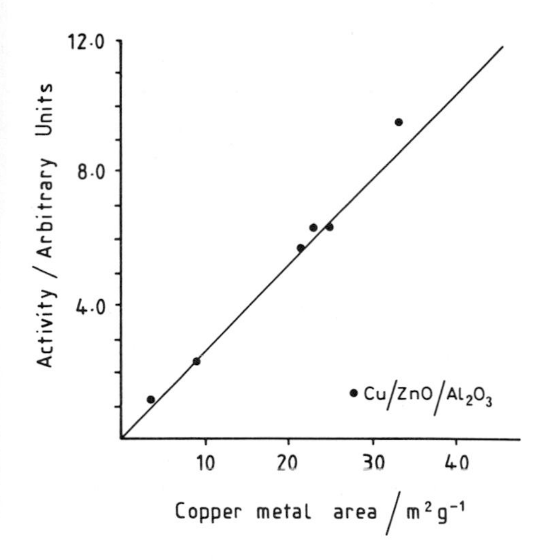

Figure 8.2 Linear dependence of catalytic activity on metal surface area (see text).

will itself vary in a fixed-bed reactor from top to bottom, since the extent of oxidation will in turn be governed by the composition of the ambient gas. It is also likely that the oxidized copper microcrystallites – which are so small as not to give rise to evidence by X-ray diffraction of a separate metallic copper phase – consist of O(ad) and OH(ad), and these surface moieties are known to be capable of interchange in the presence of H_2 and H_2O.

8.1.4 The Role of Oxide Supports: Is There Anything Special About ZnO?

It would be wrong to argue that the existence of a correlation between surface area of metallic copper and activity for methanol synthesis totally rules out a role for the supporting oxides in the overall mechanism of reaction. If the adsorbed complex involved in the rate-determining step is a very finely divided Cu^0 species in close proximity to ZnO (or to ZnO/Al_2O_3 interface), then a linear dependence between surface area and activity will still be obtained. Experiments in which methanol synthesis was performed with copper allied, separately, to the oxides of magnesium, manganese, zinc and aluminum reveal, somewhat surprisingly, no unique role for either ZnO or Al_2O_3 in governing activity for methanol synthesis. The active surface area measurements strongly indicate that only Cu^0/Cu^I, and probably only Cu^0, is implicated in the rate-determining step of synthesis; and that *any* oxide is about equally effective in promoting copper surface area and, hence, in promoting activity in the synthesis reaction. One important role served by the ZnO is that it mops up any sulphur impurity, thereby enhancing the longevity of the copper catalyst.

8.1.5 Views on the Mechanism of the Reaction

At present, there is no universally agreed view apropos the mode of action of the catalyst, of the individual elementary steps at the catalyst surface, and of the precise location of these steps. However, accumulated evidence compiled from temperature-programmed desorption (TPD), temperature-programmed reaction (TPR), IR, dynamic adsorption (e.g. by frontal chromatography) and deuterium labelling tells us the following facts.

1. Hydrogen is preferentially adsorbed dissociatively on the partially oxidised copper component of the catalyst.
2. Carbon dioxide is adsorbed on the partially reduced ZnO component of the catalyst, most significantly as CO_2^-.
3. Hydrogen and carbon dioxide interact at the Cu/ZnO interface to produce a formate species on the copper component of the catalyst. This formate is subsequently hydrogenated through methoxy to methanol, leaving a partially oxidised copper.
4. The role of the CO is to keep the copper in a more highly reduced (more active) state than can be achieved by hydrogen reduction alone, through either an Eley–Rideal or a Langmuir–Hinshelwood mechanism.

A diagrammatic representation of the overall mechanism is shown in Fig. 8.3.

The evidence is strong that very finely divided copper (too finely divided to exhibit its own X-ray diffraction pattern), in the zero oxidation state, Cu^0, is the crucial element in the reaction sphere, but this Cu^0 is in close proximity to an oxide surface site. Atomic hydrogen forms at the Cu^0, and a formate, HCO_2, is formed close to the interface at the oxide site. In all probability, the Cu^0 is itself produced by reduction of the Cu^I/ZnO solid solution with the CO, CO_2 and H_2 in the feed gas.

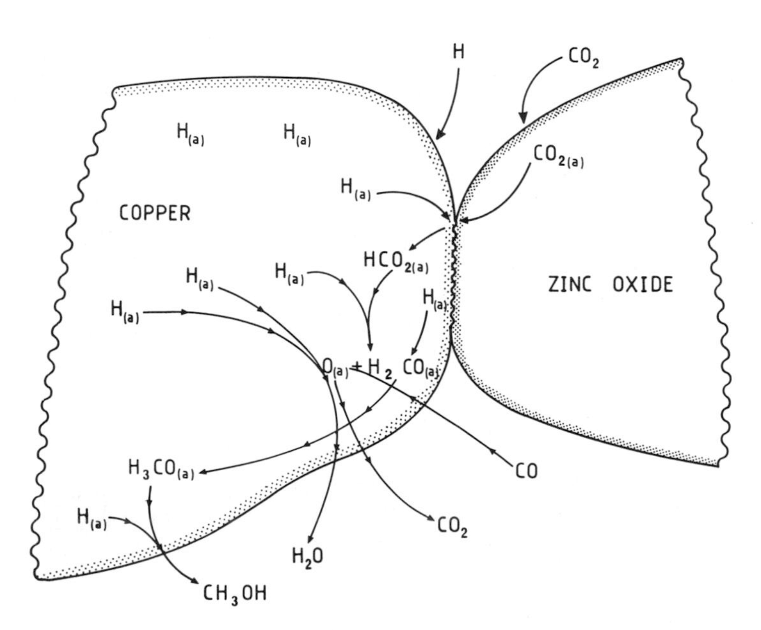

Figure 8.3 Schematic of overall mechanism for the methanol synthesis. Subscripts (a) indicate adsorbed species. (After K. C. Waugh, M. S. Spencer, G. C. Chinchin, quoted in J. R. Jennings and M. V. Zuigg, **1985** – see Section 8.12.1).

We noted earlier (Fig. 8.2) that the activity of a range of different $Cu/ZnO/Al_2O_3$ catalysts versus surface area of the copper metal (measured in situ) showed a linear dependence. We conclude therefore that the turnover frequency is identical for all copper metal areas, indicating that all, or a fixed fraction, of the Cu^0 is active. Since it is improbable that a fixed fraction of a varying area should be active, it is very likely that all the Cu^0 areas are active. This, in turn, suggests that the proposed mechanism is not quite complete because it involves the support (ZnO) as adsorber of the CO_2 and one would not expect the formate formed on the Cu^0 at the Cu/ZnO interface to be mobile. By and large, however, for the CO/CO_2H_2 feedstocks used in the ICI process, the steps represented in Fig. 8.3 yield a self-consistent picture.

It is to be emphasized that, under other conditions, Cu–ZnO catalysts function differently in transforming H_2/CO mixtures to methanol. There is little doubt that aldehydic or hydroxycarbenoic intermediates are kinetically significant in the synthesis, in that their disappearance from the surface coincides with the build-up of methanol product. Under these conditions it seems that CO, rather than CO_2, is the principal source of carbon for methanol. On yet other metal-supported methanol synthesis catalysts – such as palladium supported on a suitable oxide (e.g. MgO or SiO_2) – which are intrinsically less prone to poisoning by sulphur impurities than the copper-based ones, there is a correlation between activity in synthesis and the quantity of cationic (i.e. unreduced) palladium present in the as-prepared catalyst. The influence of the support in the palladium-catalysed reactions can be crucial, for it is known that a given syn-gas feed can be converted to methanol by a palladium catalyst supported on one type of SiO_2, and to methane, with a high selectivity, when supported on another, not too dissimilar, type of SiO_2.

For many of these catalysts there is evidence to corroborate the notion proposed by Frost that their activity derives from the increase in the number of defects in the oxide support brought about by the stabilisation of the defects by the vicinal metal.

8.1.6 Process Conditions, Reaction Configurations and Kinetics

The original process developed in Germany during the 1920s, in association with Fischer–Tropsch hydrocarbon synthesis (Section 8.2), employed a pre-reduced zinc oxide–chromia (Cr_2O_3) catalyst at high pressures between 20 and 30 MPa and in the temperature range 350–400 °C. One may draw analogies between the methanol synthesis and both ammonia synthesis and sulphur dioxide oxidation. Firstly, product formation in each of these reactions is limited by thermodynamic equilibrium; secondly each of the reactions is strongly exothermic ($-91\ kJ\ mol^{-1}$ for CH_3OH synthesis at 298 K); and thirdly, the reaction causes a net decrease in the number of moles. Moderate temperatures and high pressures should therefore favour product formation. On the other hand, too low a temperature would result in too low a reaction rate, so that, to achieve an acceptable rate, a moderate temperature (above *ca* 350 °C) must be selected, thus limiting the conversion to product because of the reaction exothermicity. Adequate cooling is necessary to prevent too large a temperature increase along the catalyst bed. Although high pressure favours product formation, this, in turn, demands high-strength construction materials for the reactor and consequent high capital costs. Furthermore, such high reaction pressures will, depending on upstream processing conditions, require compression of the feed gases with correspondingly high operating costs. The relatively recent development of a catalyst capable of achieving respectable conversions at much lower pressures (*ca* 5–10 MPa) and temperatures of about 250 °C has permitted construction of reactors which operate at these more modest pressures. The catalyst employed for the low-pressure process is the copper–zinc oxide one promoted by alumina as described above.

The high-pressure methanol synthesis reactor consists of three or more catalyst beds arranged in series and contained within a high-pressure shell which incorporates intermediate cooling between each bed and also a heat exchanger at the product exit. Figure 8.4 is a diagram of the usual type of high-pressure reactor. The feed gases pass upwards through a heat exchanger at the bottom of the reactor and thence downwards through the first catalyst bed at the top of the reactor. There is no more than a 25 °C temperature rise along each bed. Too great a temperature rise would, on the one hand, result in loss of activity and, on the other hand, cause the reaction to approach equilibrium too closely: hence the rate would get too slow (as for other exothermic equilibrium reactions, the rate of reaction first increases with temperature, passes through a maximum and then decreases towards zero at true thermodynamic equilibrium). Cooling of each bed is effected either by indirect means, using a heat exchange coil between each of the catalyst beds, or directly by injecting some of the recycled cooled feed gas between each stage of the series of catalyst beds (cold-shot cooling). The molar ratio H_2/CO in the mixture of fresh feed and recycled reacted gas at the entrance to the first bed is usually high (above *ca* 5:1) and CO_2 is sometimes added. The large proportion of H_2 (and some CO_2) helps to increase the thermal conductivity and suppress any tendency for excessive local heating within catalyst beds. The carbon dioxide may also react with hydrogen endothermically (in the reverse of the water-gas shift reaction), thus compensating for excess hydrogen and simultaneously serving to minimize large temperature rises due to reaction.

Since the development of the low-pressure (*ca* 5–10 MPa) catalytic process for the synthesis of methanol, it has been the practice to employ a rather different reactor configuration. Because the low-pressure process catalyst has a shorter life (it is more susceptible to poisoning and sinters) than the $ZnO{-}Cr_2O_3$ catalyst, the reactor is designed so that spent catalyst can be easily discharged. The ICI process employs a single bed of catalyst with quench coolers arranged at two or more levels in the catalyst bed. Such provision of gas cooling by injection also enhances an even distribution of gas within the reactor.

Nattal has reported a detailed study of the methanol synthesis kinetics for a $ZnO{-}Cr_2O_3$ catalyst. In the temperature range 330–390 °C a rate expression of the form

$$R = \frac{p_{CO}p_{H_2}^2 - \left[\frac{p_{CH_3OH}}{K}\right]}{(1 + K_1 p_{CO} + K_2 p_{H_2} + K_3 p_{CH_3OH})^3}$$

is claimed to be successful in predicting reaction rates for the high-pressure commercial process and for the purposes of reactor design. The equation, however, must be regarded as empirical in spite of the fact that a Langmuir–Hinshelwood mecha-

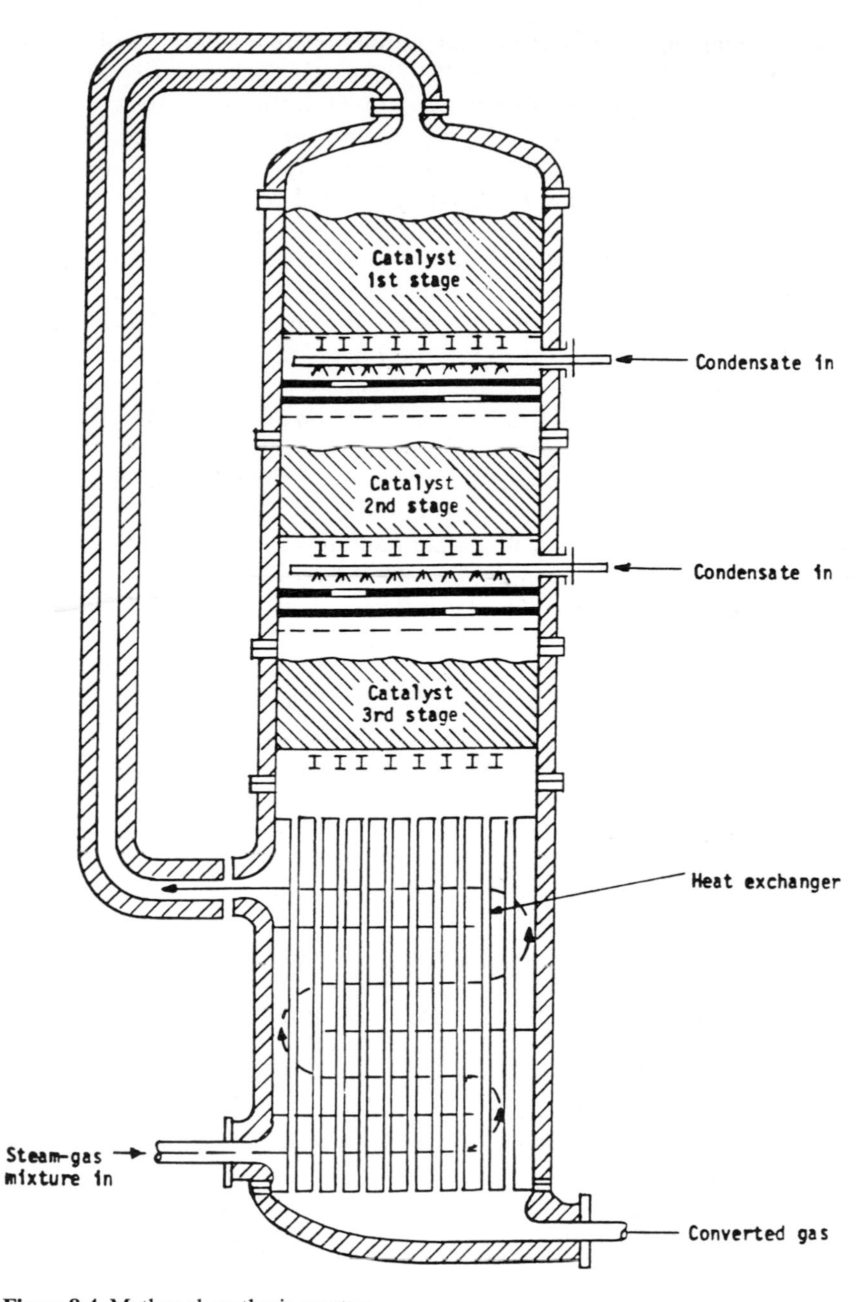

Figure 8.4 Methanol synthesis reactor.

nism (in which a statistically unlikely termolecular surface reaction is assumed rate-limiting) yields the same form of equation as was deduced experimentally.

8.2 Fischer–Tropsch Catalysis

The economic and political fortunes of nations have determined the emergence and decline and subsequent re-emergence of the Fischer–Tropsch synthesis as an industrially important process for the manufacture of hydrocarbons and oxygenated hydrocarbons from mixtures of CO and H_2, (syn-gas). Prior to World War II, Germany, who lacked reliable supplies of oil but possessed extensive deposits of coal, built a number of plants capable of generating syn-gas from coal, so that liquid fuels and other desirable products could be derived by catalytic synthesis using the discoveries published in the early 1920s. Nine plants in Germany produced, prior to World War II, about 16 000 barrels per day of liquid fuels from coal (1 barrel (bbl) ≡ 42 US gallons ≡ 0.159 m^3).

All the following reactions are examples of a Fischer–Tropsch synthesis:

$$\left.\begin{aligned} &n\,CO + 2n\,H_2 \rightarrow (CH_2)_n + n\,H_2O \\ &n\,CO + (2n+1)\,H_2 \rightarrow C_nH_{2n+2} + n\,H_2O \\ &n\,CO + 2n\,H_2 \rightarrow C_nH_{2n+1}OH + (n-1)\,H_2O \end{aligned}\right\} \tag{17}$$

$$\left.\begin{aligned} &2n\,CO + n\,H_2 \rightarrow (CH_2)_n + n\,CO_2 \\ &2n\,CO + (n+1)\,H_2 \rightarrow C_nH_{2n+2} + n\,CO_2 \\ &(2n-1)\,CO + (n-1)\,H_2 \rightarrow C_nH_{2n+1}OH + (n-1)\,CO_2 \end{aligned}\right\} \tag{18}$$

Reactions (17) and (18) are all accompanied by a substantial fall in standard free energy (Fig. 8.5); they are all exothermic, so that formation of the products is fav-

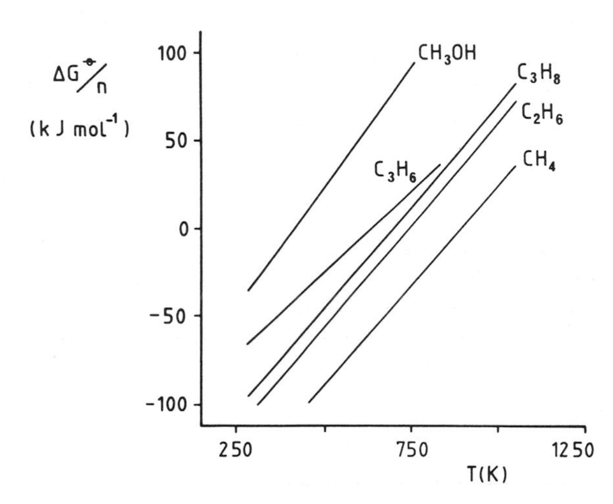

Figure 8.5 Free energy of formation plots for some simple hydrocarbons and methanol.

oured at low temperatures. Coal gasification, whether effected endothermically by steam alone (Eq. (19)), or exothermically by deliberate addition of O_2 to sustain the reaction (Eq. (20)), can be made to yield syn-gas mixtures of required CO/H_2 ratio (enriched if necessary with H_2) by utilizing the water-gas shift reaction (Eq. (15)). CO/H_2 mixtures can also be formed, as described in Section 8.1, by steam-reforming of CH_4 (Eq. (1)) and this is the basis of syn-gas production from natural gas.

$$C + H_2O \rightarrow CO + H_2 \qquad (19)$$

$$5\,C + O_2 + 3\,H_2O \rightarrow 3\,CO + CO_2 + H_2 + CH_4 \qquad (20)$$

The use of $CO–H_2$ mixtures for synthesis is not new – Sabatier and Senderens produced CH_4 using a nickel catalyst, a process now known as methanation:

$$CO + 3\,H_2 \rightarrow CH_4 + H_2O \quad \Delta H^{\ominus}_{298} = -207\,\mathrm{kJ\,mol^{-1}} \qquad (21)$$

Fischer–Tropsch plants built in Germany for the production of hydrocarbons employed a cobalt catalyst in fixed-bed reactors. (With ruthenium as the catalyst, the product is a high-molecular-weight wax, chemically identical with polyethylene.) During the early 1950s there were some unsuccessful attempts in the USA to employ fluidized-bed reactors to convert syn-gas (from natural gas) into petrol (gasoline). In due course petroleum supplies throughout the world became plentiful and interest in the Fischer–Tropsch synthesis waned, except in South Africa: in the 1950s the coal reserves there, as in Germany decades earlier, were used to generate syn-gas which, in turn, yielded waxes and gasoline (via either fixed-bed or entrained fluidized-bed reactors). The first commercial plant, known as SASOL I (South African Synthetic Oil Ltd.), was commissioned in 1955. SASOL II, an entrained fluidized-bed plant (25 atm, 330 °C) producing mainly gasoline and diesel fuels, was conceived at the time when world oil prices rose sharply in the wake of the Yom Kippur war in 1973. It was commissioned in 1980.

A typical product distribution of the liquid fractions of the entrained-bed SASOL plant is shown in Table 8.2.

There is considerable interest industrially in modern variants of the Fischer–Tropsch synthesis for very many reasons, mostly to do with the evolution of new methods of generating petrochemicals from natural gas ('C_1 chemistry') or biomass, oil sands, oil shales and coal, all of which are plentiful, rather than from oil, world supplies of which are diminishing. It would be commercially advantageous if syn-gas

Table 8.2 Product distribution of the liquid fraction from SASOL II.

Fraction	$C_5–C_{10}$ [%]	$C_{11}–C_{18}$ [%]
Olefins	70	60
Paraffins	13	15
Oxygenated compounds	12	10
Aromatics	5	15

conversion via Fischer–Tropsch synthesis could be targeted to produce C_2–C_4 olefins selectively or, alternatively, C_{10}–C_{20} paraffins which could then be utilized in the manufacture of detergent or reformed to produce gasoline free of aromatics. It would also be useful if catalytic conversion of syn-gas went to gasoline directly, i.e. with benzene and toluene comprising most of the product.

We shall see later (Section 8.2) that methanol can readily be converted to gasoline using shape-selective zeolitic acid catalysts such as ZSM-5. Economically, however, this is not ideal, as it first requires the conversion of syn-gas to methanol. ZSM-5 catalysts can also convert the hydrocarbon products of Fischer–Tropsch synthesis to gasoline. Even more attractive, bearing in mind the growing unease about use of benzene, is the conversion of reactants to yield branched olefins and other non-carcinogenic high-octane hydrocarbons.

8.2.1 Mechanistic Considerations

Since Fischer–Tropsch synthesis yields a range of products, often of high molecular weight, it has been likened to polymerization:

$$\left.\begin{aligned} &R_n + m\,(-CH_2-) \rightarrow R_{n+m} \\ &\text{or} \\ &R_n + m\,(CO + 2\,H_2) \rightarrow R_{n+m} + m\,H_2O \end{aligned}\right\} \qquad (22)$$

where $-CH_2-$, or a multiple of it, is the monomer unit – which is not, however, present initially. Given that this unit has to be generated, we perceive that Fischer–Tropsch reactions entail the individual processes of propagation followed by termination. And, indeed, just as for polymerization in general, these hydrocarbon syn-

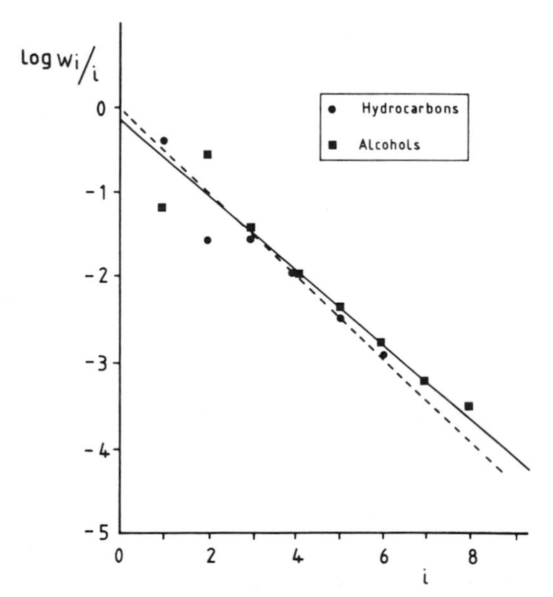

Figure 8.6 Schultz–Flory plot for the formation of homologous hydrocarbons and alkanols over a particular catalyst (see text and Eq. (23)).

theses from syn-gas often obey the so-called Schultz–Flory chain-length statistics:

$$\log(w_i/i) = A + B_i \tag{23}$$

where w_i is the weight fraction of product, i is the carbon number, and A and B are constants depending on temperature, the catalyst and other reaction conditions such as the CO/H_2 ratio and total pressure (see Fig. 8.6). We shall use this equation, along with other items of relevant information culled from different sources, to call into question some previously formulated mechanisms for the reactions, and to underpin what is now believed to be the valid one.

8.2.1.1 Does Synthesis Proceed via Hydroxymethylene Intermediates?

In the mid-1950s it was postulated that a hydroxymethylene intermediate played a crucial role in Fischer–Tropsch synthesis, and until recently the favoured mechanism of the synthesis centred on this intermediate. In essence, it entails formation of C–H bonds prior to the rupture of the C–O bond. Schematically, the overall process would consist of the steps shown in Scheme 8.1.

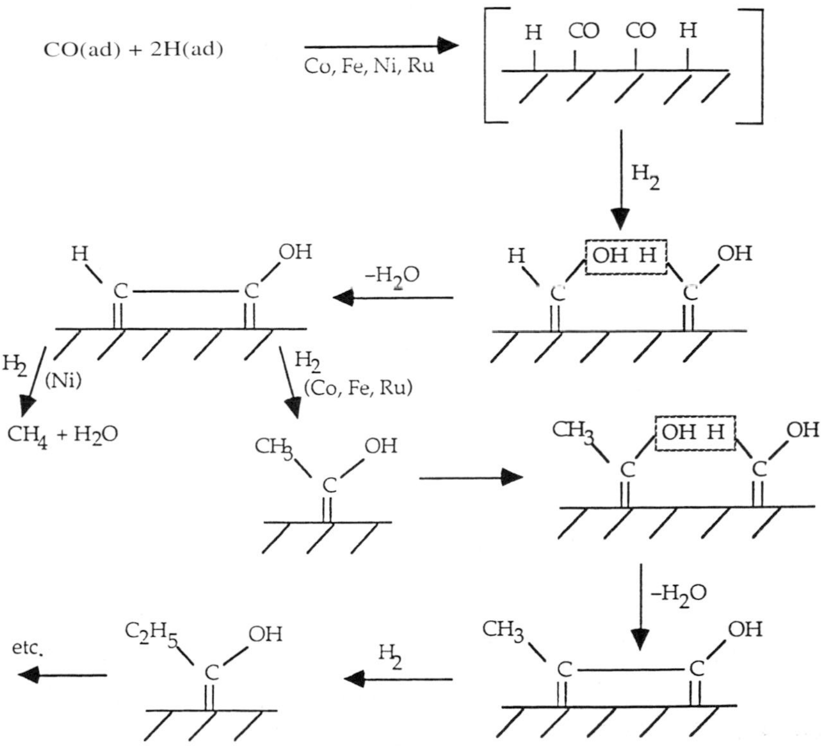

Scheme 8.1 Proposed Fischer–Tropsch mechanism involving a hydroxymethylene intermediate.

This mechanism is no longer felt to be true, partly because dynamic IR studies by D. L. King show no signs of the presence of the hydroxylmethylene surface intermediates, and because a more plausible, alternative mechanism, given in Schemes 8.2 to 8.9 below can account for the experimental facts. Even so, there could well be situations in which the surface-catalysed reduction of CO by H_2 proceeds via hydroxylmethylene intermediates; quantum mechanical calculations support their formation.

8.2.1.2 What of the CO Insertion Mechanism?

Largely by analogy with homogeneous catalytic reactions involving organometallic species, it has been argued by Olive and Muetterties that C–C bonds are formed in Fischer–Tropsch synthesis by insertion of adsorbed CO into metal alkyl groups at the catalyst surface: schematically the situation envisaged is according to Scheme 8.2. Again, this mechanism, like the hydroxymethylene one, has lost favour because there is cogent evidence that dissociation of CO is a prerequisite for Fischer–Tropsch synthesis.

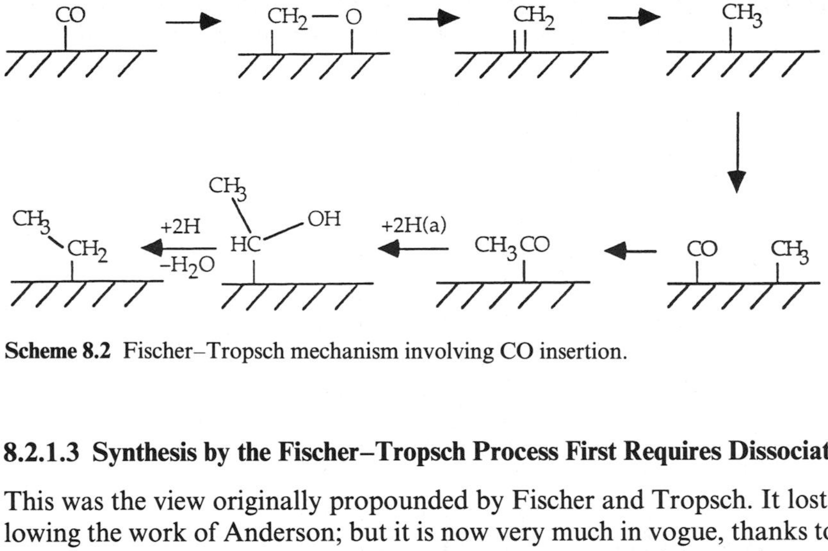

Scheme 8.2 Fischer–Tropsch mechanism involving CO insertion.

8.2.1.3 Synthesis by the Fischer–Tropsch Process First Requires Dissociation of CO

This was the view originally propounded by Fischer and Tropsch. It lost favour following the work of Anderson; but it is now very much in vogue, thanks to the efforts of many. Roberts, Joyner and McNicol produced powerful experimental evidence, especially from model studies using UPS and XPS, that all metal catalysts that are active in the Fischer–Tropsch process also dissociate CO when the gas is adsorbed, even at quite low temperatures. Subsequent experiments, notably by Sachtler and Ponec and by Brady and Pettit, confirmed that CO is first reduced to its elemental state and later converted, via CH(ad) to CH_2(ad). HREELS and ^{13}C NMR studies by Bonzel and Bell confirm that these surface intermediates do indeed form.

The results of Brady and Pettit are particularly interesting, as they lead to a simple mechanistic picture for the production of hydrocarbons in Fischer–Tropsch (FT) synthesis. Diazomethane (CH_2N_2) was chosen as a reactant, the argument being that

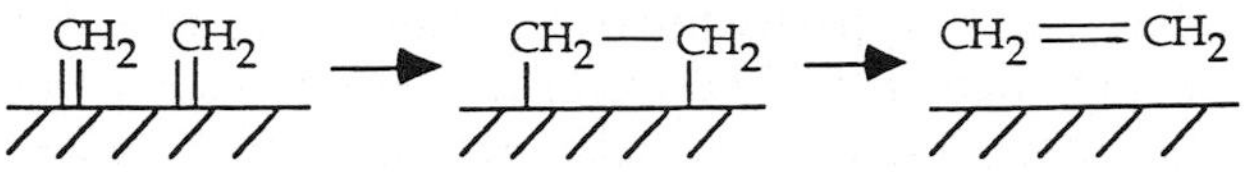

$CH_2\ CH_2 \rightarrow CH_2{-}CH_2 \rightarrow CH_2{=}CH_2$

Scheme 8.3 Reaction of CH_2 fragments at transition metal surface.

bound CH_2 groups would be formed immediately upon contact with a surface. Brady and Pettit studied the behaviour of CH_2N_2 alone, a mixture of CH_2N_2 and H_2 and a mixture of CH_2N_2, H_2 and CO. When CH_2N_2, diluted with an inert gas, is passed over surfaces of a number of transition metals (palladium, iron, cobalt, rutherium, copper) at 1 atm (25 to 200 °C), essentially only ethylene was detected as a gaseous product. From this it was concluded that the reaction of $=CH_2$ fragments at the surface leads merely to dimerization followed by desorption (Scheme 8.3). There is no polymerisation. However, when $CH_2N_2 + H_2$ mixtures were passed over these same metal catalysts, hydrocarbons were produced (Fig. 8.7) ranging from $i = 1$ to $i = 18$ and higher, depending upon the precise conditions of temperature and H_2 pressure. The chain length (i) decreased with increasing H_2 pressure. When a $CH_2N_2 + H_2$ mixture was passed over a typical FT catalyst (e.g., cobalt on kieselguhr) at 210 °C and 1 atm pressure, the product distribution (Fig. 8.8) closely mirrors that obtained when CO and H_2 mixtures are used under identical conditions.

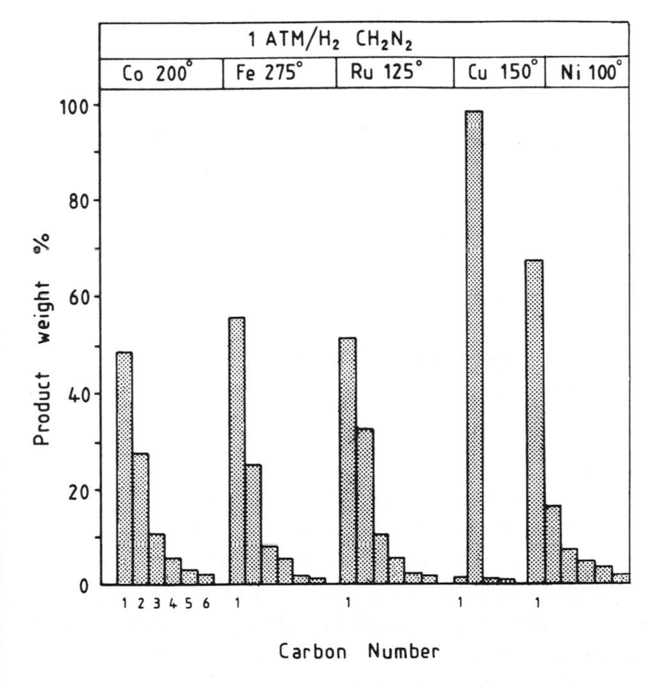

Figure 8.7 Hydrocarbon production from diazomethane and hydrogen mixtures over a series of metallic catalysts. Temperatures are in °C. (After R. C. Brady III, R. Pettit, J. Amer. Chem. Soc. **1980**, *102*, 6181.)

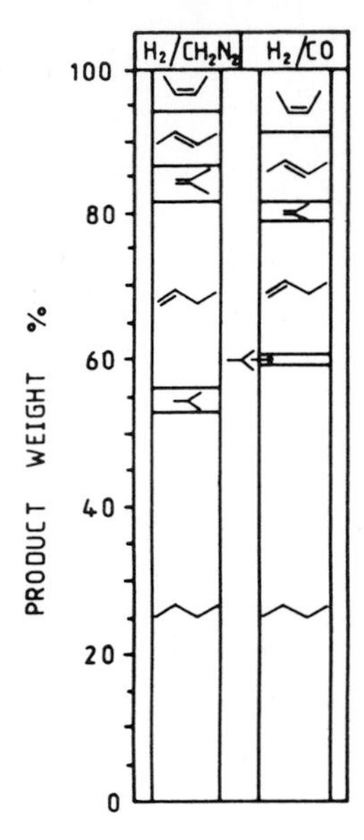

Figure 8.8 Comparison of four-carbon distribution obtained with H_2/CH_2N_2 and H_2CO over a 39% cobalt/kieselguhr catalyst at 210 °C and 1 atm pressure. (After R. C. Brady III, R. Pettit, J. Amer. Chem. Soc. **1980**, *102*, 6181.)

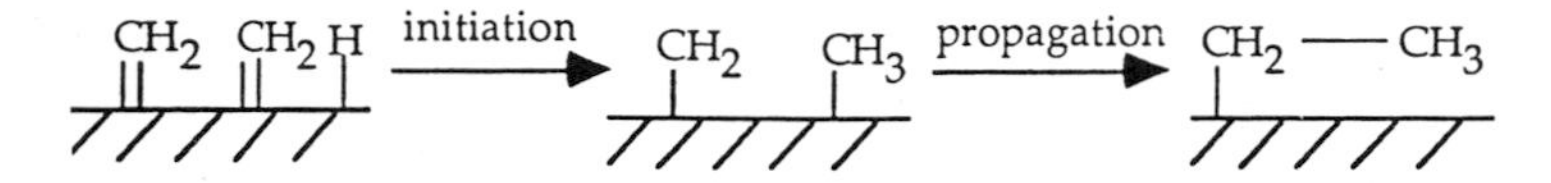

Scheme 8.4 Mechanism of polymerization of adsorbed CH_2.

There must, therefore, be a common mechanism; and we conclude that surface $=CH_2$ fragments are capable of being polymerized in the presence of H_2. Further, we argue that a chemisorbed H atom must be responsible for the *initiation* of polymerization (Scheme 8.4). Copper does not readily dissociate H_2, so that it is incapable of initiating the polymerization. This is how one interprets the exceptional behaviour of copper seen in Fig. 8.7. Propagation simply entails $=CH_2$ insertion, in much the same way as it is known to occur in organometallic chemistry (see Keim, 1991):

$$L_nM(=CH_2)(-CH_3) \longrightarrow L_nM-CH_2-CH_3$$

Termination of the growing chain at the metal surface can occur either by addition of H atoms (terminal reduction) to produce an alkane molecule or by a β-hydrogen abstraction leading to an olefin (Scheme 8.5).

The effect of adding CH_2N_2 to a $CO + H_2$ mixture (Fig. 8.9) is to boost spontaneously the surface concentration of $=CH_2$ groups, thereby casing a shift in the product distribution to longer chain lengths (higher i).

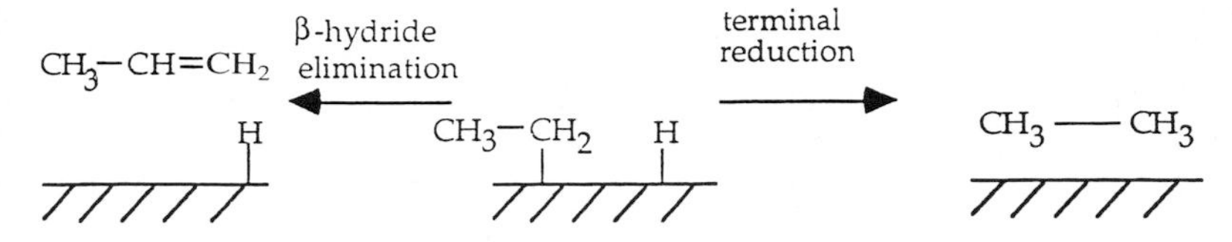

Scheme 8.5 Termination of polymerization at metal surfaces.

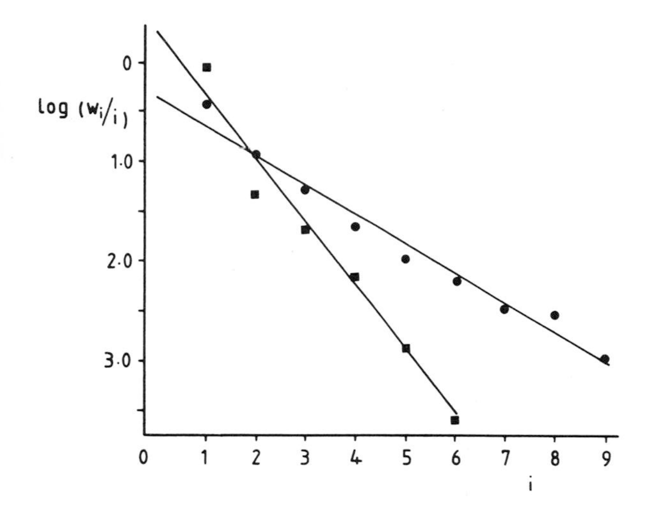

Figure 8.9 Least-squares plot of log (w_i/i) versus i for the distribution of oligomers produced when CO and H_2 (■) and CO + H_2 + CH_2N_2 (●) are passed over a cobalt catalyst.

8.2.1.4 Schultz–Flory Statistics

We can now see how Schultz–Flory statistics (Eq. (23)), governing chain length distribution, follow inexorably from the above picture. Let k_p be the rate coefficient for the propagation and, for simplicity, let k_t, the rate coefficient for termination, embrace both the β-hydride elimination and the terminal reduction steps. Furthermore, assume that these coefficients k_p and k_t are independent of chain length (i). In Scheme 8.6, R_1, R_2, etc., are surface alkyl groups containing 1,2, etc., carbon atoms and P_1, P_2, etc., are the respective gaseous hydrocarbon products. If, further, we include within k_p and k_t the surface concentrations of the reacting partner (e.g. $=CH_2$ for propagation and –H for one of the terminations) – so that these are strictly pseudo rate coefficients – and we write $k_p/(k_p + k_t) = \alpha$, a constant that is dependent on temperature and the other conditions of the synthesis, then the rate of propagation, R_p, and the rate of termination, R_t are:

$$R_p = k_p\theta_{R_n}\, R_t = k_t\theta_{R_n} \qquad (24)$$

where θ_{R_n} is the surface concentration of the chain with n carbon atoms. It follows that the distribution of the product concentrations (c_p) is governed by:

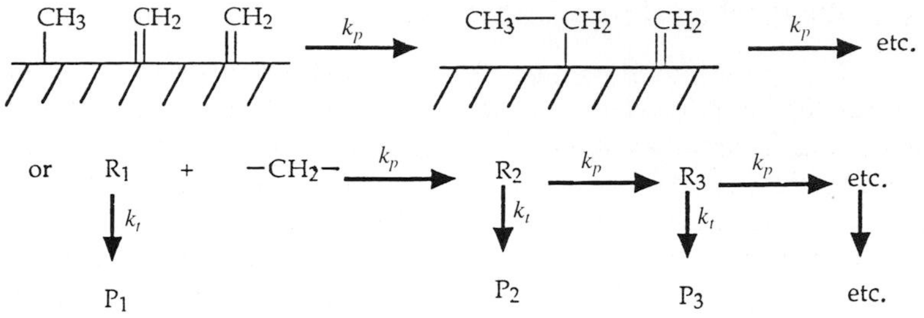

Scheme 8.6 Kinetics of polymerization.

$$c_{\mathrm{P}_{n+1}} = \alpha c_{\mathrm{P}_n} \tag{25}$$

A succession of i steps will therefore occur with probability α^{i}, i.e. the chance of obtaining a product P_i is proportional to α^i, or, in other words,

$$(w_i/i) = K\alpha^i \tag{26}$$

where K is a proportionality constant. This is synonymous with Eq. (23), which accounts for the observed distribution of chain lengths.

8.2.1.5 Other Possible Mechanisms

The fact that our favoured mechanism leads naturally to Schultz–Flory statistics must not be taken as absolute proof for that mechanism, plausible as it is on other grounds. Other mechanisms could be formulated and also comply with Eqs. (23) and (26). For example Maitlis, on the basis of ^{13}C labelling experiments and by analogy with the decomposition paths followed by dinuclear rhodium organometallic complexes (such as $[\{C_5Me_5Rh(\mu\text{–}CH_2)(CH{=}CH_2)\}]$), has proposed a new mechanism for Fischer–Tropsch synthesis of alkenes. Chain propagation is postulated to occur via reaction of surface methylenes with vinyl or alkenyl surface species as shown in Scheme 8.7(a). The surface alkenyls then combine with surface H atoms in the termination step to release the alkene product (Scheme 8.7b).

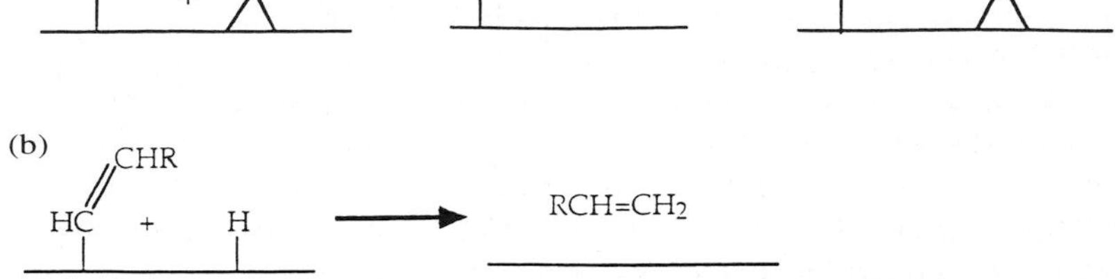

Scheme 8.7 Fischer–Tropsch mechanism according to Maithis. (a) propagation; (b) termination.

8.2.2 Fine-Tuning the Fischer–Tropsch Process

With plentiful supplies of syn-gas to hand, considerable effort is nowadays expended in developing catalysts that can direct the conversion of $CO + H_2$ mixtures to a narrow range or to particular types of products. There are two points to bear in mind. First, one may wish to circumvent somehow the restrictions imposed by Schultz–Flory statistics by, for example, effecting a short cut-off at certain values of the chain length i. The use of rutherium-doped zeolites and other catalysts containing ruthenium seems to work to this effect, but for reasons that are not clear. Second, by judicious choice of catalyst, the production of oxygenated hydrocarbons (alkanols and aldehydes) or of alkenes can be selectively favoured over that of the alkanes. Yields of 'oxygenates' can be quite significant on promoted rhodium and iron catalysts; the general principles underlying the reasons why these are selectively formed have been analysed by Sachtler. It is known, for example, that the oxides of manganese and iron are effective promoters of rhodium for the formation of C_2 oxygenates; and that incorporation of certain oxides such as ZrO_2, La_2O_3 and CeO_2 can, under specific conditions, favour ethanol or higher-alcohol production. Precisely why these catalysts and promoters function as they do remains deeply enigmatic. But there are hopes that, with our recently increased understanding of the action of promoters and poisons, our ability to design the Fischer–Tropsch catalyst of choice will be improved. Specifically, we need to manipulate α (Eq. (25)) to our advantage. Low values of α signify strong termination and hence many C_1 and C_2 entities in the products. Higher values of α signify more dominant propagation and, hence, broader product distributions. If, for example, we need to make synthetic crude oil from syn-gas, α has to be in the range 0.72–0.78 so as to generate chains of product molecules up to C_{20}.

To register significant progress we must appreciate the factors that favour certain pathways over other possible ones. Taking a cue from the organometallic analogues, one recognizes that the following reaction:

$$L_n \cdot M{-}CH_2{-}CH_3 \longrightarrow L_n \cdot M \leftarrow \begin{matrix} CH_2 \\ \| \\ CH_2 \end{matrix}$$

may take place under certain circumstances. We see here some scope – by modifying L and M in the homogeneous situation, and hence the nature of the metal surface in the heterogeneous one – for favouring alkene over alkane production. Likewise, we recognize that, although CO insertion followed by reduction is *not* the preferred path for production of alkanes in Fischer–Tropsch synthesis, CO insertion *is* the most likely route for the production of primary alcohols and aldehydes, the surface intermediates being rich in acyl groups, as evidenced by in-situ IR studies. Sachtler has chronicled the reasons for believing that the chain growth mechanisms on rhodium catalysts are identical for hydrocarbons and oxygenates and occur on the same sites for the three groups of products (Scheme 8.8). It is clear that the chain propagation parameters reach a constant value for carbon numbers greater than 3, the deviations for very short chains being in agreement with chemical experience.

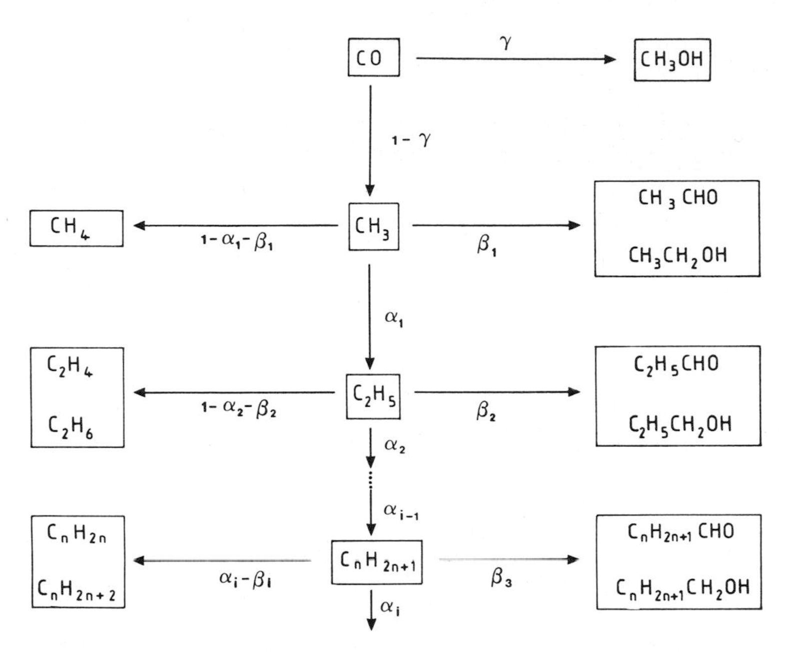

Scheme 8.8 Kinetic model for CO/H_2 reactions over rhodium catalysts with following values for the parameters of α, β and γ (after W. M. H. Sachtler, *Angew. Chem., Int. Ed. Engl.* **1992**, *31*, 1039):

$\alpha_1 = 0.14$ $\beta_1 = 0.44$ $\gamma = 0.065$

$\alpha_2 = 0.44$ $\beta_2 = 0.36$

$\alpha_3 = 0.36$ $\beta_3 = 0.21$

$\alpha_i = 0.36$ $\beta_i = 0.21$

If we focus on metallic catalysts, M, we can expect, bearing in mind the overall sequence of reactions shown in Scheme 8.9, that only a relatively small number of the elements in the Periodic Table will satisfy the conditions required for Fischer–Tropsch synthesis, since the catalyst must:

(a) possess the ability to dissociate adsorbed CO at temperatures above about 250 °C;
(b) have the right balance of strengths for the adsorbed bonds M–C and M–O, relative to the bonding in CO; and
(c) facilitate hydrogenation of M–C and M–O surface species.

The metals situated between the stepped lines in the portion of the shown Perodic Table in Scheme 8.10 are effective in dissociating CO under the typical conditions of Fischer–Tropsch synthesis. They also facilitate formation of tilted M–O bonds (of the kind described by Tamaru) as required in the reaction sequence in Scheme 8.10.

The M–O bonds formed on, for example, iron, cobalt and nickel surfaces are readily reduced by H(ad) (or by CO(ad)), leading to the liberation of gaseous H_2O and CO_2, thereby cleaning the surface and making it again accessible for more CO to

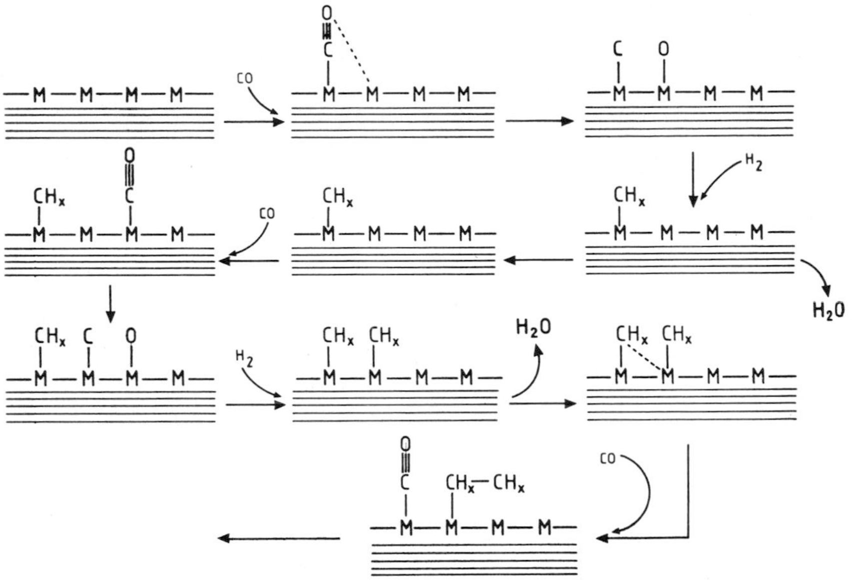

Scheme 8.9 Schematic illustration of sequence of steps occurring at the surface of a Fischer–Tropsch catalyst.

be adsorbed. This, in turn, repeats the cycle of dissociation and subsequent reduction to $=CH_2$. The system must, however, be delicately poised. Coverage of the catalyst surface by the nascent adsorbed carbon, which is reducible to $=CH_2$, must be just right. Extensive formation of linked chains of carbon or of embryonic graphite will poison the catalysis, as hydrogenation and chain propagation will then be thwarted. On the other hand, a modest population of surface precursor CO(ad) could serve a valuable purpose in that it will isolate the fruitful $=CH_2$ growth elements from isolated bound carbon atoms. Use of the powerful 'transient kinetic' approach (see Section 3.10.1) should prove illuminating in disentangling the mode of operation of

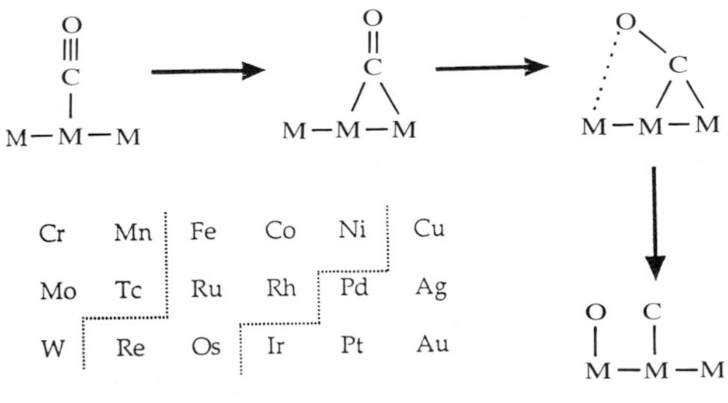

Cr	Mn	Fe	Co	Ni	Cu
Mo	Tc	Ru	Rh	Pd	Ag
W	Re	Os	Ir	Pt	Au

Scheme 8.10 Possible mechanism of CO dissociation at a metal (M) surface.

existing catalysts and in evolving new ones; promising results have already been reported by Mims.

8.2.3 Practical Fischer–Tropsch Catalysts and Process Conditions

It is not a coincidence that some industrial Fischer–Tropsch reactors function with an iron-based catalyst. We recall that dissociation of CO to elemental surface carbon is a pre-requisite. We shall see in Section 8.3 that ammonia synthesis has dissociation of the isoelectronic N_2 molecule as a pre-requisite. Just as surface elemental carbon is progressively hydrogenated in Fischer–Tropsch synthesis, so is there progressive hydrogenation of surface elemental nitrogen in the ammonia synthesis. It is well known that iron favours the latter; it is, therefroe no surprise that it can also facilitate the former.

To illustrate the kind of recipe that is rather rigidly followed in order to arrive at a successful catalyst, we describe the somewhat prescriptive procedures associated with a prominent commercial process. The fixed-bed SASOL reactors employ a precipitated catalyst prepared by adding a hot solution of iron and copper nitrates (20 : 1 Fe/Cu) to a hot solution of sodium carbonate (Na_2CO_3). The precipitate is washed with water to eliminate sodium and a solution of K_2SiO_3 is added to the slurry. Nitric acid (HNO_3) is introduced to remove some of the excess potassium and the slurry is finally filtered, partially dried, extruded and further dried to a water content of less than 10 wt.%. It is thought that the presence of the resulting silica stabilizes the high-area Fe_2O_3. The pore structure of the catalyst is largely determined by the conditions during initial precipitation of the metal oxide. The porosity is also partly controlled by the drying procedure, and not least by the addition of a low-surface-tension liquid prior to drying. The prepared catalyst is normally pre-reduced with H_2 under mild conditions so as to generate high surface area of metal. Reduction of Fe_2O_3 is facilitated by the presence of copper, which permits the low-temperature reduction that ensures maximal areas and thus suppresses crystal growth. The promoting influence that potassium exerts on the catalyst is no doubt akin to the same influence that potassium has on ammonia synthesis catalysts (see Section 8.3): it facilitates dissociation of CO which, in turn, ensures production of high-molecular-weight products.

The SASOL I and SASOL II plants mentioned earlier each employ Lurgi gasifiers which are capable of producing some $(35–55) \times 10^3\ m^3\ h^{-1}$ (expressed at STP) of raw syn-gas from coal by reaction with steam and oxygen at about 30 bar pressure. By controlling the amounts of steam and oxygen supplied, the gasifiers can be operated as thermally self-supporting units, the exothermic reaction of oxygen with coal supplying the heat requirements for the endothermic steam gasification of carbon. The exit gases issuing from the gasifiers are cooled, thus removing excess water, phenols and tar oils by condensation. Ammonia produced during gasification dissolves in the condensate. The composition of the raw syn-gas following the removal of water, phenol and ammonia is approximately 30% H_2, 30% CO, 29% CO_2, 9% CH_4 and the balance is H_2S, nitrogen and argon with trace quantities of CH_3SH, CS_2, COS and light naphthas. All the sulphur-containing gases and the napthas are removed by washing, in several stages, with cold methanol. Most of the CO_2 is absorbed simultaneously. The purified syn-gas which is fed to the Fischer–Tropsch reactors now

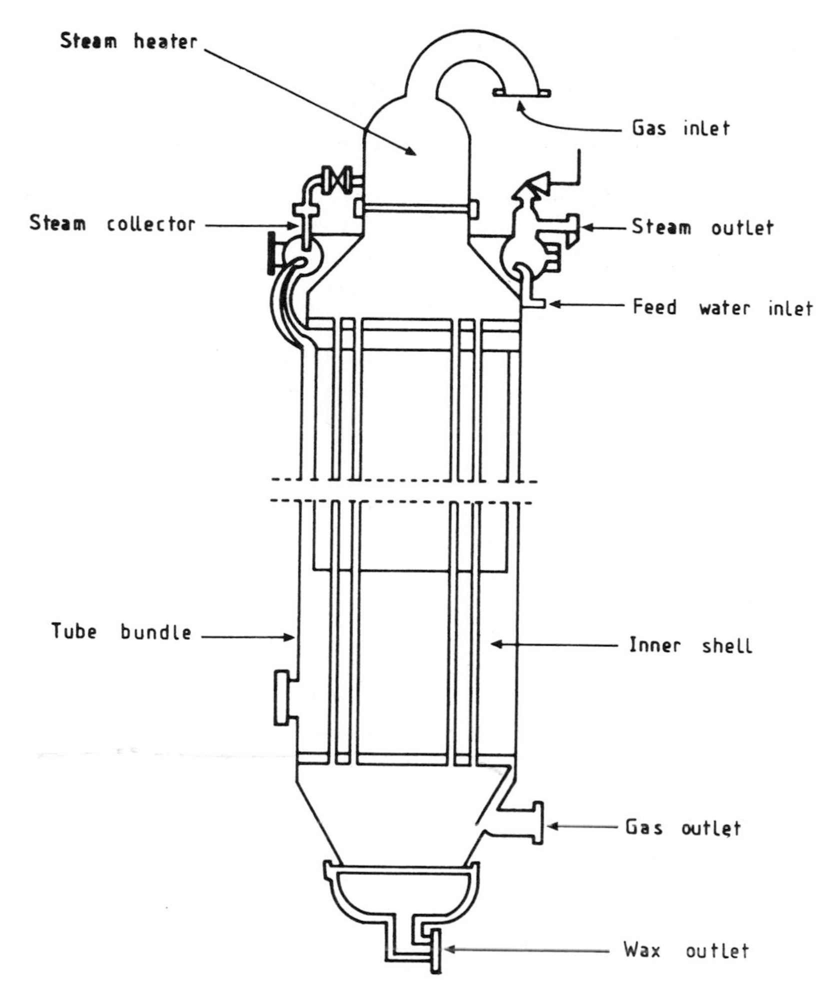

Figure 8.10 Outline of multitubular shell assembly (see text).

has a composition which is approximately 43% H_2, 43% CO, 12% CH_4 and 1% CO_2, the balance being nitrogen and argon.

Two types of catalytic reactors are used in the SASOL plants, each of which gives a different product spectrum on account of the different gas–solid contacting arrangements. Thus a multitubular shell assembly (Fig. 8.10) operating at 220–350 °C and 30 bar pressure and containing an iron catalyst packed within the tubes, produced a broad range of hydrocarbon products ranging from C_1 to C_{35} and above, about 42% of the product being medium and hard waxes and 32% gasoline and diesel fuels. A high gas velocity (relative to fixed beds) of 500 volumes of feed per unit catalyst volume per hour ensures turbulent flow conditions in the bed which, in turn, result in high rates of heat transfer from the catalyst bed to the surrounding shell containing boiling water, the temperature of which is controlled by regulating the

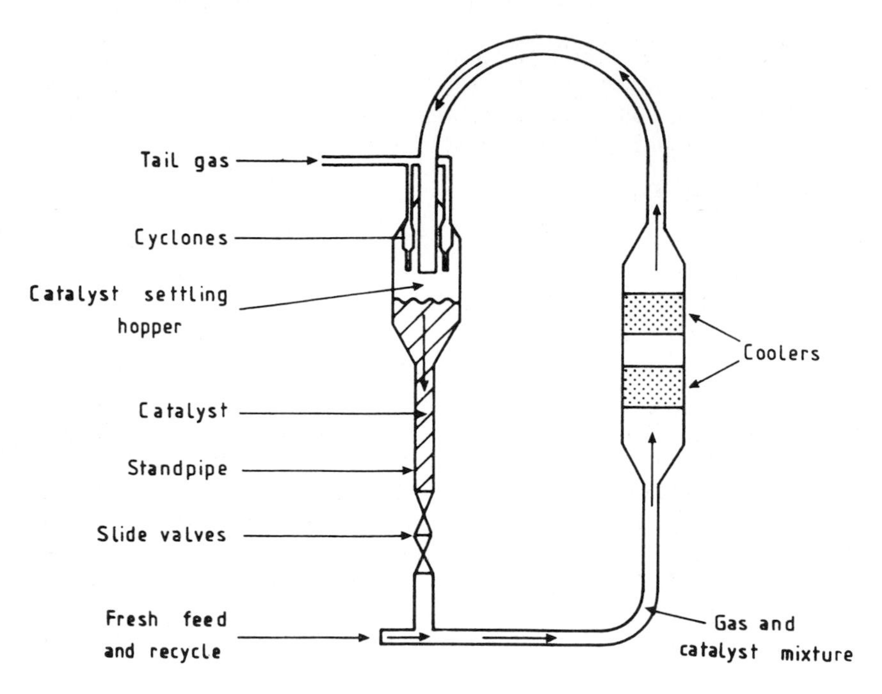

Figure 8.11 Illustration of a typical entrained fluidized-bed reactor.

pressure. As a result of this design configuration and operating procedure, the reactor operates (to a first approximation) isothermally. On the other hand, an entrained fluidized-bed reactor (Fig. 8.11) produces mainly gasoline (C_5–C_{11} paraffins) together with some lower-molecular-weight homologues, but only relatively small quantities of waxes. The iron catalyst used is finely divided and in the circulating fluidized-bed arrangement the catalyst flows downwards through a standpipe to meet this incoming feed introduced at gas velocities of about $100\,000\,m^3\,h^{-1}$. The finely divided catalyst particles are entrained by the gas flow and pass upwards through the reaction zone which contains heat exchangers to remove the heat of reaction. The catalyst and gas disengage in the settler above the standpipe, the gaseous product passing through cyclones to be collected and undergo further processes while the fine particles are returned to the standpipe. Some product gas is allowed to return with the particulate phase to be recycled through the reaction zone. The unit operates at 20–25 bar and the reaction zone temperature is 320–360 °C. Conversions of the carbon monoxide and carbon dioxide feed amount to some 85%.

8.2.4 Reductive Coupling of Two CO Ligands Forming Acetylene from Syn-Gas

In view of acetylene's central importance in the chemical industry, it would be a matter of great practical convenience if it could be prepared straightforwardly from

$CO + H_2$ mixtures. Thermodynamically, the prospects of doing so are exceedingly bleak if water is eliminated as a by-product (Eq. (17)); but they are real if CO_2 is the by-product. It can be shown that, at 1 atm pressure and 100 °C, some 25% of the CO may, in principle, be converted to C_2H_2 upon CO_2 elimination:

$$4\,CO + H_2 \rightarrow C_2H_2 + 2\,CO_2 \tag{27}$$

Laboratory tests indicate that it should indeed be possible to achieve this desirable end, and that a route of some promise is one involving reductive coupling of CO to yield multiply bonded carbon–carbon fragments.

Metals of the transition, lanthanide and actinide series can certainly convert CO ligands to multiply bonded C–C fragments; and alkali metals favour the production of entities such as $(M^+)(-OC{\equiv}CO-)(M^+)$ and its oligomers. Recent work by Lippard et al. has led to the identification of the factors that favour reductive coupling, a necessary step in generating C_2H_2 bound to the catalyst: (i) an electron-rich centre; (ii) Lewis acid coordination of the heteroatoms of the ligands to be coupled; (iii) a molecular geometry in which the orbitals involved in the coupling reaction are appropriately aligned; and (iv) a high coordination number. With these strategic points in mind the complex shown in Fig. 8.12 was identified as the product of the reductive coupling of CO in the high-coordinate metal carbonyl $[Ta(CO)_2(dmpe)_2Cl]$, where dmpe stands for 1,2-bis(dimethylphosphino)ethane. The bis(trimethylsiloxy)ethyne

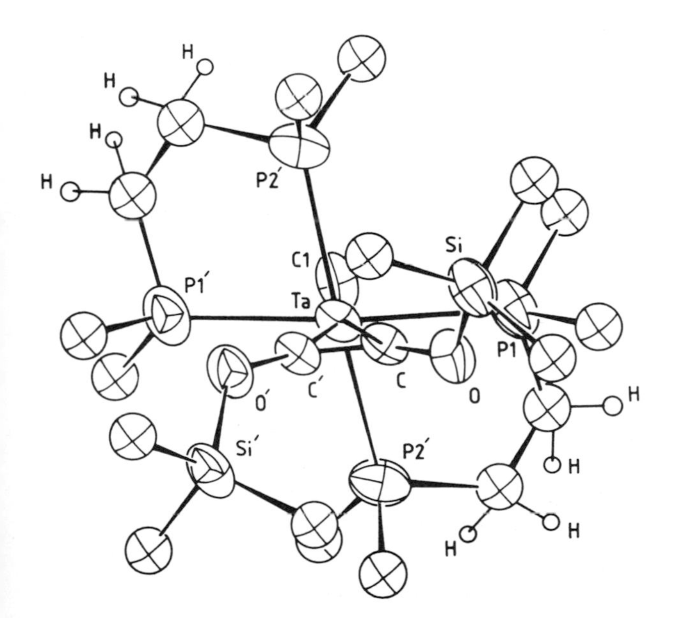

Figure 8.12 Structure of $[Ta(Me_3SiOC{-}COSiMe_3)(dmpe)_2\,Cl]$, where dmpe stands for 1,2-bis(dimethylphosphion)ethane. This embryonic coordinated acetylene is formed by reductive coupling of two CO ligands. (After S. J. Lippard, P. A. Bianconi, I. D. Williams, M. P. Engeler, *J. Amer. Chem. Soc.*, **1986**, *108*, 311.)

ligand constitutes a new type of coordinated product derived from CO, and its formation represents a novel synthetic route to an acetylene diether.

Formation of a bound ethyne moiety, essential as it is, does not take us far enough: to achieve a catalytic, as distinct from a stoichiometric, reaction the ethyne needs to be liberated and a continual turnover of CO must be effected at the catalytic centre. Jones, Schlögl and Thomas discovered that a catalyst generated by partially de-intercalating a (first-stage) sandwich compound C_9FeCl_3 with potassium naphthalenide converts a 3 : 1 mixture of H_2/CO (1 atm total pressure) with *ca* 20% efficiency and *ca* 95% selectivity into C_2H_2 at 100 °C. The catalyst itself, not surprisingly, is polyphasic and multicomponent. On the basis of X-ray diffraction, Mössbauer spectroscopy, thermogravimetry, energy-dispersive X-ray emission microanalysis and electron microscopy it was argued that the active catalytic entity probably consists of potassium-promoted Fe_2O_3 supported on a residual intercalate of iron chloride–graphite. The interaction of the minute particles of Fe_2O_3 with the still-intercalated graphite is the prime characteristic which distinguishes the performance of this catalyst from that of a physical mixture of the identified ingredients, which is only slightly active and lacks selectivity. The enhanced interaction between the Fe_2O_3 and the de-intercalated graphite (compared with pure graphite as support) probably results from the lowering of the Fermi level (see Chapter 5) as a direct consequence of residual intercalation: this electronic effect extends throughout the entire support. Subsequent experiments show that the conditions of preparation of this catalyst are rather more critical than was first thought: this is no doubt attributable to the need to achieve the requisite degree of electronic modification of the graphite, which then controls the availability of electrons for reductive coupling at the active site.

8.2.5 Methanation, Steam Reforming and Water-Gas Shift Reactions

Central to the question of syn-gas chemistry are other reactions that entail either the formation or consumption of methane and carbon dioxide. These are particularly important commercially; special catalysts have been developed to facilitate the attainment of the appropriate equilibria.

8.2.5.1 Methanation

The production of methane from syn-gas, by the reverse of reaction (1) (see Eq. (21)), is of industrial significance whenever coal and biomass are plentiful and when the price of natural gas soars. This very reaction is also relevant whenever the need arises to eliminate low concentrations of CO in a mixture with H_2. Traces of CO are to be avoided in some hydrogenation reactions because it tends to poison the hydrogenation catalyst. Moreover, in substitute natural gas (SNG), the CO content must not exceed about 0.1%; otherwise it constitutes a health hazard. At one time, especially in ammonia synthesis, in which the iron catalyst is prone to poisoning by CO, vestigial traces of this gas were removed by refrigeration involving distillation and absorption in a liquid-nitrogen bath. Nowadays, however, alumina-supported nickel catalysts are invariably employed; but, in ammonia synthesis, most of the residual

CO is removed beforehand, by high-temperature and low-temperature water-gas shift converters (Section 8.2.5.3).

The effluent concentration from the low-temperature shift converter and absorption unit amounts to between 0.5 and 1.0% of CO and rather smaller concentrations of CO_2. These levels of CO and CO_2 are too high to be tolerated by the ammonia synthesis catalyst and are reduced to a few parts per million concentration by passage over the nickel/Al_2O_3 catalyst at 300–350 °C and pressures appropriate to downstream requirements (anything ranging from 0.1 to 30 MPa or more). Gas flows are typically 1000–2000 volumes of feed per volume of catalyst bed per hour at atmospheric pressure or 20 000 volumes per volume per hour at the higher pressures. The catalyst is thus capable of tolerating a wide variety of conditions. The precise mechanism by which CO_2 is converted to methane is the subject of controversy, but is stoichiometrically represented as

$$CO_2 + 4\,H_2 \rightarrow CH_4 + 2\,H_2O \tag{28}$$

The methanation reactions are very exothermic and a temperature rise of approximately 57 °C is expected for each 1% of CO in the feed gas. The feed gas temperature must consequently be regulated to a pre-calculated level so as to avoid excessive temperature rise in the methanation reactor. Exit temperatures should not exceed 440 °C, otherwise the equilibria become unfavourable. For purposes of CO and CO_2 removal from effluent gases containing low concentrations of these oxides, the nickel content of the catalyst is about 20–25 wt.% (expressed as NiO). Typical surface areas in the oxide form are 30–80 $m^2\,g^{-1}$.

So far as SNG production is concerned, concentrations of CO, CO_2 and H_2 are much greater than the few per cent present from low-temperature shift converters. Very large temperature increases would result, therefore, if the catalysed reaction were allowed to proceed in a reactor without some recycle of the reactor exit gases. A high recycle ratio of partially cooled product gas to feed gas is thus employed when synthesizing methane for use as a gaseous fuel. Pressures are usually between 10 and 30 bar and inlet temperatures are a little in excess of the ignition temperature (which for these operating conditions is about 200 °C). Nickel carbonyl formation is also avoided by operation above 200 °C. A rather lower proportion of nickel is present in a methanation catalyst (*ca* 10–15 wt.%) than is present in a CO-removal catalyst.

Single-stage adiabatic fixed-bed reactors, with or without recycle, are generally used for methanation or CO-removal processes. The catalyst is fairly stable and has an expected life of several years. When conditions within a methanation reactor are perturbed so as to initiate carbon deposition, deactivation of the catalyst occurs by the formation of carbidic precursors leading to undesirable carbon deposition.

8.2.5.2 Steam Reforming

The catalytic processes known as steam reforming yield gas mixtures containing H_2, CH_4, CO and CO_2 in various proportions, depending upon the application for which the products are required. Steam-reforming plants may use either natural gas (predominantly CH_4) or naphtha (light petroleum distillate boiling at *ca* 150 °C and consisting chiefly of saturated hydrocarbons) as feedstock. The term 'steam reforming' is rather unfortunate and is really a misnomer: it should not be confused with catalytic reforming. The relevant reactions are:

$$CH_4 + H_2O \rightleftharpoons CO + 3\,H_2 \tag{1}$$

the water-gas shift:

$$CO + H_2O \rightleftharpoons CO_2 + H_2 \tag{29}$$

the reverse of Eq. (17):

$$C_nH_{2n+2} + n\,H_2O \rightarrow n\,CO + (2n+1)\,H_2$$

and the methanation process, the reverse of Eq. (1). One of the principal uses of the steam reforming process is in the preparation of syn-gas for methanol synthesis, and for hydrogenation reactions (after the CO has been removed), as in the synthesis of ammonia from natural gas. Depending upon the end-use, operating conditions in a steam reforming plant are adjusted so as to optimize the composition of the emergent gas. As the net reforming reactions are endothermic, heat from a furnace fired by the combustion of natural gas or naphtha is necessary to sustain the overall progress.

The choice of operating conditions within a steam reformer must be reached by due consideration of the thermodynamics and the material balance. Moreover, it is particularly important that the two equilibria

$$CO \rightleftharpoons C + CO_2 \tag{30}$$

and

$$CH_4 \rightleftharpoons C + 2\,H_2 \tag{31}$$

be kept well to the left, so that the deposition of carbon, which is deleterious to the catalyst, is avoided. This is usually achieved by employing enough steam to react with the hydrocarbon in the feed to ensure sufficiently high CO_2-to-CO and H_2-to-CH_4 ratios so that carbon is not formed. Although appropriate conditions can be predicted thermodynamically, the form of carbon which is produced (known as Dent carbon) is more active than graphite. Predictions based on free energy changes for the carbon monoxide and methane disproportionations may thus be erroneous, leading to underestimates of the steam ratio required. Equilibrium considerations aside, carbon could be formed at the catalyst surface and in pores if reactions leading to its formation (such as the methane disproportionation) are intrinsically faster than gasification reactions such as

$$H_2O + C \rightarrow CO + H_2 \tag{32}$$

resulting in its disappearance. These kinetic obstacles to the selection of appropriate equilibrium conditions can be overcome by the addition of promoters to the catalyst.

For the purpose of ammonia production, two successive catalytic steam reformers are necessary. In the primary reformer, the principal equilibria are estabilished, resulting in CO, CO_2, H_2 and CH_4 formation. In the secondary reformer, the equilibria are adjusted so that only small quantities of CH_4 remain in the synthesis gas mixture. Although CH_4 is inert so far as the ammonia synthesis is concerned, the amount of CH_4 in the recycle gas in the ammonia synthesis loop will accumulate and a purge from the loop is necessary. Thus any CH_4 remaining in the secondary reformer effluent influences the purge flow rate and the consequent loss of synthesis gas. Air is also introduced in the secondary reformer so that N_2 is present in the emergent gases in the appropriate stoichiometric ratio for synthesis. Oxygen present in the air serves to provide some of the heat consumed in the forming reactions by the combustion of a portion of the hydrogen and hydrocarbons present.

Temperatures for a steam reformer lie in the range from 400 °C to as high as 875 °C. For SNG production, temperatures are usually 400–550 °C, but ammonia synthesis requires the gas mixture to be lean in CH_4 and a temperature of about 800 °C is desirable. Operating pressures are often dictated by the downstream requirements rather than by thermodynamic equilibria. Thus, in ammonia synthesis plants the steam reformer operates at about 3 MPa (30 bar). Although this choice appears to contradict the dictates of the equilibria for a relatively low operating pressure, nevertheless costly compression requirements prior to the ammonia converter are reduced if some compression is arranged in the reformer feed and advantage is taken of the volume increase in the reforming gases on reaction.

CH_4 or naphtha feeds to a steam-reforming plant must be free of hydrogen sulphide and organic sulphur compounds as these are poisons to the steam-reforming catalyst. Consequently, they are removed by processes such as hydrodesulphurisation, adsorption or absorption. Before 1965, nickel supported on a calcium aluminosilicate base was popular, but its use was discontinued because volatilisation of silica in steam tended to occur and its subsequent deposition caused fouling of the heat-exchanger surfaces. Eventually, steam-reforming catalysts were prepared using the stable α-Al_2O_3 as a support for nickel present at about 12–20 wt.%. This proved to be rugged, having mechanical strength and long life. Potassium compounds or those of other alkali metal are sometimes added as promoters and these accelerate reaction, thereby reducing the degree of loss of activity by carbon deposition. A catalyst containing 32 wt.% NiO, 14 wt.% CaO and 54 wt.% Al_2O_3 with a low silica content is satisfactory up to 30 bar and 859 °C for a natural gas feed. For naphtha feeds, the incorporation of alkali metals (K_2O, 7 wt.% and MgO, 13 wt.%) suppresses carbon formation. Even so, if naphthas containing more than about 2 mol% of olefins are used for steam reforming, extensive carbon deposition may occur, especially if the steam-to-hydrocarbon ratio is insufficiently high. A catalyst preparation reported as suitable for use in secondary steam reforming (when air is introduced and the operating temperature is consequently higher) contains no alkali additions and nickel (analysed as NiO) is present at 18 wt.% on a calcium aluminate base. The amount of nickel present in the catalyst used in the secondary reformer is less than that in the catalyst for the primary reformer because a lower activity is now tolerable, the mean operating temperature being increased by virtue of addition of air.

The life of steam-reforming catalysts may be as high as several years. However, catalyst activity declines gradually and, concomitant with this decrease, the reformer operating temperature is increased to compensate for reduced reaction rate (see Section 7.3.3). In turn, albeit after a fairly long operating life, hot regions (referred to as 'hot spots' in the parlance of process technology) develop along the catalyst tubes, caused by localized decrements in the rate of endothermic reactions which absorb heat from the firing furnace. Ultimately, metallurgical failure may result, with drastic consequences. Plant shutdowns during normal maintenance periods are therefore arranged, either to regenerate the catalyst with steam or to replace it entirely. Sulphur contents in reformer feeds must generally not exceed 0.2 ppm, otherwise irreversible poisoning of the catalyst results.

So far as the operating conditions of steam-reforming reactors are concerned, current practice is to use temperatures in the range 850–950 °C and pressures of 3–3.5 MPa. As already noted, these are chosen to conform with downstream processing requirements. Consequently, temperatures are maintained at this high level to ensure satisfactory reaction rates. Temperatures, and therefore pressures, are limited to this range by the metallurgical properties of the tubes containing the catalyst. These material strength limitations have led to a reactor design which incorporates large amounts of catalyst in banks of vertical parallel tubes of relatively small diameter (in the approximate range 0.07–0.15 m) and heights as much as 12 m. Pressure drops in these tubes will be from 0.1–0.5 MPa, depending on tube length and catalyst pellet size.

Steam necessary for the production of synthesis gas is formed endothermically from feed water. Steam-reforming reactors are designed in such a way as to minimize steam requirements, not only for the reforming reactions themselves but also for other downstream utilities in integrated plant, as is the practice in ammonia production. Arising from these pragmatic needs, the primary reformer has evolved into a design which consists of two principal sections. The first is a firebar containing burners to supply radiant energy from the combustion of fuel gas, and the banks of vertical catalyst-packed tubes within which the hydrocarbon–steam mixture flows and reacts. The second is a vertical converter along which the hot synthesis gas product passes and which contains a number of heat exchangers by means of which other process heat and steam requirements can be met. High temperatures are maintained in the firebox so that heat fluxes of the order of 60 kW m^{-2} can be supplied to the reformer furnace tubes. About 50% of the available heat is provided by radiant heat in the firebox, the other half being drawn from the flue gas, which supplies sensible heat to the feed gas through a heat exchanger. The furnace walls are lined with refractory bricks.

As the performance of the primary reforming furnace is usually critical to the successful operation of any downstream plant, the furnace is provided generously with instrumentation, especially for checking catalyst tube temperatures, where excessively hot regions may develop. Such a local hot region may result from several reasons. It may indicate poor burner adjustment, excessive heating from flame impingement on the catalyst tube wall, too high a heat flux or insufficient draught through the firebox, resulting in poor heat distribution. Alternatively, a hot spot may arise from conditions in the process stream rather than in the furnace; for example, a coked catalyst, channelling of reactants due to poor catalyst distribution, a blocked tube, or an inactive catalyst will all result in local overheating in the process stream, where insufficient heat will be removed by a suppressed endothermic reaction. Careful control of conditions can avoid many of these unwelcome situations from arising and a programme of regular maintenance and occasional shutdown is usually necessary.

8.2.5.3 Water-Gas Shift Reaction

This is the reaction represented by Eq. (29); it has been well known ever since it gained popularity when the production of domestic and industrial fuel in the 19th century was from the pyrolysis of coals. Nevertheless, catalysts for the reaction in the temperature range 250–550 °C were not developed until it became economically viable to generate ammonia from natural gas and naphthas (Section 8.3).

As we shall see, common practice in ammonia technology is to employ a high-temperature shift converter, containing an iron oxide catalyst, followed by a low-temperature shift converter containing a copper and zinc oxide catalyst. The purpose of both water-gas shift converters is concomitantly to reduce the amount of CO in the synthesis gas from the secondary steam reformer (Section 8.2.5.2) in an ammonia processing plant and to increase the proportion of hydrogen in the synthesis gas. A typical gas mixture emerging from the secondary reformer might contain about 9% CO and 6% CO_2 (the composition will depend upon the feedstock and process conditions employed to generate the synthesis gas); the steam required to convert all of this CO to CO_2 in a single-stage reactor would be excessive and certainly uneconomic (e.g., if 98% of the CO were converted to CO_2 at 260 °C, the steam requirement would amount to 2.4 kg m^{-3} of feed gas). Consequently, the water-gas shift reaction is allowed to take place in two stages with 85–90% conversion in the

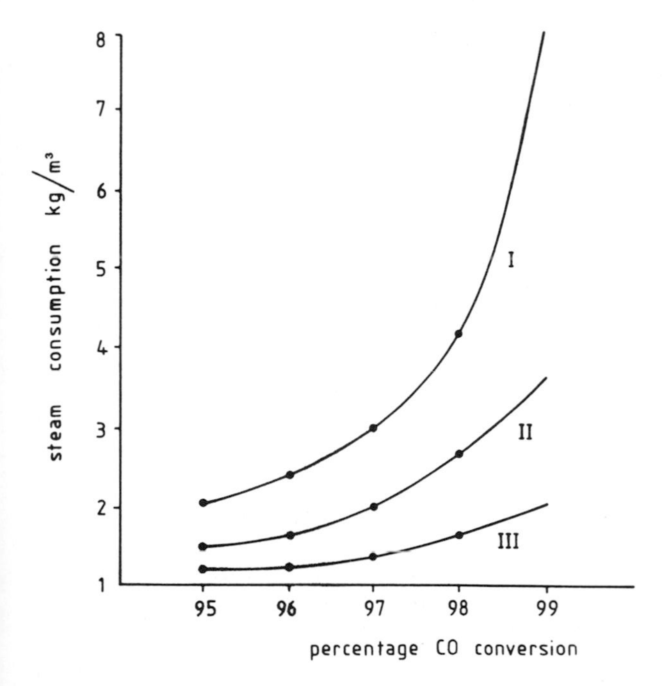

Figure 8.13 Steam requirements for water-gas shift reactions. Curve I, single-stage converter; II, two-stage converter with equal extent of conversion for each stage; III, two-stage converter with equal enthalpy change for each stage.

first stage (operated at 300–500 °C to achieve an acceptable reaction rate) and the balance occurring in one or two succeeding stages at a lower temperature favouring complete conversion. The steam requirement is seen, from Fig. 8.13, to be considerably less for a two-stage conversion with equal change of enthalpy between stages than for either a single-stage conversion or a two-stage conversion with equal extent of conversion between stages.

To match the process requirement for both a high-temperature and a low-temperature shift converter, two forms of catalyst were developed. The high-temperature shift catalyst is designed to operate between temperatures of 350 and 550 °C and reduce the CO content, for example, produced in primary and secondary reformers (Section 8.2.5.2) from about 9% to between 1.5 and 4.0%. The high-temperature shift catalyst is invariably based on iron oxide with some chromium oxide added as a textural promoter. Normally, the catalyst is unsupported and is prepared by precipitation from a ferric salt and subsequently tableted or pelleted. The iron is usually in the form of Fe_2O_3 when supplied by manufacturers, but is reduced within the shift converter to the Fe_3O_4 phase and is active in this form. The composition of the catalyst is in the region of 55 wt.% of iron and 5–6 wt.% of Cr. It is a fairly rugged catalyst and has been employed in processes ranging from atmospheric pressure to 30 bar with gases containing as much as 60% CO and 5×10^3 ppm H_2S. However, the catalyst is pyrophoric in its active form and when exposed to air (e.g. for reactor maintenance) it has to be stabilized by oxidation with an inert gas containing a low concentration of oxygen. If sufficient steam is not employed for the shift conversion, the catalyst can become partially deactivated by gradual reduction to iron, which would catalyse the formation of carbon from carbon monoxide. Excess steam as diluent also serves to control the temperature rise, due to exothermic reaction within the high-temperature shift converter.

A kinetic rate equation established for a commercial catalyst, similar in some respects to the catalyst described above, is

$$R = \frac{k(p_{CO}p_{H_2O} - kp_{CO_2}p_{H_2})}{1 + k_{CO}p_{CO} + k_{H_2O}p_{H_2O} + k_{H_2}p_{H_2} + k_{CO_2}p_{CO_2}} \tag{33}$$

and obviously reflects Langmuir–Hinshelwood kinetics (see Chapter 2 and Section 7.2.3), with each of the reactants and products responsible to some extent for retardation. Such an equation could arise from a mechanism which involves the interaction between adsorbed CO and adsorbed H_2O (either in dissociated or undissociated form). A power-law (Section 7.2.3) type of rate equation has also been proposed:

$$R = kp_{CO}{}^{0.9}p_{H_2O}{}^{0.25}p_{CO_2}{}^{0.6} \tag{34}$$

which leads to an apparent activation energy of 114 kJ mol^{-1}.

The low-temperature shift catalyst includes oxides of zinc, copper, and chromium. Its function is to reduce the CO content emerging from the high-temperature shift converter from about 1.5–4.0% down to as little as 0.5% or less. To achieve such a decrease, thermodynamic considerations dictate that the temperature should be in the region of 200–300 °C. Catalyst preparations, consisting of wt.% 30 CuO,

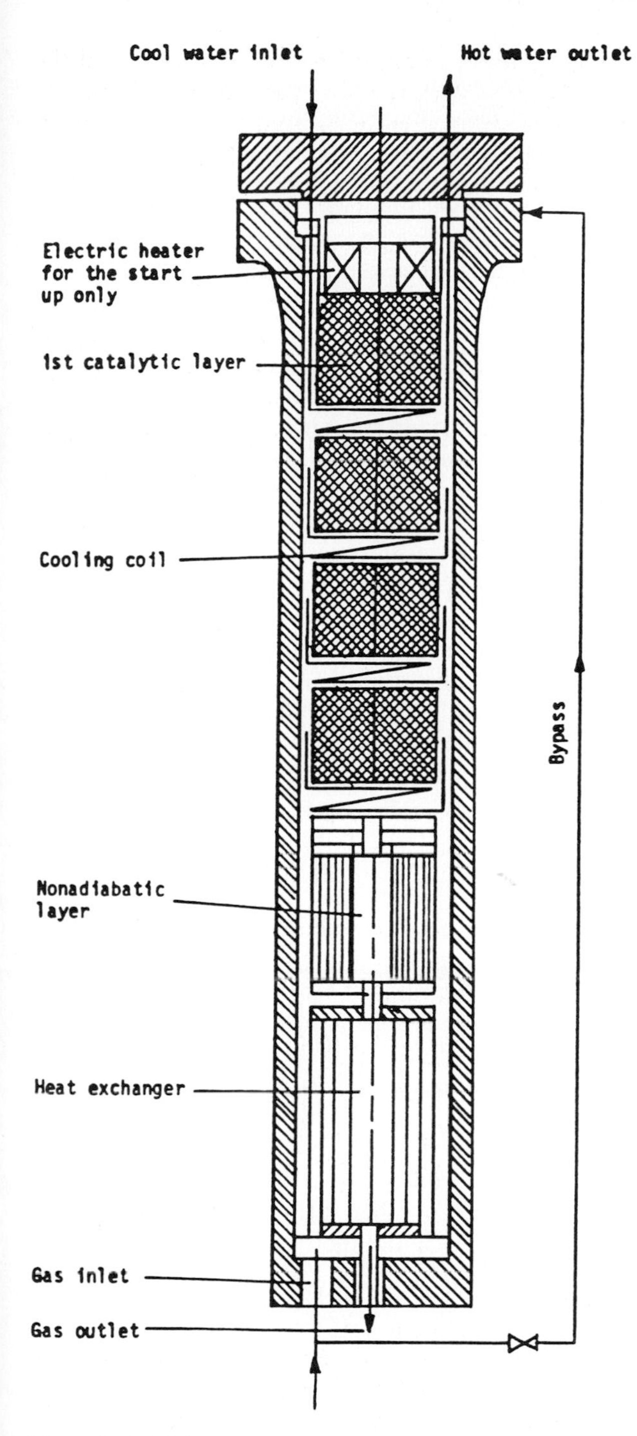

Figure 8.14 High-temperature shift converter.

45 ZnO and 13 Al_2O_3 (Cr_2O_3 is sometimes substituted for Al_2O_3), and which are reduced in H_2 prior to use, effect this reduction in CO concentration at these low temperatures, albeit at an acceptable rate. The active form of the catalyst is metallic copper. It is desirable that the copper crystallites be as small as possible to sustain stability and activity, so carefully controlled catalyst preparation is necessary, usually by alkaline precipitation from a solution of aluminum, zinc and copper salts. The catalyst becomes very pyrophoric following use and must be removed from the catalytic reactor under a blanket of nitrogen. It is susceptible to irreversible poisoning by the compounds of sulphur and chlorine, which must be absent from the feed to the low-temperature shift converter.

The high-temperature shift converter is usually constructed in the form of a two- or three-stage catalyst bed, each bed supported on a perforated tray, and contained within a stainless steel shell fabricated to withstand the high pressures usually demanded in ammonia and petrochemical processing. The feed mixture (e.g. a mixture of CO, CO_2, H_2O, H_2, N_2 and a small amount of CH_4 if the feed comes from a secondary reformer in an ammonia plant) is first passed upwards through a heat exchanger (a composite part of the reactor) which transfers heat from the hot exit gases to the incoming feed. The hot feed is then transported to the first catalyst bed incorporated at the top of the reactor, and passes downwards through this bed and subsequently to the second and third beds. The steam requirement for the reaction can be reduced by injection of hot water condensate into the gas stream between each catalyst bed; this also serves to control the temperature rise across each bed due to exothermic reaction. Figure 8.14 is a diagrammatic representation of the high-temperature shift converter.

8.3 The Synthesis of Ammonia

The current annual world capacity for the industrial production of ammonia exceeds 100 million tons, enough to provide nourishment, through its use as agricultural fertilizer, for 5×10^{11} man years. Remarkably, the catalyst employed for the synthesis of ammonia from H_2 and N_2 is almost identical to that formulated in a German patent in 1910, a composite of iron (predominantly) and the oxides of aluminum, calcium and potassium, which serve as promoters. It was Fritz Haber who first demonstrated, on 2 July 1909 in Karlsruhe, the catalytic viability of ammonia production by the high-pressure conversion of H_2 and N_2. He obtained 90 g of ammonia per hour. But the discovery of the promoted iron catalyst – still arguably the best commercial one even though more than 100 000 other formulations have by now been investigated – was made by Bosch, Mittasch and their co-workers in the laboratories of Badische Anilin und Soda Fabrik (BASF) in 1909. By 1917 more than 60 000 tons of ammonia were produced annually by means of what is nowadays termed the Haber–Bosch process.

Ironically, the first experiments with iron catalysts at BASF proved disappointing; but quite encouraging yields of ammonia were obtained using a Swedish magnetite from Gällivare. It was quickly recognized that the presence of minor impurities in the magnetite was crucial. This prompted systematic testing of a range of iron composites, and led to the realization – a turning point in heterogeneous catalysis since it is of widespread validity – that, in Mittasch's words, 'The winning catalyst is a multicomponent system.' One of the most widely used ammonia synthesis catalysts these days (BASF-S6-10) is prepared from Fe_3O_4 (magnetite) which is fused with a few per cent of K_2O, Al_2O_3 and CaO, and subsequently reduced. As we shall see later, the source of H_2 for modern ammonia synthesis plants is usually natural gas; originally the H_2 came from the steam gasification of coke.

It is obvious that the reaction

$$3\,H_2(g) + N_2(g) \rightleftharpoons 2\,NH_3 \qquad -\Delta H_{773K} = 109\,\text{kJ mol}^{-1} \tag{35}$$

is favoured, thermodynamically, at high pressures and low temperatures. In practice, to ensure that rates of conversion are sufficiently rapid, quite elevated temperatures (typically 450 °C) are used and pressures are generally in the region of 100–300 bar.

8.3.1 Catalyst Promoters are of Two Kinds

A variety of studies, to which we shall allude below, suggests that the Al_2O_3 functions as a 'structural' promoter and preserves a high area in the active iron catalyst formed from the magnetite during reduction in the reactant mixture. A similar role seems to be served by CaO. Neither of these oxides, unlike the Fe_3O_4, is reduced to the metal under reaction conditions. Neither is K_2O; but its role is thought to be very different. The heat of adsorption of molecular nitrogen is locally increased at the iron surface in the vicinity of adsorbed potassium, and there is a simultaneous lowering of the activation barrier for dissociation of N_2. K_2O, therefore, exercises more of an electronic or chemical rather than a textural or physical influence. The evidence for subscribing to these views, relating to the two distinct modes of action of promoters, is given later (Section 8.3.3.4).

8.3.2 Kinetics of the Overall Reaction: The Temkin–Pyzhev Description

It is observed experimentally that the rate of production of ammonia, over a wide range of conditions during the synthesis, may be expressed thus:

$$\frac{dp_{NH_3}}{dt} = kp_{N_2}\left[\frac{p^3{}_{H_2}}{p^2{}_{NH_3}}\right]^i \tag{36}$$

with values of i ranging between 0.50 and 0.67. It is possible to understand how Eq. (36) holds good without specifying whether N_2 is bound to the catalyst in a molecular or dissociated state. To do so, we first note that, at constant temperature, the rate of adsorption, R_a, of N_2 may be expressed by the Elovich equation (see Chapter 2):

$$R_a = k_a p_{N_2} \exp(-g\theta_N) \tag{37}$$

Likewise, the rate of desorption, R_d, is:

$$R_d = k_d \exp(h\theta_N) \tag{38}$$

Here k_a, k_d, g and h are kinetic constants and θ_N is the fraction of surface covered by nitrogen (without implying that the adsorbed state is atomic). Now suppose that nitrogen adsorption is the rate-determining step in the synthesis of ammonia. This means that the adsorbed nitrogen, which we designate N_2(ad) (without implying that it is bound in a molecular form) is not in equilibrium with N_2(g) during the overall reaction (Eq. (35)), but, instead, is in equilibrium with the gas phase H_2 and NH_3, i.e.

$$N_2(ad) + 3\,H_2(g) \rightleftharpoons 2\,NH_3(g) \tag{39}$$

If we denote the pressure of nitrogen which would be in equilibrium with the instantaneous pressure of hydrogen and ammonia by $p^*_{N_2}$, then

$$p^*_{N_2} = \frac{1}{K}\frac{p^2_{NH_3}}{p^3_{H_2}} \tag{40}$$

where K is the thermodynamic equilibrium constant for reaction (39). In effect, $p^*_{N_2}$ is the pressure of nitrogen which would be in equilibrium with the adsorbed nitrogen. At equilibrium, $R_a = R_d$, so that

$$k_a p_{N_2} \exp(-g\theta_N) = k_d \exp(h\theta_N) \tag{41}$$

Hence,

$$\theta_N = \frac{1}{(g+h)} \ln\left(\frac{k_a}{k_d} p_{N_2}\right) \tag{42}$$

where θ_N and p_{N_2} refer to equilibrium values. But, during the actual synthesis reaction, the fraction of surface covered by nitrogen, θ^r_N is in equilibrium with $p^*_{N_2}$. From equations (40) and (42) we obtain:

$$\theta^r_N = \frac{1}{(g+h)} \ln\left(\frac{k_a}{k_d} p^*_{N_2}\right) \tag{43}$$

This result is synonymous with the Temkin isotherm, $\theta = (1/a) \ln c_0 p$ (compare Section 2.6.2.3).

Substituting from Eq. (40):

$$\theta^r_N = \frac{1}{(g+h)} \ln \frac{k_a}{k_d} \frac{p^2_{NH_3}}{K p^3_{H_2}} \tag{44}$$

Since, as stated earlier, the rate of nitrogen adsorption is rate-determining, we may now write from Eqs. (37) and (44):

$$\begin{aligned}
\frac{dp_{NH_3}}{dt} &= k_a p_{N_2} \exp(-g\theta'_N) = k_a p_{N_2} \exp\left[-\frac{g}{(g+h)} \ln\left(\frac{k_a}{k_p} \frac{p^2_{NH_3}}{K p^3_{H_2}}\right)\right] \\
&= k_a \left(\frac{k_a}{k_p}\right)^{-g/(g+h)} K^{g/(g+h)} p_{N_2} (p^3_{H_2}/p^2_{NH_3})^{g/(g+h)} \\
&= k p_{N_2} \left(\frac{p^3_{H_2}}{p^2_{NH_3}}\right)^i
\end{aligned} \tag{45}$$

where the constants k and i in Eq. (36) now have their component parts identified. Clearly $i = g/(g+h)$.

We note, in passing, that a similar treatment is valid for ammonia decomposition:

$$\frac{-\mathrm{d}p_{\mathrm{NH_3}}}{\mathrm{d}t} = K_\mathrm{d} \exp(h\theta^r_\mathrm{N}) \tag{46}$$

from which we can extract more meaningful interpretations to observed rate equations such as:

$$\frac{-\mathrm{d}p_{\mathrm{NH_3}}}{\mathrm{d}t} = kp^x_{\mathrm{NH_3}} p^y_{\mathrm{H_2}} \tag{47}$$

where, usually, x is a small positive number less than unity and y ranges from -0.7 to -1.5.

Valuable as the Temkin–Pyzhev description in particular, and global expressions for rates of reactions in general, are – especially in chemical engineering contexts – such relationships do not permit us to deduce the nature of the primary steps in the catalytic synthesis of ammonia. Overall kinetic formulations, by themselves, afford little insight into the detailed mechanisms of heterogeneously catalyzed reactions. To glean the relevant information we rely more on spectroscopic, diffraction and other specific studies of the catalyst, the salient features of which we assess in the following sections.

8.3.3 The Surface of Iron Catalysts for Ammonia Synthesis Contain Several Other Elements: But is the Iron Crystalline?

Figure 8.15, taken from the work of Ertl, shows part of the X-ray photoelectron spectrum from a typical unreduced commercial catalyst. Even without taking cognizance of variations in the respective photoelectric cross-sections and electron escape

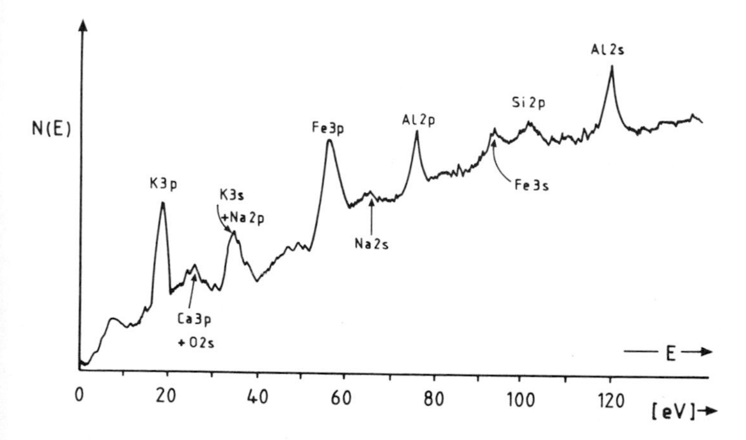

Figure 8.15 X-ray photoelectron spectrum from the surface of an unreduced commercial catalyst for ammonia synthesis. (After G. Ertl, *Angew. Chem. Int. Ed. Engl.*, **1990**, *29*, 1219.)

Table 8.3 Composition [atom%] of a promoted iron catalyst in per cent (after G. Ertl, D. Brigge, R. Schlögl, M. Weiss, *J. Catal.*, **1983**, *79*, 359).

	Fe	K	Al	Ca	O
Bulk composition	40.5	0.35	2.0	1.7	53.2
AES					
Before reduction	8.6	36.1	10.7	4.7	40.0
After reduction	11.0	27.0	17.0	4.0	41.0

depths, it is clear that the surface region of the initial catalyst has only a rather low iron content but quite high concentrations of other elements such as potassium, aluminum, silicon, etc., the nominal bulk concentrations of which are fairly low and which therefore exhibit a marked tendency for surface segregation. For the BASF S6-10 catalyst, Auger electron spectroscopic (AES) analysis yields the results shown in Table 8.3.

High-temperature (400 °C) treatment in the H_2/N_2 reaction mixture reduces the Fe_3O_4 originally present in the metallic state (as is clearly proved by the shift to lower binding energy of the core electron, Fe $2p_{3/2}$); but the valence state of the other cations (Al^{3+}, Ca^{2+} and K^+) remains unaffected. Accumulated experimental evidence points to the development of a porous iron catalyst, the high surface area resulting from the fact that Al_2O_3 and, to a lesser degree, CaO act as a structural promoter and prevent the sintering of small particles of α-Fe. Views taken in a scanning electron microscope, as well as 'element maps' obtained either by electron-induced X-ray of Auger electron emission, support this interpretation, although some care is needed since most post-mortem studies of this kind tend to be carried out after the cooled, 'expired' catalyst has been exposed to air.

It has been assumed, partly on the basis of ex-situ X-ray diffraction studies, that individual spinel units ($FeAl_2O_4$) are built into the active α-Fe catalyst and that these units cause internal strain giving rise to the formation of so-called paracrystals. The tacit assumption has also been made, recognizing that reduction of the original catalyst increases considerably the surface concentration of iron, that metallic iron is the crucial catalytically active surface component (and therefore that clean iron single-crystal surfaces are legitimate starting points for model studies).

Although in-situ X-ray powder diffractometric studies by Rayment et al., carried out with high pressures of either H_2 or H_2/N_2 mixtures, did at one time suggest that the active state of the promoted iron catalyst is largely non-crystalline, re-investigations by the same workers indicated that the crystalline α-Fe phase is the dominantly active component of the catalyst, with the (111) face being prominent. One of the intriguing facts pertaining to ammonia synthesis is that fresh catalysts and those that have been in continuous use industrially for 14 years exhibit (Fig. 8.16) essentially the same catalytic activity. This remarkable longevity, it has been suggested, stems perhaps from the quasi-liquid nature of the catalyst, since such material would be capable of constant regeneration of surface atoms so that traces of catalyst poison could be removed from the interface at which the synthesis ensues.

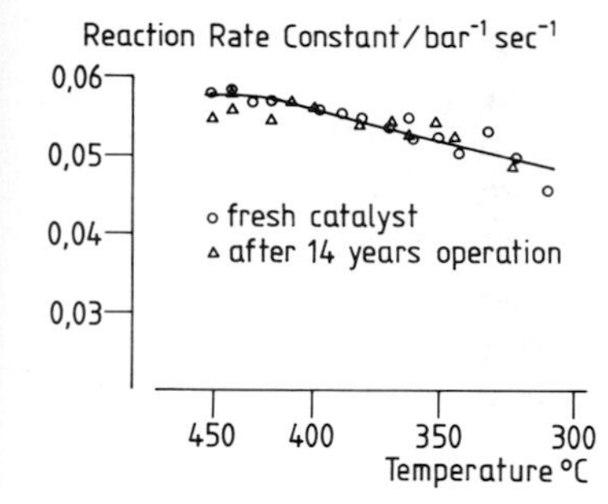

Figure 8.16 BASF catalyst S6–10 for the synthesis of ammonia. (After B. Trimm, *Proc. 8th Intl. Congr. Catalysis, Berlin 1984*, Verlag Chemie, Weinheim, **1984**, Vol. 1, p. 7.)

8.3.3.1 Does Ammonia Synthesis Proceed via Atomically or Molecularly Adsorbed Nitrogen?

We now seek to settle the issue touched upon earlier but which, in the Temkin–Pyzhev description for example, requires no specific answer, i.e. whether an atomic mechanism or a molecular one more appropriately describes the elementary processes at the catalyst surface. This topic has been extensively debated – many years have passed since Horiuti raised the question as to whether ammonia synthesis proceeds via atomic nitrogen or a hydrazine intermediate.

Recognizing the difficulty of directly monitoring the adsorbed phase under catalytic conditions (e.g. 100 bar, 450 °C), Ertl and co-workers devised an ingenious post-mortem method of answering this question. First they established the thermal stability (indicated by the temperatures T_{des} above which desorption of the respective species occurs if the sample is kept in vacuo) of the reactants:

$$H_2 \overset{K_1}{\rightleftharpoons} 2H_a \qquad T_{des} < 200\,°C \tag{48}$$

$$N_2 \overset{K_2}{\rightleftharpoons} N_2(ad) \qquad T_{des} < -100\,°C \tag{49}$$

$$N_2(ad) \overset{K_3}{\rightleftharpoons} 2N(ad) \qquad T_{des}\ 450\,°C \tag{50}$$

$$NH_3 \rightleftharpoons NH_3(ad) \qquad T_{des} < 100\,°C \tag{51}$$

From these data, it is evident that if an iron surface is operated in a H_2/N_2 mixture above 200 °C and the gas phase is subsequently pumped away while the surface is still hot, surface analysis (by AES or XPS, for example) will yield the concentration of N(ad) which was present *under working* (catalytic) *conditions*, whereas all the other species are eliminated.

Clearly, under working conditions, both atomically and molecularly adsorbed nitrogen may be present on the iron. Ammonia formation could proceed via either

$$N_2(ad) + 6\,H(ad) \rightarrow 2\,NH_3 \quad \text{(molecular route)} \tag{52}$$

or

$$N(ad) + 3\,H(ad) \rightarrow NH_3 \quad \text{(atomic route)} \tag{53}$$

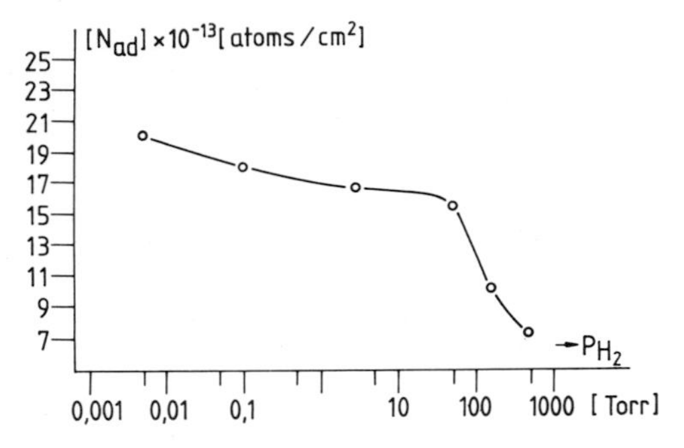

Figure 8.17 Variation of the surface concentration of atomic nitrogen ($[N_{ad}]$ in the diagram; N_a in the text), in arbitrary units, of a Fe (111) surface with the H_2 pressure after treatment in N_2/H_2 mixtures with a constant N_2 pressure of 150 Torr at 580 K. (After G. Ertl, *Catalysis*, **1983**, *4*, 210.)

A decision can be readily reached by recording the surface concentrations, N(ad), under the conditions just described. Bear in mind that N(ad) is formed via $N_2(g) \rightleftharpoons N_2(ad) \rightarrow 2\,N(ad)$.

(a) If the synthesis proceeds through the molecular route, the surface would necessarily be saturated with N(ad) under steady-state conditions, since there would exist no channel through which this species could be removed.

(b) If, on the other hand, synthesis proceeds through the atomic route, then, according to the above-mentioned reaction sequence, the steady-state concentration of N(ad) follows from

$$\frac{d[N(ad)]}{dt} = 0 = k_3[N_2(ad)] - k_0[N(ad)][H(ad)]^r$$

or

$$= k_3 k_2 p_{N_2} - k_0 K_1 [N(ad)] p_{H_2}^q$$

$$[N(ad)]_{stat} = k \frac{p_{N_2}}{p_{H_2}^q} \quad (54)$$

where q is an unknown (variable) exponent including the reaction order r and the shape of the adsorption isotherm relating [H(ad)] to P_{H_2}. It follows that, at a given temperature and partial pressure of nitrogen the stationary surface concentration of atomic nitrogen should decrease with increasing pressure of hydrogen.

Figure 8.17 shows the results obtained with a single-crystal iron surface at 310 °C, the N(ad) concentration being determined by AES. The data unambiguously favour the atomic route. Moreover, in a stoichiometric mixture (150 : 450 Torr of N_2/H_2) the concentration of N(ad) is seen to have fallen to quite low values, a result to be expected if the formation of N(ad) is the slowest step in the entire reaction sequence.

8.3.3.2 How and Where are the Reactant Gases Adsorbed at the Catalyst Surface?

Bound Nitrogen Forms a 'Surface Nitride'

Plausible arguments, rehearsed already in preceding sections, lead us to expect two forms of bound nitrogen: the molecular and the atomic. The Lennard-Jones diagram

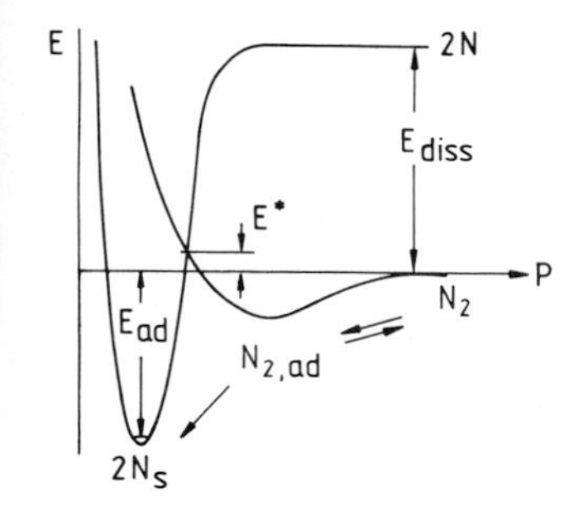

Figure 8.18 Potential-energy diagram for dissociative adsorption of nitrogen on iron. The activation energy E^* varies with surface structure as well as coverage. The adsorption energy E_{ad} is *ca* 200 kJ mol^{-1}. (After G. Ertl, *Angew. Chem. Int. Ed. Engl.*, **1990**, *29*, 1219.)

depicting their relative energies can be readily constructed (Fig. 8.18). To distinguish the relative amounts of the molecular from the atomic form of the bound state, however, a combination of UV-induced photoelectron spectroscopy (UPS) and XPS as well as HREELS is employed. The work of Roberts, Mason, Ibach, Ertl and others clearly reveals the preference for molecularly bound nitrogen at low temperatures and atomic nitrogen at high temperatures. This statement is valid both for the surfaces of single-crystal and polycrystalline iron, and by extrapolation, for the non-crystalline variety also. But most of our quantitative knowledge of the dynamics, siting and energetics of adsorption of nitrogen has emerged from model systems using the high-symmetry faces {111}, {110} and {100} of single-crystal samples pre-cleaned in ultrahigh vacuum. We now summarize this knowledge, recognizing that it may emphasize idealized situations.

First, we note that dissociative nitrogen chemisorption is a very slow process, in line with all that has been stated earlier: it is the chemical rate-determining step in synthesis. The initial sticking coefficient lies in the range 10^{-8} to 10^{-6}. Next, we find that the initial activation energy for dissociative nitrogen adsorption is rather low($<$30 kJ mol^{-1} at zero coverage). The small sticking coefficient is therefore ascribable chiefly to an unusually small pre-exponential factor.

This smallness of the pre-exponential factor arises partly because of a small sticking coefficient (*ca* 10^{-2}) into the molecular state, and partly because of an even smaller probability of N_2(ad) dissociating instead of desorbing again (*ca* 10^{-4}). (It must be remembered that N_2 is isoelectronic with CO: it is expected by analogy, and is often found in practice, to be adsorbed 'end on'.)

With increasing coverage the activation energy, E^* (Figure 8.18), increases to a value of about 50–80 kJ mol^{-1} (depending upon the crystallographic face) at the sub-monolayer coverage at which steady-state synthesis will proceed. This increase of activation energy is the root-cause for the validity of the Elovich equation (Eq. (37)), as the latter holds good under these circumstances. It is also the reason why the Temkin–Pyzhev description, given above, fits aptly the kinetics of ammonia synthesis.

Taking a cue from LEED studies, and in particular, the analysis of I–V data, for atomically bound nitrogen on Fe {100} faces, it is seen (Fig. 8.19) that the N(ad) atoms occupy fourfold hollow sites, their plane being 0.27 Å above the plane of the topmost Fe atoms. This structure, especially the disposition of the N atoms, is quite similar to the geometry of the (002) plane of bulk Fe_4N. And although this *bulk* phase cannot for thermodynamic reasons be formed under the prevailing conditions

(a)

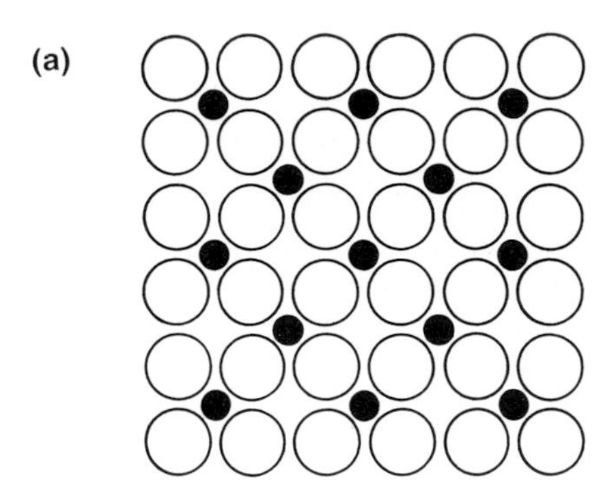

(b)

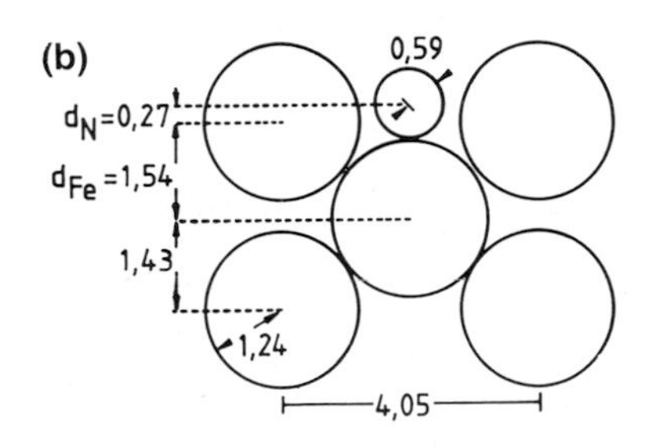

Figure 8.19 Structural model for N/Fe {100}. (a) Top view: open circles denote Fe atoms. (b) Cross-section along the {110} plane; distances in Å. (After G. Ertl, *Angew. Chem. Int. Engl.*, **1990**, *29*, 1219.)

for ammonia synthesis, we note the pedagogically revealing point that atomically adsorbed nitrogen forms a kind of surface nitride.

Surface Hydrogen Atoms Occupy Rows of Adsorption Sites

Like nitrogen, hydrogen is dissociatively adsorbed but with a fairly high initial sticking coefficient (*ca* 0.1) and with adsorption energies between about 60 and 100 kJ mol^{-1}, depending upon the orientation of the single crystal face. Under the conditions of ammonia synthesis the adsorbed hydrogen atoms are very mobile and their concentration is sub-monolayer in magnitude, the precise value being governed by the prevailing adsorption–desorption equilibrium (which is calculable). LEED analysis points to the 'long-bridge' model for the ordered array of H atom rows shown in Fig. 8.20. It is also noteworthy that the bond holding the H atoms to the surface is predominantly covalent.

The picture that emerges of the partly covered iron catalyst surface is that the N atoms are comparatively static and the H atoms are mobile; but, at temperatures greater than *ca* 400 °C, the surface concentration of the latter is small and unlikely to inhibit the adsorption of nitrogen.

We also need to enquire briefly into the siting at the iron surface, and the ease of liberation from it, of the NH_3. Below room temperature the NH_3 is adsorbed intact, and rather loosely (with an adsorption energy of about 70 kJ mol^{-1}), covering one-sixth of the surface. It is bound to the surface through coupling of the lone-pair electrons on the nitrogen to the metal. At more elevated temperatures (heating up to *ca* 500 K), chemical transformations occur in a stepwise fashion, i.e.

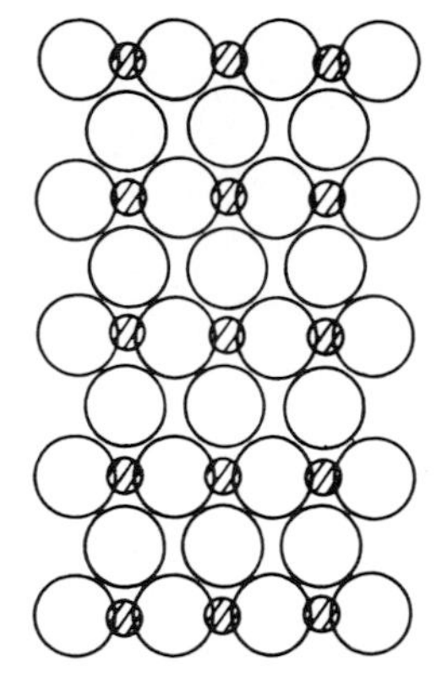

Figure 8.20 Structure model for H/Fe {110} at $\theta = 0.5$ (small hatched circles denote H). (After G. Ertl, *Catalysis*, **1983**, *4*, 210.)

$$NH_3(ad) \rightarrow NH_2(ad) + H(ad) \rightarrow NH(ad) + 2\,H(ad) \rightarrow N(ad) + 3\,H(ad)$$

Evidence for the formation of $NH_2(ad)$ is based on Grunge's work on isotopic exchange with deuterium, and on similarities of the UPS data to those from hydrazine, (N_2H_4). The presence of NH(ad) is deduced from UPS, SIMS and HREELS studies, and the adsorbed N is identified in UPS studies.

Under steady-state conditions, therefore, NH_3 is both formed and decomposed, otherwise the equilibrium $3\,H_2 + N_2 \rightleftharpoons 2\,NH_3$ could not be achieved. $^{15}N/^{14}N$ studies had also demonstrated the reality of this dynamic equilibrium.

8.3.3.3 A Potential-Energy Diagram Illustrating How the Overall Reaction Leading to Ammonia Synthesis can be Constructed

On the basis of all the facts that we have summoned, we may now summarize the sequence of steps, and the respective transitory surface species, involved in the catalysis:

$$H_2 \rightleftharpoons 2\,H(ad)$$

$$N_2 \rightleftharpoons N_2(ad) \rightleftharpoons 2\,N(ad)$$

$$N(ad) + H(ad) \rightleftharpoons NH(ad)$$

$$NH_2(ad) + H(ad) \rightleftharpoons NH_2(ad)$$

$$NH_2(ad) + H(ad) \rightleftharpoons NH_3(ad)$$

$$NH_3(ad) \rightleftharpoons NH_3(ad)$$

An appropriate potential-energy diagram, constructed by Ertl and by Bowker and Waugh, encapsulating the energetics known to be associated with each of these steps, is shown in Fig. 8.21. The magnitude of the activation energy for dissociative nitrogen adsorption, E^*, is essentially determined, as explained earlier, by the steady-state surface concentration of atomic nitrogen. Under conditions well removed from equilibrium (negligible concentration of NH_3), the reaction can schematically be de-

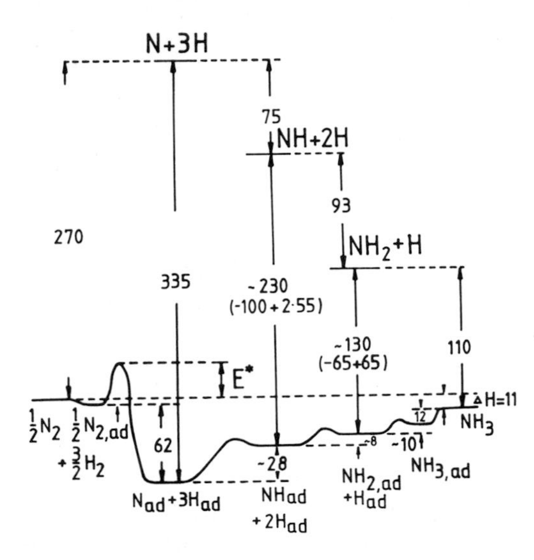

Figure 8.21 Potential-energy diagram illustrating the progress of the catalytic synthesis of NH_3 on iron. (After G. Ertl, *Catalysis*, **1983**, *4*, 210.)

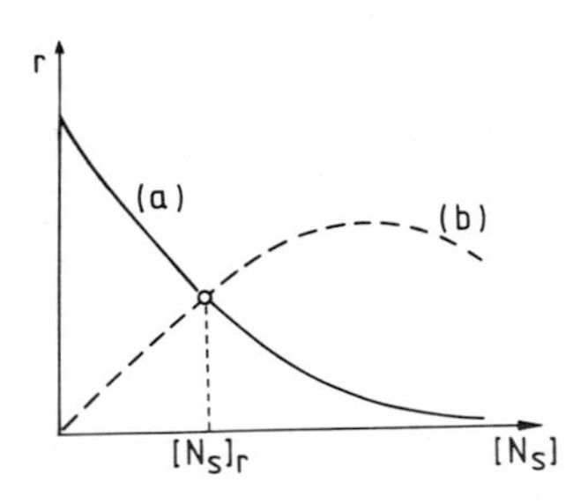

Figure 8.22 Schematic variation of the rate of dissociative nitrogen chemisorption (a) and of the rate of hydrogenation of adsorbed atomic nitrogen (b) with the surface concentration. $[N_s]_2$ is the surface concentration under steady-state conditions. (After G. Ertl, *Catalysis*, **1983**, *4*, 210.)

scribed by the following two events:

Step (a): $N_2 \rightarrow 2\,N(ad)$

Step (b): $N(ad) \rightarrow NH_3$

Figure 8.22 qualitatively summarizes the resulting situation. Curve (a) decreases approximately exponentially with increasing coverage, N(ad) (owing to the linear rise in the activation energy – recall the Elovich equation). The rate of step (b), on the other hand, increases with increasing N(ad) until inhibition of hydrogen adsorption causes a decrease again. Steady-state is reached at the point where the curves intersect. We see from Figure 8.22 why we have only a modest coverage of the catalyst surface, by N(ad), even if the pressure of the gas is very high.

8.3.3.4 How Potassium Serves as an Electronic Promoter

Now that we have demonstrated that the key rate-determining step in ammonia synthesis is the dissociative adsorption of N_2, and therefore that the magnitude of

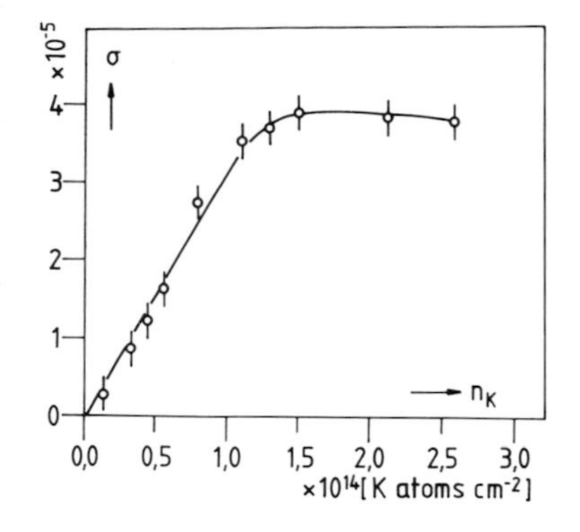

Figure 8.23 Variation of the initial sticking coefficient for dissociative N_2 adsorption, σ, with concentration of potassium n_K on an Fe {100} surface at 430 K. (After G. Ertl, S. B. Lee, M. Weiss, *Surf. Sci.*, **1981**, *108*, 357.)

E^*, the activation energy (see Figs. 8.18 and 8.21) is all-important, we can also rationalize the special role of potassium as promoter, thanks to the efforts of Ertl, Somorjai, Comsa and Topsøe et al. First we note the dramatic enhancement in the magnitude of the sticking coefficient for dissociative nitrogen adsorption with increasing surface coverage of potassium (Fig. 8.23). At a concentration of 1.5×10^{14} K atoms cm^{-2} at 430 K, the sticking coefficient of N_2 increases by a factor of 300. In addition, the presence of potassium causes the N_2 to be more tightly held to the iron surface: the adsorption energy is increased (from 30 kJ mol^{-1} to 45 kJ mol^{-1}) while simultaneously E^* is lowered by about 10 kJ mol^{-1}. These two effects are correlated: the deepening of the potential-energy well of the Lennard-Jones N_2 adsorption curve brings down the point of intersection of the curves for molecular and atomic adsorption.

Adsorption of potassium on iron surfaces also causes a pronounced diminution of the work function, which effect enhances the 'back-donation' of metallic electrons to the antibonding N_2 2π level, thus strengthening the Fe–N_2 bond. Molecular orbital calculations also indicate a net electron transfer from the metal to N_2, so that this ligand is an electron acceptor. Since the work function minimum caused by potassium adsorption reaches similar values for Fe(111), (100) and (110) it now becomes clear why differences between these high-symmetry faces seen in the absence of potassium are levelled off by its presence.

8.3.4 The Technology of Ammonia Synthesis

Since the mid-1920s NH_3 synthesis technology has developed into a reliable body of knowledge extending from catalyst formulation and manufacture to plant design and fabrication. Several secondary processes are involved in ammonia manufacture because the feedstock, which nowadays generally includes natural gas, must be manipulated in such a way as to ensure that a 3 : 1 stoichiometric mixture of H_2 and N_2 is fed to the reactor. Figure 8.24 is a diagram of a typical process arrangement. A mixture of preheated natural gas and steam is fed at about 36 bar to the tubes of a so-called primary reforming furnace, in which the natural gas (mostly CH_4) and steam are converted by a nickel catalyst to a mixture of CO, CO_2 and H_2 by the reactions:

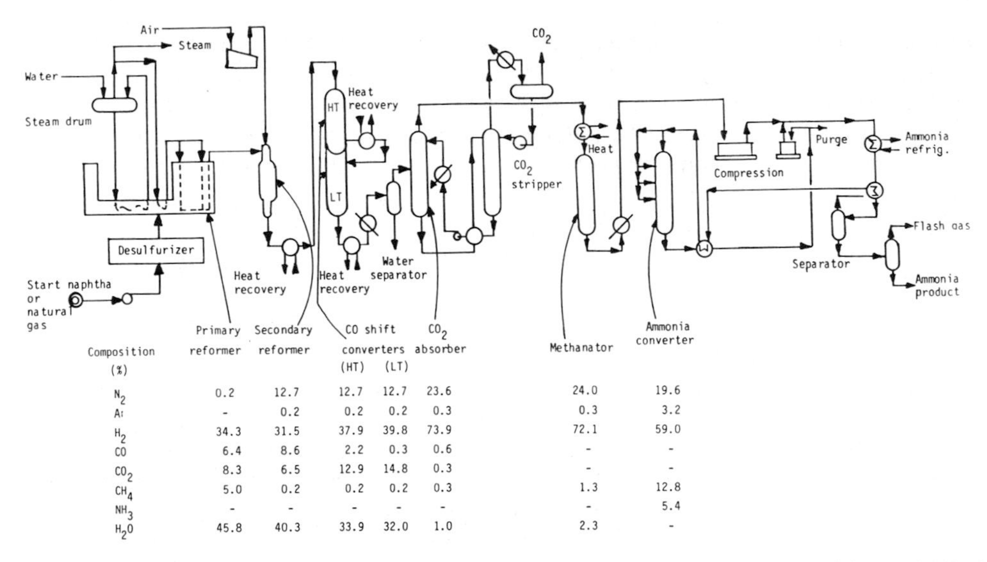

Composition (%)	Primary reformer	Secondary reformer	CO shift converters (HT)	(LT)	CO_2 absorber	Methanator	Ammonia converter
N_2	0.2	12.7	12.7	12.7	23.6	24.0	19.6
Ar	-	0.2	0.2	0.2	0.3	0.3	3.2
H_2	34.3	31.5	37.9	39.8	73.9	72.1	59.0
CO	6.4	8.6	2.2	0.3	0.6	-	-
CO_2	8.3	6.5	12.9	14.8	0.3	-	-
CH_4	5.0	0.2	0.2	0.2	0.3	1.3	12.8
NH_3	-	-	-	-	-	-	5.4
H_2O	45.8	40.3	33.9	32.0	1.0	2.3	-

Figure 8.24 Typical arrangement of plant for ammonia processing. A typical composition [vol. %] of the exit gases from each stage is indicated in the Table.

$$CH_4 + H_2O \rightleftharpoons CO + 3\,H_2$$
$$CH_4 + 2\,H_2O \rightleftharpoons CO_2 + 4\,H_2 \tag{1}$$

Both these steam-reforming reactions (Section 8.2.5) are endothermic and the necessary sensible heat (i.e. that required to raise the feed from ambient to reaction temperature) and heat of reaction are provided by burning natural gas, with the flames from the burners impinging on the outside of vertical tubes containing the nickel catalyst for the reforming reactions. Excess heat contained by the exit gases from the primary reformer is used to preheat feed gas, air, steam and boiler feed water used in other parts of the process. The function of the secondary steam reformer, where essentially the same reforming reactions occur, is twofold. First, any unconverted methane is reformed, producing additional H_2. Second, sufficient compressed air is introduced to the secondary reformer so that enough N_2 will be present for reaction with hydrogen in the NH_3 converter situated downstream, and so that a portion of the H_2 present will be burnt by the O_2 to provide the necessary heat requirements for the reforming reactions.

After the hot exit gases from the secondary reformer are cooled (providing heat for generation of some of the steam required in both reformers) they enter a shift converter in which most of the carbon monoxide present in the mixture reacts over an appropriate catalyst to produce more hydrogen by the water-gas shift reaction,

$$CO + H_2O \rightleftharpoons H_2 + CO_2 \tag{29}$$

Two successive shift converters (see Section 8.2.5) are necessary, one operating at the relatively high temperature between 325 and 500 °C and the other at 250 °C. An iron oxide catalyst (in the form of Fe_3O_4 in the reactor) is employed in the high-temperature shift converter where, because of the mildly exothermic nature of the reaction, the temperature increases and limits the conversion. More steam, sufficient both for the reaction and as a diluent preventing excessive temperature rise, must be added to the exit gases from the secondary steam reformer before entering the high-temperature shift converter. Modern practice involves cooling the exit gases from the high-temperature shift converter before entering a low-temperature shift converter containing an alumina-supported copper catalyst which converts the remaining carbon monoxide to hydrogen. Before the development of the low-temperature shift catalyst, residual CO was removed by an absorption process.

The next stage of the ammonia process is the removal of CO_2 by absorption in a basic solution containing a dissolved homogeneous catalyst. Thus, following exit from the shift converters, cooling and condensation of excess steam, the mixture – now consisting of N_2, H_2, CO_2, some residual CO and inert Ar – is passed upwards through a tower countercurrent to the hot basic solution fed to the top of the tower. Good contact between gas and liquid is achieved by an inert packing in the tower, or by the use of perforated sieve plates. The total pressure in this absorber is usually about 40 bar but will depend on upstream conditions. The basic solution is regenerated in a second tower by steam stripping at about 1 bar.

The gas mixture emerging from the absorption tower, however, still contains trace quantities of CO and CO_2, both of which are poisons to the ammonia synthesis catalyst. They are removed in a catalytic reactor (the methanator) which contains an alumina-supported nickel catalyst and which operates at about 300–350 °C. The CO is removed here by the reverse of the steam-reforming reaction (see Section 8.2.5):

$$CO + 3\,H_2 \rightarrow CH_4 + H_2O$$

Methane (present only in small amounts) is inert and the steam generated in the reaction is removed by condensation. Following this methanation reaction the CO content in the synthesis gas does not usually exceed 5 ppm.

Methanation is the penultimate stage in the ammonia synthesis process and the gas mixture emerging from the methanator contains N_2 and H_2 in stoichiometric proportions, and a small quantity of inert CH_4 and argon. These are compressed to about 150–300 bar and mixed with unreacted gases recycled from the exit of the synthesis reactor which operates between 400 and 550 °C. A common form of reactor configuration is a tubular type of construction which is self-cooled by the incoming feed gas. A quench of cooled recycle gas is introduced at one or more positions along the reactor to prevent excessive temperature rise.

From what has been outlined above, it is clear that the process engineering that is necessary is critical to economic production. There is much interaction between the variables in the process streams, especially where heat is exchanged and recovered and where gas compositions and pressures have to be carefully manipulated. A typical inventory of the exit gases from each stage is indicated in Fig. 8.24.

8.3.4.1 Reactor Configurations are Important Industrially

A number of different reactor configurations have been employed industrially to effect ammonia synthesis. Bearing in mind that the reaction is quite exothermic, ad-

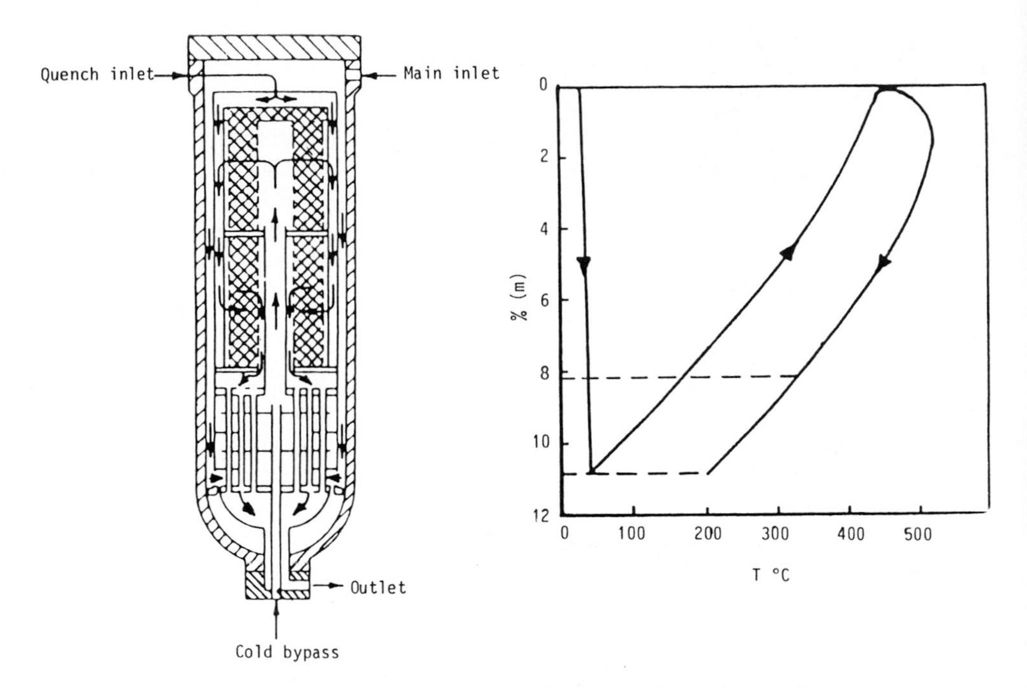

Figure 8.25 Shell-and-tube type of ammonia synthesis reactor with accompanying temperature profile.

vantage may be taken of heat exchange between relatively cold incoming feed and the reacting synthesis mixture. A multitubular shell type of configuration is shown in Fig. 8.25 in which the catalyst is loaded in the shell and the cool inlet gases pass along the tubes countercurrent to the reacting gas mixture. The temperature profile to be expected in such a multitubular shell configuration is juxtaposed with the sketch of the reactor arrangement. The relatively colder feed is warmed by heat extracted from the reaction as it passes upwards along the tubes, while the reacting gases passing downwards through the shell are cooled by the colder feed gas. Construction of this type of converter creates difficult mechanical design problems because of the high pressures employed for synthesis and the consequent requirements for a thick shell.

An alternative reactor configuration is illustrated in Figure 8.26, where heat is removed by quenching the hot reacting gases by a portion of the cold feed. The incoming feed, at about 400 °C, reacts in contact with the catalyst contained in the shell of the reactor. When the temperature reaches approximately 550 °C the reacting mixture is quenched and reaction is again allowed to proceed in the second section of the converter. It is usually necessary to quench the reacting gases a second time before the gases emerge from the converter, a 95–98% approach to chemical equilibrium being achieved.

The pressure drop in both types of reactor described can be as high as 9 bar. The recycled gases have to be recompressed to the normal operating pressures of about

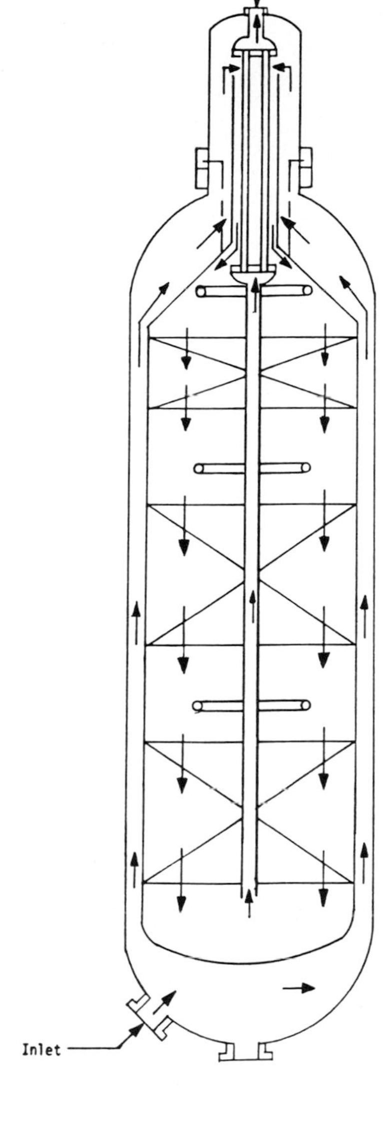

Figure 8.26 Quench type of ammonia synthesis reactor.

150 bar; such recompression is costly. A radial-flow converter sketched in Fig. 8.27 avoids such large pressure drops. In this type of reactor the shell is mounted horizontally and a gas transfer line feeds the incoming synthesis mixture of N_2 and H_2 to the first of three catalyst beds. The gases pass downwards through this catalyst bed and the direction of flow is thus normal to the direction of flow at inlet and exit. A bed of larger cross-section can therefore be employed and, because the pressure drop in such a configuration is lower than it would be in the other types of ammonia synthesis reactor described, smaller catalyst particles can be utilized with a consequent increase in catalyst effectiveness. The emergent gases from the first bed are mixed

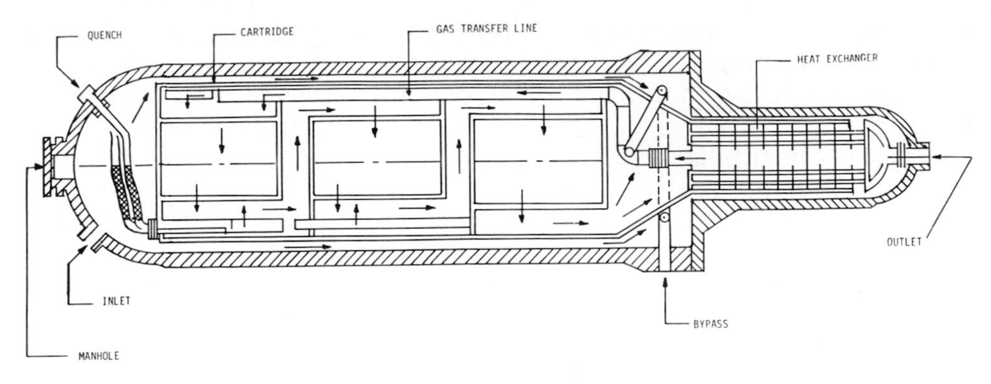

Figure 8.27 Horizontal ammonia synthesis reactor.

with a portion of the cold feed, which acts as a quench, before inlet to the second bed and finally to a third bed. A shell-and-tube single-pass heat exchanger with baffles forms an integral part of the reactor outlet to allow heat exchange between cold incoming feed and hot exit gases.

8.4 Oxidation of Ammonia: Stepping Towards the Fertilizer Industry

Ever since ammonia was first produced by the Haber process in the early years of the 20th century, its catalytic oxidation to nitric oxide has been the basis of the nitric acid manufacturing industry. As much as three-quarters of the production of nitric acid is consumed by the world fertilizer industry. Approximately 10^{10} kg of ammonium nitrate, required for the fertilizer and explosives industry, are produced in the USA per annum. This chemical is prepared by the reaction of ammonia with nitric acid; the latter is generated by addition of water to nitric oxide, the product of oxidation of the former. (Another important fertilizer used almost as extensively, ammonium sulphate, is formed from ammonia and sulphuric acid.)

Industrial plants producing nitric acid employ platinum or platinum alloy gauze as a catalyst for the reaction

$$4\,NH_3 + 5\,O_2 \rightarrow 4\,NO + 6\,H_2O \qquad -\Delta H_{298} = 226\,\text{kJ mol}^{-1}$$

Rather less than the stoichiometric ratio (14.4%) of ammonia to air is fed to the converter to avoid operating near the explosion limit of ammonia in air, which is 16% at atmospheric pressure. The operating conditions for the catalytic converter depend on the downstream process conditions, which are arranged so that the nitric oxide produced is oxidized in the gas phase to nitrogen dioxide by the homogeneous reaction

$$NO + \tfrac{1}{2}O_2 \rightarrow NO_2$$

As this latter reaction occurs rapidly under compression, it is not unusual for the ammonia–air mixture fed to the catalytic converter to be at about 5 bar or above. The reactant gases are passed downwards through the reactor, which operates at about 900 °C.

Ammonia oxidation is exceedingly exothermic and very fast. Consequently the relatively slower transport of reactant through the gas phase and the stagnant boundary layer adjacent to the solid catalyst surface (as well as the corresponding return of product to the gas phase) determines the overall rate of conversion. Such mass-transfer-limiting conditions demand that the total extent of surface area should be available at the external periphery rather than in the interior structure of the catalyst. Therefore, rather than disperse the platinum catalyst over a porous inert support (which would cause an unwelcome pressure drop), the catalyst is normally in the form of finely woven platinum gauze, about 30 or 40 layers, each about 5 mm thick, of which constitute a 4 m-diameter pad of catalyst. Only the first two or three layers of the gauze act catalytically until such time as the platinum metal has become inactive due to surface oxidation and platinum loss through vaporization of the relatively more volatile PtO_2 formed during reaction. The reaction zone consequently moves to successive layers of gauze when the first few layers have lost their activity. Gauzes are then removed from the top of the pad and fresh ones inserted at the bottom of the pad. The total loss of platinum amounts to about 200 mg per 1000 kg of nitric acid produced in a plant.

Because the pressure drop through the pad is so low, it is necessary to arrange for a uniform distribution of gas flow over the cross-section of a gauze layer and throughout the pad. Poor gas distribution through the pad would, for example, cause any unreacted ammonia to react with product nitric oxide to form N_2, thus reducing the overall efficiency. Because so much heat is generated during reaction and because of the rapid gas flow through the catalytic converter (less than 1 ms contact time is adequate) there is a delicate balance between convective heat loss and heat generation by reaction. Referring to the previous sketch of a heat loss line superimposed upon a heat generation curve (see Fig. 7.18), then clearly, to achieve the stable operating conditions at A and to obtain high conversion at a high reaction temperature, the incoming reactant gases must be ignited, otherwise little or no conversion would occur, as at C (the intersection at B represents an unstable condition). Such ignition is usually initiated by means of a flame: once the gas temperature is above B, sufficient heat is generated by reaction to overcome the loss of heat by convective flow and for the reaction to stabilize and be self-sustaining at the operating point A.

The following example demonstrates how to estimate the exit gas temperature from an ammonia-oxidation converter and thereby assess the gauze temperature.

Example

A mixture of 11% NH_3 (by volume) in air preheated to 60 °C is passed at a mass velocity of 0.49 kg m^{-2} s^{-1} through a platinum gauze pad to effect oxidation. The density of the preheated mixture is 0.4 kg m^{-3} and the available active catalyst surface is 1.2 m^{2} per square metre of gauze. The reaction is mass-transfer-limited and has an exothermic heat of reaction 226 kJ mol^{-1}. Once ignition is initiated, a conversion of

87% is achieved. Estimate (a) the mass-transfer coefficient, (b) the exit gas temperature from the converter and (c) the surface temperature of the catalyst gauze.

The molar heat capacity of the gas mixture may be taken as $31.8\,J\,mol^{-1}\,K^{-1}$. Mass- and heat-transfer correlated data predict that the interphase heat-transfer coefficient is $1.12\,J\,m^{-2}\,s^{-1}\,K^{-1}$.

Solution

Assuming the gases are in plug flow through the gauze pad (no radial or longitudinal dispersion of mass), a steady-state mass balance for the reactant ammonia gives

$$-u_l \frac{dc}{dz} = R$$

where u_l is the linear gas velocity, the concentration c is in units of moles per unit volume and R, the reaction rate, in moles per unit volume per unit time. The effectiveness factor η is taken as unity as the reaction occurs entirely at the exterior surface of the catalyst. As the reaction is mass-transfer-limited,

$$R = ah_D c$$

where a is the available interfacial area per unit volume, and h_D the interfacial mass coefficient. Combining these two equations and integrating,

$$-\ln \frac{c}{c_0} = \frac{ah_D z}{u_l}$$

where c_0 is the inlet concentration of ammonia.

From the data given, az is the interfacial area per unit gauze area available for mass transfer, which is $1.2\,m^2\,m^{-2}$. Also $u_l = \rho u_l/\rho = (0.49\,kg\,m^{-2}s^{-1})/(0.4\,kg\,m^{-3}) = 1.225\,m\,s^{-1}$. As the conversion is 87% then c/c_0 is 0.13 and thus, on substitution,

$$h_D = \frac{1}{1.47} \ln \frac{1}{0.13} = 1.39\,m\,s^{-1}$$

From a heat balance over the pad,

$$\bar{\rho} u_l \bar{c}_p \frac{dT}{dz} = (-\Delta H)R$$

(which is equivalent to Eq. (75) in Chapter 7 with $\eta = 1$ but with units expressed differently with R in dimensions of mol $L^{-3}\,T^{-1}$, $\bar{\rho}$ in mol L^{-3} and the linear velocity, rather than volumetric velocity, in $L\,T^{-1}$). Integration from an inlet temperature T_0 yields

$$T - T_0 = \frac{(\Delta H)}{-\rho \bar{c}_p}(c_0 - c) = \frac{(\Delta H)}{-\bar{c}_p} y_0 \left(1 - \frac{c}{c_0}\right)$$

where y_0 $(= c_0/\rho)$ is the mole fraction of ammonia in the feed. Substituting values given, the gas temperature at the pad exit is

$$T = 60 + \frac{226 \times 10^3}{31.8} \times 0.11 \times 0.87 = 740\,°\mathrm{C}$$

Now in the steady state (when thermal equlibrium has been established) the heat transferred from solid phase to gas phase is balanced by the heat generated by reaction, so

$$h(T_s - T) = (-\Delta H)\mathrm{R} = (-\Delta H)h_\mathrm{D}c$$

Hence

$$T_s - T = \frac{h_\mathrm{D}}{h}(-\Delta H)\frac{c}{c_0}\bar{\rho}y_0$$

in which c has been replaced by $c\bar{\rho}y_0/c_0$ on the extreme right-hand side. The molar density $\bar{\rho} = 0.4/17 = 0.023\,\mathrm{mol\,m^{-3}}$. Substituting numerical values gives

$$T_s = 740\,°\mathrm{C} + \frac{1.39\,\mathrm{m\,s^{-1}}}{1.12\,\mathrm{J\,m^{-2}\,s^{-1}\,K^{-1}}} \times 226 \times 10^3\,\frac{\mathrm{J}}{\mathrm{mol}} \times 0.13 \times 0.023\,\frac{\mathrm{mol}}{\mathrm{m^3}} \times 0.11$$

$$= 740 + 100 = 840\,°\mathrm{C}$$

8.5 In-Situ Catalytic Reaction and Separation

Advances in research and development have been driven, since the mid-1980s, by the demands of legislation and competitive commerce. Accordingly, there has been a desire to reduce material inventories, conserve and reduce energy demand, integrate process units, and maintain and improve high standards of safety. The ubiquitous use of catalysts in industrial processes has provided opportunities for such objectives as a result of one compelling and pervasive principle – that a heterogeneous catalyst has, of necessity, to adsorb a reactant in order to function as a catalyst. Such a property also provides potential for the separation of components.

Three possibilities of process innovation are discussed which involve in-situ catalytic reaction and separation. In each case products and reactants are, to a greater or lesser extent, separated in the same process unit where catalytic reaction takes place. The three process operations are:

1. catalytic distillation,
2. pressure swing reaction and
3. catalytic membrane processes.

Each of these process operations requires different configurations and employs a variety of process conditions. They are therefore considered in turn.

8.5.1 Catalytic Distillation

Chemical reaction occurring during distillation was recognised by Backhaus as early as 1921. Since that time reactive distillation has been discussed by Hofmann, Sharma and more recently by Doherty. The first commerical catalytic distillation units were

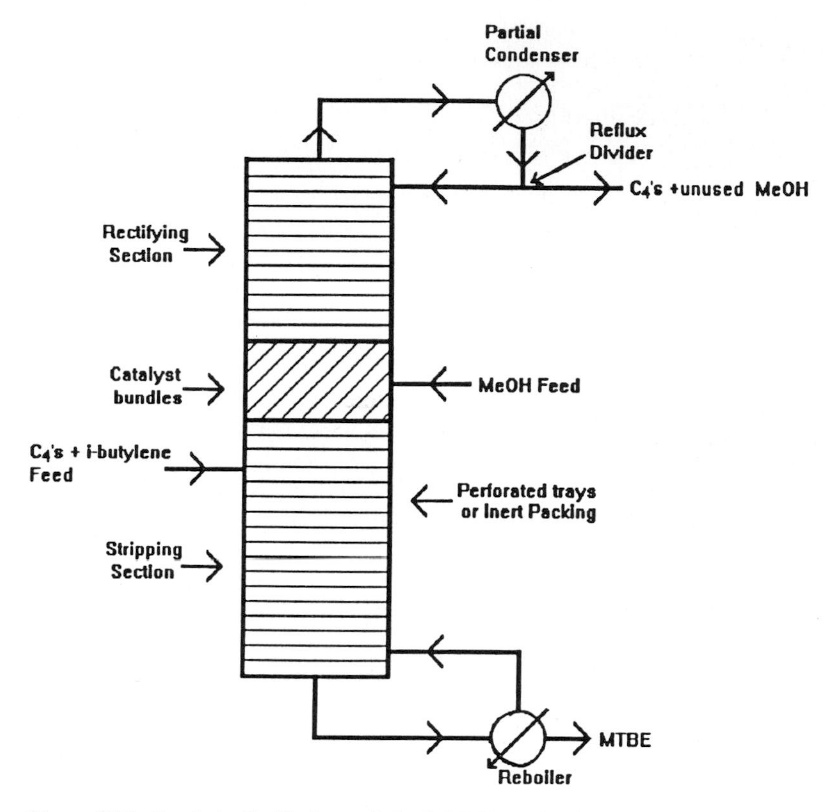

Figure 8.28 Catalytic distillation unit for MTBE production.

developed by Chemical Research and Licensing and CD Technology in conjunction with Lummus, and are producing speciality and value-added products.

As a result of legislation to reduce and eventually eliminate the emission of lead from vehicle exhausts, the production of alternative antiknock additives to petroleum spirit was realized by means of catalytic distillation. Thus the formation of methyl *t*-butyl ether (MTBE, an octane enhancer) was achieved in a single column containing an acidic resin catalyst which converts a feed of methanol and C_4 hydrocarbons (containing the reactant isobutene) to the desired product. The C_4 hydrocarbon feed enters the column immediately beneath the catalyst section situated at mid-tower level. Methanol is fed directly into the catalyst section so that the least volatile component product of the catalytic reaction, MTBE, is recovered as bottom product from the tower. Spent C_4 hydrocarbons together with any isobutene and excess methanol are recovered from the top of the tower. Figure 8.28 is an illustration of the operating arrangement.

The catalytic reaction between propylene and benzene over an acid resin catalyst is a second example of the principle of catalytic distillation. The catalyst is placed at the top of the tower so that the product cumene, less volatile than any of the other components, is removed from the reaction zone as soon as it is formed, thus max-

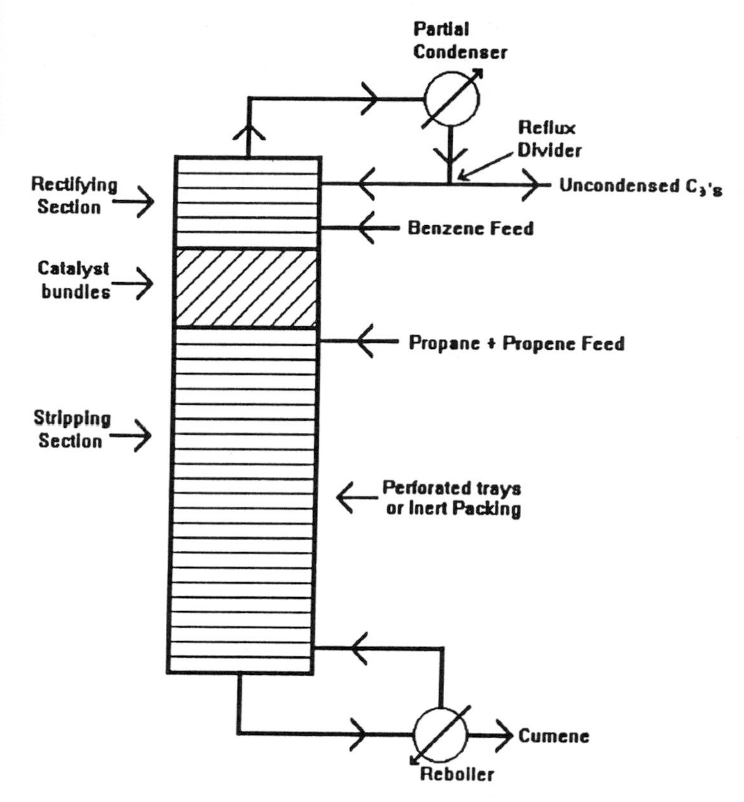

Figure 8.29 Catalytic distillation unit for cumene production.

imizing conversion by disturbance of the thermodynamic equilibrium position (Le Chetalier's principle). The feed, a mixture of propane and propylene, is fed directly beneath the catalyst section while the second reactant, benzene, is fed above the catalyst section. Benzene thus flows down to the catalyst section where it reacts with the upward flowing feed of C_3 hydrocarbons. Unused propane distils into the reflux drum and is vented. Figure 8.29 illustrates the tower arrangement.

It is clear from these two examples that catalytic reaction and separation of components by distillation occur within a single operating unit. Heat evolved from the catalytic reaction is utilized to manage the energy requirements of the column, while high conversions are possible by removal of product from the reaction zone. The position of the catalyst in the tower is determined by the relative volatilities of product and feed components and the requirement that product is removed from the reaction zone as soon as it is formed. The catalyst material usually consists of bundles of woven cloth containing pockets within which the catalyst resin resides. Stoichiometric mixtures can be employed, thus avoiding excess of reactant and eventual recovery. A further advantage accrues from the conditions of reaction which often circumvent the formation of azeotropes. Instead of representing the phase diagram in terms of mole fractions of liquid and vapour at given temperatures, Doherty

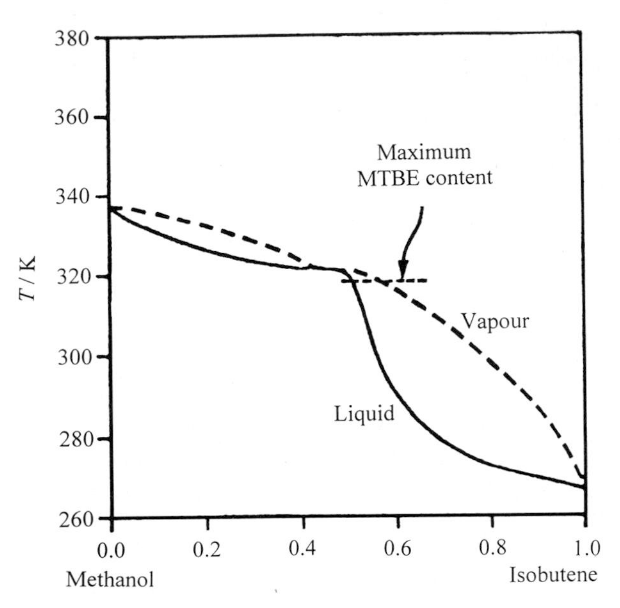

Figure 8.30 Temperature–composition diagram for ternary system MTBE–methanol–isobutene. With permission from M. F. Doherty, G. Burgud, *Trans. Inst. Chem. Eng.* **1992**, *70A*, 448.

makes use of transformed co-ordinates which are the equivalent amounts of liquid and vapour present in an equilibrium mixture

$$\nu_A A + \nu_B \rightleftharpoons \nu_T C$$

Thus, for the liquid phase

$$X_A = \frac{x_A - (\nu_A/\nu_B)x_C}{1 - \nu_T x_C} \text{ and } X_B = \frac{x_B - (\nu_B/\nu_C)x_C}{1 - \nu_T x_C} \tag{55}$$

and for the vapour phase

$$Y_A = \frac{y_A - (\nu_A/\nu_B)y_C}{1 - \nu_T y_C} \text{ and } Y_B = \frac{y_B - (\nu_A/\nu_B)y_C}{1 - \nu_T y_C} \tag{56}$$

where x_i and y_i are mole fractions (of component i) of the liquid and vapour phases respectively, and ν_i are stoichiometric coefficients; ν_T is the sum of all stoichiometric coefficients.

Figure 8.30, for example, shows that, in an MTBE–methanol–isobutene mixture, one reactive azeotrope is formed in close proximity to the maximum composition of MTBE. Formation of the reactive azeotrope can be avoided by judicious choice of reaction temperature.

Modelling the performance of a catalytic distillation system is complex, though straightforward. Material balances for components together with energy balances are formulated at three stages of the tower – the partial condenser, the reboiler and the non-equilibrium stage containing catalyst. There are, of course, additional con-

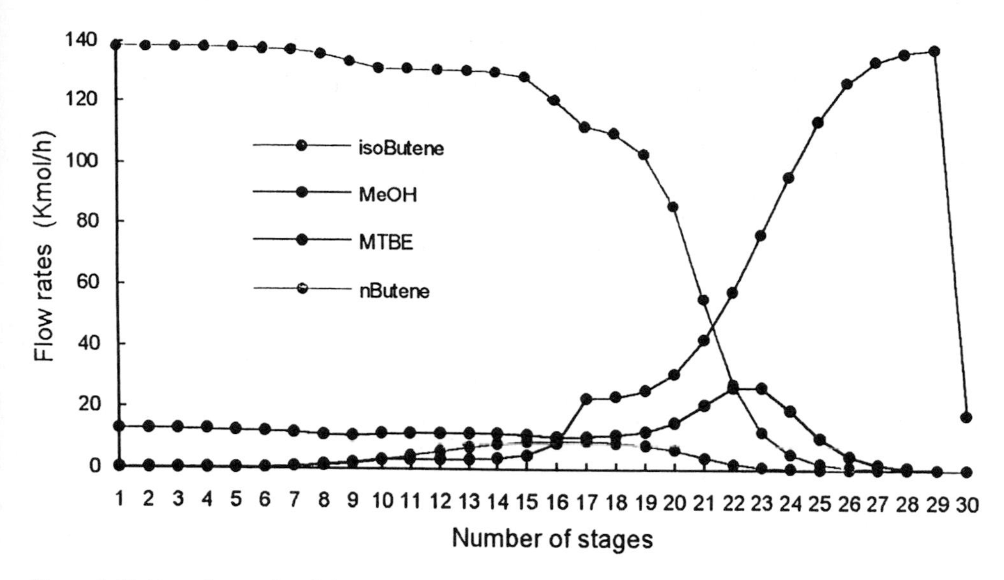

Figure 8.31 Experimental and simulated temperature profiles for the distillation of a ternary mixture of MTBE–methanol–isobutene (After Y. Zheng, X. Xu, *Trans. Inst. Chem. Eng.* **1992**, *70A*, 465.)

straints imposed by the rate of catalytic reaction and the reaction stoichiometry to consider, in addition to the usual rates of mass and heat transfer between vapour and liquid phases in the non-equilibrium stage. Although the model incorporates assumptions such as a steady state, perfect vapour and liquid mixing, no mass- or heat-transfer resistance between the liquid and catalyst phase, and uniform rates of mass transfer across the column diameter at any given height, there is, for a case simulated by Zheng and Xu and shown in Fig. 8.31, a remarkable agreement between the experimental and theoretical temperature profiles in the column.

8.5.2 Pressure Swing Reaction

The technique of pressure swing adsorption for the separation of components of a gaseous mixture is well established although, to date, no such process operation involving catalytic reaction has been introduced into industrial practice. For gas separations, the two columns packed with a suitable solid adsorbent operate sequentially in stages. While one of the columns is first pressurized, then product is recovered, and subsequently the column is depressurized and then purged, the other column in tandem is sequentially depressurized, purged and pressurized, and finally product is recovered. Thus one cycle is completed before the next similar cycle is immediately commenced. A continuous flow of product results from this cycle of operations, a mixture being separated by preferential adsorption of the more strongly adsorbed component during pressurization and preferential desorption of the least adsorbed component during depressurization. Many variations of this procedure exist in practice. There have been attempts to reduce cycle times from minutes to seconds in order to reduce material inventories and achieve smaller configurations. In principle at least, rapid cycle times can be realized in a single, rapidly pulsed, tightly packed column. Furthermore, catalytic reaction of gaseous components and simultaneous separation can be achieved in a single, rapidly pulsed column tightly packed with adsorbent as well as catalyst. The pressure variations at top and bottom of the column depend on the tightness of packing as well as on the rapidity of pulsing in and out of the column. It was shown experimentally that enhanced yields and separation of product can be obtained by employing a rapidly pulsed pressure swing reactor. In contrast to the expectations of separation by rapidly pulsed pressure swing adsorption during which product is always exhausted at the top end of the column and enriched in the most strongly adsorbed component (the least strongly adsorbed component emerging from the bottom of the column), considerable departure from such behaviour occurs and is dependent on the interactive nature of adsorption and reaction as well as column conditions, catalyst activity and adsorption capacity. Kirkby and Morgan (1994) demonstrated that there are considerable differences in product separation when pressure swing reaction is allowed to occur in a dual-column operation employing variations of the simple sequence of pressurization, product recovery and depressurization stages. Enhanced product separations for the simple cycle (consisting of pressurization, product recovery and depressurization steps), the purge cycle (when a purge step is added to the simple cycle) and the backfill cycle (when product is reintroduced to the adsorbent bed instead of the purge step) are all possible, given a correct choice of operating conditions. Pressure swing reaction has not yet been introduced as an industrial operation.

Modelling the rapid-pulsed pressure swing reactor involves time-variant boundary conditions. Boundary conditions on the sets of equations (representing mass conservation of components in the adsorption–reaction column, the equilibrium constraint and the kinetic rate equation) must account for conditions at the feed end of the column during the time when feed is introduced as a pulse, during any delay period when both feed and exhaust valves are closed, and during the period when product is exhausted from the top of the column. At the bottom end of the column flow is in a downward direction when feed is introduced to the top of the column, but is reversed when the exhaust valve (at the top of the column) is opened. Both these conditions must be properly represented. Vaporciyan and Kadlec (1989) showed that separation of product is enhanced when the principal reactant is strongly adsorbed.

8.5.3 Catalytic Membrane Processes

In principle there is no reason why certain catalysts cannot be incorporated in a membrane structure to effect catalytic reaction, the component products and reactants being separated in the normal manner simultaneously with reaction. Removal of the products of an equilibrium reaction through the membrane structure would thus, by Le Chetalier's principle, considerably enhance conversion beyond the expected equilibrium value. Furthermore, it should be possible to remove an undesired product occurring as the result of a side reaction, and therefore to improve the selectivity of reaction in favour of the principal reaction and product.

Materials which have been used for porous membranes include alumina, silica, titania, silicon carbide and nitride, and composites such as mullite and cordierite. These materials can be configured as flat plates, tubes or as parallel multichannel monoliths. Catalysts which have been incorporated in some of these materials are noble metals, nickel, molybdenum sulphide and ferric oxide.

Research in this field of catalytic membranes is relatively new and publications are not numerous. Three examples illustrate some of the concepts involved and their application. Further advances in materials are required before catalytic membranes are employed in the process industries.

An example of a membrane which enhances conversion of a reactant to a desired product is the dehydration of cyclohexane to yield the product benzene. The membrane is a palladium tube packed with an alumina-supported platinum catalyst. Hydrogen produced in the course of reaction is able to permeate the palladium tube and can be swept away by a flowing carrier gas. The conversion of cyclohexane is thus enhanced as a result of the disturbance of the equilibrium between cyclohexane and benzene. This system has been modelled by Itoh and Govind (1989) by considering the conservation of mass on either side of the palladium membrane. In addition to the usual reaction, convection and mass-transfer (for a one-dimensional representation) or radial-diffusion (for a two-dimensional representation) terms in the differential equation representing the steady-state flow of components through the packed tube, the loss of hydrogen by permeation must be included as a term. Outside the tube in which reaction occurs, hydrogen is convected by the inert carrier gas sweeping through the annulus and, of course, there is a gain of hydrogen content due to its permeation through the membrane wall. Figure 8.32 compares the results of computation with experimental observation. Agreement is good and it is also clear that the normal equilibrium conversion is far exceeded.

A second example of catalytic reaction producing hydrogen which is allowed to permeate into an annulus is the endothermic conversion of butene to butadiene. This occurs within a palladium tube packed with an alumina-supported platinum catalyst as in the previous example, but heat to support the reaction is derived by allowing the hydrogen permeate to react catalytically with oxygen within the annulus. The

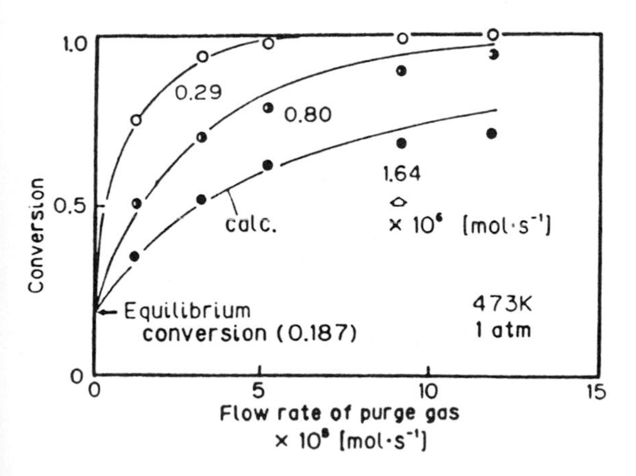

Figure 8.32 Comparison of experimental and calculated results (points and curves, respectively) for the conversion of cyclohexane to benzene in a catalytic membrane reactor. After N. Itoh, R. Govind, *AI Chem. E.* Symp. Ser. **1989**, *36*, 70.

oxygen within the annulus could flow either cocurrent or countercurrent with respect to the flow of reactants in the inside tube. Flow in the annulus countercurrent to reactant flow in the packed tube gives the larger conversion.

Although in the above two examples the catalyst is entirely separate from the membrane, there are clearly some applications where it would be possible to incorporate the catalyst within the membrane pores. Asymmetrical deposition of catalyst would be preferable to uniform deposition because it is desirable to maximize the concentration of product on the permeate side of the membrane. A large concentration gradient is thus encouraged to develop which assists the removal of the product by flowing inert gas.

An interesting and potentially useful application of a catalytic membrane reactor would be to provide hydrogen for a fuel cell delivering power to an electrically driven vehicle, thus avoiding formation of noxious exhaust gases. Methanol can be converted to hydrogen by direct decomposition:

$$CH_3OH \rightarrow CO + 2\,H_2$$

by steam reforming:

$$CH_3OH + H_2O \rightarrow CO_2 + 3\,H_2$$

or by oxidative dehydrogenation:

$$CH_3OH + \tfrac{1}{2}O_2 \rightarrow CO_2 + 2\,H_2$$

The decomposition reaction is endothermic and high temperatures would be necessary to achieve reasonable conversions. Steam reforming is mildly endothermic but requires a source of steam, and therefore heat, which is disadvantageous. Oxidative dehydrogenation, however, is moderately exothermic and requires only a supply of air. Furthermore, heat generated by reaction could be recovered and used as internal heating for the vehicle. The reactant methanol may be regarded as a convenient mode of transporting hydrogen from refinery to fuel cell. Suitable catalysts for the oxidative dehydrogenation of methanol are ZnO and Cr_2O_3 with the addition of a precious metal such as platinum or palladium. Such a catalyst mixture can be prepared as a washcoat for deposition on the channel walls of a monolith fabricated from a porous ceramic material. The principle of operation of such a catalytic membrane unit can be seen by reference to Fig. 8.33. Methanol vapour and air are fed to

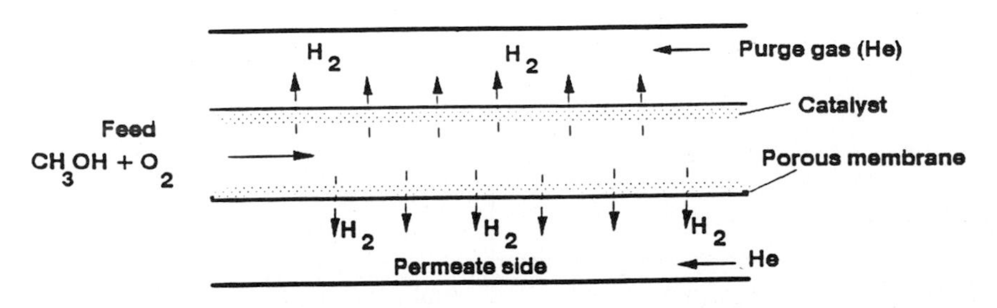

Figure 8.33 Representation of a catalytic membrane reactor.

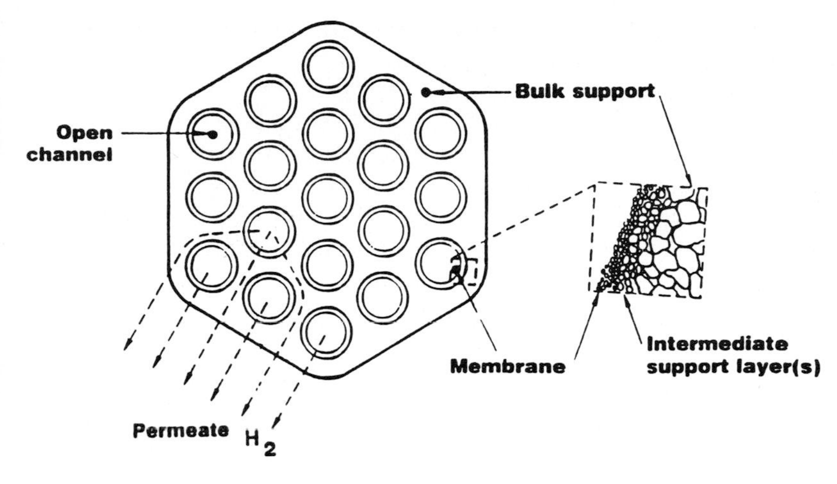

Figure 8.34 Cross-section of a catalytic membrane unit.

the catalyst-loaded monolith channels, where reaction occurs on the channel walls (the diagram shows only a single monolith channel). Hydrogen, a product of reaction, would permeate through the porous walls of the monolith by Knudsen diffusion (provided the diameters of the pores of the porous membrane walls are considerably less than the mean free path of the diffusing gas) and the rate of diffusion, by Graham's law, would be some threefold greater than that of any other product (CO as well as CO_2 could be an indirect product of reaction arising from any reactant interactive equilibria, and would have to be removed by downstream oxidation to CO_2). Figure 8.34 is an illustration of a cross section of a monolithic catalytic membrane. Such a unit would need to be approximately $2 \times 10^{-3}\,m^3$ for a fuel (methanol) consumption of $2\,g\,s^{-1}$ (equivalent to a vehicle travelling at $100\,km\,h^{-1}$ and consuming 1 litre per 10 km as indicated by the following example.

Example

How large should a catalytic membrane device (fabricated in the form of a monolith with parallel channels of circular cross-section) be to produce hydrogen (by the catalytic oxidative dehydrogenation of methanol) to feed to a fuel cell driving the vehicle at an average of $100\,km\,h^{-1}$ and consuming a litre of fuel (methanol) per 10 km travelled?

Solution

If a vehicle travels at $100\,km\,h^{-1}$ and consumes 1 litre of petrol (assume octane) every 10 kilometres, this is equivalent to consuming fuel at a rate of $2\,g\,s^{-1}$. Now the heat of combustion of hydrogen is about 2.68 times the heat of combustion of octane, so the requirement for hydrogen supply to a fuel cell would be at a rate of less than $1\,g\,s^{-1}$. Thus the required H_2 consumption is $0.5\,mol\,s^{-1}$, equivalent to $0.25\,mol\,s^{-1}$ of CH_3OH.

If the rate of reaction is diffusion-limited,

$$\text{rate} = k_g a \Delta c$$

where k_g is the mass-transfer coefficient (replacing a diffusion coefficient per unit length), a the area available for mass transfer and Δc the concentration driving force. Assume that the partial pressure of methanol in the monolith device is 0.5 bar at (say) 500 K. Thus $\Delta c = 12\,mol\,m^{-3}$. If the Sherwood number describing the mass-transfer/diffusion mechanism for molecular transport is 3 (laminar flow in the monolith channels), then

$$\frac{\text{rate}}{\Delta c} = k_g a = \left(\frac{\text{Sh} \cdot D}{d}\right)(\pi d L n)$$

where the monolith channel diameter is d, L the length and n the number of channels. Taking the diffusion coefficient $D = 10^{-5}\,m^2\,s^{-1}$ therefore gives $nL = 220\,m$.

A typical monolith might have channels of 2 mm diameter so the volume of the catalyst monolith would be

$$V = \frac{\pi d^2}{4} nL = \frac{\pi}{4}(2 \times 10^{-3}\,m)^2 \times 220\,m = 7.5 \times 10^{-4}\,m^3 = 0.8\,L$$

8.6 Automobile Exhaust Catalysts and the Catalytic Monolith

Stimulated largely by legislation in California, the car industry began seriously to seek ways of minimizing pollution from auto-exhausts in the late 1960s. The US Clean Air Act of 1970 led gradually to the introduction of catalytic mufflers (auto-exhaust catalysts), the initial objective being to reduce the emission of CO and unburnt hydrocarbons (C_xH_y). The so-called three-way catalyst (TWC) has been in use since 1979. Its name reflects the simultaneous treatment by the catalyst (see Section 1.2.2.1 and below) of the two reducing pollutants, CO and C_xH_y, and the oxidizing pollutant, oxides of nitrogen, NO_x. Three-way catalysis is possible provided the fuel/air ratio, termed λ, in the gas mixture is stoichiometric. At $\lambda < 1$ the activity for NO reduction is high, but not for the oxidation of CO and C_xH_y. At $\lambda > 1$ the reverse is the case. Hence, a special sensor-governed and electronic control system has been developed in motor vehicles that are filled with auto-exhaust catalysts so as to guarantee the desired gas composition. Indeed, the control system is rather more critical than the catalyst itself, which contains rhodium and platinum as key constituents.

An auto-exhaust catalytic converter typically contains 1–2 g of platinum, and 0.2–0.3 g of rhodium. Its introduction (from the USA and Japan initially) has influenced legislation in other countries, including Europe and South America, with the result that projected sales of catalysts are set to rise sharply (Fig. 8.35). Introduction of TWCs has also had a profound impact on the composition of petrol (gasoline). The TWCs are prone to poisoning by small amounts of impurities, the lead in petrol being particularly detrimental. Apart from the gradual phase-out of lead additives

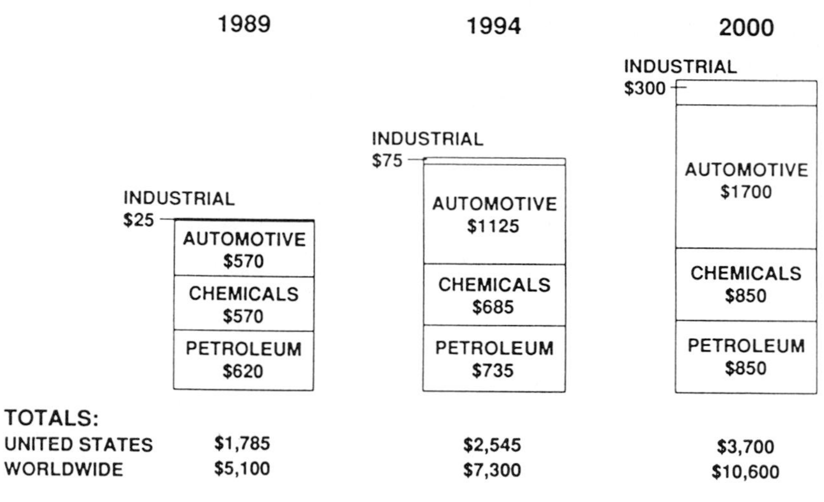

Figure 8.35 Projected sales of catalysts in the USA (in millions of constant dollars) (After J. A. Cusumano in *Perspectives in Catalysis* (J. M. Thomas, K. I. Zamaraev, eds.), Blackwells, Oxford, **1992**, p. 1.)

for this as well as health reasons, TWCs stimulated the development of industrial chemical (also catalytic) processes which produce high-octane-number compounds. Hence, cracking and reforming became even more important commercial catalytic processes. Moreover, oxygenates such as MTBE (methyl *t*-butyl ether), which have good octane numbers and also facilitate the complete combustion of CO and fuel, have already become essential constituents of the petrol. An unpredicted consequence of the introduction of the auto-exhaust catalyst is the decrease in the number of suicides: very little CO is liberated from a car exhaust fitted with a TWC.

8.6.1 The Three-Way Catalyst (TWC)

Current TWCs are quite complicated chemically. They consist of a thin layer of a porous material coated on the channel walls of a ceramic (usually cordierite; see Section 1.2.2.1) honeycomb-shaped body with the channels axially orientated in the direction of flow of the exhaust gas. The porous material is composed largely (70–80%) of a high-surface-area γ-Al_2O_3, the balance being a carefully selected mixture of oxides, each one added for a specific purpose. These are mainly rare-earth oxides, usually CeO_2, or alkaline-earth oxides such as BaO. In some formulations NiO is used to getter H_2S. Only a small fraction of the 'washcoat' weight (1–2%) consists of the noble metals rhodium, platinum and/or palladium. Other aluminas (e.g. α-Al_2O_3) or ZrO_2 may also be used as supports for the noble metals. The reducible CeO_2 (which has the fluorite structure) can function as a source of oxygen (when more is needed to ensure the right value of λ); and the oxidizable BaO may, contrariwise, mop up oxygen, thereby becoming peroxidic.

The principal reactions occurring on the auto-exhaust catalyst (TWC) are:

Oxidation: $2\,CO + O_2 \rightarrow CO_2$

$$\text{'HC'} + O_2 \rightarrow CO_2 + H_2O$$

Reduction: $2\,CO + 2\,NO \rightarrow 2\,CO_2 + N_2$

$$\text{'HC'} + NO \rightarrow CO_2 + H_2O + N_2$$

Whereas platinum (and also palladium) are good for the oxidations, rhodium comes into its own in being able to cope with NO_x elimination much more efficiently than the other metals.

8.6.2 Why is Rhodium in the Auto-Exhaust Catalyst?

It is well known that rhodium metal has water-gas shift catalytic activity and resistance to sintering and to sulphur poisoning. Useful as these facts are in the context of auto-exhaust catalysis, the key virtue that rhodium possesses is its reactivity towards NO; it was Meguerian et al. of Amoco who first realized this fact. Subsequent work, notably by Lambert and Bridge (1984) in Cambridge and by Nieuwenhuys in Leyden, showed that NO dissociates readily on Rh{111}, and that N_2 is desorbed from the rhodium surfaces from *ca* 200 to 300 °C. By contrast, platinum and palladium are less able to dissociate NO than rhodium, the adsorption in these cases being non-dissociative, with NO molecules being desorbed intact upon heating, as shown by Gorte, Schmidt and Gland.

However, facile dissociation of NO, although a necessary first step in NO reduction, is not the only feature of importance pertaining to rhodium. Fisher, Oh, Goodman and Shelef have shown that, in comparing rhodium and palladium, there appears to be very little difference in activity for reduction of NO by CO in the absence of O_2, but dramatically different behaviour in the oxidation of CO for mixtures of NO and O_2. It seems that NO is better able to compete with O_2 for CO over rhodium than over palladium since the addition of NO inhibits the $CO + O_2$ reaction over rhodium, but not over palladium. This behaviour is thought to reflect a high N atom concentration on the rhodium surface under reaction conditions, which, in turn, inhibits O_2 adsorption.

From pioneering and independent infrared studies by Lunsford (1976) (Texas) and Rochester (Dundee), one major difference in behaviour between NO adsorbed on supported rhodium and that on supported platinum and palladium is that only on supported rhodium is the dinitrosyl species observed. Sites able to adsorb dinitrosyls represent a locus for the most important event in the reduction of NO, namely the pairing of the N atoms to yield N_2.

Approximate molecular orbital (extended Hückel; see Section 5.4.6) computations by Hoffmann et al. of 'adhesion' and NO reduction in the TWC, modelled by a monolayer of either rhodium, palladium or platinum on the (0001)O and (0001)Al faces of α-Al_2O_3 prove illuminating. First, it transpires that only the aluminium-exposed (0001) face of α-Al_2O_3 is stable with 'adhered' Rh: the (0001)O face is not. Second, the Fermi level of the composite systems varies dramatically, as seen from Fig. 8.36. And as we see from Fig. 8.37, the position of the Fermi level crucially

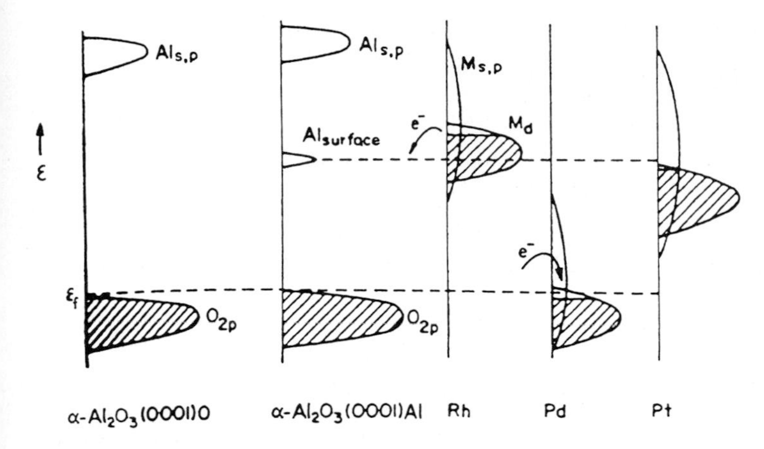

Figure 8.36 Qualitative illustration of charge transfer between metal and ceramic (α-Al_2O_3). The upper broken line indicates the position of the aluminium surface states. (After T. R. Ward, P. Alemany, R. Hoffmann, *J. Phys. Chem.*, **1993**, *97*, 7691.)

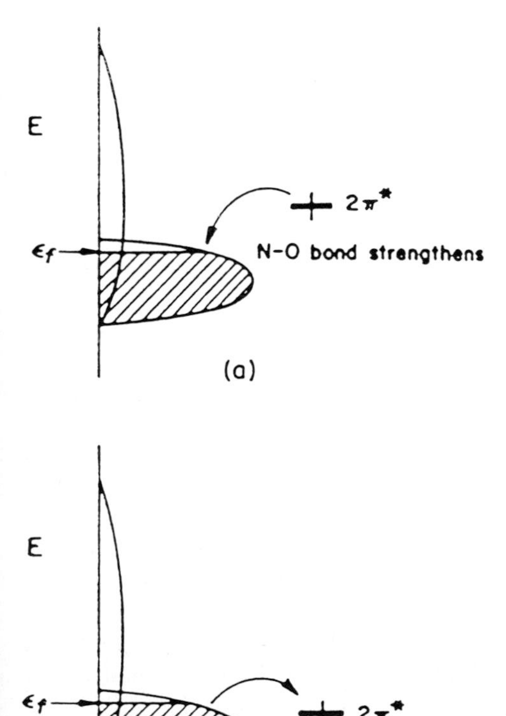

Figure 8.37 Effect of Fermi level of metal in the automobile exhaust catalyst on NO adsorption. (After R. Hoffmann, *Solids and Surfaces: A Chemists View of Bonding in Extended Structures*, VCH, Weinheim, **1988**.)

affects what happens to NO when it is adsorbed on the metal surface. Dissociative chemisorption of NO is favoured by the rhodium surface, this being a direct consequence of the position of the Fermi level vis à vis the $2\pi^*$ molecular orbital of the adsorbed NO. For rhodium the Fermi level is located above, for palladium and platinum below, the antibonding $2\pi^*$ orbital. Weakening of the NO bond ensues on rhodium, strengthening on palladium and platinum.

Another quantum mechanical conclusion is that on all three metals, the coupled product N_2O_2 gives rise to a strong N–N bond and a weakened N–O bond compared with that of a free or adsorbed NO molecule. Shelef and Graham (1994) point out that the N–N and N–O overlap populations are closer to those of a hyponitrite structure of rhodium than on palladium or platinum, however, making rhodium the most likely to reduce adsorbed NO to N_2 via the dimeric NO intermediate.

Armed with these bonding and mechanistic insights, attempts have already been made to replace the expensive rhodium by other active TWC constituents. One strategy suggested is to replace rhodium by a complex site such as Pt(Pd)–Mo, with the noble metal adsorbing the reductant and the base metal splitting the NO or providing sites for its pairwise adsorption.

There has been enormous interest in the possibility of jettisoning the noble metals by using copper-ion-exchanged ZSM-5, which the Japanese worker Iwamoto reported (in 1990) to possess attractive features for auto-exhaust catalysts. One disadvantage from which these catalysts suffer is the gradual breakdown of the zeolitic structure as a result of the leaching of aluminium ions from the framework by the water in the exhaust.

8.6.3 The Catalytic Monolith

Exhaust emission from the modern automobile internal combustion engine has stimulated the development of a catalytic combustion chamber which is designed to oxidize unburnt hydrocarbons and CO to carbon dioxide and water, and reduce NO_x to N_2. It is relevant to recall that, were it not for severe spatial constraints and variable operating conditions occasioned by a frequently changing engine (motor car) speed and exhaust temperature, a shallow fixed-bed tubular reactor containing a mixture of noble metals and promoted iron oxide would serve the purpose. However, pressure drop in the reactor must be very small if the engine performance is not to be affected adversely. Longitudinal dispersion of mass in a shallow packed bed is also significant and impinges seriously on reactor performance. For these reasons, a monolith reactor, i.e. the TWC, consists of an integral bundle of ceramic tubes, the walls of which are coated with catalyst. Figure 8.38 is a sketch of some typical monolith reactors. The honeycomb of channels has regular cross-sections, but the shape of the channels may be circular, hexagonal, square or triangular, depending on the design and mode of construction. The catalyst is usually dispersed on the channel walls by passing a slurry of high-area alumina, previously impregnated with catalyst, through the structure and subsequently calcining it in a furnace. (Metal catalyst supports have also been developed, especially by the Finnish company Kemira, in the form of open-mesh wire structures or staggered layers of metallic screens.) Monolith

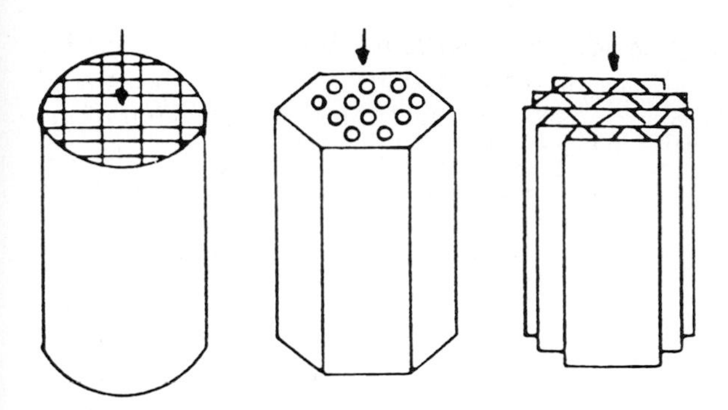

Figure 8.38 Typical monolith reactors.

channel diameters are typically 1.5 mm with a catalyst coating of 0.03 mm on walls 0.25 mm thick. The monolith is usually contained in a stainless steel cylinder about 130 mm long and 130 mm in diameter, fitted into an automobile exhaust system by connecting cones.

8.6.4 Catalytic Monoliths may be Used in Several Applications

Another application of the catalytic monolith is in gas turbine engines. If a catalytic monolith is employed to cause combustion so that substantial rates of burning are achieved at temperatures lower than those at which nitrogen forms undesirable oxides in a conventional combustion chamber, then combustion should be smooth and stable, with very little NO formation. It has been demonstrated that an assembly in which the main combustion reactions occur within the monolith section is a satisfactory means of achieving stable combustion without the emission of NO_x. The catalyst serves to sustain the overall combustion process (which occurs both heterogeneously and homogeneously within the monolith channels), thus minimizing the tendency for unstable combustion that is prevalent in combustors of conventional design. Other applications of the catalytic monolith are worth considering for domestic heating purposes (where direct heat transfer would be a great advantage) and for the industrial processing of rapid catalytic gas reactions.

The above applications are of commercial importance, not least because the pressure drop across the reactor is quite small. Thus catalytic monoliths may well have other potential applications in the future. For example, a sufficiently high reaction rate, under normal circumstances, may only be achievable by using very fine catalyst particles which would cause an unacceptably large pressure drop across a packed tubular reactor. To obviate such a problem, recourse to a fluidized-bed reactor would be the solution. However, another alternative might be the catalytic monolith. Indeed, it has been demonstrated experimentally that a single-channel reactor with catalyst deposited on the tube wall will effectively convert a mixture of CO and H_2 to methane (the methanation reaction).

8.6.5 Rate Characteristics of Catalytic Combustion Processes

It is of interest to plot the reaction rate as a function of bulk gas temperature for the catalytic monolith. Figure 8.39 shows the type of relationship to be expected. Immediately after ignition (sometimes referred to as 'light-off') occurs, the bulk gas temperature is relatively low and the overall combustion rate is controlled by the rate of chemical reaction at the catalyst surface (region A). Mass and heat transfer between the catalyst surface and the bulk gas are rapid compared with chemical reaction, so that the surface temperature and concentration of components is essentially the same as in the bulk gas. As the gas flow through the monolith is very rapid, the system behaves as though it were adiabatic and the reaction rate increases exponentially as the gas temperature increases as a result of chemical reaction at the walls of the monolith channels. Within a short distance downstream from the entrance to the monolith, the temperature will have increased sufficiently for mass

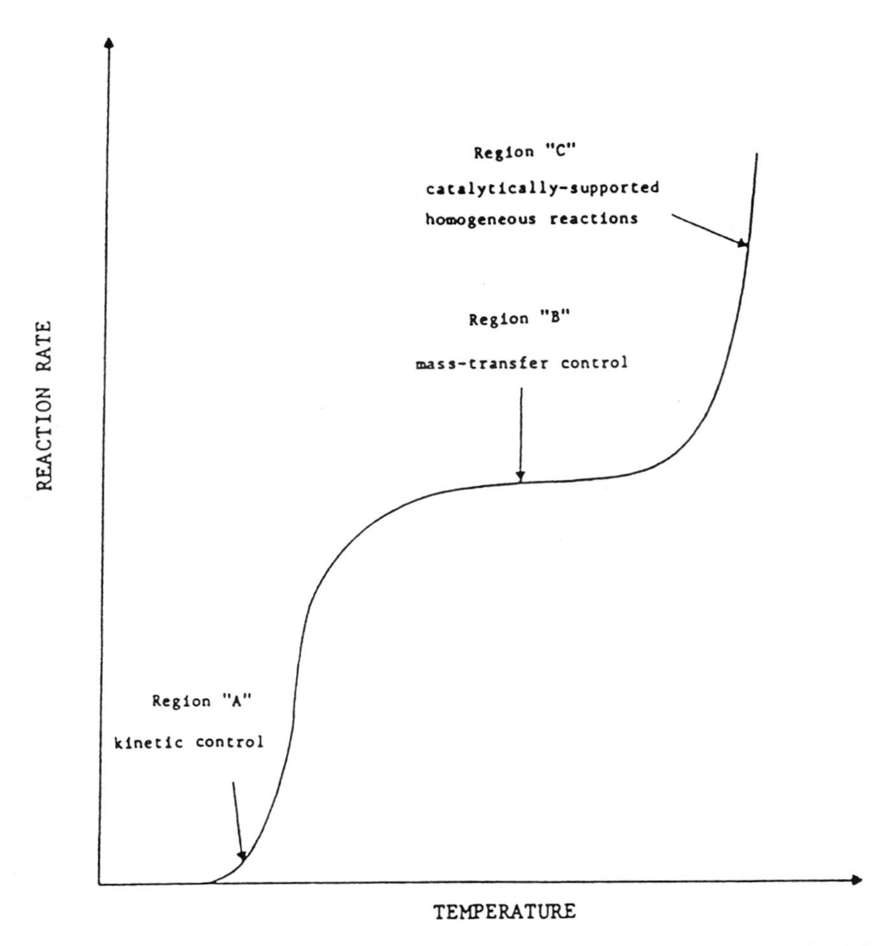

Figure 8.39 Reaction rate as a function of bulk gas temperature for a catalytic monolith reactor.

transfer from bulk gas to catalyst surface to be the dominant rate process (region B). The concentration of limiting reactant is almost zero at the catalyst surface under these conditions because the reactant is consumed so rapidly by the surface reaction. The surface temperature by now has reached values close to the adiabatic flame temperature and exerts an influence on the temperature of the bulk gas phase, which becomes sufficiently hot to initiate homogeneous combustion. Thus the phenomenon of catalytically supported thermal combustion occurs (region C) and the overall reaction rate is much greater than the maximum rate attainable by mass-transfer-limited catalytic reaction.

8.6.6 Combustion Reactions in a Catalytic Monolith Differ from Those Occurring in a Homogeneously Operated Combustor

Chain reactions in which hydrocarbon, hydrogen and hydroxyl free radicals are formed are characteristic of homogeneous combustion processes. Upper and lower limits of flammability are also a feature of non-catalytic combustion reactions, the rate of chain propagation exceeding the rate of chain termination within these limits, thus causing an infinitely fast reaction rate and a corresponding release of energy in the form of heat and light – the flame. Details of chain initiation, propagation and termination account for mechanisms by which both partially and fully oxidized products of combustion are formed. These are well understood, and help us, for present purposes, to distinguish between combustion occurring mainly in the gaseous phase and catalytic combustion occurring entirely at the surface of a catalyst. The following elementary scheme, formulated for methane, serves to illustrate the complexity of oxidation in the gas phase. At temperatures below 550 °C oxidation proceeds by an initiation reaction:

$$CH_4 + O_2 \rightarrow CH_3 + HO_2$$

Propagation of the chain reaction occurs by reactions such as:

$$CH_3 + O_2 \rightarrow CH_2O + OH$$

$$HO_2 + CH_4 \rightarrow H_2O_2 + CH_3$$

Chain branching is thought to be caused by formaldehyde, arising as a product of the propagation reactions, reacting with O_2:

$$CH_2O + O_2 \rightarrow CHO + HO_2$$

and the chain is terminated by the formation of CO:

$$CHO + O_2 \rightarrow CO + HO_2$$

and destruction of either HO_2 or OH radicals at the surface of the vessel in which combustion takes place. Such schemes account for the major products of combustion, explosion phenomena and the observed reaction kinetics. At higher temperatures CO_2 is a product, formed by a radical reaction such as:

$$CO + OH \rightarrow CO_2 + H$$

The combustion of higher hydrocarbons is represented by even more complex schemes involving initiation, propagation, branching and termination reactions. In addition to aldehydes, peroxides and CO, other products such as alcohols and olefins are formed. For the purpose of describing combustion in operating process systems, it is more convenient to represent reactions by overall stoichiometry than to list all the possible elementary reaction steps. Each stoichiometric equation is then associated with an empirical kinetic equation. For example, the gas-phase combustion of propane has been represented by the following three stoichiometric reaction steps:

$$2\,C_3H_8 + O_2 \rightarrow 3\,C_2H_4 + 2\,H_2O$$

$$C_2H_4 + 2\,O_2 \rightarrow 2\,CO + 2\,H_2O$$

$$2\,CO + O_2 \rightarrow 2\,CO_2$$

Empirical rate equations to each of the above steps can be deduced by curve-fitting techniques, matching the observed rate of disappearance of reactants with kinetic expressions containing unknown parameters (exponents of concentration, activation energy and temperature-independent factor). Activation energies of 40–60 kJ mol^{-1} have been assigned to these overall reactions using such a procedure.

The mechanism of catalytic combustion, on the other hand, is quite different from gas-phase oxidation. Conversion occurs at the catalyst surface and involves the chemisorption of reactants. Because of difficulties associated with the identification of chemisorbed species under dynamic reaction conditions, much of the evidence for the mechanism of catalytic combustion has been obtained indirectly. Catalysts which are effective for the oxidation of CO and hydrocarbons, constituents of exhaust gases from internal combustion engines, are rhodium, platinum, palladium, and some metal oxides. At their surfaces, CO and hydrocarbons are oxidized by a reduction–oxidation (redox) mechanism. In the case of CO, non-dissociative adsorption takes place on the noble metals.

With hydrocarbons it is probable that hydrogen abstraction and oxygen insertion occur in a sequence of elementary surface reactions of the type:

$$CH_4(g) \rightarrow CH_3(ad) + H(ad)$$

$$CH_3(g) \rightarrow CH_2(ad) + H(ad)$$

$$CH_2(ad) + O(ad) \rightarrow CO(ad) + H_2O$$

$$CO(ad) + O(ad) \rightarrow CO_2$$

Metal oxide supports (such as CeO_2) act as a medium for the transfer of oxygen and its state of oxidation continually changes as the surface is supplied with and depleted of oxygen. On the noble metals, as described in Sections 8.6.1 and 8.6.2, competitive adsorption between CO and O_2 occurs, the latter being dissociatively bound to the noble metals, rhodium and platinum.

For hydrocarbons, catalytic reaction tends to occur with greater ease as the molecular weight increases. It is thought that the dissociative adsorption of the hydrocarbon at the metal surface is, in most cases, the rate-determining step.

It is convenient, for the purpose of process simulation, to represent the overall catalytic oxidation as either one or two stoichiometric reactions with associated empirical kinetic equations. For propane the single-step stoichiometry is:

$$C_3H_8 + 5\,O_2 \rightarrow 3\,CO_2 + 4\,H_2O$$

while two-step stoichiometry is written:

$$2\,C_3H_8 + 7\,O_2 \rightarrow 6\,CO + 8\,H_2O$$

$$2\,CO + O_2 \rightarrow 2\,CO_2$$

For automobile exhaust applications it is desirable to reduce within the catalyst monolith the small amounts of NO formed in the internal combustion engine, in addition to oxidizing the residual CO and C_xH_y. If it were possible to maintain a stoichiometric balance between CO, H_2 and C_xH_y on the one hand and O_2 and NO on the other, then even a non-selective catalyst would equilibriate the mixture to produce CO_2 and water. In practice, it is difficult to maintain carburation such that a stoichiometric mixture of oxidizing and reducing gases is always present. A degree of selectivity is therefore important and this is achieved by means of the three-way catalyst described in Section 8.6.1.

8.6.7 Simulation of the Behaviour of a Catalytic Monolith is Important for Design Purposes

Prediction of the performance of a monolith contained within the exhaust manifold of a motor vehicle is exceedingly difficult. Variable inlet conditions caused by changing engine loads mean that different performance models are applicable, depending on the conditions obtaining at the reactor inlet. Driving the vehicle at high speed (e.g. motorway or turnpike driving) produces sufficiently high exhaust-gas velocities for gas-to-solid mass-transfer limitations to be prevalent. Conversion in the reactor may be estimated approximately by assuming that plug flow obtains through the monolith channels (the Reynolds number for exhaust-gas flow through the monolith would be about 400 for motorway driving and such conditions would result in laminar flow with a parabolic velocity profile, producing a poorer conversion than would be calculated assuming plug flow). The reactor equation for plug flow in a channel (compare Eq. (73) in Chapter 7 for a packed-bed reactor) is, from Eq. (89) in Chapter 7,

$$-u_l \frac{\mathrm{d}c}{\mathrm{d}z} = \mathrm{R} \tag{57}$$

where u_l is the linear superficial gas velocity, and concentration and reaction rate are measured on a volume basis. For gas-to-catalyst surface mass-transfer limitation the rate may be written similarly to Eq. (88) in Chapter 7:

$$\mathrm{R} = h_{\mathrm{D}}a(c - c_{\mathrm{i}}) \approx h_{\mathrm{D}}ac \tag{58}$$

(in view of the comparatively low interface concentration $(c - c_{\mathrm{i}})$ will be approximately equal to c). Combining these equations and integrating:

$$\frac{c}{c_i} = \exp(-h_D aL/u_l) \tag{59}$$

where L is the length of the monolith channel (cf. Eq. (90) in Chapter 7). Equation (59) above provides an estimate of the monolith conversion efficiency for those conditions when gas-to-solid mass transfer is rate-controlling. The mass-transfer coefficient h_D can either be estimated by means of independent experiments in the absence of chemical reaction or alternatively correlated according to Gilliland in terms of dimensionless Sherwood ($Sh = h_D d/D$), Schmidt ($Sc = \mu/\rho D$) and Reynolds ($Re = \rho u_l d/\mu$) numbers.

For conditions such as an engine starting from normal ambient temperatures or for driving at low and intermediate speed, interphase mass transfer does not limit the overall conversion rate in the monolith. A relationship which has been used with some limited success for predicting performance is the plug-flow equation with a longitudinal dispersion term added to account for mass dispersion in the direction of flow. An alternative model accounting for dispersion of both mass and heat within a shallow bed of catalyst particles is a cascade of well-stirred tank reactors. The number of mixing cells (the well-stirred tank) is set equal to the ratio of channel length to particle diameter L/d_p.

Use of the various applications of catalytic monoliths described above in conjunction with the combustion chambers of stationary or mobile turbines is of particular significance. For example, when a catalyst monolith is used to promote stable combustion as the turbine is providing maximum thrust (conditions for aircraft take-off illustrate such a need), it is of considerable advantage to the designer to be able to predict performance. A comprehensive mathematical model of the catalyst monolith must be constructed if performance is to be simulated effectively.

It is beyond the scope of this chapter to discuss all the details of a satisfactory model of the catalyst monolith. It is nevertheless instructive to consider the principles involved and the component elements of a comprehensive model. Figure 8.40 illustrates the various effects to be taken into account. Both physical and chemical changes occur within a monolith, and are interactive. We note that momentum, heat and mass are transferred in the axial and radial directions, so that a two-dimensional

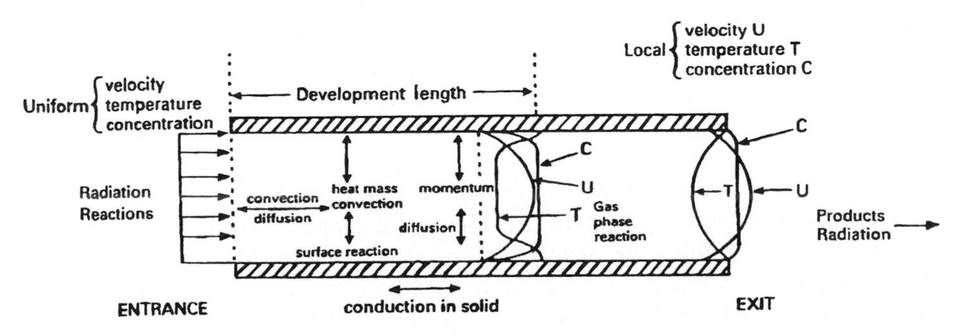

Figure 8.40 Chemical and physical effects which should be accounted for in a comprehensive model of a catalytic monolith.

model with axial symmetry will nearly always produce more accurate predictions than a one-dimensional model. The physical and chemical effects which occur are:

1. convective flow of reactants and products in the direction of flow through a channel;
2. diffusion and dispersion of components in both axial and radial directions;
3. momentum changes resulting from viscous flow;
4. convective heat transfer in the direction of flow;
5. conduction of heat along and through the walls of the monolith;
6. heat radiation from catalyst surface to gas phase and from hot zones within the gas phase to cooler zones;
7. catalytic reaction at the monolith wall; and
8. homogeneous reaction in the gas phase.

It is obvious that a comprehensive model incorporating all of these effects would be very complex. Accordingly, a number of simplifying assumptions have been made by various authors to reduce the computational problems to manageable proportions.

Computed temperature and concentration profiles for the oxidation of propane–air mixtures in a single-channel monolith agree qualitatively with observed experimental results. A two-dimensional model has been used by several authors to describe the continuity of mass and heat (taking into account the mass-transfer limitation from fluid to solid). Some authors have also recognized that the fuel may be converted to product by both catalytic and homogeneous oxidation. They conclude that at 400–500 °C most of the hydrocarbon fuel is converted to CO_2 and H_2O at the catalyst monolith surface, although the importance of homogeneous gas-phase oxidation becomes increasingly important at higher temperatures. The rate of oxidation is strongly mass-transfer-controlled at high temperatures.

Phang treated a monolith as an adiabatic entity and computed, for a single channel, the conversion of propane as a function of inlet temperature for two inlet velocities and a given initial fuel-to-air ratio. The model included all the effects enumerated above, but with the constraint that no heat is transferred to the surroundings. A developing velocity profile was taken into account and satisfactory agreement between predicted and experimentally measured conversions was obtained.

A complex one-dimensional model taking into account all of the steps, except radiation, has recently been formulated to interpret events occurring during the combustion of propane in a catalytic monolith. For engine exhaust applications where the monolith is encased within a cylindrical container, adiabatic operation may be assumed. The most elementary models assume that a single channel is representative of the bundle of symmetrical channels.

Flow within the channels is such that laminar rather than turbulent flow occurs (a Reynolds number of about 400 is typical). The various physical and chemical effects which should be accounted for when constructing an appropriate mathematical model are illustrated in Fig. 8.40. It has been noted that the thermal response time to temperature perturbations of the solid monolith is several orders of magnitude slower than response to gas-phase concentration and temperature changes. Thus it is possible to assume that a quasi-steady state exists in the gas phase, so that the conservation equation for reactant is

$$d(u_l c)/dz = R_h(c, T) - h_D a(c - c_s) \tag{60}$$

where R_h is the net rate of reaction in the homogeneous phase (a function of gas-phase concentration and temperature) and c_s is the concentration of reactant at the channel wall. A steady-state material balance at the channel wall simply asserts that the rate of mass transfer equals the reaction rate at the surface, so

$$h_D a(c - c_s) = R_s(c_s, T_s) \tag{61}$$

where R_s is the rate of surface reaction and is a function of concentration and temperature at the channel wall. A steady-state energy balance can also be written for the gas phase:

$$\rho c_p d(u_l T)/dz = (-\Delta H)R_h(c, T) + ha(T_s - T) \tag{62}$$

For the solid phase the energy conservation is, as explained, an unsteady-state balance. It is written

$$\rho c_{ps} \partial T_s/\partial t = (-\Delta H)R_s(c_s, T_s) - ha(T_s - T) + k\partial^2 T_s/\partial x^2 \tag{63}$$

This is a partial differential equation rather than an ordinary differential equation. The term on the left-hand side represents the accumulation of heat by the solid, and the first and second terms on the right-hand side describe, respectively, the rate of evolution of heat by the surface reaction and the rate of heat transfer between solid and gas phases. The last term on the right-hand side of this equation is the heat flux due to thermal conduction along the monolith walls. Some authors have also included a radiation term, but this is not necessary for sufficiently low operating temperatures. Two additional conditions are required: the equation of continuity, $d(\rho u_l)/dz = 0$, must be satisfied; and an equation describing the overall stoichiometry of reaction is needed. Boundary conditions at $z = 0$ will be $c = c_0$, $T = T_0$ (the entrance conditions) and $\partial T_s/\partial z = 0$ for all t. The initial temperature condition for all values of z would also be specified. At the exit to the monolith the condition $\partial T_s/\partial z = 0$ would also be satisfied for all t as there is no further possibility of reaction beyond the monolith exit.

The above set of non-linear ordinary and partial differential equations can be solved numerically using special numerical techniques. Figures 8.41(a) and 8.41(b) compare some experimental results (obtained by feeding a mixture of propane and air in at the inlet of the monolith) with predictions derived from the above model. Deliberate ramp changes to reactant concentration and temperature show that, although concentration and temperature trends are predicted quite well, even a complex model such as that described does not account quantitatively for all of the events that occur. The main deficiency of the model is its failure to account for heat transfer between channels and consequently to allow for differing extents of reaction in individual channels. Furthermore, it is doubtful whether experimental conditions were adiabatic. More recent work which takes into account channel interactions and radial effects in a two-dimensional model shows that radiation has a significant effect

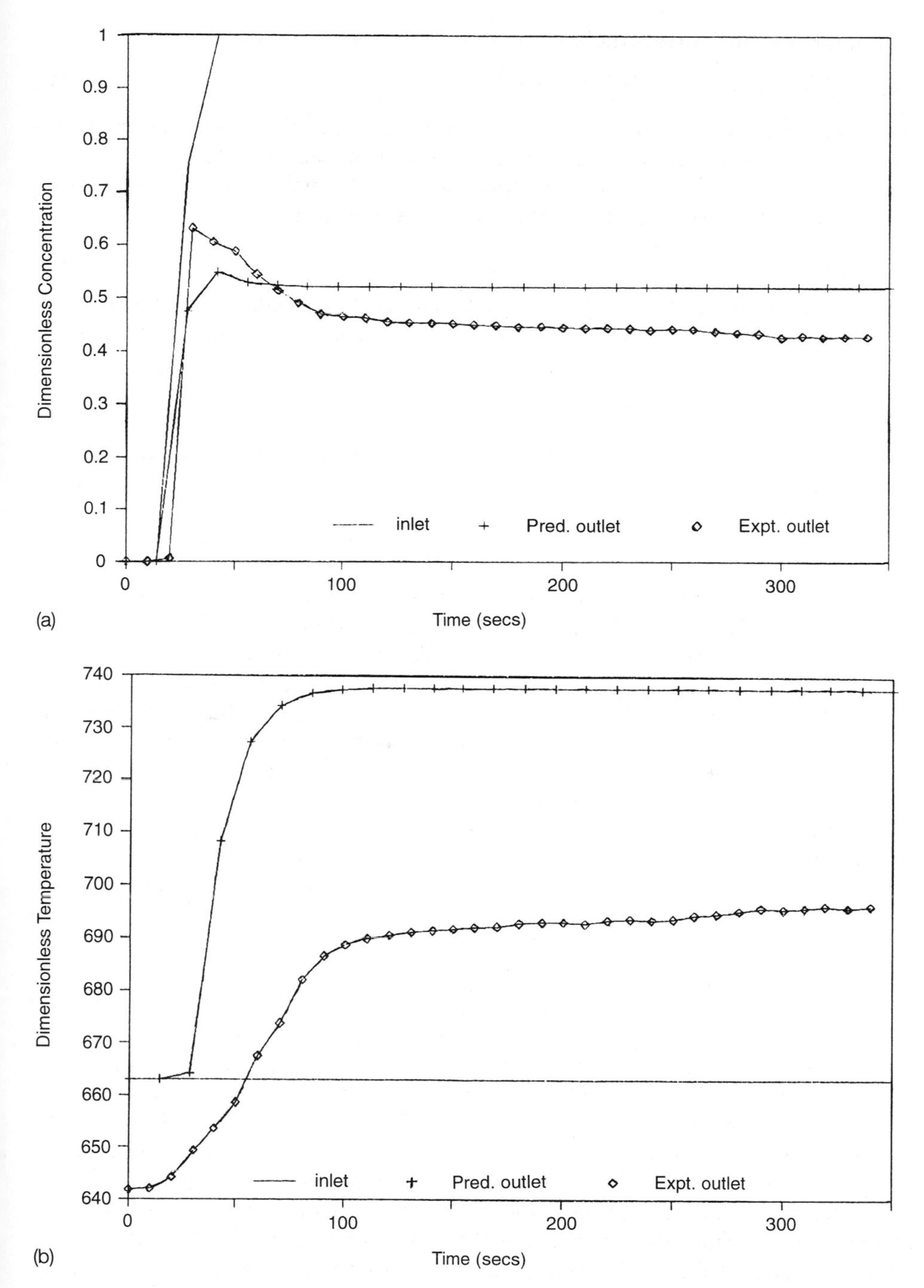

Figure 8.41 Comparison of experimental results with model predictions for propane oxidation in a catalytic monolith; (a) concentration profiles, (b) temperature profiles.

on monolith performance at high temperatures of operation. The depth of catalyst deposit on the channel walls is also important as it can influence the effectiveness factor of the catalyst layer. Two-dimensional models, however, suffer from the disadvantage that quite steep concentration and temperature gradients have to be computed close to the channel walls and this introduces the possibility of computational errors. Despite these difficulties, a recent study by Hayes et al. (1995) of CO oxidation in a tube-wall reactor containing catalyst deposited on the walls in the form of a washcoat confirms that both intraphase diffusion (within the catalyst layer) and interphase mass transport (mass transport from the bulk gas phase to the tube wall) is important. Computed results incorporating both intraphase diffusion and interphase mass transport in a two-dimensional model agree remarkably well with results obtained experimentally.

8.7 Photocatalytic Breakdown of Water and the Harnessing of Solar Energy

As much energy from the Sun reaches the Earth's surface in one hour as the world currently consumes as fossil fuel in a year. The challenge to devise satisfactory means of harnessing solar radiation is therefore irresistible. Were it properly accomplished, it would solve forever most of man's energy needs; it would also open up new avenues for chemical synthesis. Present-day efforts to convert solar energy into fuel or chemical feedstocks devolve upon discovering appropriate catalysts for the following reactions.

Generation of O_2 from water:

$$2\,H_2O \rightarrow O_2 + 4\,H^+ + 4\,e^- \quad E_r = +0.82\,V \tag{64}$$

where E_r *is the redox potential (with respect to the normal hydrogen electrode in aqueous solution) at neutral pH;*
Reduction of water to H_2:

$$2\,H^+ + 2\,e^- \rightarrow H_2 \quad E_r = -0.41\,V \tag{65}$$

Simultaneous generation of H_2 and O_2 from water:

$$2\,H_2O \xrightarrow{h\nu} 2\,H_2 + O_2 \tag{66}$$

and a comparable set of polyelectronic processes for the *photo-assisted reduction of* CO_2, e.g.

$$CO_2 + 2\,H^+ + 2\,e^- \rightarrow CO + H_2O \quad E_r = -0.52\,V \tag{67}$$

$$CO_2 + 2\,H^+ + 2\,e^- \rightarrow HCOOH \quad E_r = 0.61\,V \tag{68}$$

$$CO_2 + 4\,H^+ + 4\,e^- \rightarrow HCHO + H_2O \quad E_r = -0.48\,V \tag{69}$$

To effect reaction thermally requires 4.92 eV; but if water is split using a combination of the dielectronic reduction (reaction (65)) and the tetraelectronic oxidation

(reaction (64)), the free energy required per electron is only 1.23 eV, i.e. the sum of the redox potentials, taking regard of signs, for reactions (64) and (65).
In photosynthesis, the reaction which takes place,

$$CO_2 + H_2O \rightarrow (CH_2O)_n + O_2 \tag{70}$$

requires about the same energy (4.96 eV) as that for reaction (66). Yet Nature succeeds (admittedly not very efficiently) in driving this reaction with photons of the red light in which much of the energy of solar radiation resides (1 Einstein of photons with $\lambda = 100$ nm is equivalent to 1.24 eV). Photosynthesis proceeds by a series of interconnected steps involving, *inter alia*, light absorption, electron transfer and the separation of charge. In artificial photosynthesis, which is what – in a sense – we seek to effect, success is achieved by adopting a similar stepwise approach.
It became apparent quite early in the study of light-induced water-splitting reactions that it is prudent to utilize a photosensitizer (PS). Donors (D) or acceptors (A), or both, are also required; and the choice of a catalyst, heterogeneous or homogeneous, to assist electron transfer or to mediate dark reactions often turns out to be crucial.

8.7.1 Oxygen Generation by Photo-induced Oxidation of Water

To drive reaction (64), which is the thermodynamically most favourable (tetraelectronic) process for the photogeneration of O_2 from water, we first require the photoproduction of an oxidized species of redox potential greater than +0.82 V. We also require a redox catalyst, designated C_{ox}, for reasons which will become clear shortly. Oxidative quenching of the excited state *PS of a sensitizer PS by an electron accepter A leads to generation of PS^+:

$$^*PS + A \rightarrow PS^+ + A^- \tag{71}$$

Recombination of the charge-separated species PS^+ and A^- is prevented or minimized if A^- rapidly undergoes further irreversible transformation (such as spontaneous decomposition). Reaction of the one-electron oxidant PS^+ with water may then proceed to yield O_2 in the presence of a suitable redox catalyst capable of mediating the reaction:

$$4\,PS^+ + H_2O \xrightarrow{C_{ox}} 4\,PS + 4\,H^+ + O_2 \tag{72}$$

This scheme is summarized, in general terms, in Fig. 8.42(a), where the sensitizer PS follows a catalytic cycle while the acceptor A and water are consumed.
We note that the redox potentials of the couples involved must obey the relations:

$$O_2/H_2O(+0.82) < PS^+/PS \quad \text{and} \quad PS^+/{}^*PS < A/A^-$$

These are the general principles. The practical task is to find a suitable PS/A/C_{ox} system which obeys these principles and functions satisfactorily.
It transpires that $Ru(bipy)^{3+}{}_3$ (bipy: bipyridyl) is a sufficiently strong oxidant and may be photogenerated by oxidative quenching of $^*Ru(bipy)^{2+}{}_3$ by the complex $Co(NH_3)_5Cl^{2+}$, the latter functioning as an electron acceptor (Fig. 8.42(b)). The

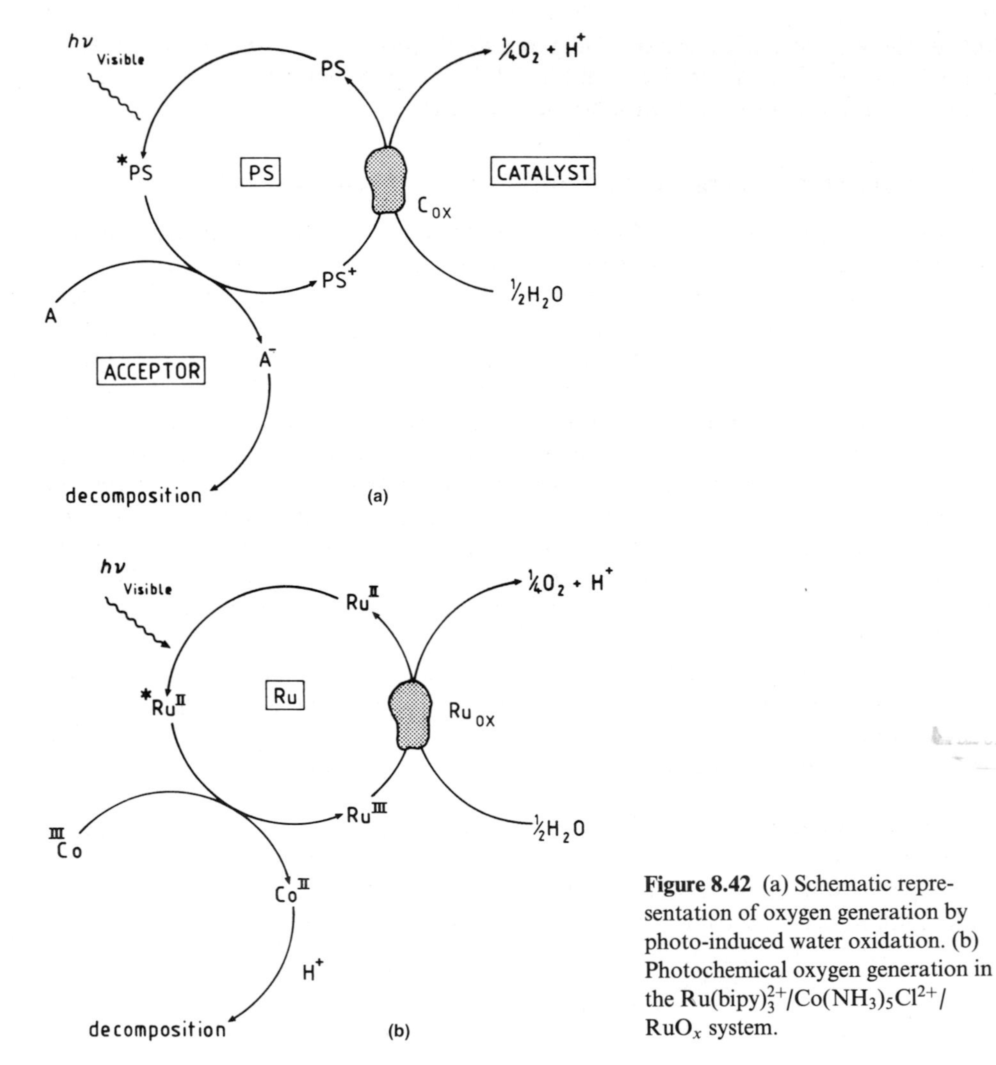

Figure 8.42 (a) Schematic representation of oxygen generation by photo-induced water oxidation. (b) Photochemical oxygen generation in the $Ru(bipy)_3^{2+}/Co(NH_3)_5Cl^{2+}/RuO_x$ system.

$^*Ru(bipy)^{2+}{}_3$ is generated by irradiation of the starting material $Ru(bipy)^{2+}{}_3$, the photosensitizer PS. Non-stoichiometric RuO_2 (best symbolized RuO_x), in the form of a powder, is a good catalyst for the thermal reduction of $Ru(bipy)^{3+}{}_3$ by water with consequent evolution of oxygen.

Effectively, the ruthenium complex undergoes a catalytic cycle, while the Co^{III} complex and water are consumed. The overall process of O_2 generation therefore involves the sacrificial consumption of the cobalt complex:

$$Co(NH_3)_5Cl^{2+} + 4\,H^+ + \tfrac{1}{2}H_2O \rightarrow Co(aq)^{2+} + 5\,NH_4^+ + Cl^- + \tfrac{1}{4}O_2 \qquad (73)$$

The successful choice of RuO_x as a redox catalyst was prompted by the fact that RuO_x anodes show high electrocatalytic activity (i.e. low overvoltage) for O_2 evolution in the electrolysis of water. The exact mechanism of the oxidation step is at present obscure. One possibility is that the Ru^{III} complex changes the catalyst by injecting holes and oxidizing surface-bound hydroxyl groups.

8.7.2 Hydrogen Generation by Photo-induced Reduction of Water

To drive reaction (65) above, which is the thermodynamically most favourable (dielectronic) process for the photogeneration of hydrogen from water, we first require the photoproduction of a reduced species R^- at a redox potential for R/R^- of less than -0.41 V. Again, as with oxygen production, we shall need a redox catalyst, C_{red}, to facilitate the dielectronic step, reaction (65). The R^- species may be formed by electron transfer from the excited state of a suitable photosensitizer such that the redox potential of the excited state, $PS^+/{}^*PS$, is more negative than that of the R/R^- couple. Effectively, substance R serves as a relay between PS and the catalyst, C_{red}, providing a means for intermediate storage of electrons. Finally, just as with the strategy outlined above for oxygen photogeneration, fast recombination processes (this time between PS^+ and R^-, for example) need to be suppressed. This may be realized by using an electron donor D which allows fast back-conversion of PS^+ into PS and is consumed in the process by a fast, irreversible decomposition of the oxidized D^+ species so formed.

The entire system (Fig. 8.42) thus consists of a photosensitive species PS, a relay species R, an electron donor D, and a catalyst C_{red}. In the scheme shown in Fig. 8.43, PS and R follow catalytic cycles, while D and H^+ are consumed, so that the light-stimulated production of H_2 from water is achieved sacrificially, just as that of O_2 is.

Lehn (1982) and his co-workers were led to choose the $Ru(bipy)^{2+}{}_3$ complex as a photosensitizer, not only because it strongly absorbs visible light, but also because it possesses the appropriate redox properties and, in addition, it is known to undergo facile light-induced electron-transfer reactions. The choice of an Rh^{III} complex as a

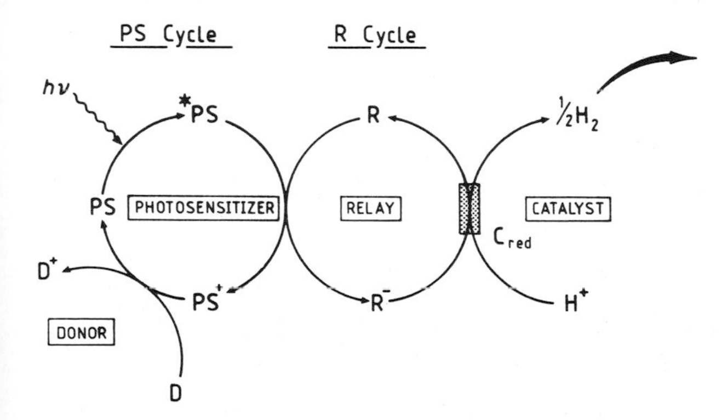

Figure 8.43 General diagram of a photosensitizer/relay/donor/catalyst, $PS/R/D/C_{red}$ system for hydrogen generation by photochemical reduction of water. The oxidized species D^+ decomposes rapidly.

relay species, apart from its having appropriate redox properties, was governed by the knowledge that Rh^{III} complexes participate in two-electron exchanges by interconversion of Rh^{III} and Rh^{I}, a fact of some interest in view of the dielectronic nature of reaction (65). (It has been known since the 19th century that finely divided or colloidal platinum is an efficient redox catalyst; indeed, Faraday's preparation in 1857 of colloidal platinum has recently been shown by Harriman *et al.* (1987) to exhibit the catalytic properties desired in photostimulated hydrogen evolution from water.) The final component chosen by Lehn was the sacrificial donor triethanolamine (TEOA), although many other species, such as ethylenediamine tetra-acetic acid (EDTA), could serve equally.

Several other viable systems for the photo-induced, stepwise conversion of water to hydrogen have been described. Thus Porter and his colleagues have used EDTA, zinc tetramethylpyridylporphyrin (ZnTMPyP), methylviologen (MV) and platinum as donor, photosensitizer, relay and catalyst, respectively. Grätzel (1983) and Kiwi used EDTA, $Ru(bipy)^{2+}{}_3$, MV^{2+} and platinum.

8.7.3 Simultaneous Generation of Hydrogen and Oxygen by Catalysed Photolysis of Water

The two systems described above permit the separate photogeneration of either H_2 or O_2 with consumption of an electron donor or acceptor, respectively, the function of which is to compete with, and to predominate over, recombination reactions. These processes represent the reductive and oxidative components of a complete water-splitting system. To be able to generate H_2 and water simultaneously (reaction (66)), conditions must be found in which the recombination reactions between charge-separated species are minimized in the absence of deliberately added trapping materials. In practice, several schemes may be envisaged to meet these kinetic requirements. In one scheme, a membrane permeable to electrons and protons would separate two half-cells of the oxidative and reductive type based on the separate systems discussed above. An electrochemical variant of this system is also possible in which electrons are led through an external circuit via electrodes, and protons permeate through the membrane. Another scheme employs a single-compartment system combining a suitable photosensitizer with two catalysts, one for water reduction, C_{red} (e.g. Pt), the other for water oxidation, C_{ox} (e.g. RuO_x). Positive results have already been reported for systems of this type. There have been many other ways proposed, and demonstrated, for the simultaneous production of H_2 and O_2 from water photolytically. Some use microemulsions or monolayer assemblies.

Yet another scheme, of demonstrated viability, involves the generation of an electron–hole (e^- and h^+) pair in a semiconductor by irradiation (see Section 1.2.2.2 and Chapter 5). Here the sensitizer is adsorbed on to a semiconductor of colloidal dimensions, and no electron relay is required (Fig. 8.44(a)). The excited state of the sensitizer injects an electron into the conduction band of the semiconductor (Fig. 8.44(b)), where it is channelled to a catalytic site for H_2 evolution. A second ultrafine catalyst, co-deposited on to the colloidal particle, mediates O_2 generation from PS^+ and H_2O, thereby regenerating the original form of the sensitizer.

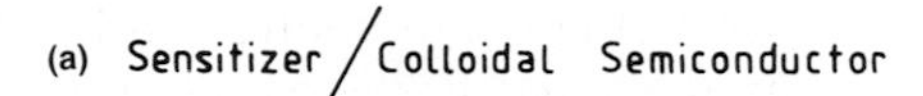

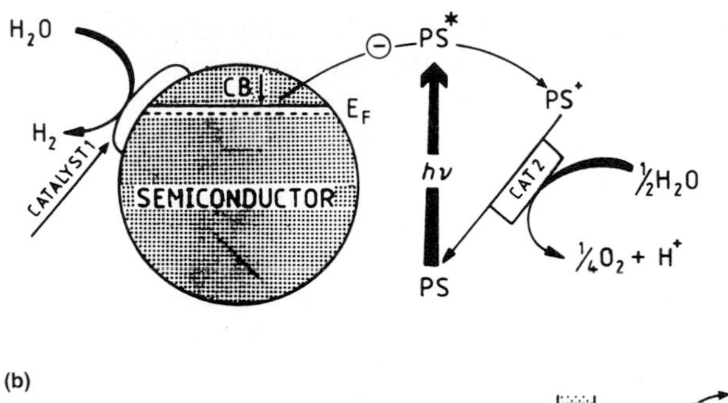

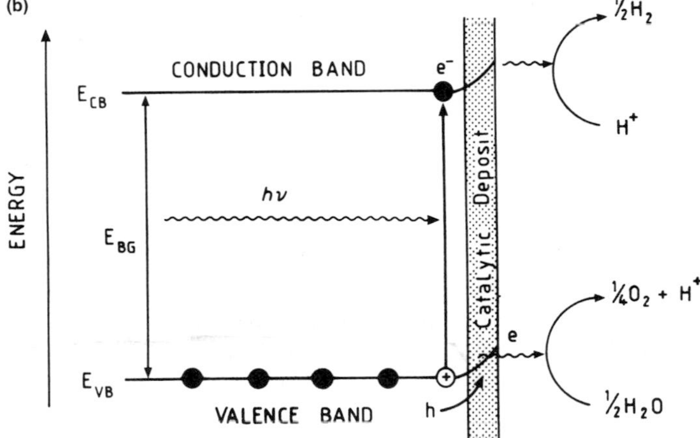

Figure 8.44 (a) Schematic representation of water photolysis by irradiation of a semiconductor material on which a metallic catalyst has been deposited. (b) The water molecule is respectively reduced and oxzidized by the electron/positive hold pair, e^-/h^+, produced by ejection of an electron from the valence band into the conduction band under irradiation with light of energy equal to or higher than that of the gap between the two bands (cf. Fig. 1.6).

Taking this approach a stage further and using a semiconductor such as $SrTiO_3$ as a photoactive solid, and co-depositing (by prior photochemical or thermal treatment) on to its exterior surface a metallic catalyst (Fig. 8.44), photolysis can be effected, to yield H_2 and O_2 simultaneously, by UV irradiation of a colloidal suspension of the coated semiconductor in water. $SrTiO_3$ has the required properties (see Fig. 5.64) in regard to the positions of the edges of the conduction and valence bands vis à vis the redox potentials of H^*/H_2 and O_2/OH. It is also chemically very inert. However, it suffers from the drawback of being sensitive only to UV light, owing to its rather large band gap (3.2 eV). Of the metals that may be co-deposited, rhodium is superior to most others (e.g. ruthenium, iridium, palladium or platinum). XPS measurements reveal that, for the most efficacious catalytic arrangements, some co-existent Rh^0 and Rh^{III} states are required, the implication here being that the photoactive step of the reaction may involve both band-gap excitation of the semiconductor supports and direct UV excitation of the deposited Rh^{III} species. The rationale therefore is that, in the band-gap excitation process, the catalytic deposit facilitates the reaction of the photoproduced electrons and holes with water: the Rh^{III} species catalyses oxygen formation and Rh^0 mediates in hydrogen generation. In other words:

$$4h^+ + 2\,H_2O \xrightarrow{Rh^{III}} O_2 + 4\,H^+ \tag{74}$$

and

$$2\,e^- + 2\,H^+ \xrightarrow{Rh^0} H_2 \tag{75}$$

Several other semiconductors could, in principle, serve as viable candidates for the support medium in the photo-induced splitting of water as the quantitative data show (see Fig. 5.64). Light-harvesting units based on the principle of a colloidally dispersed and coated semiconductor have the advantage over the simple sensitizer relay system in that photo-induced charge separation and redox catalysis are concentrated in a very small and confined reaction volume. Amongst other considerations, with this strategy all the water-splitting events can take place on a single colloidal semiconductor particle, thereby eliminating the need for bulk-phase diffusion of the reactants. On the other hand, there are the disadvanages associated with the tiny-bubble large-field problem and the hazard that hydrogen and oxygen bubbles would be in a two-to-one mixture. Grätzel (1983) has pointed out, however, that modified TiO_2 supports complying with all the criteria needed to generate hydrogen and oxygen simultaneously could also serve as adsorbents for oxygen during the photolysis. In addition to the finite adsorption that would occur spontaneously, electrons injected into the conduction band can be captured by juxtaposed O_2 converting it to O_2^- which is strongly bound to a TiO_2 surface.

Documented examples of heterogeneous systems that will photocatalyse the decomposition of water include platinized TiO_2 powder, platinized monocrystals of $SrTiO_3$–$KTaO_3$, and NiO–$SrTiO_3$ powders, which are suitable for the photodecomposition of water vapour, as is 'raw' $SrTiO_3$.

8.7.4 Other Photochemical Methods of Harnessing Solar Energy

There are many other reactions besides the photolysis of water which could in principle be employed to capture solar radiation. The basic notion, common to all photochemical processes for the conversion and storage of solar energy, hinges on the use of light energy for driving a chemical reaction in the direction opposite to its spontaneous one, thus storing energy. The total energy content of the products exceeds that of the starting reagents. This is the strategy underlying both the production of chemical feedstocks (via syn-gas) and the storage of energy in endothermic reactions (such as $CH_4 + CO_2 \rightarrow 2\,CO + 2\,H_2$) outlined in Chapter 1, where Russian and Israeli work was described.

The key problem here is to discover and select photochemical transformations

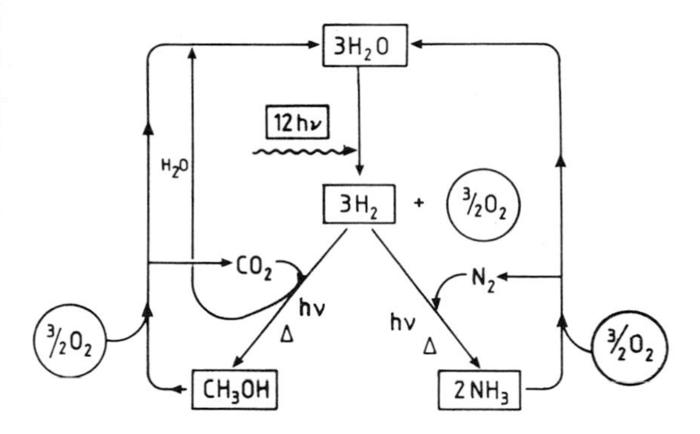

Figure 8.45 Schematic cyclic process for the production of fuels and chemicals by reactions utilizing hydrogen produced by water photolysis.

which would be catalytic (undergoing numerous reaction cycles without fatigue) and have a high storage yield. Promising results have recently been obtained on the photo-induced reduction of CO_2 to CO by a new catalytic system employing Co^I as a crucial element in a cycle, the overall reduction being

$$CO_2 + 2\,H^+ + 2\,e^- \rightarrow CO + H_2O \qquad (76)$$

The water-gas shift reaction

$$H_2O + CO \rightarrow H_2 + CO_2 \qquad (2)$$

can also be driven at a platinized TiO_2 surface under UV illumination. In the long term, there are real prospects of solar energy being utilized photochemically to produce fuels and chemical feedstocks, all derived from water, according to the scheme shown in Fig. 8.45.

8.7.5 Catalysis and Photoelectrochemistry: Photocatalysis and Photoelectrosynthesis

To appreciate how these four topics interdigitate with one another and with the topics discussed earlier, we first recall the main routes for the utilization of solar energy. Albery's block diagram (Fig. 8.46) is useful for this purpose. In a photogalvanic cell, the electron is separated from the hole associated with it, and the cell produces energy in the form of electrical power. In such a cell, the power is stored in a battery or used directly for electrochemical synthesis. An alternative strategy uses solar radiation for the photoelectrolysis of water. In this approach, the energy captured from solar radiation is stored (as H_2) for later use as a fuel or chemical reactant in synthesis. The colloidal electrode cell uses no plate electrode but rather finely dispersed electrodes (see Fig. 8.44). There are, however, a number of distinct types of so-called 'photoelectrochemical' cells which figure eminently in solar energy conversion, and it is useful to summarize the basic principles upon which they operate.

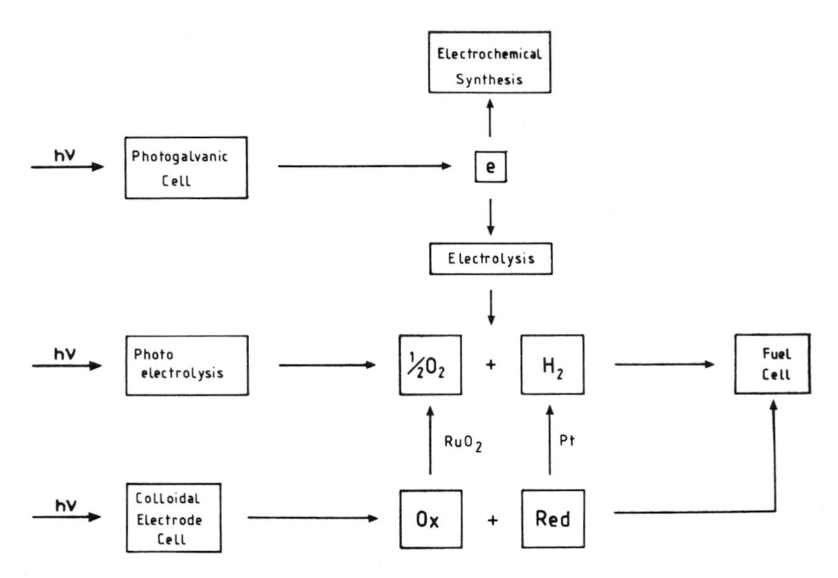

Figure 8.46 Possible routes for the use of solar energy. (After W. J. Albery, *Acc. Chem. Res.* **1982**, *15*, 142.)

8.7.5.1 The Principles

When a semiconductor electrode (typified by ZnO, TiO_2, CdSe or GaAs) is immersed in a liquid electrolyte solution, a charge-transfer process will, in general, take place and, at equilibrium, after the electrochemical potential of the solution (E_{redox}) and the potential of the electrode E_f (the Fermi level – see Chapter 5) have equalized, a relatively thick space-charge layer forms in the semiconductor. For reasons analogous to those responsible for the existence of a Schottky barrier at the junction of a semiconductor and a metal (see Section 5.4.4), band bending occurs when semiconductors form a junction with an electrolyte. In n-type semiconductors the bands bend upwards; in p-type they bend downwards (Fig. 8.47). The equilibration that leads to band bending deprives the semiconductor of some of the majority charge carriers. (The space-charge layer is also termed to depletion region.)

The band bending has an important consequence. Photogenerated minority carriers (holes for n-type, electrons for p-type) are driven to the surface that is exposed to the liquid and are available there for redox reactions. The hole for the n-type semiconductors has an oxidizing power no greater than that represented by E_{VB}, and the electron for the p-type materials (as photocathodes) has a reducing power equal to E_{CB}. Figure 8.48 illustrates the elementary processes, energetics and circuits in the so-called 'photovoltaic' (or 'regenerative') photoelectrochemical cell. Here an electrical current flows through a load in an external circuit *without* net changes occurring in the electrolyte solutions, i.e. the reduction at the photoelectrode is the reverse of that at the counterelectrode. In a solar cell of this kind, the objective is to optimize the product of output voltage E_V and photocurrent.

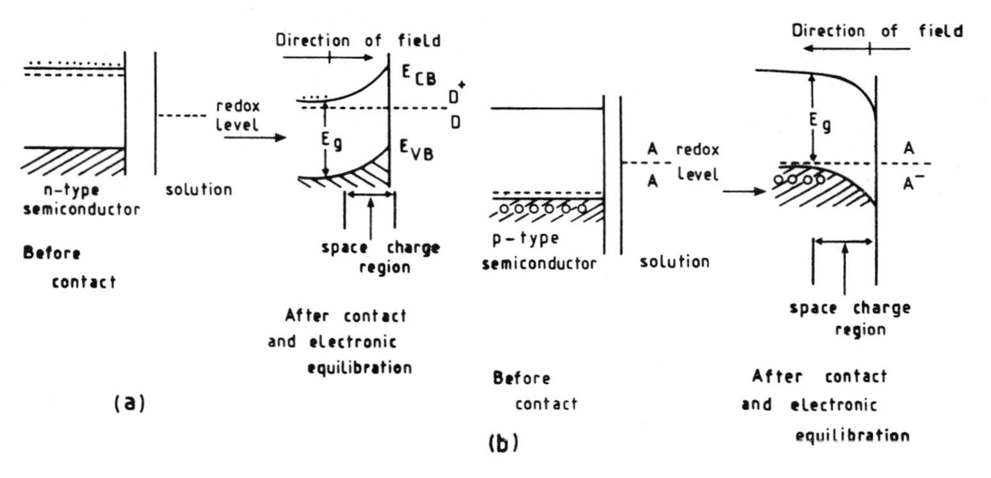

Figure 8.47 When n-type semiconductors are brought into contact with an electrolyte solution the bands bend upwards (a); in p-type semiconductors under similar circumstances the bands bend downwards (b).

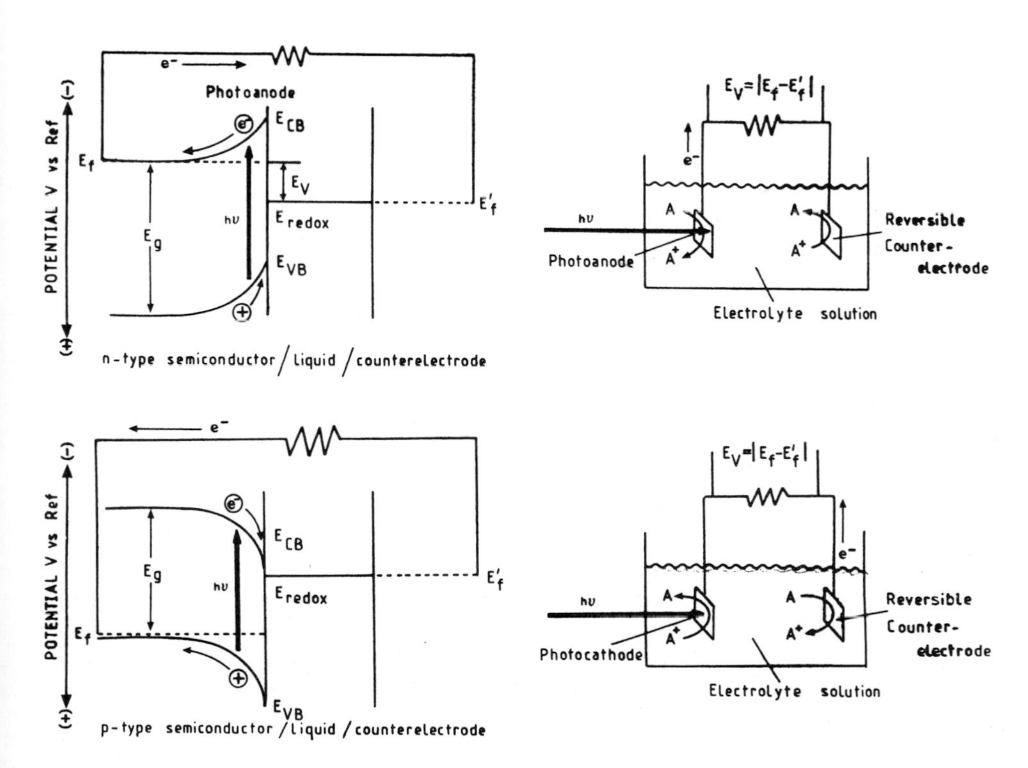

Figure 8.48 Illustration of elementary processes, energetics and circuits in a photovoltaic (or regenerative) photoelectrochemical cell for the conversion of light to electricity (see text).

In photoelectrochemical cells where there is a *net chemical reaction* we may have two distinct situations. In the 'photoelectrosynthetic' cell, the overall reaction that takes place is driven in an uphill direction against the free-energy gradient, so that the fraction of the light used for this purpose is stored as chemical energy in the products. The 'photocatalytic' cell (sometimes termed the photoelectrocatalytic cell), on the other hand, uses light to drive a thermodynamically feasible reaction (i.e. down the free-energy gradient). Here, the radiant energy is not stored as chemical energy but is used instead to overcome the energy of activation of the process – hence the term 'catalytic'.

Before any photoelectrochemical cell can operate satisfactorily, some important practical problems have to be overcome. Take, for example, the situations depicted in Fig. 8.48. The hole generated at the photoanode, as well as effecting oxidation of the species at the interface, can be involved in competing processes such as recombination of the electron–hole pair and photocorrosion or photoanodic decomposition. To be specific, n-type CdSe photoanodes will tend to corrode according to

$$CdSe + 2\,h^+ \rightarrow Cd^{2+} + Se^0 \tag{77}$$

If the solution with which it is in contact contains S^{2-} ions, reprecipitation may occur:

$$Cd^{2+} + S^{2-} \rightarrow CdS \tag{78}$$

The sustained operation of a semiconductor liquid-junction cell (a photovoltaic cell) therefore depends upon the balance of competing interfacial charge-transfer reactions. To be effective, photo-oxidation must dominate over recombination and photocorrosion. One effective way of securing the right balance is by rendering ineffective the surface states (which are inevitably present) which tend to facilitate recombination of electron–hole pairs. Methods of doing this have been evolved by Heller, who used the right choice of anions in the electrolyte solution (e.g. Se^{2-} ions for a sulphide photoanode).

Wrighton (1979) and co-workers circumvented the harmful recombination of electron–hole pairs and other side reactions in an ingenious fashion using a derivatized electrode, which, in a sense, is a tailor-made, optimized electrocatalyst. In their optimization of p-type silicon for the photochemical splitting of water to produce H_2, they first oxidized the surface very slightly to create an oxide layer a few tenths of an Ångstrøm thick, and derivatized the resulting silanol groups with the bromide salt of a dual trifunctional silane. Photons of light strike the crystalline p-type silicon segment of the electrode and the electrons so generated move (see Fig. 8.47), because of band bending, to the surface. They tunnel quantum-mechanically through the thin oxide layer and are captured by the 4,4′-bipyridinium (paraquat) moieties embedded in the silane polymer that forms the second layer of the electrode coating. This reduction step intercepts the normal recombination of the electrons and the holes that remain in the silicon. The silane polymer layer also has embedded within it microcrystals of platinum which catalyse the evolution of H_2:

$$H_2O + e^- \rightarrow \tfrac{1}{2}H_2 + OH^- \tag{79}$$

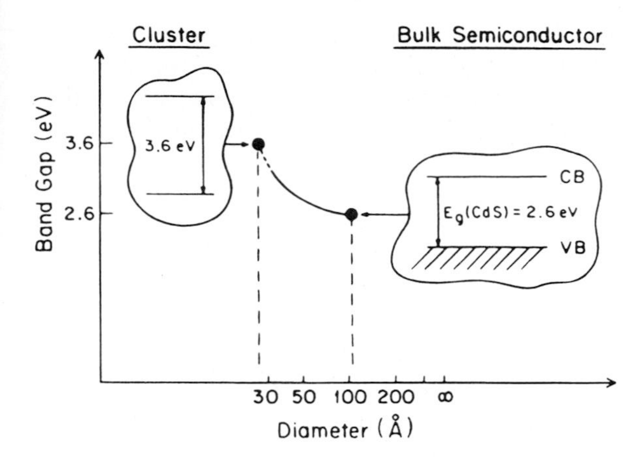

Figure 8.49 Illustration of the quantum size effect (QSE) on the band gap of a semiconductor (After A. L. Linsebigler, G. Lee, J. T. Yates, Jr., *Chem. Rev.* **1995**, *95*, 735.)

Oxidation of those paraquat entities that were earlier reduced as a result of the tunnelling releases electrons to the platinum, where they are used to facilitate reaction (79).

Another important general principle in the whole domain of photocatalysis and light-harvesting is the so-called 'quantum size effect' (QSE). These occur for semiconductor particles (Q-particles) of the order of 10–100 Å in diameter. Anomalies arise when the size of the semiconductor particle is comparable with the de Broglie wavelength of the charge carriers in the semiconductor. Brus and others in 1984 gave the necessary quantum chemical interpretation. The e^-–h^+ pairs produced in Q-particles are confined in a potential well of small geometrical dimensions. The electron and the hole do not experience the electronic delocalization that they do in a bulk semiconductor with its conduction band (CB) and valence bands (VBs), as described in Chapter 5. Confinement produces a quantization of discrete electronic states and, more importantly, increases the effective band gap of the semiconductor. Such effects, illustrated in Fig. 8.49, profoundly influence the photocatalytic properties of the solid. Whereas a bulk semiconductor may not have quite the right characteristics to straddle both the H_2/H^+ and O_2/OH^- couples (see Fig. 5.44), minute particles of the same semiconductor may do so. Huge changes in E_g (see Fig. 8.49) arise from the QSE. Whereas PbS in bulk has a band gap of 0.4 eV, particles of this material of diameter *ca* 13 Å have a gap of *ca* 2.4 eV.

As Henglein (1989), Nozik (1993) and others have shown, the increase in band gap E_g for comminuted particles is observed experimentally as a blue shift in the absorption and emission spectra of the semiconductor. Brus and Henglein have shown that the energy shift in the band gap, ΔE, as a function of particle size is given by:

$$\Delta E = \frac{\hbar_2 \pi^2}{2R_2}\left(\frac{1}{m^*_{\mathrm{c}}} + \frac{1}{m^*_{\mathrm{h}}}\right) - \frac{1.786e^2}{\varepsilon R} - 0.248E_{\mathrm{RY}} \tag{80}$$

where R is the particle radius, m^*_c and m^*_h are the effective masses for the electrons and holes, ε is the permittivity, and E_{RY} is the effective Rydberg energy (given as $e^4/2\varepsilon^2h^2(1/m^*_c + 1/m^*_h)$). ΔE gives the blue shift in the band gap of the particle and determines the transition energy (E_{TE}), the effective band gap, of the minute particle:

$$E_{TE} = \Delta E + E_g^{bulk} \tag{81}$$

8.7.5.2 Practical Examples

The first demonstrated example of sustained photoelectrolysis of water using a photoanode connected, via an external load, to an inert material was reported by Fujishima and Honda in 1972. Using the set-up schematized at the top right-hand side of Fig. 8.50, with n-type TiO_2 as a photoanode and platinum as a counterelectrode, both immersed in water, these workers showed that when the TiO_2 is irradiated with near-UV light, O_2 bubbles off from the irradiated electrode and hydrogen is evolved at the platinum electrode (see Fig. 8.50). Experiments with $D_2{}^{18}O$ show that the reactions involved are:

$$H_2O \xrightarrow{2h\nu(2h^+)} \tfrac{1}{2}O_2 + 2\,H^+ \tag{82}$$

and

$$2\,H^+ \xrightarrow{2e^-} H_2 \tag{83}$$

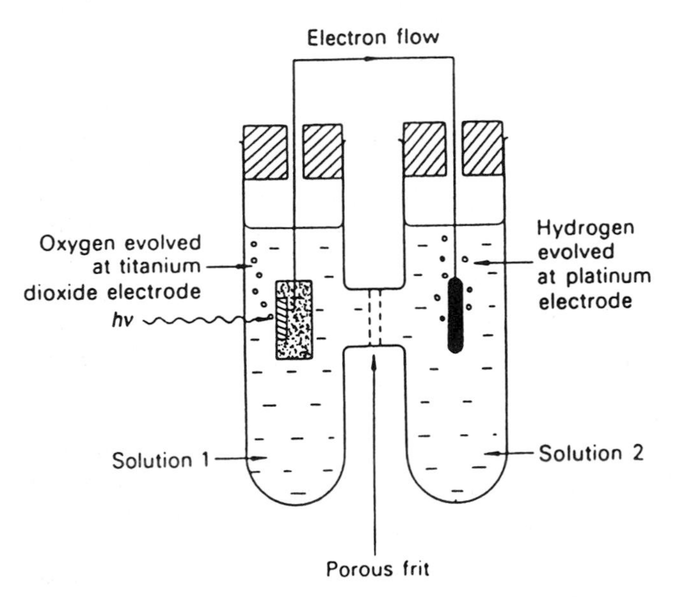

Figure 8.50 The Honda–Fujishima cell for the assisted photoelectrolysis of water. (After M. Archer, *Proc. R. Inst. GB* **1995**, *66*, 97.)

Proof that the TiO_2 anode is a photocatalyst (in effect, a sensitizer operating via its photogenerated electrons and holes) comes from the fact that its mass is the same before and after prolonged photoelectrolysis. Current densities of *ca* 0.5 A cm^{-2} for oxygen evolution were sustained in alkaline solution, using UV-laser excitation for water oxidation. However, the chemistry represented by Eqs. (82) and (83) cannot be effected simply by short-circuiting TiO_2 and platinum and illuminating with photons of energy greater than 3.0 eV (the band gap of TiO_2). Rather, it is found that a power supply in series in the external circuit, providing *ca* 0.2 V of driving force, is required. The need to deliver extra driving force, even though the band gap (3.0 eV) greatly exceeds the energy required to split water (1.23 eV), stems from the fact that E_{CB} is more positive than the H_2 evolution potential by about 0.2 V (see Fig. 5.44). $SrTiO_3$, in view of its band gap and the fact that this straddles the H_2 and O_2 evolution potentials, is a better photoanode which, upon short-circuiting to a platinum electrode and exposure to radiation, photodissociates water. Unfortunately $SrTiO_3$ has too large a band gap to be useful in a solar cell, but it is instructive to recall that this cell, with $SrTiO_3$ as the photoanode, has a conversion efficiency (of monochromatic UV light to stored chemical energy) of 25–30%; by definition it is a photoelectrosynthetic cell, the water being 'driven' thermodynamically uphill within it, via the agency of adsorbed radiation at the anode, to yield stored energy in the oxygen and hydrogen.

Photoelectrosynthetic cells utilizing p-type GaP electrodes have been shown by Bard and others to be capable of reducing CO_2 to formaldehyde and methanol. As with Fujishima and Honda's set-up for producing O_2 and H_2 from water, in these cells also an external electrical bias had to be provided so that, in reality, we are dealing here with photo-assisted electrolyses. Other chemically useful products, such as Cl_2 (now prepared industrially in a manner demanding electrical energy), have been generated by photo-oxidation of chloride using TiO_2 electrode:

$$2\,H^+ + \tfrac{1}{2}O_2 + 2\,Cl^- \rightarrow H_2O + Cl_2 \tag{84}$$

An elegant photoelectrosynthetic means now exists for executing the reduction of aqueous cupric solutions so as to generate O_2 and metallic copper:

$$Cu^{2+} + H_2O + h\nu \rightarrow Cu^0 + \tfrac{1}{2}O_2 + 2\,H^+ \quad \Delta G^\ominus = 1.71\ \mathrm{kJ\,mol^{-1}} \tag{85}$$

Preferential deposition of the Cu^0 occurs on the unilluminated side of a photoactive TiO_2.

One of the attractions of photoelectrosynthesis is that it offers a strategy, which can be achieved provided the right catalysts are developed, of converting inexpensive, readily available materials (H_2O, CO_2, N_2 or CO) into useful fuels. Another, more dramatic, illustration of the merit of photoelectrosynthesis is Bard's demonstration of amino acid synthesis (glycine, alanine, serine, aspartic acid, glutamic acid) from CH_4, NH_3 and H_2O in contact with irradiated suspensions of platinized TiO_2:

$$\begin{aligned} &NH_3 + 2\,CH_4 + 2\,H_2O \xrightarrow{h\nu} Pt/TiO_2 + H_2NCH_2CO_2H + 5\,H_2 \\ &\Delta G^\ominus = 13.2\ \mathrm{kJ\,mol^{-1}} \end{aligned} \tag{86}$$

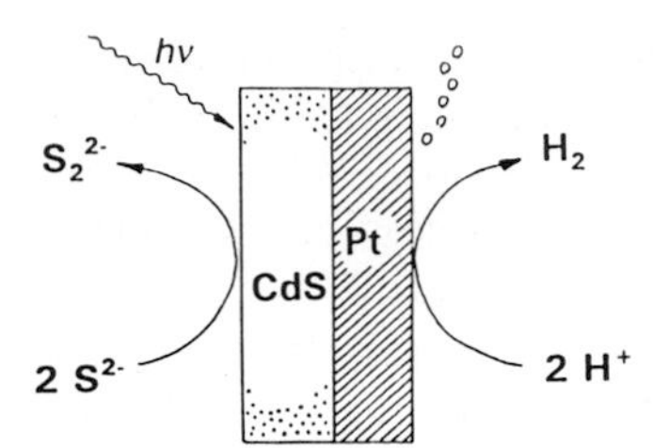

Figure 8.51 Schematic of Nozik's (1993) H_2-evolving chemical diode. (After M. Archer, *Proc. R. Inst. GB* **1995**, *66*, 97.)

Products of intermediate functionality, such as CH_3OH, C_2H_5OH and CH_3NH_2, were also generated in this way.

Of the numerous photocatalytic processes that have recently been discussed, the photo-Kolbe reaction is among the most interesting:

$$CH_3COOH(l) \xrightarrow{h\nu} \tfrac{1}{2}C_2H_6(g) + CO_2(g) + \tfrac{1}{2}H_2(g)$$
$$\Delta G^{\ominus} = -1.05\,\mathrm{kJ\,mol^{-1}} \tag{87}$$

This reaction is induced by long-wavelength UV irradiation of an n-type TiO_2 photoanode immersed in acetonitrile and some other reagent. When the same irradiation was conducted on platinized TiO_2 (anatase) powder in aqueous acetic acid, methane became the major product. The electron–hole pairs generated when light of energy greater than the band gap (≥ 3.0 to $3.2\,\mathrm{eV}$) is absorbed by the TiO_2 do not readily recombine, presumably because of the bending of bands caused by the equilibria $H + e^- \rightleftharpoons H(ad)$ and $2\,H^+ + 2\,e^- \rightleftharpoons H_2$. The low-lying holes thereby created lead to the oxidation of CH_3COO^-, thus initiating the Kolbe process. The rapid decomposition of the CH_3CO_2 radical to the CH_3 radical and CO_2 prevents any reverse reaction. The subsequent processes, judging by the production of CH_4, are reductive. A simple H_2-producing chemical diode, based on a junction of CdS and platinum, is shown in Fig. 8.51.

8.7.5.3 The Prospects

Although photochemical and photocatalytic processes are likely to be less promising as commercial prospects than photovoltaic cells in harnessing solar radiation, there is likely to be continued interest in devising more effective systems. To maximize utilization of solar energy by tailored semiconductors, one needs a solid which must satisfy several requirements. Such a solid has to have:

1. an appropriate band gap, E_g (preferably $1.5\,\mathrm{eV} < E_g < 2.5\,\mathrm{eV}$);
2. suitably positioned conduction band and valence band edges (otherwise known as flat-band potentials) where recombination of e^-–h^+ pairs in its bulk is negligible;
3. few grain or sub-grain boundaries (which are notorious e^-–h^+ recombination centres); and
4. the requisite electrochemical stability.

Great efforts are currently being made to design solids which result in the best combination of these desiderata. J. M. White et al. found that minute particles of ZnS, overlaid by a coating of CdS, all supported on high-area SiO_2, show remarkable activity for H_2 generation from water under illumination with visible light in the absence of a platinum catalyst. All this capitalizes on the quantum size effects (QSE) described above. The opening out of E_g often succeeds in better positioning of the flat-band potentials. As a consequence, reductions that cannot occur with bulk semiconductors can take place with the comminuted analogue. Thus it has recently been shown that photoevolution of H_2 occurs from aqueous colloids of HgSe (particle diameter < 50 Å) and that CO_2 reduction to formic acid from CO_2-saturated aqueous solutions occurs with CdSe colloidal catalysts of less than 50 Å diameter.

Another promising approach to better photocatalysts is deliberately to dope a parent structure with progressive, and quite large, amounts of a second material. Bi_2O_3 is known to be capable of forming a large variety of structures that are subtly related to that of the progenitor Bi_2O_3; Harriman et al. (1988) have found that when it is modified by adding Nb_2O_5 so as to form ordered solid solutions, it yields band-gap changes that augur well for designing required photocatalysts. The 31 : 1 Bi_2O_3/Nb_2O_5 solid solution, for example, freely converts aqueous solutions of 4-chlorophenol into CO_2 and Cl^- ions under visible light, thus demonstrating its potential utility in harnessing solar radiation for treatment of aqueous effluents. Yet a further approach to novel design of electrodes of possible use in the harnessing of solar energy is to adapt and fine-tune micro- and meso-porous hosts as well as clays and the structures based on pillared clays.

It has to be recognized that, in photoelectrochemical (PEC) cells based on single junctions of semiconductors with solutions or metals, photopotentials are rarely above 0.6–0.8 V. This rather low driving force limits the range of possible photo-electrosynthetic reactions that may be carried out in PEC cells without the application of an external field. Alternatively, the photoactive junction can be a metal semiconductor interface (Schottky barrier – see Section 5.4.4) with the faradaic reaction occurring at the metal, sometimes bearing an appropriate electrocatalyst. For example, PECs with the following junctions have been described: Au/n-GaP and Pt/n-GaAs. Such cells have the advantage that the semiconductor is protected from the solution and shows photopotentials that are independent of the redox potential of the species in solution. However, the reported potentials are again usually below *ca* 0.6 V. Related PECs involve photopotentials that arise at a p–n semiconductor junction, again protected from the solution by a metal overlayer. The most highly developed systems of this type are probably the p-Si/n-Si junctions separated from the solution by suitable noble metal overlayers that are used in the Texas Instruments (TI) Solar Energy System. The photopotential developed at these junctions is about 0.55 V. These junctions are produced on small (0.2 mm diameter) silicon spheres that are embedded in glass and backed by a conductive matrix to form arrays in contact with a solution. The TI system operates cyclically and provides electric power from the fuel-cell reaction $H_2 + Br_3^- \rightarrow 2\,HBr + Br^-$ ($E = 1.05$ V) when the sun is not shining, and is recharged by the reaction $2\,HBr + Br^- \rightarrow H_2 + Br_3^-$ when it is shining. In effect, therefore, the overall conversion of solar energy in this TI cell provides an open-circuit voltage of 1.05 V.

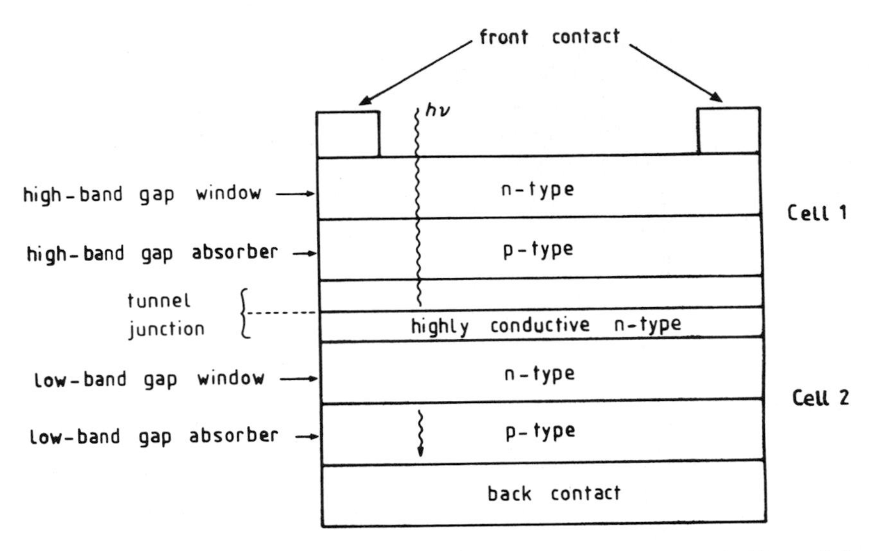

Figure 8.52 In a two-terminal cascade cell, high-energy protons are absorbed in a cell that is tuned to high energies, and low-energy protons in a cell tuned to low energies.

Future effort will doubtless focus on ways of generating higher output voltages by connecting a number of junctions in series or by coupling PECs in the appropriate fashion. Particularly interesting configurations would entail multilayer devices with several photoactive junctions, especially if semiconductors with different band gaps are utilized. This is precisely the strategy that has recently proved so successful in the fashioning of solid-state photovoltaic devices using so-called 'tandem' or 'cascade' cells (Fig. 8.52). The range of materials available for the exploitation of photovoltaic cells – amorphous silicon, $CuInSe_2$, CdTe and GaAs – are already most appropriate in the tailoring of devices with variable band gaps. To pursue the development of multijunction PECs, new kinds of junctions will be needed. But Bard and his co-workers have already achieved significant success and have demonstrated that PEC arrays can be used in the photoassisted bromination of phenol and in the chlorination of cyclohexene.

Grätzel and co-workers in Lausanne have devised an ingenious photovoltaic cell where an inorganic dye is fired on to an optically transparent base, the anode being TiO_2 (see Fig. 8.53). An I_2/I^- couple in solution is oxidized at the dyed electrode, and reduced at the other electrode. This photovoltaic cell produces electric power rather than a chemical fuel. Grätzel's cells have an efficiency greater than 10% in artificial light, surpassing the performance of amorphous silicon, and only about one-fifth of its cost. They are under commercial development.

More will undoubtedly be heard in future of the junction semiconductor of the kind fashioned in the Boreskov Institute in Novosibinsk, Russia. These are already in use as a means of eliminating H_2S from inland seas in the former Soviet Union. The band bending at the heterojunction of this device (Fig. 8.54) serves to separate the e^-–h^+ pair for productive photocatalytic purposes.

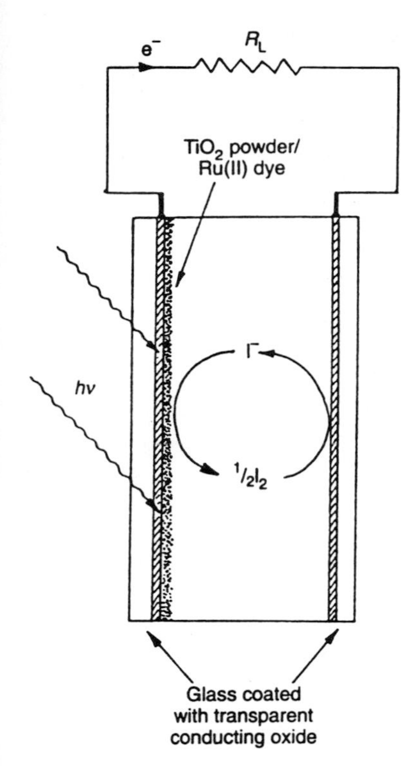

Figure 8.53 The Grätzel electrochemical photovoltaic cell. The anode consists of a high-surface-area TiO_2 paste fired on to an optically transparent base and coated with an orange ruthenium dye. The solution contains an I_2/I^- redox couple dissolved in a non-aqueous solvent. (After M. Archer, *Proc. R. Inst. GB* **1995**, *66*, 97.)

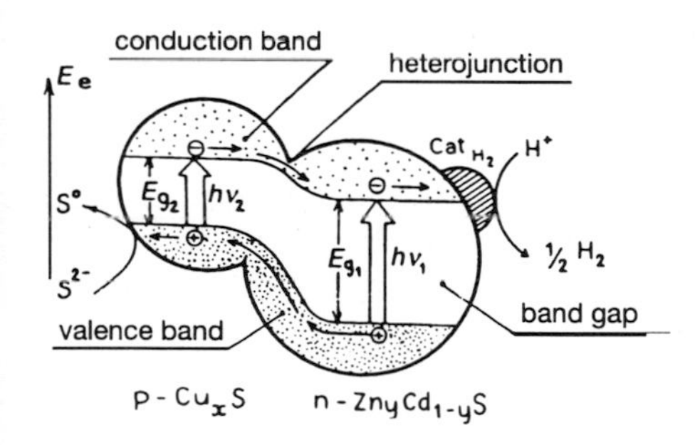

Figure 8.54 Energy diagram for the photo-separation of charges and scheme for the catalytic process in a suspended semiconductor particle with the microjunction $Cu_xS/Zn_y\,Cd_{1-y}S$. E_{g_1} and E_{g_2} are the widths of the band gaps of the semiconductor phases and $h\nu_1$ and $h\nu_2$ the corresponding photon energies required to promote electrons to the conduction band. (After J. M. Thomas, K. I. Zamaraev, *Angew. Chem.*, **1994**, *106*, 316.)

$$R^1N{=}NR^2 + R^3H \xrightarrow[\text{MeOH}]{h\nu/\text{MS (M= Zn,Cd)}} R^1N(R^3)NHR^2 \quad \mathbf{1}$$

1a, **1d**: $R^1 = R^2 =$ Ph; **1b**: $R^1 = R^2 = p$-Tolyl; **1c**: $R^1 =$ Ph, $R^2 = t$Bu

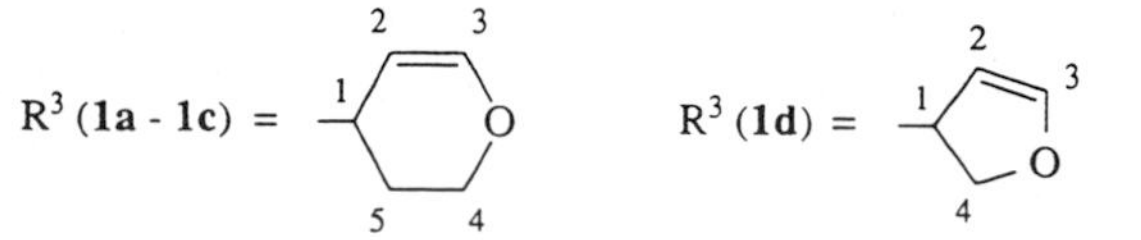

R^3 (**1a** - **1c**) = [ring: positions 1, 2, 3, O, 4, 5] R^3 (**1d**) = [ring: positions 1, 2, 3, O, 4]

Scheme 8.11

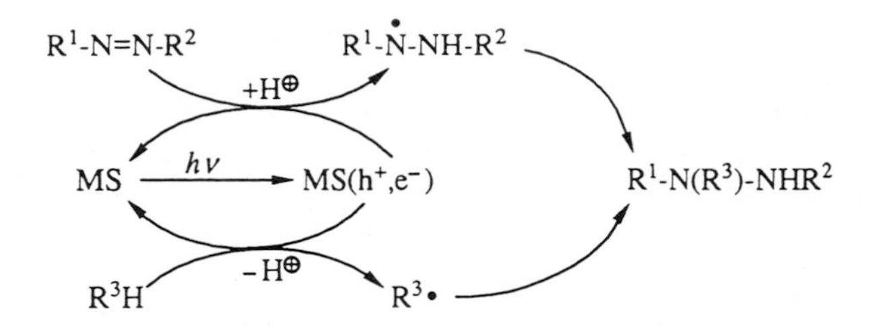

M= Zn, Cd

Scheme 8.12 Postulated photocatalysis cycle.

Kisch et al. at Erlangen, pursuing the themes of Fox in Texas and de Mayo in Ontario, have recently shown that suspensions of semiconductors such as TiO_2, ZnS and CdS have much merit as photocatalysts in the synthesis of organic compounds. Potentially useful preparative reactions include oxidations, reductions, cyclodimerizations, isomerizations and photoaddition of cyclic enol ethers to 1,2-diazenes. Scheme 8.11 summarizes the nature of the photoadditions recently reported, and Scheme 8.12 sketches the postulated photocatalytic cycle involving ZnS or CdS semiconductor catalysts.

8.8 Catalysis Using Microporous or Mesoporous Solids and Modified Clays: Its Growing Role in the Petroleum Industry and Clean Technology

On numerous occasions in this monograph zeolites and other molecular sieves have been highlighted as catalysts of growing commercial importance and also as ideal model catalysts, a situation altogether rare for catalytic solids in general. Molecular sieve solids (typified by aluminosilicates, ALPOs, SAPOs and MeALPOs, already described in Chapter 5) have, effectively, three-dimensional surfaces. Teeming with micropores, cages and channels, often interconnected, most if not all of the atoms of the bulk are also at the surface, and are accessible to reactants (and products) that are of a size capable of diffusing from or towards the exterior surfaces.

Microporous crystalline catalysts based on naturally occurring zeolites (Section 5.2.7) form an open network possessing a porosity such that 30–50% of the solid is void space. About one-quarter of the elements of the Periodic Table can now be incorporated into the framework structure of this kind of solid, which have very high areas (typically 400–500 $m^2\ g^{-1}$) with over 99%, depending on the crystallite size, of the surface area inside the bulk. By controlling, during or after synthesis, the number of heteroatoms in the framework, finely adjustable concentrations of active sites may be placed in a spatially uniform fashion within the inner surface of the catalyst. The precise location and environment of the active site – be it a proton loosely attached to a framework oxygen (as in Colour Plate 8) or a redox element such as titanium or cobalt (as in Colour Plate 6) – can be determined experimentally, especially by X-ray absorption spectroscopy.

In-situ studies of bound reactants (or inhibitors) within the channels and cavities of the catalysts may be followed quantitatively by IR spectroscopy, solid-state NMR, X-ray diffraction, X-ray absorption and fluorescence and a range of other techniques. In addition, as we saw in Section 5.3, computational methods are very productive and revealing for such catalysts.

Owing to their pivotal role in the processing of petrochemicals and the refining of oil, zeolites are manufactured on a massive scale: as long ago as 1980, 40 000 tons per annum were produced. Five years later, the world-wide consumption of zeolitic cracking catalysts alone was close to 200 000 tons per annum. In one day, a typical fluidized catalytic cracker unit consumes a ton of fresh zeolite. Each day, over 10 million barrels of oil are catalytically cracked throughout the world, a quarter of a million barrels in South Wales alone in the refineries of Amoco, BP, Gulf and Texaco.

The proven commercial catalytic uses of zeolites take advantage of two distinct features of the structural properties of this expanding class of solid:

1. the facility with which they favour the production of carbo-cations (alkylcarbenium) ions; and
2. their ability to function in a shape-selective sense.

According to IUPAC rules of nomenclature, the ions that were formerly called carb*o*nium ions are to be designated carb*e*nium ions. The term carbo-cation, which has gained increasing currency, covers both carbenium ions and the positively charged ions that contain so-called pentacoordinated carbon. The term 'carbonium ion' is still widely used, however.

In essence, the high and adjustable acidity, as well as the other characteristic features of the zeolite structure, loom large in governing the type of catalysis that these solids can sustain. However, zeolites are frequently used in a bifunctional mode: finely dispersed noble metals constitute one function and serve to facilitate dehydrogenation ⇌ hydrogenation equilibria. The Brønsted acidity of the zeolite itself con-

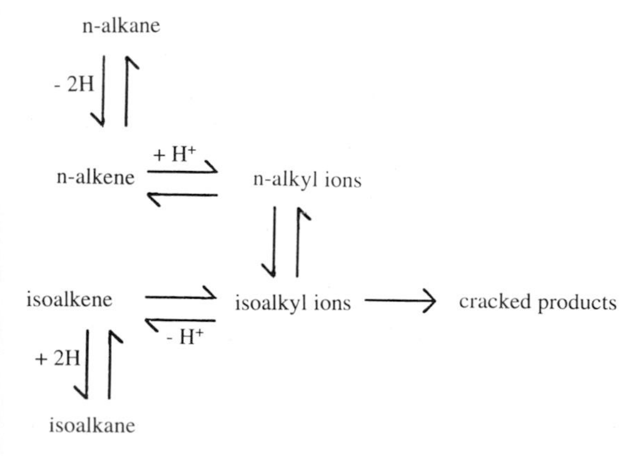

Scheme 8.13 Bifunctional behaviour of zeolites.

stitutes the second function. We therefore see how, in broad terms, bifunctional behaviour manifests itself (Scheme 8.13).

We also see how zeolites take on a pivotal role, not only in cracking but in alkylation and in the more modest rearrangements associated with isomerization. Thus, alkylcarbenium ions are formed from alkenes by reversible protonation:

$$\mathrm{>C{=}C<} + H^+ \rightleftharpoons R_2^+$$

and from alkanes by hydride transfer:

$$R_1\text{–}H + R_2^+ \rightleftharpoons R_1^+ + R_2\text{–}H$$

The alkylcarbenium ions can then undergo three possible types of reaction:

Rearrangement: $R_1^+ \rightleftharpoons R_2^+$

β-Cleavage: $R_1^+ \longrightarrow R_2^+ + C{=}C$

Addition to olefin: $R_1^+ + \mathrm{>C{=}C<} \longrightarrow R_2^+$

Indeed, 'acid catalysis' of this kind can proceed on non-zeolitic solids. In some of the classic work of Ipatieff and Pines, siliceous solids laced with concentrated mineral acids functioned by providing both a ready source of protons and a porous solid to facilitate separation of product. 'Solid phosphoric acid' (SPA), the term used to describe a mixture of kieselguhr and H_3PO_4 (see Section 1.2), was used to oligomerize propylene, the trimers being useful blending agents in petrol (gasoline) and the tetramers in the manufacture of detergents. During World War II, SPA was the catalyst employed to manufacture cumene, a crucial component of high-grade fuel for aircraft.

The supreme advantages that zeolitic acid catalysts possess compared with SPA or silica–alumina gels are their crystallinity (hence their sharply defined pore-size distribution) and their thermal stability, which permit a repeated regeneration by high-temperature oxidation of carbonaceous by-products that gradually poison a fresh catalyst. There is also the favourable dimensions of their apertures which control diffusion into and out of the microporous catalyst. This leads us, in turn, to shape-selective aspects of catalysis (see Section 1.2.1, Figs. 1.1 and 1.2 and Colour Plate 1). We should note here the importance of diffusion in catalysis by zeolites.

Although much remains to be discovered about diffusion in zeolites, thanks to the illuminating work of Weisz we may note how dramatically the diffusivity varies as a function of pore size (see Fig. 8.55). Zeolites with pore diameters in the range 4–9 Å exhibit a region of diffusivity beyond the so-called regular and Knudsen regions, and this is termed the configurational regime. In this region, where molecules diffuse through pores of near-molecular dimensions, there is a huge spread (ten orders of magnitude) of diffusivities. Subtle differences in molecular structure exert a profound effect on the diffusivity. For example, chain length of an alkane as well as decreased carbon–carbon bond mobility (as in going from an alkane to an alkene) can have a marked effect. Weisz found that, in certain zeolites, *n*-2-butene has a diffusivity some

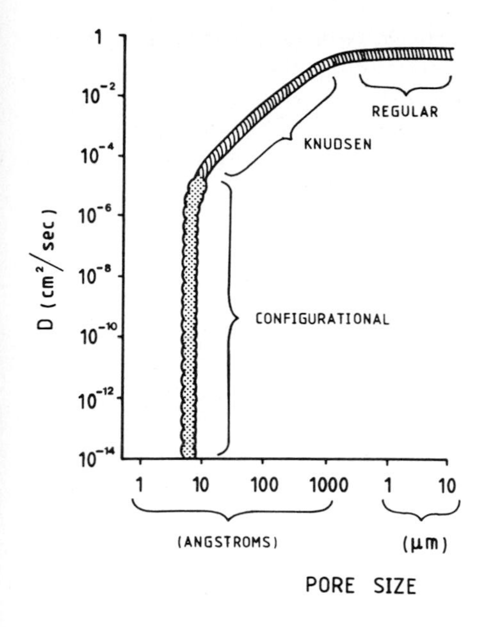

Figure 8.55 Dependence of magnitude of diffusivity upon pore dimension (After P. B. Weisz, *Pure Appl. Chem.*, **1980**, *52*, 2091.)

10^5 times that of *n*-butane. There is an immediate kinetic consequence of retarded diffusion, and the well-known Thiele modulus concept (see Section 4.6) expresses in quantitative terms the diminshed reaction rates that are associated with a fall in diffusivity. The other side of the diffusivity coin is shape-selectivity. A bulky molecule which cannot gain access to the active sites within a zeolite owing to the peculiarity of its shape has, by definition, a vanishingly small diffusivity (see Fig. 8.56).

Combining all the attributes, we may conveniently divide catalytic behaviour of zeolites into three categories:

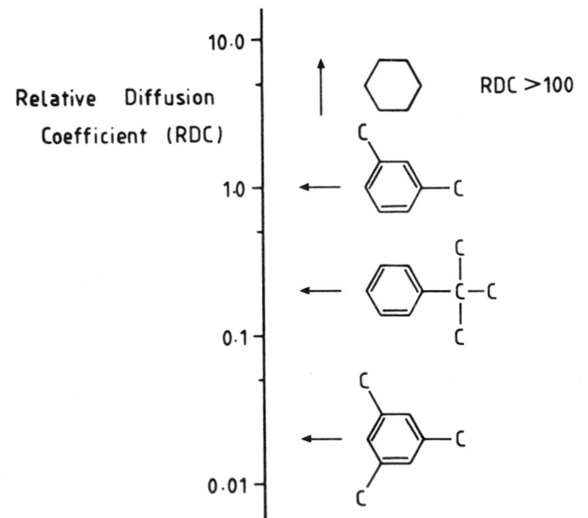

Figure 8.56 Relative diffusion coefficients for molecules of differing shape. (After P. B. Weisz, *Pure Appl. Chem.*, **1980**, *52*, 2091.)

1. reactions that rely predominantly on Brønsted acidity of the zeolite;
2. acid-catalysed reactions in which considerations of shape-selectivity dominate; and
3. non-acid reactions and/or redox reactions.

8.8.1 Activity of Zeolitic Catalysts

Silica–alumina gels served as the main catalytic cracking catalysts until they were supplanted in the mid-1960s by zeolites. The gels were not sufficiently active; they lacked selectivity; they were also thermally unstable (and hence not readily thermally regenerated); and they tended to acquire carbonaceous layers (i.e. to suffer from coking) during use. Zeolites proved superior on all scores, and soon became the catalysts of choice for cracking and hydrocracking, generally as dual-function catalysts after the incorporation of finely divided zero-valent platinum into the intrazeolitic cavities. La^{3+}–Y or Ca^{2+}–Y or simply H^+–Y zeolites are all very active solid acid catalysts (see below). A figure of merit for the catalytic activity of cracking catalysts is the so-called α-value defined as

$$\alpha = \frac{\text{activity of the particular cracking catalyst at 540 °C}}{\text{activity of a standard silica gel at 540 °C}}$$

The activity is synonymous with the first-order rate coefficient, k, for n-hexane cracking (if necessary, extrapolated to 540 °C), where

$$k = \frac{1}{t}\ln\left(\frac{1}{1-x}\right) \tag{88}$$

and x is the fraction of n-hexane cracked after a time t.

Recognizing that the key role of the acid zeolitic catalyst in cracking and hydrocracking is to produce H^+ ions which then generate the prerequisite alkylcarbenium ions, we can gain further insights into the mechanism of the catalysis by correlating changes in IR spectra with catalytic performance. Take first H^+–Y catalysts. There are two O–H stretching frequencies that are of especial significance, since they refer to the bonds that are ruptured on ionization, thereby yielding detached protons. The fact that NH_3 gas wipes out these two frequencies, when introduced to H^+–Y, shows the corresponding OH groups each to be accessible to small (basic) molecules. Only the higher-frequency peak is eliminated when pyridine is introduced, indicating that the OH group responsible for this *ca* 3650 cm^{-1} frequency protrudes into the supercage. X-ray structural analysis on H^+–Y shows precisely where the two OH bonds are situated. That associated with the lower frequency of *ca* 3550 cm^{-1} points inwards to the D6R (see Fig. 5.34) and is therefore inaccessible to molecules as bulky as pyridine or quinoline. The OH bond associated with the higher frequency of 3650 cm^{-1} is the one that plays the catalytic role, as may be seen from the slow decline in its band intensity with operation during cumene cracking. The intensity of the 3550 cm^{-1} frequency, however, remains essentially invariant, at least at temperatures below 325 °C. Corroboration of this mechanistic view came early on from the elegant work of Turkevich and Ono, who found that catalytic activity during cumene cracking correlated with the number of Brønsted acid sites determined by quinoline poisoning. Indeed, the number of active sites measured by poisoning (1×10^{21} g^{-1}) is within 15% of the number of available OH groups in the H–Y zeolite computed from the known structure.

Recalling that polyvalent cations generate Brønsted acid sites in a zeolite (see Fig. 1.18), we gain additional insight into the nature of the sites for catalytic cracking in Ca^{2+}–Y. With increasing replacement of Na^+ by Ca^{2+}, catalytic activity begins to appear at the point at which Ca^{2+} has filled in all the type I sites (16 per unit cell). We know that Ca^{2+} preferentially occupies sites of type I, and until all such sites are occupied, none of the Ca^{2+} ions can appear in the main pore structure (i.e. in the supercages) where they generate protons from available water as a result of cation hydrolysis: $Ca^{2+} + H_2O \rightarrow (CaOH)^+ + H^+$. Such hydrolysis is not feasible as long as all the Ca^{2+} ions are buried within the hexagonal prisms that circumscribe sites of type I. The parallel between the intensity of OH stretch frequencies and catalytic activity confirms the catalytic importance of accessible surface OH groups.

Differences in catalytic activity between zeolites and other acid catalysts depend, *inter alia*, on the ease or difficulty of forming carbenium ions from the hydrocarbon reactant. Hexane is difficult to ionize, so that its cracking is several orders of magnitude faster over H^+ zeolites than over silica–alumina gels. But with, say, xylene as reactant, since carbenium ions can form quite readily on each type of catalyst, the differences in rates of reaction of such aromatic hydrocarbons between zeolitic and other acid catalysts is rather small. It has recently been proposed by Haag and colleagues that the kinds of catalytic activity associated with the so-called superacids are mirrored on some zeolites. When, for example, ethylene combines with methane, ethane or propane over $TaF_5 \cdot HF$, the addition of the ethyl cation (i.e. methylcarbenium ion) produces propanc, *n*-butane and 3-methylpentane respectively:

$$H_2C{=}CH_2 + H^+ \rightarrow H_2C{-}CH_2{}^+$$

$$H_3C{-}CH_2{}^+ + CH_4 \rightarrow C_3H_8 + H^+$$

$$H_3C{-}CH_2{}^+ + n\text{-}C_4H_{10} \rightarrow H_3C{-}CH_2{-}\underset{}{\overset{\displaystyle CH_3 \atop |}{CH}}{-}CH_2{-}CH_3 + H^+$$

Such processes are no ordinary carbenium ion reactions. They are generally believed to proceed via structures (either as short-lived intermediates or as transition states) that contain so-called 'two-electron three-centre bonds', structures in which two electrons provide a bond between two H atoms and one C atom or between two C atoms and one H atom:

$$\gt C{-}H + D^+ \rightleftharpoons \left[\gt C \begin{matrix} \diagup H \\ | \\ \diagdown D \end{matrix} \right]^+ \rightleftharpoons \gt C{-}D + H^+$$

$$\begin{array}{c}\diagdown\quad\diagup\\ -C{-}C- \\ \diagup\quad\diagdown\end{array} + H^+ \rightleftharpoons \left[\begin{array}{c}\diagdown\qquad\diagup\\ -C{-}{-}{-}C- \\ \diagup\ \diagdown\ \diagup\ \diagdown \\ H\end{array}\right]^+ \rightleftharpoons \begin{array}{c}\diagdown\\ -C{-}H\\ \diagup\end{array} + \begin{array}{c}\diagup\\ C^+{-}\\ \diagdown\end{array}$$

'Pentacoordinated' carbon atoms, mentioned earlier, are implicated here.

8.8.2 Shape-Selective Zeolitic Catalysts

Although faujasitic zeolites (typically H^+–Y and La^{3+}–Y) function well in the cracking of hydrocarbons and in other types of acid catalysis, they do suffer from a few inherent disadvantages. Despite their superiority over silica–alumina gels, they nevertheless tend to 'coke' too readily; the product distributions they yield are not sufficiently selective for a wide enough range of reactions; they lack stability at really high temperatures; and they are rather too hydrophilic (i.e. oleophobic), which means they tend to repel the very hydrocarbons which we seek to convert catalytically.

Ideally, we require zeolitic catalysts that are more hydrophobic and more thermally stable, and that possess more shape-selectivity so as to permit ingress of a wide range of important reactant molecules (methanol, benzene, toluene, *n*-hexane) but that, at the same time, prevent egress and formation of larger undesirable molecules such as naphthalene and other polynuclear hydrocarbons which are responsible for coking and hence poisoning. Such shape-selective catalysts are already with us, the most renowned being ZSM-5. (However, the first laboratory-scale shape-selective catalysts, introduced by Weisz and Frillette in 1960, were based on zeolite A. A bifunctional (Pt^0) Ca^{2+}–A zeolite catalysed the combustion in O_2 of *n*-butane, *n*-butene and propane but left 99% of the isobutane in the mixture untouched.)

With the evolution of the work of Weisz, Derouane and others, three categories of shape-selective catalysis have now been identified:

1. *Reactant selectivity* occurs when only some of the reactant species can pass through the catalyst pores to the active sites. The remaining species are too large to diffuse through the pores and so they do not react. Most of the examples quoted so far fall into this category (see Figure 1.1 and Colour Plate 1).
2. *Product selectivity* occurs when, among all the product species formed within the pores, only those with small enough dimensions can diffuse out and appear as observed products. Bulky products, if formed, are either converted to less bulky ones or eventually deactivate the catalyst by blocking the pores. The observed seven-fold enhancement of hydrogenation of *trans*-2-butene over *cis*-2-butene over a (Pt^0) Ca^{2+}–A zeolite catalyst is a good example of product selectivity.
3. *Restriced transition-state selectivity* occurs when certain reactions are prevented because the corresponding transition state would require more space than is available at the active sites in the cavities. Here, reactions requiring smaller sizes of transition states proceed unimpaired; and neither reactant nor potential product species are prevented from diffusing through the pores.

Examples of all three types of selectivity have been identified in reactions of industrial significance, but it can occasionally be difficult to pinpoint the origin of the

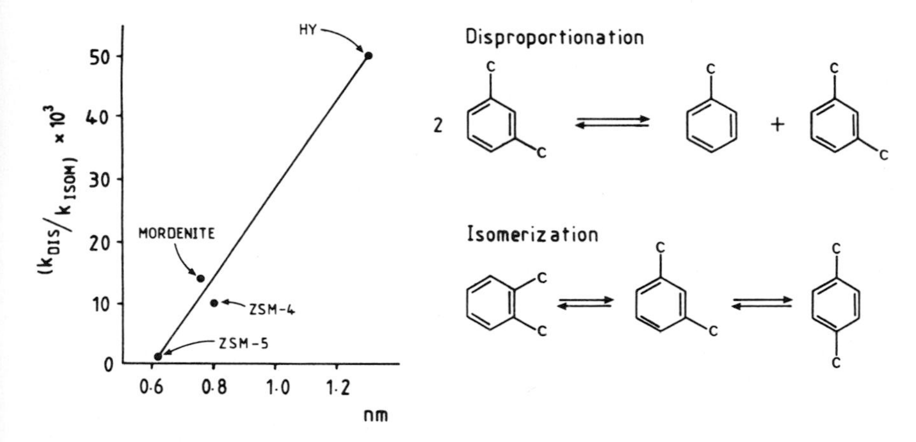

Figure 8.57 An example of restricted transition-state shape-selectivity illustrated by the ratio of isomerization to disproportionation rate coefficients as a function of pore diameter (After W. O. Haag, R. M. Lago, P. B. Weisz, *Nature*, **1984**, *309*, 589.)

selectivity under a given set of conditions. ZSM-5, the most prominent member of the so-called pentasil family (see Colour Plate 1 and Fig. 1.2) of second-generation shape-selective catalysts, exhibits all three types of selectivities categorized above. The overall advantages of these pentasils (ZSM-11 is another) is that they are rather readily prepared in highly siliceous and therefore hydrophobic forms (Si/Al ratios of $10–10^4$).

Good examples of reactant selectivity can be seen when selective hydrogenation over Pt^0/ ZSM-5 is compared with the same hydrogenations over Pt^0/Al_2O_3. Thus at 275 °C over Pt^0/AL_2O_3, both hex-1-ene and 4,4-dimethyl-hex-1-ene are about equally converted, but over Pt^0-ZSM-5, there is more than a hundred-fold preference for the conversion of the hexene over the 4,4-dimethyl-hex-1-ene.

A particularly elegant example of how restricted transition states in shape-selective catalysts can be turned to good preparative advantage is summarized in Fig. 8.57. In the course of xylene isomerization, disproportionation may, in general, also take place, provided the appropriate transition state can be reached. Disproportionation involves a large 'bimolecular' intermediate, whereas isomerization entails a rather less demanding 1,2-methyl shift. We therefore expect, and do indeed find, that the ratio of the rate coefficients k_{DIS}/k_{ISOM} increases with increasing diameter of the zeolitic cavity.

The build-up of coke as an undesirable by-product during the conversion of alkanes, alkenes, alkanols and naphthanes over solid acid catalysts was mentioned earlier. Restricted transition-state selectivity is responsible for the much-diminished rate of coking over ZSM-5 and other (e.g. theta-1) second-generation shape-selective catalysts. The formation of bulky, coke-precursor species (such as naphthalenes and other polynuclear aromatics) is greatly reduced as the diameter of the intracrystalline pores decreases.

Restricted transition states also account for the distribution of aromatics observed (Fig. 8.58) in the Mobil process for the conversion of methanol to gasoline (MTG)

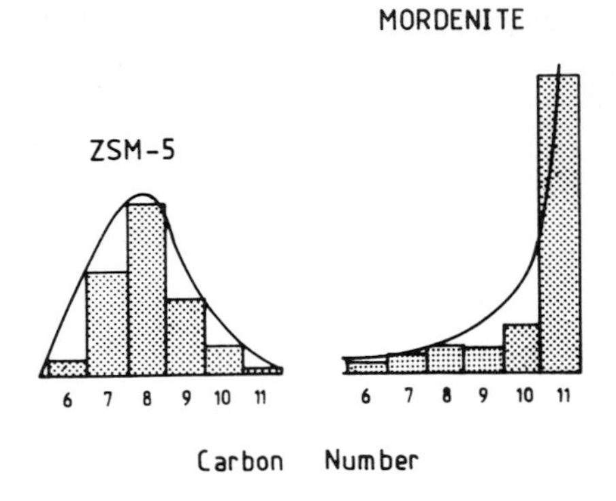

Figure 8.58 The catalysed dehydration of methanol over ZSM-5 yields products that constitute petrol (C_6 to C_9 aromatics) whereas over the larger-pore mordenite naphthalenic and higher aromatics are produced.

over ZSM-5 or to light olefins (ethylene to butene) (MTO) over solid acids such as SAPO-34:

$$n\,CH_3OH \xrightarrow[-H_2O]{[H^+]} (-CH_2-)_n$$

The alkylation of toluene with methanol to form *p*-xylene selectively is a good example of product selectivity. The various possible alkyl aromatic products have markedly different diffusivities in ZSM-5. For example, the diffusion coefficients of *p*-xylene and *o*-xylene in ZSM-5 at 315 °C are, respectively, *ca* 10^{-7} and 10^{-10} $cm^2\,s^{-1}$. We see, therefore, why, even though the equilibrium distribution of *p*-, *m*- and *o*-xylenes is 23 : 51 : 26, a ZSM-5 catalyst for the reaction

$$CH_3OH + C_6H_5{-}CH_3 \xrightarrow[-H_2O]{[H^+]} CH_3{-}C_6H_4{-}CH_3 + H_2O$$

can result in a product distribution of 46 : 36 : 18.

8.8.3 New Microporous Crystalline Catalysts

The era when heteroatoms (such as Zn or Ga in two- or three-valent form respectively, or Fe and Co in their II and III oxidation states) may be relatively routinely 'placed' in the framework sites of an ever-expanding range of known zeolite structures has long since dawned. Work by Vaughan, Ono, Inui, Ruren Xu and Chen, Jacobs, Davis and Zones and others has yielded many fascinating new catalysts, some of which are particularly efficient in dehydrocyclization of light alkanes such as propane and butane, these being largely (Brønsted) acid-catalysed reactions. Equally, thanks to pioneering efforts at the Italian laboratories of Enichem, redox elements, especially Ti ions, may be incorporated into microporous silicate networks of well-known structures such as silicalite I (the siliceous analogue of ZSM-5) and silicalite II (the siliceous analogue of ZSM-11), thereby yielding TS-1 and TS-2 respectively (see Notari, 1996).

By ringing adroit structural changes such as these, as well as by doing parallel cation (i.e. extra-framework) substitution, numerous novel catalysts may be fashioned and fine-tuned to meet particular chemical ends. Table 8.4 includes just one-

Table 8.4. Acid-catalysed reactions giving hydrocarbon products.

Reaction	Zeolite catalyst	Temperature [°C]
1 Ethylbenzene from benzene and ethylene 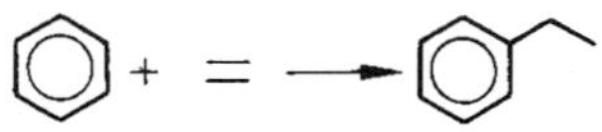	ZSM-5	400
2 Xylenes production, isomerization (including ethylbenzene) and disproportionation 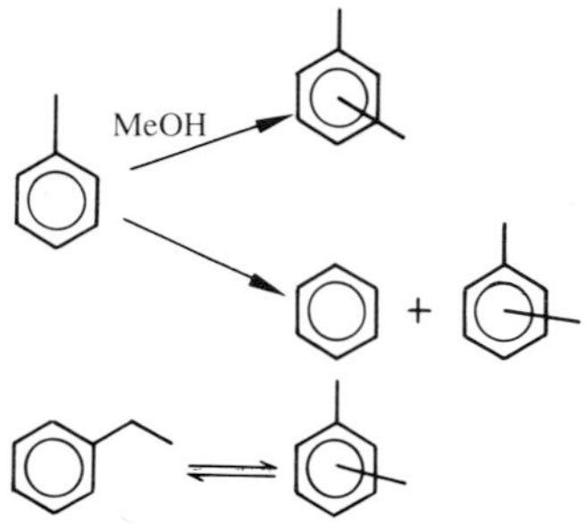 (MeOH)	Variously modified ZSM-5	300–600
3 *p*-Ethyltoluene from toluene plus ethylene 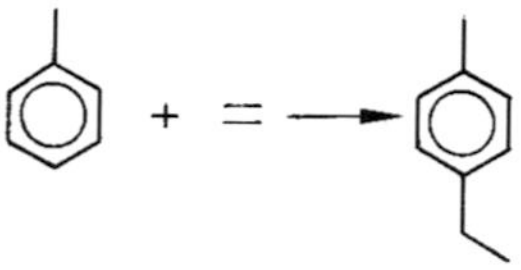	ZSM-5	400
4 New routes to aromatics: paraffins, olefins, alcohols, etc. 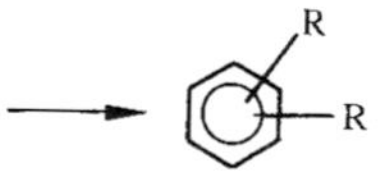 → R-substituted aromatics	ZSM-5	400–600
5 Cumene from benzene plus propylene 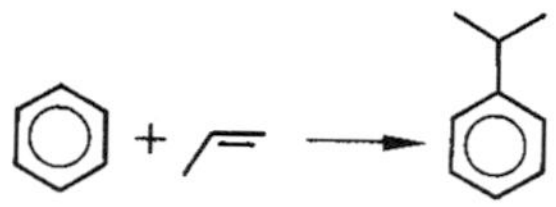	H-Y	180
6 Detergent alkylate from benzene plus C_6–C_{14} olefins 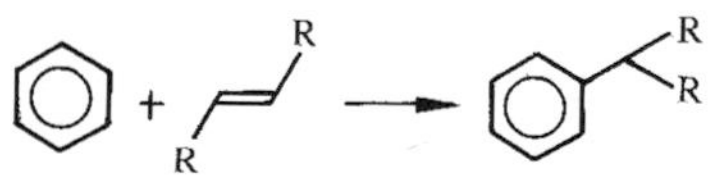	RE-Y or H-mordenite	150–210
7 Vinylcyclohexene from butadiene 	Cu-Y	95

fifth of the compendium of reactions of proven commercial significance prepared in 1992 by John, Clark and Maxwell.

Numerous acid-catalysed reactions involving functionalized hydrocarbons, bifunctional catalysts with an acidic (zeolitic) support, and examples of the use of neutralized or basic zeolitic supports are also enumerated by John et al. (1992). van Bekkum, Corma, Holderich, Jacobs, Weitkamp and others have enormously added to the list of new zeolite-catalysed reactions. It has also to be borne in mind that new microporous structures are continually being reported: the mazmorites (part-mazzite, part-mordenite) of Vaughan and Leonowicz (see Fig. 8.59) and the new zeolites SSZ-33, SSZ-37, CIT-1 and MCM-22 are a mere fraction of those recently reported, all potentially powerful frameworks for imaginative manipulation, by substitution of selected cations into extra-framework or framework sites, as well as 'ship-in-bottle' constructions of the kind pioneered by Herron and Stucky and recently exploited by Sachtler and Corma.

As both Chapter 5 and the *Atlas of Zeolite Structures* (therein quoted) make clear, a very large family of ALPOs, GAPOs, FAPOs etc. is also available for dextrous exploitation by the catalyst designer. A recent striking example is DAF-4 (see Bar-

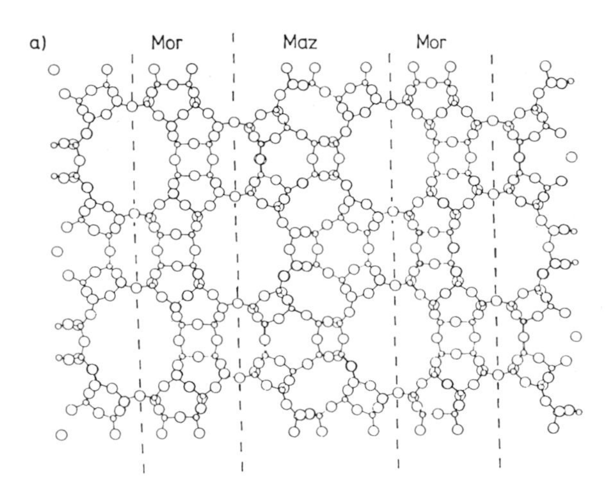

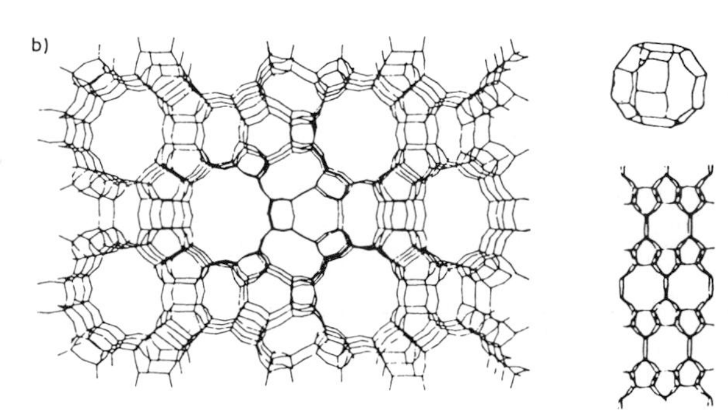

Figure 8.59 (a) Projection of ECR-1 zeolite structure along the *a*-axis showing the component sheets of mordenite (Mor) and mazzite (Maz). Large spheres denote oxygen atoms, small ones T-atoms (Si or Al). (b) Perspective view of the ECR-1 framework (left). The top right inset shows the gmelinite cage as observed in mazzite; the bottom right inset shows a layer of the mordenite structure. (After J. M. Thomas, *Angew. Chem., Int. Ed. Engl.* **1988**, *27*, 1673.)

rett et al.) which is a solid acid catalyst, designed *de novo*, to convert methanol to light alkenes.

8.8.4 Some Case Studies of In-Situ Monitoring of Catalysis with ZSM-5

Zecchina and his colleagues have carried out an elegant IR study of the oligomerization of ethylene and propylene using fast FTIR spectroscopy. Oligomerization was seen directly to proceed through: (i) formation of short-lived hydrogen-bonded precursors by interaction of the alkene with the bridging hydroxyls, Brønsted acid sites; (ii) a protonation step, and (iii) a chain-growth step. Scheme 8.14 shows the sequence. Because of the constraints imposed by the zeolite superstructure, isomerization of the oligomerizing chain inevitably takes place, forming isomers such as **1**–**4**. FTIR measurements as a function of time for the oligomerization of propylene on (H)ZSM-5 are in line with the mechanism depicted in Scheme 8.15.

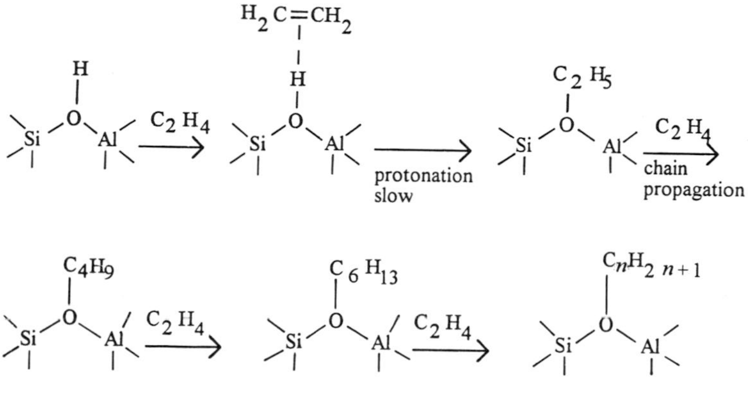

Scheme 8.14 Oligomerization of ethylene.

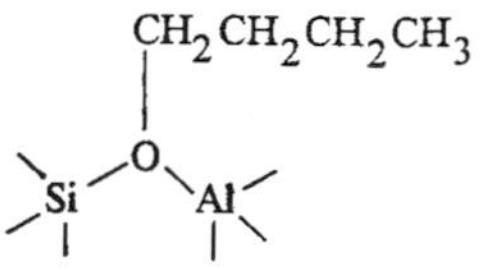

$CH_2CH_2CH_2CH_3$ **1**

$CH_2CH(CH_3)_2$ **2**

$CH_3CHCH_2CH_3$ **3**

$CH(CH_3)_3$ **4**

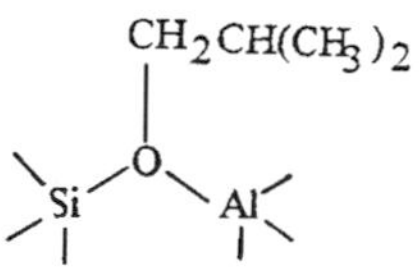

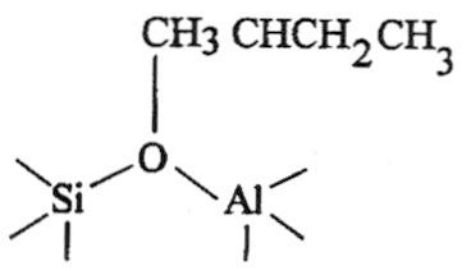

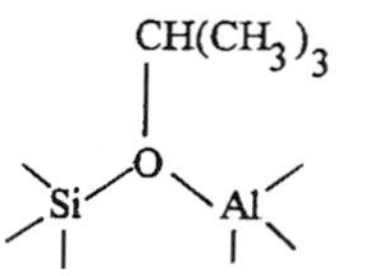

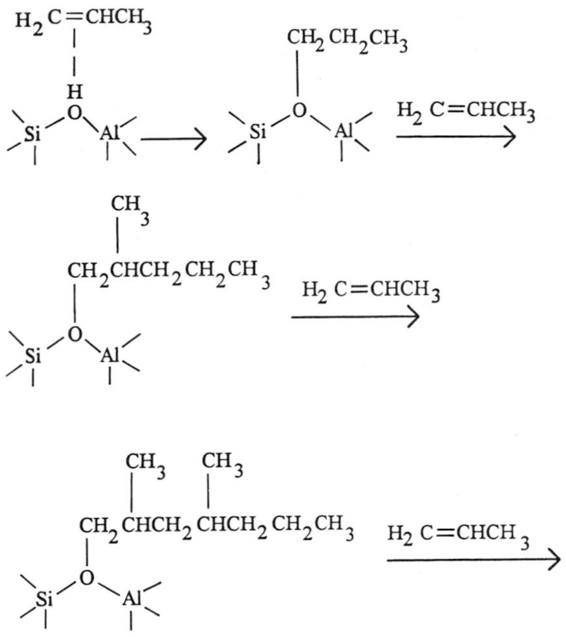

$H_2C{=}CHCH_3$ ··· H–O(Si)(Al) → $CH_2CH_2CH_3$–O(Si)(Al) $\xrightarrow{H_2C=CHCH_3}$

$CH_2CH(CH_3)CH_2CH_2CH_3$–O(Si)(Al) $\xrightarrow{H_2C=CHCH_3}$

$CH_2CH(CH_3)CH_2CH(CH_3)CH_2CH_2CH_3$–O(Si)(Al) $\xrightarrow{H_2C=CHCH_3}$

Scheme 8.15 Oligomerization of propylene.

The relative strength of the hydrogen bonds in the ethylene–OH and propylene–OH π complexes (precursors) was estimated by Zecchina et al. on the basis of the downward shift of both the ν(OH) and ν(C=C) frequencies (-389 and $-11\,cm^{-1}$ for ethylene and -539 and $-19\,cm^{-1}$ for propylene). For both molecules, the protonation of the precursors is the rate-determining step of the oligomerization process. All this is a beautiful illustration of how, with acid zeolitic catalysts, where the active sites are uniformly distributed throughout the bulk (see Thomas 1988), a clean and quantitative study may be made of a catalytic process in which the nature and number of the active sites are well defined, where the precursors can be clearly identified, and where the propagation step is also clearly pinpointed.

Using a combination of solid-state (one- and two-dimensional) multinuclear NMR, FTIR and a flow microreactor fitted to a chromatograph, Zamaraev and Thomas (1996) have elucidated the detailed mechanisms of catalytic dehydration of the four isomeric butanols on H–ZSM-5 (they also compared reactions on a crystalline aluminosilicate acid catalyst with an essentially similar set of active sites embedded in an amorphous solid). Diffusion coefficients of the reactant could be directly measured, and atomic scrambling (followed using isotopic labels) could be directly observed. Among other things, they found that pore confinement in H–ZSM-5 kinetically favours the formation of linear rather than branched C_4 carbenium ions. For *t*-BuOH (butanol), the *t*-butyl silyl ether (TBSE) could be identified. It is a rather stable species under reaction conditions (decomposing only above 373 K). At 296 K, TBSE behaves as a side-intermediate species, through which only a small fraction of *t*-butanol dehydrates. The main reaction stream by-passes TBSE, and seems to proceed through the *t*-buyl carbenium ion as the key intermediate.

The catalytic conversion of methanol to hydrocarbons (MTG and MTO) in the gasoline boiling range using ZSM-5 at *ca* 370 °C has been monitored by ^{13}C MASNMR (by Grey et al., Klinowski and Anderson, and Haw et al.). This technique probes directly the role of the active site in shape-selective catalytic reactors on zeolites in situ. The kind and quantity of chemical species present inside the catalyst particle may be directly monitored. Such information is usefully compared with the composition of the gaseous products to yield new insights into reaction pathways in molecular sieves and to assist in the design of new shape-selective catlysts. Twenty-nine different organic species were identified in the adsorbed phase and the fate of these species during the course of the reaction could be monitored. Moreover, one may unequivocally distinguish between mobile and immobile species.

8.8.5 A Rationally Chosen Zeolitic Catalyst

Acidic mordenite is the favoured catalyst to replace the environmentally harmful $AlCl_3$ now used for the alkylation of naphthalene by propylene to yield 2,6-di-isopropylnaphthalene, which is an important building-block precursor for a number of speciality polyesters and liquid-crystal polymers (see Fig. 8.60). The $AlCl_3$ catalyst yields both the 2,6- and 2,7-disubstituted isomers as well as undesirable tri- and tetra-substituted products.

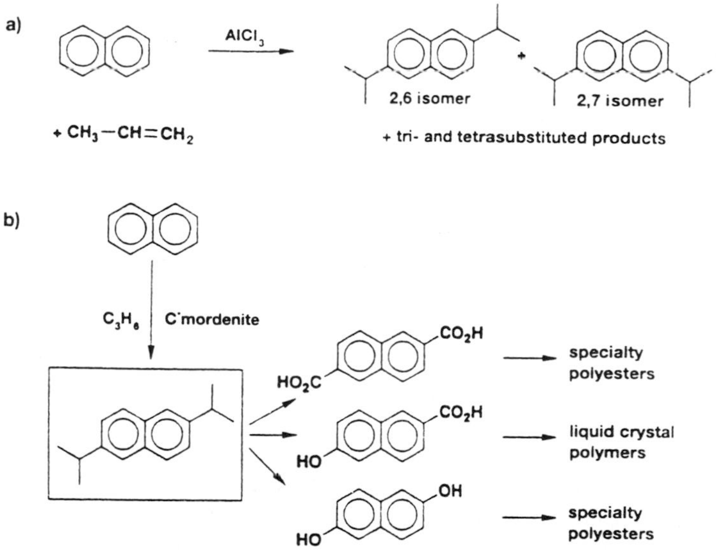

Figure 8.60 Alkylation of naphthalene by propylene catalysed by (a) $AlCl_3$ to give a mixture of products, and (b) mordenite to give 2,6-di-isopropylnaphthalene clearly; the latter can be converted to a number of monomers for specific polymers. (After J. M. Thomas, K. I. Zamaraev, *Angew. Chem.*, **1994**, *106*, 316.)

Casumano (1992) has pointed out that the choice acidic mordenite as a replacement for the environmentally harmful $AlCl_3$ (Friedel–Crafts) catalyst for the production of 2,6-di-isopropylnaphthalene was much influenced by computer modelling of the type described in Chapter 5.

Microporous, molecular-sieve catalysts are prepared via the agency of an appropriately chosen organic template which exerts a structure-directing influence on crystals that grow from a gel medium. After growth, the crystals are calcined in oxygen to burn away the template. Very recently (see Lewis et al. (1996) and Barrett et al. (1996) *de novo* design of the template molecule has proved possible for the production of a desired catalyst possessing specific pore dimensions and strategically located Brønsted acid sites inside the pores. A MeALPO catalyst, to be precise a Co $AIPO_4$ microporous solid known as solid DAF-4 (Davy Faraday number 4), which has a framework structure close to that of the mineral levyne, has been rationally designed in this way. It is efficient in catalytically converting methanol to ethene and propene.

8.8.6 New Mesoporous Catalysts

Mesoporous siliceous solids with channel apertures from 25 to 100 Å have opened up new possibilities in heterogeneous catalysis. The large-diameter (*ca* 30 Å) channels of the MCM-41 mesoporous silicas (see Section 5.2.7.2 and cover photo) permit, in principle, the direct grafting of complete metal complexes and organometallic moieties on to the inner walls of these high-surface-area (typically $>800\,m^2\,g^{-1}$) solids. This opens routes to the preparation of novel catalysts consisting of large concentrations of accessible, well-spaced and structurally well-defined active sties. For example, Maschmeyer et al. (1995) have shown how a titanocene-derived catalyst precursor anchored to the inner walls of MCM-41 serves as a powerful catalyst for the epoxidation of cyclohexene; this is a much more active than a comparable titanium-containing MCM-41 catalyst in which the titanium is part of the framework.

van Bekkum and co-workers in Delft, Corma in Spain and Pinnavaia in the USA have demonstrated how effectively the well-defined mesoporous MCM-41 silicas may be readily harnessed to effect selective oxidations (using in-built titanium ions and H_2O_2 as the key agents of conversion) of quite bulky molecules. Colour Plate 9 illustrates how van Bekkum et al. used MCM-41 to utilize shape-selectivity in such processes as the Fries reaction. Functionalized arenes such as 2,6-dimethylbenzoic acid may be anchored to the pendant Si–OH groups of the walls of the mesopores. The Fries reaction, using resorcinol, generates a benzophenone derivative from the parent arene. Figure 8.61 illustrates how Maschmeyer et al. produced, from a titanocene derivative, the powerful epoxidation catalyst for substrates such as pinene as well as smaller entities like cyclohexene.

Even before the arrival in 1992 of the MCM-41 family of sharply defined mesoporous siliceous solids, however, there existed controlled-pore glass (e.g. CPG-240), in which there is a sharply defined pore size of *ca* 24 Å. Imaginative use was made of such material, by Davis in the USA, in the immobilization of important homogeneous catalysts (as water-soluble complexes) on the inner hydrophilic surfaces of such glass. In effect, the controlled-pore glass functions as an integral part of what

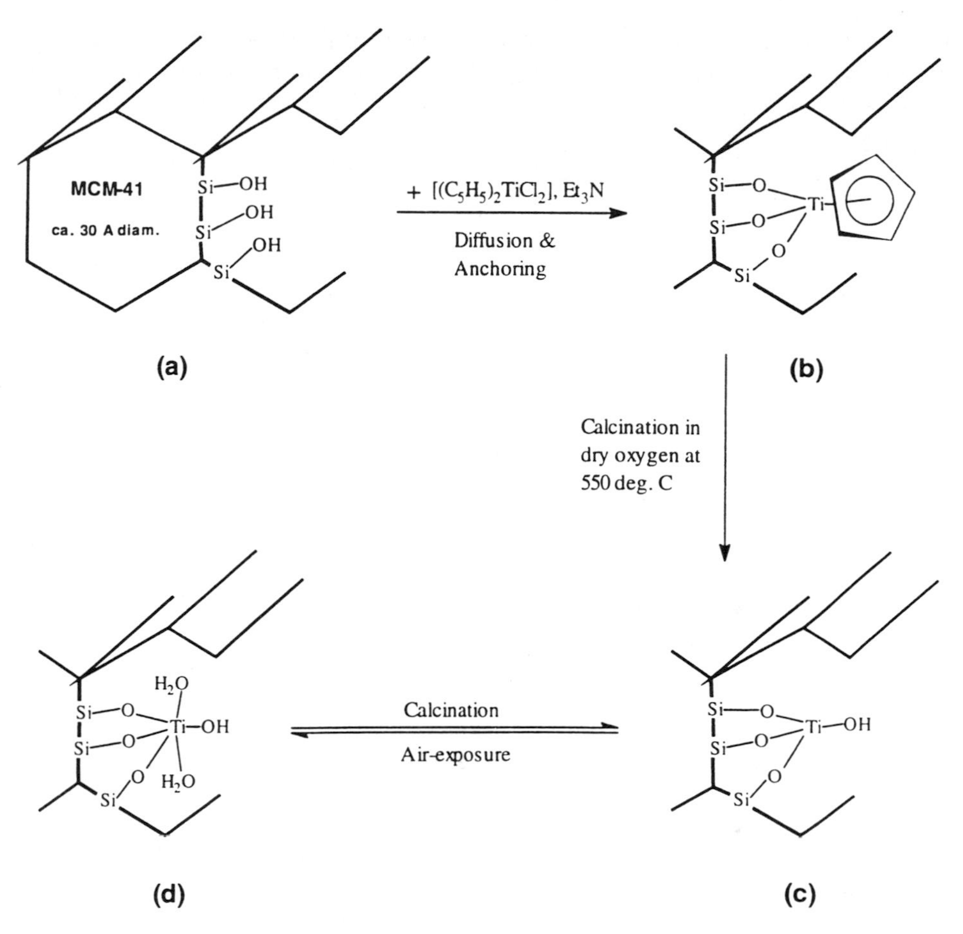

Figure 8.61 Schematic representation of the preparation of grafted titanocene-derived catalysts. The support (**a**), following the anchoring reaction, yields (**b**), which upon calcination yields the active catalyst (**c**). The bottom part shows the reversible reaction of the resulting Ti–MCM-41 epoxidation catalyst when exposed to a wet atmosphere. (After T. Maschmeyer, F. Rey, G. Sauhar, J. M. Thomas, *Nature*, **1995**, *378*, 159.)

are known as supported aqueous-phase (SAP) catalysts. When the glass is impregnated with $HRh(CO)(P[m\text{-}C_6H_4SO_3Na]_3)_3$ (otherwise called $HRh(CO)(TPPTS)_3$, where TPPTS stands for triphenylphosphine trisulphonate), a typical SAP catalyst is produced that is very effective in hydroformylation:

$$CH{=}CH \quad + CO + H_2 \rightarrow -CH(CHO)CH_2- + -CH_2CH(CHO)-$$

the cornerstone of the so-called ‘oxo’ process used commercially to produce carbonyl compounds such as alkanols, acids and esters.

The water content of the $HRh(CO)(TPPTS)_3$-based SAP catalysts greatly influences their performance. For example, when 1-heptene is hydroformylated, the turnover frequency increases by more than a hundred-fold. This increase is believed to originate in the increased mobility of the immobilized inorganic complex.

8.8.7 Clays and Other Solid Acid Catalysts

Among the many interesting properties of clays are their large capacities for cation exchange and the ability to take into their interlamellar regions water and a wide range of other molecules (alkanols, amines, amino acids, nitriles, ketones and many types of hydrocarbons). These microcrystalline materials are cheap and plentiful. In 1982, the annual worldwide use of clays, not including that of China and the countries of the former Soviet Union, exceeded 6 million tonnes ($>6 \times 10^9$ kg). How appropriate it would be, therefore, if clays functioned as efficient commercial heterogeneous catalysts. They do in any case figure eminently as precursors in the production of certain zeolitic catalysts and also of ceramic monoliths for auto-exhaust catalysts.

In the early days of oil refining, clays, notably those based on montmorillonite, were utilized by Houdry for the catalytic cracking of large hydrocarbon molecules. In due course they were supplanted by silica–alumina gels which, in turn, were superseded by zeolitic cracking catalysts. Of late, however, there has been renewed interest in clays as viable broad-spectrum catalysts for organic synthesis. They are particularly good – if care is taken to extract the full chemical advantage of the special microenvironment that exists in their interlamellar, rather than exterior, surfaces – at producing organic materials now in heavy demand, e.g. methyl *t*-butyl ether (MTBE) and ethyl acetate, which is extensively used as a multipurpose solvent. The potential of clays, when suitably tailored and modified, for isomerization and cracking is also considerable and there is abundant scope for catalyst design with these sheet silicates. Their significance biochemically, and in prebiotic processes, is also substantial.

The key structural features of clays which are relevant to discussions of their activity and selectivity are that:

1. a wide range (almost innumerable) of organic intercalates can be formed;
2. the original interlamellar, neutralizing cation can be readily replaced, as desired, by one (or possibly more) of a whole range of inorganic and organic cations;
3. they exhibit strong acidity, which is usually of the Brønsted type, partly because of the influence of the strong internal electrostatic fields (*ca* 10^6 V cm^{-1}) exerted on the interlamellar water (which generates protons by dissociation) or because of the additional influence of certain hydrated interlamellar cations, notably Al^{3+}. Cation hydrolysis, just as with strongly polarzing cations in zeolites, yields free protons:

$$M^{n+}(H_2O)_m \rightarrow [M(H_2O)_{m-1}OH]^{(n-1)+} + H^+ \tag{89}$$

The interlamellar ions present in naturally occurring vermiculites and smectite clays may be readily exchanged for potent acidic ones. These are H_3O^+ (or $H_5O_2{}^+$) ions formed by gentle and repeated acid washing or by hydrolysis according to Eq. (89). The following is a selection of the organic reactions catalysed by acidic clays:

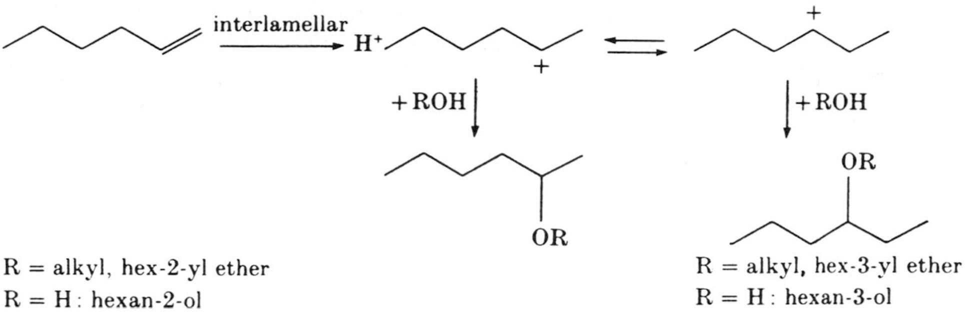

interlamellar H^+

+ROH

+ROH

OR

OR

R = alkyl, hex-2-yl ether
R = H: hexan-2-ol

R = alkyl, hex-3-yl ether
R = H: hexan-3-ol

Scheme 8.16 Ethers and alkanols are readily prepared using acidic clay catalysts. (After J. M. Thomas, *Phil. Trans.* **1990**, *A333*, 173.)

(a) conversion of primary amines or alkanols to secondary ones;
(b) isomerization, alkylation, and cyclizations;
(c) hydration, alkylation and acylation of alkenes to form alkanols, ethers and esters;
(d) dehydration of alkanols, with the formation of ethers, alkenes and naphthenes;
(e) dimerizations, oligomerizations and polymerizations;
(f) decarboxylations and polycondensations (e.g. of peptides and amino acids); and
(g) cracking and hydrocracking of hydrocarbons.

Scheme 8.16 (Thomas 1990) summarizes the essence of ether and alkanol production via acidic clay catalysts.

When used at temperatures above about a few hundred degrees Celsius, the layers of a clay catalyst tend to collapse, thereby hampering the entry of reactant species into the interlamellar spaces and nullifying their catalytic properties. If, however, an appropriate bulky cation is inserted into the spaces beforehand, the individual layers are kept apart by their pillaring effect, thus conserving access to, and the catalytic capacity of, the interior surface of the clay. The multiply charged aluminium oxy-hydroxide $[Al_{13}O_4(OH)_{24}(H_2O)^{7+}{}_{12}]$ analogue of the Keggin ion (see Fig. 5.29) functions as precursor pillars. Pillared clays have far greater stability than ordinary, acidic clays and have found application as catalysts in oil refining (e.g. in durene production by the alkylation of 1,2,4-trimethylbenzene by methanol), conversion of methanol to hydrocarbons and the esterification of alkanols.

Apart from pillared catalysts fabricated from naturally occurring clays, which, because of the presence of impurities, may lead to less than optimal solid acids, it is also possible to synthesize smectite clays, especially beidellite, in highly purified form.

Pillared clays are under intensive investigation as supports for cobalt- and ruthenium-based bimetallic catalysts of potential use in Fischer–Tropsch synthesis and in other petrochemical reactions.

The heteropolyacids described in Chapter 5 (see Colour Plate 3) are powerful catalysts for a whole range of transformations that have been particularly thoroughly explored by the Japanese workers Misono and Ono and by Haber at al. in Poland.

Metal phosphonates, prepared from metal tetrahalides by treatment with phosphoric acids by Dines and others, are layered solids consisting of an interlamellar microenvironment (rich in acidity) conducive to catalysis of organic species. These materials are not exceptionally acidic but have been used to catalyse isomerizations that are not very proton-demanding.

There is currently a wide-ranging search for new solid acid catalysts because of their importance in coping with the transformations that are central to the drive for cleaner fuels and a cleaner environment. Materials derived from phosphatoantimonates, which are crystalline, and sulphuric-acid-leached ZrO_2, which is not, are two typical examples.

Before we leave the subject of clay catalysts it is of interest to allude to their possible role in prebiotic synthesis.

8.8.7.1 Clays and their Possible Role in Replication, Evolution and the Origin of Life

Ever since Bernal's analysis of the physical basis of life, the manner in which prebiotic synthesis of biopolymers took place, involving condensation and dehydration of the active monomers believed to be present in the primaeval ocean has, along with several other relevant issues, been extensively debated. This is a vast topic, which lies outside the scope of this book. We should, however, note that the interlamellar spaces of clay catalysts can serve as favourable microenvironments within which highly specific syntheses such as the formation of amino acids, ATP and polypeptides as well as the evolution of polypeptides and proteins have been demonstrated to occur. There are other features of the chemistry of clays (smectites) that are relevant here. Thus, certain purines and pyrimidines (such as thymine), which are not normally intercalated into montmorillonite from aqueous solution, are taken up in the presence of other molecules, such as adenine, with which they may 'pair' in the interlamellar region. Such co-intercalation is not observed with non-corresponding pairs of purines and pyrimidines.

Of additional relevance is the fact that the information capacity of a single sheet-silicate layer in which isomorphous substitution has occurred is enormous, and that individual sheets may readily be (infinitely) separated by dilution and re-assembled by removal of water. In effect, referring to the first point, whenever Al^{3+} substitutes tetrahedrally for Si^{4+}, or Mg^{2+} octahedrally for Al^{3+}, a Lewis base site is created. Such sites are distributed in the sheets, but they each give rise to a Lewis acid site in the interlamellar region. (A comparable argument can encompass Brønsted sites.) A layer, therefore, may be viewed as a primitive model of a multienzyme complex. Weiss has argued that the transfer of these catalytic properties from a matrix layer on to the replica would transform the intercalating synthesis into a true replication. Cairns-Smith has likewise outlined how tacticity in a biopolymer formed in an interlamellar environment could be pre-ordained by the information encoded in the charge distributed in the silicate sheet. As far as the ease of infinite separation and re-assembly of individual sheets is concerned, the point here is that labile molecules (rich in carbon, hydrogen and nitrogen, and either formed by electrical discharges or templated from molecules that arose initially from extraterrestrial sources) may be securely accumulated and stored as clay intercalates, and that one cycle of drought and rain could provide the external conditions necessary for one cycle of replication. Cairns-Smith has speculated on the role of sheet silicates – the kandites (kaolinite and dickite) or the smectites – as crystal genes. In this regard, charge distribution within a layer, as well as polytypism in the stacking of layers, may be the key to the encoding and promulgation of information.

8.9 Catalytic Processes in the Petroleum Industry

The petroleum industry embraces primary petroleum refinery operations and a variety of processes, downstream from refinery production, which produce a whole host of consumer products. Many of the chemical operations require a catalyst. To illustrate the nature of these processes a few are selected here, such as reforming, cracking and hydrotreating, which are typical of the industry.

At the petroleum refinery a sequence of operations, including distillation, reforming and cracking, are required to meet the demands for aviation and motor fuels as well as the large number of other commodity products which can be produced from the refinery output. Initially crude oil feedstock, which may emanate from a variety of worldwide sources, is distilled either at atmospheric pressure or under vacuum conditions, depending on the products required. From the top of the atmospheric distillation column products such as C_5 and C_4 hydrocarbons emerge which form liquid petroleum gas (LPG). From a side stream near the top of the column, products in the boiling range 50–200 °C (referred to as straight-run naphtha) can be withdrawn. Higher-boiling products (boiling at about 250–300 °C), known as gas oil, are withdrawn from the middle of the column and are further processed in cracking units. Both naphtha and gas oil distillation products lead, after further treatment, to motor fuels or petrol (gasoline). From the bottom of the atmospheric distillation column relatively high-molecular-weight products emerge which are the precursor to fuel oils. Figure 8.62 is an example of the route to gasoline products obtained from the initial distillation of crude oil at atmospheric pressure.

Vacuum distillation, on the other hand, increases the proportion of middle-boiling distillates which, on further treatment, lead to lube oils and asphalt. The use of vacuum distillation avoids excessive cracking of the long-chain hydrocarbons in the feedstock. The lube oils and asphalt are products which are withdrawn from the bottom of a vacuum distillation unit, while the product from the top of the column is a gas oil requiring further processing.

8.9.1 Catalytic Reforming

The purpose of catalytic reforming is to improve the octane number (a measure of the antiknocking properties of a motor fuel) of the light products from the crude distillation unit. Light straight-run naphtha (boiling range 45–80 °C) from the crude distillation is first desulphurized and subsequently directly blended with refinery stocks to form gasoline. The heavier straight-run naphtha (boiling range 80–200 °C) is first desulphurized and then fed to a catalytic reformer unit, the products from which are blended with gasoline to form a high-octane-number gasoline.

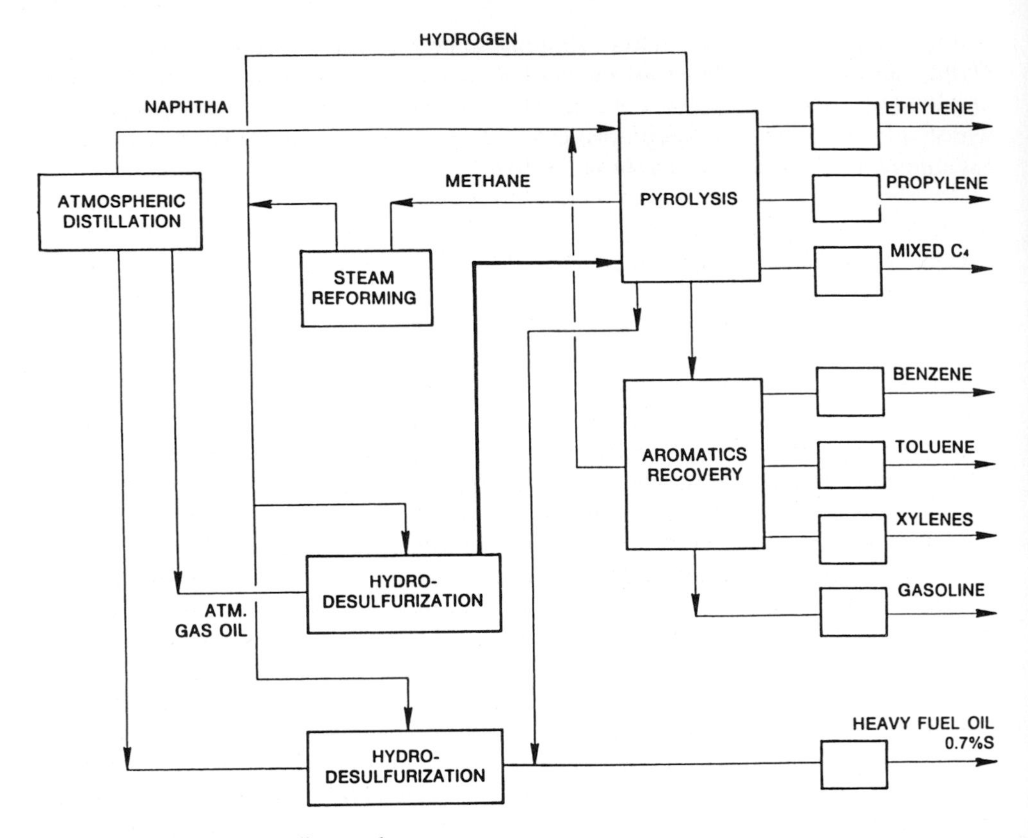

Figure 8.62 Route to gasoline products.

Some of the chemical reactions which occur in the presence of a catalyst and which contribute towards achieving a high-octane-number fuel are endothermic; a few are exothermic. Overall, it is necessary to supply heat prior to the feed passing through each catalytic reformer unit. During reforming, saturated hydrocarbons are converted to branched-chain isomers and also dehydrogenated to aromatics:

$$n\text{-}C_6H_{14} \longrightarrow CH_3CH(CH_3)CH_2CH_3 \quad \Delta H = -5.9\,\text{kJ mol}^{-1}$$

$$n\text{-}C_6H_{14} \longrightarrow C_6H_6 + 4H_2 \quad \Delta H = +266.5\,\text{kJ mol}^{-1}$$

Similar reactions occur for homologous hydrocarbons. The dehydrogenation reactions contribute much more than any other reactions to the overall endothermic nature of reforming. Cyclopentanes present in the products from the crude distillation unit are hydroisomerized in catalytic reformers:

$$C_5H_9\text{-}CH_3 \longrightarrow C_6H_6 + 3H_2 \quad \Delta H = +250.6\,\text{kJ mol}^{-1}$$

Other reactions, such as hydrodealkylation and hydrocracking, depending on the composition of naphtha emerging from the distillation unit and ultimately the source of the crude oil, also occur. Although detailed chemical analysis of the feed to a reformer unit may indicate as many as a hundred identifiable constituents, the principal catalytic reforming reactions are isomerization of saturated hydrocarbons to branched-chain alkanes, dehydrocyclization to aromatics and dehydrogenation of cycloalkanes (referred to as naphthenes) to aromatics.

Although the precise formulation and preparation of many commercial catalysts is proprietary information, it is nevertheless clear that the most active catalysts consist of platinum dispersed over the internal surface area of Al_2O_3. Many reforming reactions, particularly those involving dehydrogenation where C–H bonds are broken, are structure-insensitive and therefore largely independent of metal crystallite size. However, the isomerization of methylcyclopentane and the dehydrocyclization of heptane, both of which involve the breaking of C–C bonds, are structure-sensitive. The extent to which dispersion of platinum on the porous support is important thus depends on the prevalent reforming reactions taking place. For refinery operations where the nature and source of the feedstock often change, the practice is to employ catalysts with the platinum dispersed as crystallites about 10^{-9} m in size at a loading of between 0.1 and 2 wt% of platinum. Both the platinum and the porous alumina support, generally in the form of either η- or γ-Al_2O_3, play a catalytic role during reforming, the metal function catalysing hydrogenation–dehydrogenation reactions and the acidic function (the η- or γ-Al_2O_3) encouraging isomerization and hydrogenolysis reactions (see also Section 6.1.2 regarding the bifunctional role of some catalysts). It is clear that the aged and calcined Al_2O_3 support is usually impregnated with a platinum compound, such as chloroplatinic acid, and then reduced in H_2. Vestigial traces of chlorine (or fluorine if a salt containing fluorine is used) help to control the extent of the acid catalysing function, suppressing the otherwise excessive acid function of a catalyst prepared without traces of halogen. In addition to platinum, a second metal such as iridium, rhenium or germanium is incorporated. The so-called bimetallic multifunctional catalysts are generally considered to be superior to the single-metal supported catalyst, the second metal attenuating undesirable coke formation.

Configurations of the reforming reactor are designed for either radial flow or downward flow of the vapour-phase reactants. The radial-flow reactor, sketched in Fig. 8.63, accepts the reactants at the flanged inlet where a baffle diverts the flow into an annulus formed by a cylindrical wire mesh grid containing the catalyst and through the centre of which is a perforated pipe which leads the reformed products to the outlet. Because the flow of reactants and products is in a radial direction, the vapours pass through only a relatively small thickness of the catalyst bed, producing a pressure drop of between only 0.15 and 0.31 bar (15–30 $kN\,m^{-2}$). The outside wall of the reactor may be either hot or cool, depending on design. If the reactor wall is hot (the sketch in Fig.8.63 is a hot-wall type of radial-flow reactor) the heat insulation is exterior to the reactor shell, whose temperature is between 425 and 550 °C. The downflow reactor consists of a cylindrical shell containing the bed of catalyst pellets down through which flow the reactant vapours. Good gas distribution is essential and is provided by an inlet distributor fabricated from a perforated sheet of metal. Inert ceramic spheres are placed immediately below the distributor plate to assist gas distribution. Just above the whole catalyst bed are baskets whose

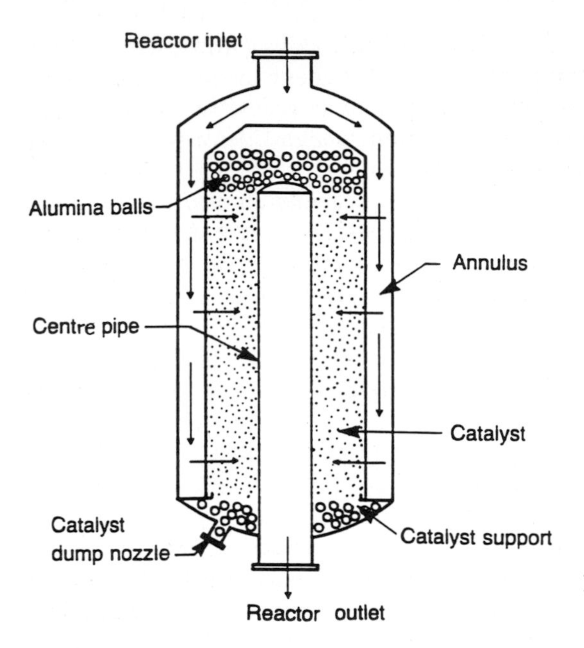

Figure 8.63 Radial-flow reforming reactor.

purpose is to collect unwanted scale and particulate fines which would otherwise tend to block bed voids and cause maldistribution of gas and a consequent high pressure drop which, under normal operating conditions, should be between 0.35 and 0.70 bar. Total residence time through a reformer is usually less then 25 s.

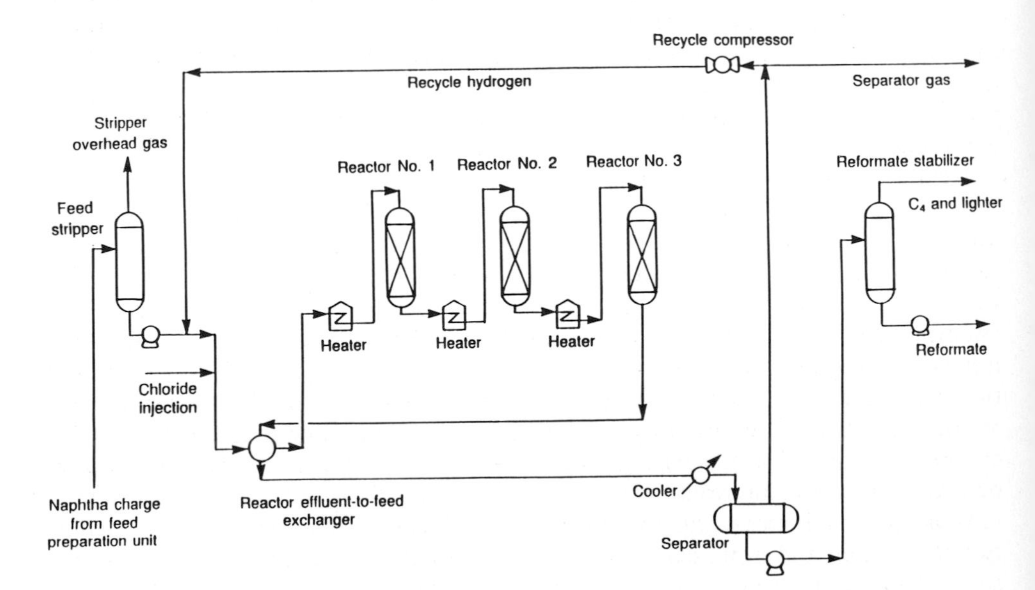

Figure 8.64 Arrangement of reactors in a reforming unit.

Temperature drops of about 100 °C occur in reforming units, occasioned by the overall endothermicity of the reforming reactions. This causes a reduction in reaction rate and octane number. To overcome such potential difficulties, three or four reactors are arranged in series with feed heaters prior to each reactor. The reformer operating temperatures are in the region of 500 °C. Figure 8.64 shows how the reactors are arranged in a reforming unit. To prevent coke formation during reforming, it is important to maintain a sufficient partial pressure of H_2. The feed to the first reactor, having been treated with H_2 (hydrotreating) to remove any poisons present such as organic sulphur and nitrogen compounds, is combined with a recycle hydrogen stream which has to be compressed to the reformer operating pressure. Pressures at which modern reformers operate are about 4–10 bar. Too high a pressure militates against net H_2 production and high octane number while too low a pressure encourages undesirable coking reactions.

8.9.2 Catalytic Cracking

The worldwide demand for petrol (gasoline) has led to refinements in the process of catalytic cracking so that refineries may obtain better yields of gasoline than ever before from a wider variety of refinery feedstocks than were formerly available. The whole object of catalyst cracking is to break down high-molecular-weight constituents of gas oils (which cannot be directly reformed, in contrast to naphthas) into lower-molecular-weight components which can then be blended with reformer products to give commercially saleable gasoline. Upstream of the catalytic cracking unit, however, the gas oils and heavier fractions from the primary distillation units have to be hydrotreated to remove undesirable components such as organic compounds containing sulphur and nitrogen which would otherwise poison the cracking catalyst. Indeed, it is the development of high-activity robust catalysts for oil cracking which has facilitated such huge increases in gasoline production in the last four decades. The catalyst used in modern catalytic cracking reactors is a solid acid – the now-ubiquitous zeolite class of silica–alumina catalysts previously described in this and earlier chapters.

Catalytic cracking involves the rupture of carbon–carbon bonds. The reactions occurring are endothermic and therefore favoured by high temperatures. Paraffins are cracked to give lower-molecular-weight paraffins and olefins. Cycloparaffins (naphthenes) crack to form olefins which subsequently react further with cycloparaffins by a hydrogen-transfer mechanism to form aromatics. Alkyl aromatics undergo dealkylation and also side-chain scission during cracking. Secondary reactions occurring include isomerization, alkyl group transfer from a dialkyl aromatic to benzene and homolgues, condensation of olefins and disproprtionation of olefins. While the principal reactions are not equilibrium-limited under catalytic cracking conditions, the secondary reactions are limited by compratively small thermodynamic equilibrium constants.

World demand for gasoline is now geared to the product with the lowest percentage of benzene (and other carcinogenic constituents). Gasoline obtained by the traditional cracking of oil, described above and below, contains too much benzene. There are already multinational oil companies that produce gasoline from natural

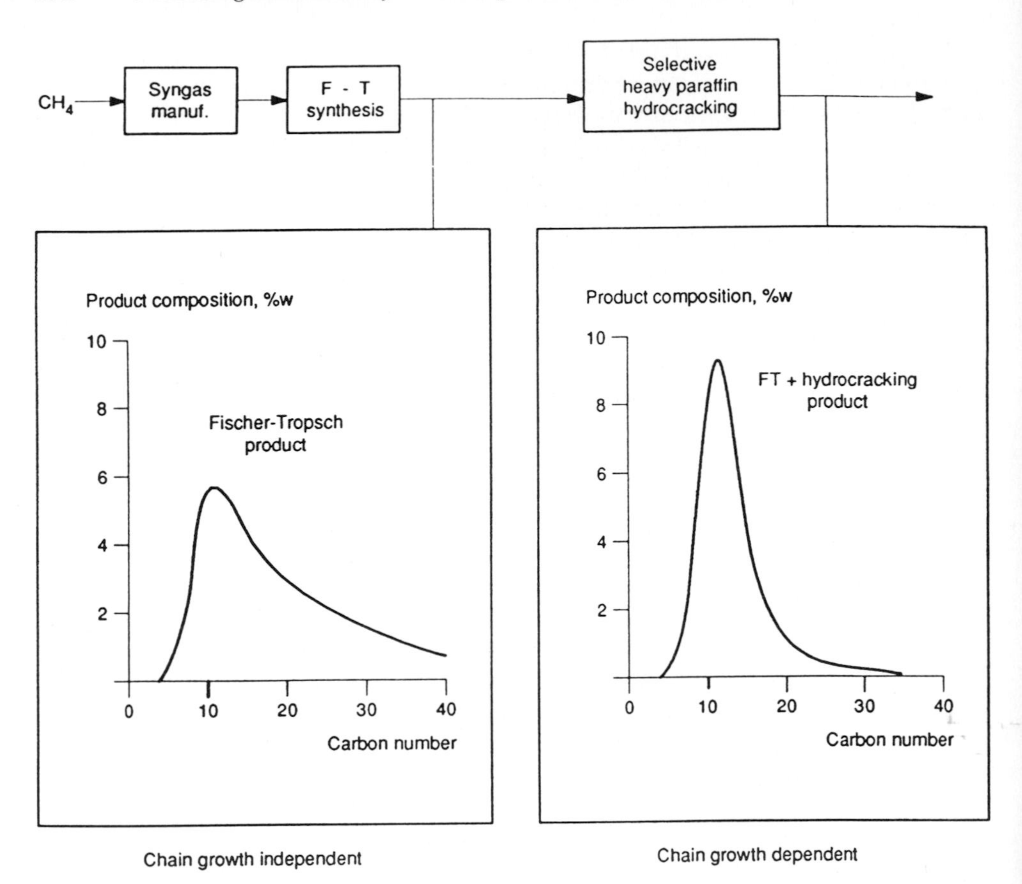

Figure 8.65 Process scheme for the so-called Shell middle distillate synthesis (SMDS) process. (After I. E. Maxwell, J. E. Naber, K. P. de Jong, *Appl. Catal.* **1995**, *68*, 153.)

gas, by first generating 'syn-gas' ($CO + H_2$ mixtures) which is then converted by Fischer–Tropsch synthesis to hydrocarbons that have large fractions of waxy product. This, in turn, is cracked, via an appropriate zeolitic catalyst, to the molecular weight distribution required for gasoline. This strategy is embodied in the *S*hell *m*iddle *d*istillate *s*ynthesis (SMDS) now in operation in Malaysia. The principles of the operation are shown in Fig. 8.65. No aromatics are generated in this way; an added bonus is that the gasoline is also devoid of deleterious sulphur impurities, since the starting material is CH_4.

8.9.2.1 Cracking Reactions

The mechanism by which catalytic cracking occurs is well established. Acidic properties of the zeolite catalysts enable the heteropolar rupture of a C–H bond in the reactant hydrocarbon. There are two possible ways in which the C–H bond can break: a carbenium ion may be formed:

$$\begin{matrix}\diagdown \\ -C-H \\ \diagup\end{matrix} \rightarrow \begin{matrix}\diagdown \\ -C^+ \\ \diagup\end{matrix} + H^-$$

or a carbanion may result from the bond breakage:

$$\begin{matrix}\diagdown \\ -C-H \\ \diagup\end{matrix} \rightarrow \begin{matrix}\diagdown \\ -C^- \\ \diagup\end{matrix} + H^+$$

By far the most common occurrence, because the least energy is required, is the formation of alkyl carbenium ions. With an increase in the number of H atoms attached to the carbon atom from which the hydride ion is abstracted, the energy required to form a carbenium ion increases. On the other hand, the stability of the carbo-cation decreases in the order tertiary > secondary > primary ion. The reacting hydrocarbon acts as a weak base in the presence of the acidic zeolite catalyst. The reactant accepts a proton from the hydrocarbon:

$$CH_3{-}CH{=}CH_2 + HX \rightleftharpoons CH_3{-}C^+{-}HCH_3 + X^-$$

the more stable secondary alkyl carbenium ion being formed in preference to its primary counterpart. If Lewis acidity (L denoting the Lewis acid) is inferred when a paraffin forms a carbo-cation by hydride ion abstraction, we have:

$$RH + L \rightleftharpoons LH^- + R^+$$

Once the ion has been formed by the interaction of the reactant hydrocarbon with the acidic zeolite catalyst, cracking occurs (C–C bond breakage) at the bond located in the β position to the carbon atom carrying the positive charge. For example, a straight-chain hydrocarbon would initially yield an olefin and a primary carbenium ion which immediately rearranges to a more stable secondary ion:

$$RCH_2\overset{+}{C}HCH_2(CH_2)_2R' \rightleftharpoons RCH_2CH{=}CH_2 + \overset{+}{C}H_2CH_2R'$$

is followed by

$$\overset{+}{C}H_2CH_2R' \rightleftharpoons CH_3\overset{+}{C}HR'$$

Continuation of this sequence of cracking leads to propylene:

$$\overset{+}{C}H_2CH_2CH_3 + X^- \rightleftharpoons CH_2{=}CHCH_3 + HX$$

A difficulty arising from any cracking mechanism is the possibility that the ultimate product could be carbonaceous material in the form of coke which blocks the active catalytic sites of the catalyst and prevents further reaction. For this reason, the catalyst has to be frequently regenerated with steam to burn off any carbonaceous deposits causing deleterious performance.

8.9.2.2 Cracking Catalysts

As outlined in Chapter 5 and the preceding sections of Chapter 8, rare-earth (RE)-exchanged faujasitic zeolites are at present the cracking catalysts of choice. Properties range from hydrophilic to hydrophobic, depending on the Si/Al ratio in the crystalline structure. Hydrophilic properties predominate when the ratio is small and a transition to hydrophobic characteristics occurs at an Si/Al ratio of between about 8 : 1 and 10 : 1. Among the many naturally occurring aluminosilicates, only erionite and mordenite are important as cracking catalysts.

8.9.2.3 The Catalytic Cracking Reactor

Before the advent of very active zeolite cracking catalysts (when silica–alumina was the only available cracking catalyst), fluidized-bed reactors coupled with catalyst regenerators (in which the catalyst was reactivated with air at elevated temperatures) were employed for cracking oil feedstocks. The more active zeolite catalysts, however, only require a fluid residence time of about 5 s to achieve good conversions to gasoline products. The fluid catalytic cracking (FCC) unit shown in Fig. 8.66 consists of a riser tube (the reactor) coupled to a catalyst disengager section which, in turn, leads directly into a two-stage regenerator. Preheated gas oil enters the base of the riser tube, within which regenerated catalyst is being conveyed upwards by the gas oil feed. Products, including gasoline, are disengaged from the upward-flowing catalyst pellets and further refined in a downstream fractionating unit. The catalyst, now falling freely under the action of gravity, is treated with steam to strip away volatile residues and then passes into the two-stage fluidized-bed regenerator where the coke is burned off the catalyst with steam and air. The regenerated catalyst then

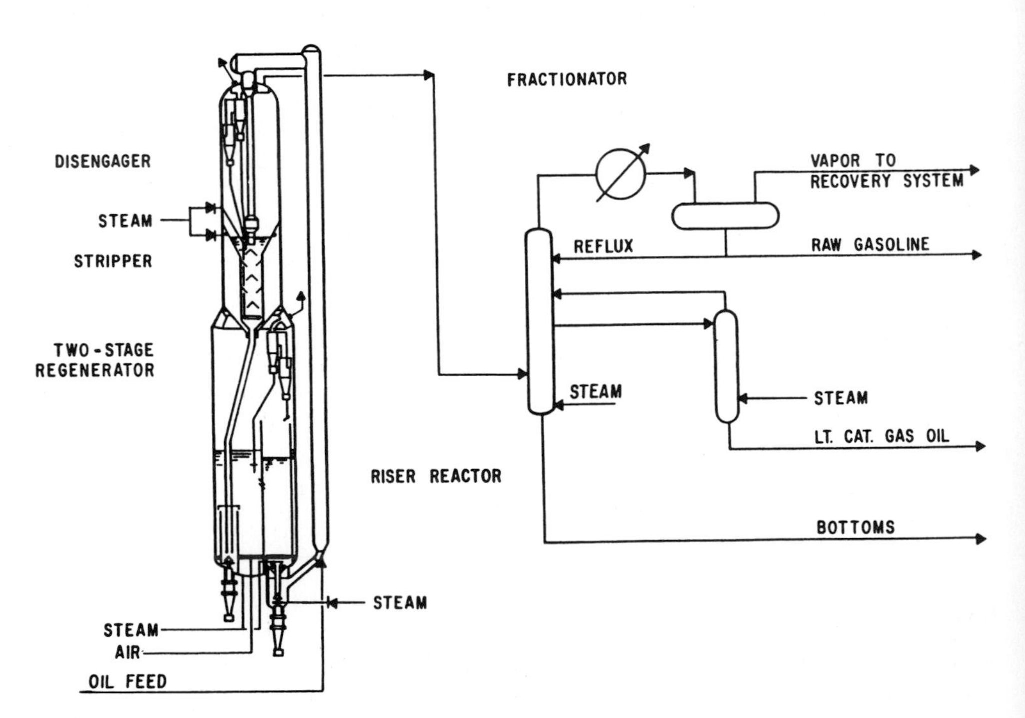

Figure 8.66 Riser tube and regenerator unit for catalytic cracking.

rejoins the gas oil feed and is further conveyed through the reactor riser tube. Conditions within the riser tube (which is categorized as a transport line reactor) are between 2 and 3 bar pressure and 475 °C at the top of the riser tube and 550 °C at the bottom. The regenerator operates between about 3 and 4 bar pressure and between 670 and 760 °C. The velocity of reacting gases in the riser tube is about 0.6–0.8 m s^{-1} while the denser catalyst particles travel upwards at about one-third of the gas velocity.

The performance of riser-tube reactors is dependent on the rate at which the zeolite catalyst activity declines as a result of coke deposits occasioned by overcracking of the gas oil in the riser tube. Weekman has analysed the events occurring in the riser tube, a concise account of which has been described by Gates et al. (1979), and concludes that the gas oil cracks via a second-order reaction producing the required gasoline fraction and other cracked products including coke which cause catalyst deterioration. If the oil feed is designated as A, the desired gasoline product as B and the overcracked products as C, then the simple kinetic scheme shown in Scheme 8.17 leads to an adequate representation of the gasoline yield.

The gas oil feed is generally a mixture of several components and it has been found that the overall rate of cracking can be represented by second-order kinetics, even though single components of the feed, taken separately, crack according to first-order kinetics. Thus, the overall cracking rate is taken to be the linear sum of the rates of cracking of individual components (each reacting independently and at their own rates) and the resultant rate curve is found to approximate to second-order kinetics. Gasoline, on the other hand, has fewer components than the gas oil feed and cracks further to coke and its precursors by a first-order rate process. For isothermal conditions and plug flow through the riser tube, Scheme 8.17 can be summarized by the two mass conservation equations,

$$-u\mathrm{d}a/\mathrm{d}z = \xi(k_1 + k_3)a^2 = \xi k_0 a^2 \tag{90}$$

and

$$u\mathrm{d}b/\mathrm{d}z = \xi k_1 a^2 - \xi k_2 b \tag{91}$$

where a and b are the weight fractions of A and B respectively at a fractional position z along the length of the riser reactor and u is the gas velocity, assumed here for simplicity as constant. The function ξ, considered to be equivalent for each reaction, is a time-dependent catalyst activity. The first of these equations may be independently integrated if ξ is regarded as constant, and yields

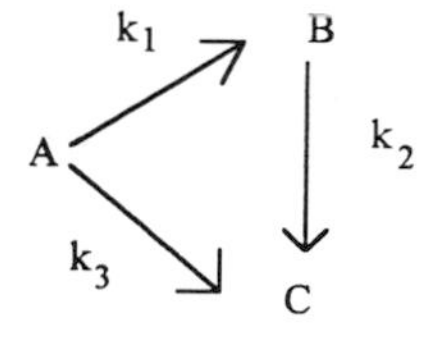

Scheme 8.17 Cracking of gas oil (A) to gasoline (B) and overcracked products (C).

$$a = 1/(1 + K_0 z) \tag{92}$$

where K_0 has been written for $\xi k_0/u$. The yield of B (i.e. b) may be found by first dividing these two equations and then multiplying through by the appropriate integrating factor. The analytical solution gives

$$b = K_1 K_2 \mathrm{e}^{-k_2/a} \left\{ \frac{\mathrm{e}^{K_2}}{K_2} - \frac{a\mathrm{e}^{K_2/a}}{K_2} + \mathrm{Ei}\left(\frac{K_2}{a}\right) - \mathrm{Ei}(K_2) \right\} \tag{93}$$

where K_1 and K_2 are written for $\xi k_1/u$ and $\xi k_2/u$ respectively and $\mathrm{Ei}(p)$ is the tabulated exponential integral of the argument p defined by

$$\int_{-\infty}^{p} (\mathrm{e}^{v}/v)\,\mathrm{d}v$$

in which v is a dummy variable. The time-dependent function of the residual catalyst activity, originally proposed by Voorhies, is

$$\xi = (1 - \beta\gamma)\mathrm{e}^{-\alpha t} \tag{94}$$

where γ is the weight fraction of coke on the catalyst, β and α are empirical constants and t is a measure of the time the catalyst has been in contact with the vapour. The factor $(1 - \beta\gamma)$ in this latter equation is the fraction of original catalyst activity remaining, which is, according to plant experience, linearly dependent on the coke content of the catalyst. Voorhies regarded catalyst decay as slow relative to the vapour residence time in the reactor and this assumption enabled a solution of the rate equations in terms of gas oil conversion and catalyst residence time for different reactor configurations. Laboratory-scale experiments emulating a moving catalyst bed reactor (within the catalyst and vapour in plug flow, which is similar to the riser-tube reactor described) provided empirical data relating to the reaction velocity constants, so that the effects of conversion on gasoline yield and of the gasoline/gas oil cracking ratio on the maximum gasoline yield could be predicted. As the temperature within the riser tube is not constant, the reactor behaving approximately as an adiabatic unit, the problem is really more complex than outlined. An energy conservation equation for the riser tube would necessarily have to be coupled with the mass conservation equations before an explicit solution to the problem could be obtained.

8.9.3 Hydrotreating

Hydroprocessing is designed to enhance the product quality of refinery streams by the use of H_2. Typical processes include hydrocracking (sometimes known as hydrorefining), hydroisomerization and hydrotreating (desulphurization, denitrogenation and demetallization processes). Hydrocracking processes range from the treatment of low-value feedstocks containing sulphur, nitrogen and heavy metals, i.e. ones which would not be suitable for catalytic reforming and cracking units. Hydrocracking is also a suitable process for converting feeds which are high in aromatic content. Hydrorefining processes operate at moderate conditions of temperature and high pressures (300–500 °C and 40–200 bar). The variety of products emerging from a hydrorefining plant include gasoline, kerosene, middle-distillate fuels and feedstocks for downstream petrochemical processes. A dual-function catalyst (acidic and hydrogenation catalytic functions), such as a platinum-impregnated silica–alumina or zeolite, facilitates the many isomerization and hydrogenation reactions which occur. In the Shell Hysomer process, for example, platinum on mordenite is used.

Two fixed-bed reactors in series are employed in the hydroprocessing of the heavy feedstocks. The first reactor is really a hydrotreating (as opposed to hydrorefining) reactor packed with a catalyst such as a cobalt–molybdenum composite. The purpose of the first reactor is to convert organic sulphur and nitrogen compounds to H_2S and NH_3 respectively, thus removing the possibility of contaminating the hydrorefining catalyst with organic sulphur and nitrogen compounds which are poisons to the dual-function catalyst.

The hydrotreating reactor is operated at between 300 and 400 °C and at 3.5–70 bar pressure, the liquid feed flowing downwards cocurrently with hydrogen. The hydrotreated product, now sulphur- and nitrogen-free, is fed together with additional H_2 to the hydrorefining reactor in series with the hydrotreating reactor. The cracked and hydrogenated products then enter gas–liquid separators, where unreacted H_2 is flashed off and recycled to the reactors; the liquid products are fed to a fractionating column, from the top of which light C_4–C_6 hydrocarbons are produced and naphthas and jet or diesel fuels are taken off from sidestreams. Figure 8.67 illustrates a typical flow diagram of a hydrocracking plant.

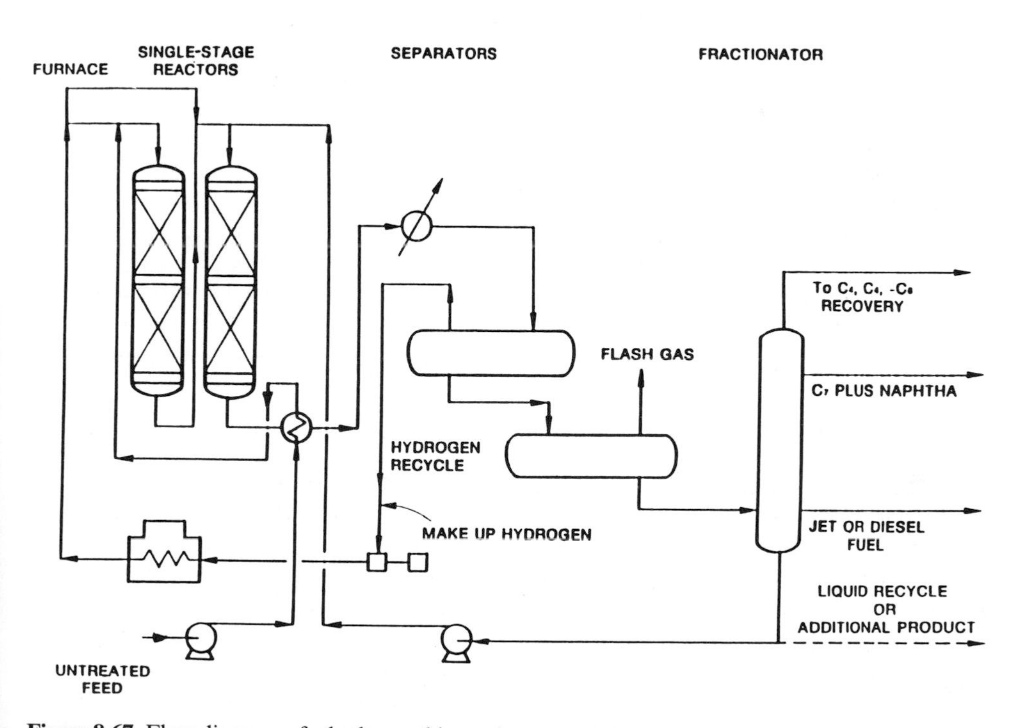

Figure 8.67 Flow diagram of a hydrocracking unit.

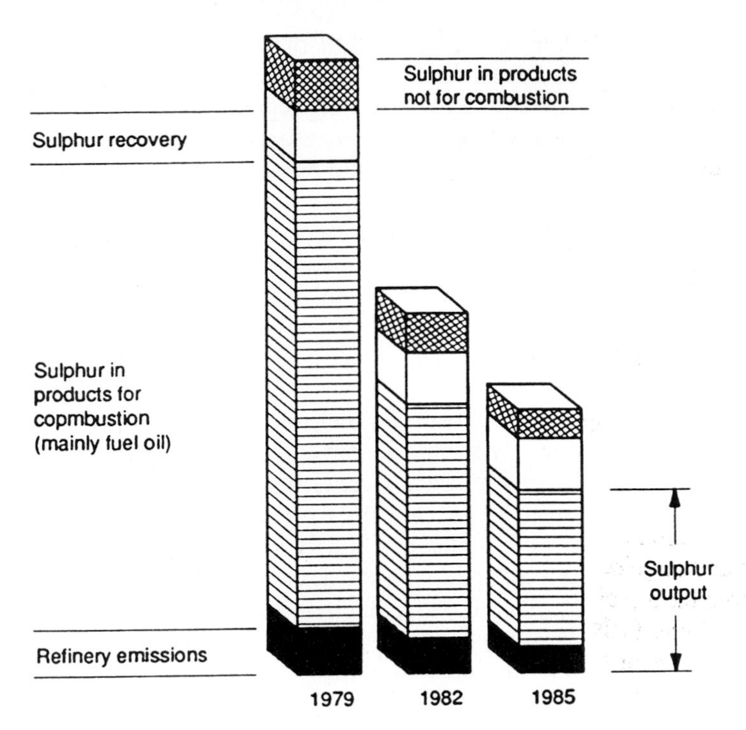

Figure 8.68 Trends in Western European sulphur emissions. (After I. E. Maxwell, J. E. Naber, K. P. de Jong, *Appl. Catal.* **1994**, *113*, 153.)

Desulphurization of feedstocks is important for three reasons: firstly, to prevent sulphur-containing organics from contaminating supported platinum reforming catalysts; secondly, to improve the colour and stability of products such as gasoline from catalytic cracking units, thus providing gasoline products low in sulphur content (referred to as a sweetening process); thirdly, to remove organic nitrogen compounds from feedstreams to catalytic cracking units lest the basic nitrogen compounds neutralize the acidic catalyst. Trends to reduce the S content of feedstocks are illustrated in Fig. 8.68. The reactions which occur during hydrotreating processes are the formation of H_2S from sulphur-containing organic compounds (mercaptans, thiols, disulphides and thiophenes) and NH_3 from organic nitrogen compounds (for example, aliphatic amines, pyrrole and pyridine). Catalysts are usually either an alumina-supported mixture of cobalt and molybdenum oxides or supported nickel and tungsten oxides. In the reactor environment these non-stoichiometric oxides are converted to their sulphide forms. Indeed, before refinery streams are fed to a hydrotreating unit, the catalyst is deliberately sulphided. Hydrogen sulphide product is then in sufficient quantity to prevent reduction of the catalyst to the inactive metal. Much speculation has occurred about the structural form of sulphided hydrodesulphurization catalysts. Some authors have proposed a monolayer of S^{2-} ions superimposed over a second layer containing O^{2-}, Mo^{3+} and Mo^{5+} and stabilized by Co^{3+} promoter ions contained in sublayers of the underlying support. Other

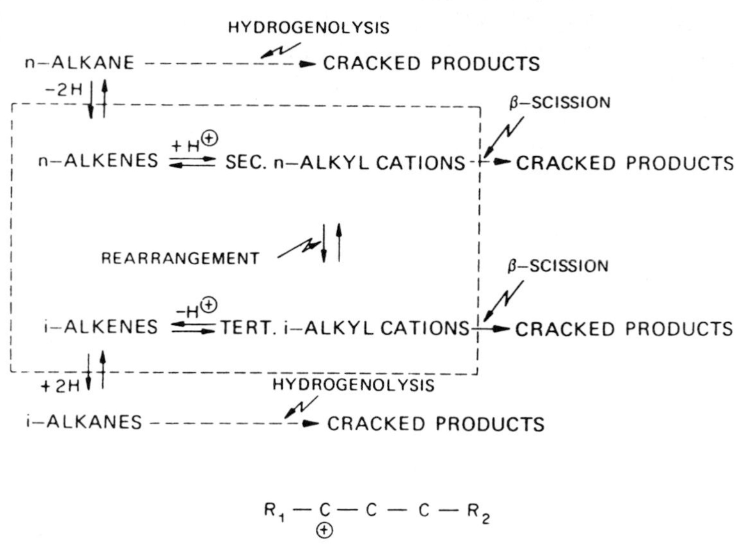

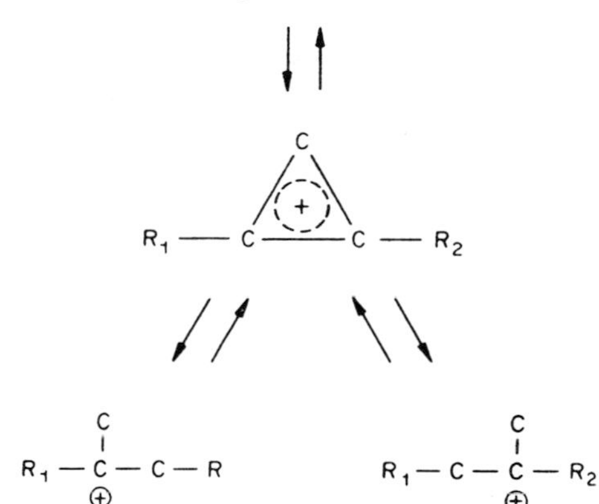

Figure 8.69 Mechanism of the hydro-isomerization reaction. (After I. E. Maxwell, C. S. John, D. M. Clark, in *Perspectives in Catalysis* (J. M. Thomas, K. I. Zamaraev, eds.) Blackwells, Oxford) **1992**, 387.)

authors, notably Chianelli, have suggested that intercalation of cobalt or nickel occurs at the edges of non-stoichiometric structures of MoS_2 or WS_2.

Hydro-isomerization is a typical acid-catalysed reaction, and the mechanism is believed to be as shown in Fig. 8.69. Carbenium ions are formed through protonation of alkenes; such ions can then rearrange, and desorb as isomers, or they can be cracked through β-scission. The intermediate in the rearrangement is thought to be a protonated cyclopropane species, as indicated in the figure. This would explain why direct C_4 isomerization does not occur in these systems. The mechanism depicted here also explains the bifunctional (hydrogenation and acidic) requirements of the catalyst. Weitkamp et al. have contributed significantly to the understanding of the hydrocracking and hydro-isomerization of C_6–C_{10} hydrocarbons.

Kinetics of desulphurization reactions of model compounds such as a thiophene can be represented quite well by Langmuir–Hinshelwood-type rate expressions. For such experimental measurements it has been shown that the rate of desulphurization is first order with respect to H_2, less than first order with respect to the reactant being desulphurized (a term in reactant concentration or partial pressure appears in the denominator of the rate expression) and inhibited by H_2S. However, although the rate of desulphurization for a mixed compound feed such as a middle distillate is first order in H_2 and inhibited by H_2S, it is approximately second order with respect to the total sulphur feed concentration. This phenomenon was referred to in Section 8.9.2.3 on catalytic cracking, and was explained by the rate of total sulphur conversion being the resultant of two or more reaction rates of first order with respect to single component concentrations. Rates of desulphurization have been shown not to be influenced by mass-transfer limitation through a liquid-phase reactant but considerable intraphase resistance is offered to the liquid and gaseous reactants within the catalyst pores. Effectiveness factors of 0.6, for example, have been reported in the literature.

Design of the reactor required to effect efficient hydrodesulphurization is a not-inconsiderable engineering problem. The feed to be desulphurized may be vapour, a mixture of liquid and vapour, or entirely liquid, depending on the source of the feed. When the source is a heavy residual oil, liquid and hydrogen are passed upward through an ebullient suspension of small catalyst particles. For a lighter oil the liquid feed trickles downwards through a fixed bed of catalyst pellets cocurrently with hydrogen: this type of reactor is commonly known as a trickle-bed reactor. A gas oil or middle distillate might be entirely in the vapour phase and passed with hydrogen downwards through fixed beds of particles, often in two stages so that there is an opportunity to remove the heat of exothermic reaction by cooling with H_2 between stages (cold-shot cooling). H_2S and unreacted H_2 emerging from these reactors is normally separated from liquid product, and the gaseous stream further scrubbed to remove H_2S. The residual H_2 is recycled to the reactor to preserve a moderate partial pressure of H_2 in the reactor and to minimise expense. Considerable operating difficulties can arise in poorly designed trickle-bed reactors. Poor contacting of the liquid with catalyst and channelling within the bed can result from inefficient liquid distribution at the reactor entrance. Potentially high pressure drops resulting in a necessary shutdown of the hydrogen recycle compressor and the whole unit can occur if the bed becomes clogged with fine catalyst dust occasioned by careless handling of the catalyst when filling the reactor. Similarly, crust formation on top of the catalyst bed can result from the liquid feed carrying in foreign material from corroded pipes or unwanted salt. Either a guard bed upstream of the trickle-bed reactor or wire mesh baskets placed on top of the catalyst bed can minimize the latter problem. Deactivation of catalyst activity can also occur by metals, such as vanadium or nickel present in heavy oil feeds, being deposited in the pores of the catalyst. An upstream guard bed containing a catalyst such as manganese which demetallizes the feed to the trickle-bed reactor is therefore often incorporated in the plant for heavy residual oil feeds. Loss of catalyst activity resulting from coke deposition is not, however, permanent. The reactor can be taken temporarily offstream so that the coke can be burned off the catalyst with air or steam at elevated temperatures.

8.10 The Role of Catalysis in Energy-Related Environmental Technology

In Chapter 1 (Section 1.4) we dealt with some intellectual and practical challenges for catalysis in the 21st century, and earlier in this chapter we dealt in depth with the key scientific and engineering aspects of the auto-exhaust catalyst. Elsewhere in this text we have illustrated the growing demands for clean technology, zero-waste processes and chemical operations which, through novel deployment of catalysts, will diminish the need and vaporization for industrial solvents. Tables 1.4 and 1.5 bring into sharp focus the exciting challenges that now face experts in catalysis.

In ending our text we dwell briefly on energy-related technologies. Worldwide, significant continued growth is expected in terms of energy consumption, and fossil fuels are expected to remain very important well into the foreseeable future. There is a marked difference in terms of energy growth between the developed countries in Europe and the Americas on the one hand, and the developing countries on the other. By about 2010, some 30 new oil refineries are expected to be built in the Far East, China and India. A graphic picture of the burgeoning developments in these countries in seen in a retrospective analysis of what has happened in South Korea

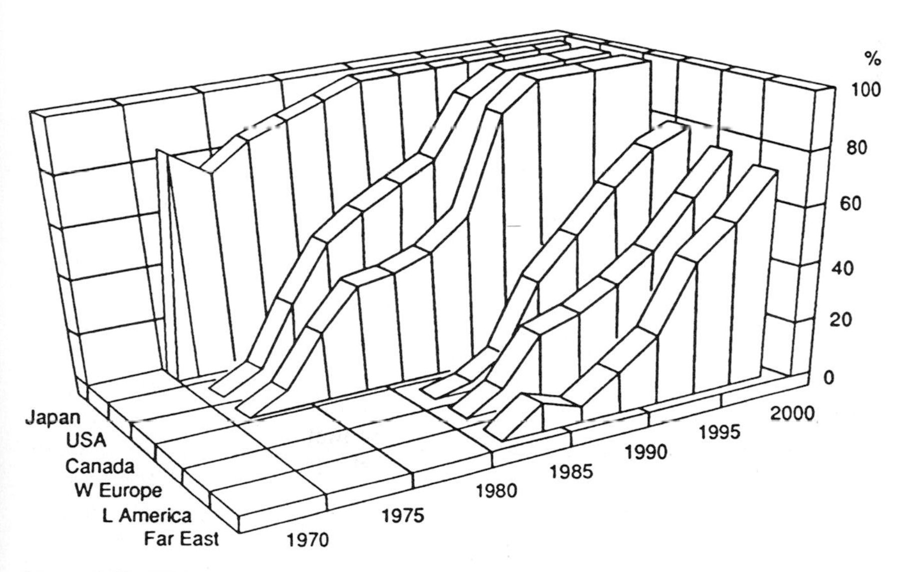

Figure 8.70 Global penetration of unleaded gasoline. (After I. E. Maxwell, J. E. Naber, K. P. de Jong, *Appl. Catal.* **1994**, *113*, 153.)

since 1965. At that time, that country refined some 16 000 barrels of oil per day – it had the capacity to cope with 35 000 barrels. However, by 1995, 1.6 million barrels were being processed, all requiring cracking and reforming catalysts, and others, for the operation.

In general, proven oil reserves are slowly eroding whereas reserves of natural gas continue to grow quite rapidly. As oil becomes progressively more scarce, the quality of the crude oil – as reflected, for example, in the hydrogen content – tends to deteriorate. In terms of trends for the products of the petrochemical industry, it is expected that factors such as requirements for reduced emissions, a shift towards middle distillates (recall Fig. 8.65) and improved combustion performance will almost certainly require that the hydrogen content of transportation fuels will steadily increase. These trends will lead to a 'hydrogen gap', which in turn will require more hydroprocessing in refineries and a concomitant need for increased production of hydrogen in the future.

Environmental legislation will continue to strive for lower emission levels, particularly for SO_2, NO_x and volatile organic compounds (VOCs). Within Western Europe since 1979, for example, as we saw in Fig. 8.68, there has been a very significant reduction in emissions of sulphur, both in the transportation fuel finally produced and in the gas streams of the refinery.

Perhaps the most dramatic change consequent upon changes in environmental legislation relates to the removal of lead from petrol (gasoline). Initially, pressure for its elimination arose from the toxicity of the emissions of lead from motor vehicles. Later, however, the fact that lead poisons the auto-exhaust catalyst became a key factor. Figure 8.70 speaks for itself.

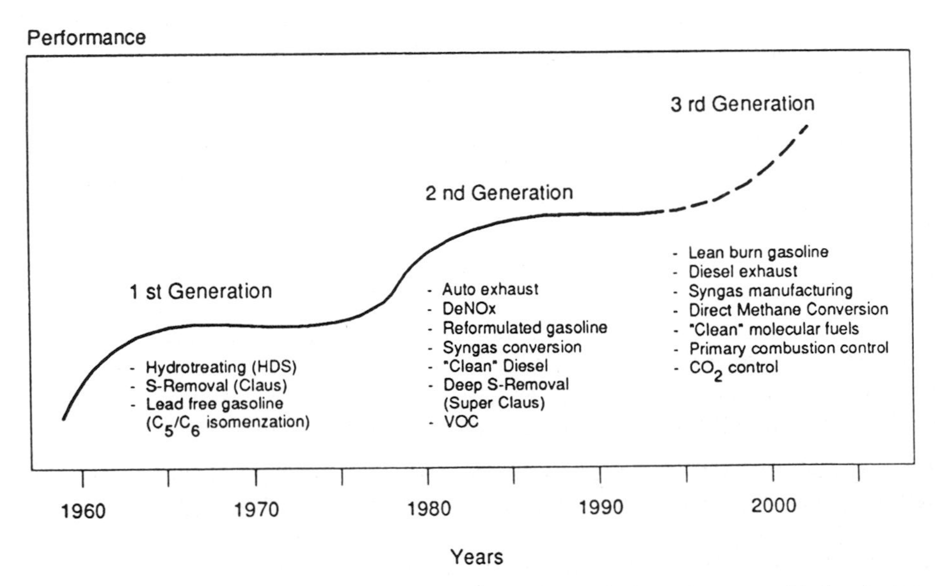

Figure 8.71 A broad-brush picture of developments in catalytic environmental technologies over a 50-year period. (After I. E. Maxwell, J. E. Naber, K. P. de Jong, *Appl. Catal.* **1994**, *113*, 153.)

Workers at KLSA (Shell) Amsterdam have represented the broad trends in the developments in catalytic environmental technologies by plotting ‘performance’ as a function of time, as in Fig. 8.71. In the 1970s and 1980s there was a thrust towards the application of catalysis in petrochemicals, particularly for the production of bulk polymer intermediates. It was in this period that the Mobil Company developed and commercialized a number of processes for the isomerization disproportionation and alkylation of aromatics utilizing their novel shape-selective zeolite ZSM-5 as a catalyst. Table 1.3 details other major catalytic advances, both prior to 1970 and after 1990. Unquestionably, in progressing from second- to third-generation catalysts (Fig. 8.71), environmental applications loom large.

To paraphrase Dr Johnson’s words of wisdom quoted in our Preface,

‘No textbook of a living subject such as catalysis can be perfect, since while it is hastening to publication, some catalysts, concepts and techniques are budding and some falling away.’

8.11 Problems

1 ‘The direct application of inorganic membrane reactors in the process industry is still limited because of a large number of technical and economic drawbacks.’ This is a quotation from G. Saracco et al. (*J. Membrane Sci.* **1994**, *94*, 105). In the light of what you have read in Chapter 8, and after consulting the cited article, highlight the key points that justify the above statement.

2 In the phenol–acetone condensation reaction to yield bisphenol A, catalysed by H^+–Y zeolites (see P. B. Venuto, P. S. Landis, *Adv. Catal.* **1968**, *18*, 259), what form of kinetics would you expect? (Consider the effect of transport limitations in your answer.)

3 Several workers (see A. Kukuoka et al., *Appl. Catal.* **1989**, *50*, 295) have considered the possibility of effecting shape-selective olefin hydroformylation. What are the principal problems with such an approach?

4 Quasi-equilibrium between catalytic intermediates and reactant or product molecules is a common assumption. As process conditions vary, this assumption may cease to be valid because of kinetic coupling between steps in the catalytic cycle. After first consulting the work of J. Halpern (*Pure Appl. Chem.* **1983**, *55*, 99) on the asymmetric hydrogenation of a prochiral alkene, show how kinetic coupling may manifest itself by substantial changes in selectivity.

[*Hint*: it is advisable first to read the work of M. Boudart and G. Djéga-Mariadassou on a typical two-step catalytic reaction: *Catal. Lett.* **1994**, *29*, 7.]

5 The accompanying set of data pertain to the question of the constituents of gasoline. Explain (a) the meaning of the acronyms used, and (b) the significance of the numerical values given.

	RON	MON	Volatility (RVP) [psi]
MTBE	118	100	8–10
ETBE	118	102	3–5
TAME	111	98	1.5
i C_5/C_6	75–102	75–102	6–27
Alkylate	92–97	90–94	3.5–4.5

6 Caprolactam is the raw material for Nylon-6:

O NH polymerization N H O n nylon 6

The key intermediate in its manufacture is cyclohexane.

Explain:

(a) how caprolactam is obtained catalytically from benzene by the so-called cumene process;

(b) how, traditionally, caprolactam is obtained from the oxime of cyclohexanone; and

(c) how the solid acid catalyst SAPO-11 promises to supplant the traditional process.

[See J. A. Cusumano, *Chem. Technol.* **1992**, 486.]

7 Monochloronaphthalenes are used in the manufacture of dyes, fungicides, insecticides, wood preservatives and as ingredients of certain cleaning agents. Lewis acid catalysts ($AlCl_3$, $FeCl_3$ and $SbCl_5$) have traditionally been used for producing 1-chloronaphthalene along with polychlorinated naphthalenes. It has been proposed that K^+ ion-exchanged zeolite L and zeolite beta are superior catalysts (A. P. Singh, S. B. Kumar, *Catal. Lett.* **1994**, *27*, 171). How would you set about conducting the appropriate experiments to test this proposition?

8 Debates on the mechanism of the Fischer–Tropsch catalysis still rage on the question of whether CO must be dissociatively adsorbed initially. Consult the report by W. H. Weinberg et al. (*J. Am. Chem. Soc.* **1993**, *115*, 4381) on their spectroscopic identification of formyl intermediates during the hydrogenation of CO on the Ru(0001) surface. How does their mechanism tally that that described in Section 8.2.1?

9 'Temperature-programmed desorption (TPD) and related techniques have played an important part in establishing the mechanism of the Cu/ZnO-catalysed synthesis of methanol.' Critically assess the veracity of this claim in the light of the recent report (Fujita, Ito, Takezawa, *Catal. Lett.* **1995**, *33*, 67) that the TPD peak of CO, usually assigned to a zinc formate surface species, arises from zinc methoxide.

10 Using the results quoted in the table and other relevant ones, elaborate upon the principles of shape-selective catalysis.

Hydrogenation catalyst (at 305 °C)	Hydrogenation [%]			
	1-Hexene	4,4-Dimethylhexene	Styrene	2-Methylstyrene
0.5 wt. % Pt on Al_2O_3	40	42	52	53
Pt on ZSM-5	95	< 1	49	< 2

11 It has been reported (R. Burch, S. Scine, *Catal. Lett.* **1994**, *27*, 177) that several metal/ZSM-5 catalysts (Pt, Rh, Co and Cu/ZSM) are effective in reducing NO to N_2 in the presence of H_2. Which measurements would you carry out to ascertain whether the reduction proceeds via a simple redox mechanism?

12 In the photocatalytic reduction of CO_2 and Rh/TiO_2 (F. Solymosi, I. Tombácz, *Catal. Lett.* **1994**, *27*, 61) doping of the TiO_2 with W^{6+} ions allegedly enhances the activity. Propose a mechanism that accounts for this result.

13 Several competing mechanisms have been proposed for the catalytic skeletal isomerization of *n*-butene to 2-methylpropylene (isobutene). γ-Al_2O_3, defective tungsten oxide WO_{3-x}, and acidic zeolites are but three types of catalyst for this reaction. (See V. Ponec et al., *Catal. Lett.* **1994**, *27*, 113 and H. Mooiweer, W. H. J. Stork, B. C. H. Krutzen, in *Zeolites and Related Microporous Materials* (ed. J. Weitkamp), Elsevier, Amsterdam, **1994**, p. 2327). Shell, Amsterdam Laboratories, have proposed the mechanism shown in Scheme 8.18 over an H^+ form ferrierite (FER) catalyst. How would you test the authenticity of this mechanism?

14 Isobutene (2-methylpropene), under appropriate conditions, can be catalytically isomerized as indicated in Scheme 8.19.

From the thermodynamic compilations of Kilpatrick et al. (*J. Res. Natl. Bur. Stds.* **1946**, *37*, 163; *ibid.* **1946**, *36*, 559) compute the equilibrium concentration of each isomer and plot the temperature dependence over the range 400–1200 K. (Compare your results with those quoted by V. R. Choudhary, *Chem. Industry Developments, Incorporating CP and E*, July **1974**, 32, where the many commercial processes of isobutene are discussed.)

15 Discuss points of physicochemical interest relating to the following statements.
(a) Spectroscopic evidence together with isotopic labelling experiments and the

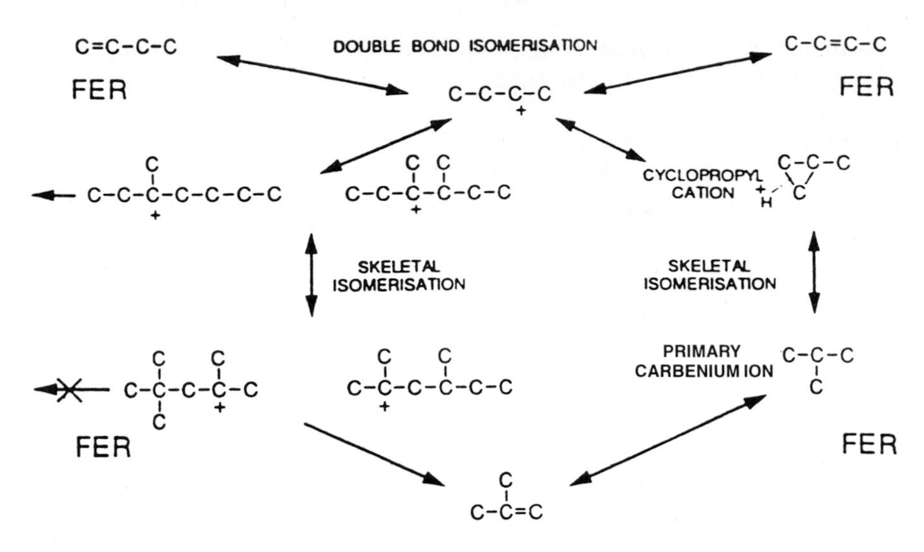

Scheme 8.18

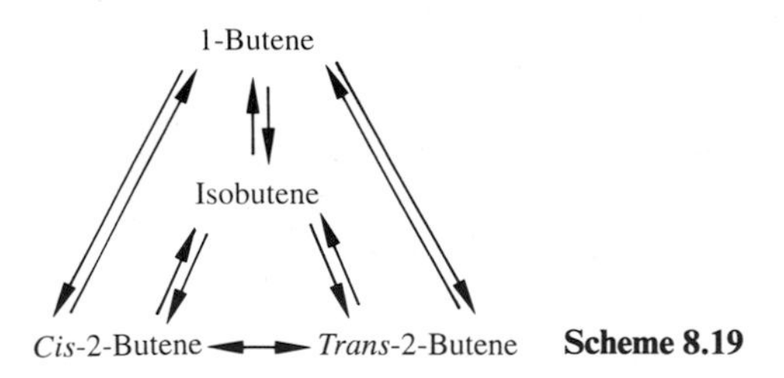

Scheme 8.19

applicability of Schultz–Flory plots reveal that Fischer–Tropsch synthesis, at iron and cobalt surfaces, is essentially a polymerization process preceded by the dissociation of CO.

(b) The nature and role of 'active sites' in heterogeneous catalytis by metals have been clarified by studies of stepped surfaces.

(c) When H–ZSM-5, a porous crystalline catalyst containing channels of diameter *ca* 5.5 Å, is poisoned by exposure to the vapour of 4-methylquinoline, the rate of catalytic cracking of *n*-hexane is almost unchanged, whereas that of 2,2-dimethylbutane is markedly diminished compared with the rate of the corresponding unpoisoned reaction.

(d) The principles governing the mode of action of shape-selective zeolitic catalysts, such as those used for alkylation of benzene or synthesis of petrol from methanol, are well understood.

16 (a) Describe the Mars–Van Krevelen mechanism and outline its applicability to the heterogeneously catalysed oxidation of propylene.

(b) Explain why irradiation of some semiconducting oxide electrodes in aqueous solution causes oxygen to be evolved, and how this phenomenon can lead to photodissociation of water by means of visible light.

17 It has been reported by H. Knözinger and P. Stolz (*Ber. Bunsenges. Phys. Chem.* **1971**, *75*, 1055) that 4-methylpyridine is more strongly bound to the Lewis acid sites of alumina than pyridine. But 2,4,6-trimethylpyridine is quantitatively displaced by pyridine. Why is this so? And what light do these facts shed on the feasibility of using substituted pyridines as probes for the acidic sites at the surfaces of solid catalysts?

18 The removal of nitrogen from organonitrogen compounds is a key process in the refining of petroleum (see B. C. Gates, J. R. Katzer, G. C. A. Schuit, *Chemistry of Catalytic Processes*, McGraw-Hill, New York, **1979**). After first consulting T. C. Ho, *Catal. Rev. Sci. Eng.* **1988**, *30*, 177, describe the test reactions and other relevant features of this process.

19 'Propylene is a vitally important feedstock in the chemical industry.' Explain why. After consulting the following reviews: Y. Moro-oka, W. Ueda, *Catalysis* **1994**, *11*, 223 (Royal Society of Chemistry); H. H. Kung, *Adv. Catal.* **1994**, *40*, 1; summarize the most successful oxidative dehydrogenation catalysts and the most likely mechanism for the conversion of propane to propylene.

20 From first principles, making use of the Gibbs free energy G, the enthalpy H, the entropy S and the heat capacity c_p, show that the standard free energy change $\Delta G^{\ominus}(T)$ as a function of absolute temperature T may be written

$$\Delta G^{\ominus} = a + bT \log_{10} T + cT$$

for any reaction.

Hence, from the data given below, calculate the decomposition pressure of PdO at 1000 K, 750 K and 500 K. What significance do the results of this calculation have for the performance of motor vehicle afterburners (monolith or flat-bed reactors) which contain palladium oxide as a component of a three-way catalyst (TWC)?

Data: to calculate $\Delta G^{\ominus}$ in $\mathrm{J\,mol^{-1}}$ use the following values for the constants:

$$a = 93784\,\mathrm{J\,mol^{-1}}; b = 24.07\,\mathrm{J\,mol^{-1}}K^{-1}; c = -154.5\,\mathrm{J\,mol^{-1}\,K^{-1}}$$

21 An ammonia synthesis reactor is operated in such a manner that advantage is taken of the exothermicity of the reaction to heat the incoming feed. Using the composite reactor–heat exchanger assembly shown in Fig. 8.72, draw up a computational scheme to ascertain the mass of catalyst necessary to employ so that a specified extent of the equilibrium synthesis reaction:

$$N_2(g) + 3\,H_2(g) \rightleftharpoons 2\,NH_3(g)$$

may be achieved.

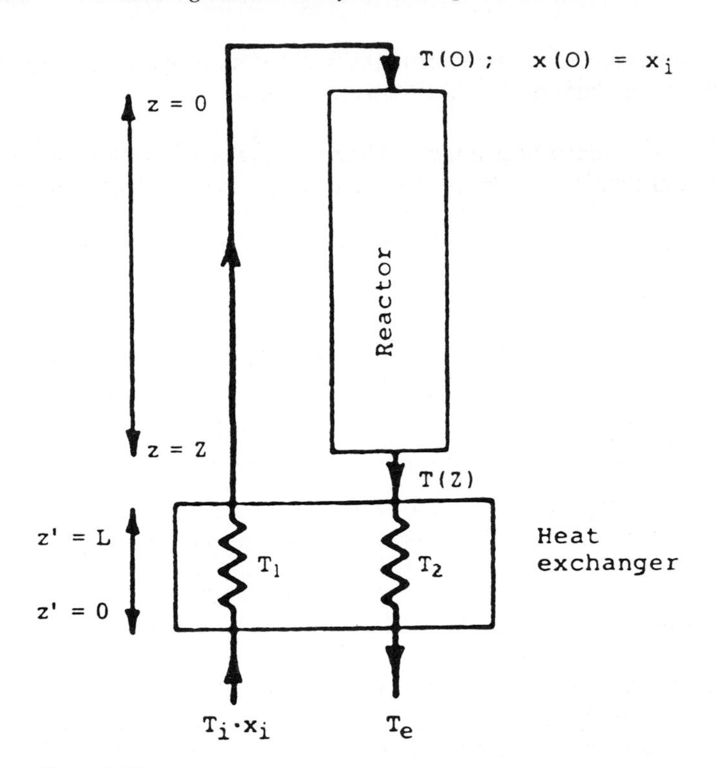

Figure 8.72

You need only write the steady-state material and energy balances and, in outline only, suggest a method of solution. State the assumptions made at each stage of your answer.

22 Estimate the number of catalyst gauze layers required to obtain a 99.8% conversion to nitric oxide of an 11% (by volume) mixture of ammonia in air at 1 bar total pressure fed upwards though the gauzes at $0.4\,\mathrm{kg\,m^{-2}\,s^{-1}}$ and an inital temperature of 60 °C. The platinum gauzes are available from manufacturers in the form of finely woven wire mesh with an available surface area of $1.19\,\mathrm{m^2}$ per square metre of a single gauze layer. It is known that the reaction is mass-transfer-controlled, the value of the mass-transfer coefficient being $1.37\,\mathrm{m\,s^{-1}}$. In practice the layers of gauze form a catalyst pad and are contained in a reactor shell of about 4 m diameter.

What temperature will the gauze have reached at the location where the conversion is 90%? Comment on the answer you obtain.

23 Describe, with the aid of practical examples, the mode of operation, construction and limitations in use of (a) an adiabatic and (b) a non-isothermal chemical reactor. Provide a mathematical model which accounts for their operational characteristics and give an analysis of their predicted behaviour.

24 The rate of the catalytic dehydrogenation of ethyl benzene (molecular weight 106 g mol^{-1}) has been reported in terms of the partial pressures of reactant and products,

$$R = k\left(P_E - \frac{P_S P_H}{K}\right)$$

where the subscripts E, S and H refer to ethylbenzene, styrene and hydrogen respectively. The specific rate constant was given as

$$\log_{10} k = -\frac{4770}{T} + 1.54$$

with k in kmol s^{-1} atm^{-1} (kg catalyst)$^{-1}$ and temperature T in Kelvin units. The equilibrium constant K is tabulated as a function of temperature.

T [K]	673	773	873	973
K [atm]	1.7×10^{-3}	2.5×10^{-2}	0.23	1.4

Using a one-dimensional reactor model with no longitudinal dispersion (see Chapter 7), write down all the numerical steps necessary to organize the data for the purpose of calculating the amount of catalyst necessary to produce 10^3 kg of styrene per day. The reactor is to operate adiabatically and consists of vertical tubes 1.2 m in diameter, packed with catalyst pellets. Specifications and data are as follows:

Feed temperature = 625 °C
Bulk density of catalyst bed = 140 kg m^{-3}
Average pressure in reactor tubes = 121 kPa
Heat of reaction = 1.4×10^5 kJ kmol^{-1}
Feed rate of ethylbenzene = 1.7×10^{-3} kmol s^{-1}
Feed rate of steam used to supply heat for reaction = 34.0×10^{-3} kmol s^{-1}
Average heat capacity of reactant and product vapours = 2.8 kJ kg^{-1}

In your answer, it will be sufficient to:
(a) produce a numerical relation between the rate of conversion per unit reactor length and reaction rate;
(b) express the reaction rate in terms of conversion and the parameters k and K;
(c) find a numerical relation which gives temperature as a function of conversion at any point along the reactor length; and
(d) outline briefly how these relations can be organized to yield the information required.

25 Give a detailed description of any one industrial reactor assembly employed to effect a heterogeneous catalytic reaction of commercial importance. Briefly outline the operating characteristics of the reactor you choose and, by the use of equations, provide an analysis by means of which this could be designed and the performance predicted.

8.12 Further Reading

8.12.1 Synthesis of Methanol

M. Bowker, R. A. Hadden, J. N. K. Hyland, K. C. Waugh, *J. Catal*, **1988**, *109*, 263.

D. B. Clarke, A. T. Bell, *J. Catal.* **1995**, *154*, 314.

G. F. Froment, K. B. Bischoff, *Chemical Reactor Analysis and Design*, Wiley, New York, **1979**, pp. 191–192, 562–571.

J. P. Hindermann, G. J. Hutchings, A. Kiennemann, *Catal. Rev. – Sci. Eng.* **1993**, *35*, 1.

J. R. Jennings, M. V. Twigg, in *Selected Developments in Catalysis* (Ed.: J. R. Jennings), Blackwells, Oxford, **1985**, p. 102.

K. Klier, C. W. Young, J. G. Nunan, *Ind. Eng. Chem. Fundam.*, **1986**, *25*, 36.

E. K. Poels, V. Ponec, in *Catalysis* **1983**, *6*, 196.

C. N. Satterfield, *Heterogeneous Catalysis in Practice*, 2nd Ed., McGraw-Hill, New York, **1991**, Chapter 10.

A. Takeuchi, J. R. Katzer, *J. Phys. Chem*, **1981**, *85*, 937.

K. C. Waugh, *Catal. Today*, **1992**, *15*, 51.

8.12.2 Fischer–Tropsch and Related Catalysis

A. T. Ashcroft, A. K. Cheetham, J. S. Foord, M. L. H. Green, C. P. Grey, A. J. Murrell, P. D. F. Vernon, *Nature (London)* **1990**, *344*, 319.

R. C. Brady III, R. Pettit, *J. Amer. Chem. Soc.*, **1980**, *102*, 6181.

D. A. Dowden, *Catalysis* **1978**, *2*, 1.

M. E. Dry, *Catal. Today* **1990**, *6*, 183.

R. A. Fiato, E. Iglesia, G. A. Somorjai (Eds.) *Top. Catal.* **1995**, *2* (special issue on Fischer–Tropsch and methanol synthesis).

E. Iglesia, S. Reyes, R. J. Madon, S. L. Soled, *Adv. Catal.* **1993**, *39*, 221.

W. Jones, R. Schlögl, J. M. Thomas, *J. Chem. Soc. Chem. Comm.*, **1984**, 464.

P. M. Maitlis, H. C. Long, R. Quyoum, M. L. Turner, Z-Q. Wang, *J. Chem. Soc. Chem. Comm.*, **1996**, 1.

V. Ponec, W. A. van Barneveld, *Ind Eng., Prod. Res. Dev.* **1979**, *18*, 26.

J. R. H. Ross, in *Catalysis* **1984**, *7*, 1.

J. R. Rostrup-Nielsen, 'Catalytic steam reforming', in *Catalysis: Science and Technology* (Eds.: J. R. Anderson, M. Boudart), Vol. 5, Springer, Heidelberg, **1984**, p. 1.

W. M. H. Sachtler, in *Chemistry and Chemical Engineering of Catalytic Processes* (Eds.: R. Prins, G. C. A. Schuit), NATO ASI No. 39, Sijthoff and Noordhoff, Alphen aan den Rijn, **1980**, p. 583.

W. M. H. Sachtler, *Proc. 8th Int. Congr. Catalysis*, Berlin **1984**, Vol. 1, VCH, Weinheim, **1984**, p. 151.

C. N. Satterfield, *Heterogeneous Catalysis in Practice*, 2nd Ed., McGraw-Hill, New York, **1991**, Chapter 10.

B. W. Wojciechowski, *Catal. Rev. – Sci. Eng.* **1988**, *30*, 629.

8.12.3 Synthesis of Ammonia

J. A. Dumesic, H. Topsøe, M. Boudart, *J. Catal.* **1975**, *37*, 513.

G. Ertl, in *Catalysis: Science and Technology* (Eds.: J. R. Anderson, M. Boudart), vol. 4, Springer, Heidelberg, *1983*, p. 257.

G. Ertl, *Angew. Chem., Int. Ed. Engl.* **1990**, *29*, 1219.

J. W. Geus, K. C. Waugh, *in Catalytic Ammonia Synthesis: Fundamentals and Practice* (Ed.: J. R. Jennings), Plenum, New York, **1991**, p. 61; see also S. R. Tennison, *ibid.*, p. 113.

P. E. Højlund–Nielsen, in *Catalytic Ammonia Synthesis: Fundamentals and Practice* (Ed.: J. R. Jennings), **1991**, Plenum, New York, pp. 285–302.

O. J. Quartulli, G. A. Wagner, *Hydrocarbon Proc.* Dec. **1978**, p. 71.

T. Rayment, R. Schlögl, J. M. Thomas, G. Ertl, *Nature*, **1985**, *315*, 311.

S. R. Tennison, see Geus and Waugh (1991).

H. Topsøe, M. Boudart, J. K. Nørskov (Eds.), *Top. Catal.* **1994**, *1* (special issue on ammonia synthesis and beyond).

8.12.4 Catalysis and Separations

M. F. Doherty, G. Burzud, *Trans. Inst. Chem Eng.* **1992**, *A70*, 448.

V. M. Gryaznov, *Platinum Met. Rev.* **1992**, *36*, 70.

N. Itoh, R. Govind, *AIChE Symp. Ser.* **1989**, *85*, 10.

N. F. Kirkby, J. E. P. Morgan, *Trans. I. Chem. E.*, **1994**, *72A*, 541.

D. Sanfilippo (Ed.), *The Catalytic Process from Laboratory to the Industrial Plant* (Italian Chem. Soc. 3rd Seminar on Catalysis, 1994) Italian Chemical Soc., **1994**; see, in particular, R. Trotta, I. Miracca on 'Synthesis and decomposition of MTBE', pp. 276–287.

G. Saracco, V. Specchia, *Cat. Rev. – Sci. Eng.* **1994**, *36*, 305.

R. Trotta, I. Miracca, see Sanfilippo (1994).

G. G. Vaporciyan, R. H. Kadlec, *AIChE J.* **1989**, *35*, 831.

8.12.5 Automobile Exhausts and the Catalytic Monolith

G. J. K. Acres, '*Perspectives in Catalysis*' (Eds.: J. M. Thomas, K. I. Zamaraev), Blackwells, Oxford, **1992**, 359.

J. Ahola, T. Maunulai, T. Salmi, H. Haario, M. Härkönen, M. Lyoma, V. J. Pohjola, Proc. 7th Nordic Sympos. on Catalysis, Turku, **1996**, paper P4-2 Finnish Chem. Societies, Turku, Finland.

J. N. Armor, *Chem. Maths.* **1994**, *6*, 730.

C. J. Bennett, R. E. Hayes, S. T. Kolaczkowski, W. J. Thomas, *Proc. R. Soc. London, Ser. A* **1992**, *439*, 465.

B. J. Cooper, S. A. Roth, *Platinum Met. Rev.* **1991**, *35*, 178.

J. A. Cusumano, *Chemtech* **1992**, *22*, 482.

G. B. Fisher, S. H. Oh, C. L. Di Maggio, S. J. Schmieg, D. W. Goodman, C. H. F. Peden, in *Proc. 9th Int. Congr. Catalysis* (1988), Vol. 3, Chemical Institute of Canada, Ottawa, **1988**, p. 1355.

S. E. Golunski, H. A. Hatcher, R. R. Rajaram, T. J. Truex, *Appl. Catal. B. Env.*, **1995**, *5*, 367.

R. D. Hawthorn, *AIChE Symp. Ser.* **1974**, *70*, 428.

B. Harrison, A. F. Diwell, C. Hallett, *Platinum met. Rev.* **1988**, *32*, 73.

R. E. Hayes, S. T. Kolaczkowski, W. J. Thomas, J. Titiloye, *Proc. R. Soc. London, Ser. A* **1995**, *448*, 321.

R. H. Heck, J. Wei, J. R. Katzer, *AIChE J.* **1976**, *22*, 477.

M. Iwamoto, *Stud. Surf. Sci. Catal.* **1990**, *54*, 121.

R. M. Lambert, M. E. Bridge, *in Chemistry and Physics of Solid Surfaces and Heterogeneous Catalysis* (Eds.: D. A. King, D. P. Woodruff), Vol. 3, Elsevier, Amsterdam, **1984**, pp. 83–101.

R. W. McCabe, R. K. Usmen, Proc. 11th Intl. Congr. Catalysis (1996), (Eds.: J. W. Hightower, W. N. Delgass, E. Iglesia, A. T. Bell), **1996**, Elsevier, Amsterdam, 355–368.

M. Shelef, G. W. Graham, *Catal. Rev. – Sci. Eng.* **1994**, *36*, 433.

K. Taylor, in *Catalysis: Science and Technology* (Eds.: J. R. Anderson, M. Boudart), Vol. 5, Springer, Heidelberg, **1984**, pp. 119–270.

T. R. Ward, P. Alemany, R. Hoffmann, *J. Phys. Chem.* **1993**, *97*, 7691.

J. Wei, *ACS Adv. Chem. Ser.* **1978**, *148*, 1.

L. C. Young, B. A. Finlayson, *AIChE J.* **1976**, *22*, 331.

8.12.6 Photocatalytic Breakdown of Water and the Harnessing of Solar Energy

W. J. Albery, *Acc. Chem. Res.* **1982**, *15*, 142.

M. Archer, *Proc. R. Inst. GB* **1995**, *66*, 97.

M. Archer, J. R. Bolton, *J. Phys. Chem.* **1990**, *94*, 8028.

A. J. Bard, M. A. Fox, *Acc. Chem. Res.* **1995**, *28*, 141.

L. E. Brus, *J. Chem. Phys.*, **1984**, *80*, 4403.

J. Cunningham, P. Sedlak, *J. Photochem. Photobiol. A* **1994**, *77*, 255.

A. Fujishima, K. Honda, *Nature (London)* **1972**, *37*, 238.

M. A. Fox, M. T. Dulay, *Chem. Rev.*, **1993**, *93*, 341.

M. Grätzel, *Energy Resources Through Photochemistry and Catalysis*, Academic Press, New York, **1983**. See also Archer, **1995**.

A. Harriman, J. M. Thomas, Zhou Wuzong, D. A. Jefferson, J. Solid State Chem., **1988**, *72*, 126.

A. Harriman, J. M. Thomas, G. R. Millward, *New Journal of Chemistry*, **1987**, *11*, 757.

A. Henglein, *Chem. Rev.* **1989**, *89*, 1861.

J. M. Lehn, *Comm. Pontif. Acad. Sci., III* **1982**, *28*, 1.

A. L. Linsebigler, G. Lee, J. T. Yates, Jr., *Chem. Rev.* **1995**, *95*, 735.

A. Nozik, in *Photocatalytic Purification and Treatment of Water and Air* (Eds.: D. F. Ollis, H. Al-Ekabi), Elsevier, Amsterdam, **1993**, p. 93.

T. N. Obee, R. T. Brown, *Environ. Sci. Techn.*, **1995**, *29*, 1223.

E. Pelizzetti, N. Serpone (Eds.), *Photocatalysis Fundamentals and Applications*, Wiley, New York, **1990**.

J. K. Thomas, *Chem. Rev.*, **1993**, *93*, 301.

J. M. Thomas, *Pure & Applied Chem.*, **1988**, *60*, 1517.

J. M. White, D. D. Beck, *J. Phys. Chem*, **1984**, *88*, 174.

M. S. Wrighton, *Acc. Chem. Res.* **1979**, *12*, 303.

K. I. Zamaraev, M. I. Khramov, V. N. Parmon, *Catal. Rev. – Sci. Eng.* **1994**, *36*, 617.

K. I. Zamaraev, *11th Intl Cong. Catalysis (1996)* (Eds.: J. N. Hightower, W. N. Delgass, E. Iglesia, A. T. Bell) Elsevier, Amsterdam, **1996**, pp. 35–50.

8.12.7 Catalysis Using Microporous or Mesoporous Solids

M. W. Anderson, J. Klinowski, *Nature*, **1989**, *339*, 200.

D. M. Antonelli, J. Y. Ying, *Angew. Chem., Int. Ed. Engl.* **1995**, *34*, 2013.

J. N. Armor, chapter in Izumi et al. (1995).

A. T. Bell, chapter in Izumi et al. (1995).

S. Bordiga, G. Ricchiardi, G. Spoto, D. Scarano, L. Carnelli, A. Zecchina, C. O. Areán, *J. Chem. Soc. Farad. Trans.*, **1993**, *89*, 1843.

A. Corma, *Chem. Rev.* May **1995**, *95*, 559.

J. A. Cusumano, chapter in Thomas and Zamaraev (1992), pp. 1–34.

W. E. Farneth, R. J. Gorte, *Chem. Rev.* May **1995**, *95*, 615.

W. F. Hölderich, in *New Frontiers in Catalysis* (Eds.: L. Guczi, F. Solymosi, P. Tetenyi) (Proc. of 10th Int. Congr. Catalysis), Akademiai, Budapest **1993**, p. 127.

Y. Izumi, H. Arai, M. Inamoto (Eds.), *Science and Technology in Catalysis 1994*, Kodansha, Tokyo, **1995**.

V. Gruver, Y. Hong, G. Panov, J. J. Fripiat, *11th Intl. Cong. Catal. 1996* (Eds.: J. W. Hightower, W. N. Delgass, E. Iglesia, A. T. Bell), Elsevier, Amsterdam, 741.

Y. Izumi, K. Urabe, M. Onaka, *Zeolite, Clay and Heteropoly Acids in Organic Reactions*, VCH, Weinheim, **1992**.

C. S. John, D. M. Clark, I. E. Maxwell, chapter in Thomas and Zamaraev (1992), pp. 387–430.

J. Klinowski, *Chem. Rev.* **1991**, *91*, 1459.

J. H. Lunsford, *Catal. Rev. – Sci. Eng.* **1976**, *12*, 137.

I. E. Maxwell, *11th Intl. Congr. Catal. (1996)* (Eds.: J. W. Hightower, W. N. Delgass, E. Iglesia, A. T. Bell), Elsevier, Amsterdam, pp. 1–30.

W. M. H. Sachtler, Z. Zhang, *Adv. Catal.* **1993**, *39*, 129.

J. M. Thomas, *Angew. Chem. Int. Ed. Eng.* **1988**, *27*, 1673.

J. M. Thomas, *Angew. Chem., Int. Ed. Engl.* **1994**, *33*, 913.

J. M. Thomas, K. I. Zamaraev (Eds.), *Perspectives in Catalysis*, Blackwell IUPAC, Oxford, **1992**.

H. van Bekkum, E. M. Flanigen, J. C. Jansen (Eds.) *Introduction to Zeolitic Science and Practice*, Elsevier, Amsterdam, **1991**.

R. A. van Santen, G. J. Kramer, *Chem. Rev.* May **1995**, *95*, 637.

D. E. W. Vaughan, *Studies in Surf. Sci. Catal.*, **1989**, *494*, 95.

P. B. Weisz, *Pure and Appl. Chem.*, **1980**, *52*, 2091.

K. I. Zamaraev, chapter in Thomas and Zamaraev (1992), pp. 35–66.

K. I. Zamaraev, J. M. Thomas, *Adv. Catal.*, **1996**, *41*, 335.

8.12.8 Catalysis Using Clays and Pillared Clays

G. Alberti, chapter in Clearfield (1982).

A. Baiker, *11th Intl. Congr. Catalysis* (Eds.: J. W. Hightower, W. N. Delgass, E. Iglesia, A. T. Bell) Elsevier, Amsterdam, **1996**, pp. 51–61.

J. A. Ballantine, J. H. Purnell, J. M. Thomas, *J. Mol. Catal.* **1984**, *27*, 157.

R. M. Barrer, *Pure Appl. Chem.* **1989**, *61*, 1903.

P. A. Barrett, J. M. Thomas, R. H. Jones, G. Sankar, C. R. A. Catlow, I. J. Shannon, *J. Chem. Soc. Chem. Commun.*, **1996**, in press.

R. Burch (Ed.), *Catal. Today* **1988**, *2*, 187 (special issue).

F. Cavani, F. Trifirò, A. Vaccari in *Catal. Today* **1991**, *11*, 173 (special issue).

S. T. Ceyer, *Science*, **1990**, *249*, 133.

J. Chen, J. M. Thomas, *J. Chem. Soc. Chem. Commun.*, **1994**, 603.

A. Clearfield (Ed.), *Inorganic Ion Exchange Materials*, CRC Press, Baton Rouge, FL, **1982**.

A. Clearfield, chapter in Clearfield (1982).

M. E. Davis, *Acc. Chem. Res.*, **1993**, *26*, 3.

E. G. Derouane, Z. Gabelicia, P. A. Jacobs, *J. Catal.*, **1981**, *70*, 238.

F. Figueras, chapter in Mitchell (1990).

W. Jones, chapter in Mitchell (1990).

W. Keim, in *Advances in Catalyst Design* (Eds.: M. Graziani, C. N. R. Rao) World Scientific, Singapore, **1991**, 345.

D. W. Lewis, C. R. A. Catlow, J. M. Thomas, D. J. Willock, G. J. Hutchings, *Nature*, **1996**, in press.

R. F. Lobo, M. E. Davis, *J. Am. Chem. Soc.*, **1995**, *117*, 3766.

T. Maschmeyer, F. Rey, G. Sankar, J. M. Thomas, *Nature*, **1995**, *378*, 159.

I. V. Mitchell (Ed.), *Pillared Layered Structures*, Elsevier, Amsterdam, **1990**.

J. A. Moulijn, P. W. N. M. Van Leeuwen, R. A. Van Santen (Eds.), *Catalysis: An Integrated Approach to Homogeneous, Heterogeneous and Industrial Catalysis*, Elsevier, Amsterdam, 1993.

B. Notari, *Adv. Catal.*, **1996**, *41*, 253.

T. J. Pinnavaia, *Science* **1983**, *220*, 365.

G. Poncelet, chapter in Mitchell (1990).

J. H. Purnell, chapter in Mitchell (1990).

C. M. Quinn (Ed.), Identification of Intermediate Species in Catalytic Reactions, *Catalysis Today*, **1992**, *12*, 339–505.

E. Ruiz-Hitzkey, chapter in Mitchell (1990).

J. Turkevich, Y. Ono, *Adv. Catal.*, **1969**, *20*, 135.

P. Venuto, *Microporous Mater.*, **1994**, *2*, 297.

P. B. Weisz, V. J. Frilette, *J. Phys. Chem.*, **1960**, *64*, 382.

Y. Xu, C. P. Grey, A. K. Cheetham, J. M. Thomas, *Catal. Lett.*, **1990**, *4*, 251.

D. T. B. Zennakoon, W. Jones, J. M. Thomas, *J. Chem. Soc. Faraday Trans.* 1., **1986**, *82*, 3081.

S. I. Zones, M. E. Davis, *Current Opinion in Solid State & Materials Science*, **1996**, *1*, 107.

J. M. Thomas, *Phil. Trans. Roy. Soc.*, **1974**, *A277*, 251.

J. M. Thomas, in *Intercalation Chemistry* (Eds.: M. S. Whittingham, A. J. Jacobson) Academic Press, New York, **1982**, p. 55.

J. M. Thomas, *Phil. Trans. Roy. Soc.*, **1990**, *A333*, 173.

J. M. Thomas, *Faraday Discuss. Chem. Soc.*, **1995**, *100*, C9.

H. van Damme, chapter in Mitchell (1990).

D. E. W. Vaughan, in *Zeolites: Facts, Figures, Future* (Eds.: P. A. Jacobs, R. A. van Santen), Elsevier, Amsterdam, **1989**, p. 95.

8.12.9 Other Solid Acid Catalysts

K. Bruckman, T. Haber, E. M. Setwicka, *Faraday Discuss. Chem. Soc.*, **1989**, *87*, 173.

N. Essayem, S. Kieger, G. Coudarier, J. C. Vedrine, in *Proc. 11th Intl. Congr. Catalysis, Baltimore* (Eds.: J. W. Hightower, E. Iglesia, W. N. Delgass, A. T. Bell) **1996**, pp. 591–600.

I. V. Kozhevnikov, *Catal. Rev. – Sci. Eng.* **1995**, *37*, 311.

M. Misono, in *New Frontiers in Catalysis* (Eds.: L. Guczi, F. Solymosi, P. Tetenyi), Elsevier, Amsterdam, **1993**, p. 69.

M. Misono, T. Okuhara, *Chem. Tech.*, **1993**, *23*, 23.

Y. Ono, in *Perspectives in Catalysis* (Eds.: J. M. Thomas, K. I. Zamaraev), Blackwells/IUPAC, **1992**, p. 431.

R. Schlögl, T. Ilkenhaus, B. Herzog, T. Braun, *J. Catal.* **1995**, *153*, 275.

J. M. Thomas, G. N. Greaves, G. Sankar, P. A. Wright, J. Chen, L. Manchese, *Angew. Chem. Int. Ed. Engl.*, **1994**, *33*, 1871.

8.12.10 Catalytic Processes in the Petroleum Industry

G. J. Antos, A. M. Aitani, J. M. Panera (ed.) *Catalytic Naphtha Reforming*, Dekker, New York, **1995**.

J. Armor, *Appl. Catal. B Environmental* **1992**, *1*, 221.

A. T. Bell, L. L. Hegedus (ed.) *Catalysis under Transient Conditions*, **1982**, ACS Symposium 178, Washington.

A. Chauvel, G. Lefebre, *Petrochemical Processes* Vol. 1, Gulf Publishing, Technip, Paris, **1989**.

B. H. Davis, *Proc. 10th Intl. Congr. Catalysis. Budapest* (Eds.: L. Guczi, F. Solymosi, P. Tetenyi) Akademiai Kiado, Budapest, **1992**, pt. A, p. 889.

B. C. Gates, J. R. Katzer, G. C. A. Schuit, *Chemistry of Catalytic Processes*, McGraw-Hill, New York, 1979, Chapters 1, 3, 5.

A. O. I. Krause, L. K. Rihko, *Ind. Eng. Chem. Res.* **1995**, *34*, 1172.

M. J. Ledoux, in *Catalysis* **1984**, *7*, 125.

I. E. Maxwell, J. E. Naber, *Catal. Lett.* **1992**, *12*, 105.

I. E. Maxwell, J. E. Naber, K. P. de Jong, *Appl. Catal.* **1994**, 113, 153.

P. C. H. Mitchell, *Catalysis* **1981**, *4*, 175.

M. Baerns, J. R. H. Ross in *Perspectives in Catalysis* (eds. J. M. Thomas, K. I. Zamaraev), Blackwell, Oxford, **1992**, 315.

R. Prins, in *Chemistry and Chemical Engineering of Catalytic Processes* (Eds.: R Prins, G. C.A. Schuit), NATO ASI No. 39, Sijthoff & Noordhoff, Alphan aan den Rijn, **1980**, p. 389.

R. Prins, 'Hydrodesulphurisation, hydrodenitrogenation, hydrodeoxygenation and hydrodechlorination', in *Handbook of Heterogeneous Catalysis* (Eds.: G. Ertl, H. Knözinger, J. Weitkamp), VCH, Weinheim, **1996**.

J. T. Richardson, *Principles of Catalyst Development*, Plenum, New York, **1989**.

J. J. Rooney in *Elementary Reaction Steps in Heterogeneous Catalysis* (Eds.: R. W. Joyner, R. A. van Santen) Kluwer, Dordrecht, **1993**, 51.

C. N. Sattersfield, *Heterogeneous Catalysis in Practice*, 2nd Ed., McGraw-Hill, **1991**, Chapter 9.

S. T. Sie, N. M. G. Senden, H M. H. van Wechem, *Catal. Today* **1991**, *8*, 371.

J. H. Sinfelt, *Bimetallic Catalysts: Discoveries, Concepts, Applications*, Wiley, New York, **1983**.

B. Sowerby, S. J. Becker, J. L. Belcher, *J. Catal.* **1996**, *161*, 377.

H. Topsøe, B. S. Clausen, F. E. Massoth, *Hydrotreating Catalysis: Science and Technology*, Springer, Berlin, **1996.**

B. C. Wiegand, C. M. Friend, *Chem. Rev.*, **1992**, *92*, 491.

Index